Reactive Oxygen Species in Metabolic Inflammation

Sumel Ashique • Neeraj Mishra
Editors

Reactive Oxygen Species in Metabolic Inflammation

Springer

Editors
Sumel Ashique
Department of Pharmaceutical Technology
Bharat Technology
Uluberia, West Bengal, India

Neeraj Mishra
Amity Institute of Pharmacy
Amity University Madhya Pradesh
Gwalior, Madhya Pradesh, India

ISBN 978-3-032-11751-9 ISBN 978-3-032-11752-6 (eBook)
https://doi.org/10.1007/978-3-032-11752-6

This Springer imprint is published by the registered company Springer Nature Switzerland AG
The registered company address is: Gewerbestrasse 11, 6330 Cham, Switzerland

Preface

Reactive oxygen species (ROS)-mediated inflammation plays a central role in the development and progression of major metabolic disorders, including obesity, diabetes, and non-alcoholic fatty liver disease (NAFLD). Excessive ROS generation disrupts redox homeostasis, leading to oxidative stress that fuels chronic, low-grade inflammation. This imbalance impairs metabolic signaling, promotes insulin resistance, and triggers mitochondrial dysfunction, ultimately establishing a self-perpetuating cycle of metabolic deterioration. Understanding these complex molecular interactions is essential for devising effective therapeutic strategies to restore cellular balance and prevent disease progression.

This book, *Reactive Oxygen Species in Metabolic Inflammation*, offers a comprehensive exploration of the intricate relationship between oxidative stress, inflammation, and metabolism. It provides in-depth coverage of critical signaling pathways, including NF-κB and Nrf2, and examines the roles of mitochondria, antioxidant defense mechanisms, and cellular stress responses in metabolic regulation. In addition, the book highlights emerging therapeutic approaches—such as natural antioxidants, postbiotics, and pharmacological modulators—that aim to counteract ROS-driven metabolic dysfunction. By integrating molecular insights with translational perspectives, this volume serves as a valuable resource for researchers, clinicians, and healthcare professionals committed to addressing the global burden of metabolic diseases through innovative, evidence-based interventions.

Uluberia, West Bengal, India — Sumel Ashique
Gwalior, Madhya Pradesh, India — Neeraj Mishra

Contents

Chapter 1
ROS-Mediated Cellular Signaling in Metabolic Disorders

Mohini Mondal, Sanket Pramanik, Mineshkumar Patel, Ankit Diwan, Ranit Kumar Samanta, Shubham Panja, Arpan Maity, and Raj Pramanik

1.1 Introduction

Reactive oxygen species, or ROS, are molecules that contain oxygen and are very reactive, like superoxide anion, hydrogen peroxide, and hydroxyl radicals. Even though they are usually seen as harmful by-products of normal cell processes, they also play an important role as signaling molecules that help control many bodily functions. In conditions such as obesity, diabetes, and metabolic syndrome (MetS), there is often an imbalance between the amount of ROS produced and the body's ability to counteract them with antioxidants. This imbalance can cause oxidative stress, which leads to harmful effects on the body [1]. Metabolic conditions represent a collection of health issues encompassing type 2 diabetes mellitus (T2D), cardiovascular disease (CVD), and similar ailments that healthcare providers identify through specific biochemical, clinical, and metabolic measurements. These metabolic conditions can alternatively be characterized by the presence of excess weight combined with high blood pressure, increased fasting blood sugar levels, triglycerides, and low-density lipoprotein (LDL) cholesterol, along with decreased

M. Mondal (✉) · S. Pramanik · R. K. Samanta · S. Panja · A. Maity · R. Pramanik
Department of Pharmaceutical Technology, Bharat Technology, Uluberia, West Bengal, India

M. Patel
Department of Pharmacology & Pharmacy Practice, Saraswati Institute of Pharmaceutical Sciences, Gandhinagar, Gujarat, India

A. Diwan
Department of Pharmaceutical Chemistry, Sagar Institute of Pharmacy and Technology Gandhinagar, Bhopal, Madhya Pradesh, India
e-mail: ankitdiwan@sistec.ac.in

S. Ashique, N. Mishra (eds.), *Reactive Oxygen Species in Metabolic Inflammation*,
https://doi.org/10.1007/978-3-032-11752-6_1

high-density lipoprotein (HDL) cholesterol levels in the blood. The steady rise in obesity rates (BMI $\geq$30 kg/m^2) and associated metabolic illnesses presents a significant obstacle for developed nations [2]. These conditions can alternatively be characterized by the presence of excess weight combined with high blood pressure; increased fasting blood sugar levels, triglycerides, and low-density lipoprotein (LDL) cholesterol; along with decreased high-density lipoprotein (HDL) cholesterol in blood plasma [3]. The steady rise in obesity rates (BMI $\geq$30 kg/m^2) and associated metabolic conditions presents a significant challenge in developed nations [4]. Metabolism is controlled by chemical reactions, and these reactions help to synchronize (1) the amounts of substances that enzymes use in reactions that only go one way and (2) the levels of small molecules like adenosine monophosphate (AMP) and adenosine diphosphate (ADP) that influence reactions that cannot easily be reversed. This creates a steady, balanced state ($-\Delta$GATP) that stays stable for a long time. Thermodynamic regulation can be demonstrated through the ATP-generating portion of glycolysis. Specifically the oxidation of glyceraldehyde-3-phosphate to pyruvate through the irreversible reaction catalyzed by pyruvate kinase (PK) depends mainly on ATP consumption rates. When ATP consumption rates change, this creates an imbalance between ATP usage and production. The resulting shift in [ATP]/[ADP][Pi] ratios, through near-equilibrium reactions before PK, modifies the concentrations of ADP and phosphoenolpyruvate (PEP), which serve as substrates for PK. These alterations in ADP and PEP levels adjust flow through PK in ways that restore balance between ATP production and consumption. These reactions happened in the earliest living things and are believed to have started how energy is used in living organisms. As evolution added more metabolic processes, more control systems were needed to fit the new processes into existing metabolism while keeping the body's balance steady. When gluconeogenesis was introduced, it required more control over PK to prevent unnecessary cycles; PK needs to be stopped during gluconeogenesis because the enzyme uses energy (ATP), which reduces the amount of glucose made and makes the process less efficient [5, 6]. Metabolism is a system of chemical reactions that happen inside all living cells to reside. These reactions are part of many connected processes within cells that help provide the energy needed for everything the cell does. The importance of these processes is shown by how they are found in different types of life, like animals, plants, fungi, and bacteria. In more complex organisms, these reactions happen in the cytosol and mitochondria, where glucose and fatty acids are the main sources of energy for animal cells. These metabolic processes follow certain paths that are either meant to make the most of energy or use less energy. Metabolism includes two types of chemical reactions: one that builds up complex molecules (called anabolism) and one that breaks them down (called catabolism). This section explains the basic ideas of how energy is used and made, along with the important chemical pathways that are needed for life (Fig. 1.1) [1–4].

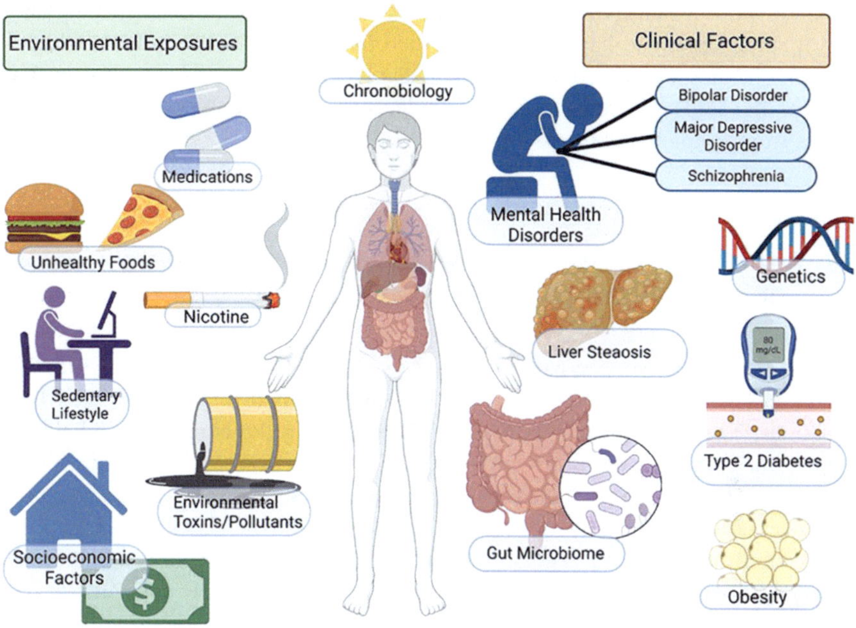

Fig. 1.1 Introduction to metabolic disorders and associated factors responsible for metabolic symptoms. (Copyright © 2022 Gangopadhyay, Ibrahim, Theberge, May and Houseknecht. This is an open-access article distributed under the terms of the Creative Commons Attribution License (CC BY) [1])

1.2 Key Metabolic Pathways: Glycolysis, TCA Cycle, and Beta-Oxidation—The Basics of Metabolism

Most people think about metabolism when they think about eating food or drinking something, or maybe they look at complicated pictures showing how thousands of chemicals connect in the body. But understanding metabolism is really important because it helps us understand life itself, and scientists have been studying it for over 150 years. The famous Nobel Prize winner Hans Krebs got interested in metabolism through his teacher, Professor Franz Knoop, who first discovered how fatty acids are broken down in the body. Krebs also figured out two important processes in metabolism, the citric acid cycle and the urea cycle, which are big parts of how the body works. These chemical processes are about making and breaking down big molecules, and they can be divided into two main groups: catabolism and anabolism. Catabolism is the process where the body breaks down big molecules into smaller ones, such as carbon dioxide, water, and ammonia. Anabolism is when the body builds up complex molecules like nucleic acids, proteins, sugars, and fats. The body needs a constant supply of energy to keep working, which helps with things like moving muscles, moving ions and other substances through the body, and making complex molecules like glycogen. Anabolism, which is also called constructive metabolism, is when living cells build up complex molecules from simpler ones, and it needs energy to happen. It works the opposite way of catabolism, which is when molecules are broken down. Examples of anabolism include making proteins from amino acids, creating DNA from nucleotides, and building complex carbohydrates from simpler sugar units.

1.3 Role of Enzymes, Hormones, and Genetic Regulation in Metabolism

Enzymatic metabolism is a natural process used by all living things. Enzymes help speed up chemical reactions inside the body and are part of many essential body functions. They make reactions happen faster, helping the body reach a balanced state. If the enzyme system does not work properly, it can cause issues in one or more organs. Many diseases are linked to the enzyme system, which is why doctors often check for signs of illness by looking at it. Hormonal signaling is important for keeping the body's tissues and metabolism stable. Research has shown that insulin and estrogen signals play a big role in how mitochondria work, how cells change, and how the body uses nutrients. Insulin helps control nutrients and mitochondria through a chain of signals involving IRS-PI3K-Akt-FoxOs and PGC1α, which is a key process that helps cells with important tasks like using energy, growing, and handling stress that involves a chain of molecular steps. This process starts when insulin receptor substrate (IRS) proteins trigger phosphatidylinositol 3-kinase (PI3K). PI3K then activates Akt, which is also called protein kinase B or PKB. Akt controls forkhead box O (FoxO) proteins, which in turn manage the genes that help the cell function properly. Estrogen helps manage protein breakdown and mitochondria through its receptors like ERα, ERβ, and GPER. Insulin and estrogen signals both affect Sirt1, mTOR, and PI3K to control cell cleaning and mitochondrial health. When these signals are not working right, it can lead to metabolic diseases. It is known that cell metabolism is closely linked to other processes inside the cell, like controlling how genes are expressed. Metabolic pathways supply the building blocks needed for gene expression and also provide ATP, which is the main energy source for these processes. But the metabolic environment can change a lot because taking in nutrients is not a steady process. Because of this, cells need to constantly adjust their gene expression to match the changing levels of metabolites and energy available [2].

1.3.1 Etiology and Pathophysiology of Metabolic Disorders

Metabolic disorders are a group of health problems that happen when the body's normal way of processing food and energy is disrupted. This can be caused by problems with enzymes, imbalances in hormones, or a mix of genetic and environmental factors. Some common examples of metabolic disorders are obesity, type 2 diabetes, metabolic syndrome, high levels of unhealthy fats in the blood, and nonalcoholic fatty liver disease. Metabolic syndrome, also known as syndrome X or obesity syndrome, is a group of health issues that can lead to serious problems like type 2 diabetes, heart disease, and other related illnesses. It is characterized by having too much fat around the waist, high blood pressure, trouble using insulin correctly, high blood sugar, high levels of a type of fat called triglycerides, and low levels of good cholesterol, which is called HDL. Insulin resistance is a major part of metabolic syndrome and is connected to health problems such as heart disease and fatty liver. Metabolic syndrome develops due to both genetic and lifestyle reasons, and it causes inflammation which increases the risk of heart disease. Studies show that the risk of getting metabolic syndrome can start at a young age and is influenced by factors like childhood trauma, social and environmental conditions, and inflammatory diseases. A large number of people have metabolic syndrome, with around 25% of the global population affected. It is especially common in developed countries where people often do not exercise much and eat a lot of high-calorie foods. Inborn errors of metabolism are diseases that you are born with. They happen when there are changes in genes that make proteins needed for metabolism. Most of these conditions are passed down through family in an autosomal recessive way. Sometimes, they can be inherited in an autosomal dominant or X-linked pattern. Other factors like the environment, how genes are regulated, and the microbes in the body may also play a role in how these disorders develop [3].

1.4 Epidemiology

Inborn errors of metabolism happen in about 1 out of every 2500 babies born. These disorders are very different from each other, so they can affect people in various ways, including how common they are, how they show up in the body, and how they are passed down through families. For example, some disorders, like mitochondrial diseases, are passed from a mother to all of her children. Other disorders might not always cause symptoms, or they might only affect certain people based on their gender. These disorders can either completely stop an enzyme from working or only partially affect it. While newborn screening tests are used to find many of these disorders early on, the tests can vary between different states and even different hospitals. These tests can detect 8–50 different diseases, but there are actually thousands of possible inborn errors of metabolism. Sometimes, the tests might miss a disorder if the baby is tested too early, before the body has built up enough of the specific chemicals needed for diagnosis. Also, if a baby has received a blood transfusion, this can affect the results. Because of these factors, even though these disorders are often thought of as problems that happen in newborns, about half of all cases appear later in life, and some people are not diagnosed until they are adults [4].

1.4.1 Pathophysiology

Some people are born with conditions that affect how their body processes carbohydrates, proteins, fats, or glycogen. Normally, when we eat, our body breaks down food into glucose, which is the main source of energy, and other by-products that are later removed from the body. If we consume more glucose than the body needs, it stores the extra as glycogen in the liver and muscles to use later when food is not available. When the body needs energy and glycogen is low, it makes new glucose from amino acids through a process called gluconeogenesis. If that is not enough, the body uses fatty acid oxidation to create fuel for the Krebs cycle. StatPearls has detailed information on specific inborn errors of metabolism, but it is still useful to think about these diseases as a group. Some of these diseases affect how the body breaks down food, which can lead to low blood sugar, called hypoglycemia. Others stop the body from getting rid of harmful chemicals, which can cause poisoning-like symptoms, such as high ammonia levels in the blood, known as hyperammonemia. Some diseases stop the body from properly using fatty acids, which can also cause low blood sugar and acidosis. Glycogen storage disorders usually do not show up suddenly and will not be covered in this discussion. Hormonal problems, especially insulin resistance, are linked to many health issues like type 2 diabetes and metabolic syndrome. When the body's cells do not respond properly to insulin, it leads to insulin resistance, which can cause blood sugar levels to rise. This issue can come from a mix of genetic factors, how you live your life, and problems with your hormones. Hormonal imbalances, especially when insulin is not working properly, can have a big effect on your health. It is important to understand how your hormones, your lifestyle, and your genes all work together to prevent and manage conditions like type 2 diabetes and metabolic syndrome [5].

1.5 Major Types of Metabolic Disorders

1.5.1 Carbohydrate Metabolism Disorders

Carbohydrates have two main roles in the body. One is helping to build and keep the structure of cells, like in DNA and some proteins. The other is giving the body energy, especially in the form of glucose. If the body has trouble handling carbohydrates, it can cause health issues. The symptoms of these problems can be similar, which makes it hard to tell them apart. Glucose is the main energy source for the body. When blood glucose levels get too low, it can lead to symptoms such as weakness, tiredness, and shaking. If the body does not use glucose, it stores it as glycogen in the liver and muscles. This can cause the liver to get bigger or lead to weak muscles and low muscle tone. Doctors use certain markers to diagnose and follow these conditions. These markers depend on which part of the body's sugar-processing system is not working right. This often involves checking the activity of a specific enzyme or looking at levels of chemicals that build up because of the problem [6].

1.5.2 Diabetes Mellitus and Glycogen Storage Diseases

Glycogen storage diseases, also called glycogenoses, are genetic disorders that affect how the body breaks down carbohydrates. These diseases can happen at any age, from birth to adulthood. They usually happen because the body lacks certain enzymes needed to break down glycogen, which causes an unusual buildup of glycogen in the liver or in the muscles used for movement. This makes it hard for the body to release glucose from glycogen, leading to low blood sugar and weakness during physical activity. Over time, this can cause serious health problems [7].

1.5.3 Lipid Metabolism Disorders

Lipids are organic compounds that do not dissolve in water but do dissolve in other types of solvents. They are made from fatty acids and sometimes include alcohol or phosphate groups, and they come in different types like triglycerides, phospholipids, and steroids. These substances act as energy stores in the body and help keep the body warm. They are also important parts of cell membranes and help send messages in the body. The body needs different kinds of lipids to keep its different parts working properly. Keeping the right amount of lipids in the blood is important for staying healthy. When there is too much lipid in the blood, it can build up in the walls of arteries, which can lead to problems inside the blood vessels. Conditions like diabetes, alcoholism, kidney disease, an underactive thyroid, liver disease, and stress can cause high levels of lipids. Extra lipids can stick to the walls of blood vessels, and the buildup of fat can lead to various atherosclerosis issues, such as stroke or heart attack. Lipid metabolism is important for several functions in the body, like storing energy, controlling hormones, sending nerve signals, and moving fat-soluble nutrients. Lipids are a good source of energy because they contain a lot of calories—9 calories per gram, which is more than protein or carbohydrates. The body can store up to 100,000 calories in the form of fat without eating for 30–40 days, as long as there is enough water. These lipids are stored in cells throughout the body, especially in a type of connective tissue called adipose tissue. They also help protect organs like the spleen, liver, heart, and kidneys from getting damaged. Lipids in the blood are processed by the liver and then sent to different parts of the body to keep the right levels. The liver plays a major role in how the body uses and stores lipids. It can store extra fat as a backup. When there is too much energy over a long time, the unused energy is saved as triglycerides in fat cells and liver cells. This process connects with the citric acid cycle, the urea cycle, and other important chemical cycles. Fatty acids are broken down through a process called oxidation, which creates a lot of ATP and uses oxygen. Glycolipids, which are made from a molecule called sn-glycerol-3-phosphate, are mainly found in the liver and fat tissue. Proteins in our diet are important because they give us the essential amino acids that our bodies need. Our bodies can make some amino acids on their own, but others, called essential ones, must come from food. Humans use 21 different amino acids, and one of them, selenocysteine, is special. It does not follow the usual way other amino acids are made, but it plays a big role in several important body functions like dealing with oxidative stress, sending signals inside cells, regulating thyroid hormones, helping proteins fold correctly, and keeping selenium levels balanced in the body [8, 9].

1.6 Risk Factors of Metabolic Disorders

1.6.1 Lifestyle

Lifestyle plays a big part in how risk factors for metabolic syndrome develop, and changing certain habits can help lower the chances of getting this disease. People with metabolic syndrome often weigh too much and do not move around much. A study looked at 8800 adults who were 25 years or older over a period of about 7 years. It found that every extra hour spent watching TV increased the risk of dying from heart-related problems by 18%. Also, not being physically active or not moving enough can make it harder for the body to clear out triglycerides and manage blood sugar levels. Many past reviews have looked at how lifestyle changes can affect metabolic syndrome. These studies focused on things like eating better, cutting down on calories, and reducing fats, carbs, and dairy. They also included exercise, like dancing, aerobic exercises, yoga, and running. A review of 6 studies showed that waist size decreased by about 4.9 cm in 643 people. Another review of 9 studies found that triglyceride levels dropped by 11% in 797 people, and fasting blood sugar levels went down by 5% in 816 people. All of these studies showed that after making lifestyle changes, the number of people with metabolic syndrome went down by 39%.

1.6.2 Diet

Diet plays a big part in how MetS develops. In a study that looked at 773 people, it was found that people who ate more meat-based diets had a higher chance of having MetS compared to those who ate vegetarian meals, with 47.55% of the meat eaters affected. This was because vegetarian diets were linked to lower levels of sugar and fats in the blood. Also, eating a lot of foods high in saturated fats over time can cause fat to build up in the liver, muscles, and pancreas, leading to insulin resistance. However, another study from Taiwan looked at a larger group of people and found that being a vegan did not lower the risk of MetS compared to people who followed other vegetarian or nonvegetarian diets.

1.6.3 Cardiovascular Disease (CVD)

Several studies showed that atherosclerosis increased the chances of developing MetS by 2–2.9 times, and CVD made MetS happen 14.6 times more often after 30 years. Recently, a group of 1516 teenagers was studied to see how CVD connects with MetS. The results found that high systolic blood pressure was strongly linked to a higher risk of MetS, while lower blood pressure and glucose levels were connected to a lower risk. Another study looked at 200 people with knee osteoarthritis to see how osteoarthritis, CVD, and MetS are related. About 84.8% of these patients met all the criteria for MetS. So finding and treating these conditions early could help prevent the growing problem of MetS.

1.6.4 Smoking

Smoking is often seen as a major risk factor for MetS. People who smoke have a higher chance of developing MetS compared to those who do not smoke, about 1.07–1.66 times higher. Smoking can lower insulin sensitivity, raise the risk of heart problems, and increase triglyceride levels. It also causes belly fat in both boys and girls during their teenage years. Girls who smoked ten cigarettes a day as teens ended up with a waist size that was nearly 3.4 cm bigger when they became adults. A study looked at the occurrence of MetS in workers who smoked ($n = 2136$) over a year. The results showed that current smokers were 6.3% more likely to develop MetS than those who did not smoke.

1.6.5 Alcohol

Strong alcohol consumption (>20 times) was correlated with low serum HDL cholesterol, increased serum triglycerides, high waist size, and incidence of hyperinsulinemia and MetS. In contrast, slight, moderate consumption of alcohol (1–19 times a month) reduced insulin levels, serum lipids, and incidence of METS ($n = 8.125$).

1.6.6 Inadequate Sleep

It is also assumed that long periods of inadequate sleep are a risk factor for metabolic syndrome. Not sleeping well enough can cause both mental and physical stress. It also increases the chances of becoming obese or developing diabetes. Similarly, obstructive sleep apnea has been found to reduce good HDL levels of cholesterol levels and increase the mirror of glucose, weight, insulin resistance, and risk of heart disease. All of these factors increase the likelihood that metabolic syndrome has developed to 87% (based on 61 participants) compared to the control group where metabolic syndrome was increased by 35% (based on 3 participants).

1.6.7 Stress

Stressful life events involving money, work, relationships, health, and housing were rated by a random group of people aged 18–78 years in Western Finland (a total of 3407 people). The study found that people who experienced very stressful events related to money, work, or health, or who had at least three stressful events in any of these areas within the past year, had higher levels of triglycerides, larger waist size, and higher body mass index (BMI). These factors increased the chances of developing metabolic syndrome (MetS) by 78.69%, compared to those who did not experience stress (Table 1.1) [10–14].

Table 1.1 Major risk factors associated with metabolic syndrome

Category	Risk factor	Impact	Ref
Genetic/family history	Family history of diabetes, hypertension, or heart disease	Inherited susceptibility increases overall risk	[10]
Age	≥45 years (MetS and T2DM risk rises with age)	Metabolic efficiency and insulin sensitivity decline with age	[10]
Ethnicity	Higher risk in certain groups	African American, Hispanic/Latino, Native American, and South Asian populations	[10]
Obesity/overweight	Particularly abdominal (visceral) obesity	Strongest predictor of insulin resistance and MetS	[11]
Physical inactivity	Sedentary lifestyle reduces insulin sensitivity	Linked to weight gain and poor glucose metabolism	[11]
Unhealthy diet	High intake of refined carbs, sugars, and saturated fats	Promotes obesity, dyslipidemia, and insulin resistance	[11]
Smoking	Contributes to inflammation and insulin resistance	Also increases cardiovascular and cancer risk	[11]
Excess alcohol use	Can raise triglycerides and blood pressure	Also contributes to liver dysfunction (e.g., NAFLD)	[12]
Sleep disorders	Poor sleep and sleep apnea	Alters glucose metabolism and increases appetite-regulating hormone disruption	[12]
Chronic stress	Elevated cortisol levels	Leads to weight gain and worsens glucose/lipid metabolism	[13]
Hormonal imbalances	PCOS, thyroid dysfunction, Cushing's syndrome	Can directly cause or exacerbate metabolic disturbances	[13]
Medications	Corticosteroids, antipsychotics, beta-blockers	May cause weight gain, insulin resistance, or lipid changes	[14]

1.7 Disorders of Nucleic Acid/Nucleotide Metabolism

Nucleic acids, such as DNA and RNA, are built from smaller parts called nucleotides. Nucleotides are the fundamental building blocks of nucleic acids and also have other important functions in the body. There are two main kinds of nitrogenous bases in nucleotides: purines and pyrimidines. Metabolism is the process that includes all the chemical reactions in the body, like creating and breaking down nucleotides [15].

1.7.1 Hyperlipidemia, Gaucher's Disease, and Fatty Liver Disease

Lipid storage diseases, also called lipidoses, are a group of inherited illnesses where harmful fats, called lipids, build up in different parts of the body. This happens when some people do not make enough of the enzymes needed to break down lipids, or they make enzymes that do not work correctly [16, 17]. Over time, having too much fat in the cells can harm tissues and organs, especially the brain, nerves, liver, spleen, heart, and bone marrow. One example is Niemann-Pick disease Type C, which happens when someone cannot break down cholesterol and other fats properly. This causes too much cholesterol to build up in the liver and spleen and too much of other fats in the brain [18, 19].

1.7.2 Gaucher Disease

Gaucher disease occurs when the body lacks a specific enzyme known as glucocerebrosidase. Without enough of this enzyme, fatty materials accumulate in certain areas of the body, such as the brain, spleen, liver, kidneys, lungs, and bone marrow. Individuals with Gaucher disease might be more prone to infections. Also, people who have Gaucher disease or carry the gene for it are more likely to develop Parkinson's disease or other related health issues. Gaucher disease is more common in certain groups, such as the Ashkenazi Jewish population. There are three common types of Gaucher disease. *Type 1*, also called the nonneuronopathic type, usually does not harm the brain. However, people with this type often experience lung issues and may also have kidney problems, along with an enlarged liver and spleen. Type 1 is the most common form of Gaucher disease in the United States and Europe. *Type 2*, known as acute infantile neuropathic Gaucher disease, causes serious brain damage. It typically starts within 3 months after birth. Kids with Type 2 often do not live past the age of 2. *Type 3*, referred to as the chronic neuronopathic form, can start at any time during childhood or even in adulthood. People with Type 3 Gaucher disease often have a swollen liver and spleen and usually face brain-related issues like seizures and memory problems. These symptoms are not as serious as in Type 2, and they usually get worse slowly over time. People with Type 3 may not live as long as others. Type 3 is the most common kind of Gaucher disease globally, but it is less common in the United States compared to Type 1 [20–24].

1.7.3 Amino Acid Metabolism Disorders

1.7.3.1 Phenylketonuria (PKU) and Maple Syrup Urine Disease

Phenylketonuria (PKU) and maple syrup urine disease (MSUD) are genetic conditions that affect how the body processes certain amino acids. PKU happens when the body lacks an enzyme called phenylalanine hydroxylase, which causes phenylalanine to build up in the body. MSUD occurs when there is a shortage of the BCKAD complex, leading to a buildup of branched-chain amino acids like leucine, isoleucine, and valine, along with their related ketoacids. If not properly managed through diet and other treatments, both conditions can cause severe health issues.

Maple syrup urine disease (MSUD) is a rare genetic disorder that makes it hard for the body to break down certain amino acids. It is often tested for in newborns, but doctors have to be very sure about the condition because it can lead to serious, long-lasting damage to the brain. Some of the key reasons people with MSUD might not survive are brain swelling, seizures, loss of awareness, breathing difficulties, and other severe metabolic problems. It is very important to recognize and treat MSUD early to reduce the risk of serious illness and death. This course is made for healthcare workers and explains the causes, signs, ways to diagnose, and methods to manage MSUD. Doctors and other healthcare staff will learn how to spot the condition early, manage diet, handle sudden health problems, and care for patients over time. This helps them give better care and improve patients' health and daily living [25–28].

1.7.4 Nucleic Acid Metabolism Disorders

1.7.4.1 Lesch-Nyhan Syndrome and Gout

Lesch-Nyhan syndrome is a condition that someone is born with, caused by a lack of an enzyme called hypoxanthine-guanine phosphoribosyltransferase or HPRT. This enzyme is part of a process that helps recycle certain chemicals called purines. Normally, the enzyme helps turn guanine and hypoxanthine into other important chemicals, but without it, these substances build up and turn into uric acid. The effects of this enzyme shortage can vary depending on how little of the enzyme is present. When the enzyme activity is less than 15%, it means the condition is quite severe. The main signs of Lesch-Nyhan syndrome include high levels of uric acid in the blood, problems with brain development and growth, unusual movements that happen without control, and harmful behaviors that hurt oneself. This information explains how to diagnose and treat Lesch-Nyhan syndrome and how different healthcare professionals work together to help people with this condition [29–31].

1.7.5 Other Inherited Metabolic Disorders

1.7.5.1 Lysosomal Storage Diseases and Mitochondrial Diseases

Lysosomal storage diseases and mitochondrial diseases are two types of inherited metabolic disorders. Lysosomal storage diseases happen when the body cannot properly break down waste materials inside lysosomes, causing these materials to build up and damage cells. Mitochondrial diseases affect the mitochondria, which are the energy centers of cells, and lead to problems with making energy. Lysosomal storage diseases are a type of inherited condition where certain substances build up in cells of different organs because the lysosomes are not working correctly. This buildup causes problems in those organs and can lead to serious illness and even death. While each of these diseases is rare on its own, together they affect many people [32].

1.7.5.2 Severe Combined Immunodeficiency (SCID)

Severe combined immunodeficiency (SCID) is a very rare and serious genetic condition that causes the body's immune system to not work properly. Because of this, people with SCID are very likely to get serious infections. Babies with SCID often have repeated, severe, and long-lasting infections that are hard to treat. If they do not get treatment, most children with SCID do not live past the age of 2 [33].

1.7.5.3 Adenylosuccinate Lyase (ADSL)

Adenylosuccinate lyase (ADSL) deficiency is a rare genetic condition that impacts the body's ability to produce certain chemicals known as purine nucleotides. It occurs when the body lacks a crucial enzyme called adenylosuccinate lyase, which is necessary for making a substance named AMP. Without enough of this enzyme, specific chemicals build up in the body, leading to a variety of symptoms, mostly affecting the nervous system [34–36].

1.8 Clinical Manifestations

When the body's hormone system is not working properly, common symptoms like feeling tired, changes in weight, and issues with the nervous system can occur. The nervous system and the hormone system are closely connected, and both help the body stay balanced. If the hormone system is not working correctly, it can lead to problems such as headaches, muscle issues, and serious brain problems, which in some cases can result in a coma. It is important to notice these nervous system symptoms when someone has a hormone disorder. This article looks at how different hormone disorders can affect the nervous system in children. Thinking about hormone disorders as a possible cause of nervous system problems is helpful. Finding and treating hormone imbalances early can help improve nervous system symptoms quickly. Understanding how the hormone system and the nervous system work together, along with knowing how nervous system problems happen because of hormone disorders, can help doctors diagnose and treat these conditions faster and better. Common clinical manifestations of neurological disorders include fatigue, changes in weight, and various neurological issues. These symptoms can significantly impact a person's daily life and require prompt medical attention. Fatigue is a very common symptom across various neurological conditions. It can manifest as physical tiredness (e.g., weakness in limbs) or cognitive fatigue (e.g., slowed processing). Neurological disorders can alter appetite, metabolism, and feeding ability, resulting in weight loss or weight gain. Endocrine dysfunctions, which may present with neurological manifestations, can likewise influence body weight. Common neurological symptoms include muscle weakness, impaired coordination, sensory abnormalities, and disturbances in vision, taste, smell, or hearing [37–39].

1.9 Current Management Strategies

1.9.1 Lifestyle and Dietary Interventions

Dietary lifestyle changes are very important in preventing diseases and managing long-term health issues. There is strong evidence showing that bad eating habits are linked to higher rates of obesity and noncommunicable diseases like heart disease, diabetes, high blood pressure, and kidney disease. Even though certain diet plans work well, they are not always used regularly in medical care. This activity focuses on including dietary changes in everyday medical care to prevent diseases and improve long-term health. It also includes important diet tips for conditions like high blood pressure, type 2 diabetes, and obesity. It talks about diets such as the Mediterranean diet, the dietary approaches to stop hypertension (DASH) diet, low-fat and low-carb diets, and intermittent fasting. It also explains how sodium and potassium impact heart and metabolism and how eating too many ultra-processed foods can lead to more diseases. There is also practical advice on how to use these diet strategies in real medical situations. Eating a balanced diet is very important for staying healthy and feeling good. Good eating habits can have a big effect on your health, both physically and mentally. Proper nutrition can improve your mood, help your body organs work better, and reduce the risk of long-term illnesses like cancer, diabetes, and heart disease. Pharmacological treatment, also known as pharmacotherapy or drug therapy, refers to the use of medications to treat, manage, or prevent diseases, alleviate symptoms, or improve a patient's overall health. It involves administering drugs, often prescribed by healthcare professionals, to target specific biological pathways or processes in the body [40, 41].

1.9.2 Emerging Research and Therapeutic Advances

1.9.2.1 Role of Microbiota in Metabolic Health

It is commonly understood that an unhealthy way of living is a big risk for metabolic diseases. In recent years, more and more research has shown that the gut microbiome and the chemicals it makes also play a big role in the start and progress of many metabolic diseases, such as obesity, type 2 diabetes, nonalcoholic fatty liver disease, and heart disease. There are many microorganisms living in the digestive system, which is a major place where the body gets energy. These microbes can process food into useful chemicals, making them important in connecting the gut microbiome with the body. The inwards of the gut microbiome is influenced by genes, the body's immune system, and what someone eats. The ability of the gut microbiome to affect metabolism and the immune system makes it important for health and disease. Because of this, focusing on the gut microbiome and the ways it works could lead to effective treatments for many metabolic diseases in the future. This review will cover how the gut microbiome affects the body's metabolism and how the chemicals made by the gut microbiome are linked to the development of various metabolic diseases. It will also look at new developments in using the gut microbiome to improve metabolic conditions [42–44].

1.9.2.2 Targeted Therapies and Personalized Medicine

New treatments for cancer, such as tyrosine kinase inhibitors and monoclonal antibodies, have made it possible for many people with cancer that has spread to live longer. These treatments are different from older methods like chemotherapy and radiation, which are known to damage cells and cause side effects because they create harmful substances called reactive oxygen species. However, how oxidative stress affects these newer treatments is still being researched. ROS are affected by these treatments, which are used for solid tumors and certain blood cancers that show that there is growing evidence about how these drugs affect

ROS, which is important for understanding drug resistance and finding better ways to treat cancer. In the past 10 years, there has been a lot of progress in creating new treatments for cancer. However, radiotherapy and traditional chemotherapy are still the main ways to treat most types of cancer. Chemotherapy is especially important in adjuvant treatment, which is used after surgery. Right now, only trastuzumab is approved for HER2-positive early breast cancer, and imatinib is approved for Kit-positive gastrointestinal stromal tumors (GISTs). In cases where cancer has spread to other parts of the body, targeted cancer therapies have greatly improved survival rates. This has been seen in cancers like HER2-positive breast cancer, kidney cancer (RCC), GISTs, skin cancer (melanoma), and colon cancer (CRC). In some trials for advanced cancers, survival rates have leveled off even after long periods of follow-up. Even today, the general belief is that it is very rare to cure metastatic cancer, and eventually, cancer cells become resistant to the drugs used. Reactive oxygen species, or ROS, are a group of very active molecules like singlet oxygen, superoxide, hydroxyl radical, and hydrogen peroxide. These molecules help the body function normally but can also play a role in causing cancer. In cancer tissues, the level of ROS is often higher than in healthy tissues. This can happen due to factors like increased metabolism, active cancer genes, and poor blood flow, which can lead to areas with less oxygen. However, some research shows that cancer cells may boost the production of antioxidant enzymes to fight the extra-oxidative stress, as mentioned in other studies. When there is too much ROS, it can be reduced by both enzyme-based antioxidants like superoxide dismutase, catalase, glutathione, and peroxiredoxins and nonenzyme antioxidants like vitamins E and A. Certain enzymes, such as thioredoxin-1 and glutathione, can help proteins that are sensitive to changes in the redox environment by reducing their cysteine residues. The production of these antioxidant proteins is controlled by a protein called NFE2L2. This protein becomes active during oxidative or electrophilic stress and is released from its inhibitor, Keap1. The success of traditional cancer chemotherapy often depends on the ability of the treatment to increase ROS levels, leading to higher oxidative stress that the cancer cells cannot handle, which can result in cell death either through apoptosis or necrosis. Chemotherapy can also have harmful effects because too much ROS can damage healthy tissues. Some chemotherapy drugs, such as anthracyclines, can hurt the heart, while platinum-based drugs may damage the kidneys. Around half of cancer patients get radiation therapy during their treatment. Both how well radiation therapy works and the side effects it causes are connected to the ROS that are made when ionizing radiation interacts with the body. The main way radiation therapy works is through the reaction of OH• with the DNA in the target tissue. Targeted cancer therapy mainly uses monoclonal antibodies, small-molecule tyrosine kinase inhibitors (TKIs), and, more recently, immunotherapies. Targeted therapy is to focus on specific pathways that are overactive in cancer cells. In theory, this should lead to better cancer cell death with fewer side effects. However, the drugs can cause specific and sometimes dangerous side effects, so their use is often limited in real-world medical settings. Initially, targeted therapies were seen as promising treatments with single targets, but as they have been used more widely, we have learned more about how they work and how cancer can become resistant to them, with reactive oxygen species (ROS) possibly playing a big role [33, 34, 36, 45–47].

1.9.3 *Role of Nanomedicine and Novel Drug Delivery*

In the area of biomedicine, drug delivery systems that respond to certain signals have become more common. These systems can release medicine exactly where it is needed in the body, which can improve treatment results and lower unwanted side effects. Reactive oxygen species, or ROS, are a natural by-product of cell metabolism. While ROS help regulate many body processes, too much of them can cause oxidative stress, which is linked to diseases like cancer, inflammation, and atherosclerosis. Because of this, drug delivery systems that respond to ROS have gained a lot of attention. These systems release medicine only in cells or tissues that produce a lot of ROS. The way these systems work depends on how they are designed. The medicine can be released through changes in the carrier's solubility, breaking down of the carrier, or breaking apart of a special link that holds the medicine [47–52].

1.10 Conclusion

A multidisciplinary approach to understanding how reactive oxygen species (ROS) work in cells brings together knowledge from areas like cell biology, biochemistry, molecular biology, and even computer modeling. This helps scientists learn how ROS act as signals that control different cell activities. It is important because ROS can be both harmful waste from metabolism and useful signals that help regulate how cells grow, change, move forward, and even die. ROS are created through many actions inside and outside the cell, and they have gained attention as new types of signal molecules that play a role in cell growth, change, movement, and death. In normal body systems, the production of ROS is balanced with antioxidant systems that protect cells from high levels of harmful ROS. When there is too much ROS and not enough antioxidants, it causes oxidative stress, which is also called "oxidative distress." This condition happens when ROS production outpaces the body's ability to handle them, leading to too much damage to important molecules in the cell. ROS like the superoxide anion and hydrogen peroxide (H2O2) are usually seen as harmful by-products of breathing in oxygen. However, more research shows that they also help maintain a balance of oxidation and reduction (redox homeostasis) in cells and are involved in many important signaling processes [53–57]. Reactive oxygen species (ROS) are types of oxygen-related free radicals that have a big and complicated role in the development and making worse of metabolic diseases. A little to moderate amount of ROS is necessary for normal cell communication, which helps the body deal with infections and control metabolism.

However, when there is too much ROS, it causes oxidative stress, which disrupts the balance inside cells. In conditions like obesity, type 2 diabetes, metabolic syndrome, and nonalcoholic fatty liver disease, high levels of ROS lead to issues such as insulin resistance, problems with mitochondria, damage to fats, and ongoing inflammation. These problems affect key organs like the liver, fat tissue, pancreas, and muscles. Oxidative stress also harms DNA, proteins, and cell membranes, creating a cycle of damage. So ROS are not just by-products of metabolism but are actively involved in causing these diseases, making them important targets for prevention and treatment. While ROS can be harmful, causing oxidative stress and cell damage linked to diseases like diabetes, heart disease, and brain disorders, they also help with normal body functions such as controlling metabolism, inflammation, and immune responses. Future research should focus on understanding the balance between producing and removing ROS, identifying specific targets of ROS in different metabolic diseases, and developing treatments that can control ROS signals for better outcomes [66–70].

1.11 Future Directions in Research and Clinical Practice

Using advanced techniques like redox proteomics and metabolomics can help find specific targets and pathways that are sensitive to reactive oxygen species (ROS) in tissues involved in metabolism. Also, antioxidants that respond in different parts of the cell, such as mitochondria and the endoplasmic reticulum, can help us better understand how ROS contribute to disease development. Instead of relying on non-selective antioxidants such as vitamins C and E, greater emphasis should be placed on targeted redox modulators, including mitochondria-directed antioxidants (e.g., MitoQ and SkQ1), agents that activate Nrf2 to enhance endogenous antioxidant defenses, and the development of organ- or cell-specific antioxidant strategies that preserve physiological ROS signaling while preventing oxidative stress–induced damage. More research is needed on how ROS affect the metabolism of immune cells in conditions like metabolic disorders, for example, how ROS influence the way macrophages behave in inflamed fat tissue. Understanding how oxidative stress and immune responses affect each other could lead to new ways to treat long-term inflammation seen in obesity and diabetes [56, 58–61].

1.12 Clinical Practice Directions

Leveraging redox profiling and genetic testing to design personalized antioxidant therapies for metabolic disorders, with treatment strategies tailored to individual oxidative stress levels and antioxidant utilization capacity. Using ROS modulators along with standard metabolic drugs like metformin and GLP-1 agonists to make treatments more effective. Controlling redox processes can help reduce drug resistance and improve insulin sensitivity and how the body handles fats. Evaluating and promoting dietary patterns, such as the Mediterranean or plant-based diets, that are rich in naturally occurring antioxidants. Planning exercise routines that carefully manage ROS production to support healthy signaling without causing harm, using a method called hormesis-based exercise therapy. Creating easy-to-use tools or tests that can check overall oxidative stress levels in a patient's blood, right at the health-care facility. Checking the balance of oxidation and reduction in the body regularly might help doctors decide the best treatment for patients with diabetes, nonalcoholic fatty liver disease, and metabolic syndrome. There is a need for carefully planned studies that compare different patient groups to check if new antioxidant medicines are safe and effective. Studies that track patients over time are important to understand how treatments that focus on redox processes influence the progression of diseases and the frequency of complications. Recent advances in targeting specific cancer cells have improved treatment options for solid tumors and lymphomas. However, these treatments are often not effective in the long term because cancer cells can become resistant, leading to treatment failure in later stages of the disease. There is a strong need for better methods to predict how well these treatments will work, to avoid harmful side effects and unnecessary costs [62–65].

References

1. Gangopadhyay A, Ibrahim R, Theberge K, May M, Houseknecht KL (2022) Non-alcoholic fatty liver disease (NAFLD) and mental illness: mechanisms linking mood, metabolism and medicines. Front Neurosci 16:1042442
2. Alberti KG, Zimmet P, Shaw J (2006) Metabolic syndrome—a new world-wide definition. A consensus statement from the international diabetes federation. Diabet Med 23(5):469–480
3. Finkelstein EA, Khavjou OA, Thompson H, Trogdon JG, Pan L, Sherry B et al (2012) Obesity and severe obesity forecasts through 2030. Am J Prev Med 42:563–570
4. Bhurosy T, Jeewon R. Overweight and obesity epidemic in developing countries: a problem with diet, physical activity, or socioeconomic status? Sci World J 2014;2014:964236
5. Leitner DR, Fruhbeck G, Yumuk V, Schindler K, Micic D, Woodward E et al (2017) Obesity and type 2 diabetes: two diseases with a need for combined treatment strategies — EASO can lead the way. Obes Facts 10:483–492
6. Rawal LB, Tapp RJ, Williams ED, Chan C, Yasin S, Oldenburg B (2012) Prevention of type 2 diabetes and its complications in developing countries: a review. Int J Behav Med 19:121–133
7. Baron S, Poast J, Richardson CJ, Nguyen D, Cloyd M (2000) Oral transmission of human immunodeficiency virus by infected seminal fluid and milk: a novel mechanism. J Infect Dis 181(2):498–504
8. Ameer F, Scandiuzzi L, Hasnain S, Kalbacher H, Zaidi N (2014) De novo lipogenesis in health and disease. Metabolism 63(7):895–902
9. Berg JM, Tymoczko JL, Stryer L (2002) The first step in amino acid degradation is the removal of nitrogen. In: Biochemistry
10. Bishop SH, Campbell JW (1965) Arginine and urea biosynthesis in the earthworm Lumbricus terrestris. Comp Biochem Physiol 15(1):51–71
11. Brown GW Jr (1962) Comparative biochemistry of urea synthesis IV.[14C] urea synthesis by liver slices of the metamorphosing tadpole. Biochim Biophys Acta 60(1):185–186

12. Chirala SS, Jayakumar A, Gu ZW, Wakil SJ (2001) Human fatty acid synthase: role of interdomain in the formation of catalytically active synthase dimer. Proc Natl Acad Sci 98(6):3104–3108
13. DeBerardinis RJ, Chandel NS (2016) Fundamentals of cancer metabolism. Sci Adv 2(5):e1600200
14. Van Esse HP, Reuber TL, van der Does D (2020) Genetic modification to improve disease resistance in crops. New Phytol 225(1):70–86
15. Frayn KN, Evans R (2019) Human metabolism: a regulatory perspective. Wiley-Blackwell
16. Gartner V, McGuire PJ, Lee PR (2015) Child neurology: medium-chain acyl-coenzyme a dehydrogenase deficiency. Neurology 85(4):e37–e40
17. Gibala MJ, Young ME, Taegtmeyer H (2000) Anaplerosis of the citric acid cycle: role in energy metabolism of heart and skeletal muscle. Acta Physiol Scand 168(4):657–665
18. Beckwith J (2011) The operon as paradigm: normal science and the beginning of biological complexity. J Mol Biol 409(1):7–13
19. Klar AJ, Halvorson HO (1974) Studies on the positive regulatory gene, GAL4, in regulation of galactose catabolic enzymes in Saccharomyces cerevisiae. Mol Gen Genet MGG 135(3):203–212
20. Johnston SA, Hopper JE (1982) Isolation of the yeast regulatory gene GAL4 and analysis of its dosage effects on the galactose/melibiose regulon. Proc Natl Acad Sci 79(22):6971–6975
21. Escalante-Chong R, Savir Y, Carroll SM, Ingraham JB, Wang J, Marx CJ, Springer M (2015) Galactose metabolic genes in yeast respond to a ratio of galactose and glucose. Proc Natl Acad Sci 112(5):1636–1641
22. Venturelli OS, Zuleta I, Murray RM, El-Samad H (2015) Population diversification in a yeast metabolic program promotes anticipation of environmental shifts. PLoS Biol 13(1):e1002042
23. Bennett MR, Pang WL, Ostroff NA, Baumgartner BL, Nayak S, Tsimring LS, Hasty J (2008) Metabolic gene regulation in a dynamically changing environment. Nature 454(7208):1119–1122
24. Liu GY, Sabatini DM (2020) mTOR at the nexus of nutrition, growth, ageing and disease. Nat Rev Mol Cell Biol 21(4):183–203
25. Rafalski VA, Mancini E, Brunet A (2012) Energy metabolism and energy-sensing pathways in mammalian embryonic and adult stem cell fate. J Cell Sci 125(23):5597–5608
26. Paik JH, Ding Z, Narurkar R, Ramkissoon S, Muller F, Kamoun WS, Chae SS, Zheng H, Ying H, Mahoney J, Hiller D (2009) FoxOs cooperatively regulate diverse pathways governing neural stem cell homeostasis. Cell Stem Cell 5(5):540–553
27. Renault VM, Rafalski VA, Morgan AA, Salih DA, Brett JO, Webb AE, Villeda SA, Thekkat PU, Guillerey C, Denko NC, Palmer TD (2009) FoxO3 regulates neural stem cell homeostasis. Cell Stem Cell 5(5):527–539
28. Zhang X, Yalcin S, Lee DF, Yeh TY, Lee SM, Su J, Mungamuri SK, Rimmelé P, Kennedy M, Sellers R, Landthaler M (2011) FOXO1 is an essential regulator of pluripotency in human embryonic stem cells. Nat Cell Biol 13(9):1092–1099
29. Herzig S, Shaw RJ (2018) AMPK: guardian of metabolism and mitochondrial homeostasis. Nat Rev Mol Cell Biol 19(2):121–135
30. Stenton SL, Kremer LS, Kopajtich R, Ludwig C, Prokisch H (2020) The diagnosis of inborn errors of metabolism by an integrative "multi-omics" approach: a perspective encompassing genomics, transcriptomics, and proteomics. J Inherit Metab Dis 43(1):25–35
31. Marques EP, Wyse AT (2019) Creatine as a neuroprotector: an actor that can play many parts. Neurotox Res 36(2):411–423
32. Guerrero RB, Kloke KM, Salazar D (2019) Inborn errors of metabolism and the gastrointestinal tract. Gastroenterol Clin 48(2):183–198
33. Canton M, Le Gall D, Feillet F, Bonnemains C, Roy A (2019) Neuropsychological profile of children with early and continuously treated phenylketonuria: systematic review and future approaches. J Int Neuropsychol Soc 25(6):624–643

34. Ferreira CR, Cassiman D, Blau N (2019) Clinical and biochemical footprints of inherited metabolic diseases. II. Metabolic liver diseases. Mol Genet Metab 127(2):117–121
35. Huang D, Zhou Q, Chao YQ, Zou CC (2019) Clinical features and genetic analysis of childhood sitosterolemia: two case reports and literature review. Medicine (Baltimore) 98(15):e15013
36. Saudubray JM, Garcia-Cazorla A (2018) An overview of inborn errors of metabolism affecting the brain: from neurodevelopment to neurodegenerative disorders. Dialogues Clin Neurosci 20(4):301–325
37. Ferreira CR, Hoffmann GF, Blau N (2019) Clinical and biochemical footprints of inherited metabolic diseases. I Movement disorders. Mol Genet Metab 127(1):28–30
38. Rutten MG, Rots MG, Oosterveer MH (2020) Exploiting epigenetics for the treatment of inborn errors of metabolism. J Inherit Metab Dis 43(1):63–70
39. Pearson TS, Pons R, Ghaoui R, Sue CM (2019) Genetic mimics of cerebral palsy. Mov Disord 34(5):625–636
40. Williams C, van der Meij BS, Nisbet J, Mcgill J, Wilkinson SA (2019) Nutrition process improvements for adult inpatients with inborn errors of metabolism using the i-PARIHS framework. Nutrit Dietet 76(2):141–149
41. Hicks J, Wartchow E, Mierau G (2011) Glycogen storage diseases: a brief review and update on clinical features, genetic abnormalities, pathologic features, and treatment. Ultrastruct Pathol 35(5):183–196
42. Özen H (2007) Glycogen storage diseases: new perspectives. World J Gastroenterol: WJG 13(18):2541
43. Kanungo S, Wells K, Tribett T, El-Gharbawy A (2018) Glycogen metabolism and glycogen storage disorders. Ann Transl Med 6(24):474
44. Sulashvili N, Beglaryan M, Gorgaslidze N, Gabunia L, Sulashvili M, Abuladze N, Giorgobiani M (2024) The scientific discussion of genetic and biochemical mechanism of glycogen storage disorders, some clinical aspects and pharmacotherapy management challenges in general. ББК 52.8 А 43. 29
45. Dimou A, Tsimihodimos V, Bairaktari E (2022) The critical role of the branched chain amino acids (BCAAs) catabolism-regulating enzymes, branched-chain aminotransferase (BCAT) and branched-chain α-keto acid dehydrogenase (BCKD), in human pathophysiology. Int J Mol Sci 23(7):4022
46. Huang D, Zhou Q, Chao YQ, Zou CC (2019) Clinical features and genetic analysis of childhood sitosterolemia: two case reports and literature review. Medicine 98(15):e15013
47. Fu R, Sutcliffe D, Zhao H, Huang X, Schretlen DJ, Benkovic S, Jinnah HA (2015) Clinical severity in Lesch–Nyhan disease: the role of residual enzyme and compensatory pathways. Mol Genet Metab 114(1):55–61
48. Gutin PH, Prados MD, Phillips TL, Wara WM, Larson DA, Leibel SA, Sneed PK, Levin VA, Weaver KA, Silver P, Lamborn K (1991) External irradiation followed by an interstitial high activity iodine-125 implant “boost” in the initial treatment of malignant gliomas: NCOG study 6G-82-2 gliomas: NCOG study 6H-82-2. Int J Radiat Oncol Biol Phys 21(3):601–606
49. Fu R, Chen CJ, Jinnah HA (2014) Genotypic and phenotypic spectrum in attenuated variants of Lesch–Nyhan disease. Mol Genet Metab 112(4):280–285
50. Mannstadt M, Clarke BL, Vokes T, Brandi ML, Ranganath L, Fraser WD, Lakatos P, Bajnok L, Garceau R, Mosekilde L, Lagast H (2013) Efficacy and safety of recombinant human parathyroid hormone (1–84) in hypoparathyroidism (REPLACE): a double-blind, placebo-controlled, randomised, phase 3 study. Lancet Diab Endocrinol 1(4):275–283
51. Fu R, Ceballos-Picot I, Torres RJ, Larovere LE, Yamada Y, Nguyen KV, Hegde M, Visser JE, Schretlen DJ, Nyhan WL, Puig JG (2014) Genotype–phenotype correlations in neurogenetics: Lesch-Nyhan disease as a model disorder. Brain 137(5):1282–1303
52. Andermann A, Blancquaert I, Beauchamp S, Déry V (2008) Revisiting Wilson and Jungner in the genomic age: a review of screening criteria over the past 40 years. Bull World Health Organ 86(4):317–319

53. Doucet BP, Jegatheesan D, Burke J (2013) Late diagnosis of Lesch-Nyhan disease variant. Case Rep 2013:bcr2013201997
54. Cameron JL, Pitt HA, Yeo CJ, Lillemoe KD, Kaufman HS, Coleman J (1993) One hundred and forty-five consecutive pancreaticoduodenectomies without mortality. Ann Surg 217(5):430
55. Torres RJ, Puig JG (2007) Hypoxanthine-guanine phosophoribosyltransferase (HPRT) deficiency: Lesch-Nyhan syndrome. Orphanet J Rare Dis 2(1):48
56. Bianco JA, Appelbaum FR, Nemunaitis J, Almgren J, Andrews F, Kettner P, Shields A, Singer JW (1991) Phase I-II trial of pentoxifylline for the prevention of transplant-related toxicities following bone marrow transplantation [published erratum appears in Blood 1992 Jun 15; 79 (12): 3397][see comments]. Blood 78:1205
57. Anderson LT, Ernst M (1994) Self-injury in Lesch-Nyhan disease. J Autism Dev Disord 24(1):67–81
58. Schroeder SR, Oster-Granite ML, Berkson G, Bodfish JW, Breese GR, Cataldo MF, Cook EH, Crnic LS, DeLeon I, Fisher W, Harris JC (2001) Self-injurious behavior: gene–brain–behavior relationships. Ment Retard Dev Disabil Res Rev 7(1):3–12
59. Podebrad F, Heil M, Reichert S, Mosandl A, Sewell AC, Böhles H (1999) 4, 5-Dimethyl-3-hydroxy-2 [5H]-furanone (sotolone)—the odour of maple syrup urine disease. J Inherit Metab Dis 22(2):107–114
60. Steinmetz KA, Potter JD (1996) Vegetables, fruit, and cancer prevention: a review. J Am Diet Assoc 96(10):1027–1039
61. Scaini G, Morais MO, Galant LS, Vuolo F, Dall'Igna DM, Pasquali MA, Ramos VM, Gelain DP, Moreira JC, Schuck PF, Ferreira GC (2014) Coadministration of branched-chain amino acids and lipopolysaccharide causes matrix metalloproteinase activation and blood–brain barrier breakdown. Mol Neurobiol 50(2):358–367
62. Casas AI, Nogales C, Mucke HA, Petraina A, Cuadrado A, Rojo AI, Ghezzi P, Jaquet V, Augsburger F, Dufrasne F, Soubhye J (2020) On the clinical pharmacology of reactive oxygen species. Pharmacol Rev 72(4):801–828
63. Zhang H, Gomez AM, Wang X, Yan Y, Zheng M, Cheng H (2013) ROS regulation of microdomain Ca2+ signalling at the dyads. Cardiovasc Res 98(2):248–258
64. Sena LA, Chandel NS (2012) Physiological roles of mitochondrial reactive oxygen species. Mol Cell 48(2):158–167
65. Giorgio M, Trinei M, Migliaccio E, Pelicci PG (2007) Hydrogen peroxide: a metabolic by-product or a common mediator of ageing signals? Nat Rev Mol Cell Biol 8(9):722–728
66. Liochev SI (2013) Reactive oxygen species and the free radical theory of aging. Free Radic Biol Med 60:1–4
67. Rhee SG (2006) H2O2, a necessary evil for cell signaling. Science. 312(5782):1882–1883. Bae YS, Oh H, Rhee SG, Do Yoo Y. Regulation of reactive oxygen species generation in cell signaling. Mol Cells. 2011 Dec 1;32(6):491–509
68. Matsuzawa A, Ichijo H (2005) Stress-responsive protein kinases in redox-regulated apoptosis signaling. Antioxid Redox Signal 7(3–4):472–481
69. Dröge W (2002) Free radicals in the physiological control of cell function. Physiol Rev 82:47
70. Finkel T (2011) Signal transduction by reactive oxygen species. J Cell Biol 194(1):7

Chapter 2
Introduction to Reactive Oxygen Species and Inflammation

Sui Sien Leong, Teck Chwen Loh, and Hooi Ling Foo

2.1 Introduction

Reactive oxygen species (ROS) and inflammation are fundamental components of cellular defence and homeostasis. ROS are unstable and highly reactive molecules naturally produced from cellular metabolism that include free radicals, such as superoxide anion (O_2^-), hydroxyl radical (•OH) and non-radical molecules like hydrogen peroxide (H_2O_2) [1]. These chemically reactive molecules containing oxygen are generated as by-products of normal cellular metabolism. Under balanced conditions, these molecules participate in various physiological processes,

S. S. Leong (✉)
Department of Animal Sciences and Fishery, Faculty of Agricultural and Forestry Sciences, Universiti Putra Malaysia Sarawak, Bintulu, Sarawak, Malaysia

Lactic Acid Bacteria Biota Technology Research Program, Research Laboratory of Probiotics and Cancer Therapeutics, UPM-MAKNA Cancer Research Laboratory (CANRES), Institute of Bioscience, Universiti Putra Malaysia, UPM Serdang, Selangor, Malaysia
e-mail: leongsuisien@upm.edu.my

T. C. Loh
Lactic Acid Bacteria Biota Technology Research Program, Research Laboratory of Probiotics and Cancer Therapeutics, UPM-MAKNA Cancer Research Laboratory (CANRES), Institute of Bioscience, Universiti Putra Malaysia, UPM Serdang, Selangor, Malaysia

Department of Animal Science, Faculty of Agriculture, Universiti Putra Malaysia, UPM Serdang, Selangor, Malaysia

H. L. Foo (✉)
Lactic Acid Bacteria Biota Technology Research Program, Research Laboratory of Probiotics and Cancer Therapeutics, UPM-MAKNA Cancer Research Laboratory (CANRES), Institute of Bioscience, Universiti Putra Malaysia, UPM Serdang, Selangor, Malaysia

Department of Bioprocess Technology, Faculty of Biotechnology and Biomolecular Sciences, Universiti Putra Malaysia, UPM Serdang, Selangor, Malaysia
e-mail: hlfoo@upm.edu.my

S. Ashique, N. Mishra (eds.), *Reactive Oxygen Species in Metabolic Inflammation*,
https://doi.org/10.1007/978-3-032-11752-6_2

such as immune defence, cell signalling and tissue repair. These species are mainly produced during mitochondrial respiration and by specialised enzymes, particularly in immune cells during host defence [2]. All living organisms have a series of antioxidant defence mechanisms mediated via various enzymes, proteins and vitamins. In tightly regulated amounts, ROS serve critical signalling functions, which modulate gene expression, regulating cellular growth and aiding in microbial clearance during infection [3]. However, when ROS production overwhelms the body's antioxidant defences, oxidative stress occurs, disrupting cellular integrity and triggering damaging cascades.

Inflammation is a complex biological response triggered by injury, infection or other harmful stimuli like pathogens, toxins or tissue damage [4]. It serves as a protective mechanism to eliminate the source of injury and initiate healing. It involves the activation of immune cells, release of cytokines and chemokines and changes in blood flow to the affected tissue [4, 5]. While acute inflammation is typically self-resolving and beneficial for repair, chronic or dysregulated inflammation can contribute to a wide array of diseases, including metabolic disorders, cardiovascular conditions and neurodegenerative diseases [5]. Notably, ROS induce oxidative stress that is deeply involved in both initiating and sustaining inflammatory responses that aim to protect and the healing process. They influence the activation of transcription factors and promote the expression of inflammatory mediators.

The interaction between ROS and inflammation is bidirectional and context dependent. On the one hand, inflammation stimulates ROS production as part of the host's defence mechanism. On the other hand, excessive and prolonged or unresolved oxidative stress can activate and amplify inflammatory signalling, creating a self-perpetuating cycle of tissue damage and immune activation. Likewise, inflammation itself can further enhance ROS production, creating a feedback loop that can escalate tissue damage. This delicate balance between the beneficial and harmful effects of ROS and inflammation highlights their dual roles in health and disease. This vicious cycle is especially evident in metabolic inflammation, where prolonged oxidative and inflammatory stress contributes to insulin resistance, endothelial dysfunction and organ impairment [6]. Understanding how ROS and inflammation are generated, how they interact and under what conditions they shift from protective to pathological roles is crucial for identifying potential targets in disease prevention and therapy.

This chapter introduces the basic concepts of ROS and inflammation, including how they form and interact. It provides a foundation of understanding for their dual roles in health and disease, setting the platform for deeper exploration in later chapters on their involvement in metabolic inflammation.

2.2 Reactive Oxygen Species (ROS)

Reactive oxygen species (ROS) are a group of highly reactive molecules derived from molecular oxygen (O_2), primarily produced during normal aerobic metabolism within cells. ROS molecules are commonly referred to as oxygen radicals due to

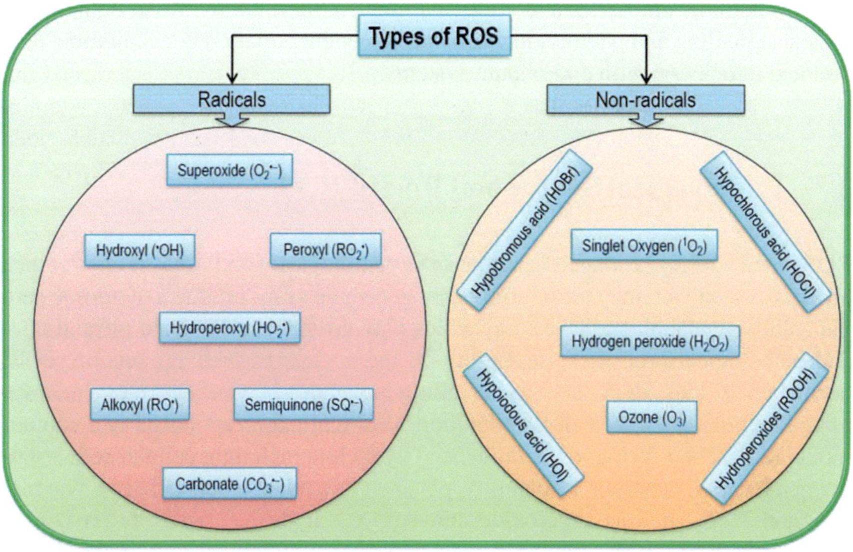

Fig. 2.1 Types of reactive oxygen species [14]

their strong oxidative potential and, in many cases, the presence of unpaired electrons that make them chemically unstable [7]. ROS are broadly classified into two main categories: free radicals and non-radical oxidants, as summarised in Fig. 2.1. Free radical ROS are species that contain one or more unpaired electrons, which makes them particularly reactive with cellular components such as nucleic acids, proteins and lipids [8], including the superoxide anion (O_2^-), hydroxyl radical (•OH), nitric oxide (•NO) and lipid-derived radicals [9]. On the other hand, non-radical ROS, although lacking unpaired electrons, are still capable of oxidising biological molecules. These include hydrogen peroxide (H_2O_2), peroxynitrite ($ONOO^-$) and hypochlorous acid (HOCl) [10].

Some non-radical ROS species are more commonly found in plants, such as hydroperoxides (ROOH), hypochlorous acid (HOCl) and excited carbonyls (RO*) [11]. Additionally, reactive oxygen intermediates (ROI), which are formed during the incomplete reduction of molecular oxygen, are also included within the broader classification of ROS. The ROI consists of singlet oxygen ($^1O_2$) and ozone (O_3), which are highly reactive due to their distinct electronic structures [7, 12]. While atmospheric oxygen typically exists in a less reactive ground state known as triplet oxygen ($^3O_2$), this form can be converted into more reactive species when it absorbs energy from processes, such as electron leakage from the mitochondrial electron transport chain (ETC), ultraviolet-B exposure or ionising radiation [13]. Furthermore, additional oxidants, such as hypobromous acid (HOBr) and hypoiodous acid (HOI), and radicals like carbonate radical ($CO_3^{\bullet-}$) and semiquinone ($SQ^{\bullet-}$) are sometimes classified under the broader umbrella of ROS, due to their similar oxidative behaviour and involvement in redox biology [13]. These diverse forms of ROS, each with unique chemical properties and biological implications, contribute to both physiological signalling and pathological damage, depending on their concentration, location and interaction with antioxidant systems.

2.3 Physiological Source and Role of ROS

ROS are not solely produced in response to pathological conditions or environmental stressors. In fact, they are continuously generated as by-products of normal aerobic cellular metabolism. These molecules play critical roles in both physiological and pathological processes, depending on the context of their production. Under normal conditions, ROS serve as signalling molecules in a variety of essential cellular functions such as cell proliferation, differentiation, apoptosis and immune modulation. Their formation is influenced by factors including cellular redox state, oxygen tension, metal ion availability, pH and antioxidant levels [15, 16]. The most common ROS, such as superoxide anion ($O_2{\bullet}^-$), hydrogen peroxide (H_2O_2) and hydroxyl radical (HO•), are generated primarily in mitochondria during oxidative phosphorylation, when electrons leak from the electron transport chain (ETC) and reduce molecular oxygen (O_2) prematurely. Other enzymatic sources, including NADPH oxidases (NOX), xanthine oxidase and cytochrome P450 enzymes, also contribute to intracellular ROS levels [17].

Superoxide ($O_2{\bullet}^-$) is one of the primary reactive oxygen species (ROS) generated in biological systems. It is formed through the monovalent (one-electron) reduction of molecular oxygen (O_2), a process that typically occurs during mitochondrial respiration or enzymatic reactions [18]. This process adds a single electron to O_2, creating a highly reactive anion radical, as shown in the equation below:

$$O_2 + e- \rightarrow O_2 \bullet -$$

This reaction typically occurs in the mitochondrial electron transport chain (ETC), particularly at complexes I and III, where electrons may leak and reduce molecular oxygen prematurely [19]. Superoxide can also be generated by NADPH oxidase (NOX) enzymes, especially in immune cells during oxidative bursts [15], as well as by xanthine oxidase and cytochrome P450 enzymes under stress conditions [20]. Normally, cells counterbalance $O_2{\bullet}^-$ levels using superoxide dismutase (SOD), which converts it into hydrogen peroxide (H_2O_2), a less reactive but still potent signalling molecule [21] as shown in the equation below:

$$2O_2 \bullet - + 2H+ \rightarrow H_2O_2 + O_2$$

However, if not adequately controlled, $O_2{\bullet}^-$ can interact with other molecules, like H_2O_2, to form highly reactive hydroxyl radicals (HO•) via Fenton or Haber-Weiss reactions, or with nitric oxide (•NO) to form peroxynitrite ($ONOO^-$), both of which can severely damage proteins, lipids and nucleic acids [22]. The balance between

ROS generation and detoxification defines whether ROS act as signalling agents or inducers of oxidative stress. H_2O_2 serves as a critical second messenger involved in modulating kinase signalling pathways and transcription factors under physiological conditions [7]. However, its intracellular and extracellular concentrations must be tightly regulated, as elevated levels can lead to oxidative damage and cell dysfunction [23].

The hydroxyl radical (HO•) is one of the most reactive species within the reactive oxygen species (ROS) family, capable of inflicting immediate and severe damage to cellular components, due to its high reactivity and short half-life. Unlike other ROS, HO• is typically not formed through enzymatic pathways but rather via non-enzymatic chemical reactions involving hydrogen peroxide (H_2O_2) and transition metals, such as iron and copper [24]. HO• is primarily generated through the Fenton reaction, in which ferrous iron (Fe^{2+}) reacts with hydrogen peroxide, as shown in the equation below:

$$Fe^{2+} + H_2O_2 \rightarrow Fe^{3+} + OH^- + HO\bullet$$

This reaction breaks down H_2O_2 into a hydroxide ion and the highly reactive hydroxyl radical [7]. The process is often sustained by the Haber-Weiss reaction, where superoxide ($O_2\bullet^-$) reduces Fe^{3+} back to Fe^{2+}, as shown in the equation below:

$$O_2\bullet - + Fe^{3+} \rightarrow O_2 + Fe^{2+}$$

HO• has a high affinity for electron-rich biomolecules, such as sulphur-containing amino acids and DNA bases, and thus contributes significantly to oxidative damage [25]. Additionally, in the presence of chloride ions and the enzyme myeloperoxidase (MPO), H_2O_2 can be converted to hypochlorous acid (HOCl), a potent oxidant used by neutrophils to kill pathogens during immune responses [26]. Other oxygen-derived radicals, such as hydroperoxyl (HOO•) and peroxyl (ROO•) radicals, are primarily involved in lipid peroxidation. These radicals initiate chain reactions that oxidise polyunsaturated fatty acids into unstable lipid hydroperoxides, which decompose into secondary products that act as biomarkers for oxidative stress and are implicated in protein and DNA damage [27, 28].

Nitric oxide (•NO), although not a ROS, interacts extensively with ROS and plays diverse physiological roles in immune regulation, vascular homeostasis and neurotransmission [29]. At optimal levels, •NO exhibits anti-inflammatory and antithrombotic properties [30]. However, excessive production or interaction with ROS, like $O2\bullet,^-$ leads to the formation of $ONOO^-$. This potent oxidant contributes to endothelial dysfunction and tissue injury, particularly in ageing and cardiovascular diseases [31].

ROS production can be significantly amplified under certain stress conditions. Environmental factors such as ultraviolet radiation, air pollutants and exposure to heavy metals, as well as internal challenges like inflammation, infection and ischemia, are known to elevate ROS levels dramatically [32]. During immune responses, phagocytes such as neutrophils and macrophages undergo an oxidative burst, intentionally producing high levels of ROS to destroy pathogens and infected cells. This can be enhanced by applying probiotics, which influence ROS levels, both by directly lowering ROS levels and by modulating the host's antioxidant defence systems [33, 34]. While beneficial in host defence, prolonged or excessive ROS production can overwhelm antioxidant systems and lead to oxidative stress, a key contributor to chronic diseases, including neurodegeneration, cancer and metabolic disorders. Thus, ROS are generated both under normal metabolic conditions and during pathological stress. Their impact on cells is dictated by the balance between their formation and the antioxidant capacity of the cell. At controlled levels, they function as integral components of intracellular signalling and defence, but when dysregulated, they act as agents of cellular damage. Thus, ROS possess a dual identity: essential for homeostasis yet potentially harmful when in excess, reflecting their complex role in health and disease [17, 35].

2.4 Inflammation

Inflammation is the body's innate and immediate response to harmful stimuli such as pathogens, physical injury, toxic chemicals and radiation. It serves as a critical defence mechanism that aims to eliminate the initial cause of cell injury, clear out damaged cells and initiate tissue repair [36]. This immune response begins as the body detects an irritant and activates a series of cellular and molecular events to contain and remove the threat [37]. Inflammation is essential for maintaining tissue homeostasis and survival, making it vital to the health of living organisms. Clinically, inflammation is characterised by five cardinal signs: heat, redness, swelling, pain and loss of function. These symptoms reflect a complex cascade of local immune, vascular and cellular responses [38]. Increased blood flow to the affected area leads to redness and warmth, whereas vascular permeability causes fluid to accumulate, resulting in swelling. Pain is triggered by the release of pro-inflammatory mediators that stimulate nerve endings, and the loss of function arises from a combination of tissue damage, swelling and pain-induced immobility [39].

Inflammation can be classified into three types based on the duration and nature of the immune response: acute, subacute and chronic inflammation. Acute inflammation is the body's immediate, short-term reaction to injury or infection and is typically marked by redness, swelling, heat, pain and occasionally loss of function [5]. It is initiated rapidly following tissue damage, microbial invasion or exposure to harmful chemicals [38]. In this phase, soluble mediators such as cytokines, chemokines and acute-phase proteins are released, which facilitate the recruitment of neutrophils and macrophages to the affected area, playing a central role in the innate immune defence [40]. This response is crucial in protecting against pathogens, including *Mycobacterium tuberculosis*, protozoa, fungi and other parasites [33].

Acute inflammation typically resolves within a few days to 2 weeks. However, if the immune response persists, it may progress to subacute inflammation, which represents a transitional phase lasting between 2 and 6 weeks. Should the condition remain unresolved beyond 6 weeks, it evolves into chronic inflammation, characterised by the infiltration of T lymphocytes and plasma cells at the site of inflammation [41]. If left uncontrolled, even acute inflammation can become harmful, potentially leading to systemic complications such as septic shock [42].

Chronic inflammation is a prolonged and often maladaptive immune response that can last for months or even years, contributing to a wide range of diseases [41]. The severity and impact of chronic inflammation depend on the nature of the initial insult and the body's capacity for healing. Persistent inflammation can lead to progressive tissue damage and organ dysfunction. According to the World Health Organisation (WHO), chronic inflammatory diseases represent one of the most significant threats to global health [41]. Alarmingly, three out of five people worldwide die from conditions linked to chronic inflammation, including stroke, heart disease, cancer, chronic respiratory conditions, diabetes and obesity [43, 44]. Common disorders associated with chronic inflammation include cardiovascular disease, diabetes, arthritis, allergies and pulmonary diseases [45]. Chronic inflammation can be further divided into two distinct types. The non-specific proliferative type, which is characterised by the formation of granulation tissue composed of mononuclear cell infiltration (lymphocytes, macrophages and plasma cells), along with the proliferation of fibroblasts, blood vessels and epithelial cells. Examples include inflammatory polyps, such as those found in the nasal cavity or cervix and lung abscesses. The second type, granulomatous inflammation, features well-defined nodular lesions called granulomas. These structures consist mainly of activated macrophages transformed into epithelioid cells, often surrounded by lymphocytes. Within granulomas, macrophages may fuse to form multinucleated giant cells, such as Langhans, Aschoff, Reed–Sternberg and tumour giant cells. Granulomatous inflammation is further subclassified into two categories: foreign body granulomas, formed in response to non-immunogenic irritants like silica (silicosis), and infectious granulomas, resulting from chronic infections such as tuberculosis and leprosy [41].

The inflammatory pathway is a complex cascade involving four primary components: inducers, sensors, mediators and effectors [46]. The process begins with inducers, which may be infectious agents (such as bacteria or viruses) or non-infectious triggers like physical injury, toxins or signals released by damaged or necrotic cells. Sensors recognise these inducers, typically pattern recognition receptors (PRRs) expressed by immune and non-immune cells, which then activate downstream responses [47]. Once activated, mediators, including cytokines, chemokines, prostaglandins and other signalling molecules, are released to orchestrate and regulate the inflammatory response. These mediators help recruit immune cells to the site of injury, induce pain and promote tissue healing. They also trigger the activation of effectors, which are the cellular and tissue elements that execute the physiological responses necessary to eliminate the threat and initiate repair [42]. The coordinated interaction among these components ensures that the inflammatory response is tailored to the type and severity of the stimulus. Ultimately, the key objective of inflammation is to eliminate the harmful agent, repair damaged tissues and re-establish homeostasis [46].

2.5 When ROS Become Harmful

ROS, while essential in moderate amounts for physiological signalling and host defence, can become harmful when their production exceeds the body's antioxidant capacity. Under normal conditions, ROS play a protective role by eliminating pathogens and regulating cellular processes, such as apoptosis and gene expression. Redox is a crucial biochemical reaction in living organisms that involves oxidation and reduction. Redox homeostasis plays a key role in maintaining cellular health, whereas a redox imbalance can lead to ROS generation exceeding the ROS-scavenging ability, defined as oxidative stress [48]. However, when ROS generation becomes excessive or persistent, it disrupts redox homeostasis, resulting in oxidative stress, a condition linked to the onset and progression of many chronic diseases [7].

Oxidative stress arises when there is an imbalance between the generation of ROS and the body's antioxidant defence systems. The physiological levels of ROS play essential roles in cell signalling, including the regulation of proliferation, differentiation and protection against malignant transformation. Excessive ROS accumulation can overwhelm antioxidant defences and trigger detrimental effects [45]. Elevated ROS levels can cause direct damage to critical biomolecules, such as DNA, proteins and lipids, resulting in genetic mutations, enzyme inactivation and cellular membrane disruption [49]. These molecular insults have been linked to a broad spectrum of pathological conditions, including sepsis, ageing, obesity, cancer, diabetes, depression and neurodegenerative diseases [50]. The cellular damage induced by high ROS levels not only disrupts normal cellular processes but also activates cell death pathways, such as apoptosis or necrosis, depending on the severity and context of the oxidative insult [35]. When oxidative stress becomes chronic, it contributes to a sustained inflammatory state and accelerates the progression of several degenerative and chronic diseases. For instance, persistent oxidative stress has been identified as a significant contributing factor in the onset and advancement of neurodegenerative disorders like Alzheimer's and Parkinson's disease, where neuronal cells are particularly vulnerable to oxidative injury [51]. Moreover, oxidative stress plays a significant role in carcinogenesis by promoting DNA mutations, genomic instability and dysregulation of tumour suppressor pathways.

In metabolic disorders such as obesity and type 2 diabetes, excessive ROS interfere with insulin signalling and exacerbate inflammatory responses. Similarly, in cardiovascular diseases, oxidative stress contributes to endothelial dysfunction, atherosclerosis and myocardial injury [52]. These findings underscore the importance of maintaining redox homeostasis to preserve cellular integrity and prevent disease progression. ROS and inflammation are closely interconnected, where ROS serve both as initiators and modulators of the inflammatory response. In the early stages of inflammation, ROS are produced primarily by activated immune cells, such as neutrophils and macrophages, in response to microbial invasion or tissue damage. This generation of ROS is a key component of the innate immune defence, as these molecules aid in destroying invading pathogens and clearing cellular debris [53]. In this context, ROS act as beneficial agents that help restore tissue homeostasis by eliminating harmful stimuli [34]. Beyond their microbicidal function, ROS also act as important signalling molecules within inflammatory pathways. Specifically, ROS modulate intracellular signalling cascades through redox-sensitive targets, influencing the activation of transcription factors, like nuclear factor-kappa B (NF-κB) and activator protein-1 (AP-1), which regulate the expression of pro-inflammatory genes [54]. This signalling function is essential for the controlled activation of inflammatory responses. However, when inflammation becomes chronic or unresolved, ROS production persists and leads to pathological consequences.

Persistent ROS generation acts not only as a trigger but also as a mediator that sustains inflammatory processes and tissue injury [55]. Under prolonged oxidative stress, ROS begin to damage essential cellular components, such as lipids, proteins and nucleic acids. This oxidative damage contributes to cell dysfunction and death, creating a vicious cycle in which ROS exacerbate inflammation and inflammation further amplify ROS production. This self-perpetuating loop intensifies tissue injury, particularly through the continuous activation of redox-sensitive transcription factors, which drive the expression of inflammatory cytokines [15, 46]. The failure to regulate this balance between ROS generation and antioxidant defence mechanisms is a hallmark of chronic inflammatory diseases. At this stage, ROS shift from being protective agents to damaging mediators that perpetuate pathological inflammation. Ultimately, this dysregulated ROS-inflammation axis plays a critical role in the development of several chronic diseases. Persistent oxidative stress and inflammation are linked to the pathogenesis of cardiovascular diseases, neurodegenerative disorders like Alzheimer's disease and various types of cancer [56]. As such, the transition of ROS from beneficial signalling molecules to harmful agents underscores the importance of tightly regulating oxidative stress in inflammatory conditions.

Interventions that target excessive ROS production or restore redox balance may therefore offer promising therapeutic strategies in preventing or managing inflammation-related diseases. The transition of ROS from protective to pathological is highly context dependent. Factors such as the duration of ROS exposure, cellular antioxidant status and the site of production determine their biological outcome. The biological effects of ROS are not fixed but instead highly dependent on the cellular context and microenvironment. Several factors influence this transition. First, the duration of ROS exposure is critical; short bursts of ROS generation are typically well tolerated or even beneficial, whereas prolonged or chronic ROS production leads to sustained oxidative stress and cellular injury. Second, the antioxidant capacity of the cell, comprising enzymatic systems (superoxide dismutase, catalase, glutathione peroxidase) and non-enzymatic antioxidants (vitamins C and E, glutathione), plays a vital role in neutralising excess ROS [24]. When antioxidant defences are overwhelmed or impaired, ROS can accumulate and initiate harmful oxidative reactions. Third, the subcellular site of ROS generation greatly influences their impact [57].

ROS produced in the mitochondria can disrupt mitochondrial membrane potential and bioenergetics, leading to apoptosis, whereas ROS generated near the nucleus may result in DNA damage and genomic instability [57]. Conversely, localised ROS at the plasma membrane or in phagosomes often participate in signalling or microbial killing without harming the host cell. Collectively, these determinants indicate that ROS function on a spectrum, with beneficial effects at physiological levels, but induce harmful outcomes when produced excessively, persistently or in vulnerable cellular compartments. Understanding this context dependency is crucial for designing targeted antioxidant therapies that mitigate oxidative damage without disrupting essential redox signalling pathways. While acute, localised ROS production is vital for pathogen clearance and wound healing, chronic systemic ROS exposure leads to cellular dysfunction and disease progression [58]. Therefore, maintaining redox balance is crucial; therapeutic strategies aimed at modulating ROS levels either by enhancing antioxidant defence or controlling ROS production are increasingly being explored for managing inflammatory and oxidative stress-related diseases [59, 60]. Types and sources of ROS and their roles in health and disease are summarised in Table 2.1.

Table 2.1 Types and sources of reactive oxygen species and their roles in health and disease

ROS type	Primary cellular sources	Physiological roles (health)	Pathological roles (disease)
Superoxide anion ($O_2^{\bullet -}$)	Mitochondrial electron transport chain (complex I and III), NADPH oxidase	Precursor for other ROS; signalling molecule for immune response and cell differentiation	Initiates lipid peroxidation; promotes inflammation and cellular damage in neurodegeneration and cardiovascular disease [35]
Hydrogen peroxide (H_2O_2)	Mitochondria, peroxisomes, superoxide dismutase (SOD) activity	Regulates gene expression, cell proliferation and apoptosis at low levels	DNA and protein oxidation promote tumour progression, insulin resistance and endothelial dysfunction [47, 50]
Hydroxyl radical ($\bullet OH$)	Fenton reaction (Fe^{2+} + H_2O_2), UV radiation, ionising radiation	No known beneficial physiological role due to extreme reactivity	Causes severe DNA strand breaks, lipid peroxidation and protein modification, leading to mutagenesis and cell death [50]
Peroxynitrite ($ONOO^-$)	Reaction of nitric oxide (NO) with superoxide anion ($O_2^{\bullet -}$)	May act in an immune response at low concentrations	Nitration of proteins; mitochondrial dysfunction; implicated in neurodegeneration and inflammation [52]
Singlet oxygen ($^1O_2$)	Photosensitisation, phagocytosis, mitochondrial respiration	Involved in immune defence and photodynamic therapy	Oxidises lipids and proteins; contributes to chronic inflammation and photo ageing [50]

2.6 Conclusion

Reactive oxygen species (ROS) play a complex and dualistic role in biological systems. At physiological levels, ROS function as essential signalling molecules involved in cell proliferation, differentiation and immune responses. However, when their production exceeds the buffering capacity of the antioxidant defence system, oxidative stress arises—leading to damage of critical biomolecules, such as lipids, proteins and nucleic acids. This imbalance is closely linked to the onset and progression of numerous pathological conditions, including neurodegenerative diseases, cardiovascular disorders, metabolic syndromes and cancer. Furthermore, the interplay between ROS and inflammation amplifies cellular damage. While ROS are initially produced as part of the immune response to eliminate pathogens, sustained ROS generation during chronic inflammation can exacerbate tissue injury and promote disease progression. This shift from protective to harmful roles underscores the importance of redox balance in maintaining cellular homeostasis. Understanding the sources, types and mechanisms of ROS provides valuable insight into their physiological relevance and pathological implications. Continued research on redox biology, including therapeutic strategies targeting ROS generation and antioxidant pathways, holds promise for the development of interventions aimed at mitigating oxidative stress-related diseases.

References

1. Collin F (2019) Chemical basis of reactive oxygen species reactivity and involvement in neurodegenerative diseases. Int J Mol Sci 20(10):2407. https://doi.org/10.3390/ijms20102407
2. Murphy MP (2009) How mitochondria produce reactive oxygen species. Biochem J 417(1):1–13. https://doi.org/10.1042/BJ20081386
3. Herb M, Schramm M (2021) Functions of ROS in macrophages and antimicrobial immunity. Antioxidants 10(2):313. https://doi.org/10.3390/antiox10020313
4. Chen L, Deng H, Cui H, Fang J, Zuo Z, Deng J, Li Y, Wang X, Zhao L (2017) Inflammatory responses and inflammation-associated diseases in organs. Oncotarget 9(6):7204–7218. https://doi.org/10.18632/oncotarget.23208
5. Zhang X, Wu X, Hu Q, Wu J, Wang G, Hong Z, Ren J (2019) Lab for trauma and surgical infections. Mitochondrial DNA in liver inflammation and oxidative stress. Life Sci 236:116464. https://doi.org/10.1016/j.lfs.2019.05.020
6. Caturano A, D'Angelo M, Mormone A, Russo V, Mollica MP, Salvatore T, Galiero R, Rinaldi L, Vetrano E, Marfella R, Monda M, Giordano A, Sasso FC (2023) Oxidative stress in type 2 diabetes: impacts from pathogenesis to lifestyle modifications. Curr Issues Mol Biol 45(8):6651–6666. https://doi.org/10.3390/cimb45080420
7. Sies H (2017) Hydrogen peroxide as a central redox signaling molecule in physiological oxidative stress: oxidative eustress. Redox Biol 11:613–619. https://doi.org/10.1016/j.redox.2016.12.035
8. Chandimali N, Bak SG, Park EH, Lim HJ, Won YS, Kim EK, Park SI, Lee SJ (2025) Free radicals and their impact on health and antioxidant defenses: a review. Cell Death Discov 11(1):19. https://doi.org/10.1038/s41420-024-02278-8

9. Radi R (2018) Oxygen radicals, nitric oxide, and peroxynitrite: redox pathways in molecular medicine. Proc Natl Acad Sci USA 115(23):5839–5848. https://doi.org/10.1073/pnas.1804932115
10. Phaniendra A, Jestadi DB, Periyasamy L (2015) Free radicals: properties, sources, targets, and their implication in various diseases. Indian J Clin Biochem 30(1):11–26. https://doi.org/10.1007/s12291-014-0446-0
11. Kapoor D, Sharma R, Handa N, Kaur H, Rattan A, Yadav P, Gautam V, Kaur R, Bhardwaj R (2015) Redox homeostasis in plants under abiotic stress: role of electron carriers, energy metabolism mediators and proteinaceous thiols. Front Environ Sci 3:13. https://doi.org/10.3389/fenvs.2015.00013
12. Waszczak C, Carmody M, Kangasjärvi J (2018) Reactive oxygen species in plant signaling. Annu Rev Plant Biol 69:209–236. https://doi.org/10.1146/annurev-arplant-042817-040322
13. Nathan C, Ding A (2010) SnapShot: reactive oxygen intermediates (ROI). Cell 140(6):951–951.e2. https://doi.org/10.1016/j.cell.2010.03.008
14. Hasanuzzaman M, Bhuyan MHMB, Zulfiqar F, Raza A, Mohsin SM, Mahmud JA, Fujita M, Fotopoulos V (2020) Reactive oxygen species and antioxidant defense in plants under abiotic stress: revisiting the crucial role of a universal defense regulator. Antioxidants 9(8):681. https://doi.org/10.3390/antiox9080681
15. Forrester SJ, Kikuchi DS, Hernandes MS, Xu Q, Griendling KK (2018) Reactive oxygen species in metabolic and inflammatory signaling. Circ Res 122(6):877–902. https://doi.org/10.1161/CIRCRESAHA.117.311401
16. Gulcin İ (2025) Antioxidants: a comprehensive review. Arch Toxicol 99(5):1893–1997. https://doi.org/10.1007/s00204-025-03997-2
17. Sies H (2020) Oxidative stress: concept and some practical aspects. Antioxidants (Basel). 9(9):852. https://doi.org/10.3390/antiox9090852
18. Fujii J, Homma T, Osaki T (2022) Superoxide radicals in the execution of cell death. Antioxidants. 11(3):501. https://doi.org/10.3390/antiox11030501
19. Zorov DB, Juhaszova M, Sollott SJ (2014) Mitochondrial reactive oxygen species (ROS) and ROS-induced ROS release. Physiol Rev 94(3):909–950. https://doi.org/10.1152/physrev.00026.2013
20. Lushchak VI (2014) Free radicals, reactive oxygen species, oxidative stress and its classification. Chem Biol Interact 224:164–175. https://doi.org/10.1016/j.cbi.2014.10.016
21. Wang Y, Branicky R, Noë A, Hekimi S (2018) Superoxide dismutases: dual roles in controlling ROS damage and regulating ROS signaling. J Cell Biol 217(6):1915–1928. https://doi.org/10.1083/jcb.201708007
22. Brown DI, Griendling KK (2015) Regulation of signal transduction by reactive oxygen species in the cardiovascular system. Circ Res 116(3):531–549. https://doi.org/10.1161/CIRCRESAHA.116.303584
23. Di Marzo N, Chisci E, Giovannoni R (2018) The role of hydrogen peroxide in redox-dependent signaling: homeostatic and pathological responses in mammalian cells. Cell 7(10):156. https://doi.org/10.3390/cells7100156
24. Kurutas EB (2016) The importance of antioxidants which play the role in cellular response against oxidative/nitrosative stress: current state. Nutr J 15(1):71. https://doi.org/10.1186/s12937-016-0186-5
25. de Almeida AJPO, de Oliveira JCPL, da Silva Pontes LV, de Souza Júnior JF, Gonçalves TAF, Dantas SH, de Almeida Feitosa MS, Silva AO, de Medeiros IA (2022) ROS: basic concepts, sources, cellular signaling, and its implications in aging pathways. Oxidative Med Cell Longev 2022:1225578. https://doi.org/10.1155/2022/1225578
26. Hawkins CL (2020) Hypochlorous acid-mediated modification of proteins and its consequences. Essays Biochem 64(1):75–86. https://doi.org/10.1042/EBC20190045
27. Ayala A, Muñoz MF, Argüelles S (2014) Lipid peroxidation: production, metabolism, and signaling mechanisms of malondialdehyde and 4-hydroxy-2-nonenal. Oxidative Med Cell Longev 2014:360438. https://doi.org/10.1155/2014/360438

28. Yadav DK, Kumar S, Choi EH, Chaudhary S, Kim MH (2019) Molecular dynamic simulations of oxidized skin lipid bilayer and permeability of reactive oxygen species. Sci Rep 9(1):4496. https://doi.org/10.1038/s41598-019-40913-y
29. Rocha BS (2021) The nitrate-nitrite-nitric oxide pathway on healthy ageing: a review of pre-clinical and clinical data on the impact of dietary nitrate in the elderly. Front Aging 2:778467. https://doi.org/10.3389/fragi.2021.778467
30. Man MQ, Wakefield JS, Mauro TM, Elias PM (2022) Regulatory role of nitric oxide in cutaneous inflammation. Inflammation 45(3):949–964. https://doi.org/10.1007/s10753-021-01615-8
31. Jomova K, Raptova R, Alomar SY, Alwasel SH, Nepovimova E, Kuca K, Valko M (2023) Reactive oxygen species, toxicity, oxidative stress, and antioxidants: chronic diseases and aging. Arch Toxicol 97(10):2499–2574. https://doi.org/10.1007/s00204-023-03562-9
32. Miazek K, Beton K, Śliwińska A, Brożek-Płuska B (2022) The effect of β-carotene, tocopherols and ascorbic acid as anti-oxidant molecules on human and animal in vitro/in vivo studies: a review of research design and analytical techniques used. Biomolecules 12(8):1087. https://doi.org/10.3390/biom12081087
33. Leong SS, Korel F, Lingoh AL, Sarbini SR, Toh SC, Abit LY, Wong SC (2023) Current probiotics application for aquaculture feed: a review. Borneo Sci J 44(2). https://doi.org/10.51200/bsj.v44i2
34. Leong SS, Loh TC, Foo HL (2025) Postbiotics and gut microbiota: gut health and oxidative stress modulation. In: Ashique S, Ghazanfar S, Mojgani N, Policarpo Tonelli FM (eds) Postbiotics in metabolic disorders, 1st edn. CRC Press, Boca Raton, pp 121–131. https://doi.org/10.1201/9781003607779
35. Schieber M, Chandel NS (2016) ROS function in redox signaling and oxidative stress. Curr Biol 24(10):R453–R462. https://doi.org/10.1016/j.cub.2014.03.034
36. Medzhitov R (2010) Inflammation 2010: new adventures of an old flame. Cell 140:771–776. https://doi.org/10.1016/j.cell.2010.03.006
37. Ferrero-Miliani L, Nielsen O, Andersen P, Girardin S (2007) Chronic inflammation: importance of NOD2 and NALP3 in interleukin-1β generation. Clin Exp Immunol 147:227–235. https://doi.org/10.1111/j.1365-2249.2006.03261.x
38. Hannoodee S, Nasuruddin DN (2025) Acute inflammatory response. StatPearls Publishing, Treasure Island (FL)
39. Zigterman BGR, Dubois L (2022) Ontsteking en infectie: de cellulaire en biochemische processen [Inflammation and infection: cellular and biochemical processes]. Ned Tijdschr Tandheelkd 129(3):125–129. Dutch. https://doi.org/10.5177/ntvt.2022.03.21138
40. Germolec DR, Shipkowski KA, Frawley RP, Evans E (2018) Markers of inflammation. Methods Mol Biol 1803:57–79. https://doi.org/10.1007/978-1-4939-8549-4_5
41. Pahwa R, Goyal A, Jialal I (2023) Chronic inflammation. StatPearls, Treasure Island (FL)
42. Medzhitov R (2008) Origin and physiological roles of inflammation. Nature 454(7203):428–435. https://doi.org/10.1038/nature07201
43. Tsai DH, Riediker M, Berchet A, Paccaud F, Waeber G, Vollenweider P, Bochud M (2019) Effects of short- and long-term exposures to particulate matter on inflammatory marker levels in the general population. Environ Sci Pollut Res Int 26(19):19697–19704. https://doi.org/10.1007/s11356-019-05194-y
44. Barcelos IP, Troxell RM, Graves JS (2019) Mitochondrial dysfunction and multiple sclerosis. Biology (Basel) 8(2):37. https://doi.org/10.3390/biology8020037
45. Deepak P, Axelrad JE, Ananthakrishnan AN (2019) The role of the radiologist in determining disease severity in inflammatory bowel diseases. Gastrointest Endosc Clin N Am 29(3):447–470. https://doi.org/10.1016/j.giec.2019.02.006
46. Bellanti F, Coda ARD, Trecca MI, Lo Buglio A, Serviddio G, Vendemiale G (2025) Redox imbalance in inflammation: the interplay of oxidative and reductive stress. Antioxidants (Basel) 14(6):656. https://doi.org/10.3390/antiox14060656

47. Suresh M, Li B, Murreddu MG, Gudima SO, Menne S (2021) Involvement of innate immune receptors in the resolution of acute hepatitis b in woodchucks. Front Immunol 12:713420. https://doi.org/10.3389/fimmu.2021.713420
48. Lee B, Ghode P, Ong D (2019) Redox regulation of cell state and fate. Redox Biol 25:101056. https://doi.org/10.1016/j.redox.2018.11.014
49. An X, Yu W, Liu J, Tang D, Yang L, Chen X (2024) Oxidative cell death in cancer: mechanisms and therapeutic opportunities. Cell Death Dis 15(8):556. https://doi.org/10.1038/s41419-024-06939-5
50. Poprac P, Jomova K, Simunkova M, Kollar V, Rhodes CJ, Valko M (2017) Targeting free radicals in oxidative stress-related human diseases. Trends Pharmacol Sci 38(7):592–607. https://doi.org/10.1016/j.tips.2017.04.005
51. Pizzino G, Irrera N, Cucinotta M, Pallio G, Mannino F, Arcoraci V, Squadrito F, Altavilla D, Bitto A (2017) Oxidative stress: harms and benefits for human health. Oxidative Med Cell Longev 2017:8416763. https://doi.org/10.1155/2017/8416763
52. Olufunmilayo EO, Gerke-Duncan MB, Holsinger RMD (2023) Oxidative stress and antioxidants in neurodegenerative disorders. Antioxidants (Basel). 12(2):517. https://doi.org/10.3390/antiox12020517
53. Andrade B, Jara-Gutiérrez C, Paz-Araos M, Vázquez MC, Díaz P, Murgas P (2022) The relationship between reactive oxygen species and the cGAS/STING signaling pathway in the inflammaging process. Int J Mol Sci 23(23):15182. https://doi.org/10.3390/ijms232315182
54. Liu J, Han X, Zhang T, Tian K, Li Z, Luo F (2023) Reactive oxygen species (ROS) scavenging biomaterials for anti-inflammatory diseases: from mechanism to therapy. J Hematol Oncol 16(1):116. https://doi.org/10.1186/s13045-023-01512-7
55. Chelombitko MA (2018) Role of reactive oxygen species in inflammation: a minireview. Moscow Univ Biol Sci Bull 73:199–202. https://doi.org/10.3103/S009639251804003X
56. Yu W, Tu Y, Long Z, Liu J, Kong D, Peng J, Wu H, Zheng G, Zhao J, Chen Y, Liu R, Li W, Hai C (2022) Reactive oxygen species bridge the gap between chronic inflammation and tumor development. Oxidative Med Cell Longev 2022:2606928. https://doi.org/10.1155/2022/2606928
57. Juan CA, Pérez de la Lastra JM, Plou FJ, Pérez-Lebeña E (2021) The chemistry of reactive oxygen species (ROS) revisited: outlining their role in biological macromolecules (dna, lipids and proteins) and induced pathologies. Int J Mol Sci 22(9):4642. https://doi.org/10.3390/ijms22094642
58. Brieger K, Schiavone S, Miller FJ Jr, Krause KH (2012) Reactive oxygen species: from health to disease. Swiss Med Wkly 142:w13659. https://doi.org/10.4414/smw.2012.13659
59. Hayyan M, Hashim MA, AlNashef IM (2016) Superoxide ion: generation and chemical implications. Chem Rev 116(5):3029–3085. https://doi.org/10.1021/acs.chemrev.5b00407
60. Usui N, Kobayashi H, Shimada S (2023) Neuroinflammation and oxidative stress in the pathogenesis of autism spectrum disorder. Int J Mol Sci 24(6):5487. https://doi.org/10.3390/ijms24065487

Chapter 3
Enzymatic and Nonenzymatic Mechanisms of ROS Generation in Metabolic Disorders

Krishnendu Adhikary, Krishnendu Ganguly, Supriya Chaudhary, Sabina Yasmin, and Md Yousuf Ansari

3.1 Introduction

3.1.1 Overview of Reactive Oxygen Species (ROS)

ROS are highly reactive molecules formed on the partial reduction of oxygen, a requirement of life, during cellular metabolism [1]. The species include free radicals such as superoxide anion and hydroxyl radical and non-radicals such as hydrogen peroxide [2]. ROS are essential components of cellular homeostasis in normal physiological states. ROS are signal messengers, controlling cell growth, immune defense, and stress adaptation [3]. Because of their dualistic nature, ROS cause phenomenal damage to proteins, lipids, and DNA when formed in excess or when antioxidant defense is compromised [4]. Disequilibrium between generation and clearance is called oxidative stress [5]. The organism has enzymic antioxidants such as superoxide dismutase, catalase, and glutathione peroxidase and nonenzymic defenses such as vitamins C and E, which collectively ensure redox homeostasis [2]. In metabolic disorders such as diabetes, obesity, and cardiovascular disease, this fine balance is compromised. Hyperglycemia, lipid overloading, and mitochondrial dysfunction cause increased formation of ROS, overloading protective mechanisms

K. Adhikary · K. Ganguly
Department of Medical Laboratory Technology, Paramedical College Durgapur, Durgapur, West Bengal, India

S. Chaudhary
Department of Neuroscience, Drake University, Des Moines, IA, USA

S. Yasmin (✉)
Department of Pharmaceutical Chemistry, College of Pharmacy, King Khalid University, Abha, Saudi Arabia
e-mail: sahussain@kku.edu.sa

M. Y. Ansari
Ibne Seena Pharmacy College, AMES group of Institutions, Shahabad, Hardoi, Uttar Pradesh, India

S. Ashique, N. Mishra (eds.), *Reactive Oxygen Species in Metabolic Inflammation*,
https://doi.org/10.1007/978-3-032-11752-6_3

[6]. Chronic oxidative stress not only causes tissue damage but also increases inflammation and insulin resistance, creating a vicious cycle which results in disease exacerbation [5].

Outside of pathology, ROS are not merely toxic by-products. They are required at regulatable levels for physiological signaling and defense against infectious pathogens [7, 8]. The issue is how to define their protective and pathologic properties as a function of concentration, location, and duration of exposure [9]. An understanding of the role of enzymatic and nonenzymatic reactions in ROS generation is thus the key to unravel the complexity of metabolic disorders [2]. Such an understanding not only defines disease mechanisms but also discloses therapeutic possibilities to restore redox homeostasis without abandoning the necessary signaling roles of ROS [3, 4].

3.1.2 ROS and Redox Homeostasis

ROS are unavoidable by-products of aerobic metabolism, which are continuously produced in cells by mechanisms like mitochondrial respiration, enzyme activity, and immune defense [1, 4]. The former understanding was that ROS are simple toxic molecules, but with the newer philosophy we can comprehend that ROS have a dual role: at low levels, ROS are important signal messengers that control the growth, repair, and defense mechanisms, but at toxic levels, ROS induce oxidative stress and cellular damage [2, 8]. Control between these two opposing events is maintained by the redox homeostasis of the body [9].

Redox homeostasis is the tightly regulated equilibrium between ROS production and the antioxidant defense system [3, 4]. Redox homeostasis relies on enzymic antioxidants, including superoxide dismutase, catalase, and glutathione peroxidase, and on the nonenzymic antioxidants glutathione, vitamin C, vitamin E, and other diet-derived phytochemicals [2, 5]. These defenses collectively detoxify surplus ROS, repair oxidative damage, and preserve cellular function. Given redox homeostasis is preserved, ROS have a useful function in such processes as immune surveillance, wound repair, and metabolic adaptation to stress [8]. Redox homeostasis disturbance by chronic hyperglycemia, lipid excess, mitochondrial dysfunction, or environmental stressors, however, leads to excess ROS accumulation [6]. Oxidative stress inactivates cellular macromolecules, induces chronic inflammation, and deranges signal cascades [5]. In metabolic disease, e.g., diabetes and obesity, failure to preserve redox balance not only accelerates tissue damage but also sustains insulin resistance and vascular disease [4]. Redox homeostasis is thus not a static condition but a dynamic process, constantly adjusting to metabolic need and environmental stress [3]. Appreciation of the way ROS have both a useful function in physiological signaling and an adverse action in pathological damage is testament to the wisdom of therapeutic approaches that seek to restore balance instead of abolition of ROS [8]. Preservation of this precarious balance is necessary in order to preserve cells and provide long-term metabolic health.

3.1.3 Importance of ROS in Metabolic Disorders

ROS are the primary causative and pathogenic factors in the development of the majority of metabolic disorders [4, 6]. Although these molecules are a natural by-product of aerobic metabolism, their level tends to abnormally rise in metabolic disorders like diabetes, obesity, metabolic syndrome, and cardiovascular disease [2, 8]. Because metabolic disorders cause derangement of normal energy homeostasis and hence cause mitochondrial dysfunction, abnormal glucose metabolism, and enhanced fatty acid oxidation—all of which increase ROS formation [1, 5].

In type 2 diabetes, for example, hyperglycemia triggers hyperproduction of ROS through glucose auto-oxidation, protein glycation, and activation of mitochondrial electron transport [4, 6]. Thus induced oxidative stress is cytotoxic to pancreatic β cells, otherwise susceptible due to impaired antioxidant defense [10]. This interferes with insulin secretion and hastens disease progression [8]. Similarly, obesity with excess adipose tissue triggers inflammatory cytokine release and mitochondrial stress, resulting in an upsurge in ROS production [11]. This increases not only local adipose inflammation but also systemic insulin resistance [5].

Cardiovascular metabolic disease complications also have direct correlations with ROS imbalance [4, 12]. Elevated ROS augment endothelial dysfunction, lipid peroxidation, and vascular inflammation and lead to atherosclerosis [11, 13]. Oxidative protein and DNA damage also disturbs cellular repair mechanisms, further amplifying tissue damage [5]. Although ROS toxic effects are established, it is interesting to observe that ROS are not only toxic by-products. Under physiological conditions, they regulate insulin signaling, vascular tone, and adaptive stress responses [8]. The problem arises when their production overloads antioxidant defenses, tipping toward oxidative stress [6]. ROS are thus both biomarkers and mediators of metabolic disease [4]. Their importance is partly to be credited for the fact that metabolic disease is often to be implicated in chronic inflammation and multi-organ damage [6]. ROS targeting by lifestyle, dietary antioxidants, or pharmacologic interventions is a potential tool to restore redox balance and arrest disease progression [2, 3] (Fig. 3.1).

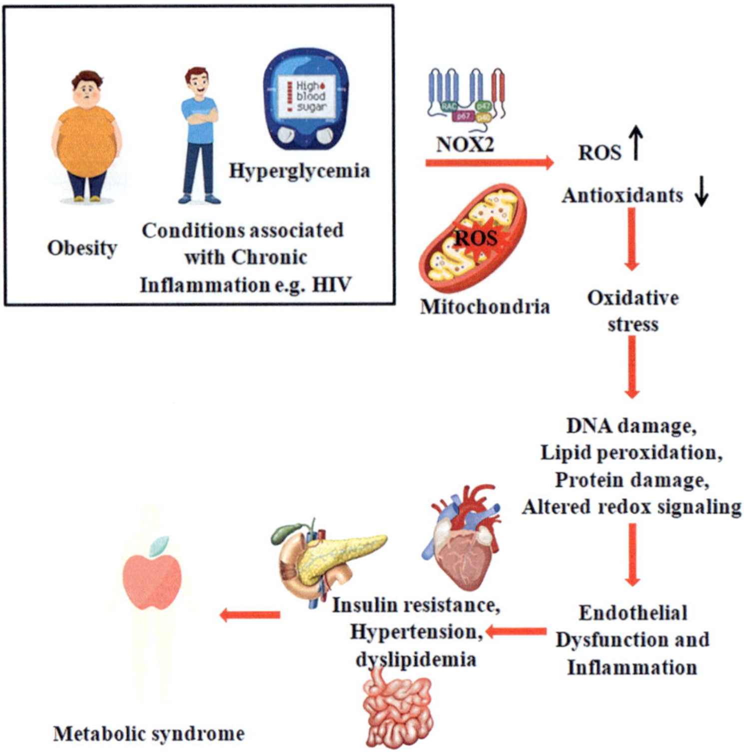

Fig. 3.1 In pathological states such as obesity, persistent inflammation, and elevated blood glucose, there is an overproduction of reactive oxygen species (ROS). This heightened ROS generation results from the activation of various enzymes located in the cytoplasm, cell membrane, and mitochondria. When ROS levels rise and antioxidant defenses are overwhelmed, oxidative stress develops. This oxidative stress damages cellular structures and disturbs redox homeostasis, leading to the buildup of irreversible oxidative by-products. These changes contribute to endothelial dysfunction, which in turn triggers insulin resistance, high blood pressure, abnormal lipid profiles, and ultimately the development of metabolic syndrome. (ROS, reactive oxygen species; NOX2, nicotinamide adenine dinucleotide phosphate (NADPH) oxidase family of enzymes)

3.2 Pathophysiological Role of ROS in Metabolic Disorders

3.2.1 Obesity and Adipose Tissue Dysfunction

Obesity is the prevalent metabolic disorder in which ROS are the pathologic drivers of the disease process [4, 11]. Fat tissue was classically described as an inert fat depot, but it is now a dynamic endocrine organ that secretes hormones, cytokines, and adipokines to regulate metabolism [14]. In obesity, excess accumulation of fat tissue leads to cellular stress, hypoxia, and low-grade chronic inflammation, all of which increase ROS production [6, 11]. At the cellular level, hypertrophic adipocytes have mitochondrial damage and increased fatty acid oxidation, both of which are ROS-producing processes of high intensity [15]. Immune cell invasion, in the form of macrophages, also causes oxidative stress by producing superoxide and other reactive species during inflammation [16]. This chronic ROS burden deconstructs redox homeostasis and initiates a cascade of pathologic effects on adipose tissue function [3, 5].

ROS directly modulate the signaling of adipocytes by acting on insulin receptor pathways and hence decrease insulin sensitivity [6, 17]. This generates an insulin-resistant condition in adipose tissue as well as skeletal muscle and liver, resulting in systemic metabolic derangement [4, 11]. Oxidative stress also inhibits secretion of adipokines, lowering protective adiponectin and increasing pro-inflammatory cytokines such as TNF-α and IL-6 [5, 16]. These actions generate a chronic inflammatory condition, resulting in tissue damage and metabolic derangement. Besides local action, ROS-mediated adipose dysfunction also has systemic implications. Plasma free fatty acids and inflammatory mediators from deranged adipose tissue cause endothelial damage, hepatic steatosis, and conversion to type 2 diabetes and cardiovascular disease [6, 8]. ROS are thus initiators and maintainers of adipose tissue dysfunction in obesity. Such mechanisms make therapeutic interventions amenable to individualization to restore redox balance, improve adipose tissue health, and prevent obesity-related metabolic derangement complications [2, 3].

3.2.2 Type 2 Diabetes and Insulin Resistance

ROS are the regulators of pathogenesis and progression of insulin resistance and type 2 diabetes mellitus (T2DM) [6, 8]. Under homeostasis, low-concentration ROS serve as signaling molecules modulating glucose transport and insulin sensitivity [2, 3]. Metabolic stress due to recovery from hyperglycemia, elevated free fatty acids, and obesity generates ROS in excess, shifting the balance toward oxidative stress, suppressing insulin signaling cascades and pancreatic β-cell function, two characteristics of T2DM [4, 17]. In peripheral tissues such as skeletal muscle and liver, ROS suppress insulin receptor function by causing serine phosphorylation of insulin receptor substrates (IRS) [18]. This posttranslational modification blocks downstream signaling by the PI3K-Akt pathway, blocking glucose transporter (GLUT4) translocation and glucose uptake [19]. Concurrently, excess ROS activate stress-sensitive kinases JNK and NF-κB, further augmenting insulin resistance and inflammatory signaling [4, 18]. Oxidative stress, therefore, not only prevents metabolic signaling but also triggers a pro-inflammatory response that augments insulin resistance [8].

The pancreas is more vulnerable to ROS damage due to inadequate antioxidant defense [10, 20]. Mitochondrial damage, DNA damage, and β-cell apoptosis by chronic oxidative stress induce defective insulin secretion [4, 21]. Defective insulin secretion and defective insulin action in target organs are a vicious cycle, and this is the etiology of the onset and development of T2DM [8]. Oxidative stress is also the pathogenesis of diabetic complications such as nephropathy, retinopathy, and neuropathy through endothelial damage and induction of vascular dysfunction [22, 23]. They are the systemic cost of ROS over glucose metabolism per se. In summary, ROS are mediators and amplifiers of insulin resistance and T2DM. Redox balance targeting with antioxidants, lifestyle intervention, and pharmacologic agents is a promising intervention to maintain β-cell function, increase insulin sensitivity, and prevent disease progression [2, 6].

3.2.3 Metabolic Syndrome and Cardiovascular Risk

Metabolic syndrome, a collection of disorders of central obesity, insulin resistance, dyslipidemia, and hypertension, is directly associated with accelerated cardiovascular disease (CVD) [24, 25]. Oxidative stress, triggered by a relative imbalance between the production of ROS and antioxidant defenses, is the driving factor of this association [3, 4]. ROS is a family of powerful mediators between metabolic derangement and vascular injury and secondary cardiovascular complications [26]. In metabolic syndrome, increased adiposity and insulin resistance trigger chronic low-grade inflammation and mitochondrial dysfunction, both of which are the source of increased production of ROS [6, 11]. Hypercirculating glucose and lipids also perpetuate oxidative stress by triggering auto-oxidation, lipid peroxidation, and the mitochondrial electron transport chain [4, 8]. Such chronic overproduction of ROS is cytotoxic to vascular endothelium, a critical regulator of vascular tone and homeostasis [23, 26].

ROS lower the bioavailability of the vasodilator nitric oxide (NO) by inducing its premature degradation to peroxynitrite [26, 27]. This induces endothelial dysfunction, impaired vasodilation, and increased vascular stiffness—earlier described markers of atherosclerosis [8]. ROS also trigger antioxidant-independent oxidation of low-density lipoprotein (LDL) to foam cells and arterial wall plaque [4, 28]. These reactions are the basis of the pathogenesis of coronary artery disease, stroke, and other cardiovascular events [26, 29]. In combination with vascular injury, oxidative stress amplifies hypertension by modulating renal sodium regulation, activation of the renin–angiotensin system, and induction of vascular smooth muscle cell growth [30, 31]. The combination of these changes provides a direct mechanistic link between metabolic syndrome and atherogenic cardiovascular risk [6, 24]. Of specific interest, whereas ROS at physiological concentrations modulate vascular signaling and adaptive responses, chronic oxidative stress tips the balance toward pathological remodeling [3, 32]. This underscores the critical importance of redox homeostasis to cardiovascular health. Therapeutic interventions that block the generation of ROS or increase antioxidant defense may be of incalculable benefit in preventing the cardiovascular burden of metabolic syndrome [2, 26] (Fig. 3.2).

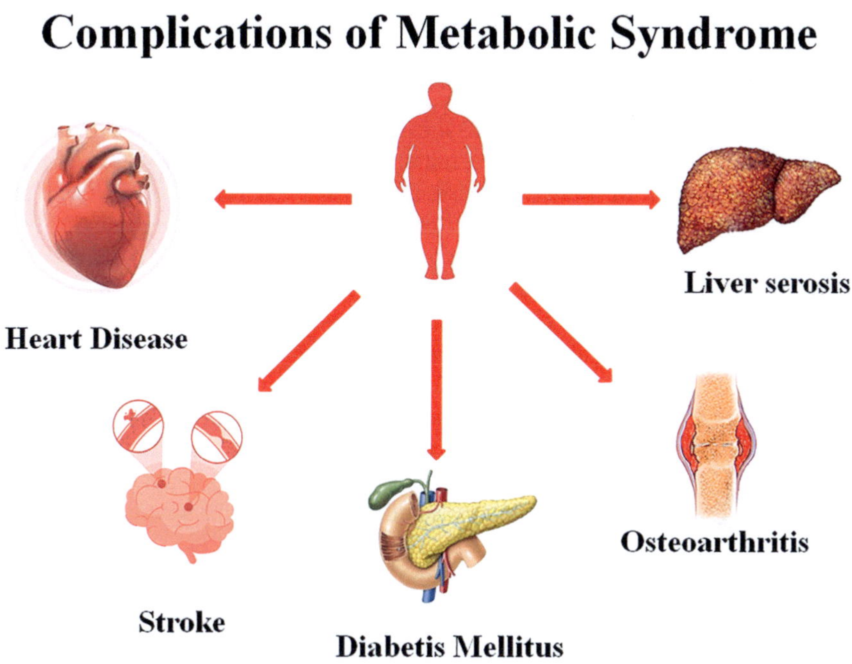

Fig. 3.2 Complications of metabolic syndrome: heart failure, brain stroke, diabetes mellitus, osteoarthritis, and liver cirrhosis occur due to metabolic syndrome

3.3 Enzymatic and Nonenzymatic Mechanisms of ROS Generation in Metabolic Disorders

3.3.1 Enzymatic Sources of ROS

Reactive oxygen species (ROS) from metabolic sources have a major impact on the pathophysiology of metabolic illnesses, such as type 2 diabetes, overweight, and nonalcoholic fatty liver disease. Nicotinamide adenine dinucleotide phosphate (NADPH) oxidases (NOX family) are important contributors to the production of superoxide and hydrogen peroxide in response to metabolic stress [33]. When nutritional conditions change, mitochondrial respiratory chain complexes—particularly complexes I and III—also generate large levels of ROS, which exacerbates insulin resistance and oxidative damage [34]. Additionally, hyperglycemia often results in an upregulation of xanthine oxidase (XO) activity, which raises ROS and contributes to cardiovascular dysfunction in endothelial cells [35]. Another enzymatic source is disconnected endothelial nitric oxide synthase (eNOS), which worsens metabolic inflammation by switching from manufacturing nitric oxide to superoxide in response to oxidative stress or tetrahydrobiopterin shortage [36]. All of these enzymatic sources work together to increase oxidative stress, sustain low-grade inflammation, and hasten the development of metabolic diseases [37].

3.3.2 NADPH Oxidase Family

The enzymes of the NADPH oxidase (NOX) family have been discovered as crucial participants in the process of oxidative stress in metabolic disorders such as diabetes type 2, obesity, and insulin resistance. Initially discovered in phagocytes, the tissue-specific forms of all of these multi-subunit enzymes (NOX1, NOX2, NOX4, and NOX5) are now recognized to actively produce reactive oxygen species (ROS) as signaling molecules in non-phagocytic cells [38]. Oxidative stress is linked to persistent low-grade inflammation in adipose tissue because ROS produced from NOX4 stimulate adipocyte enlargement and pro-inflammatory cytokine release [39]. Similarly, oxidative alteration of insulin receptor protein substrates and downstream kinases has been linked to NOX2 activation in the liver and skeletal muscle, which disrupts insulin signaling [40]. Moreover, the dysfunctional endothelial system and hypertension often seen in metabolic syndrome are caused by the overexpression of NOX1 and NOX5 in cerebral endothelial cells [41]. Therefore, targeting certain NOX isoforms is being investigated as a therapeutic approach to lessen problems in metabolic illnesses caused by oxidative stress [42].

3.3.3 Mitochondrial Electron Transport Chain

The breakdowns of the electron transport chain in mitochondria (ETC), a major source of intracellular reactive oxygen species (ROS), have a significant impact on the emergence of metabolic diseases, including type 2 diabetes mellitus, insulin resistance, and overweight and obesity [43]. Electron leakage, mostly from complexes I and III, causes increased superoxide generation under nutrient-excessive situations. This affects insulin signaling and encourages pathways of inflammation [34]. By decreasing ATP availability and raising oxidative stress, impaired mitochondrial linkage efficiency and faulty oxidative phosphorylation significantly worsen lipid formation in nonalcoholic fatty liver disease [44]. Additionally, long-term mitochondrial stress triggers mitophagy and the unfolded protein response (UPR-mt); while initially protective, this activation can lead to systemic metabolic failure and cellular senescence [45]. According to recent research, addressing mitochondrial biogenesis and redox equilibrium may help reduce damage caused by oxidation and metabolic issues [46] (Fig. 3.3).

3.3.4 Xanthine Oxidase

Xanthine oxidase (XO), a vital enzyme involved in purine metabolism, has emerged as a significant contributor to the pathophysiology of metabolic illnesses due to its capacity to generate reactive oxygen species (ROS). Elevated XO activity has been

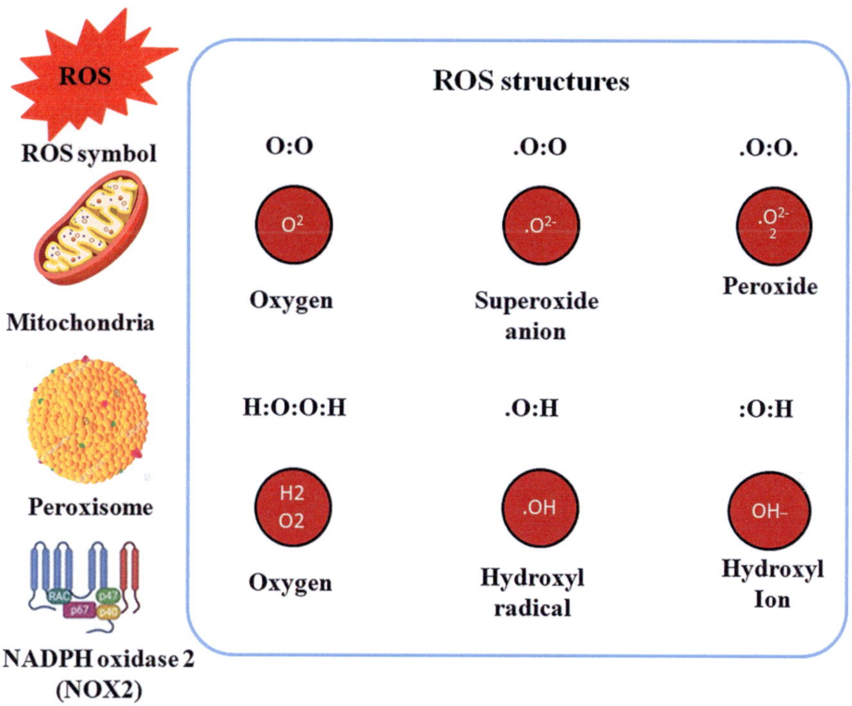

Fig. 3.3 Composition and origins of reactive oxygen species: ROS, reactive oxygen species; NADPH, nicotinamide adenine dinucleotide phosphate; NOX2, NADPH oxidase

linked to the three primary features of type 2 diabetes and the metabolic syndrome: insulin blockage, poor functioning of endothelial cells, and chronic inflammation [47, 48]. According to recent research, XO-derived ROS may also worsen oxidative stress by increasing lipid peroxidation and decreasing nitric oxide bioavailability, which might worsen cardiovascular issues that are frequently linked to metabolic illnesses [49, 50]. It has been shown that pharmacological suppression of XO with drugs like febuxostat or allopurinol can reduce metabolic risk factors, improve vascular function, and mitigate oxidative damage [51]. All of these results point to XO's critical involvement in the development and course of metabolic diseases.

3.3.5 Uncoupled Nitric Oxide Synthase (NOS)

By shifting from the manufacture of nitric oxide (NO) to the production of superoxide, uncoupled nitric oxide synthase (NOS) contributes significantly to the development of metabolic diseases, exacerbating the effects of oxidative stress and endothelial cell dysfunction [52, 53]. In conditions like obesity, insulin resistance, and type 2 diabetes, elevated oxidative stress and deficits in essential substrates such tetrahydrobiopterin (BH4) promote NOS uncoupling. Vascular problems and a reduction in NO bioavailability result from this [54, 55]. Uncoupled endothelial nitrogen oxides (eNOS), which disrupts glucose metabolism and raises inflammatory signaling, is another factor that links metabolic syndrome and vascular dysfunction [56]. Two treatment strategies that target NOS recoupling—antioxidants and BH4 supplementation—have shown promise in improving metabolic outcomes and restoring function to endothelial cells [57]. These findings all demonstrate the pathogenic significance of disconnected NOS in metabolic disorders.

3.3.6 Cytochrome P450 Enzymes

Cytochrome P450 (CYP) enzymes are now generally recognized as important modulators in the start and progression of metabolic disorders, in addition to their conventional roles in drug metabolism [58, 59]. Several CYP isoforms, such as CYP2E1, CYP4A, and CYP1A, increase oxidative stress and worsen vascular dysfunction, liver steatosis, and insulin resistance by producing reactive oxygen species (ROS) [60, 61]. For instance, CYP2E1 overexpression has been linked to metabolic disorder and nonalcoholic fatty liver disorder, or NAFLD, via increasing inflammatory signaling and lipid peroxidation in overweight and type 2 diabetes [62]. Further connecting the activity of CYP enzymes to cardiometabolic problems are aberrant CYP-derived eicosanoids, such as EETs and 20-HETE, which affect the equilibrium of glucose and vascular tone [63]. According to recent studies, CYP pathway targeting may help lower oxidative stress and inflammation in metabolic disorders [64]. All things taken into account, CYP enzymes function as both physiological regulators and amplifiers of oxidative stress, which contribute significantly to the pathogenesis of metabolic diseases.

3.4 Nonenzymatic Sources of ROS

Nonenzymatic producers of reactive oxygen species (ROS) are the primary cause of the oxidative stress observed in metabolic illnesses such as diabetes, obesity, and metabolic syndrome (Table 3.1) [65, 66]. These sources, which independently produce ROS from traditional enzyme-mediated processes, include lipid peroxidation, advanced glycation end-product (AGE) production, and auto-oxidation of glucose [67, 68]. For example, glucose experiences self-oxidation and nonenzymatic glycation processes in hyperglycemic circumstances, resulting in AGEs that interact with their cellular receptors (RAGE) to further stimulate the formation of ROS [69]. Similarly, nonenzymatic lipid peroxidation is encouraged by excessive free fatty acid buildup in metabolic diseases, which intensifies damage from oxidation and

Table 3.1 Enzymatic vs nonenzymatic mechanisms of ROS generation in metabolic disorders

Mechanism type	Key pathways/ sources	How it generates ROS	Context/relevance	Reference
Enzymatic mechanisms	NADPH oxidases (NOX1/NOX2/ NOX4, etc.)	Transfer electrons from NADPH to O_2 → produces superoxide (O_2•–)	Hyperglycemia and inflammatory signaling upregulate NOX → increased ROS in metabolic	Cipriano et al. [73]
	Mitochondrial electron transport chain (complexes I and III)	Electron leakage during oxidative phosphorylation → O_2•–, H_2O_2, OH•	IR, NAFLD, and obesity increase β-oxidation and mitochondrial ROS generation	Kumar et al. [74]
	Polyol pathway (aldose reductase, sorbitol dehydrogenase)	NADPH depletion reduces GSH; NADH increase → NADHoxidase→O_2•–	Hyperglycemia accelerates this in T2DM, promoting oxidative injury	Tangvarasittichai [75]
	Activation of protein kinase C (PKC) downstream of DAG	PKC upregulates NADPH oxidase → more ROS generation	Elevated DAG from hyperglycemia activates PKC → oxidative stress in diabetes	Pan et al. [76]
	Mitochondrial dehydrogenase complexes (e.g., OADHC)	Flavin sites leak electrons → ROS (O_2•–, H_2O_2)	Secondary source of mitochondrial ROS in altered metabolic states	Quinlan et al. [77]
Nonenzymatic mechanisms	Glucose auto-oxidation	Direct chemical oxidation of glucose → free radical formation	Occurs under hyperglycemic conditions, contributes to ROS pool	[78]
	Advanced glycation end-product (AGE) formation	Nonenzymatic glycation → AGE formation → AGE–RAGE interactions → ROS	Chronic hyperglycemia drives AGE formation → oxidative stress via RAGE	Twarda-Clapa et al. [79]
	Fenton reaction (H_2O_2 + Fe^{2+} → •OH)	High H_2O_2 levels + free iron → hydroxyl radicals (•OH)	Excess ROS (from enzymatic sources) and iron availability amplify damage	Fischbacher et al. [80]

signaling associated with inflammation [70]. These mechanisms exacerbate resistance to insulin and vascular problems by generating a self-reinforcing loop of metabolic malfunction and oxidative stress [71]. In order to develop treatment options that target the effects of oxidative stress in physiological illnesses, it is imperative that these nonenzymatic pathways be addressed [72] (Table 3.1).

3.4.1 Glucose Auto-Oxidation

Glucose auto-oxidation is one of the main nonenzymatic sources of reactive oxygen species (ROS), which feed the oxidative stress that causes metabolic disorders including hyperglycemia and metabolic syndrome [81, 82]. Excess sugar spontaneously oxidizes in hyperglycemic conditions, producing reactive intermediates such super oxide anions and hydrogen peroxide [83]. These ROS damage intracellular DNA, lipids, and protein and exacerbate the resistance to insulin and β-cell failure [84]. Moreover, advanced substances produced by glycation (AGEs), which are generated by glucose auto-oxidation, exacerbate the oxidative stress and inflammatory pathways via receptor-mediated signaling [85]. Together, these processes speed up the onset of diabetes, cardiovascular disease, and other metabolic consequences [86]. Targeting sugar auto-oxidation and the oxidative compounds that result from it is therefore one potential therapeutic strategy for the medical management of metabolic disorders [87, 88].

3.4.2 Advanced Glycation End Products (AGEs)

AGEs, also known as advanced glycation end products, are critical to the development and progression of metabolic disorders because they stimulate damage from oxidation, inflammation, and cellular dysfunction. When reducing sugars interact without enzymes with lipids, proteins, and/or nucleic acids, AGEs are produced. Under diabetics and those with the metabolic syndrome, they accumulate under hyperglycemic conditions [89, 90]. Reactive oxygen species, cell adhesion proteins, and pro-inflammatory cytokines are produced in greater quantities when AGEs connect with their individual receptor, RAGE, and activate signaling pathways such NF-κB [91] Because of this pro-inflammatory and prooxidant environment, which exacerbates insulin resistance, vascular stiffness, and endothelial dysfunction, AGEs are associated with issues including hypotension and nonalcoholic fatty liver disease [92, 93].The potential for dietary AGEs from processed foods to increase the systemic AGE load further emphasizes the importance of external as well as internal AGEs in the pathogenesis of metabolic disorders [94]. Consequently, therapeutic strategies that focus on AGE production mitigation or AGE-RAGE signaling blockage have shown promise in reducing metabolic issues [95].

3.4.3 Lipid Peroxidation

Lipid peroxidation plays an important role in the pathogenesis of metabolic disorders by promoting oxidative stress, a condition of inflammatory reactions, and cellular dysfunction [96]. In metabolic diseases like obesity, insulin resistance, and nonalcoholic fatty liver disease (NAFLD), high blood sugar levels and excess free fatty acids promote the peroxidation of polyunsaturated fatty acids in cell membranes, resulting in the production of reactive aldehydes like malondialdehyde, also known as MDA, and 4-hydroxynonenal (4-HNE) [97, 98]. These dangerous lipid peroxidation products can change proteins and DNA, interfere with the signaling of insulin pathways, and set off inflammatory cascades that exacerbate metabolic dysfunction [99, 100]. Furthermore, mitochondrial dysfunction and increased ROS generation have been connected to lipid peroxidation, resulting in a vicious loop that hastens the course of illness [101]. Additionally, recent research indicates that lipid peroxidation-targeting treatment strategies, including pharmaceutical medications, lifestyle changes, and antioxidants, may help reduce oxidative damage and enhance the metabolic results [102]. All of these results highlight the harmful role that lipid peroxidation plays in metabolic diseases.

3.4.4 *Iron and Metal-Catalyzed Oxidation*

Iron and other transition metals, particularly copper, play a crucial role in the pathophysiology of metabolic disorders by promoting the generation of reactive oxygen molecules (ROS) through Fenton and Haber-Weiss reactions [103, 104]. In cases of iron overload or dysregulated metal homeostasis, hydrogen peroxide can be nonenzymatically transformed into extremely reactive hydroxyl radicals, which damage proteins, lipids, and nucleic acids [104, 105]. This process not only results in oxidative stress but also exacerbates hepatic steatosis, insulin resistance, and β-cell dysfunction [106]—all of which are significant features of the two types of diabetes and metabolic syndromeFurthermore, iron-catalyzed lipid peroxidation results in ferroptosis, a regulated form of cell death that is becoming recognized as playing a significant role in the establishment of nonalcoholic fatty liver [107] disease (NAFLD) and cardiovascular issues [108, 109]. Recent studies have shown that metal chelators and iron metabolism modulators have therapeutic potential for reducing oxidative damage and improving physiological effects [110]. All of these results highlight the harmful role that iron and metal-catalyzed degradation play in metabolic diseases.

3.5 Molecular Mechanisms Linking Reactive Oxygen Species to Metabolic Dysfunction

The largely reactive molecules known as reactive oxygen species (ROS) appear from molecular oxygen that are substantially composed of superoxide anion, hydrogen peroxide (H_2O_2), and other oxygen-containing molecules. In normal circumstances, reactive oxygen species (ROS) have a significant part in important cellular processes similar to signaling mechanisms responsible for cell proliferation, survival, and homeostasis maintenance [111, 112]. However, when the generation of ROS goes unregulated, it leads to a state known as oxidative stress (OS). Oxidative stress has been recognized as one of the significant etiological and pathogenic factors in the development and worsening of metabolic disorders such as obesity, diabetes mellitus, and metabolic pattern [113]. Pathological consequences of ROS are due to enzymatic and nonenzymatic sources that lead to cellular redox imbalance and trigger dysfunction in the major metabolic tissues [114, 115].

In eukaryotic cells, the mitochondrial electron transport chain, specifically at complexes I and III embedded in the inner membrane, is the major physiological site of ROS generation, where partial electron leakage from the mitochondrial electron transport chain generates superoxide, which is subsequently converted into hydrogen peroxide. Within cells, reactive oxygen species are produced primarily by intrinsic enzymes, notably the mitochondrial electron transport chain and membrane-associated NADPH oxidase enzymes. However, nonenzymatic ROS production is also a significant factor in case of chronic metabolic stress [104]. When ROS production from mitochondria, NADPH oxidases, or stress-related pathways exceeds the cell's clearance capacity, it begins to induce oxidative injuries, modifying proteins, lipids, and DNA and interfering with critical signaling pathways [116, 117]. Thus, overproduction of ROS causes accumulation of ROS and ultimately oxidative stress, which is implicated in metabolic dysfunction pathogenesis by promoting insulin resistance, endothelial damage, and chronic low-grade inflammation [118]. Oxidative stress occurs when the generation of ROS exceeds the capacity of the bodies and antioxidant defenses shift in favor of ROS, overwhelming the systems designed to detoxify or buffer them. Consequently, it causes oxidative damage and disrupts redox homeostasis. It may be brought about by increased free radical production or reduced antioxidant function. The resultant oxidative environment triggers inflammatory pathways resulting in diseases like metabolic disease-associated chronic low-grade inflammation [119], though ROS is continuously produced as a natural by-product of metabolism and has beneficial functions at low levels [112].

3.5.1 Inflammatory Signaling Pathways

Oxidative stress (OS) and inflammation go hand in hand and play critical roles in the etiology of metabolic diseases. This low-grade chronic inflammation is frequently observed in many cardiovascular and metabolic diseases. Reactive oxygen species (ROS) are highly potent signaling molecules even at very low concentrations that have the capability of initiating inflammatory processes, resulting in insulin resistance, obesity, and atherosclerosis [120, 121]. Excessive generation of ROS plays a pivotal role in triggering transcription factors sensitive to redox changes, such as NF-κB and AP-1, which control immune and inflammatory responses.

Activation is responsible for increasing pro-inflammatory cytokines, which helps to exacerbate inflammation. One way is through ROS targeting IκB with its degradation and release of NF-kB. Once liberated, NF-κB is activated and moves into the nucleus, where it forms the specific DNA complexes and causes the expression of the genes that produce pro-inflammatory molecules [122]. The mechanism tends to be followed by a positive feedback mechanism, inflammation, and oxidative stress feed on one another. In addition, ROS directly induces inflammatory processes through the activation of specific signaling kinases. This intricate interaction between ROS and inflammation pathways highlights their central role in metabolic dysfunction.

3.5.1.1 NF-κB and AP-1 Activation

When ROS-mediated transcription factors are activated, there is a strong increase in pro-inflammatory expression. These transcription factors play critical roles in mediating inflammatory and immune reactions, leading to a rise in systemic pro-inflammatory cytokines. This ensuing rise in inflammatory mediators further exacerbates insulin resistance. ROS, such as hydrogen peroxide (H2O2), have the capability to directly activate the IKK kinase or oxidize specific cysteine residues on IκB. This oxidative modification, as well as phosphorylation of IκB, leads to its ubiquitination and proteasomal degradation. Once IκB separates from NF-κB, NF-κB is translocated to the nucleus. As a result it, activates many inflammatory-associated genes. By this mechanism, an auto-stimulatory positive feedback loop is generated, with NF-κB-regulated products driving their own activation, thereby exaggerating inflammation and ROS generation.

3.5.1.2 Mitogen-Activated Protein Kinases (MAPKs)

ROS are known for efficient activators of mitogen-activated protein kinases (MAPKs), a family that includes extracellular signal-regulated kinase 1/2 (ERK1/2) and p38 MAPKs. These kinases are significant regulators of cellular proliferation and have a role in hydrogen peroxide (H_2O_2)-dependent growth factor signaling. During insulin signaling, the phosphorylated serine/threonine kinases like MAPKs can impair the ability of proteins to recruit and activate downstream signaling molecules. This obstruction prevents insulin receptor substrate (IRS) proteins from associating with the insulin receptor; as a result, it hinders the signal [123].

3.5.1.3 Pro-inflammatory Cytokines

Reactive oxygen species are present which induce some pathways and result in upregulation of the synthesis and release of various pro-inflammatory cytokines. This further enhances the attraction and stimulation of immune and inflammatory cells. Some of the examples involved in this response are tumor necrosis factor-alpha (TNF-α) and interleukin-1 beta (IL-1β) This chronic condition of inflammation, or so-called inflammaging, is central to the aging process. Immune system stimulation also contributes to the metabolic syndrome by releasing these pro-inflammatory cytokines that cause insulin resistance.

3.5.1.4 Specific Organ Damage

This cascade of inflammation may lead to specific organ injury. In the liver, ROS causes membrane damage to hepatocytes, leading to cell deterioration, accumulation of collagen, fibrosis, cirrhosis, fat-mediated cell death (lipoapoptosis), and immune reactions. In the kidneys, oxidative stress initiated by ROS stimulates the release of pro-inflammatory cytokines, resulting in inflammation (nephritis) and impaired kidney function.

3.5.2 Impaired Insulin Signaling

Oxidative stress, typically augmented by chronic inflammation, is a hallmark in the pathogenesis of metabolic disorders. It disrupts insulin signaling and the trajectory of insulin resistance (IR).

3.5.2.1 Hyperglycemia as a Primary Driver

Diabetes mellitus is marked by chronic raised blood glucose levels, which over a period of years cause severe damage to different organs for a long term. This damage may result in complications such as disease of the nerves (neuropathy), disease of the eyes (retinopathy), disease of the kidneys (nephropathy), disease of the liver, and disease of the blood vessels. This chronic hyperglycemia (high blood glucose) is a major reason for ROS generation, producing deleterious molecules such as superoxide radicals, hydrogen peroxide, and hydroxyl radicals through auto-oxidation. High glucose concentrations directly enhance ROS overproduction and morphological transformation in mitochondria but also stimulate the 26S proteasome. Activation degrades GTP cyclohydrolase I, which subsequently diminishes endothelial nitric oxide synthase activity and results in the failure of normal blood vessel relaxation. Besides, high glucose stimulates protein kinase C and hexosamine pathway activity and increases advanced glycation end-product formation via complex molecular mechanisms.

3.5.2.2 Mechanisms of Insulin Resistance

Insulin resistance (IR), i.e., a reduced capacity to respond sufficiently to insulin, is mainly caused by oxidative stress from excess mitochondrial reactive oxygen species (ROS). A transient burst of low-dose hydrogen peroxide (H_2O_2) may enhance insulin signaling through inhibition of tyrosine phosphatase activity and increased phosphorylation of the insulin receptor and related proteins. Overproduction or chronic oxidative stress, however, disrupts this vital signaling pathway. The insulin signaling cascade is blocked by high levels of ROS under sustained exposure. In skeletal muscle, for instance, defective mitochondria, reduced ATP formation, and increased ROS formation are major causes of insulin resistance and obesity as well as diabetes. This is also added to by low-grade inflammation and metabolic dysregulation that are linked with obesity.

3.5.2.3 Role of Lipid Mediators and PKC

Increased levels of triglycerides, which are common in metabolic derangement, lead to the accumulation of diacylglycerol (DAG). DAG causes insulin signaling activity to decrease through the activation of heterogeneous isoforms of protein kinase C. Phosphorylation of the insulin receptor (INSR) at Thr1160 by PKC destabilizes its active form, thereby inhibiting glucose transport and phosphorylation. Diacylglycerol (DAG) generation from dihydroxyacetone phosphate also stimulates the PKC pathway, a significant mechanism in the development of cardiovascular complications via mitogen-activated protein kinase (MAPK) cascades.

3.5.2.4 Lipid Peroxidation and Cell Membrane Damage

Lipid peroxidation is the most important process involved in the induction of insulin resistance. Hydroxyl radicals and hydroperoxyl radicals cause nonspecific damage to biomolecules like neighboring organelles and the plasma membrane, leading to lipoperoxidation. The specific inhibition of glucose transporter function and the general processes of phosphorylation by damage to the plasma membrane reduce the effectiveness of insulin. Besides, high levels of hydroperoxyl radicals can also initiate a sustained chain reaction of polyunsaturated phospholipid oxidation to further impede membrane function.

3.5.2.5 Glucose Transporter Dysfunction (GLUT4)

The glucose transporter 4 (GLUT4) is the key glucose transporter that maintains the transport of glucose in insulin-sensitive tissues, skeletal muscle, and cardiac muscle. Insulin binding typically triggers a chain of signal transduction that enhances GLUT4 appearance on the plasma membrane, facilitating glucose uptake from the bloodstream. In obese and diabetic conditions, high levels of ROS are produced by mitochondrial dysfunction, protein alteration, and inactivation of antioxidant enzymes. The resultant high levels of ROS lead to insulin receptor signaling and low GLUT4 appearance on the cell membrane. Oxidative stress can also activate casein kinase-2 (CK2) to trigger, which diverts GLUT4 from the plasma membrane for degradation in lysosomes, and the result is hyperglycemia.

3.5.3 Mitochondrial Dysfunction

Mitochondria are the main source of energy production in cells and are the primary endogenous source of reactive oxygen species (ROS). Because of their critical function, mitochondrial dysfunction is closely linked with metabolic diseases, and it tends to be in a "vicious cycle" with oxidative stress.

3.5.3.1 Mitochondria as Primary ROS Generators

During oxidative phosphorylation (OXPHOS), which is essential for cells to generate adenosine triphosphate (ATP), oxygen molecules (O_2) accept electrons and are transformed into water (H_2O) as they pass through the electron transport chain (ETC). Super atmospheric superoxide is unavoidably generated in this process, which is an essential source of cellular ROS. Specifically, NADPH–ubiquinone oxidoreductase complex, also called Complex, I of the mitochondrial electron transport chain, is the predominant site for mtROS generation, one that is facilitated by extrinsic factors such as hypoxia (a state of lacking oxygen supply). Superoxide anion formation at Complex I involves electrons from the reduction of NADH to FMN or directly from reduced FMN (FMNH2) reducing oxygen. The dysfunctions of Complex III, both hereditary and acquired, also lead to an overproduction of ROS. This shows that mitochondria are the master regulators of aerobic energy production and require a carefully regulated balance between producing ROS (pro-oxidant) and systems that detoxify ROS (antioxidants). This balance is crucial to prevent oxidative damage while maintaining normal cellular function.

3.5.3.2 ROS-Induced Mitochondrial Damage

The elevated levels of ROS primarily damage mitochondria through binding with and destroying essential cellular structures like mitochondrial DNA (mtDNA), proteins, and others. This impairment causes mitochondrial dysfunction, which is evidenced by reduced mitochondrial biogenesis, disrupted membrane potential, and reduction in both the quantity and activity of oxidative proteins, all resulting from ROS accumulation in tissues and cells. This dysfunction plays a vital role in aging and most metabolic diseases, where it leads to impaired mitochondrial function, impaired oxidative capacity, and reduced ATP production. For instance, an increase in ROS causes an increased level of intracellular calcium in the mitochondria, thus further altering the membrane potential and triggering apoptosis. In diseases such as stroke, deprivation of oxygen and glucose leads to the accumulation of reducing intermediates and increased production of reactive oxygen species (ROS), which is directly responsible for mitochondrial damage.

3.5.3.3 Impact on ATP Production and Bioenergetics

The damage incurred by ROS directly impacts mitochondrial function. This results in less ATP (energy) production and further increase in the amount of ROS, thereby a self-sustaining feedback loop. Inhibited oxidative phosphorylation (OXPHOS) and defective ATP synthesis are some of the main markers for defective mitochondria. For aging, oxidative stress is the major contributor to the cumulative loss of organ function. With aging, mitochondrial enzyme activity and ROS generation are reduced. Also, cellular energy availability decreases with age in terms of elevated ATP and NADH/NAD+ ratios. Both impairments in mitochondrial function and disruptions of autophagy provoked by oxidative stress are central features of aging.

In cancer, elevated ROS not only promotes carcinogenesis and aging but also activates the JNK pathway, causing mitochondrial pathology and apoptosis [124]. Interestingly, ROS also serve as signal molecules, with low levels capable of activating AMP-activated protein kinase (AMPK). thus triggering adaptive mechanisms such as improved fatty acid oxidation, inhibition of fatty acid synthesis, and enhanced substrate uptake. However, when oxidative stress is very intense, ROS turn out to be toxic. They inhibit NAD+ levels by triggering poly-ADP-ribose polymerase (PARP). This inhibition of NAD+ then inhibits sirtuins, which are important in the regulation of mitochondrial function, antioxidant defense, cell proliferation, and metabolism.

3.5.3.4 Mitochondrial Dynamics

The regulation of proper balance between mitochondrial fission and fusion is significant for cellular health, controlling processes of the cell, and clearance of defective mitochondria. Dyshomeostasis leads to the aberrant production of ROS, altered enzymatic activity, compromised calcium homoeostasis, and overall impaired energy metabolism. The increased ROS observed with aging is partly explained by the accumulation of mutations in mtDNA, directly causing damage to mitochondrial structure and function, interfering with antioxidant defense and calcium homeostasis, and reducing the production of ATP.

3.6 Therapeutic Implications and Antioxidant Strategies

Reactive oxygen species (ROS) is a collection of chemically active molecules derived from molecular oxygen, formed as normal metabolites of aerobic respiration [125]. They consist of free radicals and non-radical derivatives and are characterized by their willingness to react with various cellular constituents. Under physiological conditions, ROS are primarily generated in the mitochondria through normal cellular metabolic processes, play essential roles in cell signaling, proliferation, and inflammation, and are necessary to maintain cellular homeostasis. However, when ROS production exceeds the capacity of the body's antioxidant defense, it leads to oxidative stress.

Overproduction of reactive oxygen species (ROS) during metabolic disorders leads to severe oxidative stress. Such disproportion involves the deleterious action of ROS overpowering the balancing effect of antioxidants. Moreover, metabolic disorders such as diabetes mellitus (DM) and obesity have a significant correlation with the overproduction of reactive oxygen species (ROS). Mitochondria are a significant source of ROS in both physiological and pathological situations. In conditions like obesity and diabetes, hyperglycemia increases metabolism through enhanced glycolysis and TCA cycle activity, resulting in greater production of electron carriers that feed into the mitochondrial electron transport chain. This results in enhanced leakage of electrons and generation of superoxide anion. Mitochondrial dysfunction, as indicated by decreased oxidative capacity, also leads to enhanced generation of ROS [126]. Excess ROS overproduction in metabolic disease generates significant oxidative stress and is therefore a significant focus for therapeutic regulation. The antioxidant strategies seek to reestablish the balance between ROS formation and the endogenous protective mechanisms of the body, thereby preventing the damaging effects of oxidative stress. These therapies are designed to restore redox homeostasis and shield cellular architecture against oxidative insult. Antioxidant strategies generally include pharmacological antioxidants, dietary and nutraceutical interventions, and selective enzyme modulation.

3.6.1 Pharmacological Antioxidants

Pharmacological antioxidant research is one of the crucial areas of research in the battle against various diseases caused by oxidative stress. Oxidative stress interferes with the normal physiological redox homeostasis, leading to the oxidative damage of metabolic pathways, nucleic acids, proteins, and phospholipids. Pharmacological antioxidants are basically specific drugs that may help our bodies fight harmful molecules called free radicals. Our bodies possess the capacity to naturally produce some antioxidants to help in maintaining free radicals in check, such as enzymatic (superoxide dismutase (SOD) and catalase (CAT)) and nonenzymatic (glutathione (GSH) and vitamin C) systems. But sometimes, especially with some diseases, we might need aid. That is where the pharmacological antioxidants take over to help in supplementing the body's endogenous defense system, thereby playing a crucial role in cellular safeguarding against oxidative damage. It must be understood that pharmacological antioxidants must be distinguished from the antioxidants obtained through nutrition. A nutritionally adequate amount of natural antioxidants from fruits, vegetables, and whole grains is highly recommended for general health as well as disease prevention. Pharmacological antioxidants are designed drugs, most often synthetically manufactured or derived from natural products, which are specifically aimed at replacing free radicals more efficiently. Pharmacological antioxidants are components of therapeutic schedules aimed at optimizing cellular health by reducing oxidative stress. Besides, drug-based antioxidants are currently well researched and used as a valuable therapeutic agent for treating various metabolic disorders involving oxidative stress. Their potential advantages are as follows.

(a) *Reducing Inflammation and Oxidative Stress*
Pharmacological antioxidants play crucial role in avoiding oxidative stress, which is a feature of most metabolic and inflammatory diseases, by ensuring redox homeostasis via neutralization of excess reactive oxygen species (ROS) and prevention of pro-inflammatory signaling cascades [127, 128].
(b) *Improving Insulin Sensitivity and Lipid Metabolism*
In persons who are obese, antioxidant supplementation can decrease body mass index (BMI) and fasting blood glucose (FBG) concentration and improve insulin resistance. The supplements also positively affect lipid metabolism by lowering atherogenic lipids.
(c) *Protection Against Complications*
Oxidative stress is tightly linked with hypertension, coronary artery disease, and cardiomyopathy. Certain conventional cardiovascular drugs, such as ACE inhibitors and calcium channel blockers, have been shown to exhibit antioxidant effects that reduce vascular oxidative stress. Antioxidant therapies are also explored to decrease the outcomes of diabetic complications, including neuropathy and renal disease [123].

Examples of Pharmacological Antioxidants and Related Strategies

(a) *Mitochondria-Targeted Antioxidants*
They include compounds such as MitoQ and MitoVitE that have been designed to live specifically in mitochondria, where most of the reactive oxygen species (ROS) production takes place. They are quite potent since they target and are able to protect against mitochondrial oxidative damage. MitoTEMPOL is another example under research for cardiovascular disease.
(b) *N-acetylcysteine (NAC)*
NAC is an antioxidant that reduces ROS both directly (by acting on hydroxyls and nitrogen dioxide) and indirectly (by providing cysteine for the formation of glutathione [129].

(c) *Coenzyme Q10 (Ubiquinone/Ubiquinol)*
The endogenous antioxidant helps mitochondrial function and improves other antioxidants' regeneration. Mitochondria-targeted CoQ10 (mitoQ) was shown to decrease the levels of ROS in ischemia/reperfusion injury and improve recovery [126].

3.6.2 Nutraceuticals and Dietary Interventions

Nutraceuticals and dietary interventions are a form of treatment of disease by modification of diet or intake of special dietary supplements. Its role is to help the body fight against free radicals and reactive oxygen species (ROS) that can destroy cells and lead to disease. Our body also has its own antioxidants that it manufactures to maintain the system in balance with ROS. In certain medical conditions, such as metabolic disorders, this inherent protection would not suffice, hence the necessity for supplementary reinforcement [125]. Nutraceuticals and dietary intervention come into the picture here and can make an important contribution by supplying exogenous antioxidants or by strengthening the endogenous antioxidant defense mechanism of the body.

3.6.2.1 Impact on Oxidative Stress and Inflammation

Certain foods and supplements possess antioxidant activity, which enables them to inactivate free radicals and other reactive species. This can prevent inflammation and promote health. Oxidative stress is when reactive oxygen species (ROS) build up within cells, which activates inflammatory response through the activation of transcription factors. With this elimination of imbalance, dietary therapy and nutraceuticals can reestablish and maintain cellular redox homeostasis, reversing oxidative stress and maintaining overall physiological function [123].

3.6.2.2 Key Nutraceuticals and Dietary Components

(a) *Vitamins*
Vitamins C and E are essential nonenzymatic antioxidants. Vitamin E (alpha-tocopherol) has been shown to interrupt lipid peroxidation and modulate cell signaling processes [124]. They play an important part in the defense against oxidative injury, especially in children with diabetes.
(b) *Glutathione (GSH)*
This tripeptide is an important nonenzymatic antioxidant that participates in the metabolism of nutrients, antioxidant protection, and regulation of cellular meta-

bolic functions, including gene expression and signal transduction. GSH supplements find application in the therapy of a range of diseases, including heart disease [125].

(c) *Polyphenols*
Polyphenols (e.g., curcumin, resveratrol, catechins) are ubiquitous in plants and have very high antioxidant activity due to their chemical structures. They have beneficial effects on various diseases, including neurodegenerative disorders, cardiovascular diseases, and inflammatory diseases.

(d) *Carotenoids*
These are also a group of nonenzymatic antioxidants that are protective against ROS.

(e) *Coenzyme Q10 (Ubiquinone/Ubiquinol)*
It is an endogenous antioxidant that participates in maintaining mitochondrial function and enhancing the regeneration of other antioxidants [126].

(f) *Dietary Habits*
Mediterranean diet and exercise regime are featured as lifestyle interventions that can reduce risk to a great extent and improve inflammatory and metabolic results by reversing excessive oxidative stress.

3.6.2.3 Selective Enzyme Modulation

It entails therapy interventions that attempt to augment the body's antioxidant defense systems, suppress particular inducers of ROS production, or fix ROS-mediated damage. They aim at the alteration of selective molecular pathways or enzymes rather than universal scavenging of ROS.

(a) *NRF2 Pathway Activation*
The NRF2 pathway regulates the expression of several antioxidant enzymes, enhancing the body's ability to inactivate oxidative stress. Oxidative stress transports NRF2 into the nucleus and activates genes for antioxidant enzymes like glutathione (GSH) and SOD, along with phase II detoxification systems.

(b) *Inhibition of ROS Production*
It attempts to reduce the production of ROS at its origin by inhibiting the activity of some ROS-generating enzymes like NOX and XO.

(c) *Antioxidant Enzyme Enhancement*
Superoxide dismutase (SOD) enzymes have a very important role in the dismutation of superoxide anions into more stable H2O26475. Low superoxide dismutase (SOD) activity raises oxidative stress, resulting in endothelial dysfunction, blood vessel damage, and atherosclerosis formation [125]. Catalase (CAT) is an enzyme that breaks down hydrogen peroxide (H_2O_2) into water and oxygen. Redox modulation regulation of ROS levels by enhanced catalase activity is essential to selectively inhibit cancer cell proliferation by redox homeostasis [126].

3.7 Conclusion and Future Scopes

Diabetic ulcers remain a major clinical challenge due to the synergistic effect of hyperglycemia, vascular insufficiency, neuropathy, chronic inflammation, and infection. Conventional interventions such as debridement, infection control, offloading, and dressings remain the mainstay but frequently insufficient in managing complex or nonhealing ulcers. Recent advances in regenerative medicine—stem cell therapy, exosome-based therapy, and bioengineered skin grafts—have been shown to be promising in inducing angiogenesis, granulation, and reepithelialization. Novel modalities of treatment such as nanotechnology-based dressings, growth factor delivery systems, negative pressure wound therapy, and hyperbaric oxygen therapy are further refining treatment protocols. Adjunctive therapy such as glycemic control, nutritional optimization, microbiome modulation, and patient education remains of paramount importance for optimal management. Next-generation studies must bring experimental therapies to cost-effective and scalable clinical practice. Genotype, epigenotype, and microbiome profiling-based personalized therapy can further improve treatment. Intelligent biomaterials, nanozyme dressings, and adaptive scaffolds can further improve healing. Artificial intelligence and digital monitoring wearables can deliver early detection, risk stratification, and personalized therapy, and microbiome-directed interventions offer a new paradigm for the treatment of infection mediated by biofilm. Central to the validation of these technologies are large multicenter trials to establish safety, cost-effectiveness, and long-term efficacy. In total, the convergence of regenerative medicine, precision technologies, and patient-centered preventive medicine has the potential to transform diabetic wound care, reduce complications, and significantly enhance the quality of life in diabetes.

References

1. Sies H, Berndt C, Jones DP (2017) Oxidative stress. Annu Rev Biochem 86:715–748. https://doi.org/10.1146/annurev-biochem-061516-045037
2. Phaniendra A, Jestadi DB, Periyasamy L (2015) Free radicals: properties, sources, targets, and their implication in various diseases. Indian J Clin Biochem 30(1):11–26. https://doi.org/10.1007/s12291-014-0446-0
3. Sies H (2020) Oxidative stress: concept and some practical aspects. Antioxidants 9(9):852. https://doi.org/10.3390/antiox9090852
4. Pizzino G, Irrera N, Cucinotta M, Pallio G, Mannino F, Arcoraci V, Squadrito F, Altavilla D, Bitto A (2017) Oxidative stress: harms and benefits for human health. Oxidative Med Cell Longev 2017:8416763. https://doi.org/10.1155/2017/8416763
5. Liguori I, Russo G, Curcio F, Bulli G, Aran L, Della-Morte D, Gargiulo G, Testa G, Cacciatore F, Bonaduce D, Abete P (2018) Oxidative stress, aging, and diseases. Clin Interv Aging 13:757–772. https://doi.org/10.2147/CIA.S158513

6. Rani V, Deep G, Singh RK, Palle K, Yadav UCS (2016) Oxidative stress and metabolic disorders: pathogenesis and therapeutic strategies. Life Sci 148:183–193. https://doi.org/10.1016/j.lfs.2016.02.002
7. Nathan C, Ding A (2010) Nonresolving inflammation. Cell 140(6):871–882. https://doi.org/10.1016/j.cell.2010.02.029
8. Sies H, Jones DP (2020) Reactive oxygen species (ROS) as pleiotropic physiological signalling agents. Nat Rev Mol Cell Biol 21(7):363–383. https://doi.org/10.1038/s41580-020-0230-3
9. Lennicke C, Cochemé HM (2021) Redox signaling and ageing: insights from Drosophila melanogaster. Free Radic Biol Med 167:321–332. https://doi.org/10.1016/j.freeradbiomed.2021.02.034
10. Robertson RP (2004) Chronic oxidative stress as a central mechanism for glucose toxicity in pancreatic islet beta cells in diabetes. J Biol Chem 279(41):42351–42354. https://doi.org/10.1074/jbc.R400019200
11. Furukawa S, Fujita T, Shimabukuro M, Iwaki M, Yamada Y, Nakajima Y, Nakayama O, Makishima M, Matsuda M, Shimomura I (2004) Increased oxidative stress in obesity and its impact on metabolic syndrome. J Clin Invest 114(12):1752–1761. https://doi.org/10.1172/JCI21625
12. Madamanchi NR, Vendrov A, Runge MS (2005) Oxidative stress and vascular disease. Arterioscler Thromb Vasc Biol 25(1):29–38. https://doi.org/10.1161/01.ATV.0000150649.39934.13
13. Förstermann U, Xia N, Li H (2017) Roles of vascular oxidative stress and nitric oxide in the pathogenesis of atherosclerosis. Circ Res 120(4):713–735. https://doi.org/10.1161/CIRCRESAHA.116.309326
14. Kershaw EE, Flier JS (2004) Adipose tissue as an endocrine organ. J Clin Endocrinol Metab 89(6):2548–2556. https://doi.org/10.1210/jc.2004-0395
15. Lee H, Lee YJ, Choi H, Ko EH, Kim JW (2013) Reactive oxygen species facilitate adipocyte differentiation by accelerating mitotic clonal expansion. J Biol Chem 284(16):10601–10609. https://doi.org/10.1074/jbc.M808043200
16. Lumeng CN, Saltiel AR (2011) Inflammatory links between obesity and metabolic disease. J Clin Invest 121(6):2111–2117. https://doi.org/10.1172/JCI57132
17. Hoehn KL, Salmon AB, Hohnen-Behrens C, Turner N, Hoy AJ, Maghzal GJ, Stocker R, Van Remmen H, Kraegen EW, Cooney GJ, Richardson AR, James DE (2009) Insulin resistance is a cellular antioxidant defense mechanism. Proc Natl Acad Sci U S A 106(42):17787–17792. https://doi.org/10.1073/pnas.0902380106
18. Hirosumi J, Tuncman G, Chang L, Görgün CZ, Uysal KT, Maeda K, Karin M, Hotamisligil GS (2002) A central role for JNK in obesity and insulin resistance. Nature 420(6913):333–336. https://doi.org/10.1038/nature01137
19. Houstis N, Rosen ED, Lander ES (2006) Reactive oxygen species have a causal role in multiple forms of insulin resistance. Nature 440(7086):944–948. https://doi.org/10.1038/nature04634
20. Lenzen S (2008) Oxidative stress: the vulnerable beta-cell. Biochem Soc Trans 36(3):343–347. https://doi.org/10.1042/BST0360343
21. Kaneto H, Matsuoka TA, Katakami N, Matsuhisa M (2005) Role of oxidative stress, endoplasmic reticulum stress, and c-Jun N-terminal kinase in pancreatic β-cell dysfunction and insulin resistance. Int J Biochem Cell Biol 37(8):1595–1608. https://doi.org/10.1016/j.biocel.2005.02.026
22. Baynes JW (1991) Role of oxidative stress in development of complications in diabetes. Diabetes 40(4):405–412. https://doi.org/10.2337/diab.40.4.405
23. Forbes JM, Cooper ME (2013) Mechanisms of diabetic complications. Physiol Rev 93(1):137–188. https://doi.org/10.1152/physrev.00045.2011
24. Grundy SM (2008) Metabolic syndrome pandemic. Arterioscler Thromb Vasc Biol 28(4):629–636. https://doi.org/10.1161/ATVBAHA.107.151092

25. Rochlani Y, Pothineni NVK, Mehta JL (2017) Metabolic syndrome: pathophysiology, management, and modulation by natural compounds. Ther Adv Cardiovasc Dis 11(8):215–225. https://doi.org/10.1177/1753944717711379
26. Madamanchi NR, Runge MS (2007) Mitochondrial dysfunction in atherosclerosis. Circ Res 100(4):460–473. https://doi.org/10.1161/01.RES.0000258450.44413.96
27. Förstermann U, Sessa WC (2012) Nitric oxide synthases: regulation and function. Eur Heart J 33(7):829–837. https://doi.org/10.1093/eurheartj/ehr304
28. Stocker R, Keaney JF (2004) Role of oxidative modifications in atherosclerosis. Physiol Rev 84(4):1381–1478. https://doi.org/10.1152/physrev.00047.2003
29. Montezano AC, Touyz RM (2012) Reactive oxygen species and endothelial function—role of nitric oxide synthase uncoupling and NOX family nicotinamide adenine dinucleotide phosphate oxidases. Basic Clin Pharmacol Toxicol 110(1):87–94. https://doi.org/10.1111/j.1742-7843.2011.00740.x
30. Rodrigo R, González J, Paoletto F, Galleano M (2011) The role of oxidative stress in the pathophysiology of hypertension. Hypertens Res 34(4):431–440. https://doi.org/10.1038/hr.2010.261
31. Harrison DG, Gongora MC (2009) Oxidative stress and hypertension. Med Clin North Am 93(3):621–635. https://doi.org/10.1016/j.mcna.2009.02.007
32. Touyz RM (2004) Reactive oxygen species, vascular oxidative stress, and redox signaling in hypertension: what is the clinical significance? Hypertension 44(3):248–252. https://doi.org/10.1161/01.HYP.0000138070.47616.9d

References for Point 3 and 4

33. Forrester SJ, Kikuchi DS, Hernandes MS, Xu Q, Griendling KK (2020) Reactive oxygen species in metabolic and inflammatory signaling. Circ Res 127(3):311–329. https://doi.org/10.1161/CIRCRESAHA.120.315386
34. García-Aguilar A, Cuezva JM (2021) The role of mitochondrial H+-ATP synthase incancer. Front Oncol 11:713381. https://doi.org/10.3389/fonc.2021.713381
35. Zhang H, Lu J, Liu X, Zhang J, Zhang W (2022) Role of xanthine oxidase in oxidative stress of metabolic disorders. Life Sci 293:120308. https://doi.org/10.1016/j.lfs.2022.120308
36. Münzel T, Daiber A, Steven S (2021) Involvement of oxidative stress in the pathogenesis of hypertension and vascular dysfunction. Antioxidants Redox Signal 34(9):682–712. https://doi.org/10.1089/ars.2020.8105
37. Di Meo S, Iossa S, Venditti P (2023) Oxidative stress and mitochondrial dysfunction in metabolic disorders: a critical update. J Transl Med 21:70. https://doi.org/10.1186/s12967-023-03879-2
38. Pendyala S, Natarajan V (2020) Redox regulation of Nox proteins. Antioxidants Redox Signal 32(9):636–655. https://doi.org/10.1089/ars.2019.7827
39. Li Y, Zhou J, Zhao L, Zhang H, Xu W (2022) NOX4-derived ROS in adipose tissue dysfunction and metabolic disorders. Redox Biol 54:102365. https://doi.org/10.1016/j.redox.2022.102365
40. Jiang L, Xu J, Li Y, Liu Y, He L (2023) NADPH oxidase 2-mediated oxidative stress impairs insulin signaling in metabolic disease. Front Endocrinol 14:1148724. https://doi.org/10.3389/fendo.2023.1148724
41. Kim JA, Wei Y, Sowers JR (2021) Role of NADPH oxidases in the pathogenesis of metabolic syndrome and hypertension. Antioxidants 10(5):752. https://doi.org/10.3390/antiox10050752
42. Wang X, Zhang J, Chen P, Liu Y, Sun J (2024) Therapeutic targeting of NOX enzymes in metabolic syndrome: advances and perspectives. Pharmacol Res 197:106556. https://doi.org/10.1016/j.phrs.2024.106556

43. Ribas V, García-Ruiz C, Fernández-Checa JC (2020) Mitochondria, cholesterol and cancer cell metabolism. Clin Transl Oncol 22(11):1990–2001. https://doi.org/10.1007/s12094-020-02388-4
44. Kim J, Wei Y, Sowers JR (2022) Role of mitochondrial dysfunction in insulin resistance. Antioxidants 11(1):48. https://doi.org/10.3390/antiox11010048
45. Zhang H, Li X, Zhang Y, Wang Y (2023) Mitochondrial unfolded protein response in metabolic disorders. J Transl Med 21:214. https://doi.org/10.1186/s12967-023-04012-1
46. Zhou T, Xue Y, Xu X, Wang X, Liu Y (2024) Targeting mitochondrial redox balance as a therapeutic approach in metabolic syndrome. Pharmacol Res 197:106562. https://doi.org/10.1016/j.phrs.2024.106562
47. Liu X, Zhang M, Liu X, Cheng Y, Zhang Y (2021) Association of xanthine oxidase with oxidative stress and insulin resistance in metabolic syndrome. Free Radic Res 55(4):389–397. https://doi.org/10.1080/10715762.2021.1886898
48. Yu H, Guo X, Xu Y, Xu L (2022) The role of xanthine oxidase in metabolic disorders and cardiovascular disease. Front Pharmacol 13:849568. https://doi.org/10.3389/fphar.2022.849568
49. Ishimoto T, Yamamoto Y, Sugimoto K, Nakagawa T, Kato T (2020) Xanthine oxidase and metabolic disorders: pathophysiological insights and therapeutic perspectives. Redox Biol 34:101567. https://doi.org/10.1016/j.redox.2020.101567
50. El-Sayed AM, El-Marasy SA, El Awdan SA, Kenawy SA (2023) Role of xanthine oxidase inhibition in the treatment of metabolic syndrome. Eur J Pharmacol 941:175541. https://doi.org/10.1016/j.ejphar.2023.175541
51. Zhang X, Li Q, Xie J, Wang Y (2021) Effects of xanthine oxidase inhibitors on metabolic parameters in patients with hyperuricemia: a systematic review and meta-analysis. Clin Ther 43(4):e91–e105. https://doi.org/10.1016/j.clinthera.2021.01.016
52. Chen L, Gao Y, Zhang Z, Zhang Y (2021) Role of nitric oxide synthase uncoupling in endothelial dysfunction and metabolic disorders. Life Sci 276:119410. https://doi.org/10.1016/j.lfs.2021.119410
53. Förstermann U, Li H (2022) Therapeutic strategies for recoupling of endothelial nitric oxide synthase in cardiovascular and metabolic diseases. Pharmacol Therap 238:108194. https://doi.org/10.1016/j.pharmthera.2022.108194
54. Gao Y, Chen X, Liu J, Xu Y (2020) The impact of NOS uncoupling on insulin resistance and metabolic syndrome. Free Radic Biol Med 156:84–92. https://doi.org/10.1016/j.freeradbiomed.2020.04.025
55. Li R, Yang T, Zhang M, Wang X (2023) Uncoupled endothelial nitric oxide synthase in metabolic disorders: mechanisms and therapeutic potential. Front Pharmacol 14:1176723. https://doi.org/10.3389/fphar.2023.1176723
56. Münzel T, Gori T, Bruno RM, Taddei S (2021) Endothelial dysfunction, oxidative stress, and uncoupled eNOS in metabolic syndrome. Eur Heart J 42(45):4596–4608. https://doi.org/10.1093/eurheartj/ehab346
57. Liu P, Zhu H, Xu Y, Chen J (2022) Tetrahydrobiopterin supplementation improves metabolic function by recoupling endothelial nitric oxide synthase. Front Endocrinol 13:838567. https://doi.org/10.3389/fendo.2022.838567
58. Nebert DW, Russell DW (2021) Clinical importance of the cytochromes P450. Lancet 398(10298):1155–1168. https://doi.org/10.1016/S0140-6736(21)01609-7
59. Tang L, Ma Y, Liu Y, Sun X (2022) Cytochrome P450 enzymes and metabolic diseases: mechanistic insights and therapeutic implications. Metabolism 128:155127. https://doi.org/10.1016/j.metabol.2022.155127
60. Wang J, Gao J, Guo X, Xu Y (2020) The role of CYP2E1 in oxidative stress, insulin resistance, and nonalcoholic fatty liver disease. Life Sci 256:117868. https://doi.org/10.1016/j.lfs.2020.117868

61. Xie T, Zhou J, Chen Y, Zhang Q (2021) Contribution of cytochrome P450 enzymes to oxidative stress in metabolic syndrome. Redox Biol 43:101979. https://doi.org/10.1016/j.redox.2021.101979
62. Zhang R, Liu P, Wang S, Li X (2023) CYP2E1-induced oxidative stress and inflammation in metabolic disorders. Free Radic Biol Med 196:202–211. https://doi.org/10.1016/j.freeradbiomed.2023.02.023
63. Kang Y, Hu J, Zhao Y, Wang X (2022) Cytochrome P450-derived eicosanoids in metabolic disorders: emerging therapeutic targets. Front Pharmacol 13:836205. https://doi.org/10.3389/fphar.2022.836205
64. Luo J, Li J, Liu H, Zhang M (2024) Targeting cytochrome P450 enzymes for the treatment of metabolic syndrome: advances and perspectives. Pharmacol Res 197:106503. https://doi.org/10.1016/j.phrs.2024.106503
65. Mao Y, Wu Y, Zhang D, Chen X (2021) Role of non-enzymatic oxidative stress in obesity and type 2 diabetes. Redox Biol 43:101887. https://doi.org/10.1016/j.redox.2021.101887
66. Zhang Y, Chen L, Wang Y, Zhao J (2023) Non-enzymatic oxidative stress pathways in metabolic syndrome: emerging insights. Front Endocrinol 14:1145623. https://doi.org/10.3389/fendo.2023.1145623
67. Singh VP, Bali A, Singh N, Jaggi AS (2020) Advanced glycation end products and diabetic complications. Korean J Physiol Pharmacol 24(1):1–14. https://doi.org/10.4196/kjpp.2020.24.1.1
68. Zhang X, Sun J (2021) Non-enzymatic ROS formation in hyperglycemia: pathways and implications. Biomed Pharmacother 136:111256. https://doi.org/10.1016/j.biopha.2021.111256
69. Ahmed N, Moinuddin, Ali A (2022) Non-enzymatic glycation and its contribution to oxidative stress in metabolic disorders. Life Sci 292:120331. https://doi.org/10.1016/j.lfs.2021.120331
70. Wang L, Zhang Q, Huang X, Li P (2021) Lipid peroxidation and its role in metabolic diseases. Oxidative Med Cell Longev 2021:6645460. https://doi.org/10.1155/2021/6645460
71. Zhao Z, Liu H, Xu Y, Zhang M (2024) Targeting non-enzymatic ROS generation to mitigate metabolic complications. Pharmacol Res 200:106761. https://doi.org/10.1016/j.phrs.2024.106761
72. Liu R, Zhang Y, Li J, Xu W (2022) Therapeutic strategies targeting non-enzymatic ROS sources in metabolic syndrome. Free Radic Res 56(9–10):890–902. https://doi.org/10.1080/10715762.2022.2106142
73. Cipriano A, Viviano M, Feoli A, Milite C, Sarno G, Castellano S, Sbardella G (2023) NADPH oxidases: from molecular mechanisms to current inhibitors. J Med Chem 66(17):11632–11655
74. Kumar A, Davuluri G, Welch N, Kim A, Gangadhariah M, Allawy A, Priyadarshini A, McMullen MR, Sandlers Y, Willard B, Hoppel CL (2019) Oxidative stress mediates ethanol-induced skeletal muscle mitochondrial dysfunction and dysregulated protein synthesis and autophagy. Free Radic Biol Med 145:284–299
75. Tangvarasittichai S (2015) Oxidative stress, insulin resistance, dyslipidemia and type 2 diabetes mellitus. World J Diabetes 6(3):456
76. Pan D, Xu L, Guo M (2022) The role of protein kinase C in diabetic microvascular complications. Front Endocrinol 13:973058
77. Quinlan CL, Goncalves RL, Hey Mogensen M, Yadava N, Bunik VI, Brand MD (2014) The 2-oxoacid dehydrogenase complexes in mitochondria can produce superoxide/hydrogen peroxide at much higher rates than complex I. J Biol Chem 289(12):8312–8325
78. González P, Lozano P, Ros G, Solano F (2023) Hyperglycemia and oxidative stress: an integral, updated and critical overview of their metabolic interconnections. Int J Mol Sci 24(11):9352
79. Twarda-Clapa A, Olczak A, Białkowska AM, Koziołkiewicz M (2022) Advanced glycation end-products (AGEs): formation, chemistry, classification, receptors, and diseases related to AGEs. Cells 11(8):1312
80. Fischbacher A, von Sonntag C, Schmidt TC (2017) Hydroxyl radical yields in the Fenton process under various pH, ligand concentrations and hydrogen peroxide/Fe (II) ratios. Chemosphere 182:738–44

81. Li P, Zhao Z, Xu Y, Zhang M (2022) Hyperglycemia-induced glucose auto-oxidation and its pathological consequences in metabolic disorders. Biomed Pharmacother 146:112569. https://doi.org/10.1016/j.biopha.2021.112569
82. Zhao Q, Wang J, Xu Y (2021) Glucose auto-oxidation and oxidative stress in diabetes mellitus: implications for therapy. Life Sci 276:119401. https://doi.org/10.1016/j.lfs.2021.119401
83. Sun X, Li W, Zhang Q, Chen H (2020) The contribution of glucose auto-oxidation to ROS generation in diabetes. Redox Biol 37:101710. https://doi.org/10.1016/j.redox.2020.101710
84. Wang L, Zhao J, Gao Y, Zhang X (2023) Mechanistic insights into glucose auto-oxidation in diabetes and metabolic syndrome. Front Endocrinol 14:1156231. https://doi.org/10.3389/fendo.2023.1156231
85. Liu J, Zhang Y, Chen Q, Sun L (2021) Oxidative stress induced by glucose auto-oxidation and its impact on metabolic syndrome. Oxidative Med Cell Longev 2021:5582345. https://doi.org/10.1155/2021/55823
86. Chen Y, Xu T, Wang Y, Liu R (2023) The role of glucose auto-oxidation in the progression of diabetic complications. Free Radic Biol Med 201:20–30. https://doi.org/10.1016/j.freeradbiomed.2023.02.011
87. Zhang R, Liu P, Wang S, Li X (2024) Targeting glucose auto-oxidation in metabolic disorders: current perspectives. Pharmacol Res 199:106721. https://doi.org/10.1016/j.phrs.2024.10672187
88. Wang L, Zhao J, Gao Y, Zhang X (2023) Pathophysiological roles of AGEs in metabolic syndrome: recent advances. Front Endocrinol 14:1167123. https://doi.org/10.3389/fendo.2023.1167123
89. Ahmed N, Moinuddin, Ali A (2022) Non-enzymatic glycation and its contribution to oxidative stress in metabolic disorders. Life Sci 292:120331. https://doi.org/10.1016/j.lfs.2021.1203316
90. Sun X, Li W, Zhang Q, Chen H (2021) AGE accumulation and its role in insulin resistance and metabolic dysfunction. Redox Biol 44:102011. https://doi.org/10.1016/j.redox.2021.102011
91. Chen Y, Xu T, Wang Y, Liu R (2023) The AGE–RAGE axis in metabolic disorders: mechanisms and therapeutic implications. Pharmacol Res 189:106699. https://doi.org/10.1016/j.phrs.2023.106699
92. Zhang R, Liu P, Wang S, Li X (2024) The contribution of advanced glycation end products to metabolic disease: mechanistic insights. Biomed Pharmacother 168:115690. https://doi.org/10.1016/j.biopha.2024.115690
93. Zhao Q, Wang J, Xu Y (2021) Advanced glycation end products and vascular complications in diabetes. Life Sci 276:119401. https://doi.org/10.1016/j.lfs.2021.119401
94. Liu J, Zhang Y, Chen Q, Sun L (2022) Dietary advanced glycation end products and their impact on metabolic health. Nutrients 14(2):312. https://doi.org/10.3390/nu14020312
95. Xu Y, Chen J, Li X, Zhang R (2023) Targeting AGE-RAGE signaling for the treatment of metabolic syndrome. Front Pharmacol 14:1198765. https://doi.org/10.3389/fphar.2023.1198765
96. Chen Y, Wang Y, Xu T, Zhang R (2023) Lipid peroxidation and its implications in insulin resistance and metabolic syndrome. Redox Biol 64:102789. https://doi.org/10.1016/j.redox.2023.102789
97. Liu P, Zhao J, Zhang Y, Xu Y (2022) Lipid peroxidation-derived aldehydes and metabolic syndrome: pathophysiological relevance. Front Endocrinol 13:856732. https://doi.org/10.3389/fendo.2022.856732
98. Zhang Q, Wang J, Xu Y (2020) Lipid peroxidation in obesity and insulin resistance: mechanistic insights. Life Sci 256:117945. https://doi.org/10.1016/j.lfs.2020.117945
99. Hu J, Zhang X, Liu Q, Sun M (2023) The role of reactive aldehydes from lipid peroxidation in metabolic diseases. Free Radic Biol Med 201:119–130. https://doi.org/10.1016/j.freeradbiomed.2023.02.0197

100. Li X, Chen Q, Zhang L, Wang S (2024) Lipid peroxidation and mitochondrial dysfunction in metabolic disorders: emerging insights. Biomed Pharmacother 170:116876. https://doi.org/10.1016/j.biopha.2024.116876
101. Zhao Z, Liu H, Xu Y, Zhang M (2021) Lipid peroxidation and mitochondrial dysfunction in the pathogenesis of metabolic disorders. Metabolism 121:154834. https://doi.org/10.1016/j.metabol.2021.154834
102. Sun X, Gao Y, Li W, Chen H (2022) Targeting lipid peroxidation in metabolic syndrome: therapeutic potential and challenges. Front Pharmacol 13:963247. https://doi.org/10.3389/fphar.2022.963247
103. Zhang Q, Liu Y, Wang J, Xu Y (2021) Iron overload and oxidative stress in metabolic syndrome: pathophysiology and therapeutic perspectives. Life Sci 273:119311. https://doi.org/10.1016/j.lfs.2021.119311
104. Wang L, Zhao J, Gao Y, Zhang X (2023) Iron-driven oxidative stress and metabolic dysfunction: recent advances. Free Radic Biol Med 201:10–22. https://doi.org/10.1016/j.freeradbiomed.2023.02.015
105. Chen Y, Wang X, Zhang Q, Zhao J (2022) Iron-catalyzed oxidative stress in the pathogenesis of metabolic syndrome. Redox Biol 53:102329. https://doi.org/10.1016/j.redox.2022.102329
106. Li R, Liu J, Xu Y, Zhang M (2024) Dysregulated iron metabolism and oxidative stress in metabolic disorders: mechanistic insights and therapeutic strategies. Biomed Pharmacother 170:116945. https://doi.org/10.1016/j.biopha.2024.116945
107. Liu P, Sun M, Wang Y, Chen H (2023) Iron homeostasis and its implications in obesity and insulin resistance. Front Endocrinol 14:1132456. https://doi.org/10.3389/fendo.2023.1132456
108. Sun X, Zhang L, Li W, Chen H (2021) Ferroptosis and metabolic diseases: emerging connections. Front Cell Dev Biol 9:733944. https://doi.org/10.3389/fcell.2021.733944
109. Zhao Z, Li H, Xu Y, Zhang M (2024) Ferroptosis as a therapeutic target in metabolic disorders. Pharmacol Res 199:106751. https://doi.org/10.1016/j.phrs.2024.106751
110. Xu Y, Chen J, Li X, Zhang R (2023) Targeting iron-catalyzed ROS generation to mitigate metabolic complications. Front Pharmacol 14:1198763. https://doi.org/10.3389/fphar.2023.1198763

References for Point 5 and 6

111. Lennicke C, Cochemé HM (2021) Redox metabolism: ROS as specific molecular regulators of cell signaling and function. Mol Cell 81(18):3691–3707
112. Juan CA, Pérez De La Lastra JM, Plou FJ, Pérez-Lebeña E (2021) The Chemistry of Reactive Oxygen Species (ROS) revisited: outlining their role in biological macromolecules (DNA, lipids and proteins) and induced pathologies. Int J Mol Sci. 22(9):4642
113. Forrester SJ, Kikuchi DS, Hernandes MS, Xu Q, Griendling KK (2018) Reactive oxygen species in metabolic and inflammatory signaling. Circ Res 122(6):877–902
114. Hajam YA, Rani R, Ganie SY, Sheikh TA, Javaid D, Qadri SS et al (2022) Oxidative stress in human pathology and aging: molecular mechanisms and perspectives. Cells 11(3):552
115. Andrés CMC, Pérez De La Lastra JM, Juan CA, Plou FJ, Pérez-Lebeña E (2024) Antioxidant metabolism pathways in vitamins, polyphenols, and selenium: parallels and divergences. Int J Mol Sci 25(5):2600
116. Masenga SK, Kabwe LS, Chakulya M, Kirabo A (2023) Mechanisms of oxidative stress in metabolic syndrome. Int J Mol Sci 24(9):7898
117. Bhatti JS, Bhatti GK, Reddy PH (2017 May) Mitochondrial dysfunction and oxidative stress in metabolic disorders — a step towards mitochondria based therapeutic strategies. Biochim Biophys Acta BBA Mol Basis Dis 1863(5):1066–1077

118. Rotariu D, Babes EE, Tit DM, Moisi M, Bustea C, Stoicescu M et al (2022) Oxidative stress—complex pathological issues concerning the hallmark of cardiovascular and metabolic disorders. Biomed Pharmacother 152:113238
119. Newsholme P, Cruzat VF, Keane KN, Carlessi R, De Bittencourt PIH (2016) Molecular mechanisms of ROS production and oxidative stress in diabetes. Biochem J 473(24):4527–4550
120. Lindsay RT, Rhodes CJ (2025) Reactive oxygen species (ROS) in metabolic disease—Don't shoot the metabolic messenger. Int J Mol Sci 26(6):2622
121. De Almeida AJPO, De Oliveira JCPL, Da Silva Pontes LV, De Souza Júnior JF, Gonçalves TAF, Dantas SH et al (2022) ROS: basic concepts, sources, cellular signaling, and its implications in aging pathways. Giustarini D, editor. Oxid Med Cell Longev. 2022(1):1225578
122. Bešlo D, Golubić N, Rastija V, Agić D, Karnaš M, Šubarić D et al (2023) Antioxidant activity, metabolism, and bioavailability of polyphenols in the diet of animals. Antioxidants 12(6):1141
123. Nuñez-Selles AJ, Nuñez-Musa RA, Guillen-Marmolejos RA (2025) Linking oxidative stress biomarkers to disease progression and antioxidant therapy in hypertension and diabetes mellitus. Front Mol Biosci 12:1611842
124. Chaudhary P, Janmeda P, Docea AO, Yeskaliyeva B, AbdullRazis AF, Modu B et al (2023) Oxidative stress, free radicals and antioxidants: potential crosstalk in the pathophysiology of human diseases. Front Chem 11:1158198
125. Jena AB, Samal RR, Bhol NK, Duttaroy AK (2023) Cellular red-ox system in health and disease: the latest update. Biomed Pharmacother 162:114606
126. Heimes F, Berendes LS, Hannibal L, Park J (2025) Antioxidant therapy in inborn metabolic diseases. Mol Genet Metab 145(4):109176
127. Tumurkhuu M, Bhat SA, Hossain MZ, Shafiq M, Hasnain MS, Nayak AK et al (2025) Editorial: cellular and molecular mechanisms in metabolic disorders: role of inflammation and oxidative stress. Front Pharmacol 16:1580553
128. Wang J, Liao B, Wang C, Zhong O, Lei X, Yang Y (2022) Effects of antioxidant supplementation on metabolic disorders in obese patients from randomized clinical controls: a meta-analysis and systematic review. Medina S, editor. Oxid Med Cell Longev. 2022(1):7255413
129. Tun S, Spainhower CJ, Cottrill CL, Lakhani HV, Pillai SS, Dilip A et al (2020) Therapeutic efficacy of antioxidants in ameliorating obesity phenotype and associated comorbidities. Front Pharmacol 11:1234

Chapter 4
ROS-Mediated Cellular Signaling Pathways in Metabolic Disorders

Sathish Kumar Gunasekaran, Mirunalini Gobinath, Parikshit Roychowdhury, Naina Mohamed Pakkir Maideen, and Gowthamarajan Kuppusamy

4.1 Introduction

Reactive oxygen species (ROS) were long thought to be harmful by-products of aerobic metabolism that killed cells and sped up aging. But this perspective has changed a lot in the previous few decades, which is completely obvious that ROS are important signaling molecules that are involved in many cellular activities at normal levels. The breakdown of this fragile redox signaling network has become a major cause of many metabolic disorders. ROS are a collection of molecules that include oxygen which are chemically reactive. They are made when molecular oxygen (O_2) breaks down naturally in the body. These types of cells play vital functions in cell signaling and keeping everything in balance. The partial reduction of O_2

S. K. Gunasekaran
Department of Pharmacology, JSS College of Pharmacy, JSS Academy of Higher Education & Research,
Mysuru, Karnataka, India

M. Gobinath
Department of Pharmacognosy & Phytopharmacy, JSS College of Pharmacy, JSS Academy of Higher Education & Research, Ooty, Tamil Nadu, India

P. Roychowdhury · G. Kuppusamy (✉)
Department of Pharmaceutics, JSS College of Pharmacy, JSS Academy of Higher Education & Research,
Ooty, Tamil Nadu, India
e-mail: gowthamsang@jssuni.edu.in

N. M. P. Maideen
College of Medicine, Mohammed Bin Rashid University of Medicine and Health Sciences, Dubai, UAE

Dubai Health, Dubai, UAE
e-mail: nmmaideen@dubaihealth.ae

S. Ashique, N. Mishra (eds.), *Reactive Oxygen Species in Metabolic Inflammation*,
https://doi.org/10.1007/978-3-032-11752-6_4

usually leads to the formation of ROS, which can be split into two main groups: free radicals and non-radicals. The first type, free radicals, are molecules that have one or more unpaired electrons. This makes them very reactive. The superoxide anion (O_2·−) is a good example of this. It is made by reducing O_2 by one electron. It is thought to be the principal ROS because it is the first one to generate and is mostly made in the mitochondria by the electron transport chain (ETC) [1]. The hydroxyl radical (·OH) is another important free radical. It is the most reactive and harmful ROS. The Fenton or Haber-Weiss reactions, which are sped up by transition metals like iron (Fe2+) or copper (Cu+), can turn hydrogen peroxide into it [2]. The second group is made up of non-radicals, which are chemically reactive entities that do not have unpaired electrons but can easily turn into free radicals. Hydrogen peroxide (H_2O_2) is the most important non-radical ROS. Superoxide dismutation, which happens on its own or with the help of superoxide dismutase (SOD) enzymes, creates this. H_2O_2 is a stable molecule that can pass across membranes, unlike charged ROS. This makes it a great signaling molecule that can move about inside cells [3]. This chapter will explain the basic ideas behind ROS biology, from their chemical structure to the complicated ways they help keep cells in balance. It will also look at how problems with ROS can lead to common diseases like obesity, type 2 diabetes, nonalcoholic fatty liver disease (NAFLD), and heart problems like atherosclerosis, as well as inherited amino acid metabolic disorders.

4.2 Concept of Oxidative Stress Homeostasis

Cells have developed a complex mechanism to regulate the generation of ROS alongside their elimination, a condition referred to as redox homeostasis. This equilibrium is essential for sustaining optimal cellular function and depends on a network of both enzymatic and nonenzymatic antioxidants. The enzymatic antioxidants form a primary line of defense. This group includes superoxide dismutase (SOD1/2/3), which convert the superoxide radical (O_2·−) to hydrogen peroxide (H_2O_2); catalase (CAT), which decomposes H_2O_2 into water and oxygen; and the glutathione peroxidase (GPx) and thioredoxin (Trx) systems, which are responsible for reducing H2O2 and others [4]. Alongside these enzymes, the cell utilizes nonenzymatic antioxidants. These are small molecules capable of scavenging free radicals, with prominent examples being glutathione (GSH), vitamin C (ascorbic acid), vitamin E (α-tocopherol), and various carotenoids. Oxidative stress occurs when this delicate balance is disrupted, either due to excessive ROS production or a deficiency in the antioxidant defense system. This imbalance leads to a shift toward a more pro-oxidative state, resulting in the indiscriminate oxidation of cellular macromolecules, including lipids, proteins, and DNA. This damage can disrupt cellular function and, if severe and protracted, result in cell death and contribute to the pathogenesis of various diseases, including metabolic disorders such as obesity, type 2 diabetes, and nonalcoholic fatty liver disease (NAFLD) [5].

4.2.1 ROS Physiology in Cellular Metabolism

ROS are not just waste products; they are important parts of how metabolism works. At low, controlled levels, they operate as second messengers, changing the activity of different signaling pathways that are important for keeping the body's metabolism in balance. One of the most well-known activities of ROS is in the signaling of insulin. When insulin binds to its receptor, NADPH oxidases (NOX) generate ROS for a short time. This burst of peroxides temporarily stops protein tyrosine phosphatases (PTPs) like PTP1B from dephosphorylating and inactivating the insulin receptor and its substrate (IRS-1). This inhibition keeps the phosphorylation cascade going, which makes the insulin signal stronger and helps the body take in more glucose [6, 7].

The difference between the signaling and harmful effects of ROS is a key idea in modern redox biology. It marks a move from the original conception of these molecules as only harmful. The amount, location, and length of time that ROS are made are the main things that determine its dual character [8]. At low, transient, and localized levels, ROS function as highly specific signaling molecules, or "eustress," essential for physiological processes. The chemical features of hydrogen peroxide are mostly what give it this signaling ability. It is relatively stable, can pass through membranes, and can selectively oxidize certain cysteine residues on target proteins, changing their function [9, 10]. This specific oxidation can be undone by cellular antioxidant systems like thioredoxin, which lets you change the signaling pathways on the fly [11]. For instance, the short-lived ROS burst that NADPH oxidase 4 (NOX4) makes is very important for insulin signaling and the differentiation of adipocytes [12, 13]. In the same way, mitochondria-derived ROS are not merely waste products; they are necessary for cells to adapt to low oxygen levels by stabilizing hypoxia-inducible factor 1-alpha (HIF-1α) and controlling autophagy [14, 15]. This controlled synthesis makes sure that ROS serve as exact second messengers, changing the pathways that control cell growth, division, and metabolic adaption [16, 17].

On the other hand, when the cell's ability to fight off ROS is overwhelmed, it goes into a state of oxidative stress, which causes broad and nonspecific damage to cells [5]. Pathological situations, including metabolic overload in obesity and diabetes, can cause this change from signaling to harm. In these cases, too much food leads to a huge rise in mitochondrial respiration and the release of ROS [18, 19]. When this happens, extremely reactive species like the hydroxyl radical (·OH) are generated, which destroy cellular macromolecules without caring who they are [20]. This causes lipid peroxidation, which damages the structure and function of membranes [21]; protein carbonylation and aggregation, which slows down enzyme activity and cellular processes [22]; and DNA damage, like the formation of 8-oxo-2′-deoxyguanosine (8-oxodG), which can cause mutations and genomic instability [23]. This broad oxidative damage messes up the signaling pathways that ROS typically controls, which directly causes cells to stop working and disease to start [24, 25]. For example, long-term oxidative stress in metabolic tissues makes insulin receptor signaling less effective, causes inflammation by turning on NF-κB, and causes apoptosis, all of which are signs of metabolic diseases [26–29]. So the most important things that decide whether ROS are helpful or harmful are the cellular context and the fragile balance of redox homeostasis [30, 31].

4.2.2 Sources of ROS in Tissue Metabolism

There are many different cellular sources of ROS, and different organelles and enzyme systems all play a role in the overall redox environment. In metabolic diseases, the mitochondria, the NADPH oxidase family, and the endoplasmic reticulum are the main places where ROS are produced. To figure out how these sources make ROS, we need to know how they work. This is important for figuring out what function they play in disease development (Fig. 4.1).

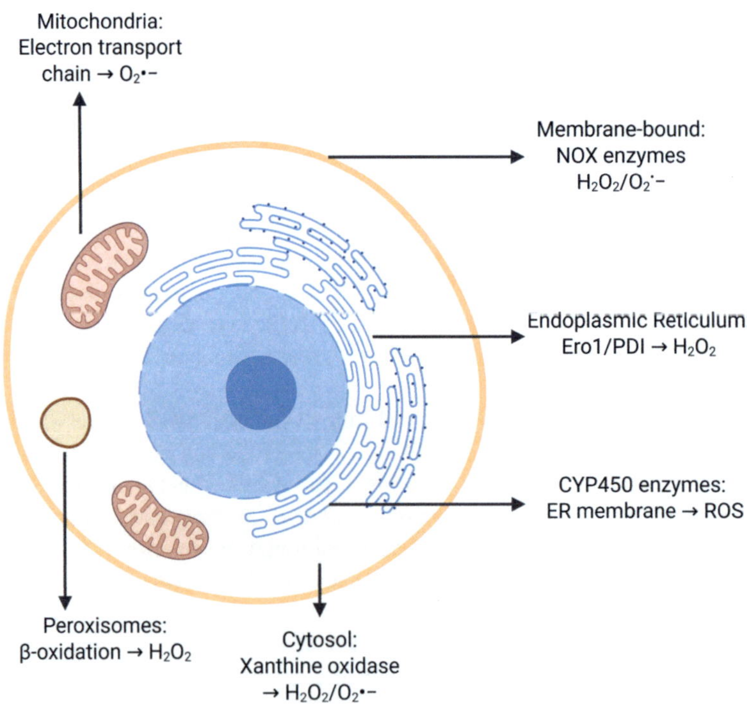

Fig. 4.1 Oxidative stress mechanisms in intracellular metabolism

4.3 Mitochondria and ROS in Metabolic Dysfunction

Mitochondria are the powerhouses of the cell, but they are also the principal source of endogenous ROS (mitoROS) [32]. While a small amount of ROS production is an unavoidable consequence of aerobic respiration, this process is significantly exacerbated in metabolic diseases.

4.3.1 Electron Transport Chain and Complex I/III-Derived ROS

The mitochondrial electron transport chain (ETC), which has four protein complexes (I–IV), is where most mitoROS are made. Electrons from NADH and FADH2 move along the ETC to molecular oxygen, which is the last electron acceptor, during oxidative phosphorylation. But a small number of electrons, about 0.1–2%, can leak out too soon and combine with oxygen to make superoxide (O2·−) [33]. Complex I (NADH: ubiquinone oxidoreductase) and Complex III (ubiquinone: cytochrome c oxidoreductase) are the main places where this electron leakage happens [34]. At Complex I, superoxide is mostly discharged into the mitochondrial matrix, often by a process termed reverse electron transfer (RET). This happens when the mitochondrial membrane potential is high and the coenzyme Q pool is very low [35]. During the Q-cycle at Complex III, superoxide can be let out into both the matrix and the space between the membranes [36].

4.3.2 Mitochondrial Dysfunction in Diabetes and Obesity

Tissues are perpetually exposed to elevated levels of glucose and free fatty acids when the body is experiencing metabolic excess, such as in obesity or type 2 diabetes. This surge of nutrients sends too many reducing equivalents (NADH and FADH2) into the electron transport chain, which is more than ATP synthase can manage [19]. When the proton gradient goes up and the mitochondrial membrane potential stays high, the flow of electrons slows down, and some of them start to leak out, which makes superoxide production go up a lot [37]. The extra mitochondrial ROS then hurt mitochondrial DNA, proteins, and lipids, starting a cycle of malfunction and even more ROS that keep going on its own [38]. This cascade is a big reason why skeletal muscle, liver, and fat tissue become resistant to insulin [39].

4.3.3 Metabolic Reprogramming and mitoROS Signaling

Beyond their damaging effects, mitoROS also act as signals to initiate adaptive responses. For instance, a mild increase in mitoROS can trigger mitochondrial biogenesis through the PGC-1α pathway to enhance respiratory capacity [40]. They are also essential for initiating mitophagy, the selective removal of damaged mitochondria, which is a critical quality control mechanism [41]. But in chronic metabolic diseases, this protective signaling systems can get overloaded or go off track. When mitoROS levels stay high, the signal can flip from helping the cell adapt to pushing it toward harmful changes, like shifting metabolism toward aerobic glycolysis (the Warburg effect) in some cells or ramping up pro-inflammatory pathways both of which can fuel disease progression [42, 43].

4.3.4 NADPH Oxidase-Mediated ROS Generation

The NADPH oxidase (NOX) family of enzymes represents the only known enzyme system whose primary function is to generate ROS [44]. Unlike mitochondria, which produce ROS as a by-product, NOX enzymes actively produce superoxide and/or hydrogen peroxide in a tightly regulated manner. Seven different NOX isoforms have been found in humans (NOX1–5 and DUOX1–2). Several of them are active in key metabolic tissues, each with their own role in health and disease. NOX1 shows up in high amounts in vascular smooth muscle cells and the lining of the colon, and it has been linked to angiotensin II-driven high blood pressure and vascular problems in diabetes [45]. The most well-known form, NOX2 (also called gp91phox), is famous for powering the phagocytic respiratory burst that helps immune cells kill pathogens. But it is not limited to immune cells; it is also present in endothelial cells, heart muscle cells, and fat cells. In these settings, NOX2 activity fuels inflammation and contributes to insulin resistance [46, 47]. In contrast, NOX4 is constitutively active and is a major source of H_2O_2 in many tissues, including the kidney, heart, and adipose tissue. It has a complex, dual role; while it is essential for physiological processes like adipocyte differentiation [12], its overexpression is strongly linked to diabetic nephropathy and cardiac fibrosis [48, 49]. Lastly, NOX5, which is activated by intracellular calcium, is implicated in endothelial dysfunction and atherosclerosis [50].

Most NOX isoforms, especially NOX1 and NOX2, do not just switch on by themselves. They need a multi-protein team to come together, including cytosolic regulators like p47phox, p67phox, p40phox, and the small GTPase Rac [44]. This assembly happens in response to triggers that are common in metabolic disorders, such as high blood sugar, advanced glycation end products (AGEs), angiotensin II, and inflammatory signals like TNF-α [51, 52]. NOX4 is different. It is generally active all the time, and its activity is controlled mostly by how much of it is made at the gene expression level [48].

The overactivity of NOX enzymes is a major cause of problems that come with diabetes. In diabetic nephropathy, the activation of NOX4 and NOX5 in podocytes and mesangial cells causes ROS-driven fibrosis and damage to the kidneys [49, 53]. NOX1 and NOX2 play a role in retinal inflammation and neovascularization in diabetic retinopathy [54]. Also, ROS that come from NOX are important in the development of atherosclerosis because they cause endothelial dysfunction, monocyte infiltration, and foam cell production in the arterial wall [55].

4.4 Endoplasmic Reticulum Stress and ROS Production

The endoplasmic reticulum (ER) is a critical organelle for protein synthesis, folding, and modification, as well as lipid biosynthesis. The ER is increasingly recognized as a significant source of ROS and a key player in the inflammatory processes that characterize metabolic diseases. The inside of the ER has a special oxidative environment that is crucial for making disulfide bonds, which are key for many secretory and membrane proteins to fold correctly. Protein disulfide isomerase (PDI) carries out this step, working in tandem with ER oxidoreductin 1 (Ero1). Ero1 passes electrons to molecular oxygen and, in the process, produces hydrogen peroxide (H2O2) as a direct by-product [56, 57]. Normally, antioxidants in the ER keep this reactive oxygen species (ROS) production in check. But when the demand for protein folding rises like when the body ramps up insulin production in response to high blood sugar, misfolded proteins can pile up, and Ero1 activity can generate more ROS than the system can handle [58].

The cell goes into a state called ER stress when unfolded or misfolded proteins stack up in the ER lumen. The unfolded protein response (UPR) is an adaptive signaling network that kicks in to bring the ER back into balance [59]. If the stress is too strong or lasts too long, the UPR can go from protecting cells to causing them to die and inflammation. ER stress and reactive oxygen species (ROS) feed into each other in a harmful loop [60]. The higher the demand for protein folding, the more ROS are produced. In turn, ROS damage the protein-folding process and ER function, which deepens the stress. Two major UPR branches, IRE1α and PERK, can switch on pro-inflammatory pathways like JNK and NF-κB [27, 61]. This interaction fuels the chronic, low-grade inflammation, or "meta-inflammation," which ties obesity to insulin resistance and type 2 diabetes [62].

Mitochondria, NOX enzymes, and the endoplasmic reticulum (ER) are important generators of reactive oxygen species (ROS), but other enzyme systems also add a lot to the oxidative load in metabolic disorders. For example, the enzyme xanthine oxidase (XO) speeds up the process of turning hypoxanthine and xanthine into uric acid. This process makes both superoxide (O2·−) and hydrogen peroxide (H2O2). XO activity is often higher in people with hyperuricemia, which is often linked to metabolic syndrome and is thought to cause problems with blood vessels [63]. In addition, various parts of cells and pathways also help make ROS. Peroxisomes, which are necessary for the β-oxidation of very-long-chain fatty

Table 4.1 ROS-generating system in metabolic tissues

ROS source	Primary location	ROS produced	Key activators in metabolic disease	Primary role in pathophysiology
Mitochondria (ETC)	Inner mitochondrial membrane	Superoxide (O2·−)	Nutrient overload (high glucose, FFAs), high membrane potential	Drives insulin resistance, β-cell failure, hepatic steatosis
NADPH oxidase 1 (NOX1)	Plasma membrane (e.g., vascular smooth muscle)	Superoxide (O2·−)	Angiotensin II, inflammatory cytokines	Vascular inflammation, hypertension
NADPH oxidase 2 (NOX2)	Plasma membrane (e.g., phagocytes, endothelial cells)	Superoxide (O2·−)	Inflammatory stimuli (LPS, TNF-α), hyperglycemia, AGEs	Adipose tissue inflammation, atherosclerosis
NADPH oxidase 4 (NOX4)	ER, mitochondria (e.g., adipocytes, kidney cells)	Hydrogen peroxide (H2O2)	High glucose, TGF-β, chronic metabolic stress (regulated by expression)	Adipocyte dysfunction, diabetic nephropathy, cardiac fibrosis
Endoplasmic reticulum (Ero1)	ER lumen	Hydrogen peroxide (H2O2)	High protein folding load (e.g., proinsulin synthesis), ER stress	ER stress-ROS crosstalk, β-cell dysfunction, hepatic inflammation

acids, make a lot of H2O2. Normally, peroxisomal catalase breaks down this substance, but when there is too much lipids, too much H2O2 might leak into the cytosol [64]. Like this, cytochrome P450 enzymes, which are mostly found in the ER, can become "uncoupled" while they are breaking down xenobiotics and endogenous chemicals. This can cause the release of superoxide and H2O2 [65]. Also, enzymes in the arachidonic acid pathway, like lipoxygenases (LOX) and cyclooxygenases (COX), help with inflammation and generate ROS as a consequence of their catalytic activity [66] (Table 4.1).

4.5 Important Signaling Pathways in ROS-Mediated Cell

Reactive oxygen species (ROS), especially hydrogen peroxide (H_2O_2), act as signaling molecules by causing specific and reversible changes to proteins after they are made, most often by oxidizing reactive cysteine thiol groups [25]. These redox changes can replace key kinases, phosphatases, and transcription factors on or off, shaping major signaling pathways. In metabolic disorders, however, constant overproduction of ROS leads to long-term, abnormal control of these pathways, which fuels inflammation, insulin resistance, and cellular damage.

4.5.1 ROS-Mediated NF-κB Activation Mechanisms

NF-κB, which stands for nuclear factor kappa-light-chain-enhancer of activated B cells, is a major regulator of inflammation, immunological responses, and cell survival. When it is not in balance, it causes the kind of long-lasting, low-level inflammation that causes numerous metabolic illnesses [67]. An inhibitory protein called IκB (inhibitor of κB) keeps NF-κB in the cytoplasm when it is not active. The IκB kinase (IKK) complex phosphorylates IκB, which causes IκB to be ubiquitinated and broken down by the proteasome. This is the usual way that NF-κB is activated. This lets NF-κB go to the nucleus and start the transcription of target genes [68]. At different amounts, ROS can turn on this pathway. H2O2 can turn on the IKK complex directly or indirectly. This could happen via oxidizing and stopping phosphatases that stop IKK from working or by turning on upstream kinases like Akt and MEKK1 [69, 70]. Also, ROS can turn on unusual types of protein kinase C (PKC), which can also add phosphate groups to the IKK complex and turn it on [71]. There is also some evidence that ROS can directly change parts of the NF-κB pathway, such as IKK and NF-κB itself, to make them more active [72].

Excess free fatty acids and hyperglycemia in metabolic organs stimulate ROS generation from mitochondria and NOX enzymes, resulting in persistent NF-κB activation [26]. The production of pro-inflammatory cytokines such as TNF-α and IL-6 in adipose tissue induces insulin resistance both locally and systemically [73]. NF-κB activation in the liver can lead to insulin resistance and nonalcoholic steatohepatitis (NASH) [74]. In pancreatic β cells, NF-κB-mediated inflammation can cause cellular malfunction and death, reducing insulin production [75].

A detrimental positive feedback loop often exists between ROS, NF-κB, and cytokine signaling. ROS-activated NF-κB stimulates the production of cytokines like TNF-α. In turn, TNF-α, upon binding to its receptor, can activate NOX enzymes, leading to further ROS production and amplification of the NF-κB signal [76]. This feed-forward mechanism perpetuates a state of chronic inflammation, which is a key driver of metabolic disease progression [77].

4.5.2 MAPK Signaling Cascades

The mitogen-activated protein kinase (MAPK) cascades are crucial signaling pathways that convert a wide range of extracellular stimuli into cellular responses, including proliferation, differentiation, and stress adaptation. The three major MAPK families, ERK1/2, JNK, and p38 are redox sensitive and play pivotal roles in metabolic regulation and disease.

Stress-activated kinases, especially JNK, play a big role in causing insulin resistance. When JNK is switched on, it can phosphorylate insulin receptor substrate 1 (IRS-1) at certain inhibitory serine sites, like Ser307 [78, 79]. This stops the insulin receptor from adding the tyrosine phosphorylation that IRS-1 needs to work properly. As a result, the PI3K-Akt pathway does not get activated, and glucose uptake is reduced. This is a common end point for different types of metabolic stress—things like reactive oxygen species, ER stress, and inflammatory cytokines can all trigger it [80].

4.5.3 ERK1/2 Pathway and Metabolic Regulation

The extracellular signal-regulated kinase (ERK1/2) pathway is typically associated with growth and proliferation. While it plays a role in mediating some of insulin's effects, its chronic activation by oxidative stress can have detrimental metabolic consequences [81]. ROS can activate the ERK pathway by oxidizing and inhibiting MAPK phosphatases (MKPs), such as MKP3, which normally restrain ERK activity [82].

4.5.4 JNK and p38 MAPK in Stress Responses

The c-Jun N-terminal kinase (JNK) and p38 MAPK pathways switch on mainly in response to cellular stress, such as oxidative stress or inflammatory cytokines [83]. Reactive oxygen species (ROS) are strong triggers for these stress kinases. One common route is through oxidation that disables upstream inhibitors. For example, glutathione S-transferase pi (GSTP) normally binds to and holds JNK in check, but oxidation can break that hold. Another example is apoptosis signal-regulating kinase 1 (ASK1), a MAPK kinase kinase that activates when its inhibitor, thioredoxin, detaches. This detachment happens after thioredoxin is oxidized [84, 85].

4.5.5 PI3K-Akt Signaling Network

The phosphoinositide 3-kinase (PI3K)-Akt (also known as protein kinase B) pathway is the central node in the insulin signaling cascade, responsible for most of the metabolic actions of insulin, including glucose transport, glycogen synthesis, and protein synthesis. It is not easy to understand how ROS controls the PI3K-Akt pathway. A short burst of ROS, like the one that happens when NOX enzymes are stimulated by insulin, can quickly boost pathway activation. The main way this is done is by stopping phosphatases that block the route from working [7]. When insulin binds to the insulin receptor, it adds phosphate groups to IRS proteins, which make docking sites for PI3K. Then, PI3K makes phosphatidylinositol (3,4,5)-trisphosphate (PIP3), which brings Akt and its activating kinase, PDK1, to the membrane. This causes Akt to be phosphorylated and turned on [86]. When Akt is turned on, it adds phosphate groups to several downstream targets to carry out insulin's effects. For example, it moves the GLUT4 glucose transporter to the cell membrane in muscle and fat tissue [87].

The lipid phosphatase and tensin homolog (PTEN) are the main negative regulators of the PI3K-Akt pathway. It does this by dephosphorylating PIP3 back to PIP2, which stops the signal [88]. PTEN is very susceptible to oxidation. H2O2 can oxidize its catalytic cysteine residue in a way that can be reversed, which makes it inactive [89]. This oxidative inactivation of PTEN takes the "brake" off the PI3K-Akt pathway, which makes insulin signaling stronger and lasts longer. This is a normal way for the body to boost the effects of insulin in the short term [90]. However, long-term oxidative stress can cause PTEN to become permanently oxidized and signal to become out of control, which can lead to diseases like cancer. On the other hand, when insulin resistance is present, other ROS-mediated mechanisms, like JNK activation, take over and stop the PI3K-Akt pathway, even though PTEN could be inhibited [91].

4.5.6 Keap1-Nrf2-ARE Antioxidant Response

In contrast to the pro-inflammatory and stress-inducing pathways, the Keap1-Nrf2 pathway is the primary cellular defense mechanism against oxidative stress. It is a master regulator of the antioxidant and cytoprotective response. The repressor protein Kelch-like ECH-associated protein 1 (Keap1) keeps the transcription factor nuclear factor erythroid 2-related factor 2 (Nrf2) at low levels by targeting it for ubiquitination and proteasomal destruction [92]. Keap1 has a lot of cysteine residues that are very reactive and act as sensors for oxidants and electrophiles. When ROS or other electrophiles modify these cysteines, Keap1 changes shape in a way that stops it from breaking down Nrf2 [93]. This lets the new Nrf2 to build up, become stable, and go to the nucleus [94]. Nrf2 heterodimerizes with tiny Maf proteins in the nucleus and binds to a DNA sequence called the antioxidant response element (ARE) in the promoter regions of its target genes [95]. This starts the transcription of more than 200 genes that safeguard cells. These include enzymes that help protect against free radicals (like SOD, catalase, and GPx), make and recycle glutathione (like GCLC, GCLM, and GSR), and get rid of toxins (like NQO1 and HO-1) [96, 97].

The Nrf2 pathway is very important for protecting against metabolic disorders because it fights oxidative stress and inflammation. Studies have indicated that activating Nrf2 can make insulin work better, lower body fat, and protect against diabetes problems such nephropathy and cardiomyopathy [98, 99]. Clinical investigations have revealed that pharmacological activators of Nrf2, including bardoxolone methyl, may be able to help people with diabetic kidney disease [100]. The Nrf2 system is an important way for cells to adapt to persistent oxidative stress that comes with metabolic diseases. It does this by strengthening the cell's own antioxidant defenses [101].

4.6 ROS-Mediated Metabolic Dysfunction

Disruption of redox homeostasis is not just a result of metabolic disease; it is a major cause of it. Chronic oxidative stress, which comes from the sources and activates the pathways mentioned above, is a direct cause of the cellular and organ dysfunction that is a hallmark of metabolic diseases. Oxidative stress is a unifying mechanism that links nutrient overload to both the failure of insulin-producing pancreatic β cells and the development of insulin resistance in peripheral tissues, the two core defects in type 2 diabetes [37].

Oxidative stress is especially harmful to pancreatic β cells. They have low quantities of important antioxidant enzymes like catalase and glutathione peroxidase, which makes them more likely to be harmed by ROS [102]. Chronic hyperglycemia makes β cells work harder to make insulin, which speeds up mitochondrial metabolism and raises mitoROS levels. This oxidative stress can make it harder for the important transcription factor PDX-1 to attach to the insulin promoter, which can affect insulin gene expression [103]. Also, ROS can start apoptotic pathways, which causes β-cell mass to slowly decrease and insulin secretion to drop, which eventually leads to the development of diabetes [104, 105].

In tissues outside of the brain, like skeletal muscle, liver, and fat, ROS-induced insulin resistance is mostly caused by the activation of signaling pathways that are susceptible to stress. ROS activate JNK and other stress kinases that add phosphate groups to IRS-1 on serine residues that stop it from working. This separates it from the insulin receptor and stops PI3K-Akt signaling further down the line [78, 79]. ROS also cause inflammation through NF-κB, which makes insulin resistance worse by making inflammatory cytokines [73, 74]. This problem with the insulin signaling cascade stops these tissues from taking in and using glucose normally, which leads to high blood sugar levels throughout the body.

Hyperglycemia alone keeps oxidative stress going in a bad way. High glucose levels can cause more ROS to be made in a number of ways: (1) by increasing flux through the mitochondrial ETC, (2) by activating NOX enzymes, (3) by encouraging the production of advanced glycation end products (AGEs), which bind to their receptor (RAGE) and start making ROS, and (4) by sending extra glucose down pathways like the polyol pathway, which uses up NADPH and weakens the glutathione antioxidant system [37, 106]. This cycle of high blood sugar and oxidative stress that keeps going is a major cause of the long-term microvascular (nephropathy, retinopathy) and macrovascular (atherosclerosis) problems that come with diabetes [107].

Obesity is characterized by the expansion of adipose tissue, which becomes a major source of ROS and pro-inflammatory signals, a state often termed "adipo-inflammation" [108]. In obesity, enlarged adipocytes develop mitochondrial dysfunction and show increased activity of NOX4, resulting in elevated ROS generation [47, 109]. This oxidative stress, together with reduced oxygen availability in adipose tissue, promotes the recruitment of macrophages. These immune cells, particularly the pro-inflammatory M1 subtype, intensify ROS production and release a range of inflammatory cytokines such as TNF-α and IL-6 [73, 110]. The outcome is a persistent, low-grade inflammatory state that contributes to both local and systemic insulin resistance. Leptin is an adipokine that controls hunger and energy balance. It can also cause oxidative stress. When people are resistant to leptin, which is prevalent in obesity, high amounts of leptin can cause different types of cells, such as endothelium and immune cells, to make more ROS. This can lead to inflammation and problems with the heart and blood vessels [111]. The interaction between leptin signaling and ROS-sensitive pathways including JAK/STAT and MAPK adds another layer of connection between obesity and metabolic problems throughout the body [112]. An increasing amount of research highlights the crucial role of gut bacteria in sustaining metabolic balance. An imbalance in this microbial community, termed dysbiosis and often linked to obesity, can compromise the integrity of the intestinal lining, leading to heightened gut permeability or "leaky gut" [113]. This weakened barrier permits the transport of microbial components, particularly lipopolysaccharide (LPS), into circulation. Upon entering circulation, LPS can attach to Toll-like receptor 4 (TLR4), which is present on the surface of immunological and metabolic cells. This interaction stimulates the generation of reactive oxygen species (ROS) via NOX-dependent mechanisms, consequently fostering an inflammatory state [114]. The gut microbiota is a crucial external factor that can influence the body's systemic redox status and has a vital role in the development and progression of metabolic disorders [115].

Cardiovascular disease is the leading cause of mortality in patients with metabolic disorders, and oxidative stress is a central pathogenic link. Oxidative stress mostly affects the endothelium, which is the inner lining of blood vessels. Excess ROS, which can come from mitochondria, NOX, and uncoupled endothelial nitric oxide synthase (eNOS), can pick up nitric oxide (NO), which is an important vasodilator and antiatherogenic chemical [116]. This decreases in NO bioavailability, along with inflammation caused by ROS, causes endothelial dysfunction. This is the first step in the formation of atherosclerosis and is marked by poor vasodilation, increased leukocyte adhesion, and a pro-thrombotic state [117, 118]. Oxidative stress is what causes atherogenesis to happen in many ways. One of the most important things that happens is that reactive oxygen species (ROS) oxidize low-density lipoprotein (LDL) cholesterol. This makes oxidized LDL (oxLDL) easy for macrophages to take up, which makes lipid-laden foam cells, which are the main sign of atherosclerotic plaques [119]. ROS also makes it easier for monocytes to get into the artery wall by increasing the production of adhesion molecules on endothelial cells. Inflammation and cell growth caused by ROS inside the plaque make it develop and become less stable, which makes it more likely to break and cause a heart attack or stroke [120, 121]. Metabolic stresses such pressure overload, ischemia, and diabetes cause a huge rise in ROS generation from mitochondria and NOX enzymes in cardiomyocytes in the heart [122]. This oxidative stress is bad for the heart because it directly damages contractile proteins, makes mitochondria stop working and run out of energy, causes cardiomyocyte death and hypertrophy, and drives interstitial fibrosis [123, 124]. This redox imbalance is a big reason why compensatory cardiac hypertrophy turns into overt heart failure.

Nonalcoholic fatty liver disease (NAFLD) is the hepatic manifestation of the metabolic syndrome, ranging from simple steatosis (fatty liver) to nonalcoholic steatohepatitis (NASH), which can progress to cirrhosis and hepatocellular carcinoma. People with NAFLD have too many free fatty acids in their liver from food and fat cells that do not respond to insulin. The increase in mitochondrial fatty acid β-oxidation leads to a big increase in mitoROS production [125]. Oxidative damage and inflammation that happen after the first-time steatosis happens make it worse. Reactive oxygen species (ROS) from mitochondria, along with endoplasmic reticulum (ER) stress and cytochrome P450 enzymes, start lipid peroxidation and cytokine synthesis (through NF-κB and JNK pathways) and direct damage to hepatocytes, which makes it easier for simple steatosis to turn into the more serious inflammatory condition of nonalcoholic steatohepatitis (NASH) [126, 127]. As in other tissues, hepatic oxidative stress is a major cause of insulin resistance. ROS-activated stress kinases (JNK, p38) and inflammatory pathways (NF-κB) disrupt insulin signaling, impairing insulin's ability to suppress hepatic glucose production [74, 128]. This failure to inhibit gluconeogenesis is a primary contributor to the fasting hyperglycemia seen in type 2 diabetes, further highlighting the central role of hepatic ROS in the pathology of metabolic disease [129]. While much of the focus on metabolic disease centers on glucose and lipid dysregulation, inborn errors of amino acid metabolism also feature oxidative stress as a key pathogenic mechanism, particularly in the brain. In disorders such as phenylketonuria (PKU) and maple syrup urine disease (MSUD), genetic defects in specific enzymes lead to the accumulation of amino acids (phenylalanine in PKU; branched-chain amino acids in MSUD) and their toxic metabolites [130]. These metabolites can directly impair mitochondrial function by inhibiting ETC complexes and the Krebs cycle, leading to increased mitoROS production and a state of chronic oxidative stress [131, 132]. Furthermore, the accumulation of these compounds can deplete the brain's antioxidant defenses, particularly by reducing the availability of glutathione (GSH) [133]. This combination of enhanced ROS generation and compromised antioxidant capacity is a major contributor to the severe neurological damage, including intellectual disability and seizures that characterizes these untreated disorders [134].

4.7 Antioxidant Defense System and Metabolic Production

Cells have a complicated, multilevel antioxidant defense system to protect themselves from the ongoing threat of oxidative harm. Both endogenous enzymes and small-molecule antioxidants make up this network, which works together to neutralize ROS and keep redox balance. This system effectiveness is a key sign of cell health, and when it is damaged or overloaded with metabolic diseases, it is a major cause of disease progression. The first line of defense against ROS is a group of highly efficient enzymes that catalyze the detoxification of specific reactive species.

Superoxide dismutases (SODs) are metalloenzymes that change the superoxide radical (O_2-) into molecular oxygen (O_2) and hydrogen peroxide (H2O2), which is less reactive. There are three primary isoforms in mammals: cytosolic copper-zinc SOD (SOD1), mitochondrial manganese SOD (SOD2), and extracellular SOD (SOD3) [135]. These isoforms are different because of where they reside and the metal they contain. Catalase, a peroxisomal enzyme that quickly turns H_2O_2 into water and oxygen, then detoxifies the result, H2O2 [136]. SOD and catalase work together to protect cells from harm caused by superoxide.

The glutathione peroxidase (GPx) enzyme family is an important way for the body to get rid of H2O2. Glutathione peroxidases (GPxs) use the tripeptide glutathione (GSH) as a reducing agent to change hydrogen peroxide (H2O2) or organic hydroperoxides into water or the alcohols that go with them [137]. This mechanism changes glutathione (GSH) into glutathione disulfide (GSSG). Glutathione reductase (GR) turns GSSG into GSH using NADPH, which keeps the cells' supply of this important antioxidant [138]. Peroxiredoxins (Prdxs) are a ubiquitous family of highly efficient, cysteine-based peroxidases that reduce H2O2, peroxynitrite, and organic hydroperoxides [139]. They are arguably the most important enzymes for detoxifying low levels of H2O2 involved in cell signaling. During their catalytic cycle, the active site cysteine of a Prdx is oxidized by peroxide to sulfenic acid. This is then regenerated by the thioredoxin (Trx) system. The oxidized Trx is, in turn, reduced by thioredoxin reductase (TrxR), another NADPH-dependent enzyme [11, 140]. This complete system is vital for controlling redox signaling and protecting against oxidative damage. In addition to enzymatic defenses, cells utilize a variety of small molecules, obtained from the diet or synthesized endogenously, which can directly scavenge ROS or participate in redox reactions.

Glutathione (GSH) is the most common nonprotein thiol in cells, and it is an important part of the body's defense against antioxidants. It may immediately get rid of free radicals, detoxify xenobiotics, and is the main co-substrate for GPx and other enzymes [141].

Vitamin C, also known as ascorbic acid, is a strong water-soluble antioxidant that may directly neutralize superoxide, hydroxyl radicals, and singlet oxygen. It is also very important for bringing back the oxidized form of vitamin E [142]. Vitamin E (α-tocopherol) is the main lipid-soluble antioxidant. It protects cell membranes from lipid peroxidation. It stops free radical damage from spreading in membranes by giving a hydrogen atom to peroxyl radicals, which breaks the chain of damage [143].

Carotenoids, like β-carotene and lycopene, are also lipid-soluble antioxidants that work well to stop singlet oxygen and peroxyl radicals from forming, especially when the oxygen partial pressure is low [144]. Many different types of polyphenols found in plants, like flavonoids (like quercetin) and stilbenes (like resveratrol), have been shown to be very good at fighting free radicals. They can work by directly getting rid of free radicals, binding transition metals that help make ROS, and maybe most crucially changing how cells communicate with one another.

Many polyphenols are strong activators of the Nrf2 antioxidant response pathway, which increases the production of protective enzymes in the body [145, 146].

The antioxidant defense system is not static but is dynamically regulated at multiple levels to respond to changes in the cellular redox environment. The primary mechanism for regulating antioxidant capacity in response to stress is the transcriptional activation of antioxidant genes. The Keap1-Nrf2 pathway is the master regulator of this response, inducing the expression of a wide array of genes that contain an antioxidant response element (ARE) in their promoters [95, 96]. Other transcription factors, such as FoxO proteins, can also contribute to the expression of antioxidant enzymes like SOD2 and catalase in response to growth factor withdrawal or stress [147]. The activity of antioxidant enzymes can also be rapidly modulated by post-translational modifications. For example, phosphorylation, acetylation, and ubiquitination can alter the activity, stability, or localization of enzymes like SOD2 and Prdxs, allowing for fine-tuning of the antioxidant response without the need for new protein synthesis [148, 149]. Antioxidant defenses are strategically localized within different subcellular compartments to deal with ROS at their site of production. For instance, SOD2 is located exclusively in the mitochondrial matrix to detoxify superoxide from the ETC, while SOD1 resides in the cytoplasm and intermembrane space. Catalase is concentrated in peroxisomes, the site of high H2O2 production from fatty acid oxidation. This compartmentalization is crucial for efficiently managing localized ROS signaling and preventing oxidative damage [150].

4.8 Therapeutic Targeting of ROS Signaling Pathways

The most direct approach to combating oxidative stress is the use of antioxidants. However, early enthusiasm based on preclinical data has been tempered by largely disappointing results from large-scale clinical trials. The strategy of using high-dose supplements of direct ROS scavengers, such as vitamin E, vitamin C, and β-carotene, has generally failed to show significant benefit in preventing cardiovascular disease or diabetes [151, 152]. This lack of efficacy may be due to several factors, including poor bioavailability, inability to reach the specific subcellular sites of pathological ROS production, and the potential for high doses to disrupt essential physiological ROS signaling, leading to adverse effects [153]. A more sophisticated approach involves the development of synthetic compounds that mimic the activity of endogenous antioxidant enzymes. SOD mimetics (e.g., tempol) and GPx mimetics (e.g., ebselen) are catalytic antioxidants that can neutralize large numbers of ROS molecules without being consumed themselves [154]. These compounds have shown promise in various animal models of metabolic and cardiovascular disease, but their clinical development has been slow, often hampered by issues of stability, delivery, and off-target effects [155, 156].

There is considerable interest in plant-derived polyphenols and other natural compounds that possess antioxidant and anti-inflammatory properties. Compounds like resveratrol (from grapes), curcumin (from turmeric), and epigallocatechin gallate (EGCG, from green tea) have been shown to ameliorate metabolic dysfunction in preclinical studies [157, 158]. Their beneficial effects are likely multifactorial, involving not only direct ROS scavenging but also the activation of protective pathways like Nrf2 and SIRT1. However, translating these findings to humans is often limited by the low bioavailability of these compounds [159]. An alternative to direct antioxidant therapy is to target the underlying metabolic dysfunction that drives ROS production or to bolster the cell's own adaptive pathways.

AMP-activated protein kinase (AMPK) is a master regulator of cellular energy homeostasis. Its activation can improve metabolic health by promoting fatty acid oxidation and enhancing insulin sensitivity. Drugs like metformin, the first-line therapy for type 2 diabetes, are known to activate AMPK, and part of their therapeutic effect is attributed to the subsequent reduction in mitochondrial ROS production [160, 161]. Newer, more direct AMPK activators are under development as potential therapies for metabolic diseases [162].

Since mitochondria are the primary source of pathological ROS in metabolic diseases, delivering antioxidants directly to this organelle is a highly attractive strategy. This has been achieved by conjugating an antioxidant moiety, such as coenzyme Q10 or vitamin E, to a lipophilic cation (e.g., triphenylphosphonium), which causes the molecule to accumulate within the negatively charged mitochondrial matrix [163]. Compounds like MitoQ have shown significant benefits in preclinical models of cardiovascular and metabolic diseases and are being explored in clinical trials [164, 165].

Sirtuin 1 (SIRT1) is an NAD+-dependent deacetylase that plays a key role in metabolic regulation, stress resistance, and longevity. It can deacetylate and activate transcription factors like PGC-1α and FoxO, leading to improved mitochondrial function and enhanced expression of antioxidant enzymes [166]. Activators of SIRT1, such as resveratrol and synthetic sirtuin-activating compounds (STACs), have been shown to improve metabolic health and protect against age-related diseases in animal models, partly by reducing oxidative stress [167, 168].

4.9 Limitations and Future Perspectives

Currently, oxidative stress is assessed by measuring the downstream products of ROS-induced damage. Commonly used plasma or urine markers include F2-isoprostanes and malondialdehyde (MDA) as indicators of lipid peroxidation, protein carbonyls for protein oxidation, and 8-hydroxy-2′-deoxyguanosine (8-OHdG) for DNA damage [22, 169]. While useful, these markers reflect systemic, cumulative damage and may lack the sensitivity to detect early or localized changes in redox signaling. Measuring the ratio of reduced to oxidized glutathione (GSH/GSSG) in blood cells can provide insight into the overall cellular redox state [170]. Future progress depends on the development of more sensitive and specific detection methods. Redox-sensitive fluorescent probes are being developed for use in cell and animal models to visualize ROS production in real-time within specific organelles [171]. The field of "redox proteomics" aims to identify specific proteins that are oxidatively modified under pathological conditions, which could yield highly specific biomarkers of disease activity [172]. Furthermore, advances in mass spectrometry are enabling more precise quantification of a wide array of oxidative damage products in clinical samples [173]. While many potential biomarkers exist, few have been rigorously validated in large-scale clinical studies. Future research must focus on correlating these biomarkers with clinical outcomes, establishing standardized measurement protocols, and defining clear reference ranges. Such validation is essential before these markers can be incorporated into routine clinical practice to guide personalized treatment decisions [174].

The translation of our understanding of ROS signaling into effective therapies has been a major challenge, marked by the failure of broad-spectrum antioxidant trials. The future lies in more targeted and nuanced approaches. The current drug development pipeline includes several promising strategies. As mentioned, Nrf2 activators like bardoxolone methyl have been investigated for diabetic kidney disease [100]. Inhibitors targeting specific ROS sources, particularly NOX isoforms like NOX1/4 and NOX2, are in various stages of preclinical and early clinical development for cardiovascular and fibrotic diseases [175]. Mitochondrial-targeted antioxidants, such as MitoQ, continue to be tested in a range of conditions, including neurodegenerative and cardiovascular diseases [165].

The key challenge is to selectively inhibit pathological ROS production without interfering with the essential physiological roles of ROS in cell signaling. The failure of non-targeted antioxidants like vitamin E highlights the danger of this "sledgehammer" approach [153]. Future therapies must be designed with high specificity for the disease-relevant ROS source (e.g., mitochondrial Complex I vs. NOX4) and the affected tissue. Issues of drug delivery, stability, and potential for off-target effects remain significant hurdles [176]. Future research will likely focus on drugging the specific protein targets of pathogenic ROS signaling. For example, developing drugs that prevent the ROS-mediated activation of JNK or NF-κB, or that protect critical proteins like PTEN from oxidative inactivation, could offer a more precise way to restore normal signaling. Another emerging area is targeting the interplay between the microbiome and host ROS production, potentially using probiotics or engineered bacteria to improve gut barrier function and reduce metabolic endotoxemia [177, 178].

While pharmacological interventions are crucial, lifestyle modifications remain the cornerstone of preventing and managing metabolic diseases. Both exercise and diet profoundly impact cellular redox balance. Physical exercise provides a fascinating example of the dual role of ROS. An acute bout of exercise transiently increases ROS production, which acts as a beneficial signaling event (a phenomenon known as mitochondrial hormesis or "mitohormesis") [179]. This ROS signal is essential for activating pathways like PGC-1α and Nrf2, leading to increased mitochondrial biogenesis and upregulation of endogenous antioxidant defenses. These adaptations are central to the health benefits of regular exercise, improving insulin sensitivity and cardiovascular function. This explains the paradoxical finding that supplementing with high-dose antioxidants can blunt some of the positive effects of exercise training [153].

4.10 Conclusions and Recommendations

The complex link between reactive oxygen species and cellular metabolism has progressed from a simple model of oxidative damage to a more sophisticated knowledge of redox signaling. It is now obvious that ROS are more than just hazardous by-products; they are basic signaling molecules whose spatial, temporal, and quantitative modulation is essential for metabolic health. Chronic dysregulation of this signaling, caused by dietary excess and sedentary lifestyle, is a common thread running through the pathogenesis of obesity, type 2 diabetes, cardiovascular disease, and NAFLD. Redox homeostasis and metabolic health are closely connected. When things are working properly, temporary and localized ROS signals are necessary for insulin to work properly, for fat cells to change, and for the body to adjust to physiological stress like exercise. A strong and flexible antioxidant defense network carefully controls these "eustress" signals. But the current Western diet has a lot of extra nutrients all the time, which throws off this delicate balance. The systemic oxidative stress that results turns ROS from important messengers into harmful agents, which causes the main pathogenic processes of inflammation and insulin resistance that link all metabolic diseases. Redox biology is always changing. The idea that subcellular compartmentalization is important is becoming more popular. "Cellular" oxidative stress is no longer enough. The individual organelle that makes it, whether it is the mitochondrion, ER, or peroxisome, sends out a different ROS signal that turns on different pathways downstream. Mitohormesis is the idea that a modest level of mitochondrial stress can cause a positive, adaptive response. This idea has big implications for understanding the benefits of exercising and eating less. Also, the idea of interorgan crosstalk, in which oxidative stress in one tissue (like adipose) can cause problems in another (like the liver or blood vessels) through circulating molecules, shows how metabolic disease affects the whole body. Despite tremendous advances, many questions remain. A significant research gap is a lack of techniques for precisely measuring and manipulating ROS in specific subcellular compartments in vivo. Developing such tools is critical for understanding the unique functions of various ROS sources in disease. The precise molecular targets of ROS that mediate pathological signaling are still being discovered, and identifying these "redox switches" opens a plethora of potential therapeutic targets. Further research into the interactions between the gut microbiome, host metabolism, and systemic redox state is an especially fascinating field. Understanding how microbial metabolites regulate host ROS generation and signaling could open totally new therapeutic possibilities.

References

1. Adam-Vizi V (2005) Production of reactive oxygen species in brain mitochondria: contribution by electron transport chain and non–electron transport chain sources. Antioxid Redox Signal 7(9–10):1140–1149
2. Halliwell B, Gutteridge JM (1992) Biologically relevant metal ion-dependent hydroxyl radical generation. An update. FEBS Lett 307(1):108–112

3. Sies H (2017) Hydrogen peroxide as a central redox signaling molecule in physiological oxidative stress: oxidative eustress. Redox Biol 11:613–619
4. Forman HJ, Maiorino M, Ursini F (2010) Signaling functions of reactive oxygen species. Biochemistry 49(5):835–842
5. Checa J, Aran JM (2020) Reactive oxygen species: drivers of physiological and pathological processes. J Inflamm Res 13:1057–1073
6. Goldstein BJ, Mahadev K, Wu X, Zhu L, Motoshima H (2005) Role of insulin-induced reactive oxygen species in the insulin signaling pathway. Antioxid Redox Signal 7(7–8):1021–1031
7. Mahadev K, Wu X, Zilbering A, Zhu L, Lawrence JTR, Goldstein BJ (2001) Hydrogen peroxide generated during cellular insulin stimulation is integral to activation of the distal insulin signaling cascade in 3T3-L1 adipocytes. J Biol Chem 276(52):48662–48669
8. Sies H (2015) Oxidative stress: a concept in redox biology and medicine. Redox Biol 4:180–183
9. Forman HJ (2016) Redox signaling: an evolution from free radicals to aging. Free Radic Biol Med 97:398–407
10. Rhee SG, Chae HZ, Kim K (2005) Peroxiredoxins: a historical overview and speculative preview of novel mechanisms and emerging concepts in cell signaling. Free Radic Biol Med 38(12):1543–1552
11. Holmgren A, Lu J (2010) Thioredoxin and thioredoxin reductase: current research with special reference to human disease. Biochem Biophys Res Commun 396(1):120–124
12. Schröder K, Wandzioch K, Helmcke I, Brandes RP (2009) Nox4 acts as a switch between differentiation and proliferation in preadipocytes. Arterioscler Thromb Vasc Biol 29(2):239–245
13. Mahadev K, Motoshima H, Wu X, Ruddy JM, Arnold RS, Cheng G et al (2004) The NAD(P)H oxidase homolog Nox4 modulates insulin-stimulated generation of H_2O_2 and plays an integral role in insulin signal transduction. Mol Cell Biol 24(5):1844–1854
14. Chandel NS, McClintock DS, Feliciano CE, Wood TM, Melendez JA, Rodriguez AM et al (2000) Reactive oxygen species generated at mitochondrial complex III stabilize hypoxia-inducible factor-1α during hypoxia: a mechanism of O2 sensing. J Biol Chem 275(33):25130–25138
15. Scherz-Shouval R, Shvets E, Fass E, Shorer H, Gil L, Elazar Z (2007) Reactive oxygen species are essential for autophagy and specifically regulate the activity of Atg4. EMBO J 26(7):1749–1760
16. Bryan HK, Olayanju A, Goldring CE, Park BK (2013) The Nrf2 cell defence pathway: Keap1-dependent and-independent mechanisms of regulation. Biochem Pharmacol 85(6):705–717
17. Collins Y, Chouchani ET, James AM, Menger KE, Cochemé HM, Murphy MP (2012) Mitochondrial redox signalling at a glance. J Cell Sci 125(4):801–806
18. Houstis N, Rosen ED, Lander ES (2006) Reactive oxygen species have a causal role in multiple forms of insulin resistance. Nature 440(7086):944–948
19. Schieber M, Chandel NS (2014) ROS function in redox signaling and oxidative stress. Curr Biol 24(10):R453–R462
20. Halliwell B (2006) Oxidative stress and neurodegeneration: where are we now? J Neurochem 97(6):1634–1658
21. Ayala A, Muñoz MF, Argüelles S (2014) Lipid peroxidation: production, metabolism, and signaling mechanisms of malondialdehyde and 4-hydroxy-2-nonenal. Oxidative Med Cell Longev 2014:1–31
22. Dalle-Donne I, Rossi R, Giustarini D, Milzani A, Colombo R (2003) Protein carbonyl groups as biomarkers of oxidative stress. Clin Chim Acta 329(1–2):23–38
23. Cooke MS, Evans MD, Dizdaroglu M, Lunec J (2003) Oxidative DNA damage: mechanisms, mutation, and disease. FASEB J 17(10):1195–1214
24. McMinimy R, Manford AG, Gee CL, Chandrasekhar S, Mousa GA, Chuang J et al (2024) Reactive oxygen species control protein degradation at the mitochondrial import gate. Mol Cell 84(23):4612–4628

25. Klomsiri C, Karplus PA, Poole LB (2011) Cysteine-based redox switches in enzymes. Antioxid Redox Signal 14(6):1065–1077
26. Evans JL, Goldfine ID, Maddux BA, Grodsky GM (2003) Are oxidative stress-activated signaling pathways mediators of insulin resistance and β-cell dysfunction? Diabetes 52(1):1–8
27. Hotamisligil GS (2010) Endoplasmic reticulum stress and the inflammatory basis of metabolic disease. Cell 140(6):900–917
28. Newsholme P, Cruzat V, Arfuso F, Keane K (2014) Nutrient regulation of insulin secretion and action. J Endocrinol 221(3):R105–R120
29. Zeeshan HMA, Lee GH, Kim HR, Chae HJ (2016) Endoplasmic reticulum stress and associated ROS. Int J Mol Sci 17(3):327
30. Ray PD, Huang BW, Tsuji Y (2012) Reactive oxygen species (ROS) homeostasis and redox regulation in cellular signaling. Cell Signal 24(5):981–990
31. Sena LA, Chandel NS (2012) Physiological roles of mitochondrial reactive oxygen species. Mol Cell 48(2):158–167
32. Turrens JF (2003) Mitochondrial formation of reactive oxygen species. J Physiol 552(2):335–344
33. Murphy MP (2009) How mitochondria produce reactive oxygen species. Biochem J 417(1):1–13
34. Brand MD (2016) Mitochondrial generation of superoxide and hydrogen peroxide as the source of mitochondrial redox signaling. Free Radic Biol Med 100:14–31
35. Chouchani ET, Kazak L, Jedrychowski MP, Lu GZ, Erickson BK, Szpyt J et al (2016) Mitochondrial ROS regulate thermogenic energy expenditure and sulfenylation of UCP1. Nature 532(7597):112–116
36. Muller FL, Liu Y, Van Remmen H (2004) Complex III releases superoxide to both sides of the inner mitochondrial membrane. J Biol Chem 279(47):49064–49073
37. Brownlee M (2001) Biochemistry and molecular cell biology of diabetic complications. Nature 414(6865):813–820
38. Shadel GS, Horvath TL (2015) Mitochondrial ROS signaling in organismal homeostasis. Cell 163(3):560–569
39. Lowell BB, Shulman GI (2005) Mitochondrial dysfunction and type 2 diabetes. Science 307(5708):384–387
40. St-Pierre J, Drori S, Uldry M, Silvaggi JM, Rhee J, Jäger S et al (2006) Suppression of reactive oxygen species and neurodegeneration by the PGC-1 transcriptional coactivators. Cell 127(2):397–408
41. Youle RJ, Narendra DP (2011) Mechanisms of mitophagy. Nat Rev Mol Cell Biol 12(1):9–14
42. Wellen KE, Thompson CB (2010) Cellular metabolic stress: considering how cells respond to nutrient excess. Mol Cell 40(2):323–332
43. Martínez-Reyes I, Chandel NS (2020) Mitochondrial TCA cycle metabolites control physiology and disease. Nat Commun 11(1):102
44. Lambeth JD (2004) NOX enzymes and the biology of reactive oxygen. Nat Rev Immunol 4(3):181–189
45. Lassègue B, San Martín A, Griendling KK (2012) Biochemistry, physiology, and pathophysiology of NADPH oxidases in the cardiovascular system. Circ Res 110(10):1364–1390
46. Inoguchi T, Li P, Umeda F, Yu HY, Kakimoto M, Imamura M et al (2000) High glucose level and free fatty acid stimulate reactive oxygen species production through protein kinase C–dependent activation of NAD (P) H oxidase in cultured vascular cells. Diabetes 49(11):1939–1945
47. DeVallance E, Li Y, Jurczak MJ, Cifuentes-Pagano E, Pagano PJ (2019) The role of NADPH oxidases in the etiology of obesity and metabolic syndrome: contribution of individual isoforms and cell biology. Antioxid Redox Signal 31(10):687–709
48. Gorin Y, Block K (2013) Nox4 and diabetic nephropathy: with a friend like this, who needs enemies? Free Radic Biol Med 61:130–142

49. Sedeek M, Callera G, Montezano A, Gutsol A, Heitz F, Szyndralewiez C et al (2010) Critical role of Nox4-based NADPH oxidase in glucose-induced oxidative stress in the kidney: implications in type 2 diabetic nephropathy. Am J Physiol Renal Physiol 299(6):F1348–F1358
50. Touyz RM, Anagnostopoulou A, Rios F, Montezano AC, Camargo LL (2019) NOX5: molecular biology and pathophysiology. Exp Physiol 104(5):605–616
51. Gao L, Mann GE (2009) Vascular NAD(P)H oxidase activation in diabetes: a double-edged sword in redox signalling. Cardiovasc Res 82(1):9–20
52. Gray SP, Di Marco E, Okabe J, Szyndralewiez C, Heitz F, Montezano AC et al (2013) NADPH oxidase 1 plays a key role in diabetes mellitus–accelerated atherosclerosis. Circulation 127(18):1888–1902
53. Jha JC, Banal C, Chow BSM, Cooper ME, Jandeleit-Dahm K (2016) Diabetes and kidney disease: role of oxidative stress. Antioxid Redox Signal 25(12):657–684
54. Al-Shabrawey M, Smith S (2010) Prediction of diabetic retinopathy: role of oxidative stress and relevance of apoptotic biomarkers. EPMA J 1(1):56–72
55. Madamanchi NR, Runge MS (2007) Mitochondrial dysfunction in atherosclerosis. Circ Res 100(4):460–473
56. Tu BP, Weissman JS (2004) Oxidative protein folding in eukaryotes: mechanisms and consequences. J Cell Biol 164(3):341–346
57. Santos CXC, Tanaka LY, Wosniak J, Laurindo FRM (2009) Mechanisms and implications of reactive oxygen species generation during the unfolded protein response: roles of endoplasmic reticulum oxidoreductases, mitochondrial electron transport, and NADPH oxidase. Antioxid Redox Signal 11(10):2409–2427
58. Malhotra JD, Kaufman RJ (2007) The endoplasmic reticulum and the unfolded protein response. In: Seminars in cell & developmental biology. Elsevier, pp 716–731. Available from: https://www.sciencedirect.com/science/article/pii/S1084952107001498
59. Walter P, Ron D (2011) The unfolded protein response: from stress pathway to homeostatic regulation. Science 334(6059):1081–1086
60. Cao SS, Kaufman RJ (2014) Endoplasmic reticulum stress and oxidative stress in cell fate decision and human disease. Antioxid Redox Signal 21(3):396–413
61. Urano F, Wang X, Bertolotti A, Zhang Y, Chung P, Harding HP et al (2000) Coupling of stress in the ER to activation of JNK protein kinases by transmembrane protein kinase IRE1. Science 287(5453):664–666
62. Özcan U, Cao Q, Yilmaz E, Lee AH, Iwakoshi NN, Özdelen E et al (2004) Endoplasmic reticulum stress links obesity, insulin action, and type 2 diabetes. Science 306(5695):457–461
63. Battelli MG, Polito L, Bolognesi A (2014) Xanthine oxidoreductase in atherosclerosis pathogenesis: not only oxidative stress. Atherosclerosis 237(2):562–567
64. Fransen M, Nordgren M, Wang B, Apanasets O (2012) Role of peroxisomes in ROS/RNS-metabolism: implications for human disease. Biochim Biophys Acta 1822(9):1363–1373
65. Zangar RC, Davydov DR, Verma S (2004) Mechanisms that regulate production of reactive oxygen species by cytochrome P450. Toxicol Appl Pharmacol 199(3):316–331
66. Chen Y, Yu CY, Deng WM (2019) The role of pro-inflammatory cytokines in lipid metabolism of metabolic diseases. Int Rev Immunol 38(6):249–266
67. Shoelson SE, Lee J, Goldfine AB (2006) Inflammation and insulin resistance. J Clin Invest 116(7):1793–1801
68. Hayden MS, Ghosh S (2008) Shared principles in NF-κB signaling. Cell 132(3):344–362
69. Kamata H, Honda SI, Maeda S, Chang L, Hirata H, Karin M (2005) Reactive oxygen species promote TNFα-induced death and sustained JNK activation by inhibiting MAP kinase phosphatases. Cell 120(5):649–661
70. Morgan MJ, Liu Z-G (2011) Crosstalk of reactive oxygen species and NF-κB signaling. Cell Res 21(1):103–115
71. Oliveira-Marques V, Marinho HS, Cyrne L, Antunes F (2009) Role of hydrogen peroxide in NF-κB activation: from inducer to modulator. Antioxid Redox Signal 11(9):2223–2243

72. Gloire G, Legrand-Poels S, Piette J (2006) NF-κB activation by reactive oxygen species: fifteen years later. Biochem Pharmacol 72(11):1493–1505
73. Weisberg SP, McCann D, Desai M, Rosenbaum M, Leibel RL, Ferrante AW (2003) Obesity is associated with macrophage accumulation in adipose tissue. J Clin Invest 112(12):1796–1808
74. Cai D, Yuan M, Frantz DF, Melendez PA, Hansen L, Lee J et al (2005) Local and systemic insulin resistance resulting from hepatic activation of IKK-β and NF-κB. Nat Med 11(2):183–190
75. Pickup JC (2004) Inflammation and activated innate immunity in the pathogenesis of type 2 diabetes. Diabetes Care 27(3):813–823
76. Riedl SJ, Salvesen GS (2007) The apoptosome: signalling platform of cell death. Nat Rev Mol Cell Biol 8(5):405–413
77. Donath MY, Shoelson SE (2011) Type 2 diabetes as an inflammatory disease. Nat Rev Immunol 11(2):98–107
78. Aguirre V, Uchida T, Yenush L, Davis R, White MF (2000) The c-Jun NH2-terminal kinase promotes insulin resistance during association with insulin receptor substrate-1 and phosphorylation of Ser307. J Biol Chem 275(12):9047–9054
79. Hirosumi J, Tuncman G, Chang L, Görgün CZ, Uysal KT, Maeda K et al (2002) A central role for JNK in obesity and insulin resistance. Nature 420(6913):333–336
80. Solinas G, Becattini B (2017) JNK at the crossroad of obesity, insulin resistance, and cell stress response. Mol Metab 6(2):174–184
81. Cagnol S, Chambard J (2010) ERK and cell death: mechanisms of ERK-induced cell death—apoptosis, autophagy and senescence. FEBS J 277(1):2–21
82. Hancock JT, Desikan R, Neill SJ (2001) Role of reactive oxygen species in cell signalling pathways. Biochem Soc Trans 29(2):345–349
83. Davis RJ (2000) Signal transduction by the JNK group of MAP kinases. Cell 103(2):239–252
84. Adler V (1999) Regulation of JNK signaling by GSTp. EMBO J 18(5):1321–1334
85. Saitoh M (1998) Mammalian thioredoxin is a direct inhibitor of apoptosis signal-regulating kinase (ASK) 1. EMBO J 17(9):2596–2606
86. Manning BD, Cantley LC (2007) AKT/PKB signaling: navigating downstream. Cell 129(7):1261–1274
87. Taniguchi CM, Emanuelli B, Kahn CR (2006) Critical nodes in signalling pathways: insights into insulin action. Nat Rev Mol Cell Biol 7(2):85–96
88. Maehama T, Dixon JE (1998) The tumor suppressor, PTEN/MMAC1, dephosphorylates the lipid second messenger, phosphatidylinositol 3,4,5-trisphosphate. J Biol Chem 273(22):13375–13378
89. Lee SR, Yang KS, Kwon J, Lee C, Jeong W, Rhee SG (2002) Reversible inactivation of the tumor suppressor PTEN by H2O2. J Biol Chem 277(23):20336–20342
90. Trinh VH, Nguyen Huu T, Sah DK, Choi JM, Yoon HJ, Park SC et al (2024) Redox regulation of PTEN by reactive oxygen species: its role in physiological processes. Antioxidants 13(2):199
91. Li YZ, Di Cristofano A, Woo M (2020) Metabolic role of PTEN in insulin signaling and resistance. Cold Spring Harb Perspect Med 10(8):a036137
92. Itoh K, Wakabayashi N, Katoh Y, Ishii T, Igarashi K, Engel JD et al (1999) Keap1 represses nuclear activation of antioxidant responsive elements by Nrf2 through binding to the amino-terminal Neh2 domain. Genes Dev 13(1):76–86
93. Suzuki T, Yamamoto M (2015) Molecular basis of the Keap1-Nrf2 system. Free Radic Biol Med 88:93–100
94. Baird L, Dinkova-Kostova AT (2011) The cytoprotective role of the Keap1-Nrf2 pathway. Arch Toxicol 85(4):241–272
95. Kensler TW, Wakabayashi N, Biswal S (2007) Cell survival responses to environmental stresses via the Keap1-Nrf2-ARE pathway. Annu Rev Pharmacol Toxicol 47(1):89–116
96. Hayes JD, Dinkova-Kostova AT (2014) The Nrf2 regulatory network provides an interface between redox and intermediary metabolism. Trends Biochem Sci 39(4):199–218

97. Magesh S, Chen Y, Hu L (2012) Small molecule modulators of Keap1-Nrf2-ARE pathway as potential preventive and therapeutic agents. Med Res Rev 32(4):687–726
98. Uruno A, Furusawa Y, Yagishita Y, Fukutomi T, Muramatsu H, Negishi T et al (2013) The Keap1-Nrf2 system prevents onset of diabetes mellitus. Mol Cell Biol 33(15):2996–3010
99. David JA, Rifkin WJ, Rabbani PS, Ceradini DJ (2017) The Nrf2/Keap1/ARE pathway and oxidative stress as a therapeutic target in type II diabetes mellitus. J Diabetes Res 2017:1–15
100. De Zeeuw D, Akizawa T, Audhya P, Bakris GL, Chin M, Christ-Schmidt H et al (2013) Bardoxolone methyl in type 2 diabetes and stage 4 chronic kidney disease. N Engl J Med 369(26):2492–2503
101. He F, Ru X, Wen T (2020) NRF2, a transcription factor for stress response and beyond. Int J Mol Sci 21(13):4777
102. Lenzen S (2008) Oxidative stress: the vulnerable β-cell. Biochem Soc Trans 36(3):343–347
103. Robertson RP, Harmon J, Tran POT, Poitout V (2004) β-cell glucose toxicity, lipotoxicity, and chronic oxidative stress in type 2 diabetes. Diabetes 53(suppl_1):S119–S124
104. Butler AE, Janson J, Bonner-Weir S, Ritzel R, Rizza RA, Butler PC (2003) β-cell deficit and increased β-cell apoptosis in humans with type 2 diabetes. Diabetes 52(1):102–110
105. Donath MY, Ehses JA, Maedler K, Schumann DM, Ellingsgaard H, Eppler E et al (2005) Mechanisms of β-cell death in type 2 diabetes. Diabetes 54(suppl_2):S108–S113
106. Giacco F, Brownlee M (2010) Oxidative stress and diabetic complications. Schmidt AM, editor. Circ Res 107(9):1058–1070
107. Forbes JM, Cooper ME (2013) Mechanisms of diabetic complications. Physiol Rev 93(1):137–188
108. Choe SS, Huh JY, Hwang IJ, Kim JI, Kim JB (2016) Adipose tissue remodeling: its role in energy metabolism and metabolic disorders. Front Endocrinol 7:30
109. Furukawa S, Fujita T, Shimabukuro M, Iwaki M, Yamada Y, Nakajima Y et al (2017) Increased oxidative stress in obesity and its impact on metabolic syndrome. J Clin Invest 114(12):1752–1761
110. Lumeng CN, Saltiel AR (2011) Inflammatory links between obesity and metabolic disease. J Clin Invest 121(6):2111–2117
111. Bouloumié A, Marumo T, Lafontan M, Busse R (1999) Leptin induces oxidative stress in human endothelial cells. FASEB J 13(10):1231–1238
112. Frühbeck G, Catalán V, Rodríguez A, Gómez-Ambrosi J (2018) Adiponectin-leptin ratio: a promising index to estimate adipose tissue dysfunction. Relation with obesity-associated cardiometabolic risk. Adipocyte 7(1):57–62
113. Cani PD, Amar J, Iglesias MA, Poggi M, Knauf C, Bastelica D et al (2007) Metabolic endotoxemia initiates obesity and insulin resistance. Diabetes 56(7):1761–1772
114. You BR, Shin HR, Han BR, Kim SH, Park WH (2015) Auranofin induces apoptosis and necrosis in HeLa cells via oxidative stress and glutathione depletion. Mol Med Rep 11(2):1428–1434
115. Jones RM (2016) The influence of the gut microbiota on host physiology: in pursuit of mechanisms. Yale J Biol Med 89(3):285
116. Förstermann U, Sessa WC (2012) Nitric oxide synthases: regulation and function. Eur Heart J 33(7):829–837
117. Incalza MA, D'Oria R, Natalicchio A, Perrini S, Laviola L, Giorgino F (2018) Oxidative stress and reactive oxygen species in endothelial dysfunction associated with cardiovascular and metabolic diseases. Vasc Pharmacol 100:1–19
118. Münzel T, Sinning C, Post F, Warnholtz A, Schulz E (2008) Pathophysiology, diagnosis and prognostic implications of endothelial dysfunction. Ann Med 40(3):180–196
119. Goldstein JL, Brown MS (2009) The LDL receptor. Arterioscler Thromb Vasc Biol 29(4):431–438
120. Libby P, Ridker PM, Maseri A (2002) Inflammation and atherosclerosis. Circulation 105(9):1135–1143

121. Stocker R, Keaney JF (2004) Role of oxidative modifications in atherosclerosis. Physiol Rev 84(4):1381–1478
122. Tsutsui H, Kinugawa S, Matsushima S (2011) Oxidative stress and heart failure. Am J Phys Heart Circ Phys 301(6):H2181–H2190
123. Moris D, Spartalis M, Spartalis E, Karachaliou GS, Karaolanis GI, Tsourouflis G et al (2017) The role of reactive oxygen species in the pathophysiology of cardiovascular diseases and the clinical significance of myocardial redox. Ann Transl Med 5(16):326
124. Forte M, Nocella C, De Falco E, Palmerio S, Schirone L, Valenti V et al (2016) The pathophysiological role of NOX2 in hypertension and organ damage. High Blood Press Cardiovasc Prev 23(4):355–364
125. Begriche K, Igoudjil A, Pessayre D, Fromenty B (2006) Mitochondrial dysfunction in NASH: causes, consequences and possible means to prevent it. Mitochondrion 6(1):1–28
126. Tilg H, Moschen AR (2010) Evolution of inflammation in nonalcoholic fatty liver disease: the multiple parallel hits hypothesis. Hepatology 52(5):1836–1846
127. Paradies G, Paradies V, Ruggiero FM, Petrosillo G (2014) Oxidative stress, cardiolipin and mitochondrial dysfunction in nonalcoholic fatty liver disease. World J Gastroenterol 20(39):14205
128. Samuel VT, Shulman GI (2012) Mechanisms for insulin resistance: common threads and missing links. Cell 148(5):852–871
129. Perry RJ, Samuel VT, Petersen KF, Shulman GI (2014) The role of hepatic lipids in hepatic insulin resistance and type 2 diabetes. Nature 510(7503):84–91
130. Blau N, Van Spronsen FJ, Levy HL (2010) Phenylketonuria. Lancet 376(9750):1417–1427
131. Mescka C, Moraes T, Rosa A, Mazzola P, Piccoli B, Jacques C et al (2011) In vivo neuroprotective effect of L-carnitine against oxidative stress in maple syrup urine disease. Metab Brain Dis 26(1):21–28
132. Schuck PF, Malgarin F, Cararo JH, Cardoso F, Streck EL, Ferreira GC (2015) Phenylketonuria pathophysiology: on the role of metabolic alterations. Aging Dis 6(5):390
133. Peiris A, Weerasinghe N, Tchikadze N, Janelidze M, Khatiashvili I (2025) Phenylketonuria associated early onset parkinsonism: a case of rapid progression following dietary nonadherence. Jr Res 3(2):42–46
134. Mahjoub S, Masrour-Roudsari J (2012) Role of oxidative stress in pathogenesis of metabolic syndrome. Caspian J Intern Med 3(1):386
135. Fukai T, Ushio-Fukai M (2011) Superoxide dismutases: role in redox signaling, vascular function, and diseases. Antioxid Redox Signal 15(6):1583–1606
136. Glorieux C, Calderon PB (2017) Catalase, a remarkable enzyme: targeting the oldest antioxidant enzyme to find a new cancer treatment approach. Biol Chem 398(10):1095–1108
137. Brigelius-Flohé R, Maiorino M (2013) Glutathione peroxidases. Biochim Biophys Acta 1830(5):3289–3303
138. Deponte M (2013) Glutathione catalysis and the reaction mechanisms of glutathione-dependent enzymes. Biochim Biophys Acta 1830(5):3217–3266
139. Rhee SG, Woo HA (2011) Multiple functions of peroxiredoxins: peroxidases, sensors and regulators of the intracellular messenger H_2O_2, and protein chaperones. Antioxid Redox Signal 15(3):781–794
140. Perkins A, Nelson KJ, Parsonage D, Poole LB, Karplus PA (2015) Peroxiredoxins: guardians against oxidative stress and modulators of peroxide signaling. Trends Biochem Sci 40(8):435–445
141. Meister A, Anderson ME (1983) Glutathione. Annu Rev Biochem 52(1):711–760
142. Padayatty SJ, Katz A, Wang Y, Eck P, Kwon O, Lee JH et al (2003) Vitamin C as an antioxidant: evaluation of its role in disease prevention. J Am Coll Nutr 22(1):18–35
143. Traber MG, Atkinson J (2007) Vitamin E, antioxidant and nothing more. Free Radic Biol Med 43(1):4–15
144. Stahl W, Sies H (2005) Bioactivity and protective effects of natural carotenoids. Biochim Biophys Acta 1740(2):101–107

145. Vauzour D, Rodriguez-Mateos A, Corona G, Oruna-Concha MJ, Spencer JP (2010) Polyphenols and human health: prevention of disease and mechanisms of action. Nutrients 2(11):1106–1131
146. Pandey KB, Rizvi SI (2009) Plant polyphenols as dietary antioxidants in human health and disease. Oxidative Med Cell Longev 2(5):270–278
147. Nakamura T, Sakamoto K (2008) Forkhead transcription factor FOXO subfamily is essential for reactive oxygen species-induced apoptosis. Mol Cell Endocrinol 281(1–2):47–55
148. Ozden O, Park SH, Kim HS, Jiang H, Coleman MC, Spitz DR et al (2011) Acetylation of MnSOD directs enzymatic activity responding to cellular nutrient status or oxidative stress. Aging (Albany NY) 3(2):102
149. Salmeen A, Barford D (2005) Functions and mechanisms of redox regulation of cysteine-based phosphatases. Antioxid Redox Signal 7(5–6):560–577
150. Okado-Matsumoto A, Fridovich I (2001) Subcellular distribution of superoxide dismutases (SOD) in rat liver: Cu, Zn-SOD in mitochondria. J Biol Chem 276(42):38388–38393
151. Bjelakovic G, Nikolova D, Gluud C (2013) Antioxidant supplements to prevent mortality. JAMA 310(11):1178–1179
152. Sesso HD, Buring JE, Christen WG, Kurth T, Belanger C, MacFadyen J et al (2008) Vitamins E and C in the prevention of cardiovascular disease in men: the Physicians' Health Study II randomized controlled trial. JAMA 300(18):2123–2133
153. Ristow M, Zarse K, Oberbach A, Klöting N, Birringer M, Kiehntopf M et al (2009) Antioxidants prevent health-promoting effects of physical exercise in humans. Proc Natl Acad Sci USA 106(21):8665–8670
154. Batinić-Haberle I, Rebouças JS, Spasojević I (2010) Superoxide dismutase mimics: chemistry, pharmacology, and therapeutic potential. Antioxid Redox Signal 13(6):877–918
155. Bianchi C, Wakiyama H, Faro R, Khan T, McCully JD, Levitsky S et al (2002) A novel peroxynitrite decomposer catalyst (FP-15) reduces myocardial infarct size in an in vivo peroxynitrite decomposer and acute ischemia-reperfusion in pigs. Ann Thorac Surg 74(4):1201–1207
156. Salvemini D, Riley DP, Cuzzocrea S (2002) SOD mimetics are coming of age. Nat Rev Drug Discov 1(5):367–374
157. Aggarwal BB, Harikumar KB (2009) Potential therapeutic effects of curcumin, the anti-inflammatory agent, against neurodegenerative, cardiovascular, pulmonary, metabolic, autoimmune and neoplastic diseases. Int J Biochem Cell Biol 41(1):40–59
158. Timmers S, Konings E, Bilet L, Houtkooper RH, van de Weijer T, Goossens GH et al (2011) Calorie restriction-like effects of 30 days of resveratrol supplementation on energy metabolism and metabolic profile in obese humans. Cell Metab 14(5):612–622
159. Manach C, Williamson G, Morand C, Scalbert A, Rémésy C (2005) Bioavailability and bioefficacy of polyphenols in humans. I. Review of 97 bioavailability studies. Am J Clin Nutr 81(1):230S–242S
160. Zhou G, Myers R, Li Y, Chen Y, Shen X, Fenyk-Melody J et al (2001) Role of AMP-activated protein kinase in mechanism of metformin action. J Clin Invest 108(8):1167–1174
161. Martin-Montalvo A, Mercken EM, Mitchell SJ, Palacios HH, Mote PL, Scheibye-Knudsen M et al (2013) Metformin improves healthspan and lifespan in mice. Nat Commun 4(1):2192
162. Hardie DG, Ross FA, Hawley SA (2012) AMPK: a nutrient and energy sensor that maintains energy homeostasis. Nat Rev Mol Cell Biol 13(4):251–262
163. Smith RAJ, Murphy MP (2010) Animal and human studies with the mitochondria-targeted antioxidant MitoQ. Ann N Y Acad Sci 1201(1):96–103
164. Gane EJ, Weilert F, Orr DW, Keogh GF, Gibson M, Lockhart MM et al (2010) The mitochondria-targeted anti-oxidant mitoquinone decreases liver damage in a phase II study of hepatitis C patients: mitoquinone and liver damage. Liver Int 30(7):1019–1026
165. Snow BJ, Rolfe FL, Lockhart MM, Frampton CM, O'Sullivan JD, Fung V et al (2010) A double-blind, placebo-controlled study to assess the mitochondria-targeted antioxidant MitoQ as a disease-modifying therapy in Parkinson's disease. Mov Disord 25(11):1670–1674

166. Brunet A, Sweeney LB, Sturgill JF, Chua KF, Greer PL, Lin Y et al (2004) Stress-dependent regulation of FOXO transcription factors by the SIRT1 deacetylase. Science 303(5666):2011–2015
167. Baur JA, Pearson KJ, Price NL, Jamieson HA, Lerin C, Kalra A et al (2006) Resveratrol improves health and survival of mice on a high-calorie diet. Nature 444(7117):337–342
168. Milne JC, Lambert PD, Schenk S, Carney DP, Smith JJ, Gagne DJ et al (2007) Small molecule activators of SIRT1 as therapeutics for the treatment of type 2 diabetes. Nature 450(7170):712–716
169. Graille M, Wild P, Sauvain JJ, Hemmendinger M, Canu IG, Hopf NB (2020) Urinary 8-isoprostane as a biomarker for oxidative stress. A systematic review and meta-analysis. Toxicol Lett 328:19–27
170. Jones DP, Carlson JL, Mody VC Jr, Cai J, Lynn MJ, Sternberg P Jr (2000) Redox state of glutathione in human plasma. Free Radic Biol Med 28(4):625–635
171. Dickinson BC, Chang CJ (2011) Chemistry and biology of reactive oxygen species in signaling or stress responses. Nat Chem Biol 7(8):504–511
172. Butterfield DA, Dalle-Donne I (2012) Redox proteomics. Antioxid Redox Signal 17(11):1487–1489
173. Perluigi M, Swomley AM, Butterfield DA (2014) Redox proteomics and the dynamic molecular landscape of the aging brain. Ageing Res Rev 13:75–89
174. Milne GL, Sanchez SC, Musiek ES, Morrow JD (2007) Quantification of F2-isoprostanes as a biomarker of oxidative stress. Nat Protoc 2(1):221–226
175. Green DE, Murphy TC, Kang BY, Kleinhenz JM, Szyndralewiez C, Page P et al (2012) The Nox4 inhibitor GKT137831 attenuates hypoxia-induced pulmonary vascular cell proliferation. Am J Respir Cell Mol Biol 47(5):718–726
176. Forman HJ, Zhang H (2021) Targeting oxidative stress in disease: promise and limitations of antioxidant therapy. Nat Rev Drug Discov 20(9):689–709
177. Quigley EMM (2012) Prebiotics and probiotics: their role in the management of gastrointestinal disorders in adults. Nutr Clin Pract 27(2):195–200
178. Suez J, Zmora N, Segal E, Elinav E (2019) The pros, cons, and many unknowns of probiotics. Nat Med 25(5):716–729
179. Ristow M, Schmeisser S (2011) Extending life span by increasing oxidative stress. Free Radic Biol Med 51(2):327–336

Chapter 5
Epigenetic Regulation of Oxygenases Involved in ROS Production in Metabolic Disorders

Neetu Sachan, Prakhar Varshney, Phool Chandra, and Pradeep Singh

5.1 Introduction

Normal cell metabolism creates ROS. During oxidative phosphorylation, mitochondria make these. Oxidation by ROS was thought to harm proteins, lipids, and DNA. They maintain metabolic homeostasis as signaling molecules [1]. Low-to-moderate ROS are secondary messengers in immunology, cell division, proliferation, and metabolism. Oxidative stress, obesity, type 2 diabetes, heart disease, and other metabolic diseases may result from antioxidant defense or ROS generation difficulties [2]. Socrates famously said, "Know thyself." Understand yourself and your role in the world. They affect glucose absorption, insulin signaling, and lipid metabolism via redox-sensitive mechanisms. High ROS may damage mitochondria, activate stress kinases, and cause inflammatory cascades, impacting metabolism. Oxidase XO, NOX, and P450 create ROS [3]. To reduce ROS, these enzymes are closely controlled transcriptionally, posttranscriptionally, and epigenetically. Epigenetic control modifies gene expression without changing DNA sequence, making it fascinating. DNA methylation, histone modification, and noncoding RNA-mediated regulation regulate gene expression in response to environmental changes, making them vital to cell function [4]. Aberrant epigenetic landscapes may activate oxidative metabolism and redox balance genes in metabolic disorders.

N. Sachan
Department of Pharmaceutical Sciences, School of Pharmaceutical & Biological Sciences, Harcourt Butler Technical University (HBTU), Kanpur, Uttar Pradesh, India

P. Varshney · P. Chandra (✉)
Teerthanker Mahaveer College of Pharmacy, Teerthanker Mahaveer University, Moradabad, Uttar Pradesh, India

P. Singh
Department of Pharmacy, College of Health Sciences, Debre Tabor University, Debra Tabor, Ethiopia

S. Ashique, N. Mishra (eds.), *Reactive Oxygen Species in Metabolic Inflammation*,
https://doi.org/10.1007/978-3-032-11752-6_5

Diabetes may demethylate NOX enzyme promoters, triggering ROS. Eliminating acetyl groups from antioxidant gene loci histones may enhance oxidative stress by limiting toxin-removing enzymes [1]. Oxygenases influence epigenetics. Cell metabolism affects prolyl hydroxylases and TET enzymes, which need oxygen and α-ketoglutarate for efficient function. These enzymes remove methyl groups from chromatin [5]. Epigenetics alter oxygenases' influence on cell redox and metabolites. Epigenetic regulation affects metabolic issues substantially. Chronic inflammation, elevated blood sugar, and free fatty acids indicate metabolic disease [6, 7]. Epigenetic drugs that balance ROS-producing enzyme expression may restore redox equilibrium and metabolic issues [8]. Epigenetic regulation, ROS, and oxygenase activity maintain metabolic balance. Several metabolic problems result from disrupting this sensitive mechanism. Epigenetic processes that regulate oxygenase production and activity may lead to new precision medicine treatments for metabolic diseases and oxidative stress [9].

ROS are reactive and may alter biological macromolecules, causing cell death. Examples of these chemicals are H_2O_2, O_2, and hydroxyl radicals. Recent paradigm shifts show that ROS regulate intracellular signaling and physiological processes and are not only harmful molecules or by-products of aerobic metabolism [10, 11]. ROS normally regulate insulin sensitivity, glucose metabolism, adipocyte development, and mitochondrial biogenesis. Low quantities are needed for physiological signaling, while high levels induce oxidative stress and metabolic issues. Concentration affects signaling [12]. Most ROS are produced by mitochondria during electron transport chain (ETC). Some electrons from complexes I and III convert oxygen to superoxide, which may increase ROS. ROS are produced by mitochondria, xanthine oxidase, uncoupled NOS, NOX, and cytochrome P450 enzymes [13]. These oxygenases are strictly controlled to minimize ROS, benefiting the organism. Obesity, T2DM, NAFLD, and metabolic syndrome disrupt this redox equilibrium. Glucose and free fatty acids cause mitochondrial dysfunction and ROS. Insulin resistance and chronic inflammation increase oxygenase enzymes and ROS [1]. Long-term oxidative stress may damage pancreatic β cells, alter transcription factors including NF-κB and FOXO, and impede insulin receptor signaling by activating JNK and IKKβ that may affect metabolism [14].

5.2 Oxygenases and ROS Production in Metabolic Disorders

Oxygenases, a large set of enzymes, indirectly affect ROS levels and oxygenate organic substrates. ROS generation is carefully controlled in the body and required for signaling functions such as immune regulation, cell division, proliferation, and metabolic management [15]. However, abnormal oxygenase activity promotes oxidative stress in ill individuals, particularly those with metabolic disorders, including obesity, insulin resistance, T2DM, and NAFLD. Oxidative stress disrupts mitochondrial function, sends pro-inflammatory and profibrotic signals, and alters cellular redox balance [16]. The most well-known oxygenase systems that create ROS in

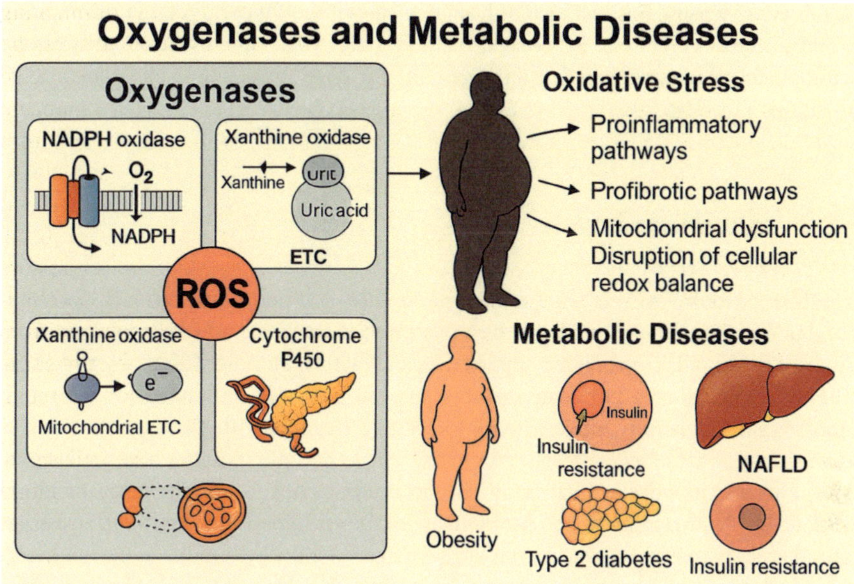

Fig. 5.1 Role of oxygenases in ROS generation and metabolic disorders: Key oxygenase systems NADPH oxidase, xanthine oxidase, mitochondrial ETC, and cytochrome P450 contribute to oxidative stress, which triggers pro-inflammatory and profibrotic pathways, mitochondrial dysfunction, and redox imbalance, ultimately leading to metabolic diseases such as obesity, insulin resistance, type 2 diabetes, and NAFLD

metabolic disorders are the NADPH oxidase (NOX) family, xanthine oxidase, mitochondrial electron transport chain (ETC), and cytochrome P450 enzymes [17] (Fig. 5.1).

5.2.1 *NADPH Oxidases (NOX Family)*

The NADPH oxidase (NOX) family of membrane-bound enzymes generates ROS, not as a result of their activity. These enzymes help move electrons from NADPH to molecular oxygen, which makes superoxide anion (O_2^-). Different isoforms of the NOX family, such as NOX1–5 and DUOX1/2, are found in different tissues and have different ways of regulating themselves and generating ROS. NOX_4 and NOX_2 are directly linked to problems with metabolism. When you are overweight, NOX_4 levels rise in fat tissue [18]. This activates redox-sensitive serine kinases (including JNK and p38 MAPK), which stop insulin receptor activation. This makes you resistant to insulin both locally and across your body. Reactive oxygen species (ROS) from NOX lower the bioavailability of nitric oxide (NO) in the vascular endothelium, which leads to endothelial dysfunction, a key sign of diabetic vasculopathy. NOX enzymes are also controlled by transcriptional and epigenetic means in metabolic diseases. For instance, it has been shown that hyperglycemia raises NOX expression via epigenetic processes, including DNA hypomethylation and histone acetylation, which makes tissue damage and oxidative stress worse in diabetes complications [15, 19].

5.2.2 *Xanthine Oxidase*

Xanthine oxidase (XO) is another crucial enzyme that helps produce ROS in metabolic situations. When it changes hypoxanthine to xanthine and subsequently xanthine to uric acid, it produces superoxide and hydrogen peroxide as by-products. XO has a normal role in purine metabolism, but ischemia, inflammation, and high blood sugar levels may make it work too much. Patients with metabolic syndrome as well as type 2 diabetes have higher levels of XO activity in their endothelial cells, liver, and adipose tissue [20]. This has been connected to inflammation of blood vessels, lipid peroxidation, and problems with the endothelium. Oxidative stress not only damages macromolecules, but it also causes the body to make adhesion molecules and pro-inflammatory cytokines. This links XO-derived ROS to the low-grade chronic inflammation that is a sign of metabolic diseases. Also, metabolic problems might be a sign of high blood uric acid levels, which is common in people who have insulin resistance [21]. Uric acid's pro-oxidant activities within cells may keep a cycle of oxidative damage going. In both preclinical and clinical studies, XO inhibitors like febuxostat and allopurinol have shown promise in lowering oxidative stress and improving metabolic markers (Fig. 5.1) [22].

5.2.3 *Mitochondrial Electron Transport Chain*

The electron transport chain (ETC), which is found in the inner mitochondrial membrane, is a significant generator of reactive oxygen species (ROS) and the main place where cells get their energy. To make adenosine triphosphate (ATP) during oxidative phosphorylation, electrons flow between complexes I and IV. But a very small number of electrons, especially those from complexes I and III, might leak out too soon and turn oxygen into superoxide. Mitochondrial ROS changes redox-sensitive signaling pathways under normal physiological conditions [23]. On the other hand, too much glucose and free fatty acids may overload mitochondria, make electrons leak out more easily, and produce too much ROS in metabolic diseases. This is sometimes called "mitochondrial oxidative stress," and it has been related to damage to liver cells, problems with insulin signaling, problems with pancreatic β cells, and problems with the heart. Mitochondrial ROS may also impair mitochondrial biogenesis and electron transport chain (ETC) function, creating a vicious cycle of oxidative damage [24]. This problem makes cellular stress worse. It often comes with changes to mitochondrial DNA and problems with mitophagy [25].

People have looked at uncoupling proteins (UCPs) and mitochondria-targeted antioxidants (such as MitoQ and SkQ1) as approaches to lower oxidative stress in mitochondria in people with metabolic disorders [26].

5.2.4 *Cytochrome P450 Enzymes*

Cytochrome P450 (CYP) enzymes are a large group of heme-containing monooxygenases that help break down medicines, foreign compounds, and natural molecules, including lipids and steroid hormones. Two examples of ROS that may be generated during these oxidative reactions are superoxide and hydrogen peroxide [27]. This happens when electrons leak out of the CYP catalytic cycle. In metabolic illnesses, including NAFLD and type 2 diabetes, changes in the expression and activity of several CYP isoforms, such as CYP2E1, CYP4A, and CYP1A2, are connected to higher levels of ROS in the liver [28]. CYP2E1 is very interesting since it causes a lot of oxidative stress, has poor coupling efficiency, and has high NADPH oxidase activity [29]. In steatotic and insulin-resistant livers, high levels of CYP2E1 expression have been associated with lipid peroxidation, mitochondrial dysfunction, and inflammatory signaling. Also, depending on the context and quantity, CYP-derived metabolites like hydroxyeicosatetraenoic acids (HETEs) and epoxyeicosatrienoic acids (EETs) may change how sensitive insulin is and how much inflammation there is, both for the better and the worse [30]. Changes in hormones and diet have also been demonstrated to modify the epigenetics of CYP enzymes, especially via histone acetylation and noncoding RNAs. These alterations might affect the body's natural oxidative equilibrium, how well it metabolizes drugs, and how well it works [31].

5.3 Epigenetic Mechanisms Regulating Oxygenases

Epigenetic regulation is necessary for modulating gene expression without affecting the main DNA sequence. Cells may react to changes in their environment and metabolism, such as diet, stress, toxins, inflammation, and other signals. These things are quite significant when it comes to metabolic diseases [32]. At the epigenetic level, oxygenases such as cytochrome P450 enzymes (CYPs), xanthine oxidase (XO), NADPH oxidases (NOXs), and mitochondrial enzymes that make ROS are rigorously regulated to keep ROS generation in check. These epigenetic mechanisms are not working properly, which leads to too much ROS production [33]. This causes oxidative stress, inflammation, and tissue dysfunction in metabolic diseases such as obesity, type 2 diabetes mellitus (T2DM), and nonalcoholic fatty liver disease (NAFLD). The main epigenetic alterations that affect the expression of the oxygenase gene are DNA methylation and demethylation, histone modifications, and noncoding RNAs. These modifications can modify how accessible chromatin is, how transcription factors bind to it, and how stable RNA is, either on their own or together. This may influence how genes that code for oxygenase are transcribed and translated [34, 35].

5.3.1 DNA Methylation and Demethylation

A methyl group is added to the 5′ position of cytosine residues in CpG dinucleotides during DNA methylation. DNMTs such as DNMT1 (maintenance methylation) and DNMT3A/B (de novo methylation) accelerate this process [36]. Gene repression is connected to promoter CpG island methylation, which hinders transcription factors and methyl-binding proteins from condensing chromatin. Metabolic disorders affect oxygenase expression by aberrant DNA methylation, as many now recognize [37]. In adipocytes and hepatocytes, fructose and inflammation cause promoter hypomethylation of the ROS-producing NOX4 gene. This boosts ROS and gene expression. These changes cause insulin resistance and low-level liver and fat tissue inflammation [38].

In insulin-resistant patients, superoxide dismutase (SOD), catalase, and GPx genes are hypermethylated. It limits toxin clearance and promotes oxidative imbalance. Dual methylation activates pro-oxidant genes and inhibits antioxidant genes, worsening metabolic disease oxidative stress [39]. Ten–Eleven Translocation (TET) enzymes demethylate DNA together. TET proteins speed DNA demethylation from 5-methyl to 5-hydroxymethylcytosine. TET enzyme activity depends on cell redox and metabolic status, since they need oxygen and α-ketoglutarate for operation. Metabolic diseases with low TET activity may make regulatory gene demethylation problematic [40]. It worsens oxygenase dysregulation. Changes in mitochondrial metabolism or ROS-induced enzyme inhibition may cause this. Constant DNA methylation and demethylation alter oxygenase enzyme expression. Abnormal NOX, XO, and CYP gene methylation increases oxidative stress and metabolic disease [41].

5.3.2 Histone Modifications: Acetylation, Methylation, and Phosphorylation

Histone proteins wrap DNA to make nucleosomes, chromatin's building blocks. N-terminal histone tails undergo multiple posttranslational modifications (PTMs), such as acylation, methylation, phosphorylation, ubiquitination, and sumoylation [42]. These modifications make DNA accessible to transcriptional machinery, affecting chromatin shape and activity. Histone tail lysine residues get acetyl groups from histone acetyltransferases (HATs). Histone acetylation starts transcription. Relaxed chromatin reduces histones' positive charge and makes it harder for them to attach to negatively charged DNA, improving transcription [43]. These acetyl groups are removed by histone deacetylases (HDACs), making transcription harder and chromatin more compact. CYP isoform and NOX enzyme promoter histone acetylation changes come from metabolic disorders. High fat and sugar consumption enhances HAT and lowers HDAC [44]. This hyperacetylates NOX4 promoter histones. The liver and endothelial tissues are oxidized by NOX4. HDAC inhibitors have different effects. Some HDAC medicines enhance insulin sensitivity and decrease inflammation, whereas others activate NOX genes to increase ROS generation. Targeting histone acetylation machinery is necessary [45].

Histone methylation may trigger transcription depending on the amino acid residue and amount. H3K4me3 is connected with active transcription, whereas H3K9me3 and H3K27me3 are related to repression [46]. HMTs and KDMs control marks. These enzymes progressively control oxygenases. In type 2 diabetes, demethylation of repressive markers such as H3K27me3 in the NOX1 promoter region causes blood vessel problems and increased NOX1 expression. Obesity and diabetes reduce free radical defense by losing activating methyl tags at antioxidant gene promoters [47]. Histone phosphorylation and chromatin rearrangement are typical stress responses. Serine and threonine residues often suffer. H2AX phosphorylation (γH2AX) is a typical sign of ROS-induced DNA damage. Even if it is not the major source of transcription alterations, histone phosphorylation may activate pro-oxidant genes in chromatin when metabolism is challenged. Histone modifications control oxygenase production flexibly and reversibly. Acetyl-CoA, S-Adenosylmethionine (SAM), and Flavin Adenine Dinucleotide (FAD) impact them. This links oxidative equilibrium, chromatin state, and metabolism [48].

5.4 Regulation by Noncoding RNAs (miRNAs, lncRNAs)

Some types of noncoding RNAs (ncRNAs) include microRNAs (miRNAs), long noncoding RNAs (lncRNAs), small interfering RNAs (siRNAs), and circular RNAs (circRNAs). These make up a large part of the human transcriptome. miRNAs and lncRNAs are two types of RNA that have become very important for controlling gene expression after transcription and epigenetically [49]. This includes controlling the expression of oxygenases, which produce reactive oxygen species. miRNAs are around 22-nucleotide-long RNAs that bind to the 3′ untranslated regions (3′UTRs) of target mRNAs and stop them from being translated or break them down. In endothelial cells, miR-25 and miR-146a stop NOX4 from being made. In diabetes, when these two are turned down, it causes oxidative stress in blood vessels and too much NOX4 to be made [50]. miR-34a, which is more active in obesity and type 2 diabetes, indirectly raises ROS levels by targeting SIRT1, which is an important regulator of mitochondrial ROS detoxification. miR-155 makes insulin resistance worse by raising inflammation and NOX2 expression in macrophages. lncRNAs are usually longer than 200 nucleotides and influence gene expression in several ways, including transcriptional interference, chromatin remodeling, and competing endogenous RNAs (ceRNAs) that soak up miRNAs [51]. lncRNA H19 helps cause insulin resistance and fat buildup in the liver by controlling the expression of the CYP enzyme and the production of ROS. lncRNA GAS5 changes oxidative stress by modulating the expression of antioxidant genes [52]. It may also change the way oxygenase works in people who are overweight. These ncRNAs use both environmental signals and epigenetic states to carefully control the expression of oxygenase and the production of ROS. Changes in their expression in metabolic illnesses establish feedback loops that make oxidative damage and metabolic dysfunction worse [53].

5.5 Epigenetic Dysregulation in Metabolic Diseases

This is because metabolic disorders like obesity, insulin resistance, type 2 diabetes mellitus (T2DM), nonalcoholic fatty liver disease (NAFLD), and related heart and metabolic problems are becoming more common. Epigenetic dysregulation has been shown to have a big effect on the pathophysiology and development of metabolic disorders. Epigenetic alterations, such as DNA methylation, histone modifications, and noncoding RNA-mediated regulation, affect gene expression in a way that may be passed down through generations [54]. Unlike mutations, epigenetic alterations are very sensitive to things like diet, exercise, inflammation, and exposure to toxins, and they may be able to be reversed. These modifications may train or reprogram metabolic pathways at different points in life, from when a person is a fetus to when they are an adult. It has even been shown that these modifications may pass on metabolic concerns to future generations [55]. It is very essential to know that the disruption of epigenetic control affects several critical genes that are involved in mitochondrial function, lipid metabolism, inflammation, oxidative stress, and glucose homeostasis [56].

5.5.1 *Obesity and Insulin Resistance*

Obesity excess body fat is a major risk factor for insulin resistance and metabolic syndrome. Adipose tissue not only stores but also secretes adipokines, inflammatory cytokines, and metabolic regulators. Epigenetic dysregulation in adipocytes and preadipocytes affects adipose tissue function, remodeling, and inflammation [57]. Lipid metabolism and adipogenesis genes have altered DNA methylation. Obesity is linked to hypomethylation of the PPARγ promoter, which regulates adipocyte differentiation and promotes adipogenesis. Obesity often causes promoter hypermethylation in genes associated with thermogenesis and mitochondrial function, such as UCP1 and PGC-1α, leading to reduced energy expenditure and fat storage [58]. Histone modifications affect adipose tissue dysfunction. Histone acetylation and methylation regulate insulin signaling pathway genes. Chronic low-grade inflammation in obese individuals' adipose tissue is connected to increased histone acetylation at inflammatory gene promoters (e.g., TNF-α and IL-6) [59]. Noncoding RNAs, including miR-143, miR-103, and miR-27a, affect genes involved in adipocyte development, insulin signaling, and fat accumulation, contributing to obesity [60]. Aberrant miRNA expression increases adipose tissue inflammation and metabolic dysfunction [61].

5.5.2 Type 2 Diabetes Mellitus

Type 2 diabetes is a long-term metabolic illness that makes it hard for the body to handle glucose and produces high blood sugar. Insulin resistance and β-cell dysfunction are two of its most common symptoms. More and more evidence is coming in to support the hypothesis that epigenetic memory is a big part of how diabetes develops and stays around, even when blood sugar levels are brought under control [62]. Hyperglycemia is an epigenetic modulator all by itself. It helps epigenetic markers build up, which turns on genes that cause inflammation and oxidative stress for a long time. This is called "metabolic memory." For example, high blood sugar levels over a long time induce H3K4 methylation and acetylation to stay at the NF-κB p65 promoter [63]. This keeps inflammatory genes active in vascular cells and makes diabetes worse. DNA methylation of key transcription factors, including Pdx1, Nkx6.1, and MafA, makes it hard for pancreatic β cells to make and release insulin. Also, oxidative stress causes histone deacetylation at insulin gene promoters, which stops insulin expression [64]. In diabetes circumstances, miRNAs like miR-375 and miR-29a are overexpressed. These miRNAs stop genes that help β cells grow and release insulin. These miRNAs not working well makes β-cell failure worse. It is fascinating that certain epigenetic changes that are connected to diabetes may be fixed with medication. The fact that drugs like metformin, resveratrol, and HDAC inhibitors may change some of the bad epigenetic marks shows that epigenetic-targeted therapy might work for type 2 diabetes [65].

5.5.3 Nonalcoholic Fatty Liver Disease (NAFLD)

NAFLD includes cirrhosis, fibrosis, NASH, and simple steatosis. Overweight, elevated cholesterol, and insulin resistance are major factors. Under metabolic stress, the liver, which breaks down lipids and carbohydrates, is particularly susceptible to epigenetic changes [66]. NAFLD is associated with global DNA hypomethylation and site-specific hypermethylation of oxidative stress and lipid oxidation-protecting genes. Methylation inhibits the CPT1A promoter, which encodes a key fatty acid oxidation enzyme. This produces hepatocyte fat. Lipogenesis genes are also regulated by histone acetylation [67]. Histone acetyltransferases like p300 increase SREBP-1c and FASN, which stimulate hepatic de novo lipogenesis. NAFLD is also linked to lower SIRT1 activity, which impairs mitochondrial biogenesis and fatty acid oxidation. NAFLD is also affected by noncoding RNAs [68]. One of the most frequent hepatic miRNAs, miR-122, is downregulated in NAFLD and linked in poor lipid metabolism. Steatotic livers have more miR-34a, which targets SIRT1 and worsens oxidative stress and liver cell destruction. lncRNAs such as H19, MEG3, and lncARSR regulate liver inflammation, lipid metabolism, and fibrogenesis via interacting with chromatin-modifying enzymes and miRNAs [69].

5.5.4 Cardiometabolic Complications

Heart disease, hypertension, atherosclerosis, and cardiovascular disease (CVD) are connected to metabolic syndrome, type 2 diabetes, and NAFLD. DNA methylation of endothelial nitric oxide synthase (eNOS) promoter regions reduces NO production, vasodilation, and vascular tone in diabetes and hypertension [70]. Endothelial cells hypomethylate pro-inflammatory genes like VCAM-1 and ICAM-1 to assist monocytes in producing atheromas. Atherosclerosis histone changes affect macrophage type and lipids. More H3K27 acetylation near pro-inflammatory cytokine promoters makes plaques unstable and foam cells probable. However, anti-inflammatory gene histone methylation is more likely to be lost, causing atherogenic inflammation [71].

MiR-33, miR-126, and miR-21 regulate endothelial cell integrity, inflammatory signaling, and cholesterol breakdown. Cardiometabolic disease-related vascular calcification, atherosclerotic plaque, and cardiac hypertrophy include these miRNAs [72]. Curcumin, quercetin, HDAC inhibitors, and miRNA mimics/inhibitors may improve heart disease epigenetics. In the genetic-environmental interaction, epigenetic dysregulation causes metabolic diseases. Epigenetic changes affect metabolic, inflammatory, and oxidative stress genes. Reverse dynamic changes [73]. Insulin resistance, obesity, T2DM, NAFLD, and cardiometabolic disorders worsen due to abnormal DNA methylation, histone changes, and noncoding RNA profiles. Epigenetic changes may be reversed, simplifying metabolic disease diagnosis, quantification, and treatment. To develop precision therapies and customized medicine, we must understand these illnesses' epigenetics [54].

5.6 Epigenetic Regulation of Oxygenases Involved in ROS Production in Metabolic Disorders

This table lists the primary oxygenase enzymes that produce reactive oxygen species (ROS) and how these enzymes are controlled epigenetically in connection with metabolic diseases. It talks about their physiological roles, how they are linked to conditions like obesity, type 2 diabetes mellitus (T2DM), nonalcoholic fatty liver disease (NAFLD), and heart problems, as well as specific epigenetic changes that affect how they work and how they are expressed, like DNA methylation, histone modifications, and noncoding RNA regulation. Understanding these regulatory pathways makes it feasible to use targeted epigenetic and antioxidant therapy to treat metabolic illnesses (Table 5.1) [74].

5.7 Therapeutic Implications and Emerging Strategies

Epigenetic medicines, often known as "epidrugs," work on the enzymes that add, read, or erase epigenetic markers. Histone acetyltransferases (HATs), histone methyltransferases (HMTs), histone deacetylases (HDACs), and DNA methyltransferases (DNMTs) are all examples of these. A big part of their therapeutic promise is changing how genes are expressed and fixing the aberrant epigenetic landscapes that come with metabolic diseases [84].

5.7.1 *Epigenetic Drugs Targeting ROS Pathways*

Metabolic illness mice recover normal methylation using FDA-approved cancer drugs like 5-azacytidine and decitabine. In diabetes and obesity, antioxidant genes are hypermethylated, while pro-oxidant genes such as NOX4 are hypomethylated, creating ROS imbalance. DNMT inhibitors may fix methylation and redox imbalance. But using them in noncancerous conditions is risky since they are not specific and may influence other things [85]. HDAC inhibitors, including vorinostat, trichostatin A, and valproic acid, may lower oxidative stress by regulating inflammation, mitochondrial biogenesis, and antioxidant defense genes. The balance between ROS generation and clearance decreases hepatic steatosis and enhances insulin action in animal models. HDAC3 inhibition enhances liver mitochondrial activity and lipid metabolism without side effects [86]. Various illnesses can be treated with EZH2 or LSD1 inhibitors. Compounds may restore oxygenase or the indirect oxidative pathway gene transcription. Stopping G9a, an HMT that has repressive tags like H3K9me2, may activate oxidative stress genes [87]. Scientists are creating nanoparticle carriers to deliver epidrugs to the liver, adipose, and pancreatic islets [88].

5.7.2 *Nutraceuticals and Lifestyle Interventions*

Nutritional epigenetics, which alters markers and ROS without medications, is becoming popular. Lifestyle and diet affect epigenetics and oxidative homeostasis. Nutritional supplements may affect antioxidants and epigenetics. For example, turmeric curcumin inhibits HDACs, activates histone acetyltransferases, and removes

Table 5.1 Epigenetic regulation of oxygenases involved in ROS production in metabolic disorders

S. no	Oxygenase enzyme	Physiological role and ROS involvement	Metabolic disorder association	Epigenetic mechanism	Reference
1.	Mitochondrial complexes I and III	Electron transport and ATP production; ROS leak under metabolic stress	Obesity, T2DM, neurodegeneration	Global histone hypoacetylation under nutrient stress, SIRT3/ PGC-1α regulated by miR-34a and miR-210, DNA methylation of mitochondrial biogenesis genes	[75]
2.	eNOS (endothelial NOS)	Produces nitric oxide (NO); dysfunctional eNOS produces superoxide instead of NO	Cardiovascular complications of diabetes	Promoter hypermethylation (reduced eNOS expression), miR-21 and miR-126 alterations, H3K27me3 enrichment in vascular tissues	[76, 77]
3.	Dual oxidase (DUOX2)	Produces H_2O_2 in mucosal and epithelial defense	Gut inflammation, metabolic endotoxemia	miR-29a/miR-155 dysregulation, altered chromatin accessibility via inflammation-induced histone marks	[78]
4.	NADPH oxidase (NOX4)	Catalyzes electron transfer to oxygen, forming superoxide (O_2^-); major ROS source in non-phagocytic cells	Obesity, T2DM, NAFLD, atherosclerosis	DNA hypomethylation of the NOX4 promoter, histone acetylation (↑ H3/H4 acetylation)	[79]
5.	Xanthine oxidase (XO)	Oxidizes hypoxanthine to xanthine and uric acid, producing superoxide and hydrogen peroxide	T2DM, hypertension, NAFLD	DNA hypermethylation of antioxidant gene promoters (e.g., SOD), histone acetylation at pro-inflammatory gene loci, miR-34a (indirect regulation via SIRT1 inhibition)	[80, 81]
6.	Cytochrome P450 (e.g., CYP2E1)	Metabolizes lipids and xenobiotics; generates ROS during uncoupled reactions	NAFLD, alcohol-induced liver disease	Promoter hypomethylation of CYP2E1 in fatty liver, miR-378 and miR-155 (dysregulated)	[82, 83]

CpG islands [89]. Nrf2 activity increases, and NOX enzymes decrease. Resveratrol from red wine and grapes enhances mitochondrial biogenesis and reduces reactive oxygen species (ROS) through activation of the NAD^+-dependent histone deacetylase SIRT1. It also reduces endothelial NOX and blood vessel oxidative stress. Cruciferous vegetables contain sulforaphane. HDAC activity diminishes, and phase II detoxification enzymes and antioxidants increase [90]. It also reduces oxidative metabolic tissue damage. EGCG is an epigenetic and antioxidant green tea catechin. Research shows that reducing DNMT and HDAC may improve liver oxidative balance and insulin sensitivity [91]. Natural compounds can safely and cheaply change oxidative stress epigenetic markers, but they need to be studied. Reducing calories increases mitochondria, SIRT1, and oxidative stress [92]. Ketogenic diets may reduce oxidative stress by increasing β-hydroxybutyrate, an HDAC inhibitor.

5.7.3 *CRISPR-Based Epigenome Editing*

CRISPR-based epigenome editing is a promising gene control technique. Epigenome editing allows precise, reversible gene expression changes. This differs from standard gene-editing methods, which permanently alter DNA sequences. These methods may precisely and targetedly modulate oxygenase gene expression by targeting promoters or enhancers like NOX4, XO, or CYP2E1 [93]. Demethylation of antioxidant response element (ARE) regions in diabetic mice may increase oxidative defense by restoring Nrf2 activity, according to proof-of-concept research. In obese mice, shutting down NOX gene promoters reduces inflammation and ROS. This method is accurate enough to modify genes in specific tissues without changing the epigenome. Exosomes, nanoparticles, and Adeno-Associated Viruses (AAVs can be used to safely and efficiently insert CRISPR-dCas9 complexes into live cells [94]. CRISPR-based epigenome editing for metabolic diseases is still in preclinical trials despite its potential. Immunological responses, long-term safety, off-target consequences, and regulatory issues must be addressed. However, its tailored, reversible, and programmable therapies may soon transform therapy [95].

5.8 Future Perspectives and Research Directions

The study of how oxidative stress and epigenetic regulation work together in metabolic illnesses is a fast-growing field of biomedical research. As our scientific understanding improves, targeting the epigenetic mechanisms that govern oxygenase expression and ROS equilibrium gives us both fresh information about how diseases start and useful targets for diagnosis, prevention, and therapy [96].

5.8.1 Deepening the Mechanistic Understanding

Even while we know a lot about individual modifications like DNA methylation and histone acetylation, we still do not know much about how different epigenetic layers work together. New tools like single-cell epigenomics and spatial transcriptomics may help us understand these complicated regulatory networks. They can show us how epigenetic changes that are particular to certain cell types cause illness and how metabolic tissues are different from one another [97].

5.8.2 Identification of Epigenetic Biomarkers

Epigenetic indicators offer a lot of potential as dynamic, noninvasive biomarkers for early identification, risk assessment, and therapy monitoring of metabolic diseases. To uncover and confirm epigenetic biomarkers that are connected to key clinical outcomes like insulin resistance, liver fibrosis, or heart problems, future research should concentrate on large-scale, multi-omics cohort studies. By combining personalized epigenetic landscape-based prediction models with machine learning techniques, they might be made faster [98].

5.8.3 Advancing Precision Epigenetic Therapies

The invention of CRISPR-based epigenome editing provides a revolutionary way to fix epigenetic alterations that cause illness in a precise and reversible way. Unlike regular epigenetics that change the global chromatin architecture, CRISPR-dCas9 fusion tools may change the promoters or enhancers of oxygenase genes like NOX4 or CYP2E1 without changing the DNA sequence itself. Future studies should work on making these tools better by reducing immune reactions, making them safer, and creating delivery methods that target particular tissues (such as lipid nanoparticles, AAVs, or exosomes). It is also worth looking into the idea of employing inducible systems that respond to metabolic signals to dynamically control genes that make ROS in real time [98]. Epigenetic plasticity means that things like diet, exercise, and changes to the microbiota may have a big impact on gene expression and oxidative stress levels. It is becoming more and more obvious that stress, environmental contaminants, and a mother's nutrition may program the risk of metabolic illnesses via epigenetic pathways. This developmental reprogramming changes how genes involved in energy metabolism, mitochondrial function, and ROS control are expressed throughout life [99].

5.9 Conclusion

Epigenetic regulation and reactive oxygen species (ROS) homeostasis interact in a complex way to affect the pathophysiology and progression of metabolic disorders like obesity, type 2 diabetes, NAFLD, and heart disease. ROS are produced by oxygenases such as NADPH oxidases (NOX family), xanthine oxidase, cytochrome P450 enzymes, and mitochondrial complexes. Epigenetic processes include DNA methylation, histone changes, and noncoding RNAs that control enzyme production. Epigenetic pathway dysregulation may cause metabolic problems, including chronic oxidative stress, low-grade inflammation, mitochondrial dysfunction, and food metabolization abnormalities. Data reveals that epigenetic alterations induce metabolic issues, not just passive reflections of illness conditions. Unregulated microRNAs and lncRNAs, mis-methylated pro-oxidant and antioxidant gene promoters, and unbalanced histone acetylation and methylation states disrupt the delicate redox balance and accelerate illness. These changes are dynamic and, most importantly, reversible since they are generally caused by environmental factors including overeating, inactivity, pollution, and childhood stress. Epigenetics is a cutting-edge therapy since it can be reversed. Epigenetic medicines like DNMT and HDAC inhibitors may limit oxygenase activity and restore redox balance. Curcumin, resveratrol, sulforaphane, and green tea polyphenols are additional bioactive nutraceuticals with intriguing epigenetic and antioxidant effects. New technologies like CRISPR-based epigenome editing may adjust gene expression with unprecedented precision without altering the DNA sequence, changing targeted medicine forever. Precision medicine in metabolic care may be possible by combining epigenetic data with individualized nutrition, exercise, and microbiota manipulation. Problems persist. Better tissue-specific delivery, long-term epigenetic therapy safety, and biomarker clinical validation are needed. The epigenetic modulation of oxygenases in metabolic diseases suggests a shift from treating symptoms to treating causes. Future medicines may be more effective, last longer, and be more patient specific by targeting ROS and oxidative stress molecular switches.

References

1. Lennicke C, Cochemé HM (2021) Redox metabolism: ROS as specific molecular regulators of cell signaling and function. Mol Cell 81:3691–3707. https://doi.org/10.1016/J.MOLCEL.2021.08.018
2. Schieber M, Chandel NS (2014) ROS function in redox signaling and oxidative stress. Curr Biol 24:R453–R462. https://doi.org/10.1016/J.CUB.2014.03.034
3. Saltiel AR, Kahn CR (2001) Insulin signalling and the regulation of glucose and lipid metabolism. Nature 414:799–806. https://doi.org/10.1038/414799A
4. Rahat B, Sharma R, Ali T, Kaur J (2023) Epigenetic regulation of gene expression: an overview of classical and recently discovered novel players. In: Perinatal and developmental epigenetics: Volume 32 in Translational epigenetics. Academic Press, pp 3–45. https://doi.org/10.1016/B978-0-12-821785-6.00010-4

5. Chen Y, Shen YQ (2025) Role of reactive oxygen species in regulating epigenetic modifications. Cell Signal 125:111502. https://doi.org/10.1016/J.CELLSIG.2024.111502
6. Lu C, Thompson CB (2012) Metabolic regulation of epigenetics. Cell Metab 16:9–17. https://doi.org/10.1016/J.CMET.2012.06.001/ASSET/406E2617-3D4E-47CD-B0D2-735A6D7B78AC/MAIN.ASSETS/GR4.JPG
7. Chen C, Wang Z, Qin Y (2022) Connections between metabolism and epigenetics: mechanisms and novel anti-cancer strategy. Front Pharmacol 13:935536. https://doi.org/10.3389/FPHAR.2022.935536
8. García-Giménez JL, Seco-Cervera M, Tollefsbol TO, Romá-Mateo C, Peiró-Chova L, Lapunzina P et al (2017) Epigenetic biomarkers: current strategies and future challenges for their use in the clinical laboratory. Crit Rev Clin Lab Sci 54:529. https://doi.org/10.1080/10408363.2017.1410520
9. Al-Awar A, Hussain S (2024) Interplay of reactive oxygen species (ROS) and epigenetic remodelling in cardiovascular diseases pathogenesis: a contemporary perspective. Front Biosci 29:398. https://doi.org/10.31083/J.FBL2911398
10. Murphy MP, Bayir H, Belousov V, Chang CJ, Davies KJA, Davies MJ et al (2022) Guidelines for measuring reactive oxygen species and oxidative damage in cells and in vivo. Nat Metab 4:651–662. https://doi.org/10.1038/S42255-022-00591-Z;SUBJMETA=1647,45,631;KWRD=BIOCHEMISTRY,BIOLOGICAL+TECHNIQUES
11. Juan CA, de la Lastra JMP, Plou FJ, Pérez-Lebeña E (2021) The chemistry of reactive oxygen species (ROS) revisited: outlining their role in biological macromolecules (DNA, lipids and proteins) and induced pathologies. Int J Mol Sci 22:4642. https://doi.org/10.3390/IJMS22094642
12. Hong Y, Boiti A, Vallone D, Foulkes NS (2024) Reactive oxygen species signaling and oxidative stress: transcriptional regulation and evolution. Antioxidants 13:312. https://doi.org/10.3390/ANTIOX13030312
13. Zhao RZ, Jiang S, Zhang L, Yu Z-B (2019) Mitochondrial electron transport chain, ROS generation and uncoupling (Review). Int J Mol Med 44:3–15. https://doi.org/10.3892/IJMM.2019.4188
14. Rains JL, Jain SK (2011) Oxidative stress, insulin signaling, and diabetes. Free Radic Biol Med 50:567–575. https://doi.org/10.1016/J.FREERADBIOMED.2010.12.006
15. Averill-Bates D (2024) Reactive oxygen species and cell signaling. Review. Biochim Biophys Acta 1871:119573. https://doi.org/10.1016/J.BBAMCR.2023.119573
16. Reddy VP (2023) Oxidative stress in health and disease. Biomedicine 11:2925. https://doi.org/10.3390/BIOMEDICINES11112925
17. Lam GY, Huang J, Brumell JH (2010) The many roles of NOX2 NADPH oxidase-derived ROS in immunity. Semin Immunopathol 32(4):415–430. https://doi.org/10.1007/S00281-010-0221-0
18. Vermot A, Petit-Härtlein I, Smith SME, Fieschi F (2021) Nadph oxidases (Nox): an overview from discovery, molecular mechanisms to physiology and pathology. Antioxidants 10:890. https://doi.org/10.3390/ANTIOX10060890
19. Takata T, Araki S, Tsuchiya Y, Watanabe Y (2020) Oxidative stress orchestrates MAPK and nitric-oxide synthase signal. Int J Mol Sci 21:8750. https://doi.org/10.3390/IJMS21228750
20. Furuhashi M (2020) New insights into purine metabolism in metabolic diseases: role of xanthine oxidoreductase activity. Am J Physiol Endocrinol Metab 319:E827–E834. https://doi.org/10.1152/AJPENDO.00378.2020
21. Higashi Y (2022) Roles of oxidative stress and inflammation in vascular endothelial dysfunction-related disease. Antioxidants 11:1958. https://doi.org/10.3390/ANTIOX11101958
22. Sautin YY, Johnson RJ (2008) Uric acid: the oxidant–antioxidant paradox. Nucleosides Nucleotides Nucleic Acids 27:608. https://doi.org/10.1080/15257770802138558
23. Nolfi-Donegan D, Braganza A, Shiva S (2020) Mitochondrial electron transport chain: oxidative phosphorylation, oxidant production, and methods of measurement. Redox Biol 37:101674. https://doi.org/10.1016/J.REDOX.2020.101674

24. González P, Lozano P, Ros G, Solano F (2023) Hyperglycemia and oxidative stress: an integral, updated and critical overview of their metabolic interconnections. Int J Mol Sci 24:9352. https://doi.org/10.3390/IJMS24119352
25. Yang Y, Wu Y, Sun XD, Zhang Y (2021) Reactive oxygen species, glucose metabolism, and lipid metabolism. In: Oxidative stress: human diseases and medicine. Springer, Singapore, pp 213–235. https://doi.org/10.1007/978-981-16-0522-2_9
26. Eisner V, Picard M, Hajnóczky G (2018) Mitochondrial dynamics in adaptive and maladaptive cellular stress responses. Nat Cell Biol 20(7):755–765. https://doi.org/10.1038/s41556-018-0133-0
27. Baggelaar MP, Maccarrone M, van der Stelt M (2018) 2-Arachidonoylglycerol: a signaling lipid with manifold actions in the brain. Prog Lipid Res 71:1–17. https://doi.org/10.1016/J.PLIPRES.2018.05.002
28. Schubert ML (2004) H2-receptor antagonists. In: Encyclopedia of gastroenterology. Academic Press, pp 395–401. https://doi.org/10.1016/B0-12-386860-2/00381-6
29. Shaik S, Dubey KD (2021) The catalytic cycle of cytochrome P450: a fascinating choreography. Trends Chem 3:1027–1044. https://doi.org/10.1016/J.TRECHM.2021.09.004
30. Bo T, Gao L, Yao Z, Shao S, Wang X, Proud CG et al (2024) Hepatic selective insulin resistance at the intersection of insulin signaling and metabolic dysfunction-associated steatotic liver disease. Cell Metab 36:947–968. https://doi.org/10.1016/J.CMET.2024.04.006
31. Abraham MJ, El Sherbini A, El-Diasty M, Askari S, Szewczuk MR (2023) Restoring epigenetic reprogramming with diet and exercise to improve health-related metabolic diseases. Biomolecules 13:318. https://doi.org/10.3390/BIOM13020318
32. Stephens KE, Miaskowski CA, Levine JD, Pullinger CR, Aouizerat BE (2012) Epigenetic regulation and measurement of epigenetic changes. Biol Res Nurs 15:373. https://doi.org/10.1177/1099800412444785
33. Cipriano A, Viviano M, Feoli A, Milite C, Sarno G, Castellano S et al (2023) NADPH oxidases: from molecular mechanisms to current inhibitors. J Med Chem 66:11632. https://doi.org/10.1021/ACS.JMEDCHEM.3C00770
34. Masenga SK, Kabwe LS, Chakulya M, Kirabo A (2023) Mechanisms of oxidative stress in metabolic syndrome. Int J Mol Sci 24(9):7898. https://doi.org/10.3390/IJMS24097898
35. Colak E, Pap D (2021) The role of oxidative stress in the development of obesity and obesity-related metabolic disorders. J Med Biochem 40:1. https://doi.org/10.5937/JOMB0-24652
36. Ying Z, Chen T (2023) Mechanisms of DNA methylation and demethylation during mammalian development. In: Handbook of epigenetics: the new molecular and medical genetics, 3rd edn, pp 11–26. https://doi.org/10.1016/B978-0-323-91909-8.00009-8
37. Stefansson OA, Sigurpalsdottir BD, Rognvaldsson S, Halldorsson GH, Juliusson K, Sveinbjornsson G et al (2024) The correlation between CpG methylation and gene expression is driven by sequence variants. Nat Genet 56:1624–1631. https://doi.org/10.1038/s41588-024-01851-2
38. Guo S, Chen X (2015) The human Nox4: gene, structure, physiological function and pathological significance. J Drug Target 23:888–896. https://doi.org/10.3109/1061186X.2015.1036276
39. Accili D, Deng Z, Liu Q (2025) Insulin resistance in type 2 diabetes mellitus. Nat Rev Endocrinol 21(7):413–426. https://doi.org/10.1038/s41574-025-01114-y
40. Zhang X, Zhang Y, Wang C, Wang X (2023) TET (ten-eleven translocation) family proteins: structure, biological functions and applications. Signal Transduct Target Ther 8:1–20. https://doi.org/10.1038/S41392-023-01537-X;SUBJMETA=1647,176,208,2210,631;KWRD=EPIGENETICS,EPIGENETICS+ANALYSIS
41. Chen S, Li Q, Shi H, Li F, Duan Y, Guo Q (2024) New insights into the role of mitochondrial dynamics in oxidative stress-induced diseases. Biomed Pharmacother 178:117084. https://doi.org/10.1016/J.BIOPHA.2024.117084
42. McGinty RK, Tan S (2015) Nucleosome structure and function. Chem Rev 115:2255–2273. https://doi.org/10.1021/CR500373H

43. Nitsch S, Shahidian LZ, Schneider R (2021) Histone acylations and chromatin dynamics: concepts, challenges, and links to metabolism. EMBO Rep 22:e52774. https://doi.org/10.15252/EMBR.202152774
44. Zhang SY, Zhang LY, Wen R, Yang N, Zhang TN (2024) Histone deacetylases and their inhibitors in inflammatory diseases. Biomed Pharmacother 179:117295. https://doi.org/10.1016/J.BIOPHA.2024.117295
45. Charidemou E, Kirmizis A (2024) A two-way relationship between histone acetylation and metabolism. Trends Biochem Sci 49:1046–1062. https://doi.org/10.1016/J.TIBS.2024.10.005
46. He S, Yiu G, Zhou P, Chen DF (2012) Epigenetic mechanisms of retinal disease. In: Retina Fifth Ed, vol 1, pp 642–651. https://doi.org/10.1016/B978-1-4557-0737-9.00033-3
47. Lopes JR, Chiba DE, Dos Santos JL (2021) HIV latency reversal agents: a potential path for functional cure? Eur J Med Chem 213:113213. https://doi.org/10.1016/j.ejmech.2021.113213
48. Rossetto D, Avvakumov N, Côté J (2012) Histone phosphorylation: a chromatin modification involved in diverse nuclear events. Epigenetics 7:1098. https://doi.org/10.4161/EPI.21975
49. Nemeth K, Bayraktar R, Ferracin M, Calin GA (2024) Non-coding RNAs in disease: from mechanisms to therapeutics. Nat Rev Genet 25:211–232. https://doi.org/10.1038/S41576-023-00662-1;SUBJMETA=2568,331,337,384,521,631;KWRD=LONG+NON-CODING+RNAS,MIRNAS,SMALL+RNAS
50. Checa J, Aran JM (2020) Reactive oxygen species: drivers of physiological and pathological processes. J Inflamm Res 13:1057. https://doi.org/10.2147/JIR.S275595
51. Kim Y, Kim OK (2021) Potential roles of adipocyte extracellular vesicle–derived miRNAs in obesity-mediated insulin resistance. Adv Nutr 12:566–574. https://doi.org/10.1093/ADVANCES/NMAA105
52. Huang H, Zhang G, Ge Z (2021) lncRNA MALAT1 promotes renal fibrosis in diabetic nephropathy by targeting the miR-2355-3p/IL6ST axis. Front Pharmacol 12:647650. https://doi.org/10.3389/FPHAR.2021.647650
53. Peterson SJ, Dave N, Kothari J (2020) The effects of heme oxygenase upregulation on obesity and the metabolic syndrome. Antioxid Redox Signal 32:1061–1070. https://doi.org/10.1089/ARS.2019.7954
54. Wondmkun YT (2020) Obesity, insulin resistance, and type 2 diabetes: associations and therapeutic implications. Diabetes Metab Syndr Obes 13:3611. https://doi.org/10.2147/DMSO.S275898
55. Al Aboud NM, Tupper C, Jialal I (2023) Genetics, epigenetic mechanism. StatPearls
56. Akanbi CA, Rotimi DE, Ojo AB, Ojo OA (2025) Endocrine-disrupting chemicals (EDCs) and epigenetic regulation in embryonic development: mechanisms, impacts, and emerging trends. Toxicol Rep 14:101885. https://doi.org/10.1016/J.TOXREP.2024.101885
57. Mohamed SM, Shalaby MA, El-Shiekh RA, El-Banna HA, Emam SR, Bakr AF (2023) Metabolic syndrome: risk factors, diagnosis, pathogenesis, and management with natural approaches. Food Chem Adv 3:100335. https://doi.org/10.1016/J.FOCHA.2023.100335
58. Samblas M, Milagro FI, Martínez A (2019) DNA methylation markers in obesity, metabolic syndrome, and weight loss. Epigenetics 14:421. https://doi.org/10.1080/15592294.2019.1595297
59. Kawai T, Autieri MV, Scalia R (2020) Adipose tissue inflammation and metabolic dysfunction in obesity. Am J Physiol Cell Physiol 320:C375. https://doi.org/10.1152/AJPCELL.00379.2020
60. Chu DT, Thi YVN, Chew NWS (2023) Histone modifications in fat metabolism and obesity. Prog Mol Biol Transl Sci 197:135–152. https://doi.org/10.1016/BS.PMBTS.2023.01.003
61. Peng J, He L (2017) IRS posttranslational modifications in regulating insulin signaling. J Mol Endocrinol 60:R1. https://doi.org/10.1530/JME-17-0151
62. American Diabetes Association Professional Practice Committee (2024) Pharmacologic approaches to glycemic treatment: standards of care in diabetes—2024. Diabetes Care 47:S158–S178. https://doi.org/10.2337/DC24-S009
63. Prasad M, Rajagopal P, Devarajan N, Veeraraghavan VP, Palanisamy CP, Cui B et al (2022) A comprehensive review on high -fat diet-induced diabetes mellitus: an epigenetic view. J Nutr Biochem 107:109037. https://doi.org/10.1016/J.JNUTBIO.2022.109037

64. Steven S, Frenis K, Oelze M, Kalinovic S, Kuntic M, Jimenez MTB et al (2019) Vascular inflammation and oxidative stress: major triggers for cardiovascular disease. Oxid Med Cell Longev 2019:7092151. https://doi.org/10.1155/2019/7092151
65. Angelescu MA, Andronic O, Dima SO, Popescu I, Meivar-Levy I, Ferber S et al (2022) miRNAs as biomarkers in diabetes: moving towards precision medicine. Int J Mol Sci 23:12843. https://doi.org/10.3390/IJMS232112843
66. Heyens LJM, Busschots D, Koek GH, Robaeys G, Francque S (2021) Liver fibrosis in non-alcoholic fatty liver disease: from liver biopsy to non-invasive biomarkers in diagnosis and treatment. Front Med (Lausanne) 8:615978. https://doi.org/10.3389/FMED.2021.615978
67. Hyun J, Jung Y (2020) DNA methylation in nonalcoholic fatty liver disease. Int J Mol Sci 21:8138. https://doi.org/10.3390/IJMS21218138
68. Kim JY, Yang AY, Kim K, Kwon HH, Leem J, Kim YA (2024) Pharmacological inhibition of p300 ameliorates steatosis, inflammation, and fibrosis in mice with non-alcoholic steatohepatitis. Heliyon 10:e30908. https://doi.org/10.1016/J.HELIYON.2024.E30908
69. Colaianni F, Zelli V, Compagnoni C, Miscione MS, Rossi M, Vecchiotti D et al (2024) Role of circulating microRNAs in liver disease and HCC: focus on miR-122. Genes (Basel) 15:1313. https://doi.org/10.3390/GENES15101313
70. Lee CH, Lui DTW, Lam KSL (2022) Non-alcoholic fatty liver disease and type 2 diabetes: an update. J Diabetes Investig 13:930. https://doi.org/10.1111/JDI.13756
71. Lee HT, Oh S, Ro DH, Yoo H, Kwon YW (2020) The key role of DNA methylation and histone acetylation in epigenetics of atherosclerosis. J Lipid Atheroscler 9:419. https://doi.org/10.12997/JLA.2020.9.3.419
72. Mori MA, Ludwig RG, Garcia-Martin R, Brandão BB, Kahn CR (2019) Extracellular miRNAs: from biomarkers to mediators of physiology and disease. Cell Metab 30:656. https://doi.org/10.1016/J.CMET.2019.07.011
73. Chen M, Li Y, Zhang M, Ge S, Feng T, Chen R et al (2024) Histone deacetylase inhibition enhances extracellular vesicles from muscle to promote osteogenesis via miR-873-3p. Signal Transduct Target Ther 9:256. https://doi.org/10.1038/S41392-024-01976-0;SUBJMETA=136,578,631,692,699,815;KWRD=BONE+REMODELLING,TRAUMA
74. Pecchillo Cimmino T, Ammendola R, Cattaneo F, Esposito G (2023) NOX dependent ROS generation and cell metabolism. Int J Mol Sci 24:2086. https://doi.org/10.3390/IJMS24032086
75. Vercellino I, Sazanov LA (2021) The assembly, regulation and function of the mitochondrial respiratory chain. Nat Rev Mol Cell Biol 23(2):141–161. https://doi.org/10.1038/s41580-021-00415-0
76. Shu X, Keller TCS, Begandt D, Butcher JT, Biwer L, Keller AS et al (2015) Endothelial nitric oxide synthase in the microcirculation. Cell Mol Life Sci 72(23):4561–4575. https://doi.org/10.1007/S00018-015-2021-0
77. Tran N, Garcia T, Aniqa M, Ali S, Ally A, Nauli S (2022) Endothelial nitric oxide synthase (eNOS) and the cardiovascular system: in physiology and in disease states. Am J Biomed Sci Res 15:153. https://doi.org/10.34297/ajbsr.2022.15.002087
78. Grasberger H, Gao J, Nagao-Kitamoto H, Kitamoto S, Zhang M, Kamada N et al (2015) Increased expression of DUOX2 is an epithelial response to mucosal dysbiosis required for immune homeostasis in mouse intestine. Gastroenterology 149:1849–1859. https://doi.org/10.1053/j.gastro.2015.07.062
79. Cipriano A, Viviano M, Feoli A, Milite C, Sarno G, Castellano S et al (2023) NADPH oxidases: from molecular mechanisms to current inhibitors. J Med Chem 66:11632–11655. https://doi.org/10.1021/ACS.JMEDCHEM.3C00770/ASSET/IMAGES/LARGE/JM3C00770_0009.JPEG
80. Kostić DA, Dimitrijević DS, Stojanović GS, Palić IR, Dordević AS, Ickovski JD (2015) Xanthine oxidase: isolation, assays of activity, and inhibition. J Chem 2015:294858. https://doi.org/10.1155/2015/294858
81. Aziz N, Jamil RT (2023) Biochemistry, xanthine oxidase. StatPearls

82. Manikandan P, Nagini S (2017) Cytochrome P450 structure, function and clinical significance: a review. Curr Drug Targets 19:38–54. https://doi.org/10.2174/1389450118666170125144557
83. Veith A, Moorthy B (2018) Role of cytochrome P450s in the generation and metabolism of reactive oxygen species. Curr Opin Toxicol 7:44–51. https://doi.org/10.1016/J.COTOX.2017.10.003
84. Miranda Furtado CL, Dos Santos Luciano MC, Da Silva Santos R, Furtado GP, Moraes MO, Pessoa C (2019) Epidrugs: targeting epigenetic marks in cancer treatment. Epigenetics 14:1164. https://doi.org/10.1080/15592294.2019.1640546
85. Uddin MG, Fandy TE (2021) DNA methylation inhibitors: retrospective and perspective view. Adv Cancer Res 152:205. https://doi.org/10.1016/BS.ACR.2021.03.007
86. Kumar V, Kundu S, Singh A, Singh S (2022) Understanding the role of histone deacetylase and their inhibitors in neurodegenerative disorders: current targets and future perspective. Curr Neuropharmacol 20:158. https://doi.org/10.2174/1570159X19666210609160017
87. Li T, Yu C, Zhuang S (2021) Histone methyltransferase EZH2: a potential therapeutic target for kidney diseases. Front Physiol 12:640700. https://doi.org/10.3389/FPHYS.2021.640700/XML/NLM
88. Majchrzak-Celinska A, Warych A, Szoszkiewicz M (2021) Novel approaches to epigenetic therapies: from drug combinations to epigenetic editing. Genes (Basel) 12:208. https://doi.org/10.3390/GENES12020208
89. Tiffon C (2018) The impact of nutrition and environmental epigenetics on human health and disease. Int J Mol Sci 19:3425. https://doi.org/10.3390/IJMS19113425
90. Ngo V, Duennwald ML (2022) Nrf2 and oxidative stress: a general overview of mechanisms and implications in human disease. Antioxidants 11:2345. https://doi.org/10.3390/ANTIOX11122345
91. Yan Z, Zhong Y, Duan Y, Chen Q, Li F (2020) Antioxidant mechanism of tea polyphenols and its impact on health benefits. Anim Nutr 6:115–123. https://doi.org/10.1016/J.ANINU.2020.01.001
92. Irshad R, Husain M (2021) Natural products in the reprogramming of cancer epigenetics. Toxicol Appl Pharmacol 417:115467. https://doi.org/10.1016/J.TAAP.2021.115467
93. Cai R, Lv R, Shi X, Yang G, Jin J (2023) CRISPR/dCas9 tools: epigenetic mechanism and application in gene transcriptional regulation. Int J Mol Sci 24:14865. https://doi.org/10.3390/IJMS241914865
94. Buendia I, Michalska P, Navarro E, Gameiro I, Egea J, León R (2016) Nrf2–ARE pathway: an emerging target against oxidative stress and neuroinflammation in neurodegenerative diseases. Pharmacol Ther 157:84–104. https://doi.org/10.1016/J.PHARMTHERA.2015.11.003
95. Fadul SM, Arshad A, Mehmood R (2023) CRISPR-based epigenome editing: mechanisms and applications. Epigenomics 15:1137–1155. https://doi.org/10.2217/EPI-2023-0281;CTYPE:STRING:JOURNAL
96. García-Giménez JL, Cánovas-Cervera I, Nacher-Sendra E, Dolz-Andrés E, Sánchez-Bernabéu Á, Agúndez AB et al (2025) Oxidative stress and central metabolism pathways impact epigenetic modulation in inflammation and immune response. Free Radic Biol Med 233:378–399. https://doi.org/10.1016/J.FREERADBIOMED.2025.04.004
97. Bird A (2024) Transgenerational epigenetic inheritance: a critical perspective. Front Epigenet Epigenom 2:1434253. https://doi.org/10.3389/FREAE.2024.1434253
98. García-Giménez JL, Ushijima T, Tollefsbol TO (2016) Epigenetic biomarkers: new findings, perspectives, and future directions in diagnostics. In: Epigenetic biomarkers and diagnostics. Academic Press, pp 1–18. https://doi.org/10.1016/B978-0-12-801899-6.00001-2
99. Hua D, Petrina N, Young N, Cho JG, Poon SK (2024) Understanding the factors influencing acceptability of AI in medical imaging domains among healthcare professionals: a scoping review. Artif Intell Med 147:102698. https://doi.org/10.1016/J.ARTMED.2023.102698

Chapter 6
Interplay Between ROS and Inflammation in Alcoholic Fatty Liver Disease

Vishal Thakur, Swati Mishra, Vijay Kumar, Anchal Sharma, Saurabh Sharma, and Chandni Sharma

6.1 Introduction

6.1.1 Alcoholic Fatty Liver Disease

AFLD is considered as one most prevalent diseases in today's world. A common reason for this condition is the long-term consumption of alcohol [1]. Various studies known are supposed to make a guess that the prevalence rate of fatty liver disease will increase by 2–3 times by the year 2030. Recent data documented from NHANES report that the occurrence rate of AFLD associated with the consumption of alcohol in adults is around 4% [2]. In fact one of the main risk factors associated with the progression of alcohol fatty liver disease is the overconsumption of alcohol over a particular limit despite of age and gender [3]. An elevated level of endotoxin because of the excessive consumption alcohol is reported in many studies. A similar study by Fukuii et al. (1991) proved that overuse of alcohol can lead to the succession of fatty liver disease. The study was performed in 85 individuals. Outcomes of the study have shown that alcohol consumption causes the generation of endotoxemia [4]. Another study by Bode et al. (1987) have also shown the similar result that consumption of alcohol results in the development of the disease due to the elevated level of endotoxins. These studies have supported the premise and the association of endotoxins with the generation of the disease [5]. External toxins are linked with the generation of ROS, which are further involved with the progression of the disease. ROS are involved with the oxidation of cellular organelles, and components

V. Thakur · S. Mishra · V. Kumar · S. Sharma · C. Sharma (✉)
Shiva Institute of Pharmacy, Bilaspur, Himachal Pradesh, India
e-mail: chandni.sharma@shivapharmacy.com

A. Sharma
Department of Pharmaceutical Sciences, Guru Nanak Dev University, Amritsar, Punjab, India

S. Ashique, N. Mishra (eds.), *Reactive Oxygen Species in Metabolic Inflammation*,
https://doi.org/10.1007/978-3-032-11752-6_6

are -SO_2 radical, -H_2O_2, -OH radical, etc. Another impact given by the excessive oxidation and inflammatory response is the progression of AFLD [6]. The liver is considered as the main organ associated with the metabolism of alcohol and also the main affected organ by alcohol-induced diseases. The succession of AFLD consists of a series of steps such as steatosis, steatohepatitis, and then cirrhosis [7].

6.1.2 Impact of ROS

All compounds are meant to be stable via the pairing of electrons; meanwhile the involvement of unpaired electron causes the generation of an extremely reactive or unstable molecule. Such molecules are considered as free radicals. The free radicals formed by oxygen are called ROS [8, 9]. Several types of free radicals are occupied in the classes of ROS such as superoxide, hydrogen peroxide, and hydrochloric acid [10]. These ROS molecules produced after alcohol metabolism [11] *via* different mechanisms such as excessive electron outburst from mitochondria during oxidative phosphorylation, over production of the same during normal cellular functions or via hyper activity of NADPH oxidase [12]. Liver cells like hepatocytes and Kupffer cells are directly involved with the generation and utilization of ROS. These free radicals are openly involved with the destruction of cells by lipid peroxidation and the breaking of DNA strands [13]. Beside these, there are various other moieties which are involved in hepatocytes stress mechanisms and damage of biomolecules viz., peroxynitrate. It is formed by the interaction of nitric oxide (an intercellular messenger for various cellular processes) with oxygen molecules. The species is linked with the oxidative as well the nitrative damage of the cells [14].

6.1.3 Fatty Liver

A major indication of AFLD is the gathering of fat in the liver by the high consumption of alcohol [1, 15]. A study has shown that a ten times thick fold of triglycerides in the liver was observed in mice after the consumption of high dose of alcohol [16]. Fatty liver is considered as the condition of hepatocytes with more than 5% of fat deposition. Initially the fat deposition in hepatocytes begins at the central vein which further progressed into the hepatic lobule and ends into the portal region. The condition is often associated with the accumulation of hepatic triglycerides [17]. Under certain cases fatty liver further leads to inflammatory response and cirrhosis [1]. Alcohol dehydrogenase is a major enzyme, associated with the metabolism of alcohol initially. The end product formed in this process in acetate which further is metabolized by the acetate dehydrogenase enzyme into a toxic product. One of the major enzymes involved in the pathogenesis of the disease is cytochrome P45 P2E1 which is responsible for the process of ethanol oxidation and finally results in to the

Table 6.1 Sources of reactive oxygen species in alcoholic fatty liver disease

Sr. No.	Source	Mechanism of ROS generation	Impact on the liver
1.	Alcohol dehydrogenase (ADH)	Converts ethanol into acetaldehyde, leading to ROS formation	Oxidative stress, lipid peroxidation
2.	Cytochrome P450 2E1 (CYP2E1)	Ethanol oxidation leads to ROS overproduction	Mitochondrial damage, DNA adducts
3.	Mitochondria	Leakage of electrons in oxidative phosphorylation	Impaired ATP synthesis, apoptosis
4.	NADPH oxidase (NOX enzymes)	Over-activation during alcohol intake	Inflammation, Kupffer cell activation
5.	Kupffer cells and immune cells	Produce ROS during inflammatory response	Cytokine release, hepatocyte injury

lipogenesis [18]. A comprehensive Table 6.1 is provided below revealing various sources of ROS in AFLD.

6.1.4 Diagnosis

AFLD can be diagnosed via different techniques such as by determining the Analgesia Nociception Index (ANI) (ALD/NAFLD), mean corpuscular volume (MCV), aspartate transaminase/alanine transaminase ratio (AST/ALT ratio), and BMI test [19].

6.2 Pathophysiology

The ordinary pathway concerned with the processing of alcohol is via acetaldehyde through enzyme alcohol dehydrogenase in the cytosol. The former compound is metabolized by the ALDH enzyme in the mitochondria. Less availability of the ALDH enzyme might result in the aggregation of acetaldehyde. The other pathways associated with the succession of disease are linked with the generation of more ROS radicals that further leads to lipid peroxidation and also impacts mitochondrial glutathione level. The process is the result of microsomal enzyme oxidation system or MEOS. The enzyme which is mainly affected via this mechanism is CYP2E1 as an effect of the long-term consumption of alcohol. The development of the former mechanism further results in elevated oxidative stress and liver diseases [20]. The two main mechanisms involved in the succession of the AFLD are related to two mechanisms of alcohol metabolism. The first one is linked with the impact of alcohol dehydrogenase enzyme, and another one is the impact of CYP2E1. Further elaboration of these mechanisms is given earlier. The amount of acetate is gradually increased by the activity of the ALDH enzyme. These two mechanisms lead to the

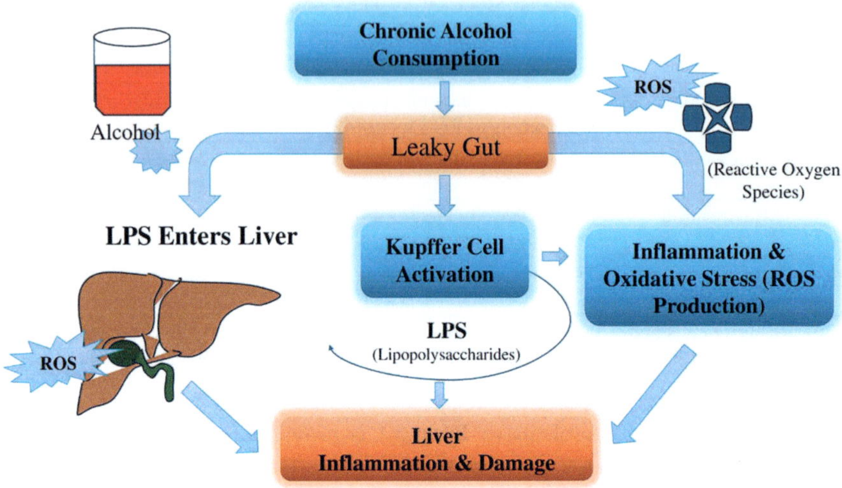

Fig. 6.1 Progression ofAFLD: role of chronic alcohol consumption, ROS, and LPS

reduction in the intensity of NAD-NADH (schematic representation depicted in Fig. 6.1). An increased level of the former results in the impact on metabolism of other complex molecules too. The consequences of the previously given mechanism are shown on the gluconeogenesis and citric acid cycle impairment and further impact the ketogenesis and synthesis of fatty acid [21]. The CYP2E1 enzyme shows its impact by producing oxidative stress that further leads to the breakdown of membrane and lipid peroxidation. The acetaldehyde produced mentioned earlier due to the oxidation of ethanol prevents the repair of alkylated nucleoproteins [22].

6.2.1 Impact of Long-Term Alcohol Consumption on the Liver

The utilization of alcohol by humans in different forms started around 11,000 years B.C [23]. The consumption of alcohol in today's world is an integral part of various cultures. The global alcohol industry contributes around 1 trillion USD revenue per year to the world economy [24]. There are various consumption patterns of alcohol by people ranging from occasional drinking habits to prolonged use of alcohol [25]. The first known impact of alcoholic liver disease is shown as hepatic steatosis, in which aggregation of fatty acids and lipids occurs in the liver [26]. Steatosis is a reversible process because the liver can regenerate its cells [27]. But if steatosis is left untreated, it can result in the development of alcoholic hepatitis in which there is a continuous immersion of Mallory substances in liver tissues [28].

6.3 Inflammatory Pathways Activated in AFLD

6.3.1 *Pattern Recognition Receptors (TLRs)*

The inflammatory cascade of AFLD typically starts when pattern recognition receptors (PRRs), an arm of the innate immunity system of the body, get activated. Among the primary receptors is the Toll-like receptor 4 (TLR4) [29]. Both chronic and acute intake of alcohol are found to relax the intestinal barrier in a condition of "leaky gut." Hence, cytotoxic molecules from microbes and in particular lipopolysaccharide (LPS), a molecule of the outer membrane component of Gram-negative bacteria, are no more restricted and gain access to the portal system from the gut with direct access to the liver [30].

In the liver, LPS principally signals through liver cells' TLR4 complex, in particular, macrophage cells referred to as Kupffer cells [31]. With the binding of LPS to TLR4, it produces a robust "danger signal," and it triggers a cascade of a cell's signaling from alcohol-associated gut issues to liver inflammatory responses. Besides microbial signals (PAMPs), TLR4 is activated similarly by molecules of DAMPs that are produced by liver cells that die or get damaged. Consequently, the damage and inflammation keep going round and round and keep intensifying continuously [32].

6.3.2 *NF-κB Pathway Activation*

Below TLR4 activation lies one of the most important regulative controls of inflammatory gene expression: the nuclear factor-kappa B (NF-κB) signaling pathway. When at rest, NF-κB proteins are kept in the cytoplasm in a complexed form with inhibitory proteins referred to as inhibitor of κB (IκB). Signaling pathway initiated upon TLR4 activation results in IκB phosphorylation followed by proteolytic degradation [33].

This activity unleashes the NF-κB dimer in order for it to translocate into the cell nucleus. In the nucleus, NF-κB complexates with specific DNA sequences in promoter areas of various genes and acts as a master switch for the induction of the transcription of pro-inflammatory cytokines, chemokines, and adhesion molecules [33, 34]. Consequently, activation of the NF-κB signal pathway is the central event underlying the conversion of the early danger signals recognized by TLR4 into a strong and prolonged inflammatory reaction at the liver level.

6.3.3 Cytokines and Chemokines (TNF-α, IL-1-β, IL-6, MCP-1)

The genes that are regulated by NF-κB synthesize various pro-inflammatory proteins. These proteins control inflammatory reactions and liver damage in AFLD.

6.3.3.1 Tumor Necrosis Factor-Alpha (TNF-α)

Activated Kupffer cells mainly produce TNF-α. TNF-α plays a role in liver injury in AFLD. TNF-α causes cell death in the liver indirectly through a procedure termed apoptosis while raising the production of destructive molecules known as reactive oxygen species (ROS), and it all adds up to cell stress and damage [35].

6.3.3.2 Interleukin-1beta (IL-1β)

IL-1β is a very strong inflammatory molecule. It is a two-step cautious procedure in making the active form of the molecule. The first utilizes a signaling molecule named NF-κB in the production of the inactive form of the molecule, pro-IL-β. The second utilizes another signaling molecule in making it active, and it is done through the inflammasome [36].

6.3.3.3 Interleukin-6 (IL-6)

Multifunction cytokine IL-6 is accountable for a significant percentage of acute inflammatory reaction and insulin resistance of hepatocytes. High IL-6 for a long duration is directly associated with severity of alcoholic liver damage and is implicated in liver regeneration as well [37].

6.3.3.4 Monocyte Chemoattractant Protein-1 (MCP-1)

It is otherwise referred to as C–C motif chemokine ligand 2 (CCL2). MCP-1 has a central role in the recruitment of immune cells by forming a chemical gradient that results in attracting the monocytes in the bloodstream toward liver inflammatory sites. Upon recruitment of the monocytes at the latter sites, they get converted into macrophages that enhance locoregional inflammation in addition to spreading hepatocellular injury [38].

6.3.4 *Role of Inflammasomes (E.G., NLRP3)*

Large multimeric protein complexes within cells, inflammasomes act as prime sites for integrating inflammation. During AFLD, the NLRP3 inflammasome was identified as an essential sensor for connecting metabolic stress/ROS with inflammation. Activation of NLRP3 inflammasome, as with its prime substrate IL-1β production, is also biphasic.

6.3.4.1 Signal 1 (Priming)

It is derived from TLR4-NF-κB signaling, which enhances NLRP3 components and pro-IL-1β levels in cell [36–38].

6.3.4.2 Signal 2 (Activation)

There are many different stimuli that reveal cells are stressed or damaged and are able to activate this signal. Key inducers of AFLD are ROS generated when metabolizing ethanol, problems with the mitochondria, and other DAMPs released when the cell is damaged.

When the NLRP3 protein is elicited once or twice, it oligomerizes and assembles the inflammasome complex. The complex continues activating the enzyme caspase-1. Cleavage of pro-IL-1β from its precursor form to the mature active form is the primary role of active caspase-1, and the mature active form is secreted from the cell. Therefore, NLRP3 inflammasome connects cell stress induced by alcohol with IL-1β's strong inflammatory response.

6.4 Molecular Interplay Between ROS and Inflammation

During the hyper-activation of the immune response to an injury induced by overconsumption of alcohol, it leads to the formation of more reactive oxygen species. Furthermore the transformation of alcohol into acetaldehyde by the activity of the enzyme alcohol dehydrogenase also results in disease progression. Another mechanism which includes the metabolism of alcohol by the activity of the enzyme P4502E1 is also associated with the overproduction of ROS. These mechanisms contribute into different types of impacts in the liver such as impairment in the endoplasmic reticulum, lysosome dysfunction, and also mitochondrial injury and finally lead to autophagy of liver cells [39].

One of the prominent factors related with the development of alcohol-induced fatty liver disease is the effect of the NAD+/NADH ratio. The factor contributes largely for the proper homeostasis mechanism of liver cells. During the oxidation of ethanol to acetaldehyde/acetate, it needs the consumption of NAD+ and results in the disruption of the NAD+/NADH ratio. This disruption leads to the inhibition of several processes related with homeostasis like synthesis of fatty acid, oxidation of fatty acid, and gluconeogenesis. This inhibition of processes results in the accumulation of fat in the liver [40, 41].

6.4.1 *Impact of Inflammation in the Progression of Alcoholic Fatty Liver Disease*

The oxidative stress induced by the consumption of alcohol explained earlier can lead to the formation of many inflammatory mediators like cytokines. The mediators are related with the generation of more immune responses in the liver and enhance inflammation [42–45]. It also involves the activation of damage-associated molecular patterns (DAMPs) which further causes the progression of inflammation by the activation of immune cells like neutrophils and macrophages [46–49]. The overconsumption of alcohol is associated with the generation of abduct protein and DNA fragments because the action of ROS is associated with the activation of adaptive immunity by T and B cells [49]. The impact of ROS-derived inflammation and the stimulation of different immune cells by alcohol liver injury are linked with the succession of AFLD [50, 51].

6.4.2 *Role of Macrophages*

The advancement of AFLD causes the generation of hepatic monocytes which further differentiates into macrophages and Kupffer cells [52, 53]. The generation of oxidative agents by the activated macrophages and Kupffer cells is done by the activity of alcohol and NOX [54, 55]. During the consumption of alcohol, the ROS are associated with the generation of various types of inflammatory mediators such as TNF-α and NF-kB [55]. This proves that oxidative stress caused by macrophage activation can lead to the progression of AFLD [55].

6.4.3 *Role of Neutrophils*

Neutrophils are key cells which are related with the succession of various immune and inflammatory diseases [56]. An important activity shown by neutrophils in the development of AFLD is done by the movement of former cells by the release of various inflammatory cytokines during liver damage done by alcohol consumption

[57, 58]. Activated neutrophils are also linked to the rise of oxidative stress by the discharge of ROS, resulting in the aggravation of the earlier disease [59, 60].

6.5 Consequences of ROS-Inflammation Cross Talk in AFLD

6.5.1 Hepatocyte Injury and Apoptosis/Necrosis

Hepatocytes constitute the liver's primary cellular components. These are subjected to maximum stress brought by the onslaught of ROS-induced inflammation [61]. Damage is demonstrated in particular through oxidative stress occasioned by ROS induced by ethanol metabolism and cytochrome P450 2E1 (CYP2E1) overload. Oxidative stress leads to the initiation of lipid peroxidation, a vicious chain reaction that targets cell membranes and destroys them in terms of function. At the same time, DNA and proteins undergo damage, which may result in programmed cell death or cell cycle arrest.

Concurrently, an inflammatory cascade caused by ROS results in the production of pro-inflammatory cytokines, such as tumor necrosis factor-alpha (TNF-α). TNF-α worsens the susceptibility of hepatocytes to oxidative injury and consequently initiates apoptosis. When the injury is intense, a form of cell death that is uncontrolled and referred to as necrosis results. Unlike apoptosis, cell death through necrosis causes the release of cellular materials containing DAMPs that stimulate the inflammatory response and consequently a vicious cycle of injury and inflammation and death.

6.5.2 Stellate Cell Activation and Fibrosis

One of the most significant outcomes of the ROS-inflammation cross talk is the activation of hepatic stellate cells (HSCs) [62]. These cells are normally present in the space of Disse where vitamin A is stored in quiescence by them. But under the condition of liver injury, HSCs are activated by a plethora of stimuli including pro-inflammatory cytokines like transforming growth factor-beta (TGF-β) and reactive oxygen radicals themselves.

Once activated, HSCs are converted into myofibroblast-like cells. Now in the form of myofibroblast-like cells, they lose vitamin A stores and commence excessive production of extracellular matrix (ECM) proteins, of which collagen is the predominant component. Profuse production and deposits of this ECM lead to scarring tissue formation, which is referred to as fibrosis. Fibrotic tissue constricts the liver, disrupts the normal architecture of the liver, and obstructs the flow of blood. Going on continuously, this slow devastation may progress to cirrhosis, a fatal condition of chronic liver disease where extreme scarring and liver failure account for the condition.

6.5.3 Altered Lipid Metabolism

The ROS-inflammation axis greatly accelerates the disequilibrium of lipid metabolism in the liver [63]. Lipid metabolism in a hepatocyte is the fine balance between the biosynthesis and degradation of fatty acids and fatty acid uptake. Oxidative stress and inflammatory cytokines interfere with this equilibrium. When oxidative stress is present, the very powerhouse of the cell, the mitochondria, is compromised. Mitochondria carry out beta-oxidation in the production of energy by breaking down fatty acids. Impaired beta-oxidation leads to elongation of fatty acids. Other cytokines of the inflammatory phase, such as specifically TNF-α, are also metabolic disruptors. They upregulate the fatty acid biosynthetic enzymes and concurrently suppress the genes of its breakdown. Double effect of increased biosynthesis facilitated by diminished breakdown characterizes the typical hallmark of steatosis: an intrahepatocellular triglyceride accumulation. Lipid accumulation is not a singular occurrence but instead a substrate for further ROS generation and further inflammation and thus initiates a vicious lipid toxicity and oxidative stress cycle (shown in the form of flowchart in Fig. 6.2).

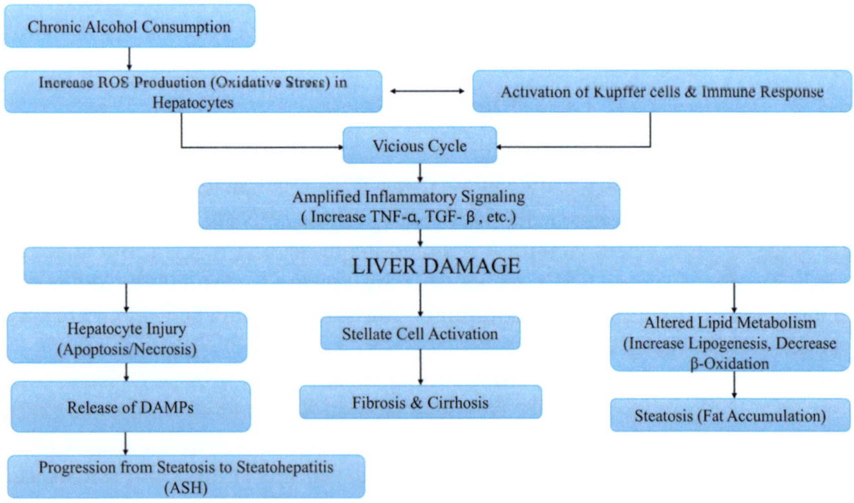

Fig. 6.2 Alcohol-induced oxidative stress leading to liver damage

6.5.4 *Progression from Steatosis to Steatohepatitis*

One of the larger concerns with ROS and inflammation acting in concert is that simple fat deposition in the liver (steatosis) progresses to the more severe and inflammatory form of alcoholic steatohepatitis or ASH [64]. Whereas liver fat accumulation is solved by the mere remedy of abstaining from drinking, alcoholic steatohepatitis is the turning point in the disease process. And it is where the relationship turns abusive. The progression in alcoholic steatohepatitis is the culmination of the combined effects of ROS-induced oxidative stress, cytokine-mediated inflammation, cell death of the hepatocytes, and the activation of the stellate cells. Alcoholic steatohepatitis represents an advanced stage in which inflammation and liver cell injury follow with the accumulation of fat. The vicious cycle of self-sustaining injury significantly increases the risk of cirrhosis and, ultimately, liver failure in which the capacity of the liver to regenerate and function is severely lost [65].

6.6 Therapeutic Implications

6.6.1 *Antioxidant Enzymes: Role of Glutathione System, Superoxide Dismutase, and Catalase*

In humans, there are different kinds of antioxidant agents that are involved in preventing liver cells from oxidative stress caused by different compounds. These antioxidant enzymes contribute to the prevention of oxidative damage to hepatocytes through different mechanisms. A combination of glutathione and enzyme glutathione peroxidase is utilized for the elimination of most free radicals. Such effect is also caused by the glutathione reductase enzyme that increases the glutathione effect of free radicals. At the same time, there are antioxidant enzymes such as superoxide dismutase which cause the decrease of impact of highly toxic oxidative agent known as superoxide radical. The final product of this process is the formation of hydrogen peroxide which is also taken into account as the oxidative agent and is responsible for the development of alcoholic fatty liver disease. The catalase enzyme activity minimizes the impact of this free radical and alleviates oxidative stress [66, 67]. The three forms of the superoxide dismutase enzyme which are manganese-SOD, copper-zinc-SOD, and free-form extracellular SOD are associated in order to contribute to the antioxidant effect and also in the enhancement of MAPK activity [68].

6.6.2 Reduction of Antioxidant Activity Due to Chronic Alcohol Consumption

Various studies have proved that the impact of alcohol on different systems of the body was produced. The damage on the stomach, pancreas, and brain was clearly observed even with the acute consumption of alcohol. In case of antioxidant activity, the long-term consumption of alcohol was also impacted by many mechanisms. Many studies show that long-term alcohol consumption can seriously impact the processing of vitamin A in the body and also its enzymatic regulation further to show its impact [69]. In short, the metabolism of vitamin A was impacted in the liver due to the interaction between retinol and ethanol which are both dependent on dehydrogenase enzymes for their further metabolism [70]. Various studies have also evidently shown that the metabolism of alcohol causes the increase in the level of NADH/NAD ratio. This increase in the NADH/NAD ratio is directly associated with the transformation of Fe 3+ to Fe 2+ ions. This reduction of Fe 3+ into Fe 2+ ions can directly generate oxidizing agents such as -OH radical from H2O2 [71]. Many studies show that there was a noteworthy decline in the level of α-tocopherol due to the impact of ascorbic acid. The former is also associated with the reduction of impacts induced due to chronic alcohol consumption such as elevation of glutathione effect on the liver [72]. A brief Table 6.2 is shown for various antioxidant defense mechanisms and impairment in AFLD.

6.7 Future Directions and Research Gaps

6.7.1 Oxidative Stress and Inflammatory Biomarkers in AFLD

The majority of monitoring and diagnosis processes are through general imaging methods and regular blood tests that frequently cannot portray the complete disease pathology. These biomarkers should level zero in the root of the issues: oxidative

Table 6.2 Comparison of antioxidant defense mechanism vs. impairments in AFLD

Sr. No.	Antioxidant system	Normal function	Impairment in AFLD
1.	Glutathione (GSH)	Neutralizes free radicals and detoxifies H_2O_2	Depleted by chronic alcohol, reduced defense
2.	Superoxide dismutase (SOD)	Converts superoxide into hydrogen peroxide	Activity reduced, increased oxidative stress
3.	Catalase	Break downs hydrogen peroxide into water and oxygen	Reduced activity leads to accumulation of ROS
4.	Vitamin E (tocopherol)	Lipid-soluble antioxidant protecting membranes	Levels reduced with alcohol use
5.	Retinoids (vitamin A)	Regulate gene expression and protect against ROS	Metabolism impaired by alcohol and leading to deficiency

stress and inflammation. Rather than focusing on overall liver injury signs, researchers wish to uncover markers that directly quantify the harmful molecules formed in the processes. Malondialdehyde (MDA), for instance. From a chemical perspective, is an oxidative stress product capable of inflicting cell membrane damage. Measuring the MDA level could permit a reasonable estimation of oxidative stress taking place inside the liver. Moreover, measuring inflammatory markers such as the C-reactive protein (CRP) could present the condition of inflammation in a more direct way [15, 73].

A set of these specific biomarkers would be wonderful. Such a "biomarker panel" could help in the early diagnosis of AFLD, even before the person feels sick, better than current methods. In addition, it would enable physicians to evaluate the response to therapy—a decrease in the levels of these harmful molecules would suggest that the treatment is effective. For instance, a decrease in MDA and CRP levels in a patient after starting a new medication would present strong evidence that the treatment is really working in getting rid of oxidative stress and inflammation.

6.7.2 Targeting Mitochondria and Redox Signaling Systems

The mitochondria are tiny power plants within a cell. They supply energy to the cells, but in AFLD, the mitochondria get damaged by alcohol and are therefore big-time producers of harmful reactive oxygen species. This damage produces a bad cycle: alcohol harms the mitochondria, which then make more ROS, causing more damage. Therefore, protecting these mitochondria seems to be a promising approach to the treatment of AFLD. A way of achieving this is with mitochondria-targeted antioxidants, which travel directly to the mitochondria and neutralize ROS in situ [74]. The second strategy is attempting to regulate the defense pathways of the cells, more commonly referred to as redox signaling pathways. The most well-known pathway is the Nrf2 pathway. Nrf2 is visualized as a master switch of the antioxidant defense system of the cell. There are genes of the Nrf2 pathway that have blueprints for potent-acting antioxidant enzymes when the master switch is activated. Researchers now look for natural and synthetic molecules that might flip this "switch" on to flood the liver with protection against free radical stress [75, 76]. The formation of AFLD can be prevented at its very origin by safeguarding mitochondria and thus enhancing the body's antioxidant defense.

6.7.3 Precision Medicine Strategies

But not all heavy drinkers develop AFLD. Genetics, lifestyle, and colonizing gut microbes vary and determine the body's reaction to booze. Therefore, a generic approach to treating it does not work on everyone. Precision medicine aims at avoiding this deadfall by taking up methods of treating that are tailored to the body of the person. Sikarin et al. continued to describe that in the future, moreover, plans exist for conducting research in the identification of genes harboring the variations that possess a potential susceptibility to AFLD. Some of them, for instance, might have a specific variation in genes that would either metabolize alcohol or manage stress. By testing the patient's genes and other biological information, a doctor would then be capable of delineating a risk profile pattern in the patient. A patient with a genetic predisposition to oxidative stress may be offered specific antioxidants, while a patient with a genetic predisposition to excessive inflammation may be offered drugs with an anti-inflammatory effect. Furthering intervention and outcomes for sufferers of AFLD are to be found in this extremely personalized approach of intervention where the clinician considers the genetics of the patient and the lifestyle of the patient [77–79].

6.8 Conclusion

Alcoholic fatty liver disease (AFLD) is a global health problem induced by the harmful interaction of reactive oxygen species (ROS) and inflammation. Extended heavy drinking overloads liver capacity to metabolize substances. These happen most commonly with the ADH and CYP2E1 systems that result in enhanced ROS production. ROS causes oxidative stress and causes liver cell (hepatocytes) injury through lipid peroxidation processes. At the same time, alcohol weakens the intestinal barrier. Thus, bacterial endotoxins from the gut, such as LPS, enter the liver. This triggers a strong inflammatory reaction through pattern recognition receptors like TLR4, activating the NF-κB pathway and initiating the release of pro-inflammatory cytokines like TNF-α and interleukins. Immune cells, i.e., macrophages and neutrophils, infiltrate the liver and contribute to inflammatory and oxidative stress response and create an injury cycle. This ROS and inflammatory destructive cycle is the major driving force behind the progression of development of AFLD from basic fat accumulation (steatosis) to advanced stage of steatohepatitis, fibrosis, and cirrhosis. The end results are liver cell death, activation of the hepatic stellate cells that lead to scarring, and disrupted lipid metabolism. This exacerbates the accumulation of fat in the liver. Future treatment methods need to target the interruption of this cycle. Possible solutions include using the body's natural antioxidant systems, developing antioxidants that target mitochondria, and finding selective biomarkers for early detection and therapy monitoring. The long-term solution will involve a personalized medicine approach tailored to the genetic and environmental factors of each individual so that we can effectively address this complex disease.

Acknowledgments We would like to express our heartfelt gratitude to Shiva Institute of Pharmacy and its management for their constant encouragement and support. We extend our sincere thanks to our respected principal, *Dr. Saurabh Sharma*, for providing the infrastructural support.

References

1. Seitz HK, Bataller R, Cortez-Pinto H, Gao B, Gual A, Lackner C, Mathurin P, Mueller S, Szabo G, Tsukamoto H (2018) Alcoholic liver disease. Nat Rev Dis Primers. 4(1):16. https://doi.org/10.1038/s41572-018-0014-7. Erratum in: Nat Rev Dis Primers 2018 Aug 28;4(1):18. doi: 10.1038/s41572-018-0021-8
2. Wong VW, Chu WC, Wong GL, Chan RS, Chim AM, Ong A, Yeung DK, Yiu KK, Chu SH, Woo J, Chan FK (2012) Prevalence of non-alcoholic fatty liver disease and advanced fibrosis in Hong Kong Chinese: a population study using proton-magnetic resonance spectroscopy and transient elastography. Gut 61(3):409–415
3. Traversy G, Chaput JP (2015) Alcohol consumption and obesity: an update. Curr Obes Rep 4(1):122–130
4. Fukui H, Brauner B, Bode JC, Bode C (1991) Plasma endotoxin concentrations in patients with alcoholic and non-alcoholic liver disease: reevaluation with an improved chromogenic assay. J Hepatol 12(2):162–169
5. Bode C, Kugler V, Bode JC (1987) Endotoxemia in patients with alcoholic and non-alcoholic cirrhosis and in subjects with no evidence of chronic liver disease following acute alcohol excess. J Hepatol 4(1):8–14
6. Sun Q, Zhang W, Zhong W, Sun X, Zhou Z (2017) Pharmacological inhibition of NOX4 ameliorates alcohol-induced liver injury in mice through improving oxidative stress and mitochondrial function. Biochim Biophys Acta (BBA)-General Subjects 1861(1):2912–2921
7. Massey VL, Arteel GE (2012) Acute alcohol-induced liver injury. Front Physiol 3:193
8. Muriel P (2009) Role of free radicals in liver diseases. Hepatol Int 3(4):526–536
9. Ježek P, Hlavatá L (2005) Mitochondria in homeostasis of reactive oxygen species in cell, tissues, and organism. Int J Biochem Cell Biol 37(12):2478–2503
10. Balazy M, Nigam S (2003) Aging, lipid modifications and phospholipases—new concepts. Ageing Res Rev 2(2):191–209
11. De Minicis S, Bataller R, Brenner DA (2006) NADPH oxidase in the liver: defensive, offensive, or fibrogenic? Gastroenterology 131(1):272–275
12. Granger DN, Kvietys PR (2015) Reperfusion injury and reactive oxygen species: the evolution of a concept. Redox Biol 6:524–551
13. Sun T, Xiao S, Wang M, Xie Q, Zhang L, Gong M, Zhang D, Zhou C (2023) Reactive oxygen species scavenging nanozymes: emerging therapeutics for acute liver injury alleviation. Int J Nanomedicine 18:7901–7922
14. Azzi A (2022) Oxidative stress: what is it? Can it be measured? Where is it located? Can it be good or bad? Can it be prevented? Can it be cured? Antioxidants 11(8):1431
15. Gao B, Bataller R (2011) Alcoholic liver disease: pathogenesis and new therapeutic targets. Gastroenterology 141(5):1572–1585
16. Hege M, Jung F, Sellmann C, Jin C, Ziegenhardt D, Hellerbrand C, Bergheim I (2018) An iso-α-acid-rich extract from hops (Humulus lupulus) attenuates acute alcohol-induced liver steatosis in mice. Nutrition 45:68–75
17. Hall P (1985) Pathological spectrum of alcoholic liver disease. In: Alcoholic liver disease: pathology and pathogenesis. Edward Arnold, London
18. Seitz HK, Moreira B, Neuman MG (2023) Pathogenesis of alcoholic fatty liver a narrative review. Life 13(8):1662
19. Dunn W, Angulo P, Sanderson S, Jamil LH, Stadheim L, Rosen C, Malinchoc M, Kamath PS, Shah VH (2006) Utility of a new model to diagnose an alcohol basis for steatohepatitis. Gastroenterology 131(4):1057–1063
20. Dunn W, Shah VH (2016) Pathogenesis of alcoholic liver disease. Clin Liver Dis 20(3):445
21. Rasineni K, Casey CA (2012) Molecular mechanism of alcoholic fatty liver. Indian J Pharmacol 44(3):299–303
22. Lieber CS (2004) Alcoholic fatty liver: its pathogenesis and mechanism of progression to inflammation and fibrosis. Alcohol 34(1):9–19

23. Liu L, Wang J, Rosenberg D, Zhao H, Lengyel G, Nadel D (2018) Fermented beverage and food storage in 13,000 y-old stone mortars at Raqefet cave, Israel: investigating Natufian ritual feasting. J Archaeol Sci Rep 21:783–793
24. Wadsworth EJ, Moss SC, Simpson SA, Smith AP (2004 Jun) Factors associated with recreational drug use. J Psychopharmacol 18(2):238–248
25. Research A, Staff CR (2018) Drinking patterns and their definitions. Alcohol Res 39(1):17
26. Steiner JL, Lang CH (2017) Alcohol, adipose tissue and lipid dysregulation. Biomolecules 7(1):16
27. Bergheim I, McClain CJ, Arteel GE (2006) Treatment of alcoholic liver disease. Dig Dis 23(3–4):275–284
28. Basaranoglu M, Turhan N, Sonsuz A, Basaranoglu G (2011) Mallory-Denk bodies in chronic hepatitis. World J Gastroenterol: WJG 17(17):2172
29. Szabo G, Bala S (2013) Alcoholic liver disease and the gut-liver axis. World J Gastroenterol 19(8):1293–1301
30. Rao RK, Seth A, Sheth P (2004) Recent advances in alcoholic liver disease I. Role of intestinal permeability and endotoxemia in alcoholic liver disease. Am. J. Physiol. Gastrointest. Liver Physiol 286(6):G881–G884
31. Uesugi T, Froh M, Arteel GE, Bradford BU, Thurman RG (2001) Toll-like receptor 4 is involved in the mechanism of early alcohol-induced liver injury in mice. Hepatology 34(1):101–108
32. Mandrekar P, Szabo G (2010) Signalling pathways in alcoholic liver disease. Nat Rev Gastroenterol Hepatol 7(10):556–566
33. Luedde T, Schwabe RF (2011) NF-κB in the liver—linking injury, inflammation, and hepatocellular carcinoma. Virchows Arch 458(4):433–445
34. Yin M, Wheeler MD, Kono H, Bradford BU, Gallucci RM, Luster MI, Thurman RG (1999) Essential role of tumor necrosis factor α in alcohol-induced liver injury in mice. Gastroenterology 117(4):942–952
35. Szabo G, Petrasek J (2015) Inflammasome activation and function in liver disease. Nat Rev Gastroenterol Hepatol 12(7):387–400
36. El-Assal O, Hong F, Kim WH, Radaeva S, Gao B (2003) IL-6-deficient mice are resistant to ethanol-induced liver injury. Hepatology 37(5):1081–1089
37. Mandrekar P, Bala S, Catalano D, Kodys K, Szabo G (2009) The opposite effects of CpG and candidal DNA on alcohol-induced inflammation are mediated by distinct toll-like receptor 9 domains. J Immunol 183(12):8123–8132
38. Petrasek J, Bala S, Csak T, Lippai D, Kodys K, Menashy V et al (2012) IL-1 receptor antagonist ameliorates inflammasome-dependent alcoholic steatohepatitis in mice. J Clin Invest 122(10):3476–3489
39. Ambade A, Mandrekar P (2012) Oxidative stress and inflammation: essential partners in alcoholic liver disease. Int J Hepatol 2012(1):853175
40. Cederbaum AI (2012) Alcohol metabolism. Clin Liver Dis 16(4):667–685
41. Donohue TM Jr (2007) Alcohol-induced steatosis in liver cells. World J Gastroenterol: WJG 13(37):4974
42. Ge X, Antoine DJ, Lu Y, Arriazu E, Leung TM, Klepper AL, Branch AD, Fiel MI, Nieto N (2014) High mobility group box-1 (HMGB1) participates in the pathogenesis of alcoholic liver disease (ALD). J Biol Chem 289(33):22672–22691
43. Cai Y, Xu MJ, Koritzinsky EH, Zhou Z, Wang W, Cao H, Yuen PS, Ross RA, Star RA, Liangpunsakul S, Gao B (2017) Mitochondrial DNA–enriched microparticles promote acute-on-chronic alcoholic neutrophilia and hepatotoxicity. JCI Insight 2(14):e92634
44. Mihm S (2018) Danger-associated molecular patterns (DAMPs): molecular triggers for sterile inflammation in the liver. Int J Mol Sci 19(10):3104
45. Hoyt LR, Randall MJ, Ather JL, DePuccio DP, Landry CC, Qian X, Janssen-Heininger YM, van der Vliet A, Dixon AE, Amiel E, Poynter ME (2017) Mitochondrial ROS induced by chronic ethanol exposure promote hyper-activation of the NLRP3 inflammasome. Redox Biol 12:883–896

46. Kany S, Janicova A, Relja B (2019) Innate immunity and alcohol. J Clin Med 8(11):1981
47. Kim A, Bellar A, McMullen MR, Li X, Nagy LE (2020) Functionally diverse inflammatory responses in peripheral and liver monocytes in alcohol-associated hepatitis. Hepatol Commun 4(10):1459–1476
48. Xu MJ, Zhou Z, Parker R, Gao B (2017) Targeting inflammation for the treatment of alcoholic liver disease. Pharmacol Ther 180:77–89
49. Mottaran E, Stewart SF, Rolla R, Vay D, Cipriani V, Moretti M, Vidali M, Sartori M, Rigamonti C, Day CP, Albano E (2002) Lipid peroxidation contributes to immune reactions associated with alcoholic liver disease. Free Radic Biol Med 32(1):38–45
50. Fialkow L, Wang Y, Downey GP (2007) Reactive oxygen and nitrogen species as signaling molecules regulating neutrophil function. Free Radic Biol Med 42(2):153–164
51. Murphy MP, Holmgren A, Larsson NG, Halliwell B, Chang CJ, Kalyanaraman B, Rhee SG, Thornalley PJ, Partridge L, Gems D, Nyström T (2011) Unraveling the biological roles of reactive oxygen species. Cell Metab 13(4):361–366
52. Italiani P, Boraschi D (2014) From monocytes to M1/M2 macrophages: phenotypical vs. functional differentiation. Front Immunol 5:514
53. Wen Y, Lambrecht J, Ju C, Tacke F (2021) Hepatic macrophages in liver homeostasis and diseases-diversity, plasticity and therapeutic opportunities. Cell Mol Immunol 18(1):45–56
54. Panday A, Sahoo MK, Osorio D, Batra S (2015) NADPH oxidases: an overview from structure to innate immunity-associated pathologies. Cell Mol Immunol 12(1):5–23
55. Kono H, Rusyn I, Yin M, Gäbele E, Yamashina S, Dikalova A, Kadiiska MB, Connor HD, Mason RP, Segal BH, Bradford BU (2000) NADPH oxidase–derived free radicals are key oxidants in alcohol-induced liver disease. J Clin Invest 106(7):867–872
56. Rosales C (2018) Neutrophil: a cell with many roles in inflammation or several cell types? Front Physiol 9:324475
57. Luk GD, Silverman AL, Giardiello FM (1987) Biochemical markers in patients with familial colonic neoplasia. Semin. Surg. Oncol 3(2):126–132
58. Kolaczkowska E, Kubes P (2013) Neutrophil recruitment and function in health and inflammation. Nat Rev Immunol 13(3):159–175
59. Nguyen GT, Green ER, Mecsas J (2017 Aug) Neutrophils to the ROScue: mechanisms of NADPH oxidase activation and bacterial resistance. Front Cell Infect Microbiol 25(7):373
60. Blüml S, Rosc B, Lorincz A, Seyerl M, Kirchberger S, Oskolkova O, Bochkov VN, Majdic O, Ligeti E, Stöckl J (2008) The oxidation state of phospholipids controls the oxidative burst in neutrophil granulocytes. J Immunol 181(6):4347–4353
61. Hepatocyte Injury and Apoptosis/Necrosis, Albano E (2006) Oxidative mechanisms in the pathogenesis of alcoholic liver disease. Proc Nutr Soc 65(3):324–331
62. Stellate Cell Activation and Fibrosis, Friedman SL (2008) Mechanisms of hepatic fibrosis. J Gastroenterol 43(3):193–206
63. Altered Lipid Metabolism, Gorden DL, Reid RA (2012) Ethanol and the liver. Compr Physiol 2(3):1629–1647
64. Progression from Steatosis to Steatohepatitis, Gao B, Bataller A (2011) Alcoholic liver disease: pathogenesis and new therapeutic targets. Gastroenterology 141(5):1572–1585
65. Role of ROS and Inflammation in Disease Progression, Mandrekar P, Szabo G (2009) Signaling pathways in alcoholic liver disease. Curr Mol Med 9(5):553–563
66. Kalyanaraman B (2013) Teaching the basics of redox biology to medical and graduate students: oxidants, antioxidants and disease mechanisms. Redox Biol 1(1):244–257
67. Chen Y, Dong H, Thompson DC, Shertzer HG, Nebert DW, Vasiliou V (2013) Glutathione defense mechanism in liver injury: insights from animal models. Food Chem Toxicol 60:38–44
68. Hashimoto N, Noda T, Kim SJ, Yamauchi H, Takigawa S, Matsuura-Endo C, Suzuki T, Han KH, Fukushima M (2010) Colored potato extracts induce superoxide dismutase-2 mRNA via ERK1/2 pathway in HepG2 cells. Plant Foods Hum Nutr 65(3):266–270
69. Leo MA, Lieber CS (1982) Hepatic vitamin A depletion in alcoholic liver injury. N Engl J Med 307(10):597–601

70. Leo MA, Lieber CS (1999) Alcohol, vitamin A, and β-carotene: adverse interactions, including hepatotoxicity and carcinogenicity. Am J Clin Nutr 69(6):1071–1085
71. Heaton MB, Mitchell JJ, Paiva M (2000) Amelioration of ethanol-induced neurotoxicity in the neonatal rat central nervous system by antioxidant therapy. Alcohol Clin Exp Res 24(4):512–518
72. Suresh MV, Kumar CS, Lal JJ, Indira M (1999) Impact of massive ascorbic acid supplementation on alcohol induced oxidative stress in Guinea pigs. Toxicol Lett 104(3):221–229
73. Cichoż-Lach H, Michalak A (2014) Oxidative stress in liver diseases. World J Gastroenterol 20(19):4563–4573
74. Nassir F, Ibdah JA (2014) Role of mitochondria in alcoholic liver disease. World J Gastroenterol 20(9):2136–2142
75. Wu D, Cederbaum AI (2009) Alcohol, oxidative stress, and free radical damage. In: Alcohol-related organ damage. Humana Press, pp 1–26
76. Kensler TW, Wakabayashi N (2010) Nrf2: friend or foe for chemoprevention? Carcinogenesis 31(1):90–99
77. Agarwal DP (2001) Genetic polymorphisms of alcohol metabolizing enzymes. Pathol Biol 49(9):703–709
78. Takamatsu M, Yamauchi M, Maezawa Y, Saito S, Maeyama S, Uchikoshi T (2000) Genetic polymorphisms of interleukin-1β in association with the development of alcoholic liver disease in Japanese patients. J Gastroenterol Hepatol 15(5):500–505
79. McClain CJ, Barve S, Deaciuc I et al (1999) Alcoholic liver disease and the gut: a clinical–molecular review. Alcohol Clin Exp Res 23(8):1378–1386

Chapter 7
Reactive Oxygen Species and Eicosanoid-Mediated Inflammation in Metabolic Disorders

Pritam Bera, Anubhab Kuila, Sagnik Samanta, and Debasmita Paul

7.1 Introduction

Metabolic disorders, like obesity, type 2 diabetes mellitus (T2DM), metabolic syndrome, and non-alcoholic fatty liver disease (NAFLD), are a global health crisis. These conditions share common pathological features, including nutrient excess, insulin resistance, and dysregulated lipid metabolism. Acute or chronic inflammatory processes usually coexist with the onset and development of these diseases, and many studies have suggested a correlation between metabolic diseases and inflammation [1].Two interconnected mediators, reactive oxygen species (ROS) and eicosanoids, form a powerful biochemical axis that sustains inflammation and metabolic dysregulation. The development of metabolic diseases is assisted by oxidative stress and chronic inflammatory conditions. Reactive oxygen species (ROS) and oxidative free radicals are produced in excess when cells and tissues undergo oxidative stress, in which the oxidative and anti-oxidative systems are out of balance [2]. Lipid-derived mediators particularly eicosanoids (prostaglandins, leukotrienes (LTs), thromboxanes (TXs), resolvins, lipoxins (LXs)), are essential in inflammation, vascular tone regulation, and insulin sensitivity [3]. A complex biological process known as oxidative stress is defined by a high level of reactive oxygen species (ROS), which affect the body's REDOX balance and subtly cause oxidative damage [4]. Oxidative stress is a condition in which the cells' and tissues' oxidative and anti-oxidative systems are out of balance, leading to an excess of reactive oxygen species (ROS) and oxidative free radicals. Overproduction of ROS may damage the proteins, lipids, and nucleic acids of the cell, resulting in cellular dysfunction. This

P. Bera · A. Kuila · S. Samanta · D. Paul (✉)
Department of Pharmaceutical Technology, Bharat Technology,
Uluberia, Howrah, West Bengal, India

S. Ashique, N. Mishra (eds.), *Reactive Oxygen Species in Metabolic Inflammation*,
https://doi.org/10.1007/978-3-032-11752-6_7

dysfunction may include changed cell signaling and cell cycle regulation, loss of energy metabolism, genetic mutations, altered cellular transport mechanisms, and a general decrease in biological activity, immune activation, and inflammation [5]. Reactive oxygen species (ROS) are produced during oxidative metabolism in the mitochondria and in response to bacterial invasion, cytokines, and xenobiotic [6]. According to recent research, ROS damage cells and, at high concentrations, cause programmed cell death (PCD). ROS can function as signaling molecules in a number of cellular processes at low concentrations [7]. They also have a signaling role in the cell and are produced specifically by enzymes like NOXes (nicotinamide adenine dinucleotide phosphate [NADPH] oxidases) [8]. ROS signaling may be involved in normal physiological processes or contribute to a maladaptive response that result in metabolic dysfunction and inflammatory signaling, depending on the source of ROS, cell type, and tissue environment [9]. Various lifestyles affect beneficial metabolic processes, leading to disorders. The body uses different lipids for a variety of metabolic processes, including hormone production, energy storage, signaling, biomarkers, energy metabolism, and structural components [10]. Lipids are diverse chemical groups that play a part in a number of processes in the body [11]. The accumulation of lipid peroxidation (LPO) products in human tissues is a major contributor to cellular and tissue dysfunction, which is a major factor in aging and the majority of diseases linked to oxidative stress and ageing [12].The three phases of LPO are initiation, propagation, and termination, and they are carried out using three different techniques [13]. Biomolecules like proteins, lipids, structural carbohydrates, and nucleic acids are harmed by free radical-mediated oxidation, free radical-independent oxidation, and non-enzymatic oxidation [14]. Eicosanoids are a broad class of bioactive lipids that are produced from polyunsaturated fatty acids (PUFA) [15]. Lipid mediators referred to as eicosanoids, which include prostaglandins, leukotrienes, and thromboxanes [16], produced through the pathways of lipoxygenase (LOX), cyclooxygenase (COX), and epoxygenase, are crucial for controlling a number of pathophysiological processes like inflammation [17]. They are crucial endogenous metabolic regulators that are produced when phospholipases, sphingomyelinases, and cytochrome P450 (CYP) hydroxylases break down phospholipids, sphingolipids, and cholesterol enzymatically and non-enzymatically [18]. The roles of ROS and eicosanoids have been studied separately; their combined impact on metabolic disorders is beginning to be researched. The rising prevalence of obesity and diabetes, understanding this oxidative–lipid axis, has profound clinical implications. Therapeutic interventions targeting these pathways—ranging from antioxidants and COX/LOX inhibitors to nutritional interventions with omega-3 fatty acids. They represent promising avenues but require evaluation of efficacy and safety. By consolidating evidence across molecular, cellular, and clinical studies, this chapter aims to highlight the central role of ROS–eicosanoid interactions in metabolic disease progression and discuss emerging therapeutic opportunities.

7.2 Reactive Oxygen Species in Metabolic Disorders

Reactive oxygen species (ROS) are oxygen-containing free radicals and non-radical oxidants that are by-products of aerobic metabolism. In healthy physiology, ROS serve critical signaling roles in regulating immune defense, vascular tone, and insulin sensitivity. However, when ROS production overwhelms antioxidant defense systems, a state of oxidative stress develops, leading to structural cell damage, activation of inflammatory pathways, and disruption of metabolic homeostasis. In the context of metabolic disorders, excessive ROS generation is a key driver of chronic inflammation, insulin resistance, β-cell dysfunction, and vascular pathology.

7.2.1 Mechanisms of ROS Production and Redox Imbalance

The mitochondrion is the primary intracellular source of ROS. In the case of ROS formation, defective oxidative phosphorylation occurs and electron transport is mismatched. From the oxidative phosphorylation one electron comes into ubiquinone (Q). When ubiquinone accepts one electron then it converts into partially reduced ubiquinone radical. This partially reduced ubiquinone radical passes one electron to an oxygen molecule, and oxygen molecules convert into superoxide radicals. The superoxide radicals are converted into hydrogen peroxide in the presence of superoxide dismutase enzymes. This hydrogen peroxide undergoes reaction with iron and forms hydroxyl radical, and can cause cellular damage. Anytime protein is transporting electrons through a system, especially with multiple transfers, there is some leakage from the system. Single electron escaping from complex-I and complex-III of the ETC and generate the superoxide radical [19, 20]. NADPH oxidases (NOX) include both cytosolic subunits (p47phox, p67phox, p40phox) and membrane-bound subunits (gp91phox, p22phox, Rac). During activation of NOX, cytosolic subunits compromise a multi-component enzyme and migrate to the plasma membrane, to dock with the membrane subunit. These multi-subunit enzymes produce superoxide anion. Superoxide has physiological functions in host defense and signaling. Sustainable oxidative stress is a result of the chronic activation of NOX enzymes in obesity and diabetes due to hyperglycemia, hyperlipidemia, and inflammatory cytokines [21]. Endoplasmic reticulum (ER) is responsible for protein folding and lipid biosynthesis. Nutrient overload induces ER stress, and the accumulation of misfolded proteins occurs. This activates the unfolded protein response (UPR). This stimulates ROS production through protein disulfide isomerase (PDI) and ER oxidoreductin-1 (ERO1). Persistent ER stress contributes to inflammation and apoptosis in hepatocytes and adipocytes. Cytosolic Xanthine oxidase, Cytochrome P450 monooxygenases are not "SYP" family proteins; they belong to the CYP (Cytochrome P450) superfamily and are primarily localized in the endoplasmic reticulum (ER). The mitochondrial ETC and the NADPH oxidase are the key enzyme classes that are used to break superoxide or the superoxide dismutase enzymes (SOD). The SOD enzyme helps to convert superoxide into hydrogen peroxide [22].

7.2.2 ROS-Mediated Activation of Inflammatory Pathways

Immune and inflammatory responses, adhesion, differentiation, proliferation, autophagy, senescence, and apoptosis are all regulated by NF-κB. Rel, RelA (p65), RelB, p50/p105, and p52/p100 are part of its family. Both canonical (IKKβ-mediated) and noncanonical (IKKα-mediated) pathways are involved in NF-κB activation. Canonical pathway involves ubiquitination, phosphorylation, and degradation of IκBα for nuclear translocation. Noncanonical pathway is activated by LTβ, CD40L, B-cell Activating Factor (BAFF), and viruses, processing p100 to p52. NF-B is influenced by reactive oxygen species through modulating IB phosphorylation and other mechanisms, which results in increased levels of antioxidant proteins [23, 24]. ROS triggers the activation of the Nucleotide-binding Oligomerization Domain (NOD)-like receptor family pyrin domain-containing 3 (NLRP3) inflammasome. Mitochondrial reactive oxygen species play a crucial role in promoting the assembly of the inflammasome complex, which consequently activates caspase-1 and causes the maturation of IL-1 and IL-18. These cytokines are responsible for the dysfunction of β-cells and amplifying chronic inflammation in metabolic tissues [23, 25]. The mitogen-activated protein kinase (MAPK) cascades, consisting of the extracellular signal-related kinases (ERK1/2), the c-Jun N-terminal kinases (JNK), the p38 kinase (p38), and the big MAP kinase 1 (BMK1/ERK5) pathway. Reactive oxygen species (ROS) initiate MAPK pathways, including JNK, p38 MAPK, and extracellular signal-regulated kinase (ERK). JNK is particularly important because its activation results in serine phosphorylation of insulin receptor substrate-1, which affects insulin signaling. The activation of MAPK mediated by ROS also increases the expression of adhesion molecules in endothelial cells, resulting in vascular inflammation [23, 26].

7.2.3 Contribution of ROS to Metabolic Dysfunction

Reduced levels of antioxidant enzymes, such as SOD and Glutathione peroxidase (GPx), in pancreatic β-cells lead to their distinct vulnerability to oxidative stress. The buildup of reactive oxygen species (ROS) in β-cells is intensified by chronic hyperglycemia and hyperlipidemia, causing mitochondrial dysfunction, DNA damage, and eventual apoptosis. The ongoing presence of reactive oxygen species (ROS) in adipocytes and macrophages is associated with obesity, saturated fatty acids, inflammation, and systemic insulin resistance [27, 28]. Decreased levels of antioxidant enzymes, of which SOD and Gpx are the most important, in the pancreatic β-cells lead to these cells a very special susceptibility to oxidative stress. The load of ROS in β-cells is aggravated by the glucose toxicity and the lipotoxicity, which are both chronic; as a result, the mitochondria become dysfunctional, DNA damage occurs, and the cells die by apoptosis. By blocking the transcription factor PDX-1, elevated ROS also affect the transcription of the insulin gene. Both impaired insulin secretion and progressive cell loss in T2DM are caused by oxidative stress [29, 30]. In endothelial cells, overproduction of ROS decreases endothelial function and triggers vascular damage, induces inflammatory responses, and promotes the expression of adhesion molecules and pro-inflammatory cytokines [31]. Hypertrophic adipocytes and infiltrating macrophages drive chemokine release (MCP-1, IL-8). It recruits more immune cells and establishes a vicious cycle of inflammation. This "inflamed adipose phenotype" produces excess free fatty acids and cytokines, worsening systemic insulin resistance.

7.3 Eicosanoids and Inflammation

The term "Eicosanoid" (from the Greek word eicosa—which means "twenty") refers to straight-chain polyunsaturated fatty acids (PUFAs) containing 20 carbons [32]. These are lipid-based signaling molecules derived from arachidonic acid (AA) and other polyunsaturated fatty acids (PUFAs). Eicosanoids in physiological systems and pathological processes play a role in inflammation and immune response [33]. Depending on their particular type and the relationship, they can act as both pro- and anti-inflammatory agents. Enzymes such as cyclooxygenases (COX) and lipoxygenases (LOX) synthesize them by converting the precursor molecule into various eicosanoids, such as lipoxins, prostaglandins, and leukotrienes.

7.3.1 Biosynthesis of Eicosanoids

The enzymatic oxidation of polyunsaturated fatty acids (PUFAs), mainly arachidonic acid (AA) and eicosapentaenoic acid (EPA), produces eicosanoids. Cyclooxygenases (COX), lipoxygenases (LOX), and cytochrome P450 (CYP) enzymes are involved in eicosanoid biosynthesis and catalyze distinct eicosanoid synthesis pathways. The LOX pathway produces leukotrienes (LTs), lipoxins (LXs), and prostacyclin (PGI2); and the COX pathway produces prostaglandins (PGs), thromboxanes (TXs), and prostacyclin [34]. The biosynthesis occurs mainly in three pathways. The cyclooxygenase pathway is responsible for the formation of all eicosanoids with ring structures, such as prostaglandins and thromboxanes. Two isoforms of the cyclooxygenase enzyme (COX-1) and (COX-2) are implicated in this pathway. COX-1 contributes to the synthesis of prostanoids. Vascular homeostasis, kidney function, platelet aggregate formation, and other regular processes are among the physiological cellular functions that COX-1 regulates. When there is inflammation and chronic disease, COX-2 increases the production of prostanoids [35]. The fatty acid is transformed into hydroperoxyeicosatetraenoic acids (HPETEs) in the lipoxygenase (LOX) pathway, which act as bridges for the production of lipoxins and leukotrienes. 5-LOX changes AA into 5-HPETE, which is then further broken down to produce leukotriene A4 (LTA4). Several leukotrienes, including leukotriene B4 (LTB4), leukotriene C4 (LTC4), and leukotriene D4 (LTD4), can be produced by enzymatic conversion of LTA4 [35]. The route involves cytochrome P450 enzymes metabolizing PUFAs, mainly AA, to create epoxyeicosatrienoic acids (EETs) and hydroxyeicosatetraenoic acids (HETEs). EETs control inflammation, blood pressure, and vasodilation, while HETEs control vascular tone and immune response [36].

7.3.2 Pro-inflammatory vs. Anti-inflammatory Eicosanoids

Eicosanoids have a dual role in inflammation, both pro-inflammatory and anti-inflammatory mediators. Pro-inflammatory eicosanoids prostaglandin E2 (PGE2), leukotriene B4 (LTB4), and thromboxane A2 (TXA2) are generated through the action of cyclooxygenase-2 (COX-2) and 5-lipoxygenase (5-LOX) enzymes on arachidonic acid [37–39]. These lipid mediators improve the body's defense mechanisms by increasing vascular permeability, facilitating immune cell migration to inflamed tissues, and inducing cardinal signs of inflammation such as pain and fever [38]. Their activation causes chronic low-grade inflammation, the reason for several metabolic and cardiovascular disorders [37]. Anti-inflammatory and pro-resolving eicosanoids stop excessive immune activation and restore tissue homeostasis. Lipoxins, resolvins, and protectins are generated by 15 lipoxygenase (15-LOX) pathways and omega-3 fatty acid-derived lipid metabolism [40, 41]. They terminate inflammation, reduce fever and pain, and promote tissue healing by dampening neutrophil infiltration while increasing macrophage-mediated clearance of cellular debris [40]. Changes from pro-inflammatory to anti-inflammatory eicosanoid production are the resolution phase of inflammation, and their therapeutic potential in chronic inflammatory conditions like obesity, type 2 diabetes, and atherosclerosis [41]. Prostaglandins (PGs) are a class of physiologically active lipid molecules known as eicosanoids that exhibit a variety of hormone-like properties [42]. Prostaglandins (PGs) are metabolites of arachidonic acid generated by the enzyme cyclooxygenase (COX).While PGs derived from COX-2 are assumed to exhibit pro-inflammatory characteristics and COX-1 is accepted to mediate cytoprotective actions on the gastrointestinal mucosa [43], PGE2 was potent at increasing blood flow, and PGD_2 produced an increase only at high doses [44]. Leukotrienes are chemottractants for neutrophils and macrophages. In obesity, leukotrienes (LTB4,

LTC4) recruit immune cells to adipose tissue, sustaining chronic inflammation and impairing insulin signaling. Elevated LTB4 also worsens hepatic inflammation in non-alcoholic fatty liver disease (NAFLD) [45]. The pathophysiology of asthma and other respiratory disorders may be particularly related to leukotrienes (LTs) C4 and B4, which are strong pro-inflammatory mediators synthesized in myeloid cells from arachidonic acid with a wide range of biologic activities, such as smooth muscle contraction, mucus hypersecretion, and leukocyte activation [46]. Leukotriene B4 stimulates neutrophil, monocyte, and eosinophil recruitment and activation [47]. LTC4 has a number of inflammatory effects, such as bronchoconstriction, vasoconstriction, increased post-capillary permeability with edema, and smooth muscle constriction [48]. Cysteinyl leukotrienes (LTC4, LTD4, LTE4) are also linked to vascular dysfunction and atherosclerosis in obese and diabetic patients. Endothelial cells and platelets produce the metabolite thromboxane from arachidonic acid. It is a smooth muscle mitogen and causes platelet aggregation and vasoconstriction [49]. TXA2 has prothrombotic qualities because it promotes platelet aggregation and activation [50]. Thromboxane B2 may be a biomarker for intestinal damage brought on by radiation therapy [51]. Elevated TXA2 contributes to prothrombotic states in diabetes and metabolic syndrome, increasing cardiovascular risk.

7.4 ROS and Eicosanoid Crosstalk

7.4.1 ROS-Mediated Lipid Peroxidation and Eicosanoid Generation

Reactive oxygen species (ROS) are important in modulating lipid signaling through their influence on polyunsaturated fatty acids (PUFAs). Lipid peroxidation is a central event linking ROS with eicosanoid generation. ROS attack PUFAs, arachidonic acid (AA), linoleic acid, and docosahexaenoic acid (DHA), generating reactive aldehydes (malondialdehyde (MDA), 4-hydroxynonenal (HNE)) and isoprostanes, which are both biomarkers and mediators of oxidative stress [52]. These lipid peroxidation products exert direct pro-inflammatory effects and provide substrates for enzymatic eicosanoid biosynthesis. Cyclooxygenases (COX), lipoxygenases (LOX), and cytochrome P450 epoxygenases use oxidized fatty acids to produce prostaglandins, leukotrienes, thromboxanes, and epoxyeicosatrienoic acids [53]. ROS activate COX enzymes by modifying cysteine residues and altering their redox state, increasing prostaglandin synthesis [54]. LOX isoforms such as 5-LOX are sensitive to oxidative environments, with ROS-driven phosphorylation increasing leukotriene biosynthesis [55]. ROS are facilitators of enzymatic eicosanoid production (Fig. 7.1).

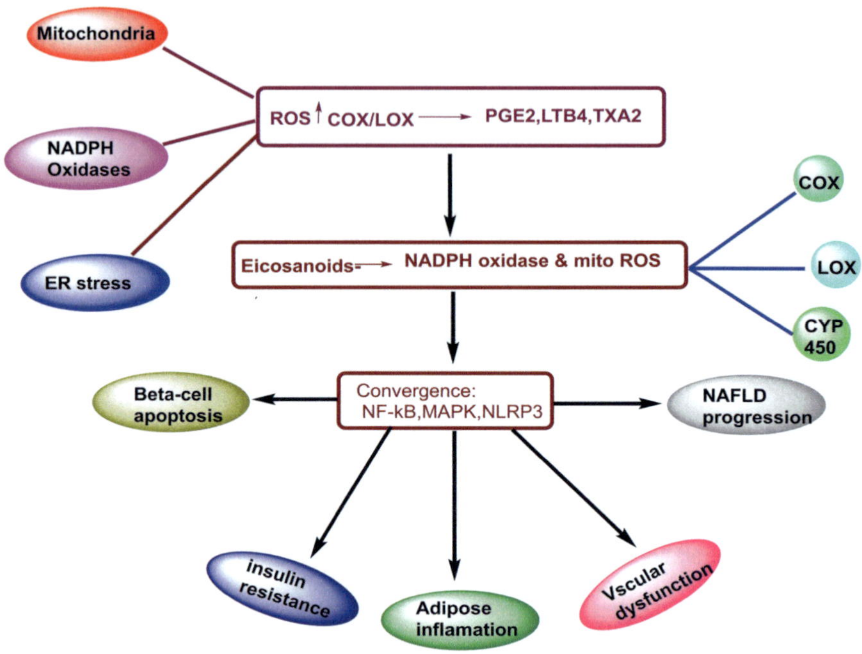

Fig. 7.1 ROS–Eicosanoid amplification loop in metabolic disorders

7.4.2 *Synergistic Role of ROS and Eicosanoids in Amplifying Inflammation*

The crosstalk between ROS and eicosanoids makes a feed forward loop that amplifies inflammatory signaling. ROS promote pro-inflammatory eicosanoids such as PGE2, LTB4, and TXA2. These eicosanoids increase ROS production by stimulating NADPH oxidase activity, mitochondrial dysfunction, and nitric oxide synthase uncoupling [56]. PGE2 activates EP receptors on macrophages and adipocytes. It increases NADPH oxidase activity and superoxide generation [38]. Leukotrienes stimulate neutrophil ROS production, perpetuating oxidative stress [39]. This bidirectional synergy plays a role in metabolic disorders. In obesity and type 2 diabetes mellitus (T2DM), excess nutrients and adipocyte hypertrophy create a pro-oxidant milieu that causes both ROS and eicosanoid accumulation. The convergence of these pathways drives chronic, low-grade inflammation, "meta-inflammation," which causes insulin resistance, β-cell dysfunction, and vascular pathology [57].

7.4.3 *Impact on Adipose Tissue Inflammation, Insulin Signaling, and Vascular Dysfunction*

Adipose tissue is a major site for ROS–eicosanoid crosstalk. ROS increases adipocyte secretion of pro-inflammatory adipokines (TNF-α, IL-6), synergizes with PGE2 and LTB4. With the help of macrophages and neutrophils, an inflamed adipose microenvironment is created [58]. LTB4 activates its receptor BLT1 on immune cells and fat cells, which impacts the JNK pathway and stops insulin receptor signaling [59]. In pancreatic β-cells, prolonged exposure to reactive oxygen species (ROS) with increased prostaglandin E2 (PGE2) inhibits insulin secretion and induces apoptosis [60]. When vascular endothelial cells are exposed to oxidative stress and thromboxane A2, impaired nitric oxide bioavailability occurs, which is a sign of endothelial dysfunction. It also increased platelet aggregation, and contributed to atherosclerosis [61]. ROS and eicosanoids constitute an integrated inflammatory axis that influences various tissues associated with metabolic syndrome, exacerbating systemic dysfunction.

7.5 Therapeutic Approaches

The interplay of reactive oxygen species (ROS) and eicosanoids forms a central axis in the pathogenesis of metabolic disorders. Targeting this axis has emerged as a therapeutic strategy. It can be categorized into antioxidants targeting ROS, COX/LOX inhibitors, and nutritional interventions. Targeting the ROS–eicosanoid axis through antioxidants, COX/LOX inhibitors and nutritional strategies hold promise for mitigating metabolic disorders. While conventional antioxidants show modest benefits, advances in mitochondria-targeted agents and phytochemical formulations are enhancing therapeutic efficacy. COX/LOX modulators and leukotriene antagonists present novel pharmacological opportunities. They represent feasible, low-risk strategies. Future research must focus on combination therapies, biomarker-driven interventions. Long-term clinical validation is needed to optimize outcomes.

7.5.1 *Antioxidants Targeting ROS*

Natural antioxidants like Vitamin C and Vitamin E are widely studied. Vitamin C scavenges ROS and activates the NF-κB pathway, which in turn attenuates oxidative damage and inflammation of pulmonary system [62]. Vitamin C inhibits phospholipase A_2 activation to reduce the generation of ROS that leads to reduced arachidonic acid release, and attenuates the synthesis of prostaglandins and leukotrienes. The beneficial effects of metabolic disorders are: reduced oxidative stress for atherosclerosis and diabetes, and better endothelial activity. Vitamin E (α-tocopherol) is the predominant antioxidant in biological systems. It protects cellular membranes through inhibition of lipid peroxidation and ROS [63]. It inhibits the formation of pro-inflammatory prostaglandins and thromboxanes by preventing the peroxidation of arachidonic acid. Valuable for metabolic abnormalities, it protects from oxidation in cardiovascular disease (CVD), insulin resistance, and obesity. A variety of synthetic antioxidants that mimic the beneficial effects of natural antioxidants are also available. NAC (N-acetylcysteine) is one of them; it is known to prevent ROS-dependent apoptosis. NAC is a synthetic precursor of intracellular glutathione (GSH) and cysteine. Its ability to scavenge free radicals either directly through the redox potential of thiols or indirectly through raising glutathione levels in cells gives it its anti-ROS effect. The role of ROS in drug-induced apoptosis is frequently verified by ROS scavengers, particularly NAC [64]. Antioxidant compounds that contain ubiquinone (MitoQ) or vitamin E (MitoVit E) that are targeted to mitochondria have been employed in succession to combat in recent years [65]. MitoQ is a ubiquinone derivative linked to triphenylphosphonium, a lipophilic cation that allows its accumulation within mitochondria via the electrochemical gradient. Ubiquinone's various benefits as an efficient antioxidant have led to the development of this moiety. As an antioxidant and an electron transporter, ubiquinone is essential to the ETC. By lowering the mitochondria's synthesis of lipid peroxyl radicals, it stops lipid peroxidation [66]. In recent research, MitoQ protects against metabolic syndrome via altering redox signaling pathways. It has been demonstrated that MitoQ reduces human NP cell apoptosis and mitochondrial dysfunction. MitoQ improves Nrf2 activity, mitigates lysosomal function damage and mitophagosome–lysosome fusion impairment, and restores mitochondrial dynamic equilibrium. MitoQ can extend the survival period of NP cells in an ROS-rich environment and facilitate the removal of defective mitochondria brought on by oxidative stress; these outcomes are in line with those of in vivo studies [67].

7.5.2 COX/LOX Inhibitors

Arachidonic acid is converted to prostaglandin by the COX enzyme, and when ROS activates Cox enzymes, pro-inflammatory prostaglandins are increased. Both anti-inflammatory steroids and nonsteroidal anti-inflammatory drugs (NSAIDs) can prevent the production of prostaglandins. Aspirin, ibuprofen, and naproxen are examples of typical NSAIDs that pharmacologically target both PGHS isozymes. NSAIDs help to reduce chronic inflammation in obesity, type 2 diabetes, and atherosclerosis [68]. LOX enzymes convert PUFAs into leukotrienes, and also LOX metabolites increase the ROS generation. LOX inhibitors inhibit LOX enzyme and reduce the production of harmful lipid mediators. Three leukotriene receptor antagonists, pranlukast, zafirlukast, and montelukast have also demonstrated good success in the treatment of asthma. These medications reduce airway constriction and mucus buildup in the lungs and airways by blocking the binding of leukotriene D4, LTC4, and LTE4 to the CysLTR1 in the lungs and bronchial tubes [69].

7.5.3 Nutritional Interventions

The presence of prostanoids of the 2-series (PGI2, PGE2, and TXA2) in cell membranes formed from arachidonic acid (AA) lipid peroxidation is inhibited by omega-3, EPA, and DHA. They activate many proteins that induce mitogenic cell signaling pathways. Evidence also showed that PUFAs can alter the fate of a number of signal transduction pathways linked to a strong anti-inflammatory and anti-proliferative effect. They can down-regulate the activities of several transcription factors, such as NF-κB, which results in a decrease in the production of pro-inflammatory cytokines (TNF-α/interleukin-1) [70]. Fish oil supplements have been shown to reduce PGE2 and 4 series LTs production by inflammatory cells in healthy human volunteers over a period of weeks to months [71]. Dietary polyphenols (or flavonoids) are effective free radical and reactive oxygen species (ROS) scavengers due to the presence of aromatic structural features, numerous hydroxyl groups, and a strongly conjugated system. They have the power to negate ROS or suppress cellular OS, allowing them to avert oxidative damages of biomolecules (lipids, proteins, DNA) and reduce tissue inflammation [72]. Polyphenols help maintain normal blood glucose levels by preserving pancreatic cells and boosting insulin production by neutralizing ROS. Polyphenols have antioxidant and anti-inflammatory properties and can reduce chronic inflammation associated with obesity and insulin resistance [73] (Table 7.1).

7.6 Challenges and Future Directions

The interaction between reactive oxygen species (ROS) and eicosanoids is a complex and dynamic regulatory network that perpetuates inflammatory processes in metabolic disorders, like obesity, type 2 diabetes mellitus (T2DM), non-alcoholic fatty liver disease (NAFLD), and cardiovascular diseases (CVDs). Advancements have been achieved in clarifying the mechanistic pathways governing ROS production and eicosanoid synthesis; numerous obstacles exist for efficacious therapeutic and diagnostic applications. The complexity arises from cellular heterogeneity, tissue-specific effects, feedback regulatory loops, and the dual functions of both ROS and eicosanoids in physiological and pathological contexts.

Table 7.1 Therapeutic approaches targeting ROS–eicosanoid axis in metabolic disorders

Category	Agents/Examples	Mechanism of action	Therapeutic implications in metabolic disorders	References
Antioxidants (conventional)	Vitamin C	Scavenge ROS; inhibit phospholipase A2 activation→↓ arachidonic acid release; attenuate prostaglandin & leukotriene synthesis	Oxidative stress in atherosclerosis & diabetes; improve endothelial function; reduce pulmonary oxidative damage	[74]
	Vitamin E (a-tocopherol)	Inhibit lipid peroxidation; block peroxidation of arachidonic acid →↓ pro-inflammatory prostaglandins & thromboxanes	Protect against CVD, insulin resistance, obesity; maintain membrane stability	[75]
Synthetic antioxidants	N-Acetylcysteine (NAC)	Precursor of glutathione; scavenge ROS directly via thiols & indirectly by raising intracellular GSH; prevent ROS-dependent apoptosis	Attenuate oxidative stress-induced apoptosis; used in drug-induced ROS toxicity	[76]
Mitochondria-targeted antioxidants	Mitochondria-targeted antioxidants	Conjugated to lipophilic cation (triphenylphosphonium) → accumulate in mitochondria; reduce lipid peroxyl radicals; restore mitochondrial redox balance	Protect against metabolic syndrome; improve Nrf2 activity; prevent mitochondrial dysfunction & apoptosis in ROS-rich environment	[77–79]
COX inhibitors	NSAIDs (aspirin, ibuprofen, naproxen)	Inhibit cyclooxygenase (COX)→↓ prostaglandins; reduce ROS-induced pro-inflammatory mediator synthesis	↓Chronic inflammation in obesity, type 2 diabetes, atherosclerosis	[80]
LOX inhibitors & leukotriene antagonists	LOX inhibitors; montelukast, zafirlukast, pranlukast	Block leukotriene biosynthesis or receptor binding (CysLTR1) → ↓ bronchoconstriction, mucus, ROS generation	Beneficial in asthma; potential use in inflammatory pathways linked with metabolic disorders	[81]
Nutritional interventions (PUFAs)	Omega-3 fatty acids (EPA, DHA)	Replace arachidonic acid in membranes → → pro-inflammatory prostanoids (PGE2, TXA2, LTs); inhibit NF-κB pathway →↓ cytokines (TNF-α, IL-1)	Anti-inflammatory & anti-proliferative; improve insulin sensitivity; lower cardiovascular risk	[82, 83]
Nutritional interventions (polyphenols)	Polyphenol-rich diets (flavonoids)	Scavenge ROS; protect biomolecules (lipids, proteins, DNA); preserve pancreatic ß-cells; ↑ insulin secretion; downregulate NF-κB	Scavenge ROS; protect biomolecules (lipids, proteins, DNA); preserve pancreatic ß-cells; ↑ insulin secretion; downregulate NF-κB	[84, 85]

7.6.1 Complexity of ROS–Eicosanoid Interplay

The interactions between ROS and eicosanoids depend a lot on the situation. Under normal conditions, ROS and eicosanoids are both important signaling molecules that help cells grow, form new blood vessels, and protect the body. During metabolic stress, their unregulated production results in persistent low-grade inflammation, commonly referred to as "metaflammation" [86]. The therapeutic approaches are complicated due to the dual roles of ROS, acting as signaling intermediates at low concentrations and as damaging agents at elevated levels. Eicosanoids exhibit functional heterogeneity: prostaglandin E2 (PGE2) and leukotriene B4 (LTB4) are primarily pro-inflammatory; lipoxins and resolvins facilitate the resolution of inflammation [86]. This duality suggests that indiscriminate inhibition of ROS or eicosanoids may not consistently produce beneficial results and potentially interfere with protective homeostatic functions. Subsequent research is needed to clarify tissue-specific interactions between reactive oxygen species (ROS) and eicosanoids. In adipose tissue, excessive ROS generation increases LTB4 production, which attracts macrophages and exacerbates insulin resistance [87]. In contrast, in hepatic tissue, ROS-induced lipid peroxidation plays a more critical role in the progression of NAFLD [88]. Single-cell transcriptomics, when integrated with lipidomics and redox profiling, may elucidate the heterogeneity of these responses.

7.6.2 Lack of Reliable Biomarkers

The lack of dependable and standardized biomarkers for assessing oxidative stress and eicosanoid signaling in individuals with metabolic disorders is a challenge. Numerous surrogate markers of oxidative stress, including malondialdehyde (MDA), 4-hydroxynonenal (4-HNE), and protein carbonyls, are commonly assessed; these indicators are deficient in specificity and sensitivity [89]. Circulating eicosanoid concentrations may not accurately represent localized tissue activity, given that numerous lipid mediators exhibit autocrine or paracrine effects with swift turnover [90]. The advancement of high-throughput lipidomic platforms has facilitated the identification of a wide array of eicosanoids and their metabolites; clinical validation and standardization across laboratories are inadequate. Subsequent investigations should concentrate on the development of composite biomarker panels that incorporate redox-sensitive molecules, eicosanoid signatures, and inflammatory mediators. These panels help classify patients into distinct groups of disease progression and treatment validation. Improvements in mass spectrometry and bioinformatics-based data integration are important for reaching this goal [87].

7.6.3 Translational Challenges in Therapy

Preclinical models showed promise, but trying to turn those results into treatments that target ROS and eicosanoids has been difficult. Vitamin E and N-acetylcysteine (NAC) are examples of broad-spectrum antioxidants that have not worked in large clinical trials. This is partly because they are not very bioavailable, do not target specific cells, or restore normal redox signaling [87]. Chronic use of nonsteroidal anti-inflammatory drugs (NSAIDs), which block cyclooxygenase (COX) enzymes, poses cardiometabolic risks [91]. Selective lipoxygenase (LOX) inhibitors and dual COX/LOX inhibitors have shown certain advantages in animal models; however, human trials are limited [92]. A promising avenue entails the utilization of mitochondria-targeted antioxidants (e.g., MitoQ, SkQ1), which regulate mitochondrial reactive oxygen species (ROS) production while preserving critical redox signaling [93]. Pro-resolving mediators like resolvins, protectins, and maresins, which come from omega-3 fatty acids, have the potential to shift the balance toward resolving inflammation instead of suppressing it [94]. Before these mediators can be used in clinical settings, problems with stability, pharmacokinetics, and large-scale synthesis must be solved.

7.6.4 Systems Biology and Network Modeling

Because of many complex feedback loops between ROS and eicosanoid pathways, systems biology methods are needed to find important regulatory nodes that can be used for treatment. Computational models that include redox dynamics, lipid mediator fluxes, and signaling cascades can help to predict how a system will react to changes. Recent developments in network pharmacology indicate that combination therapies aimed at multiple nodes (e.g., ROS scavengers in conjunction with COX/LOX modulators) demonstrate greater efficacy than monotherapies [95]. Subsequent research ought to amalgamate multi-omics data, encompassing genomics, transcriptomics, metabolomics, and lipidomics, to develop predictive models that reflect patient-specific variability.

7.6.5 Inter-Individual Variability and Precision Medicine

Patient heterogeneity constitutes a significant challenge in the formulation of universal therapeutic strategies. Genetic polymorphisms in oxidative stress-related enzymes (such as NADPH oxidase subunits, superoxide dismutase, and glutathione peroxidase) and eicosanoid-metabolizing enzymes (including COX-2, 5-LOX, and CYP450 epoxygenases) affect individual susceptibility to metabolic inflammation [96]. Lifestyle factors, including diet, microbiome composition, and physical activity, influence both redox homeostasis and lipid mediator profiles [97]. Precision medicine strategies that incorporate genetic, epigenetic, and metabolic data facilitate the customization of antioxidant or eicosanoid-targeted therapies for particular patient cohorts. People with polymorphisms in the 5-LOX promoter regions may get the most benefit from LOX inhibitors. People with mitochondrial DNA variants that affect ROS production may get the most benefit from antioxidants that target mitochondria [98]. Promising paths are starting to show up. First, modulating the gut microbiota presents a potent approach to regulate both reactive oxygen species (ROS) and eicosanoid signaling, as microbial metabolites like short-chain fatty acids (SCFAs) and secondary bile acids affect host redox balance and lipid mediator synthesis [99]. Second, researchers are looking into delivery systems based on nanotechnology to make antioxidants and lipid mediator modulators more bioavailable and target-specific tissues [100]. Third, innovative Clustered Regularly Interspaced Short Palindromic Repeats (CRISPR)-based gene-editing tools facilitate the targeted modification of redox and eicosanoid pathway genes in preclinical models, allowing for meticulous analysis of causal relationships [101].

7.7 Conclusion

Obesity, type 2 diabetes mellitus, and metabolic syndrome are all metabolic disorders that cause a long-lasting low-grade inflammation, known as metaflammation. The interaction between reactive oxygen species (ROS) and eicosanoids is the center of this disease. They create a feed-forward loop that makes oxidative stress, immune activation, and tissue dysfunction worse.

Reactive oxygen species (ROS), produced mainly by mitochondria, NADPH oxidases, and endoplasmic reticulum stress, function both as deleterious oxidants and as essential signaling molecules. They turn on redox-sensitive pathways in metabolic tissues, like NF-κB, MAPKs, and the NLRP3 inflammasome. These cascades interfere with insulin signaling, induce inflammation in adipose tissue, hinder pancreatic β-cell functionality, and exacerbate endothelial damage.

Eicosanoids, come from arachidonic acid and other polyunsaturated fatty acids, change the inflammatory environment even more. In obesity and diabetes, pro-inflammatory mediators prostaglandin E2 (PGE2), leukotriene B4 (LTB4), and thromboxane A2 (TXA2) are in charge. They cause immune cells to gather, blood vessels to stop working properly, and insulin resistance. The synthesis of anti-inflammatory and pro-resolving mediators, including lipoxins, resolvins, and epoxyeicosatrienoic acids (EETs) is inhibited. It obstructs the resolution of inflammation. The interaction between ROS and eicosanoids is crucial to the advancement of disease. ROS facilitates lipid peroxidation and augment COX/LOX activity, biasing eicosanoid profiles toward pro-inflammatory variants. Eicosanoids activate enzymes that make ROS, like NADPH oxidases, which makes oxidative stress worse. This reciprocal relationship elucidates the persistence of metaflammation in metabolic disorders and its significant correlation with cardiovascular complications, fatty liver disease, and β-cell failure. Interventions aimed at oxidative–lipid axis show a lot of promise for treatment. Traditional antioxidants have demonstrated restricted effectiveness. Mitochondria-targeted agents and NADPH oxidase inhibitors provide accurate regulation of reactive oxygen species (ROS). Pharmacological approaches COX/LOX inhibitors, thromboxane receptor antagonists, and soluble epoxide hydrolase inhibitors show promise in restoring balance to eicosanoid signaling. Nutritional strategies omega-3 fatty acids and polyphenols confer supplementary advantages by altering lipid mediator profiles to pro-resolving states. There are still big problems even though things are getting better. ROS and eicosanoids have important roles in the body and their suppression must be weighed against the risk of hurting the body's ability to fight infections and heal tissues. There are no reliable biomarkers for tracking ROS–eicosanoid changes in patients, and differences between people make treatment harder. Going forward, the field needs to use precision medicine methods that combine genetic, dietary, and microbiome data with biomarker-driven patient stratification. Future directions encompass the creation of synthetic pro-resolving mediator analogs, nanomedicine-based delivery systems, and integrative omics-driven biomarkers. These advancements in conjunction with systems biology models may facilitate the rational design of multi-targeted therapies that mitigate both oxidative and lipid-mediated inflammation. In summary, the ROS–eicosanoid axis is a key factor that causes inflammation and metabolic problems. By focusing on this interaction, we might find new ways to treat metabolic diseases, restore metabolic balance, preserve organ function, and lower the global burden of these diseases.

References

1. Chen J, Yang Y, Kong W (2022) Cross talk between inflammation and metabolic disorders. Mediators Inflamm [Internet] 2022:9821506. Available from: https://doi.org/10.1155/2022/9821506
2. Rani V, Deep G, Singh RK, Palle K, Yadav UCS (2016) Oxidative stress and metabolic disorders: pathogenesis and therapeutic strategies. Life Sci [Internet] 148:183–193. Available from: https://doi.org/10.1016/j.lfs.2016.02.002
3. Tuomisto K, Palmu J, Long T, Watrous JD, Mercader K, Lagerborg KA et al (2022) A plasma metabolite score of three eicosanoids predicts incident type 2 diabetes: a prospective study in three independent cohorts. BMJ Open Diabetes Res Care [Internet] 10(2):e002519. Available from: https://doi.org/10.1136/bmjdrc-2021-002519
4. Hotamisligil GS (2006) Inflammation and metabolic disorders. Nature 444(7121):860–867
5. Chatterjee S (2016) Oxidative stress, inflammation, and disease. In: Oxidative stress and biomaterials. Academic Press
6. Ray PD, Huang B-W, Tsuji Y (2012) Reactive oxygen species (ROS) homeostasis and redox regulation in cellular signaling. Cell Signal [Internet] 24(5):981–990. Available from: https://doi.org/10.1016/j.cellsig.2012.01.008
7. Li Z, Xu X, Leng X, He M, Wang J, Cheng S et al (2017) Roles of reactive oxygen species in cell signaling pathways and immune responses to viral infections. Arch Virol [Internet] 162(3):603–610. Available from: https://doi.org/10.1007/s00705-016-3130-2

8. Forrester SJ, Kikuchi DS, Hernandes MS, Xu Q, Griendling KK (2018) Reactive oxygen species in metabolic and inflammatory signaling. Circ Res [Internet] 122(6):877–902. Available from: https://doi.org/10.1161/CIRCRESAHA.117.311401
9. Valko M, Leibfritz D, Moncol J, Cronin MTD, Mazur M, Telser J (2007) Free radicals and antioxidants in normal physiological functions and human disease. Int J Biochem Cell Biol [Internet] 39(1):44–84. Available from: https://doi.org/10.1016/j.biocel.2006.07.001
10. Natesan V, Kim SJ (2021) Lipid metabolism, disorders and therapeutic drugs-review. Biomol Ther 29(6)
11. Benedetti A, Comporti M, Esterbauer H (1980) Identification of 4- hydroxynonenal as a cytotoxic product originating from the peroxidation of liver microsomal lipids. Biochim Biophys Acta 620:281–296
12. Negre-Salvayre A, Auge N, Ayala V, Basaga H, Boada J, Brenke R et al (2010) Pathological aspects of lipid peroxidation. Free Radic Res [Internet] 44(10):1125–1171. Available from: https://doi.org/10.3109/10715762.2010.498478
13. Catalá A (2006) An overview of lipid peroxidation with emphasis in outer segments of photoreceptors and the chemiluminescence assay. Int J Biochem Cell Biol [Internet] 38(9):1482–1495. Available from: https://doi.org/10.1016/j.biocel.2006.02.01
14. Niki E, Yoshida Y, Saito Y, Noguchi N (2005) Lipid peroxidation: mechanisms, inhibition, and biological effects. Biochem Biophys Res Commun [Internet] 338(1):668–676. Available from: https://doi.org/10.1016/j.bbrc.2005.08.072
15. Hardwick JP, Eckman K, Lee YK, Abdelmegeed MA, Esterle A, Chilian WM et al (2013) Eicosanoids in metabolic syndrome. Adv Pharmacol [Internet] 66:157–266. Available from: https://doi.org/10.1016/B978-0-12-404717-4.00005-6
16. Mayatepek E (2022) Disorders of eicosanoid metabolism. In: physician's guide to the diagnosis, treatment, and follow-up of inherited metabolic diseases. Springer International Publishing, Cham
17. Agarwal S, Reddy GV, Reddanna P (2009) Eicosanoids in inflammation and cancer: the role of COX-2. Expert Rev Clin Immunol [Internet] 5(2):145–165. Available from: https://doi.org/10.1586/1744666X.5.2.145
18. Farooqui AA. In metabolic syndrome: an important risk factor for stroke, Alzheimer disease, and depression. New York: Springer; 2013
19. Zhao RZ, Jiang S, Zhang L, Yu ZB (2019) Mitochondrial electron transport chain, ROS generation and uncoupling. Int J Mol Med 44:3–1
20. Cobley JN (2020) Mechanisms of mitochondrial ROS production in assisted reproduction: the known, the unknown, and the intriguing. Antioxidants (Basel) [Internet] 9(10):933. Available from: https://doi.org/10.3390/antiox9100933
21. Bedard K, Krause KH (2007) The NOX family of ROS-generating NADPH oxidases: physiology and pathophysiology. Physiol Rev 87:245–313
22. Hanukoglu I (2006) Antioxidant protective mechanisms against reactive oxygen species (ROS) generated by mitochondrial P450 systems in steroidogenic cells. Drug Metab Rev [Internet] 38(1–2):171–196. Available from: https://doi.org/10.1080/03602530600570040
23. Zhang J, Wang X, Vikash V, Ye Q, Wu D, Liu Y et al (2016) ROS and ROS-mediated cellular signaling. Oxid Med Cell Longev 2016
24. Bonizzi G, Karin M (2004) The two NF-κB activation pathways and their role in innate and adaptive immunity. Trends Immunol 25(6):280–288
25. Li P, Chang M (2021) Roles of PRR-mediated signaling pathways in the regulation of oxidative stress and inflammatory diseases. Int J Mol Sci 22
26. Junttila MR, Li S-P, Westermarck J (2008) Phosphatase-mediated crosstalk between MAPK signaling pathways in the regulation of cell survival. FASEB J [Internet] 22(4):954–965. Available from: https://doi.org/10.1096/fj.06-7859rev
27. Wang N, Wu Y, Bian J, Qian X, Lin H, Sun H et al (2017) Current development of ROS-modulating agents as novel antitumor therapy. Curr Cancer Drug Targets [Internet] 17(2):122–136. Available from: https://doi.org/10.2174/1568009616666160216125833

28. Yesupatham A, Saraswathy R (2025) Role of oxidative stress in prediabetes development. Biochem Biophys Rep [Internet] 43:102069. Available from: https://doi.org/10.1016/j.bbrep.2025.102069
29. Martens GA, Cai Y, Hinke S, Stangé G, Van De Casteele M, Pipeleers D (2005) Glucose suppresses superoxide generation in metabolically responsive pancreatic β cells. J Biol Chem 280(21):20389–20396
30. Tiedge M, Lortz S, Drinkgern J, Lenzen S (1997) Relation between antioxidant enzyme gene expression and antioxidative defense status of insulin-producing cells. Diabetes 46(11):1733–1742. https://doi.org/10.1016/j.bbrep.2025.102069
31. Szocs K (2004) Endothelial dysfunction and reactive oxygen species production in ischemia/reperfusion and nitrate tolerance. Gen Physiol Biophys 23:265–296
32. Beare-Rogers JL, Dieffenbacher A, Holm JV (2001) Lexicon of lipid nutrition (IUPAC technical report). Pure Appl Chem [Internet] 73(4):685–744. Available from: https://doi.org/10.1351/pac200173040685
33. Wikipedia contributors. Eicosanoid [Internet]. Wikipedia, The Free Encyclopedia. 2025. Available from: https://en.wikipedia.org/w/index.php?title=Eicosanoid&oldid=1304975914
34. Mosaad E (2020) The role (s) of eicosanoids and exosomes in human parturition. Front Physiol 11
35. Dennis EA, Norris PC (2015) Eicosanoid storm in infection and inflammation. Nat Rev Immunol [Internet] 15(8):511–523. Available from: https://doi.org/10.1038/nri3859
36. Gabbs M, Leng S, Devassy JG, Monirujjaman M, Aukema HM (2015) Advances in our understanding of Oxylipins derived from dietary PUFAs. Adv Nutr [Internet] 6(5):513–540. Available from: https://doi.org/10.3945/an.114.007732.PMC4561827
37. Mittal M, Siddiqui MR, Tran K, Reddy SP, Malik AB (2014) Reactive oxygen species in inflammation and tissue injury. Antioxid Redox Signal [Internet]. 20(7):1126–1167. Available from: https://doi.org/10.1089/ars.2012.5149
38. Ricciotti E, FitzGerald GA (2011) Prostaglandins and inflammation. Arterioscler Thromb Vasc Biol 31(5):986–1000. https://doi.org/10.1161/ATVBAHA.110.207449
39. Peters-Golden M, Henderson WR Jr (2007) Leukotrienes. N Engl J Med 357(18):1841–1854. https://doi.org/10.1056/NEJMra071371
40. Serhan CN (2014) Pro-resolving lipid mediators are leads for resolution physiology. Nature [Internet] 510(7503):92–101. Available from: https://doi.org/10.1038/nature13479
41. González-Périz A, Clària J (2007) New approaches to the modulation of the cyclooxygenase-2 and 5-lipoxygenase pathways. Curr Top Med Chem 7(3):297–309. https://doi.org/10.2174/156802607779941378
42. Bailey J (ed) (2012) Prostaglandins, leukotrienes, and lipoxins: biochemistry, mechanism of action, and clinical applications. Springer Science & Business Media
43. Morteau O (2001) Prostaglandins and inflammation: the cyclooxygenase controversy. In: Górski A, Krotkiewski H, Zimecki M (eds) Inflammation. Springer, Dordrecht. https://doi.org/10.1007/978-94-015-9702-9_6
44. Williams TJ (1979) Prostaglandin E2, prostaglandin I2 and the vascular changes of inflammation. Br J Pharmacol [Internet] 65(3):517–524. Available from: https://doi.org/10.1111/j.1476-5381.1979.tb07860.x
45. Butola LK, Dhok A, Ambad R, Kanyal D, Jha RK (2021) Leukotrienes and inflammation-a review. Indian J Forensic Med Toxicol 15(2):295–301
46. Wardlaw AJ, Hay H, Cromwell O, Collins JV, Kay AB (1989) Leukotrienes, LTC4 and LTB4, in bronchoalveolar lavage in bronchial asthma and other respiratory diseases. J Allergy Clin Immunol [Internet] 84(1):19–26. Available from: https://doi.org/10.1016/0091-6749(89)90173-5
47. Crooks SW, Stockley RA (1998) Leukotriene B4. Int J Biochem Cell Biol 30(2):173–178. https://doi.org/10.1016/B978-008055232-3.62930-7
48. Editors (2008) xPharm: the comprehensive pharmacology reference. Elsevier, p 1
49. Rucker D, Dhamoon AS (2022) Physiology, thromboxane A2. InStatPearls

50. Swartz HM, Mason RP, Hogg N (2005) Free radicals and medicine. Biomedi EPR Part A Free Radicals [Internet] 23:25–74. Available from: https://doi.org/10.1007/0-387-26741-7_353
51. Higgs GA, Moncada S, Salmon JA, Seager K (1983) The source of thromboxane and prostaglandins in experimental inflammation. Br J Pharmacol [Internet] 79(4):863–868. Available from: https://doi.org/10.1111/j.1476-5381.1983.tb10530.x
52. Swartz HM, Mason RP, Hogg N (2005) Free radicals and medicine. Biomed EPR Part A Free Radicals [Internet] 23:25–74. Available from: https://doi.org/10.1007/0-387-26741-7_353
53. Dennis EA, Norris PC (2015) Eicosanoid storm in infection and inflammation. Nat Rev Immunol 15(8):511–523. https://doi.org/10.1038/nri3859
54. Landino LM, Crews BC, Timmons MD, Morrow JD, Marnett LJ (1996) Peroxynitrite, the coupling product of nitric oxide and superoxide, activates prostaglandin biosynthesis. Proc Natl Acad Sci USA 93(26):15069–15074. https://doi.org/10.1073/pnas.93.26.15069
55. Werz O (2002) 5-lipoxygenase: cellular biology and molecular pharmacology. Curr Drug Targets Inflamm Allergy 1(1):23–44. https://doi.org/10.2174/1568010023344959
56. Ashique S, Mishra N, Mantry S, Garg A, Kumar N, Gupta M et al (2025) Crosstalk between ROS-inflammatory gene expression axis in the progression of lung disorders. Naunyn Schmiedebergs Arch Pharmacol [Internet] 398(1):417–448. Available from: https://doi.org/10.1007/s00210-024-03392-1
57. Lumeng CN, Saltiel AR (2011) Inflammatory links between obesity and metabolic disease. J Clin Invest 121(6):2111–2117. https://doi.org/10.1172/JCI57132
58. Ye J (2011) Adipose tissue vascularization: its role in chronic inflammation. Curr Diab Rep 11(3):203–210. https://doi.org/10.1007/s11892-011-0183-1
59. Li P, Oh DY, Bandyopadhyay G, Lagakos WS, Talukdar S, Osborn O, Johnson A, Chung H, Maris M, Ofrecio JM, Taguchi S, Lu M, Olefsky JM (2015) LTB4 promotes insulin resistance in obese mice by acting on macrophages, hepatocytes and myocytes. Nat Med 21(3):239–247. https://doi.org/10.1038/nm.3800
60. Robertson RP (2004) Chronic oxidative stress as a central mechanism for glucose toxicity in pancreatic islet beta cells in diabetes. J Biol Chem 279(41):42351–42354. https://doi.org/10.1074/jbc.R400019200
61. Madamanchi NR, Vendrov A, Runge MS (2005) Oxidative stress and vascular disease. Arterioscler Thromb Vasc Biol 25(1):29–38. https://doi.org/10.1161/01.ATV.0000150649.39934.13
62. Liu Z, Ren Z, Zhang J, Chuang CC, Kandaswamy E, Zhou T, Zuo L (2018) Role of ROS and nutritional antioxidants in human diseases. Front Physiol 9:477. https://doi.org/10.3389/fphys.2018.00477
63. Costa ML, Sales CH, Neto JV, Sarti FM, Rogero MM, Fisberg RM (2025) Vitamin E is associated with inflammatory biomarkers and metabolic syndrome: insights from a population study. Nutrit Res (New York, NY) 140:161–172. https://doi.org/10.1016/j.nutres.2025.06.009
64. Halasi M, Wang M, Chavan TS, Gaponenko V, Hay N, Gartel AL (2013) ROS inhibitor N-acetyl-L-cysteine antagonizes the activity of proteasome inhibitors. Biochem J [Internet] 454(2):201–208. Available from: https://doi.org/10.1042/BJ20130282
65. Bhatti JS, Bhatti GK, Reddy PH (2017) Mitochondrial dysfunction and oxidative stress in metabolic disorders – a step towards mitochondria based therapeutic strategies. Biochimica et Biophysica Acta (BBA) 1863(5):1066–1077. https://doi.org/10.1016/j.bbadis.2016.11.010
66. Oyewole AO, Birch-Machin MA (2015) Mitochondria-targeted antioxidants. FASEB J 29(12):4766–4771. https://doi.org/10.1096/fj.15-275404
67. Wang DK, Zheng HL, Zhou WS, Duan ZW, Jiang SD, Li B, Zheng XF, Jiang LS (2022) Mitochondrial dysfunction in oxidative stress-mediated intervertebral disc degeneration. Orthop Surg 14(8):1569–1582. https://doi.org/10.1111/os.13302
68. Smith WL, Murphy RC (2015) The eicosanoids. In: Elsevier eBooks [Internet], pp 259–296. Available from: https://doi.org/10.1016/b978-0-444-63438-2.00009-2
69. Wisastra R, Dekker FJ (2014) Inflammation, cancer and oxidative lipoxygenase activity are intimately linked. Cancer 6(3):1500–1521. https://doi.org/10.3390/cancers6031500

70. Kar A, Ghosh P, Patra P, Chini DS, Nath AK, Saha JK et al (2023) Omega-3 fatty acids mediated cellular signaling and its regulation in human health. Clin Nutrit Open Sci [Internet] 52:72–86. Available from: https://doi.org/10.1016/j.nutos.2023.10.004
71. Calder PC (2013) Omega-3 polyunsaturated fatty acids and inflammatory processes: nutrition or pharmacology?: Omega-3 fatty acids and inflammation. Br J Clin Pharmacol [Internet] 75(3):645–662. Available from: https://doi.org/10.1111/j.1365-2125.2012.04374.x
72. Rudrapal M, Khairnar SJ, Khan J, Dukhyil AB, Ansari MA, Alomary MN, Alshabrmi FM, Palai S, Deb PK, Devi R (2022) Dietary polyphenols and their role in oxidative stress-induced human diseases: insights into protective effects, antioxidant potentials and mechanism(s) of action. Front Pharmacol 13:806470. https://doi.org/10.3389/fphar.2022.806470
73. Mukherjee S, Chopra H, Goyal R, Jin S, Dong Z, Das T, Bhattacharya T (2024) Therapeutic effect of targeted antioxidant natural products. Discover nano 19(1):144. https://doi.org/10.1186/s11671-024-04100-x
74. Carr A, Maggini S (2017) Vitamin C and immune function. Nutrients [Internet] 9(11):1211. Available from: https://doi.org/10.3390/nu9111211
75. Traber MG, Atkinson J (2007) Vitamin E, antioxidant and nothing more. Free Radic Biol Med [Internet] 43(1):4–15. Available from: https://doi.org/10.1016/j.freeradbiomed.2007.03.024
76. Dekhuijzen PNR (2004) Antioxidant properties of N-acetylcysteine: their relevance in relation to chronic obstructive pulmonary disease. Eur Respir J [Internet] 23(4):629–636. Available from: https://doi.org/10.1183/09031936.04.00016804
77. Smith RAJ, Hartley RC, Murphy MP (2011) Mitochondria-targeted small molecule therapeutics and probes. Antioxid Redox Signal [Internet] 15(12):3021–3038. Available from: https://doi.org/10.1089/ars.2011.3969
78. Kelso GF, Porteous CM, Coulter CV, Hughes G, Porteous WK, Ledgerwood EC et al (2001) Selective targeting of a redox-active ubiquinone to mitochondria within cells. J Biol Chem [Internet] 276(7):4588–4596. Available from: https://doi.org/10.1074/jbc.m009093200
79. Snow BJ, Rolfe FL, Lockhart MM, Frampton CM, O'Sullivan JD, Fung V et al (2010) A double-blind, placebo-controlled study to assess the mitochondria-targeted antioxidant MitoQ as a disease-modifying therapy in Parkinson's disease. Mov Disord [Internet] 25(11):1670–1674. Available from: https://doi.org/10.1002/mds.23148
80. Mhmedical.com. [cited 2025 Sept 19]. Available from: https://accessmedicine.mhmedical.com/content.aspx?bookid=2189&sectionid=170271972
81. Peters-Golden M, Henderson WR Jr (2007) Leukotrienes. N Engl J Med [Internet] 357(18):1841–1854. Available from: https://doi.org/10.1056/nejmra071371
82. Calder PC (2015) Marine omega-3 fatty acids and inflammatory processes: effects, mechanisms and clinical relevance. Biochim Biophys Acta Mol Cell Biol Lipids [Internet] 1851(4):469–484. Available from: http://dx.doi.org/https://doi.org/10.1016/j.bbalip.2014.08.010
83. Kromhout D, Yasuda S, Geleijnse JM, Shimokawa H (2012) Fish oil and omega-3 fatty acids in cardiovascular disease: do they really work? Eur Heart J [Internet] 33(4):436–443. Available from: https://doi.org/10.1093/eurheartj/ehr362
84. Pandey KB, Rizvi SI (2009) Plant polyphenols as dietary antioxidants in human health and disease. Oxid Med Cell Longev [Internet] 2(5):270–278. Available from: https://doi.org/10.4161/oxim.2.5.9498
85. Vinayagam R, Xu B (2015) Antidiabetic properties of dietary flavonoids: a cellular mechanism review. Nutr Metab Lond [Internet] 12(1) Available from: https://doi.org/10.1186/s12986-015-0057-7
86. Hotamisligil GS (2006) Inflammation and metabolic disorders. Nature 444(7121):860–867. https://doi.org/10.1038/nature05485
87. Spite M, Clària J, Serhan CN (2014) Resolvins, specialized proresolving lipid mediators, and their potential roles in metabolic diseases. Cell Metab [Internet] 19(1):21–36. Available from: https://doi.org/10.1016/j.cmet.2013.10.006

88. Sanyal AJ (2005) Mechanisms of disease: pathogenesis of nonalcoholic fatty liver disease. Nat Clin Pract Gastroenterol Hepatol 2(1):46–53. https://doi.org/10.1038/ncpgasthep0084
89. Dalle-Donne I, Rossi R, Colombo R, Giustarini D, Milzani A (2006) Biomarkers of oxidative damage in human disease. Clin Chem 52(4):601–623. https://doi.org/10.1373/clinchem.2005.061408
90. Funk CD (2001) Prostaglandins and leukotrienes: advances in eicosanoid biology. Science (New York, N.Y.) 294(5548):1871–1875. https://doi.org/10.1126/science.294.5548.1871
91. Grosser T, Smyth E, FitzGerald G (2017) Pharmacotherapy of inflammation, fever, pain, and gout. In: Brunton LL, Hilal-Dandan R, Knollmann BC (eds) Goodman & Gilman's: the pharmacological basis of therapeutics, 13e. McGraw-Hill Education. https://accessmedicine.mhmedical.com/content.aspx?bookid=2189&sectionid=170271972
92. Pergola C, Werz O (2010) 5-lipoxygenase inhibitors: a review of recent developments and patents. Expert Opin Ther Pat 20(3):355–375. https://doi.org/10.1517/13543771003602012
93. Murphy MP (2014) Antioxidants as therapies: can we improve on nature? Free Radic Biol Med 66:20–23. https://doi.org/10.1016/j.freeradbiomed.2013.04.010
94. Serhan CN, Chiang N, Van Dyke TE (2008) Resolving inflammation: dual anti-inflammatory and pro-resolution lipid mediators. Nat Rev Immunol 8(5):349–361. https://doi.org/10.1038/nri2294
95. Kitano H (2002) Systems biology: a brief overview. Science (New York, N.Y.) 295(5560):1662–1664. https://doi.org/10.1126/science.1069492
96. Brownlee M (2005) The pathobiology of diabetic complications: a unifying mechanism. Diabetes 54(6):1615–1625. https://doi.org/10.2337/diabetes.54.6.1615
97. Nicholson JK, Holmes E, Kinross J, Burcelin R, Gibson G, Jia W, Pettersson S (2012) Host-gut microbiota metabolic interactions. Science (New York, N.Y.) 336(6086):1262–1267. https://doi.org/10.1126/science.1223813
98. Wallace DC (2018) Mitochondrial genetic medicine. Nat Genet 50(12):1642–1649. https://doi.org/10.1038/s41588-018-0264-z
99. Tilg H, Zmora N, Adolph TE, Elinav E (2020) The intestinal microbiota fuelling metabolic inflammation. Nat Rev Immunol 20(1):40–54. https://doi.org/10.1038/s41577-019-0198-4
100. Torchilin VP (2006) Multifunctional nanocarriers. Adv Drug Deliv Rev 58(14):1532–1555. https://doi.org/10.1016/j.addr.2006.09.009
101. Doudna JA, Charpentier E (2014) Genome editing. The new frontier of genome engineering with CRISPR-Cas9. Science (New York, N.Y.) 346(6213):1258096. https://doi.org/10.1126/science.1258096

Chapter 8
Cellular Cytokine Storm-Mediated Inflammation in Metabolic Disorders

Bratati Bandyopadhyay, Souparna Sasmal, Ayantika Kanrar, and Shreya Adhikary

8.1 Introduction

The cytokine storm (CS), also known as hypercytokinemia, has become a well-known concept in the medical field as well as in the popular press. It is used to define an unregulated and over-reactive immune reaction where massive amounts of pro-inflammatory cytokines are released. Although the term is widely used, a clear molecular definition that can distinguish CS and physiological inflammation has been difficult to establish. The major differences are the cytokines, their levels, and the thresholds that change protective inflammation into pathological hyperinflammation [1]. Initially used in the early 1990s to describe graft-versus-host disease (GvHD), the term was also later applied to a wide range of infectious diseases including influenza virus infection, variola virus, group A Streptococcus (GAS), Epstein-Barr virus (EBV)-associated hemophagocytic lymphohistiocytosis (HLH), and the severe acute respiratory syndrome coronavirus (SARS-CoV) [2]. This was further reinforced by the occurrence of highly pathogenic avian influenza (H5N1) in 2005 that underlined the role of CS in clinical practice. Recently, the COVID-19 pandemic caused by SARS-CoV-2 raised new interest in CS because of its close correlation with severe disease and death [3]. Outside of the infectious setting, hyperinflammatory states (cytokine storm) have been reported in metabolic disease, particularly obesity. In this environment, macrophages accumulate around dying adipocytes, forming crown-like structures (CLS), that promote tissue fibrosis and metabolic disturbances. The kidney and liver, among other organs, develop CLS-like pathology after injury and steatohepatitis, respectively. These findings highlight the complexity of the interplay between immunity and metabolism, through receptors of the innate immune system and lipid-binding proteins including the

B. Bandyopadhyay (✉) · S. Sasmal · A. Kanrar · S. Adhikary
Department of Pharmaceutical Technology, Bharat Technology, Jadurberia, Uluberia, Howrah, West Bengal, India

S. Ashique, N. Mishra (eds.), *Reactive Oxygen Species in Metabolic Inflammation*,
https://doi.org/10.1007/978-3-032-11752-6_8

adipocyte/macrophage fatty acid-binding proteins (FABPs) aP2 and MAL1 [4]. The absence of FABPs in animal models induces metabolic and anti-inflammatory effects similar to those of peroxisome proliferator-activated receptor (PPAR) gamma ligands. In addition, FABPs control insulin responsiveness and inflammatory pathway through the activation of c-Jun N-terminal Kinase (JNK) and Inhibitor of κB Kinase (IKK)). Removal of them confers resistance to hallmark signs of metabolic syndrome atherosclerosis, fatty liver disease, insulin resistance, and visceral obesity [5].

8.1.1 Molecular and Cellular Mechanisms of Cytokine Storm

Cytokine storm (CS) is a potentially fatal hyperinflammatory condition that occurs when the release of cytokines is uncontrolled because of an imbalance of the immune system. It starts with the activation of pattern recognition receptors (PRRs), e.g., Toll-like receptors (TLRs) by a pathogen or injury, activating downstream signaling cascades, e.g., MyD88-MAPK (Mitogen-Activated Protein Kinase) and TRIF (TIR domain-containing adaptor inducing IFN-β)-IRF3 pathways, which culminate in the activation of nuclear factor kappa B (NF-kB) and massive transcription of pro-inflammatory cytokines. This reaction is also enhanced by JAK-STAT signaling (especially through Interleukin-6 (IL-6) and IFN-γ), NLRP3 inflammasome engagement resulting in IL-1β and IL-18 maturation, and PANoptosis, a cell-death pathway that combines pyroptosis, apoptosis, and necroptosis [6]. Several immune and non-immune cells (macrophages, neutrophils (through NETs), T lymphocytes, dendritic cells, adipocytes, and Kupffer cells) also contribute to the cytokine milieu, and genetic predispositions include those of PRF1, IL-6, and NLRC4 polymorphisms. CS is clinically seen in COVID-19, sepsis, CAR-T treatment, and hemophagocytic lymphohistiocytosis (HLH) and is a powerful predictor of mortality. Among the most important cytokines, interleukin-6 (IL-6) is a versatile mediator of the acute-phase response and coagulation abnormalities; its overexpression critically contributes to tissue damage, whereas its blockade with tocilizumab mitigates inflammation but increases the risk of infections [7]. Tumor necrosis factor-alpha (TNF-α) is secreted primarily by macrophages and mediates innate and adaptive immunity, vascular and metabolic processes, and is involved in rheumatoid arthritis, Alzheimer's, atherosclerosis, and sepsis; its blockage improves inflammation and metabolic profiles. Interleukin-1beta (IL-1β), which is cleaved by caspase-1 in the inflammasome, does not only coordinate host defense but also causes pancreatic 8-cell dysfunction, insulin resistance, and type 2 diabetes [8]. Other cellular contributors include macrophages, which secrete TNF-α, IL-1β, and chemokines, exhibit M1/M2 plasticity, and contribute to systemic inflammation; T cells, which mediate adaptive immunity through Th1 and Th2 cytokines and contribute to liver damage in alcohol-related disease; adipocytes, which release pro-inflammatory mediators in obesity and recruit immune cells to form crown-like structures; and Kupffer cells, which serve as hepatic sentinels but also drive hepatitis. Existing

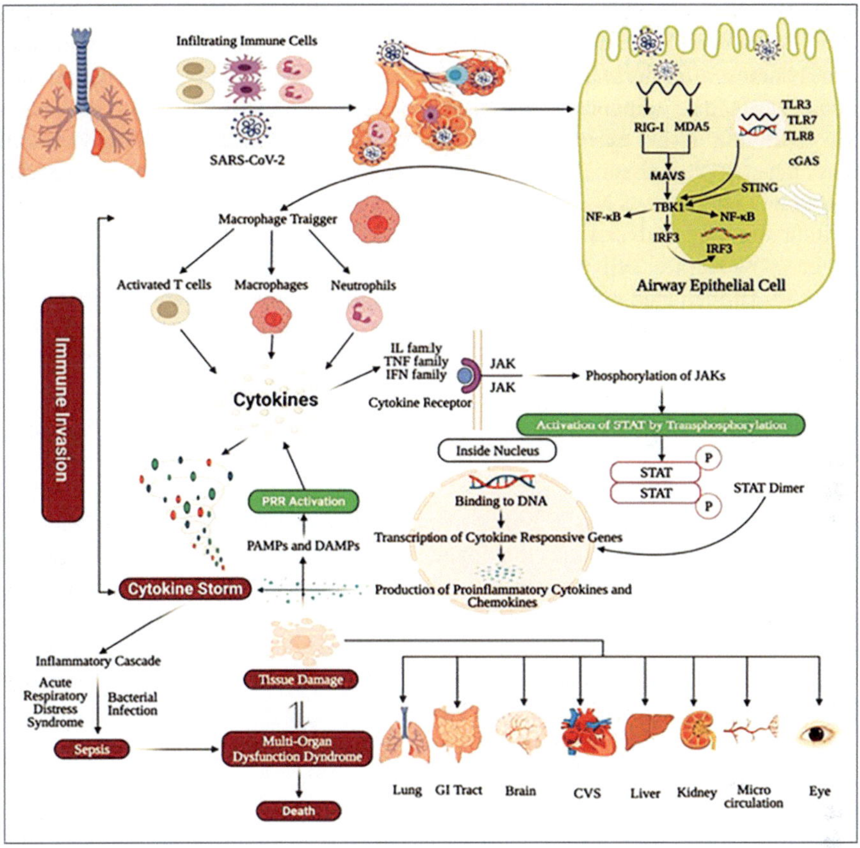

Fig. 8.1 Mechanism of cytokine storm induced by SARS-CoV-2 infection

treatment plans aim to inhibit IL-6, IL-1, and JAK pathways, with new treatment options to include stem cell-based therapy and biomarker-guided treatment regimens [9] (Fig. 8.1).

8.2 Chronic Low-Grade Inflammation in Metabolic Disorders

Chronic low-grade inflammation (metaflammation) is asymptomatic and closely connected with visceral adipose tissue (VAT) unlike acute inflammation. VAT is metabolically active producing TNF-α, IL-6, and leptin, which attracts macrophages and interferes with insulin signaling. The adaptive immune cells, such as B and T lymphocytes, also influence the activity of macrophages, which increases insulin resistance and vascular dysfunction. The outcome is the development of type 2 diabetes, atherosclerosis, and cardiovascular disease [10]. Immune regulation is actively involved in metabolic organs, such as adipose tissue, liver, skeletal muscle, and pancreas. In normal conditions, anti-inflammatory immune cells provide homeostasis (e.g., adiponectin secretion, pathogen clearance by Kupffer cells, myokine secretion during exercise). Obesity destabilizes this balance leading to adipocyte hypertrophy, tissue hypoxia, and increased pro-inflammatory adipokines/cytokines. This attracts M1 macrophages and CD8+ T cells, causing a continuation of inflammation. In liver, non-alcoholic fatty liver disease (NAFLD) and fibrosis are triggered by Kupffer cell activation and involvement of the stellate cells. In the pancreas, IL-1β-mediated inflammation kills β-cells and systemic inflammation induces skeletal muscle insulin resistance [11].

8.3 Cytokine Storm and Obesity-Related Inflammation

In obesity, the adipose tissue grows mainly by hypertrophy, not hyperplasia, and when adipocyte size surpasses the oxygen diffusion limit (~100 m), local hypoxia, necrosis, and inflammation ensue, resulting in systemic metabolic dysfunction. This hypertrophy is a chronic cause of insulin resistance, type 2 diabetes and cardiovascular disease. Enlarged adipocytes secrete monocyte chemoattractant protein-1 (MCP-1/CCL2) which attracts macrophages that surround necrotic adipocytes to form crown-like structures (CLS). These infiltrating macrophages secrete pro-inflammatory cytokines including TNF-α, IL-6, and IL-1β, which enhance local inflammation and insulin resistance; eliminating adipose tissue macrophages will increase insulin sensitivity, and thus should be therapeutic targets. Of these cytokines, IL-6 is dualistic, where adipocyte-secreted IL-6 promotes macrophage infiltration and metabolic dysfunction, and exercise-induced skeletal muscle IL-6 exhibits anti-inflammatory effects and improves insulin sensitivity, and its effect is determined by the mode of signaling—trans-signaling is pro-inflammatory, and classic signaling is anti-inflammatory [12]. Similarly, overexpression of MCP-1 in adipose tissue in obese individuals stimulates the recruitment of monocytes and the accumulation of macrophages, which lead to persistent inflammation, and its inhibition results in the enhancement of insulin sensitivity in animal models, hence, a potential therapeutic target. TNF-α, another central mediator, is a regulator of immune responses, inflammation, apoptosis, tissue remodeling, but its dysregulation leads to insulin resistance, dyslipidemia, and cardiovascular risk. In addition to metabolic disease, TNF-α is also relevant to neuroinflammation and has been found to play a role in the Alzheimer's disease, highlighting the breadth of its effects on systemic health [13].

8.4 Cytokine Storm in Type II Diabetes Mellitus

A cytokine storm is a loss of control of the excessive release of pro-inflammatory cytokines. In T2DM, this will be more easily achieved because of the underlying condition of chronic low-grade inflammation, immune dysregulation, and metabolic dysfunction. Increased cytokines, such as IL-1, IL-6, and TNF-α, exacerbate insulin resistance, poor glycemic control, tissue damage, and an increased vulnerability to infections including COVID-19 [14].

In T2DM, immune dysregulation is the cause of cytokine storm and chronic inflammation, with limited T and NK cell populations combined with sustained high levels of pro-inflammatory cytokines, an imbalance of the immune system [15]. The result is enhanced by hyperglycemia and metabolic stress, which contribute to increased ROS production in mitochondria and hypoxia-inducible factor-1 alpha (HIF-1 alpha) activity, which drives excessive production of cytokines. This forms a vicious cycle wherein excessive cytokines exacerbate insulin resistance and the dysfunction of pancreatic β-cells and predispose patients to ketoacidosis and infection-related organ failure [16]. These key inflammatory mediators directly interfere with insulin signaling; TNF-α promotes insulin receptor substrate-1 (IRS-1) serine phosphorylation and activates IKKbeta/NF-kappaB and JNK pathways, IL-1β reduces insulin sensitivity and secretion in β-cells under hyperglycemic conditions, and IL-6 activates JAK-STAT signaling to increase SOCS3 expression and inhibit GLUT-4 (Glucose Transporter) and IRS-1, leading to decreased glucose uptake [17]. Collectively, these cytokine-mediated changes result in degradation of the IRS proteins, activation of negative signaling pathways, impaired glucose transport in muscle and adipose tissue and increased hepatic glucose production. Chronic glucotoxicity, lipotoxicity, and cytokine exposure also contribute to the stress and apoptosis of β-cells as a result of ER dysfunction and oxidative stress with β-cells being particularly susceptible due to their poor antioxidant defenses. Prolonged ER stress leads to the unfolded protein response (UPR) that, when not resolved, results in apoptosis through CHOP (C/EBP Homologous, a crucial transcription factor induced by endoplasmic reticulum (ER) stress) and other pro-apoptotic proteins [18]. In the meantime, the mitochondrial ROS production and cytokine amplification through IL-1β, TNF-α, IFN-γ, and nitric oxide production exacerbate the NF-kappa B and STAT1 signaling, which leads to the mitochondrial injury and activation of apoptosis. The β-cells are killed in several ways, including the intrinsic pathway of mitochondrial apoptosis (caspase activation and release of cytochrome c), unresolved ER stress, death receptor activation by cytokines, and finally, apoptosis decreases the mass of β-cells, exacerbates hyperglycemia, and increases T2DM progression. The potential treatments to preserve β-cells include therapeutic approaches used to counter oxidative stress, ER stress, inflammation, and autophagy [19].

8.5 Cytokine-Mediated Hepatic Inflammation in NAFLD and NASH

Non-alcoholic fatty liver disease (NAFLD) and its more severe manifestation non-alcoholic steatohepatitis (NASH) are significant healthcare problems across the world and are closely associated with obesity, insulin resistance, diabetes, and metabolic syndrome. They are propagated by lipid accumulation, oxidative stress, immune activation, and cytokine-mediated hepatic inflammation. The liver-resident macrophages, Kupffer cells, are the key players in sensing lipotoxic signals and microbial products via pattern recognition receptors [20]. This stimulates NF-kB and inflammasome signaling, which results in the secretion of pro-inflammatory cytokines including TNF-α, IL-6, and IL-1β. These cytokines increase the damage to the hepatocytes, attract immune cells, and stimulate the hepatic stellate cells (HSCs), which secrete extracellular matrix proteins and contribute to fibrosis. TNF-α triggers apoptosis, insulin resistance, and inflammation by signaling through TNFR1 and TNFR2, thereby influencing immune cell recruitment and tissue remodeling [21]. IL-6 signals through the JAK/STAT3 pathway, where classic signaling promotes regeneration, while trans-signaling drives inflammation, fibrosis, and metabolic imbalance. High IL-6 levels are associated with the aggravation of steatosis and fibrosis. Pyroptosis caused by IL-1β through inflammasomes including NLRP3 (NOD-, LRR-, and pyrin domain-containing protein 3, a key sensor in the innate immune system that forms the NLRP3 inflammasome), and monocyte recruitment by chemokines such as MCP-1 maintain the injury process [22]. Cytokines also act in synergy with TGF-β to increase HSC activation and deposition of collagen. Oxidative and ER stress predispose hepatocytes to cytokine-induced injury by liberating more danger signals that sustain inflammation. This produces a loop of lipotoxicity, cytokine production, immune infiltration, and fibrosis to progress NAFLD to NASH. Therapies which target cytokines including TNF-α or IL-6 inhibitors are promising in reducing fibrosis and inflammation but have pleiotropic effects which make them difficult to treat. Such selective modulation as inhibiting IL-6 trans-signaling without blocking protective pathways could have safer effects. In sum, inflammation mediated by cytokines is a critical factor in NAFLD/NASH and an important focus of treatment development [23].

8.6 Role of Gut Microbiota in Cytokine-Driven Inflammation

The gut microbiota is pivotal in cytokine-mediated inflammation and development of chronic liver diseases including NAFLD and NASH. It is an active ecosystem that controls metabolism, immunity, and barrier health [24]. Dysbiosis or microbial imbalance decreases the number of beneficial microbes such as *Lactobacillus* and *Bifidobacterium* and increases harmful microbes such as *Enterobacteriaceae* and *Clostridium*. This imbalance decreases the production of short-chain fatty acid

(SCFA), compromises the intestinal barrier, and results in a leaky gut. Consequently, microbial products like LPS (Lipopolysaccharide), peptidoglycans, and bacterial DNA get into the portal circulation and into the liver. LPS plays a very crucial role since it stimulates TLR4 in Kupffer cells, hepatocytes, and stellate cells. This activates NF-kB and MAPK pathways, resulting in TNF-α, IL-6, IL-1β, and MCP-1 release [25]. These cytokines cause hepatocyte injury, fibrosis, and systemic inflammation. TNF-α induces cell death and insulin resistance, IL-6 stimulates JAK/STAT3 signal transduction, and IL-1β activates inflammasomes. The hepatic stellate cells are transformed into myofibroblasts, which generate collagen and fibrosis. Endothelial activation is also a mechanism of LPS recruitment of immune cells. In addition to the liver, LPS-induced cytokines modify adipose tissue by changing macrophage to pro-inflammatory M1 and interfering with adipokines, which exacerbates insulin resistance [26]. Dysbiosis also interferes with bile acid metabolism further facilitating the production of toxic products such as ethanol, which exacerbates oxidative stress. This connection is validated by animal models since germ-free or TLR4-deficient mice exhibit less liver damage. Higher levels of LPS, gut permeability, and cytokine activation are also observed in the human studies of NAFLD patients. Such therapies include probiotics, prebiotics, synbiotics, antibiotics, TLR4 antagonists, and bile acid receptor modulators, all of which enhance barrier function of the gut and reduce inflammation [27]. Nevertheless, microbiota-inflammation relations are complicated, as some metabolites, such as SCFAs and indoles, are anti-inflammatory. Therefore, it is necessary to modulate, and not suppress, the targeted modulation. Therefore, dysbiosis, endotoxemia, and LPS-TLR4 signaling present an essential connection between the gut and liver, perpetuating cytokine-induced inflammation and fibrosis in NAFLD/NASH, and is a potential target of precision therapy [28].

8.7 Clinical Implications and Biomarkers

The clinical implications of cytokine biology in metabolic disease have become an important area of translational research, with the mediators of inflammation produced in adipose tissue, liver, skeletal muscle, gut, and immune cells not only being drivers of pathophysiology, but also measurable indicators of disease onset, progression, and prognosis, making circulating cytokines highly valuable as diagnostic and prognostic biomarkers [29]. The idea that metabolic disease is synonymous with chronic low-grade inflammation—often termed "metaflammation"—has shifted the clinical focus toward the molecular fingerprints of this low-grade state of inflammation, especially involving cytokines such as tumor necrosis factor-alpha (TNF-α), interleukin-6 (IL-6), interleukin-1-beta (IL-1β), C-reactive protein (CRP, an IL-6-induced acute-phase reactant), monocyte chemoattractant [30]. The circulating cytokine profiles in clinical practice can be used both as a diagnostic feature of active tissue inflammation and as a prognostic parameter predicting the development of type 2 diabetes from subclinical insulin resistance, simple steatosis to non-alcoholic steatohepatitis (NASH), obesity to cardiovascular diseases (CVD), and neurodegenerative disorders [31]. Higher TNF-α, e.g., has been repeatedly associated with impaired insulin receptor signaling in adipose tissue and skeletal muscle by serine phosphorylation of insulin receptor substrate-1 (IRS-1), and its circulating levels are correlated with obesity severity, indices of insulin resistance (HOMA-IR), and the risk of type 2 diabetes, making TNF-α not only a mechanistic driver, but also a potential biomarker of disease activity [32]. Similarly, IL-6 exhibits dual and context-dependent clinical effects: adipose-derived IL-6 promotes hepatic Creactive protein synthesis and insulin resistance, whereas exercise-induced skeletal muscle IL-6 exerts anti-inflammatory and insulin-sensitizing effects; however, chronically elevated plasma IL-6 levels are consistently associated with the progression of metabolic syndrome, atherosclerosis, and Non-alcoholic fatty liver disease (NAFLD/NASH), making it a predictive biomarker of cardiometabolic disease [33]. Another central cytokine, IL-1β, is secreted upon inflammasome activation and acts directly on the pancreatic beta-cells by causing apoptosis and decreased insulin secretion; in patients, elevated serum IL-1β levels are associated with progression of impaired glucose tolerance to full diabetes, and blockade of IL-1 in trials significantly improves glycemic control. Another important cytokine, IL-1β, is released during the activation of inflammasome and directly interferes with the functionality of pancreatic beta-cells by increasing apoptosis and reducing insulin secretion; in patient populations, higher levels of circulating IL-1β are associated with progression to diabetes following development of impaired glucose tolerance, and in interventional trials, IL-1 antagonists (anakinra, canakinumab) have proven effective at improving glycemic control, providing further support to the biomarker role [34]. MCP-1, a chemokine highly expressed in hypertrophic adipose tissue, mediates recruitment of macrophages and the formation of the crown-like structure around necrotic adipocytes; in clinical practice, plasma MCP-1 levels are associated with visceral adiposity and predict systemic insulin resistance, whereas experimental models demonstrate that MCP-1 inhibitors improve insulin sensitivity, which reflects its prognostic and therapeutic importance. In the NAFLD and NASH scenario, Kupffer cell-derived cytokines including TNF-α, IL-6, and transforming growth factor-beta (TGF-beta) are mediators of hepatocyte damage, steatosis, and fibrosis; thus, their blood concentrations (along with secondary acute-phase proteins such as CRP and serum amyloid A) are under investigation as noninvasive markers that may replace or complement invasive liver biopsies to assess disease severity [35]. Of note, high-sensitivity CRP (hsCRP), which is not a cytokine per se but an IL-6-driven acute-phase protein, has become clinically accepted as a measure of systemic inflammation and cardiovascular risk, and its addition into risk prediction algorithms improves prognostic performance of myocardial infarction, stroke, and heart failure in obese and diabetic patients. In addition to single cytokines, panels of biomarkers that combine several inflammatory mediators are emerging to enhance the accuracy of diagnosis; e.g., a panel of TNF-α, IL-6, MCP-1, and adiponectin has been demonstrated to differentiate between metabolically healthy and unhealthy obesity, and multiplex cytokine profiling can identify patients with NAFLD who are at risk of developing advanced fibrosis [36]. A prognostic viewpoint is that longitudinal studies show that persistent high cytokine levels precede and predict metabolic degradation: IL-6 and TNF-α levels predict future development of type 2 diabetes, IL-18 levels forecast cardiovascular events independent of conventional risk factors, and MCP-1 predicts worsening of hepatic fibrosis in NAFLD [37]. In addition, inflammatory cytokines may also be used as markers of treatment response; weight-reducing surgical or lifestyle interventions are associated with marked decreases in circulating TNF-α, IL-6, and MCP-1, which correlate with improved insulin sensitivity and lipid profiles, pointing to their potential use as measures of treatment effect [38]. Moreover, clinical trials of anti-cytokine medications, including TNF-α blockers (e.g., etanercept, infliximab) and IL-1 inhibitors, and IL-6 receptor antagonists (e.g., tocilizumab) not only confirm cytokine pathways as viable targets of therapy but also indicate the utility of these molecules as biomarkers to stratify patients and predict outcomes. In cardiovascular disease, inflammatory biomarkers (IL-6, TNF-α, CRP, and soluble TNF receptors (sTNFR1 and sTNFR2)) add prognostic information about future events in addition to lipid profiles and blood pressure measurements, and thus metabolic and inflammatory risk assessment are combined [39]. In neurodegenerative diseases related to metabolic dysfunction, like Alzheimer's disease, TNF-α and IL-6 in circulation have been correlated with deteriorating cognitive performance, indicating a possible role as early markers of metabolically related neuroinflammation. The clinical utility of cytokine biomarkers is also facilitated by technological advances: multiplex immunoassays, high-sensitivity ELISA, and proteomic-based platforms can quantify dozens of cytokines simultaneously using small plasma samples, which can be used as cytokine signatures to support personalized medicine strategies [40]. Combining cytokine profiling with genomics and metabolomics is resulting in multi-omics biomarker discovery and cytokine profiles are combined with genetic risk scores, metabolite signatures, and microbiome composition to refine diagnostic and prognostic models of complex metabolic diseases [41]. Although these findings are encouraging, clinical translation has been complicated by biological variability of cytokine concentration attributed to circadian rhythm, acute infections, comorbidities, and medication use, and technical challenges including assay standardization and reproducibility. However, there are still a number of obstacles to cytokine profiling becoming a normal component of clinical practice in endocrinology, hepatology, cardiology, and metabolic medicine, but these are gradually being overcome by concerted attempts to harmonize the methodology and establish reference ranges [42]. In brief, circulating cytokines are both mechanistic and measurable biomarkers of metabolic inflammation and have diagnostic and prognostic utility in obesity, type 2 diabetes, NAFLD/NASH, and cardiovascular disease. Their prognostic value in the disease progression and severity highlights their clinical significance, and the potential of their combined use with more advanced multi-omics and precision medicine in early detection, risk stratification, and monitoring of patients with metabolic disorders is promising [43].

8.8 Future Perspectives

Although there has been progress in the study of cytokines, the role they play in chronic inflammatory disease like rheumatoid arthritis (RA) has been slow. Although early research implied the presence of infectious triggers, later research failed to support the notion [44]. However, recent findings on Toll-like receptors (TLRs) have rekindled the interest in pathogen-immune interactions as shown in Table 8.1. The TLRs serve as important sensors that discriminate between self and non-self-molecular patterns and mediate various aspects of immunity, such as cytokine gene expression. Notably, TLRs are not only stimulated by microbial products but also by endogenous tissue damage products, so they have the potential to perpetuate inflammation even in the absence of infection by pathogens [45]. This knowledge supports the feasibility of TLR-mediated cytokine dysregulation in RA and other chronic inflammatory diseases, which can become a new route to treatment. Circulating cytokines may be useful in the context of myeloproliferative neoplasms (MPNs) to characterize disease at diagnosis and may have prognostic implications

Table 8.1 Key insights into the role of Toll-like receptors (TLRs) and cytokines in rheumatoid arthritis (RA), chronic inflammation, and myeloproliferative neoplasms (MPNs), highlighting their clinical implications, prognostic value, and therapeutic potential

Focus area	Key insights	Clinical/research implications	Ref.
Role of TLRs in RA and chronic inflammation	TLRs act as sensors of microbial and endogenous ligands; regulate cytokine gene expression; sustain inflammation even without pathogens	Revive interest in pathogen-immune interactions; support hypothesis of TLR-driven cytokine dysregulation in RA; potential new therapeutic targets	[48]
Endogenous activation of TLRs	Endogenous ligands from tissue damage can activate TLRs, perpetuating sterile inflammation	Highlights role of non-infectious triggers in chronic inflammation; expands treatment strategies beyond anti-infective approaches	[49]
Cytokines in myeloproliferative neoplasms (MPNs)	Circulating cytokines may aid in disease characterization at diagnosis	Serve as diagnostic and prognostic biomarkers; help stratify patients based on cytokine profiles	[50]
Prognostic value in MPNs	Cytokine levels may predict disease progression and outcomes	Enables personalized risk assessment; guides therapeutic decision-making	[51]
Therapeutic monitoring with cytokine assays	JAK2 inhibitors shown to modulate cytokine levels in clinical settings	Cytokine assays may serve as real-time tools to monitor therapeutic efficacy and disease progression	[52]
Future therapeutic avenue	Targeting TLRs and cytokine pathways directly	May lead to innovative treatments for RA, MPNs, and other chronic inflammatory conditions	[53]

in predicting outcomes [46]. Moreover, there is clinical evidence that JAK2 inhibitors have the capability of altering circulating cytokines, and therefore, cytokine assays can be used as an effective means of monitoring therapeutic response and disease progression in real time [47](Table 8.1).

8.9 Conclusion

Cytokine-mediated inflammation is also a fundamental factor in the onset and development of various ailments such as cancer, autoimmune diseases, and metabolic diseases. In metabolic conditions like obesity, type 2 diabetes, and non-alcoholic fatty liver disease (NAFLD), the uncontrolled secretion of pro-inflammatory cytokines, commonly referred to as a cytokine storm, enhances immune dysfunction and promotes disease pathogenesis. Cytokines play a role in immune balance and tissue homeostasis under physiological conditions. But in cytokine storms, the excessive production of IL-6, IL-1β, TNF-α, and IFN-γ is released by hyperactivated immune cells i.e., macrophages, neutrophils, and lymphocytes. This unregulated cascade leads to systemic inflammation, endothelial dysfunction, and direct tissue damage exacerbating the already tenuous immune-metabolic balance of people with chronic metabolic diseases. Cytokine storms in these patients worsen insulin resistance, oxidative stress, and cellular damage in vital organs, including the pancreas, the liver, and adipose tissue, where low-grade inflammation is already present. This increases the threat of serious complications, such as cardiovascular disease, hepatic fibrosis, and metabolic crises. Moreover, immune dysregulation interferes with recovery of infections and other inflammatory insults further increasing morbidity and mortality. Due to such implications, treatment interventions aimed at cytokine signaling and inflammatory cascades are important to enhance metabolic health and clinical outcomes. These strategies can involve pharmacological inhibitors, lifestyle, probiotics, or new immunomodulatory agents that aim to bring balance back to the immune system and to preserve organ functionality. Additional studies of the molecular pathways connecting cytokine imbalance with the development of chronic diseases are necessary to create individualized treatment strategies. Cytokines are thus a biomarker and a therapeutic target-giving clinician's powerful tools to diagnose, prognose, monitor the disease, and optimize treatment across a broad spectrum of pathologies, including cancer and metabolic diseases.

References

1. Jarczak D, Nierhaus A (2022) Cytokine storm—definition, causes, and implications. IJMS 23:11740. https://doi.org/10.3390/ijms231911740
2. Nie J, Zhou L, Tian W, Liu X, Yang L, Yang X, Zhang Y, Wei S, Wang DW, Wei J (2025) Deep insight into cytokine storm: from pathogenesis to treatment. Sig Transduct Target Ther 10:112. https://doi.org/10.1038/s41392-025-02178-y

3. Di Guardo G (2025) Highly pathogenic avian influenza a (H5N1) virus: how far are we from a new pandemic? Veter Sci 12:566. https://doi.org/10.3390/vetsci12060566
4. Guria S, Hoory A, Das S, Chattopadhyay D, Mukherjee S (2023) Adipose tissue macrophages and their role in obesity-associated insulin resistance: an overview of the complex dynamics at play. Biosci Rep 43:BSR20220200. https://doi.org/10.1042/BSR20220200
5. Hotamisligil GS, Bernlohr DA (2015) Metabolic functions of FABPs—mechanisms and therapeutic implications. Nat Rev Endocrinol 11:592–605. https://doi.org/10.1038/nrendo.2015.122
6. Fan Y, Guan B, Xu J, Zhang H, Yi L, Yang Z (2023) Role of toll-like receptor-mediated pyroptosis in sepsis-induced cardiomyopathy. Biomed Pharmacother 167:115493. https://doi.org/10.1016/j.biopha.2023.115493
7. Shrestha S, Hong C-W (2023) Extracellular mechanisms of neutrophils in immune cell crosstalk. Immune Netw 23:e38. https://doi.org/10.4110/in.2023.23.e38
8. Jang D, Lee A-H, Shin H-Y, Song H-R, Park J-H, Kang T-B, Lee S-R, Yang S-H (2021) The role of tumor necrosis factor alpha (TNF-α) in autoimmune disease and current TNF-α inhibitors in therapeutics. IJMS 22:2719. https://doi.org/10.3390/ijms22052719
9. Hemmatian G, Rostamzadeh D, Baghaei K, Shabani M (2025) Monocytes and macrophages as unique cellular compartments governing non-alcoholic fatty liver disease and inflammation. AJMB. https://doi.org/10.18502/ajmb.v17i2.18560
10. Popa AD, Niță O, Caba L, Gherasim A, Graur M, Mihalache L, Arhire LI (2023) From the sun to the cell: examining obesity through the lens of vitamin D and inflammation. Meta 14:4. https://doi.org/10.3390/metabo14010004
11. Ren Y, Zhao H, Yin C, Lan X, Wu L, Du X, Griffiths HR, Gao D (2022) Adipokines, hepatokines and myokines: focus on their role and molecular mechanisms in adipose tissue inflammation. Front Endocrinol 13:873699. https://doi.org/10.3389/fendo.2022.873699
12. Horwitz A, Birk R (2023) Adipose tissue hyperplasia and hypertrophy in common and syndromic obesity—the case of BBS obesity. Nutrients 15:3445. https://doi.org/10.3390/nu15153445
13. Gkrinia EMM, Belančić A (2025) The mechanisms of chronic inflammation in obesity and potential therapeutic strategies: a narrative review. CIMB 47:357. https://doi.org/10.3390/cimb47050357
14. Sen A, Kumar D, Anand P (2025) Diabetes and COVID-19: the vicious cycle of a dual burden. Outbreak Manag Resp 1:2539083. https://doi.org/10.1080/29947677.2025.2539083
15. Berbudi A, Rahmadika N, Tjahjadi AI, Ruslami R (2020) Type 2 diabetes and its impact on the immune system. CDR 16:442–449. https://doi.org/10.2174/1573399815666191024085838
16. Berbudi A, Khairani S, Tjahjadi A (2025) Interplay between insulin resistance and immune dysregulation in type 2 diabetes mellitus: implications for therapeutic interventions, ITT. Immunotargets Ther 14:359–382. https://doi.org/10.2147/ITT.S499605
17. Chen J, Fei S, Chan LWC, Gan X, Shao B, Jiang H, Li S, Kuang P, Liu X, Yang S (2025) Inflammatory signaling pathways in pancreatic β-cell: new insights into type 2 diabetes pathogenesis. Pharmacol Res 216:107776. https://doi.org/10.1016/j.phrs.2025.107776
18. Silva Rosa SC, Nayak N, Caymo AM, Gordon JW (2020) Mechanisms of muscle insulin resistance and the cross-talk with liver and adipose tissue. Physiol Rep 8. https://doi.org/10.14814/phy2.14607
19. Aparicio-Trejo OE, Hernández-Cruz EY, Reyes-Fermín LM, Ceja-Galicia ZA, Pedraza-Chaverri J (2025) The role of redox signaling in mitochondria and endoplasmic reticulum regulation in kidney diseases. Arch Toxicol 99:1865–1891. https://doi.org/10.1007/s00204-025-04041-z
20. Vetrano E, Rinaldi L, Mormone A, Giorgione C, Galiero R, Caturano A, Nevola R, Marfella R, Sasso FC (2023) Non-alcoholic fatty liver disease (NAFLD), type 2 diabetes, and non-viral Hepatocarcinoma: pathophysiological mechanisms and new therapeutic strategies. Biomedicine 11:468. https://doi.org/10.3390/biomedicines11020468

21. Taru V, Szabo G, Mehal W, Reiberger T (2024) Inflammasomes in chronic liver disease: hepatic injury, fibrosis progression and systemic inflammation. J Hepatol 81:895–910. https://doi.org/10.1016/j.jhep.2024.06.016
22. Jee Y-M, Lee J-Y, Ryu T (2025) Chronic inflammation and immune dysregulation in metabolic-dysfunction-associated Steatotic liver disease progression: from steatosis to hepatocellular carcinoma. Biomedicine 13:1260. https://doi.org/10.3390/biomedicines13051260
23. Ma X, Huang T, Chen X, Li Q, Liao M, Fu L, Huang J, Yuan K, Wang Z, Zeng Y (2025) Molecular mechanisms in liver repair and regeneration: from physiology to therapeutics. Sig Transduct Target Ther 10:63. https://doi.org/10.1038/s41392-024-02104-8
24. Pandey H, Goel P, Srinivasan VM, Tang DWT, Wong SH, Lal D (2025) Gut microbiota in non-alcoholic fatty liver disease: pathophysiology, diagnosis, and therapeutics. World J Hepatol 17. https://doi.org/10.4254/wjh.v17.i6.106849
25. Shen Y, Fan N, Ma S, Cheng X, Yang X, Wang G (2025) Gut microbiota dysbiosis: pathogenesis, diseases, prevention, and therapy. MedComm 6:e70168. https://doi.org/10.1002/mco2.70168
26. Zhang Z, Wang J, Li H, Niu Q, Tao Y, Zhao X, Zeng Z, Dong H (2025) The role of the interleukin family in liver fibrosis. Front Immunol 16:1497095. https://doi.org/10.3389/fimmu.2025.1497095
27. Li F, McClain CJ, Feng W (2019) Microbiome dysbiosis and alcoholic liver disease. Liver Res 3:218–226. https://doi.org/10.1016/j.livres.2019.09.001
28. Zhao M, Chu J, Feng S, Guo C, Xue B, He K, Li L (2023) Immunological mechanisms of inflammatory diseases caused by gut microbiota dysbiosis: a review. Biomed Pharmacother 164:114985. https://doi.org/10.1016/j.biopha.2023.114985
29. Al-Mansoori L, Al-Jaber H, Prince MS, Elrayess MA (2022) Role of inflammatory cytokines, growth factors and Adipokines in Adipogenesis and insulin resistance. Inflammation 45:31–44. https://doi.org/10.1007/s10753-021-01559-z
30. Kuryłowicz A, Koźniewski K (2020) Anti-inflammatory strategies targeting metaflammation in type 2 diabetes. Molecules 25:2224. https://doi.org/10.3390/molecules25092224
31. Ramoni D, Liberale L, Montecucco F (2024) Inflammatory biomarkers as cost-effective predictive tools in metabolic dysfunction-associated fatty liver disease. World J Gastroenterol 30:5086–5091. https://doi.org/10.3748/wjg.v30.i47.5086
32. Okuma H, Tsuchiya K (2024) Tissue-specific activation of insulin signaling as a potential target for obesity-related metabolic disorders. Pharmacol Ther 262:108699. https://doi.org/10.1016/j.pharmthera.2024.108699
33. Rohm TV, Meier DT, Olefsky JM, Donath MY (2022) Inflammation in obesity, diabetes, and related disorders. Immunity 55:31–55. https://doi.org/10.1016/j.immuni.2021.12.013
34. Meier DT, De Paula Souza J, Donath MY (2025) Targeting the NLRP3 inflammasome–IL-1β pathway in type 2 diabetes and obesity. Diabetologia 68:3–16. https://doi.org/10.1007/s00125-024-06306-1
35. Taylor EB (2021) The complex role of adipokines in obesity, inflammation, and autoimmunity. Clin Sci 135:731–752. https://doi.org/10.1042/CS20200895
36. Fatima Z, Shakeel R, Chaudhry SAA, Qadri M, Kakar AI, Ahmad B, Ahmad T, Akilimali A (2025) The relation between C-reactive protein (CRP) and risk of incident heart failure in patients with cardiovascular disease: a narrative review. Ann Med Surg 87:4356–4361. https://doi.org/10.1097/MS9.0000000000003455
37. Mohammadi M, Gozashti MH, Aghadavood M, Mehdizadeh MR, Hayatbakhsh MM (2017) Clinical significance of serum IL-6 and TNF-α levels in patients with metabolic syndrome. Rep Biochem Mol Biol 6:74–79
38. Abulmeaty MMA, Ghneim HK, Alkhathaami A, Alnumair K, Al Zaben M, Razak S, Al-Sheikh YA (2023) Inflammatory cytokines, redox status, and cardiovascular diseases risk after weight loss via bariatric surgery and lifestyle intervention. Medicina 59:751. https://doi.org/10.3390/medicina59040751

39. Mallick R, Basak S, Chowdhury P, Bhowmik P, Das RK, Banerjee A, Paul S, Pathak S, Duttaroy AK (2025) Targeting cytokine-mediated inflammation in brain disorders: developing new treatment strategies. Pharmaceuticals 18:104. https://doi.org/10.3390/ph18010104
40. Asimakidou E, Saipuljumri EN, Lo CH, Zeng J (2025) Role of metabolic dysfunction and inflammation along the liver–brain axis in animal models with obesity-induced neurodegeneration. Neural Regen Res 20:1069–1076. https://doi.org/10.4103/NRR.NRR-D-23-01770
41. Puig-Castellví F, Pacheco-Tapia R, Deslande M, Jia M, Andrikopoulos P, Chechi K, Bonnefond A, Froguel P, Dumas M-E (2023) Advances in the integration of metabolomics and metagenomics for human gut microbiome and their clinical applications. TrAC Trends Anal Chem 167:117248. https://doi.org/10.1016/j.trac.2023.117248
42. De Assis LVM, Kramer A (2024) Circadian de(regulation) in physiology: implications for disease and treatment. Genes Dev 38:933–951. https://doi.org/10.1101/gad.352180.124
43. Zheng H, Sechi LA, Navarese EP, Casu G, Vidili G (2024) Metabolic dysfunction-associated steatotic liver disease and cardiovascular risk: a comprehensive review. Cardiovasc Diabetol 23:346. https://doi.org/10.1186/s12933-024-02434-5
44. Mueller A-L, Payandeh Z, Mohammadkhani N, Mubarak SMH, Zakeri A, Alagheband Bahrami A, Brockmueller A, Shakibaei M (2021) Recent advances in understanding the pathogenesis of rheumatoid arthritis: new treatment strategies. Cells 10:3017. https://doi.org/10.3390/cells10113017
45. Nie L, Cai S-Y, Shao J-Z, Chen J (2018) Toll-like receptors, associated biological roles, and signaling networks in non-mammals. Front Immunol 9:1523. https://doi.org/10.3389/fimmu.2018.01523
46. Mendez Luque LF, Blackmon AL, Ramanathan G, Fleischman AG (2019) Key role of inflammation in myeloproliferative neoplasms: instigator of disease initiation, progression. And symptoms. Curr Hematol Malig Rep 14:145–153. https://doi.org/10.1007/s11899-019-00508-w
47. Huang J, Zhou C, Deng J, Zhou J (2022) JAK inhibition as a new treatment strategy for patients with COVID-19. Biochem Pharmacol 202:115162. https://doi.org/10.1016/j.bcp.2022.115162
48. Kircheis R, Planz O (2024) Special Issue "The Role of Toll-Like Receptors (TLRs) in Infection and Inflammation 2.0". IJMS 25:9709. https://doi.org/10.3390/ijms25179709
49. Chen R, Zou J, Zhong X, Liu J, Kang R, Tang D (2025) DAMP signaling networks: from receptors to diverse pathophysiological functions. J Adv Res:S2090123225005776. https://doi.org/10.1016/j.jare.2025.07.047
50. Masselli E, Pozzi G, Gobbi G, Merighi S, Gessi S, Vitale M, Carubbi C (2020) Cytokine profiling in myeloproliferative neoplasms: overview on phenotype correlation, outcome prediction, and role of genetic variants. Cells 9:2136. https://doi.org/10.3390/cells9092136
51. Loscocco GG, Coltro G, Guglielmelli P, Vannucchi AM (2021) Integration of molecular information in risk assessment of patients with myeloproliferative neoplasms. Cells 10:1962. https://doi.org/10.3390/cells10081962
52. Lin X, Li X, Zhai Z, Zhang M (2025) JAK-STAT pathway, type I/II cytokines, and new potential therapeutic strategy for autoimmune bullous diseases: update on pemphigus vulgaris and bullous pemphigoid. Front Immunol 16:1563286. https://doi.org/10.3389/fimmu.2025.1563286
53. Cheung TT, McInnes IB (2017) Future therapeutic targets in rheumatoid arthritis? Semin Immunopathol 39:487–500. https://doi.org/10.1007/s00281-017-0623-3

Chapter 9
Eicosanoids and Bioactive Lipids in Modulation of Inflammation in Metabolic Disorders

Subhasundar Maji, Sangita Mishra, and Slim Smaoui

Abbreviations

NAFLD	Non-alcoholic fatty liver disease
COX	Cyclooxygenases
LOX	Lipoxygenases
ROS	Reactive oxygen species
NF-κB	Nuclear factor kappa-light-chain-enhancer of activated B cells
MAPK	Mitogen-activated protein kinase
PPARs	Peroxisome proliferator-activated receptors

9.1 Introduction

Metabolic diseases such as diabetes, obesity, and non-alcoholic fatty liver disease (NAFLD) have become major global health issues in recent decades. What makes these conditions particularly complex and dangerous is that they are not just about fat accumulation or glucose imbalance. At their core lies a silent, persistent form of inflammation that subtly rewires normal physiological processes, pushing the body into a chronic state of immune activation. This phenomenon, often termed metabolic inflammation, contributes significantly to the formation of metabolic disorders [1–3]. A major piece of this puzzle lies in a group of powerful, lipid-derived molecules known as bioactive lipids. Among them, eicosanoids stand out as some of the most influential. These signalling molecules—produced from polyunsaturated fatty acids like arachidonic acid (AA)—serve as molecular

S. Maji · S. Mishra
CSIR-Indian Institute of Chemical Biology, Kolkata, India

S. Smaoui (✉)
Univercity of SFAX, SFAX, Tunisia
e-mail: slim.smaoui@cbs.rnrt.tn

S. Ashique, N. Mishra (eds.), *Reactive Oxygen Species in Metabolic Inflammation*,
https://doi.org/10.1007/978-3-032-11752-6_9

messengers between cells. They have the remarkable ability to either amplify inflammation (as seen with prostaglandins (PGs) and leukotrienes (LTs)) or shut it down and promote healing (as with lipoxins (LXs) and resolvins (Rvs)). The balance between these opposing forces is crucial: when disrupted, it can tip tissues into a prolonged state of inflammation that underpins many metabolic conditions [4, 5]. These lipids don't act in isolation. Their production is controlled by tightly regulated enzymatic pathways, particularly cyclooxygenases (COX), lipoxygenases (LOX), and cytochrome P450 (CYP) enzymes. However, under the oxidative stress (OS) that often goes with metabolic overload—such as high-fat diets or persistent hyperglycaemia—these enzymes can behave abnormally. Reactive oxygen species (ROS), which are elevated in metabolic tissues under stress, directly influence both the production and function of eicosanoids. In this way, lipid signalling becomes entangled with oxidative signalling, creating a feedback loop that sustains inflammation [6, 7]. This chapter explores how eicosanoids and related lipid mediators regulate inflammation at the molecular level in the context of metabolic disorders. We will dive into how these lipids are synthesized, the roles they play in various tissues, and how they interact with key inflammatory and metabolic pathways such as nuclear factor kappa B (NF-κB), mitogen-activated protein kinase (MAPK), and peroxisome proliferator-activated receptors (PPARs). We will also examine how immune cells like macrophages respond to these lipid

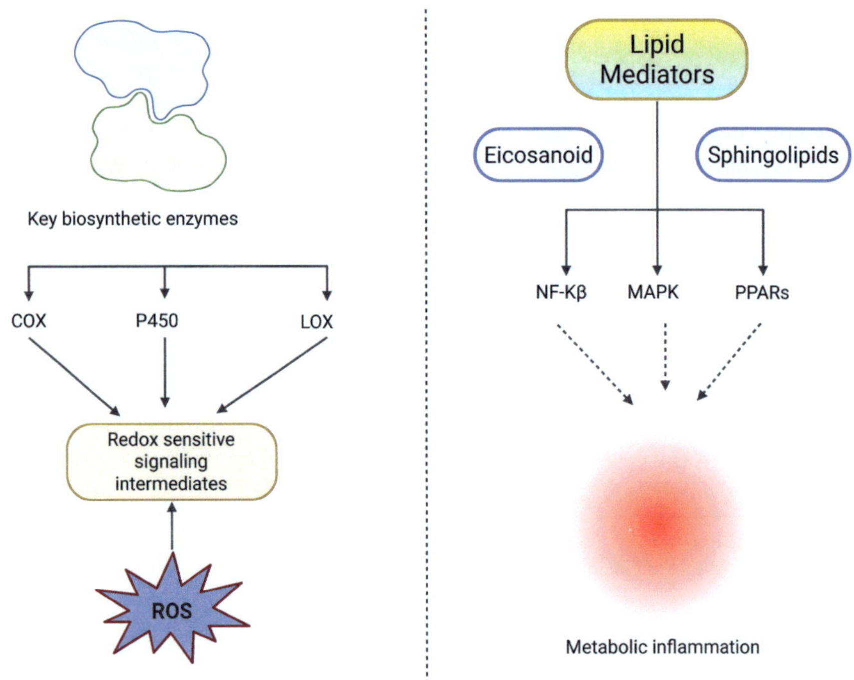

Fig. 9.1 Schematic diagram of dynamic interplay between lipid mediator networks

signals in metabolically active tissues like fat, liver, and muscle. By unpacking the role of bioactive lipids in metabolic inflammation, this chapter aims to provide a deeper understanding of their potential as biomarkers and therapeutic targets—offering new hope in the fight against some of the most pressing health issues of our time (Fig. 9.1).

9.2 Biochemistry and Biosynthesis of Eicosanoids

9.2.1 The Arachidonic Acid Pathway: Key Precursors and Enzymatic Networks

The journey of eicosanoid biosynthesis begins when cellular membranes are disrupted by inflammatory stimuli, physical trauma, or metabolic stress. This activation triggers phospholipase A2 (PLA2) enzymes to liberate arachidonic acid and additional membrane phospholipid-derived polyunsaturated fatty acids [8–10]. The type IV cytosolic PLA2 (cPLA2) is particularly important, as it specifically releases arachidonic acid from the second carbon of the glycerol backbone in phospholipids (SN2) position of membrane phospholipids under conditions of cellular activation [11, 12]. Once released, arachidonic acid becomes the substrate for three major enzymatic pathways that collectively form the arachidonic acid cascade. Each pathway is named after its initial enzyme and produces distinct families of bioactive mediators with unique biological functions.

9.2.2 Cyclooxygenase (COX) Pathway: Prostaglandins and Thromboxanes

The cyclooxygenase pathway represents one of the most clinically relevant branches of eicosanoid biosynthesis. Two major COX isoforms exist, each with distinct regulatory patterns and tissue distributions. The majority of tissues constitutively express COX-1, which has homeostatic purposes, including gastric cytoprotection, platelet aggregation, and vascular tone regulation. In contrast, COX-2 is inducible and dramatically upregulated during inflammation, making it a primary contributor to the inflammatory eicosanoid pool [13, 14]. Both enzymes convert arachidonic acid to prostaglandin H2 (PGH2), the central intermediate that serves as precursor for multiple proteinoids. Specific synthases then transform PGH2 into various prostaglandins (PGE2, PGD2, PGF2α), prostacyclins (PGI2), and thromboxanes A2 (TXA2). Each of these mediators exhibits distinct receptor-binding profiles and biological activities, from vasodilation and pain sensitization (PGE2) to platelet aggregation and vasoconstriction (TXA2) [15].

9.2.3 Lipoxygenase (LOX) Pathway: Leukotrienes and Lipoxins

The lipoxygenase pathway encompasses several distinct enzymes that insert molecular oxygen at specific positions along the arachidonic acid chain. The most extensively studied is 5-lipoxygenase (5-LOX), which produces leukotrienes—potent mediators of neutrophil recruitment and bronchoconstriction. Arachidonic acid is initially converted to 5-hydroperoxyeicosatetraenoic acid (5-HPETE) by 5-LOX, which is then transformed to leukotriene A4 (LTA4). LTA4 serves as a branch point: it can be hydrolysed to LTB4 (a powerful neutrophil chemoattractant) or conjugated with glutathione to form cysteinyl leukotrienes (LTC4, LTD4, LTE4) that cause bronchoconstriction and increased vascular permeability [16, 17]. In addition to producing various hydroxylated fatty acids, 12-lipoxygenase (12-LOX) and 15-lipoxygenase (15-LOX) are crucial for the manufacture of lipoxins. Lipoxins represent a crucial class of dual-acting mediators: they are anti-inflammatory and pro-resolving, actively promoting the termination of neutrophil infiltration and stimulating macrophage-mediated clearance of apoptotic cells [9, 15, 18, 19].

9.2.4 Cytochrome P450 Pathway: Epoxygenases and Omega-Hydroxylases

The cytochrome P450 (CYP) pathway adds another layer of complexity to eicosanoid metabolism [20]. Different CYP enzymes produce distinct classes of metabolites: CYP2C and CYP2J subfamilies function as epoxygenases, converting arachidonic acid to epoxyeicosatrienoic acids (EETs). EETs exhibit anti-inflammatory properties, promoting vasodilation and inhibiting inflammatory gene expression. Soluble epoxide hydrolase (sEH) can further break them down to produce dihydroxyeicosatrienoic acids (DiHETEs), which frequently have counteracting, pro-inflammatory effects. CYP4A and CYP4F subfamilies operate as omega-hydroxylases, producing 20-HETE as the predominant metabolite. 20-HETE is particularly relevant in liver pathology, where it accounts for 50–75% of all CYP-derived eicosanoids and shows elevated levels in cirrhosis, NAFLD, and alcoholic liver disease. This metabolite promotes hepatic stellate cell activation and liver fibrosis through TGF-β signalling pathways [9, 20, 21].

9.2.5 Regulation of Lipid Mediator Production

The cellular production of eicosanoids is tightly regulated at multiple levels. Enzyme expression varies dramatically between cell types and activation states. For example, COX-2 expression can increase 10–100-fold within hours of inflammatory stimulation, while 5-LOX activity is predominantly found in myeloid cells like neutrophils and macrophages [22]. Substrate availability represents another key regulatory mechanism. The cellular pools of arachidonic acid are not infinite, and different phospholipase A2 enzymes show specificity for different membrane locations and phospholipid species. This creates distinct pools of substrate that can be mobilized under different conditions [23, 24]. Redox regulation adds an additional layer of complexity. Many eicosanoids biosynthetic enzymes are sensitive to the local redox environment. Reactive oxygen species (ROS), which are elevated in metabolic tissues under conditions of obesity and insulin resistance, can directly modulate enzyme activity. For instance, ROS can activate certain phospholipase A2 isoforms while inhibiting others, fundamentally altering the balance between pro-inflammatory and anti-inflammatory lipid mediator production [23]. Cross-pathway interactions further complicate the regulatory landscape. A remarkable example is the interaction between COX-2 and 5-LOX pathways: COX-2 can metabolize 5-LOX-derived 5S-HETE to produce novel hemiketal eicosanoids, representing a direct biochemical crosstalk between traditionally separate pathways. Such interactions suggest that the eicosanoid network operates as an integrated system rather than parallel independent pathways [25].

9.3 Classification and Functions of Eicosanoids

9.3.1 Prostaglandins and Thromboxanes: Dual Roles in Inflammation

Prostaglandins represent the most diverse and extensively studied family of eicosanoids, with each subtype exhibiting distinct receptor-binding patterns and biological activities. PGE2, the most abundant prostaglandin in inflammatory conditions, exemplifies the complex dual nature of eicosanoid signalling. Through its four EP receptors (EP1–EP4), PGE2 can simultaneously promote inflammation by sensitizing pain receptors and causing fever, while also limiting inflammatory damage by upregulating anti-inflammatory IL-10 production in macrophages [9, 26].

Prostaglandin D2 (PGD2) functions primarily through DP1 and DP2 receptors, with dramatically different outcomes. DP1 activation promotes neuroprotection (NP) and mast cell maturation, while DP2 activation drives eosinophil recruitment and allergic responses. This receptor-specific signalling illustrates how a single lipid mediator can have opposing effects depending on the cellular context. Thromboxane A2 (TXA2) operates as a more straightforward pro-inflammatory mediator, promoting platelet aggregation, vasoconstriction, and T-cell activation through TP receptors. The balance between TXA2 and prostacyclin (PGI2) is particularly important in cardiovascular health, as PGI2 counteracts TXA2 by inhibiting platelet aggregation and promoting vasodilation [9].

9.3.2 Leukotrienes and Lipoxins: Equilibrium Between Anti-inflammatory and Pro-inflammatory Reactions

The leukotriene family consists of potent pro-inflammatory mediators that play crucial roles in acute inflammatory responses. LTB4 acts as one of the most powerful neutrophil chemoattractant known, binding to Leukotriene B_4 Receptor 1 (BLT1) receptors to promote cell recruitment and vascular leakage. The cysteinyl leukotrienes (LTC4, LTD4, LTE4) bind to CysLT1 and CysLT2 receptors, causing bronchoconstriction and vascular leakage that characterize asthmatic responses [27, 28]. Lipoxins indicate a shift in our perspective on how to reduce inflammation. Unlike other eicosanoids that either promote or simply inhibit inflammation, lipoxins actively orchestrate the resolution process. LXA4 and LXB4 bind to the Lipoxin A4 Receptor/Formyl Peptide Receptor 2 (ALX/FPR2) receptor, triggering a coordinated programme that includes:

- Cessation of neutrophil recruitment and activation.
- Enhanced macrophage phagocytosis of apoptotic neutrophils (efferocytosis).
- Stimulation of anti-inflammatory cytokine production.
- Promotion of tissue repair and regeneration.

Resolvins, protectins (PDs), and maresins (MaR) are examples of specialized pro-resolving mediators (SPMs).

The discovery of specialized pro-resolving mediators has revolutionized our understanding of inflammation resolution. Originating from the omega-3 fatty acids eicosapentaenoic (EPA) and docosahexaenoic acid (DHA), these mediators are a distinct class of bioactive lipids that have developed especially to actively stop inflammatory reactions and return tissue homeostasis [29, 30]. The E-series (derived from EPA) and D-series (derived from DHA) subfamilies of resolvins are separated. RvE1 binds to ChemR23 and BLT1 receptors, demonstrating how SPMs can reprogram traditional pro-inflammatory signalling pathways for resolution. RvD1 acts through ALX/FPR2 and GPR32 receptors, promoting macrophage phagocytosis and bacterial clearance while reducing pro-inflammatory mediator production [31, 32].

Protectins, including neuroprotection D1 (NPD1), exhibit tissue-protective functions beyond simple anti-inflammatory activity. They promote cell survival, enhance tissue regeneration, and protect against oxidative damage [33]. *Maresins* represent the most recently identified SPM family, with maresin 1 (MaR1) showing particularly potent pro-resolving activities. MaR1 binds to LGR6 receptors on macrophages, promoting their polarization toward tissue repair phenotypes and enhancing efferocytosis [34, 35]. The temporal regulation of SPM production is crucial for their biological function. SPMs are typically biosynthesized during the resolution phase of inflammation, creating a natural "resolution programme" that actively switches off inflammatory signalling and promotes tissue repair. This timing is orchestrated by changes in enzyme expression and cellular activation states as inflammation progresses.

9.3.3 Membrane-Associated and Modified Eicosanoids

Not all bioactive eicosanoids exist as free mediators. Membrane-esterified eicosanoids represent an important regulatory mechanism. Phospholipid-bound HETEs (PL-HETEs) can modulate signal transduction pathways and serve as storage forms for lipoxin precursors. Phospholipid-bound prostaglandins (PL-PGs) can reduce Toll-like receptor responses, demonstrating how membrane association can alter eicosanoid function [9, 36]. Aspirin-triggered SPMs represent a unique class of modified resolution mediators. Aspirin irreversibly acetylates COX-2, changing its enzymatic activity to produce 15R-hydroxy-fatty acids instead of prostaglandins. These aspirin-triggered precursors can be further metabolized by 5-LOX to produce aspirin-triggered lipoxins (ATLs) and aspirin-triggered resolvins (AT-RvDs), which often exhibit enhanced potency compared to their naturally occurring counterparts [37, 38]. This intricate classification system reveals eicosanoids as a sophisticated regulatory network rather than simple inflammatory mediators. Each class operates within specific temporal and spatial contexts, with receptor-specific signalling, tissue-selective expression, and coordinated interactions that collectively determine the outcome of inflammatory responses. Understanding these distinct functions is crucial for developing targeted therapeutic interventions that can selectively enhance beneficial eicosanoid signalling while suppressing pathological inflammatory pathways.

9.4 Molecular Signalling Pathways Modulated by Eicosanoids

9.4.1 NF-κB and MAPK Signalling in Lipid-Mediated Inflammation

One of the main inflammatory signalling cascades that eicosanoids influence is the nuclear factor kappa B (NF-κB) pathway. This transcription factor complex is attached to inhibitory proteins (IκB) and persists in the cytoplasm in an inactive form. IκB kinase (IKK) complexes phosphorylate and degrade IκB proteins in response to inflammatory stimulation, which permits NF-κB dimers to go into the nucleus and trigger the transcription of inflammatory genes [39–41]. Eicosanoids influence NF-κB signalling through multiple mechanisms. Reactive oxygen species (ROS), often elevated during eicosanoid biosynthesis, can directly activate NF-κB through IκBα tyrosine phosphorylation and subsequent degradation via calpain-mediated pathways. However, the relationship is complex: while some eicosanoids promote NF-κB activation, others actively inhibit it. PGE2 can suppress NF-κB activity in macrophages through EP receptor-mediated signalling, while certain lipoxins and resolvins directly interfere with NF-κB nuclear translocation [26, 42].

Mitogen-activated protein kinase (MAPK) pathways represent another critical node of eicosanoid signalling. The three major MAPK cascades—ERK, JNK, and p38—are all modulated by lipid mediators. In adipose tissue, the interaction between adipocytes and macrophages creates a paracrine signalling loop where free fatty acids activate MAPK pathways in macrophages, leading to increased TNF-α production. This TNF-α then activates MAPK signalling in adipocytes, creating a self-perpetuating inflammatory cycle [43, 44]. Crosstalk between NF-κB and MAPK pathways is particularly evident in metabolic inflammation. MAPK activation can directly phosphorylate NF-κB subunits, enhancing their transcriptional activity. Conversely, NF-κB can upregulate MAPK pathway components, creating positive feedback loops that sustain inflammatory signalling [45, 46].

9.4.2 PPARs and Eicosanoid-Dependent Transcriptional Regulation

Peroxisome proliferator-activated receptors (PPARs) represent a unique class of nuclear receptors that function as direct targets for numerous eicosanoids. Unlike membrane-bound eicosanoid receptors that trigger rapid signalling cascades, PPARs mediate slower, transcriptional responses that fundamentally reprogram cellular metabolism and inflammatory responses [47, 48]. The tissue distribution and ligand preferences of the three PPAR isotypes vary. In metabolically active organs such as the liver, heart, and skeletal muscle, PPARα is extensively expressed and binds a variety of eicosanoids, including LTB4 and 15d-PGJ2. Following ligand interaction, PPARα attaches to peroxisome proliferator response elements (PPREs) in target gene promoters and forms heterodimers with the retinoid X receptor (RXR) [49, 50]. PPARγ shows the highest expression in adipose tissue and plays crucial roles in adipocyte differentiation and insulin sensitivity. A metabolite of PGD2, 15-deoxy-Δ12,14-prostaglandin J2 (15d-PGJ2), is one of the strongest endogenous PPARγ ligands. This connection creates a direct link between cyclooxygenase-derived eicosanoids and metabolic gene regulation [50]. PPARβ/δ has the broadest tissue expression pattern and responds to various fatty acids and eicosanoids. It plays particularly important roles in skeletal muscle fatty acid oxidation and inflammatory resolution [51]. The anti-inflammatory actions of PPARs occur through multiple mechanisms. PPARs can directly upregulate anti-inflammatory genes while simultaneously interfering with NF-κB and AP-1 signalling through protein-protein interactions. This dual action—transcriptional activation of anti-inflammatory programs combined with transcriptional repression of inflammatory genes—makes PPARs powerful regulators of the inflammatory response [52].

9.4.3 Crosstalk with Insulin and Adipokine Signalling Pathways

Eicosanoids exert profound effects on insulin signalling pathways, contributing significantly to metabolic inflammation and insulin resistance. The mechanisms involve both direct interference with insulin signal transduction and indirect effects through inflammatory mediator production.

A common characteristic of lipid-mediated insulin resistance is the buildup of diacylglycerol (DAG), which triggers the activation of protein kinase C (PKC) isoforms that phosphorylate serine residues on insulin receptor substrate 1 (IRS-1) directly. This serine phosphorylation impairs the tyrosine phosphorylation necessary for normal insulin signalling, effectively blocking downstream PI3K-Akt activation. The specific fatty acid composition of DAG molecules appears crucial, with saturated fatty acid (SFA)-containing DAG species showing greater insulin resistance-promoting effects [53–55]. Ceramides represent another class of bioactive lipids that interfere with insulin signalling. Unlike DAG, which acts primarily at the receptor level, ceramides directly inhibit Akt/PKB (Akt = Protein Kinase B) activation through mechanisms involving protein phosphatase 2A (PP2A) activation. This creates a bottleneck in insulin signalling that prevents glucose uptake and glycogen synthesis [56, 57]. Phosphatidic acid (PA) adds another layer of complexity to lipid-insulin interactions. PA derived from different biosynthetic pathways exhibits opposite effects on insulin signalling; phospholipase D (PLD)-derived PA stabilizes mTORC2 complexes, promoting insulin sensitivity, while glycerol-3-phosphate acyltransferase (GPAT)-derived PA disrupts Mechanistic Target of Rapamycin Complex 2 (mTORC2) assembly, impairing insulin signalling [58]. The integration of adipokine signalling with eicosanoid pathways creates additional regulatory networks. Adiponectin, an anti-inflammatory adipokine, can be suppressed by inflammatory eicosanoids like PGE2 and LTB4. Conversely, leptin signalling can be enhanced by certain pro-resolving mediators, creating feedback loops that influence both metabolic homeostasis and inflammatory responses.

9.5 Interplay Between Eicosanoids and Reactive Oxygen Species (ROS)

9.5.1 Peroxidation of Lipids and Oxidative Stress in Metabolic Tissues

Oxidative stress (OS) is caused by the imbalanced generation of reactive oxygen and nitrogen species (ROS and RNS, respectively), which is linked to a reduction in the quantity or expression of antioxidant systems and a decline in their function [59]. It functions as a signal transduction molecule and offers cell protection when present in insignificant amounts in cells. However, when overproduced, as in the case of inflammatory tissues, they can produce additional highly reactive species that can irreversibly damage proteins, lipids, and nucleic acids [59]. Amongst them lipid peroxidation through the free radicals from oxidative stress in different metabolic tissue is our concern. Evidence suggested that hypermetabolic rate in hyperthyroidism is due to oxidative stress which helps in production of excess level of free radicals and increased level of lipid peroxidase enzyme which induces the lipid peroxidation end products [60]. Oxidative stress could be a major consequence in obesity cases. Through a few biochemical processes, including the production of superoxide anion through oxidative phosphorylation, glyceraldehyde auto-oxidation, protein kinase C (PKC) activation, and the polyol and hexosamine pathways, increased consumption of fats, carbohydrates, and saturated fatty acids—particularly trans-fatty acids—leads to increased oxidative stress. By promoting the deposition of adipose tissue, including adipocyte proliferation, differentiation, and expansion, oxidative stress also contributes significantly to the development of obesity. Through promoting lipogenesis, insulin signalling, oxidative stress, and reactive oxygen species production, TNF-α affects lipid metabolism, death of adipose tissues, and the inflammatory response. ROS production is enhanced by the activation of pro-inflammatory transcription factors and the release of inflammatory cytokines, including nuclear factor kappa B (NF-κB) and activator protein-1 (AP-1) [61]. TNF-α also promotes the release of IL-6. The stimulating impact of IL-1β generated from monocytes, which causes tissue damage, immunologic events, and infection, may increase the production of IL-6 [61]. Adipocytes, pancreatic β-cells, macrophages, and monocytes all produce IL-6, which controls inflammation and energy homeostasis by encouraging the production of pro-inflammatory cytokines [62]. The correlation between elevated serum IL-6 levels and the onset of diabetes mellitus, obesity, hypertension, and impaired glucose tolerance has been extensively shown [61]. The brain tissues of people with Alzheimer's disease (AD) exhibit higher levels of oxidative stress and lipid peroxidation, according to several studies. Lipid peroxidation in brain tissues is primarily caused by thiobarbituric acid reactive substances (TBARS), which are increased in all AD brain regions except the middle frontal gyrus and reached statistical significance in the hippocampus and pyriform cortex. In AD, TBARS is marginally significant in the amygdala when compared to control [63]. Numerous studies have demonstrated that the increased level of TBRAS causes lipid oxidation in Alzheimer's disease brains, which may be a subsequent effect of other changes such as lower antioxidant levels or raised polyunsaturated fatty acids (PUFAs) [64]. Changes in brain phospholipids, which are involved in a few membrane activities, have been demonstrated to contribute to AD [65]. The primary cause of neurodegeneration and pathomechanisms in Alzheimer's brain tissue may be a dramatic reduction in phosphatidylcholine and phosphatidylethanolamine, which are precursors of phospholipids like choline and ethanolamine. Decreased production, increased degradation, and increased lipid peroxidation can all contribute to the loss of membrane phospholipids [66]. Phospholipid depletion in the AD brain may be caused by increased free-radical-mediated lipid peroxidation, according to studies showing PUFA changes in the AD brain. When PUFA is oxidized, a substantial number of aldehydes, primarily 4-hydroxyalkenals, are formed by lipid peroxidation. These aldehydes are extremely reactive and cause oxidative stress and cytotoxicity [66].

9.5.2 ROS-Mediated Modulation of Eicosanoid Biosynthetic Enzymes

Eicosanoids are oxidized forms of 20-carbon polyunsaturated fatty acids (PUFAs) that are produced by a variety of enzymes, including cytochrome P450 (cytP450), lipoxygenase (LOX), and cyclooxygenase (COX). The primary substrate for the synthesis of eicosanoid is arachidonic acid (AA). The COX pathway produces prostaglandins (PGs) and thromboxanes (TXs), the LOX pathways produce leukotrienes (LTs) and lipoxins (LXs), and the cytP450 pathways produce epoxy, hydroxy, and dihydroxy derivatives [67]. NADPH oxidases, inflammatory cells, and mitochondrial respiration all produce superoxide, hydrogen peroxide, and hydroxyl ions, which can alter the biosynthesis of eicosanoid enzymes through transcriptional control, post-translational modifications, feedback amplifications, and other mechanisms. ROS uses the NF-kB or standard route to modify the enzymes involved in eicosanoid production. Arachidonic acid and prostaglandin are released more when the NF-kB and MAPK cascade activates the activated transcription factor activator protein 1 (AP-1) and CREB-binding protein (CBP/p300). This causes the transcription of cytosolic phospholipase A2 (cPLA2), cyclooxygenase-2 (COX-2), and membrane-bound prostaglandin E synthase 1 (mPGES-1). Prostaglandin D2 (PGD2) is produced when ROS triggers the hematopoietic-type PGD synthase (H-PGDS). It has been shown that ROS directly raises the amount of arachidonic acid by generating superoxide ions and H_2O_2. Prostaglandin D2 biosynthesis depends on ROS. By controlling hematopoietic-type PGD synthase (H-PGDS), ROS also contribute significantly to the production of this prostaglandin, albeit the exact mechanism is still unclear.

9.5.3 Redox-Dependent Regulation of Lipid Signalling Cascades

Lipid signalling such as eicosanoid, phospholipid, and sphingolipid, and fatty acid signalling are important for cellular processes, cell proliferation, apoptosis, metabolism, and migration. A complex lipid signalling network with several nodes of contact and cross-regulation is made up of enzymes such as phospholipases, prostaglandin synthase, 5-lipoxygenase, phosphoinositide 3-kinase, sphingosine kinase, and sphingomyelinase, as well as their downstream targets. Human disorders could result from an imbalance in their network [68]. The lipid signalling cascade, which in turn causes inflammation, cancer, metabolic diseases like obesity and diabetes, and other conditions, can be activated or inactivated by redox regulation such as thiol-mediated oxidation, nitrosylation, sulfenylation, and lipid peroxidation of the important enzymes like phospholipase A_2, cyclooxygenases, and lipoxygenases. A crucial component of cellular metabolism, redox metabolism connects the network of oxidative PTMs and redox processes. ROS is a harmful waste product of cells that can accelerate the irreversible and indiscriminate oxidation of DNA, proteins, and lipids [69]. Through altering the activity of transcription factors (TFs) and metabolic enzymes, as well as controlling gene expression, redox signalling may regulate cell function [69]. Numerous cell surface receptors raise intracellular H_2O_2 and have a second messenger-like effect on downstream signalling [70].

9.5.4 Adipocytes: Lipid Mediators in Adipose Tissue Inflammation

The metabolism of fat cells varies significantly depending on the anatomical location of adipose tissue, making it a heterogeneous organ. It is still unknown, though, how bioactive lipid mediators are distributed throughout various fat depots and how they relate to complicated illnesses. According to the report, a variety of specialized pro-resolving mediators (SPM) are found in human adipose tissue, including resolvin (Rv) D1, RvD2, protectin (PD) 1, lipoxin (LX) A4, and the monohydroxy biosynthetic pathway markers of RvD1 and PD1 (17-HDHA), RvE1 (18-HEPE), and maresin 1 (14-HDHA), based on LC-MS data. SC fat was also shown to include the "classic" eicosanoids prostaglandin (PG) E2, PGD2, PGF2α, leukotriene (LT) B4, 5-hydroxyeicosatetraenoic acid (5-HETE), 12-HETE, and 15-HETE [71].

Inflammation is essential for tissue wound healing and host defence against invasive infections. Higher species have developed defence systems to guarantee that the inflammatory response resolves in a particular time-limited manner since unresolved inflammation is harmful to the host [72]. Lipid mediators made from polyunsaturated fatty acids are one of the processes that promote resolution and are very effective at reducing inflammation [73]. Together with the resolvin, protectin, and maresin families, which are derived from the omega-3 fatty acids eicosapentaenoic (EPA) and docosahexaenoic (DHA) acids, lipoxins (LX), which are produced in human tissues through transcellular biosynthesis from the omega-6 fatty acid arachidonic acid (AA), are widely acknowledged as the most potent and collectively form a new genus of specialized pro-resolving mediators (SPMs). These SPMs can resolve a range of acute inflammatory reactions with little harm to the surrounding tissue [74]. According to some research, patients with perivascular disease and concurrent open foot ulcers have interestingly excessive amounts of lipid mediators with pro- and anti-inflammatory qualities in their periwound foot fat. It has been demonstrated that SPM mimetics of these naturally occurring bio templates have anti-inflammatory and anti-fibrotic properties [75]. Recent research suggests that SPMs may help diabetics recover their wounds [76] and in keeping a mouse burn wound model free of necrosis and subsequent thrombosis [77].

9.5.5 *Macrophages and Dendritic Cells: Eicosanoids in Immune Polarization*

In recent research, we have attempted to comprehend how the metabolism of free fatty acids (FAs) shapes the functional characteristics of dendritic cells (DCs) and macrophages. Numerous studies demonstrate how fatty acid metabolism regulates the activity of dendritic cells and macrophages, and how the metabolism of polyunsaturated fatty acids and their oxidation products, such as prostaglandins, leukotrienes, and pro resolving lipid mediators, modulates and impacts the biology of DCs and macrophages [78]. Macrophages are myeloid-derived innate immune cells that perform a variety of tasks. They are crucial for phagocytosing and eliminating microorganisms during an infection, promoting wound healing and tissue regeneration during the resolution phase of an inflammatory response, and preserving tissue homeostasis in noninflammatory circumstances [79, 80]. In in vitro culture, conventionally activated macrophages (also called M1 macrophages) and alternatively activated macrophages (also called M2 macrophages) are the most extensively researched macrophages; in in vivo culture, however, they are demonstrated to be pro-inflammatory and anti-inflammatory models, respectively [81]. Dendritic cells are myeloid-derived innate immune response cells, just like macrophages. Dendritic cells, which function as specialized APCs with the ability to trigger a T-cell response, are the mediators between the innate and adaptive immune responses [82]. Furthermore, because DCs are exposed to tolerizing signals throughout steady state, they also contribute to the induction of tolerance. By encouraging peripheral T-cell anergy and death, reducing effector and memory T-cell responses, and stimulating Treg development and activation, these tolerogenic DCs can induce tolerance [83, 84].

It is commonly known that exogenous SFA signalling generally causes DCs and macrophages to become more inflammatory [85]. Likewise, polyunsaturated fatty acids have an impact on dendritic cells and macrophage function. Given the mounting evidence supporting their critical role in the active resolution of inflammation, specialized pro-resolving mediators have attracted a lot of interest lately. Deregulation of SPM metabolism is already strongly linked to a number of chronic inflammatory diseases, including type 2 diabetes, Alzheimer's disease, atherosclerosis, and arthritis [86]. The effects of eicosanoids on macrophages and dendritic cells are listed below (Tables 9.1 and 9.2):

Table 9.1 Types of eicosanoids and source and effect on macrophages

Eicosanoid	Source	Effect on macrophages
Palmitic acid	Saturated fatty acid	Linked to poor wound healing, a low-grade chronic inflammatory state, and elevated production of IL-1β and IL-23 [87]
PGE_2	COX-2 pathway	Inhibits the activation and polarization of murine BM-macrophages both in vitro and in vivo by suppressing M1 cytokines such as IL-12 and TNF alpha, promoting IL-10, and converting it to an M2 profile [88]
PGD_2	COX-2 pathway	DP1/DP2 receptors, dampens inflammatory cytokines, promotes resolution
LTB_4	5-LOX pathway	An elevated pro-inflammatory signature is linked to this. By utilizing a receptor antagonist to prevent LTB4 signalling, these effects are lessened and M1 macrophage activation is increased
Lipoxins (LXA_4)	15-LOX pathway	Pro-resolving: Reduces M1 responses, promotes tissue repair
Specialized pro-resolving mediators	Oxidation of eicosanoids and polyunsaturated fatty acids	SPMs enhance the phagocytic and efferocytotic functions of macrophages and encourage the transition from M1-like to M2-like macrophages

Table 9.2 Eicosanoid's influence on dendritic cells and cell polarization

Eicosanoid	Effect on dendritic cells
PGE_2	Enhances the dendritic cell's migration and also induces IL-10 production which can lead to tolerogenic dendritic cells
PGD_2	Binds DP1/DP2 receptors which inhibits DC maturation, decreases IL-12, promotes Th2/Treg conversion
LTB_4	Enhances DC antigen uptake and pro-inflammatory cytokine production, favouring Th1 polarization
LXA_4	Inhibits DC maturation and antigen presentation which promotes resolution and tolerance
Specialized pro-resolving mediators	Specialized pro-resolving mediators bind to PPARs. By controlling their metabolic, PPAR signalling is known to contribute to reducing the inflammatory activity of DCs and macrophages [89]

9.6 Hepatocytes and Hepatic Stellate Cells: Liver Inflammation and Fibrosis

The interaction of parenchymal cells (hepatocytes) and non-parenchymal cells (such as stellate cells, kupffer cells, and liver sinusoidal endothelial cells) causes damage to the liver. Hepatocyte damage and immunological activation or hepatic stellate cell activation are the primary causes of liver injury. Liver fibrosis follows persistent inflammation in chronic liver illnesses such as viral hepatitis or non-alcoholic fatty liver disease (NAFLD). Long-term morbidity such as cirrhosis or liver cancer, as well as mortality in non-alcoholic steatohepatitis (NASH) and non-alcoholic fatty liver disease (NAFLD) are significantly influenced by liver fibrosis [90]. Hepatic stellate cells (HSCs) are activated by inflammatory reactions and can either transdifferentiate into matrix-producing myofibroblasts or modify immunological processes through chemokines and cytokines [90]. Antifibrotic substances that promote blood clotting and matrix metalloproteinases (MMP) that break down extracellular matrix (ECM) to reduce immune cell infiltration are generated during acute inflammation [91]. Phagocytes remove debris from inflammation by releasing an anti-inflammatory mediator [92], to reconstruct the extracellular matrix (ECM) and ultimately close the wound; this mechanism attracts endothelial cells, blood vessels, and activates macrophages and myofibroblasts. It has been demonstrated that the primary ECM producers in hepatic fibrogenesis are hepatic stellate cells [93]. HSCs are in charge of the ECM turnover and process and store retinoids in the resting liver. Profibrotic stimuli cause HSCs to transdifferentiate into myofibroblasts [94]. According to a recent study, HSCs may play the following role in hepatitis caused by Concanavalin A (Con A). Con A injection causes the sinusoids to produce large amounts of inflammatory cytokines, including TNF-α and IFN-γ, initially by Kupffer cells and T cells and then, in a paracrine fashion, by HSCs. Significant liver damage results from this cytokine production amplification. Simultaneously, Kupffer cells release PGD2, which inhibits inflammatory responses, acts on HSC DP1, and limits the amplification loop of hepatic inflammation. Therefore, constitutive liver inflammation may cause the liver's DP1-PGD2 pathway to malfunction.

9.7 Crosstalk Between Cell Types in Metabolic Tissues

The paracrine communication networks established through eicosanoid signalling create integrated tissue responses that extend beyond individual cell types. In adipose tissue, the adipocyte-macrophage axis represents a well-characterized example. Adipocyte-derived free fatty acids activate macrophages to produce TNF-α, which then stimulates further fatty acid release from adipocytes through hormone-sensitive lipase activation. This creates a self-perpetuating inflammatory cycle mediated by both protein and lipid signals [95]. Tissue-specific microenvironments influence eicosanoid signalling through local factors. The oxygen tension, pH, and metabolite concentrations characteristic of different tissues alter enzyme activities and substrate availabilities, creating tissue-specific eicosanoid profiles even when the same cell types are present [96]. Systemic circulation of eicosanoids and their precursors allows for inter-tissue communication. Liver-derived 20-HETE can influence cardiovascular function, while adipose tissue-derived adipokines can alter hepatic eicosanoid metabolism. This systemic integration means that metabolic dysfunction in one tissue can influence eicosanoid signalling throughout the body [97]. The temporal dynamics of cell-type-specific eicosanoid production create coordinated tissue responses. Early-phase inflammatory eicosanoids produced by resident immune cells recruit additional cell types, while late-phase specialized pro-resolving mediators coordinate the transition to tissue repair and homeostasis. Understanding these temporal patterns is crucial for developing therapeutic interventions that work with, rather than against, natural resolution processes.

9.8 Pathophysiological Relevance in Metabolic Disorders

9.8.1 Obesity-Induced Inflammation and Lipid Imbalance

Obesity fundamentally alters eicosanoid homeostasis through multiple interconnected mechanisms that create and sustain chronic low-grade inflammation. The expansion of adipose tissue beyond its healthy storage capacity triggers a cascade of metabolic and inflammatory changes that profoundly impact eicosanoid biosynthesis and signalling. Adipose tissue hypoxia represents a critical early event in obesity-induced eicosanoid dysregulation. As adipocytes hypertrophy, local oxygen supply becomes insufficient, leading to HIF-1α activation and metabolic reprogramming. This hypoxic environment favours cyclooxygenase-2 (COX-2) upregulation while suppressing enzymes required for specialized pro-resolving mediator synthesis. The result is a fundamental shift from resolution-promoting to inflammation-sustaining eicosanoid profiles [98]. Free fatty acid flux increases dramatically in obesity, providing excess substrate for inflammatory eicosanoid synthesis. Saturated fatty acids like palmitate not only serve as COX and LOX substrates but also directly activate Toll-like receptor 4 (TLR4) signalling in immune cells. This dual mechanism—increased substrate availability combined with enhanced inflammatory activation—amplifies eicosanoid-mediated inflammatory signalling [99]. The macrophage phenotype switch in obese adipose tissue creates a self-reinforcing inflammatory environment. Crown-like structures form around dying adipocytes, where M1-polarized macrophages generate high levels of PGE2, LTB4, and other inflammatory eicosanoids. These mediators not only sustain local inflammation but also contribute to systemic inflammatory burden through spillover into systemic circulation. Adipokine dysregulation in obesity creates additional feedback loops with eicosanoid signalling. Reduced adiponectin and increased leptin resistance alter cellular responses to both inflammatory and anti-inflammatory eicosanoids. TNF-α and IL-6 overproduction enhances inflammatory eicosanoid biosynthesis while suppressing resolution pathways [100, 101].

9.8.2 Insulin Resistance and Lipid Mediator Dysregulation

Insulin resistance development involves complex interactions between altered eicosanoid signalling and disrupted metabolic homeostasis. Multiple classes of bioactive lipids contribute to insulin resistance through distinct but complementary mechanisms that impair insulin signal transduction at different levels. Diacylglycerol (DAG) accumulation in insulin-sensitive tissues represents a well-characterized mechanism of lipid-mediated insulin resistance. Membrane-bound DAG species, particularly those containing saturated fatty acids, activate protein kinase C θ (PKCθ) in skeletal muscle and PKCε in liver. These kinases phosphorylate serine residues on insulin receptor substrate-1 (IRS-1), directly impairing tyrosine phosphorylation and blocking downstream PI3K-Akt signalling [102]. Ceramide-mediated insulin resistance operates through different mechanisms, primarily targeting Akt/PKB activation. Saturated fatty acids promote ceramide synthesis through the de novo pathway, leading to activation of protein phosphatase 2A (PP2A) and atypical protein kinase C ζ (aPKCζ). These enzymes directly inhibit Akt phosphorylation and activity, creating a bottleneck in insulin signalling that prevents glucose uptake and glycogen synthesis [103]. Tissue-specific patterns of lipid-mediated insulin resistance reflect different cellular environments and metabolic demands. Skeletal muscle shows sensitivity to DAG-mediated PKC activation, while hepatic insulin resistance involves more complex interactions between multiple lipid species and inflammatory pathways. Adipose tissue insulin resistance creates a vicious cycle where impaired insulin signalling enhances lipolysis, increasing free fatty acid release and further promoting inflammatory eicosanoid production [104]. Subcellular localization of bioactive lipids significantly influences their insulin resistance-promoting effects. Plasma membrane-associated DAG shows stronger correlations with insulin resistance than total cellular DAG content. Similarly, endoplasmic reticulum (ER)-localized ceramides may have different effects than mitochondrial or plasma membrane pools [54]. Temporal dynamics of lipid accumulation and insulin resistance development reveal the progressive nature of metabolic dysfunction. Acute lipid exposure initially activates compensatory mechanisms, including enhanced insulin secretion and improved clearance pathways. However, chronic lipid excess overwhelms these protective systems, leading to sustained activation of insulin resistance pathways and eventual β-cell dysfunction.

9.8.3 Non-alcoholic Fatty Liver Disease (NAFLD): Inflammatory Lipidome Shifts

NAFLD progression involves dramatic alterations in hepatic eicosanoid metabolism that both reflect and drive disease advancement. The transition from simple steatosis through steatohepatitis to fibrosis and cirrhosis is accompanied by characteristic changes in lipid mediator profiles that serve both as biomarkers and therapeutic targets. Simple steatosis represents the earliest stage of NAFLD, characterized by triglyceride accumulation in hepatocytes without significant inflammation. Even at this stage, eicosanoid biosynthetic enzyme expression begins to change, with modest increases in COX-2 and alterations in cytochrome P450 activity. 20-HETE production increases progressively, correlating with the degree of hepatic fat accumulation. Steatohepatitis development involves a fundamental shift toward inflammatory eicosanoid dominance. Prostaglandin E2 (PGE2) production increases dramatically in response to hepatocellular lipotoxicity and inflammatory cell recruitment. Leukotriene synthesis by infiltrating neutrophils and activated Kupffer cells creates a pro-inflammatory microenvironment that sustains hepatic inflammation [105]. Hepatic fibrosis progression is directly mediated by specific eicosanoids that activate hepatic stellate cells; 20-HETE emerges as a key fibrogenic mediator, directly stimulating stellate cell activation through TGF-β pathway modulation. Studies in cirrhosis patients show that 20-HETE accounts for 50–75% of total hepatic eicosanoid production, far exceeding prostaglandin and leukotriene levels [105]. Resolution deficit represents a crucial but often overlooked aspect of NAFLD pathogenesis. Specialized pro-resolving mediator synthesis decreases progressively as NAFLD advances, preventing effective resolution of hepatic inflammation. Lipoxin A4, resolvins, and protectins show reduced production in advanced NAFLD, contributing to the chronic inflammatory state. Systemic implications of hepatic eicosanoid dysregulation extend beyond liver pathology. The liver's central role in metabolism means that altered hepatic eicosanoid production influences systemic inflammatory burden and metabolic homeostasis. Circulating 20-HETE levels correlate with cardiovascular disease risk in NAFLD patients, suggesting direct systemic effects of hepatic eicosanoid overproduction. Biomarker potential of eicosanoid profiles in NAFLD has generated significant clinical interest. Multi-eicosanoid panels can distinguish between simple steatosis and steatohepatitis with high accuracy, potentially replacing invasive liver biopsies for disease monitoring; 11,12-diHETrE, dhk PGD2, and 20-COOH AA represent particularly promising biomarkers that reflect different aspects of NAFLD pathophysiology [106]. Gender and genetic influences on NAFLD-associated eicosanoid dysregulation add complexity to disease patterns. Polymorphisms in cytochrome P450 genes influence 20-HETE production capacity and correlate with fibrosis severity. Hormonal factors affect eicosanoid biosynthetic enzyme expression, contributing to observed gender differences in NAFLD prevalence and progression. Therapeutic implications of understanding NAFLD lipidome shifts are profound. Selective inhibition of 20-HETE synthesis shows promise for preventing fibrosis progression. Enhancement of specialized pro-resolving mediator production through dietary omega-3 fatty acid supplementation or direct mediator administration represents another therapeutic avenue [106]. The pathophysiological relevance of eicosanoids in metabolic disorders extends far beyond simple inflammatory mediation. These molecules serve as crucial signalling intermediates that link metabolic dysfunction with inflammatory activation, creating self-perpetuating cycles that drive disease progression. Understanding these mechanisms opens new therapeutic possibilities for breaking these cycles and restoring metabolic health through targeted manipulation of eicosanoid signalling pathways.

9.9 Therapeutic Targeting of Eicosanoids in Metabolic Inflammation

9.9.1 Pharmacological Modulators of Eicosanoid Biosynthesis

Cyclooxygenase inhibitors represent the most clinically established approach to eicosanoid modulation, with decades of experience in inflammatory disease management. Non-selective NSAIDs like aspirin and ibuprofen inhibit both COX-1 and COX-2, reducing prostaglandin production but also causing gastrointestinal complications due to impaired gastroprotective PGE2 synthesis. Selective COX-2 inhibitors were developed to maintain anti-inflammatory efficacy while reducing gastro intestinal (GI) toxicity, but cardiovascular safety concerns have limited their widespread use [107, 108]. Aspirin's unique mechanism extends beyond simple COX inhibition. Low-dose aspirin irreversibly acetylates COX-2, converting it from a prostaglandin-synthesizing enzyme to a 15R-HETE-producing enzyme. These 15R-HETE products serve as precursors for aspirin-triggered specialized pro-resolving mediators (AT-SPMs), including aspirin-triggered lipoxins and aspirin-triggered resolvins. This mechanism explains aspirin's unique cardioprotective effects and suggests that its benefits may involve active resolution enhancement rather than simple anti-inflammatory action [109–111]. Lipoxygenase modulation offers additional therapeutic opportunities; 5-LOX inhibitors like zileuton effectively reduce leukotriene production and show efficacy in asthma treatment. However, their application to metabolic disorders remains limited, partly due to the recognition that complete 5-LOX inhibition may also prevent beneficial lipoxin synthesis. Selective leukotriene receptor antagonists like montelukast provide targeted inhibition of pro-inflammatory signalling while preserving other 5-LOX product functions [112, 113]. Cytochrome P450 enzyme modulation represents an emerging therapeutic approach; 20-HETE synthesis inhibitors show promise for treating hepatic fibrosis and cardiovascular disease. Soluble epoxide hydrolase (sEH) inhibitors aim to preserve beneficial EET levels while preventing their conversion to potentially harmful DiHETE metabolites. These approaches target specific CYP pathways rather than broadly inhibiting eicosanoid synthesis [114, 115]. Dual inhibition strategies attempt to address the complexity of eicosanoid networks. COX-2/5-LOX dual inhibitors like licofelone aim to reduce both prostaglandin and leukotriene production while potentially preserving resolution pathways. Flavocoxid, a natural dual inhibitor, has shown survival benefits in sepsis models by reducing inflammatory eicosanoids while increasing lipoxin levels [116, 117].

9.9.2 Nutritional and Lifestyle Interventions Affecting Lipid Mediators

Omega-3 fatty acid supplementation represents the most extensively studied nutritional approach to eicosanoid modulation. Eicosapentaenoic acid (EPA) and docosahexaenoic acid (DHA) serve as direct precursors for specialized pro-resolving mediators, including resolvins, protectins, and maresins. Clinical studies demonstrate that omega-3 supplementation can shift eicosanoid profiles toward resolution-promoting patterns, reducing inflammatory markers and improving metabolic parameters [118, 119]. EPA supplementation specifically competes with arachidonic acid as a COX substrate, producing 3-series prostaglandins and 5-series leukotrienes that are generally less inflammatory than their arachidonic acid-derived counterparts. DHA supplementation provides substrate for D-series resolvins and protectins, directly enhancing endogenous resolution capacity. The optimal EPA:DHA ratio for metabolic benefits remains under investigation, with some evidence suggesting that higher EPA ratios may be more effective for inflammatory conditions [120–122]. Dietary fat quality profoundly influences tissue eicosanoid profiles. Saturated fat consumption increases tissue arachidonic acid levels and promotes inflammatory eicosanoid synthesis. Monounsaturated fats like those found in olive oil provide oleic acid, which can be converted to oleoylethanolamide (OEA), an endocannabinoid-related molecule with anti-inflammatory and metabolic benefits. Trans fats disrupt normal eicosanoid metabolism and should be avoided [123, 124]. Caloric restriction and intermittent fasting influence eicosanoid metabolism through multiple mechanisms. Weight loss reduces adipose tissue inflammatory burden and associated eicosanoid overproduction. Fasting periods may enhance specialized pro-resolving mediator synthesis while reducing inflammatory eicosanoid production. Exercise training complements dietary interventions by improving tissue oxidative capacity and reducing inflammatory signalling [125–127]. Polyphenol consumption can modulate eicosanoid pathways through direct enzyme inhibition and indirect effects on gene expression. Curcumin inhibits COX-2 and 5-LOX while activating PPARγ, creating a dual anti-inflammatory and pro-resolution effect. Green tea catechins can inhibit phospholipase A2 activity, reducing substrate availability for inflammatory eicosanoid synthesis. Resveratrol influences eicosanoid metabolism through SIRT1 activation and subsequent effects on inflammatory gene expression.

9.9.3 Emerging Strategies: Lipidomics, Precision Therapies, and Drug Repositioning

Lipidomics-guided therapy represents a paradigm shift toward personalized eicosanoid-targeted interventions. High-throughput lipid profiling can identify individual eicosanoid signatures that predict disease risk and therapeutic responses. Multi-eicosanoid biomarker panels show promise for early detection of metabolic dysfunction and monitoring of therapeutic interventions. Precision dosing of omega-3 fatty acids based on individual lipid profiles may optimize therapeutic outcomes. Genetic polymorphisms in eicosanoid biosynthetic enzymes influence individual responses to dietary and pharmacological interventions. Fatty Acid Desaturase (FADS1/2) gene variants affect omega-3 fatty acid metabolism and may guide personalized supplementation strategies. Specialized pro-resolving mediator therapy is emerging as a novel therapeutic approach. Synthetic resolvins, protectins, and maresins are being developed as direct therapeutic agents. These metabolically stable analogues resist degradation while maintaining biological activity. RvE1 analogues show promise for treating inflammatory diseases, while maresin analogues demonstrate organ-protective effects [128]. Drug repositioning based on eicosanoid mechanisms offers rapid therapeutic development opportunities. Metformin, widely used for diabetes treatment, influences eicosanoid metabolism through activated protein kinase (AMPK) activation and may enhance resolution signalling. Statins affect eicosanoid synthesis through effects on membrane composition and enzyme expression. Fibrates, PPARα agonists, directly modulate eicosanoid-responsive gene expression [129]. Combination therapy approaches aim to target multiple aspects of eicosanoid dysregulation simultaneously. Omega-3 fatty acids combined with COX-2 inhibitors may enhance anti-inflammatory effects while providing resolution pathway substrate. PPAR agonists combined with specialized pro-resolving mediators could address both transcriptional and post-translational aspects of inflammatory resolution [128]. Delivery system innovations are addressing challenges in eicosanoid-based therapeutics. Liposomal formulations protect sensitive lipid mediators from degradation while targeting specific tissues. Nanoparticle delivery systems can enhance bioavailability and provide sustained release of eicosanoid modulators. Prodrug approaches allow oral administration of otherwise unstable lipid mediators. Biomarker-guided monitoring enables personalized adjustment of eicosanoid-targeted therapies. Real-time lipid profiling can track therapeutic responses and guide dose optimization. Genetic testing for eicosanoid pathway variants may predict therapeutic responses and guide treatment selection. Inflammatory resolution indices provide quantitative measures of therapeutic efficacy beyond simple inflammatory marker reduction [130]. The therapeutic targeting of eicosanoids in metabolic inflammation represents a rapidly evolving field that bridges traditional pharmacology with cutting-edge precision medicine approaches. Success in this area requires understanding the complex networks of eicosanoid signalling, individual genetic and metabolic variations, and the temporal dynamics of inflammatory and resolution processes. As our understanding of these systems continues to advance, more sophisticated and effective therapeutic strategies will undoubtedly emerge, offering new hope for treating the growing burden of metabolic inflammatory diseases.

9.10 Conclusions and Future Perspectives

The comprehensive examination of eicosanoids and bioactive lipids in metabolic disorders reveals a sophisticated regulatory network that fundamentally challenges traditional views of inflammation as a simple pathological process. Rather than merely mediating tissue damage, eicosanoids orchestrate complex biological programs that encompass inflammatory initiation, immune cell recruitment, pathogen clearance, tissue protection, and active resolution. The temporal orchestration of lipid mediator networks represents a crucial paradigm shift in our understanding. Early phase eicosanoids like prostaglandins and leukotrienes serve essential protective functions by promoting immune cell recruitment and pathogen clearance. However, the failure to transition to resolution-promoting specialized pro-resolving mediators leads to chronic inflammatory states that characterize metabolic disorders. This temporal dysregulation, rather than excessive inflammatory initiation, may represent the fundamental pathological process in conditions like obesity, insulin resistance, and NAFLD [131]. Tissue-specific eicosanoid signatures emerge as critical determinants of disease progression and therapeutic response. The liver's predominance of 20-HETE production in metabolic disease contrasts with adipose tissue's complex balance of prostaglandins, leukotrienes, and resolution mediators. These tissue-specific patterns suggest that successful therapeutic interventions must account for local microenvironmental factors rather than applying uniform systemic approaches. Inter-organ communication through circulating eicosanoids creates system-wide effects that extend beyond local tissue pathology. Hepatic 20-HETE overproduction influences cardiovascular function, while adipose tissue inflammatory eicosanoids contribute to systemic insulin resistance. This recognition of eicosanoids as inter-tissue signalling molecules opens new therapeutic possibilities for targeting systemic metabolic dysfunction through localized interventions [132, 133]. The integration of metabolic and inflammatory signalling through eicosanoid pathways represents a fundamental biological principle with broad implications. Lipid metabolism intermediates like diacylglycerols and ceramides directly influence inflammatory signalling, while inflammatory eicosanoids alter metabolic enzyme activities and insulin sensitivity. This bidirectional relationship means that therapeutic interventions targeting either metabolic or inflammatory pathways may have unexpected effects on the other system.

9.10.1 Challenges and Opportunities in Therapeutic Modulation

Therapeutic complexity arises from the dual nature of many eicosanoids and their context-dependent effects. PGE2, for example, can simultaneously promote tissue damage through pain and fever induction while protecting against excessive inflammation through anti-inflammatory cytokine upregulation. Similarly, complete inhibition of 5-LOX reduces harmful leukotriene production but may also prevent beneficial lipoxin synthesis. These complexities necessitate more sophisticated therapeutic approaches than simple pathway inhibition. Individual variability in eicosanoid metabolism presents both challenges and opportunities for personalized medicine. Genetic polymorphisms in biosynthetic enzymes, particularly FADS1/2 and cytochrome P450 variants, significantly influence individual responses to dietary and pharmacological interventions. Pharmacogenomic approaches to eicosanoid-targeted therapy may optimize treatment outcomes while minimizing adverse effects. Drug development challenges include the chemical instability of many bioactive lipids and the complexity of delivery systems required for therapeutic efficacy. Specialized pro-resolving mediators are particularly susceptible to oxidation and enzymatic degradation, requiring innovative formulation approaches. However, these challenges also present opportunities for developing novel drug delivery technologies and metabolically stable analogues that retain biological activity. Biomarker development offers significant opportunities for improving therapeutic monitoring and patient selection. Multi-eicosanoid panels can provide detailed portraits of inflammatory and resolution status, enabling more precise therapeutic targeting. Real-time lipid profiling technologies may eventually allow dynamic adjustment of therapeutic interventions based on changing eicosanoid profiles.

9.10.2 Emerging Technologies in Lipid Mediator Research

Mass spectrometry advances are revolutionizing our ability to study eicosanoid networks in health and disease. High-resolution accurate mass spectrometry combined with comprehensive lipid databases now enables identification and quantification of hundreds of bioactive lipids simultaneously. Spatial lipidomics techniques are beginning to reveal tissue-specific eicosanoid distributions and their relationships to pathological processes. Systems biology approaches are essential for understanding the complexity of eicosanoid networks. Mathematical modelling of eicosanoid biosynthetic pathways can predict therapeutic outcomes and identify optimal intervention points. Network analysis reveals unexpected connections between different lipid mediator families and their roles in disease progression. Artificial intelligence applications in lipidomics research are accelerating biomarker discovery and therapeutic target identification. Machine learning algorithms can identify complex patterns in multidimensional lipid datasets that exceed human analytical capabilities. Predictive modelling based on eicosanoid profiles may enable early disease detection and prevention. Synthetic biology approaches offer possibilities for engineered biosynthesis of specialized pro-resolving mediators. Recombinant enzyme systems could provide cost-effective production of therapeutic lipid mediators. Cell-based delivery systems engineered to produce specific eicosanoids at disease sites represent an innovative therapeutic approach. Nanotechnology applications are addressing delivery challenges for lipid-based therapeutics. Lipid nanoparticles can protect sensitive mediators while providing targeted delivery to specific tissues. Smart delivery systems that respond to local inflammatory conditions could provide precise spatial and temporal control of eicosanoid release. The future of eicosanoid research and therapeutic development lies in embracing the complexity of these systems rather than seeking overly simplified solutions. Success will require interdisciplinary collaboration between lipid biochemists, immunologists, systems biologists, and clinical researchers. Most importantly, future therapeutic approaches must work with natural resolution mechanisms rather than against them, enhancing the body's endogenous capacity for inflammatory resolution and tissue repair. The potential impact of eicosanoid-targeted therapeutics extends far beyond metabolic disorders. As we better understand the role of specialized pro-resolving mediators in various diseases, applications may expand to cancer, cardiovascular disease, neurodegenerative disorders, and infectious diseases. The principles learned from studying metabolic inflammation may prove broadly applicable to the resolution of pathological inflammation across diverse disease contexts. Public health implications of this research are profound. By understanding how dietary factors, lifestyle interventions, and environmental exposures influence eicosanoid signalling, we may develop more effective prevention strategies for metabolic diseases. The integration of nutritional science with advanced lipidomics may enable precision nutrition approaches tailored to individual eicosanoid profiles. In conclusion, the field of eicosanoids and bioactive lipids in metabolic inflammation represents a remarkable convergence of fundamental biology, cutting-edge technology, and urgent clinical need. The insights gained from this research promise to transform our understanding of inflammatory diseases and provide new tools for promoting human health and treating metabolic disorders that affect billions of people worldwide.

References

1. Godoy-Matos AF, Silva Júnior WS, Valerio CM (2020) NAFLD as a continuum: from obesity to metabolic syndrome and diabetes. Diabetol Metab Syndr [Internet] 12(1):60. [cited 2025 Aug 15]. Available from: https://pmc.ncbi.nlm.nih.gov/articles/PMC7359287/
2. Blackstone RP (2016) Obesity-related diseases and syndromes: insulin resistance, type 2 diabetes mellitus, non-alcoholic fatty liver disease, cardiovascular disease, and metabolic syndrome. Obesity [Internet]:83–108. [cited 2025 Aug 15]. Available from: https://link.springer.com/chapter/10.1007/978-3-319-39409-1_5
3. Oladipupo SO, Ezenabor EH, Ojo AB, Ogunlakin AD, Ojo OA (2025) Interplay of the pathophysiological mechanisms of non-alcoholic fatty liver disease, diabetes mellitus, and inflammation: a growing threat to public health. Obes Med [Internet]. 55:100613. [cited 2025 Aug 15]. Available from: https://www.sciencedirect.com/science/article/pii/S2451847625000338
4. Jarc E, Petan T (2020) A twist of FATe: lipid droplets and inflammatory lipid mediators. Biochimie [Internet] 169:69–87. [cited 2025 Aug 15]. Available from: https://www.sciencedirect.com/science/article/pii/S0300908419303396
5. Kaffe E, Tisi A, Magkrioti C, Aidinis V, Mehal WZ, Flavell RA et al (2024) Bioactive signalling lipids as drivers of chronic liver diseases. J Hepatol [Internet] 80(1):140–154. [cited 2025 Aug 15]. Available from: https://www.sciencedirect.com/science/article/pii/S0168827823050997
6. Yoon H, Shaw JL, Haigis MC, Greka A (2021) Lipid metabolism in sickness and in health: emerging regulators of lipotoxicity. Mol Cell [Internet] 81(18):3708–3730. [cited 2025 Aug 15]. Available from: https://www.cell.com/action/showFullText?pii=S1097276521006948
7. Santos AL, Preta G (2018) Lipids in the cell: organisation regulates function. Cell Mol Life Sci [Internet] 75(11):1909–1927. [cited 2025 Aug 15]. Available from: https://link.springer.com/article/10.1007/s00018-018-2765-4

8. Khanapure SP, Garvey DS, Janero DR, Letts LG (2007) Eicosanoids in inflammation: biosynthesis, pharmacology, and therapeutic frontiers. Curr Top Med Chem [Internet] 7(3):311–340. [cited 2025 Jul 29]. Available from: https://www.eurekaselect.com/article/22773
9. Dennis EA, Norris PC (2015) Eicosanoid storm in infection and inflammation. Nat Rev Immunol [Internet] 15(8):511–523. [cited 2025 Jul 29]. Available from: https://www.nature.com/articles/nri3859
10. Zurier RB (2017) Prostaglandins, leukotrienes, and related compounds. In: Kelley and Firestein's textbook of rheumatology: volumes 1–2, 10th edn, pp 366–383.e3. [cited 2025 Jul 29]. Available from: https://www.sciencedirect.com/science/article/abs/pii/B9780323316965000243
11. Braza MKE, Dennis EA, Amaro RE (2025) Conformational dynamics and activation of membrane-associated human group IVA cytosolic phospholipase A2 (cPLA2). J Phys Chem Lett [Internet] 16(24):6059–6065. [cited 2025 Jul 29]. Available from: https://doi.org/10.1021/acs.jpclett.5c00860?ref=article_openPDF
12. Khan SA, Ilies MA (2023) The phospholipase A2 superfamily: structure, isozymes, catalysis, physiologic and pathologic roles. Int J Mol Sci [Internet] 24:1353. [cited 2025 Jul 29]. Available from: https://www.mdpi.com/1422-0067/24/2/1353/htm
13. Morita I (2002) Distinct functions of COX-1 and COX-2. Prostaglandins Other Lipid Mediat [Internet]:68–69. 165–75. [cited 2025 Jul 29]. Available from: https://www.sciencedirect.com/science/article/abs/pii/S0090698002000291
14. Chang YC, Yang SF, Huang FM, Liu CM, Tai KW, Hsieh YS (2003) Proinflammatory cytokines induce cyclooxygenase-2 mRNA and protein expression in human pulp cell cultures. J Endodont 29(3):201–204
15. Wang B, Wu L, Chen J, Dong L, Chen C, Wen Z et al (2021) Metabolism pathways of arachidonic acids: mechanisms and potential therapeutic targets. Signal Transduct Target Ther [Internet] 6(1):1–30. [cited 2025 Jul 29]. Available from: https://www.nature.com/articles/s41392-020-00443-w
16. Mashima R, Okuyama T (2015) The role of lipoxygenases in pathophysiology; new insights and future perspectives. Redox Biol [Internet] 6:297–310. [cited 2025 Jul 29]. Available from: https://www.sciencedirect.com/science/article/pii/S2213231715000968?via%3Dihub
17. Ramalho R, Guimarães C, Santareno S, Cerqueira J, Português J, Cruz J (2020) Erratum: the role of adipose tissue and macrophages in chronic inflammation associated with obesity – clinical implications (Acta Medica Portuguesa (2008) 21 (489–496). https://doi.org/10.20344/amp.1635). Acta Med Port 33(3):217
18. Wood I, Trostchansky A, Rubbo H (2020) Structural considerations on lipoxygenase function, inhibition and crosstalk with nitric oxide pathways. Biochimie [Internet] 178:170–180. [cited 2025 Jul 29]. Available from: https://www.sciencedirect.com/science/article/abs/pii/S0300908420302352
19. Serhan CN, Yacoubian S, Yang R (2008) Anti-inflammatory and proresolving lipid mediators. Annu Rev Pathol Mech Dis [Internet] 3(Volume 3, 2008):279–312. [cited 2025 Jul 29]. Available from: https://www.annualreviews.org/content/journals/10.1146/annurev.pathmechdis.3.121806.151409
20. Shoieb SM, El-Ghiaty MA, Alqahtani MA, El-Kadi AOS (2020) Cytochrome P450-derived eicosanoids and inflammation in liver diseases. Prostaglandins Other Lipid Mediat [Internet] 147:106400. [cited 2025 Jul 29]. Available from: https://www.sciencedirect.com/science/article/abs/pii/S1098882319301510?via%3Dihub
21. Alba MM, Ebright B, Hua B, Slarve I, Zhou Y, Jia Y et al (2023) Eicosanoids and other oxylipins in liver injury, inflammation and liver cancer development. Front Physiol 14:1098467
22. Khanapure S, Garvey D, Janero D, Gordon Letts L (2007) Eicosanoids in Inflammation: biosynthesis, pharmacology, and therapeutic frontiers. Curr Top Med Chem [Internet] 7(3):311–340. [cited 2025 Jul 29]. Available from: https://pubmed.ncbi.nlm.nih.gov/17305573/

23. Forrester SJ, Kikuchi DS, Hernandes MS, Xu Q, Griendling KK (2018) Reactive oxygen species in metabolic and inflammatory signaling. Circ Res [Internet] 122(6):877–902. [cited 2025 Jul 29]. Available from: https://doi.org/10.1161/CIRCRESAHA.117.311401?download=true
24. Masenga SK, Kabwe LS, Chakulya M, Kirabo A (2023) Mechanisms of Oxidative Stress in Metabolic Syndrome. Int J Mol Sci [Internet] 24(9):7898. [cited 2025 Jul 29]. Available from: https://pmc.ncbi.nlm.nih.gov/articles/PMC10178199/
25. Griesser M, Suzuki T, Tejera N, Mont S, Boeglin WE, Pozzi A et al (2011) Biosynthesis of hemiketal eicosanoids by cross-over of the 5-lipoxygenase and cyclooxygenase-2 pathways. Proc Natl Acad Sci U S A [Internet] 108(17):6945–6950. [cited 2025 Jul 29]. Available from: https://doi.org/10.1073/pnas.1019473108?download=true
26. Gervois P, Torra IP, Fruchart JC, Staels B (2000) Regulation of lipid and lipoprotein metabolism by PPAR activators. Clin Chem Lab Med [Internet] 38(1):3–11. [cited 2025 Jul 29]. Available from: https://www.degruyterbrill.com/document/doi/10.1515/CCLM.2000.002/html
27. Cuzzo B, Lappin SL (2023) Physiology, Leukotrienes. StatPearls. [cited 2025 Jul 29]. Available from: https://www.ncbi.nlm.nih.gov/books/NBK526114/
28. Kanaoka Y, Boyce JA (2014) Cysteinyl leukotrienes and their receptors; emerging concepts. Allergy Asthma Immunol Res [Internet] 6(4):288–295. [cited 2025 Jul 29]. Available from: https://doi.org/10.4168/aair.2014.6.4.288
29. Fredman G, Serhan CN (2024) Specialized pro-resolving mediators in vascular inflammation and atherosclerotic cardiovascular disease. Nat Rev Cardiol [Internet] 21(11):808–823. [cited 2025 Jul 29]. Available from: https://www.nature.com/articles/s41569-023-00984-x
30. Natami M, Hosseini SM, Khaleel RA, Addulrahman TS, Zarei M, Asadi S et al (2024) The role of specialized pro-resolving mediators (SPMs) in inflammatory arthritis: A therapeutic strategy. Prostaglandins Other Lipid Mediat [Internet]. 170:106798. [cited 2025 Jul 29]. Available from: https://www.sciencedirect.com/science/article/abs/pii/S1098882323000953
31. Basil MC, Levy BD (2015) Specialized pro-resolving mediators: endogenous regulators of infection and inflammation. Nat Rev Immunol [Internet] 16(1):51–67. [cited 2025 Jul 29]. Available from: https://www.nature.com/articles/nri.2015.4
32. Duvall MG, Levy BD (2015) DHA- and EPA-derived resolvins, protectins, and maresins in airway inflammation. Eur J Pharmacol [Internet]. 785:144. [cited 2025 Jul 29]. Available from: https://pmc.ncbi.nlm.nih.gov/articles/PMC4854800/
33. Bazan NG (2009) Neuroprotectin D1-mediated anti-inflammatory and survival signaling in stroke, retinal degenerations, and Alzheimer's disease. J Lipid Res [Internet] 50(Suppl):S400–S405. [cited 2025 Aug 3]. Available from: https://www.sciencedirect.com/science/article/pii/S0022227520306465
34. Fattori V, Pinho-Ribeiro FA, Staurengo-Ferrari L, Borghi SM, Rossaneis AC, Casagrande R et al (2019) The specialised pro-resolving lipid mediator maresin 1 reduces inflammatory pain with a long-lasting analgesic effect. Br J Pharmacol [Internet] 176(11):1728. [cited 2025 Aug 3]. Available from: https://pmc.ncbi.nlm.nih.gov/articles/PMC6514290/
35. Liu WC, Yang YH, Wang YC, Chang WM, Wang CW (2023) Maresin: macrophage mediator for resolving inflammation and bridging tissue regeneration—a system-based preclinical systematic review. Int J Mol Sci [Internet] 24(13):11012. [cited 2025 Aug 3]. Available from: https://www.mdpi.com/1422-0067/24/13/11012/htm
36. Hammond VJ, O'Donnell VB (2011) Esterified eicosanoids: generation, characterization and function. Biochim Biophys Acta [Internet] 1818(10):2403. [cited 2025 Aug 3]. Available from: https://pmc.ncbi.nlm.nih.gov/articles/PMC3740819/
37. Romano M, Cianci E, Simiele F, Recchiuti A (2015) Lipoxins and aspirin-triggered lipoxins in resolution of inflammation. Eur J Pharmacol [Internet]. 760:49–63. [cited 2025 Aug 3]. Available from: https://www.sciencedirect.com/science/article/abs/pii/S0014299915003295
38. Gilligan MM, Gartung A, Sulciner ML, Norris PC, Sukhatme VP, Bielenberg DR et al (2019) Aspirin-triggered proresolving mediators stimulate resolution in cancer. Proc Natl Acad Sci

USA [Internet]. 116(13):6292–6297. [cited 2025 Aug 3]. Available from: https://pubmed.ncbi.nlm.nih.gov/30862734/
39. Bäck M, Hansson G (2018) Basic mechanisms of atherosclerosis. In: Chronic coronary artery disease: a companion to braunwald's heart disease [Internet], pp 45–54. [cited 2025 Aug 3]. Available from: https://www.sciencedirect.com/science/article/abs/pii/B9780323428804000042
40. Mao H, Zhao X, Sun SC (2025) NF-κB in inflammation and cancer. Cell Mol Immunol [Internet] 22(8):811–839. [cited 2025 Aug 3]. Available from: https://www.nature.com/articles/s41423-025-01310-w
41. Guo Q, Jin Y, Chen X, Ye X, Shen X, Lin M et al (2024) NF-κB in biology and targeted therapy: new insights and translational implications. Signal Transduct Target Ther [Internet]. 9(1):1–37. [cited 2025 Aug 3]. Available from: https://www.nature.com/articles/s41392-024-01757-9
42. Nakajima S, Kitamura M (2013) Bidirectional regulation of NF-κB by reactive oxygen species: a role of unfolded protein response. Free Radic Biol Med [Internet] 65:162–174. [cited 2025 Aug 3]. Available from: https://www.sciencedirect.com/science/article/abs/pii/S0891584913003018
43. Moens U, Kostenko S, Sveinbjørnsson B (2013) The role of Mitogen-Activated Protein Kinase-Activated Protein Kinases (MAPKAPKs) in inflammation. Genes (Basel) [Internet] 4(2):101. [cited 2025 Aug 3]. Available from: https://pmc.ncbi.nlm.nih.gov/articles/PMC3899974/
44. Zlobin A, Bloodworth JC, Osipo C (2019) Mitogen-Activated Protein Kinase (MAPK) signaling. In: Predictive biomarkers in oncology: applications in precision medicine [Internet], pp 213–221. [cited 2025 Aug 3]. Available from: https://link.springer.com/chapter/10.1007/978-3-319-95228-4_16
45. Oeckinghaus A, Hayden MS, Ghosh S (2011) Crosstalk in NF-κB signaling pathways. Nat Immunol [Internet] 12(8):695–708. [cited 2025 Aug 3]. Available from: https://www.nature.com/articles/ni.2065
46. Gottschalk RA, Martins AJ, Angermann BR, Dutta B, Ng CE, Uderhardt S et al (2016) Distinct NF-κB and MAPK Activation Thresholds Uncouple Steady-State Microbe Sensing from Anti-pathogen Inflammatory Responses. Cell Syst [Internet]. 2(6):378–390. [cited 2025 Aug 3]. Available from: https://www.cell.com/action/showFullText?pii=S2405471216301508
47. Asgharzadeh F, Memarzia A, Alikhani V, Beigoli S, Boskabady MH (2024) Peroxisome proliferator-activated receptors: key regulators of tumor progression and growth. Transl Oncol [Internet] 47:102039. [cited 2025 Aug 3]. Available from: https://www.sciencedirect.com/science/article/pii/S1936523324001669
48. Chinetti G, Fruchart JC, Staels B (2000) Peroxisome proliferator-activated receptors (PPARs): nuclear receptors at the crossroads between lipid metabolism and inflammation. Inflamm Res [Internet] 49(10):497–505. [cited 2025 Aug 3]. Available from: https://link.springer.com/article/10.1007/s000110050622
49. Kersten S (2014) Integrated physiology and systems biology of PPARα. Mol Metab [Internet] 3(4):354–371. [cited 2025 Aug 3]. Available from: https://www.sciencedirect.com/science/article/pii/S221287781400043X
50. Pawlak M, Lefebvre P, Staels B (2015) Molecular mechanism of PPARα action and its impact on lipid metabolism, inflammation and fibrosis in non-alcoholic fatty liver disease. J Hepatol [Internet]. 62(3):720–733. [cited 2025 Aug 3]. Available from: https://www.sciencedirect.com/science/article/pii/S016882781400806X
51. Meng X, Wang L, Du YC, Cheng D, Zeng T (2024) PPARβ/δ as a promising molecular drug target for liver diseases: a focused review. Clin Res Hepatol Gastroenterol [Internet]. 48(6):102343. [cited 2025 Aug 3]. Available from: https://www.sciencedirect.com/science/article/abs/pii/S2210740124000640
52. Straus DS, Glass CK (2007) Anti-inflammatory actions of PPAR ligands: new insights on cellular and molecular mechanisms. Trends Immunol [Internet] 28(12):551–558. [cited

2025 Aug 3]. Available from: https://www.sciencedirect.com/science/article/abs/pii/S147149060700244X
53. Kolczynska K, Loza-Valdes A, Hawro I, Sumara G (2020) Diacylglycerol-evoked activation of PKC and PKD isoforms in regulation of glucose and lipid metabolism: a review. Lipids Health Dis [Internet] 19(1):1–15. [cited 2025 Aug 3]. Available from: https://lipidworld.biomedcentral.com/articles/10.1186/s12944-020-01286-8
54. Campomanes P, Zoni V, Vanni S (2019) Local accumulation of diacylglycerol alters membrane properties nonlinearly due to its transbilayer activity. Commun Chem [Internet] 2(1):1–8. [cited 2025 Aug 3]. Available from: https://www.nature.com/articles/s42004-019-0175-7
55. Eichmann TO, Lass A (2015) DAG tales: the multiple faces of diacylglycerol—stereochemistry, metabolism, and signaling. Cell Mol Life Sci [Internet]. 72(20):3931. [cited 2025 Aug 3]. Available from: https://pmc.ncbi.nlm.nih.gov/articles/PMC4575688/
56. Sokolowska E, Blachnio-Zabielska A (2019) The role of ceramides in insulin resistance. Front Endocrinol (Lausanne) [Internet]. 10:436871. [cited 2025 Aug 3]. Available from: www.frontiersin.org
57. Larsen PJ, Tennagels N (2014) On ceramides, other sphingolipids and impaired glucose homeostasis. Mol Metab [Internet]. 3(3):252–260. [cited 2025 Aug 3]. Available from: https://www.sciencedirect.com/science/article/pii/S2212877814000167
58. Zhang C, Klett EL, Coleman RA (2013) Lipid signals and insulin resistance. Clin Lipidol [Internet] 8(6):659–667. [cited 2025 Aug 3]. Available from: https://www.tandfonline.com/doi/abs/10.2217/clp.13.67
59. Vona R, Gambardella L, Cittadini C, Straface E, Pietraforte D (2019) Biomarkers of oxidative stress in metabolic syndrome and associated diseases. Oxid Med Cell Longev [Internet] 2019(1):8267234. [cited 2025 Jul 29]. Available from: https://doi.org/10.1155/2019/8267234
60. Venditti P, Balestrieri M, Di MS (1997) Effect of thyroid state on lipid peroxidation, antioxidant defences, and susceptibility to oxidative stress in rat tissues. J Endocrinol 155:151–157
61. Colak E, Pap D (2021) The role of oxidative stress in the development of obesity and obesity-related metabolic disorders. J Med Biochem [Internet] 40(1):1. [cited 2025 Jul 29]. Available from: https://pmc.ncbi.nlm.nih.gov/articles/PMC7857849/
62. Muntean C, Sasaran MO, Crisan A, Banescu C (2022) Effects of PPARG and PPARGC1A gene polymorphisms on obesity markers. Front Public Health 10:962852
63. Hajimohammadreza I, Brammer M (1990) Brain membrane fluidity and lipid peroxidation in Alzheimer's disease. Neurosci Lett [Internet] 112(2–3):333–337. [cited 2025 Aug 5]. Available from: https://pubmed.ncbi.nlm.nih.gov/2359533/
64. Free ND (2015) Radicals and oxidative stress in neurodegenerative disorders. In: Free radicals in human health and disease [Internet], pp 143–158. [cited 2025 Aug 5]. Available from: https://link.springer.com/chapter/10.1007/978-81-322-2035-0_11
65. Markesbery WR (1997) Oxidative stress hypothesis in alzheimer's disease. Free Radic Biol Med [Internet]. 23(1):134–147. [cited 2025 Aug 5]. Available from: https://www.sciencedirect.com/science/article/pii/S0891584996006296
66. Esterbauer H, Schaur RJ, Zollner H (1991) Chemistry and biochemistry of 4-hydroxynonenal, malonaldehyde and related aldehydes. Free Radic Biol Med [Internet]. 11(1):81–128. [cited 2025 Aug 5]. Available from: https://pubmed.ncbi.nlm.nih.gov/1937131/
67. Calde PC (2020) Eicosanoids. Essays Biochem [Internet]. 64(3):423–441. [cited 2025 Aug 5]. Available from: /essaysbiochem/article/64/3/423/226134/Eicosanoids
68. Wymann MP, Schneiter R (2008) Lipid signalling in disease. Nat Rev Mol Cell Biol [Internet] 9(2):162–176. [cited 2025 Aug 9]. Available from: https://www.nature.com/articles/nrm2335
69. Lennicke C, Cochemé HM (2021) Redox metabolism: ROS as specific molecular regulators of cell signaling and function. Mol Cell [Internet]. 81(18):3691–3707. [cited 2025 Aug 10]. Available from: https://www.cell.com/action/showFullText?pii=S1097276521006857
70. Veal E, Day A (2011) Hydrogen peroxide as a signaling molecule. Antioxid Redox Signal 15(1):147–151

71. Clària J, Nguyen BT, Madenci AL, Keith Ozaki C, Serhan CN (2013) Diversity of lipid mediators in human adipose tissue depots. Am J Physiol Cell Physiol [Internet] 304(12):1141–1149. [cited 2025 Aug 10]. Available from: https://doi.org/10.1152/ajpcell.00351.2012
72. Serhan CN, Brain SD, Buckley CD, Gilroy DW, Haslett C, O'Neill LAJ et al (2007) Resolution of inflammation: state of the art, definitions and terms. FASEB J [Internet] 21(2):325–332. [cited 2025 Aug 10]. Available from: https://pubmed.ncbi.nlm.nih.gov/17267386/
73. Serhan CN, Chiang N, Van Dyke TE (2008) Resolving inflammation: dual anti-inflammatory and pro-resolution lipid mediators. Nat Rev Immunol [Internet] 8(5):349–361. [cited 2025 Aug 10]. Available from: https://pubmed.ncbi.nlm.nih.gov/18437155/
74. Serhan CN (2010) Novel lipid mediators and resolution mechanisms in acute inflammation: to resolve or not? Am J Pathol [Internet] 177(4):1576–1591. [cited 2025 Aug 10]. Available from: https://pubmed.ncbi.nlm.nih.gov/20813960/
75. Börgeson E, Docherty NG, Murphy M, Rodgers K, Ryan A, O'Sullivan TP et al (2011) Lipoxin A 4 and benzo-lipoxin A 4 attenuate experimental renal fibrosis. FASEB J [Internet] 25(9):2967–2979. [cited 2025 Aug 10]. Available from: https://pubmed.ncbi.nlm.nih.gov/21628447/
76. Hellmann J, Tang Y, Spite M (2012) Proresolving lipid mediators and diabetic wound healing. Curr Opin Endocrinol Diabetes Obes [Internet]. 19(2):104–108. [cited 2025 Aug 10]. Available from: https://pubmed.ncbi.nlm.nih.gov/22374140/
77. Bohr S, Patel SJ, Sarin D, Irimia D, Yarmush ML, Berthiaume F (2013) Resolvin D2 prevents secondary thrombosis and necrosis in a mouse burn wound model. Wound Repair Regen [Internet] 21(1):35–43. [cited 2025 Aug 10]. Available from: https://pubmed.ncbi.nlm.nih.gov/23110665/
78. Almeida L, Everts B (2021) Fa(c)t checking: how fatty acids shape metabolism and function of macrophages and dendritic cells. Eur J Immunol [Internet] 51(7):1628–1640. [cited 2025 Aug 11]. Available from: https://doi.org/10.1002/eji.202048944
79. Wynn TA, Barron L (2010) Macrophages: master regulators of inflammation and fibrosis. Semin Liver Dis [Internet]. 30(3):245–257. [cited 2025 Aug 11]. Available from: https://pubmed.ncbi.nlm.nih.gov/20665377/
80. Gordon S (2003) Alternative activation of macrophages. Nat Rev Immunol [Internet]. 3(1):23–35. [cited 2025 Aug 11]. Available from: https://pubmed.ncbi.nlm.nih.gov/12511873/
81. Murray PJ, Allen JE, Biswas SK, Fisher EA, Gilroy DW, Goerdt S et al (2014) Macrophage activation and polarization: nomenclature and experimental guidelines. Immunity [Internet] 41(1):14–20. [cited 2025 Aug 11]. Available from: https://pubmed.ncbi.nlm.nih.gov/25035950/
82. Sallusto F, Lanzavecchia A (1994) Efficient presentation of soluble antigen by cultured human dendritic cells Is maintained by granulocyte/macrophage colony-stimulating factor plus interleukin 4 and downregulated by tumor necrosis factor α. J Exp Med [Internet]. 179(4):1109–1118. [cited 2025 Aug 11]. Available from: https://pubmed.ncbi.nlm.nih.gov/8145033/
83. Kryczanowsky F, Raker V, Graulich E, Domogalla MP, Steinbrink K (2016) IL-10–modulated human dendritic cells for clinical use: identification of a stable and migratory subset with improved tolerogenic activity. J Immunol [Internet] 197(9):3607–3617. [cited 2025 Aug 11]. Available from: https://pubmed.ncbi.nlm.nih.gov/27683749/
84. Li H, Shi B (2015) Tolerogenic dendritic cells and their applications in transplantation. Cell Mol Immunol [Internet]. 12(1):24–30. [cited 2025 Aug 11]. Available from: https://pubmed.ncbi.nlm.nih.gov/25109681/
85. Fritsche KL (2015) The Science of Fatty Acids and Inflammation. Advances in Nutrition [Internet]. 6(3):293S–301S. [cited 2025 Aug 11]. Available from: https://pubmed.ncbi.nlm.nih.gov/25979502/
86. Serhan CN, Levy BD (2018) Resolvins in inflammation: emergence of the pro-resolving superfamily of mediators. J Clin Invest [Internet] 128(7):2657–2669. [cited 2025 Aug 11]. Available from: https://pubmed.ncbi.nlm.nih.gov/29757195/

87. Snodgrass RG, Boß M, Zezina E, Weigert A, Dehne N, Fleming I et al (2016) Hypoxia potentiates palmitate-induced pro-inflammatory activation of primary human macrophages. J Biol Chem [Internet] 291(1):413–424. [cited 2025 Aug 11]. Available from: https://pubmed.ncbi.nlm.nih.gov/26578520/
88. Zasłona Z, Serezani CH, Okunishi K, Aronoff DM, Peters-Golden M (2012) Prostaglandin E2 restrains macrophage maturation via E prostanoid receptor 2/protein kinase A signaling. Blood [Internet]. 119(10):2358–2367. [cited 2025 Aug 11]. Available from: https://pubmed.ncbi.nlm.nih.gov/22234697/
89. Yin X, Zeng W, Wu B, Wang L, Wang Z, Tian H et al (2020) PPARα inhibition overcomes tumor-derived Exosomal lipid-induced dendritic cell dysfunction. Cell Rep [Internet] 33(3) [cited 2025 Aug 11]. Available from: https://pubmed.ncbi.nlm.nih.gov/33086073/
90. Hammerich L, Tacke F (2023) Hepatic inflammatory responses in liver fibrosis. Nat Rev Gastroenterol Hepatol [Internet] 20(10):633–646. [cited 2025 Aug 11]. Available from: https://www.nature.com/articles/s41575-023-00807-x
91. Elkington PTG, O'Kane CM, Friedland JS (2005) The paradox of matrix metalloproteinases in infectious disease. Clin Exp Immunol [Internet] 142(1):12–20. [cited 2025 Aug 11]. Available from: https://pubmed.ncbi.nlm.nih.gov/16178851/
92. Wen Y, Lambrecht J, Ju C, Tacke F (2021) Hepatic macrophages in liver homeostasis and diseases-diversity, plasticity and therapeutic opportunities. Cell Mol Immunol [Internet] 18(1):45–56. [cited 2025 Aug 11]. Available from: https://pubmed.ncbi.nlm.nih.gov/33041338/
93. Mederacke I, Hsu CC, Troeger JS, Huebener P, Mu X, Dapito DH et al (2013) Fate tracing reveals hepatic stellate cells as dominant contributors to liver fibrosis independent of its aetiology. Nat Commun [Internet]:4. [cited 2025 Aug 11]. Available from: https://pubmed.ncbi.nlm.nih.gov/24264436/
94. Riordan SM, Williams R (1999) Tolerance after liver transplantation: does it exist and can immunosuppression be withdrawn? J Hepatol [Internet] 31(6):1106–1119. [cited 2025 Aug 11]. Available from: https://pubmed.ncbi.nlm.nih.gov/10604588/
95. Su J, Song Y, Zhu Z, Huang X, Fan J, Qiao J et al (2024) Cell–cell communication: new insights and clinical implications. Signal Transduct Target Ther [Internet]. 9(1):1–52. [cited 2025 Aug 13]. Available from: https://www.nature.com/articles/s41392-024-01888-z
96. Dennis EA, Norris PC (2015) Eicosanoid storm in infection and inflammation. Nat Rev Immunol [Internet]. 15(8):511–523. [cited 2025 Aug 13]. Available from: https://www.nature.com/articles/nri3859
97. Brock TG, Peters-Golden M (2007) Activation and regulation of cellular eicosanoid biosynthesis. Sci World J 7:1273–1284
98. Engin A (2024) Adipose tissue hypoxia in obesity: clinical reappraisal of hypoxia hypothesis. Adv Exp Med Biol [Internet] 1460:329–356. [cited 2025 Aug 14]. Available from: https://link.springer.com/chapter/10.1007/978-3-031-63657-8_11
99. Boden G (2009) Free fatty acids, a major link between obesity, insulin resistance, inflammation, and atherosclerotic vascular disease. Cardiovasc Endocrinol [Internet]:61–70. [cited 2025 Aug 14]. Available from: https://link.springer.com/chapter/10.1007/978-1-59745-141-3_4
100. Gkrinia EMM, Belančić A (2025) The Mechanisms of Chronic Inflammation in Obesity and Potential Therapeutic Strategies: A Narrative Review. Curr Issues Mol Biol [Internet]. 47(5):357. [cited 2025 Aug 14]. Available from: https://www.mdpi.com/1467-3045/47/5/357/htm
101. Ebke LA, Nestor-Kalinoski AL, Slotterbeck BD, Al-Dieri AG, Ghosh-Lester S, Russo L et al (2014) Tight association between macrophages and adipocytes in obesity: implications for adipocyte preparation. Obesity [Internet] 22(5):1246–1255. [cited 2025 Aug 14]. Available from: https://onlinelibrary.wiley.com/doi/pdf/10.1002/oby.20634?msockid=2eecf218ba1b6e9821abe187bbc66f7e

102. Grabner GF, Xie H, Schweiger M, Zechner R (2021) Lipolysis: cellular mechanisms for lipid mobilization from fat stores. Nat Metab [Internet]. 3(11):1445–1465. [cited 2025 Aug 14]. Available from: https://www.nature.com/articles/s42255-021-00493-6
103. Reali F, Morine MJ, Kahramanoä Ullarl O, Raichur S, Schneider HC, Crowther D et al (2017) Mechanistic interplay between ceramide and insulin resistance. Sci Rep [Internet] 7(1):1–9. [cited 2025 Aug 14]. Available from: https://www.nature.com/articles/srep41231
104. Corcoran MP, Lamon-Fava S, Fielding RA (2007) Skeletal muscle lipid deposition and insulin resistance: effect of dietary fatty acids and exercise. Am J Clin Nutr [Internet]. 85(3):662–677. [cited 2025 Aug 14]. Available from: https://www.sciencedirect.com/science/article/pii/S0002916523279721
105. Kennedy L, Meadows V, Sybenga A, Demieville J, Chen L, Hargrove L et al (2021) Mast cells promote nonalcoholic fatty liver disease phenotypes and microvesicular steatosis in mice fed a western diet. Hepatology [Internet] 74(1):164–182. [cited 2025 Aug 14]. Available from: https://doi.org/10.1002/hep.31713
106. Masoodi M, Gastaldelli A, Hyötyläinen T, Arretxe E, Alonso C, Gaggini M et al (2021) Metabolomics and lipidomics in NAFLD: biomarkers and non-invasive diagnostic tests. Nat Rev Gastroenterol Hepatol [Internet] 18(12):835–856. [cited 2025 Aug 14]. Available from: https://www.nature.com/articles/s41575-021-00502-9
107. Ju Z, Li M, Xu J, Howell DC, Li Z, Chen FE (2022) Recent development on COX-2 inhibitors as promising anti-inflammatory agents: the past 10 years. Acta Pharm Sin B [Internet]. 12(6):2790–2807. [cited 2025 Aug 7]. Available from: https://www.sciencedirect.com/science/article/pii/S2211383522000156
108. Patrignani P, Patrono C (2015) Cyclooxygenase inhibitors: from pharmacology to clinical read-outs. Biochimica et Biophysica Acta (BBA) – Mol Cell Biol Lipids [Internet] 1851(4):422–432. [cited 2025 Aug 7]. Available from: https://www.sciencedirect.com/science/article/abs/pii/S1388198114001942
109. Patrignani P, Tacconelli S, Piazuelo E, Di Francesco L, Dovizio M, Sostres C et al (2014) Reappraisal of the clinical pharmacology of low-dose aspirin by comparing novel direct and traditional indirect biomarkers of drug action. J Thromb Haemost [Internet] 12(8):1320–1330. [cited 2025 Aug 7]. Available from: https://www.sciencedirect.com/science/article/pii/S1538783622040223
110. Lei J, Zhou Y, Xie D, Zhang Y (2015) Mechanistic insights into a classic wonder drug-aspirin. J Am Chem Soc [Internet] 137(1):70–73. [cited 2025 Aug 7]. Available from: https://doi.org/10.1021/ja5112964?ref=article_openPDF
111. Patrono C (2024) Low-dose aspirin for the prevention of atherosclerotic cardiovascular disease. Eur Heart J [Internet] 45(27):2362. [cited 2025 Aug 7]. Available from: https://pmc.ncbi.nlm.nih.gov/articles/PMC11242460/
112. Yan M, Zhang S, Li C, Liu Y, Zhao J, Wang Y et al (2021) 5-lipoxygenase as an emerging target against age-related brain disorders. Ageing Res Rev [Internet] 69:101359. [cited 2025 Aug 7]. Available from: https://www.sciencedirect.com/science/article/abs/pii/S1568163721001069
113. Sinha S, Doble M, Manju SL (2019) 5-Lipoxygenase as a drug target: a review on trends in inhibitors structural design, SAR and mechanism based approach. Bioorg Med Chem [Internet]. 27(17):3745–3759. [cited 2025 Aug 7]. Available from: https://www.sciencedirect.com/science/article/abs/pii/S0968089619306777
114. Lan R, Zhang MJ, Liu K, Meng FF, Xu XH, Wang CC et al (2024) Cytochrome P450-derived Epoxyeicosatrienoic acid, the regulation of cardiovascular-related diseases, and the implication for pulmonary hypertension. Cardiovasc Drugs Ther [Internet]:1–17. [cited 2025 Aug 7]. Available from: https://link.springer.com/article/10.1007/s10557-024-07655-0
115. Foti RS (2023) Cytochrome P450 and other drug-metabolizing enzymes as therapeutic targets. Drug Metab Dispos [Internet] 51(8):936–949. [cited 2025 Aug 7]. Available from: https://www.sciencedirect.com/science/article/abs/pii/S0090955624010201

116. Du L, Du S, Li J, Wang H (2023) Design, synthesis, and biological evaluation of dual-target COX-2/5-LOX inhibitors for the treatment of inflammation. Med Chem Res [Internet]. 32(2):218–238. [cited 2025 Aug 7]. Available from: https://link.springer.com/article/10.1007/s00044-022-02995-8
117. Rudrapal M, Eltayeb WA, Rakshit G, El-Arabey AA, Khan J, Aldosari SM et al (2023) Dual synergistic inhibition of COX and LOX by potential chemicals from Indian daily spices investigated through detailed computational studies. Sci Rep [Internet]. 13(1):1–27. [cited 2025 Aug 7]. Available from: https://www.nature.com/articles/s41598-023-35161-0
118. Bland JS (2022) Therapeutic use of Omega-3 Fatty Acids for immune disorders in search of the ideal Omega-3 supplement. Integr Med (Encinitas) [Internet] 21(5):14. [cited 2025 Aug 7]. Available from: https://pmc.ncbi.nlm.nih.gov/articles/PMC9831136/
119. Banaszak M, Dobrzyńska M, Kawka A, Górna I, Woźniak D, Przysławski J et al (2024) Role of Omega-3 fatty acids eicosapentaenoic (EPA) and docosahexaenoic (DHA) as modulatory and anti-inflammatory agents in noncommunicable diet-related diseases—reports from the last 10 years. Clin Nutr ESPEN [Internet]. 63:240–258. [cited 2025 Aug 7]. Available from: https://www.sciencedirect.com/science/article/pii/S2405457724001992
120. Nelson JR, Budoff MJ, Wani OR, Le V, Patel DK, Nelson A et al (2021) EPA's pleiotropic mechanisms of action: a narrative review. Postgrad Med [Internet] 133(6):651–664. [cited 2025 Aug 7]. Available from: https://www.tandfonline.com/doi/pdf/10.1080/00325481.2021.1921491
121. Toth PP, Chapman MJ, Parhofer KG, Nelson JR (2022) Differentiating EPA from EPA/DHA in cardiovascular risk reduction. Am Heart J Plus [Internet] 17:100148. [cited 2025 Aug 7]. Available from: https://www.sciencedirect.com/science/article/pii/S2666602222000659
122. Nelson JR, Raskin S (2019) The eicosapentaenoic acid:arachidonic acid ratio and its clinical utility in cardiovascular disease. Postgrad Med [Internet] 131(4):268–277. [cited 2025 Aug 7]. Available from: https://www.tandfonline.com/doi/pdf/10.1080/00325481.2019.1607414
123. Kinsella JE, Broughton KS, Whelan JW (1990) Dietary unsaturated fatty acids: interactions and possible needs in relation to eicosanoid synthesis. J Nutr Biochem [Internet] 1(3):123–141. [cited 2025 Aug 7]. Available from: https://www.sciencedirect.com/science/article/abs/pii/0955286390900119
124. Mathias MM, Dupont J (1989) Effects of dietary fat on eicosanoid production in normal tissues. In: Carcinogenesis and dietary fat [Internet], pp 29–52. [cited 2025 Aug 7]. Available from: https://link.springer.com/chapter/10.1007/978-1-4613-1585-8_3
125. Alfahl SO (2025) Evaluation of the effectiveness of intermittent fasting versus caloric restriction in weight loss and improving cardiometabolic health: a systematic review and meta-analysis. J Taibah Univ Med Sci [Internet] 20(2):159–168. [cited 2025 Aug 7]. Available from: https://www.sciencedirect.com/science/article/pii/S1658361225000186
126. Di Francesco A, Deighan AG, Litichevskiy L, Chen Z, Luciano A, Robinson L et al (2024) Dietary restriction impacts health and lifespan of genetically diverse mice. Nature [Internet] 634(8034):684–692. [cited 2025 Aug 7]. Available from: https://www.nature.com/articles/s41586-024-08026-3
127. Semnani-Azad Z, Khan TA, Chiavaroli L, Chen V, Bhatt HA, Chen A et al (2025) Intermittent fasting strategies and their effects on body weight and other cardiometabolic risk factors: systematic review and network meta-analysis of randomised clinical trials. BMJ [Internet]:389. [cited 2025 Aug 7]. Available from: https://pubmed.ncbi.nlm.nih.gov/40533200/
128. Natto ZS, Yaghmoor W, Alshaeri HK, Van Dyke TE (2019) Omega-3 Fatty acids effects on inflammatory biomarkers and lipid profiles among diabetic and cardiovascular disease patients: a systematic review and meta-analysis. Sci Rep [Internet] 9(1):1–10. [cited 2025 Aug 14]. Available from: https://www.nature.com/articles/s41598-019-54535-x
129. Almulhim M, Ghasemian A, Mojtaba M, Karami F, Asmaa YSA et al (2025) Drug repositioning as a promising approach for the eradication of emerging and re-emerging viral agents. Mol Divers [Internet] 123AD. [cited 2025 Aug 14]; Available from: https://doi.org/https://doi.org/10.1007/s11030e-025-11131-8

130. Zhou Y, Tao L, Qiu J, Xu J, Yang X, Zhang Y et al (2024) Tumor biomarkers for diagnosis, prognosis and targeted therapy. Signal Transduct Target Ther [Internet] 9(1):1–86. [cited 2025 Aug 14]. Available from: https://www.nature.com/articles/s41392-024-01823-2
131. Brennan E, Kantharidis P, Cooper ME, Godson C (2021) Pro-resolving lipid mediators: regulators of inflammation, metabolism and kidney function. Nat Rev Nephrol [Internet] 17(11):725–739. [cited 2025 Aug 14]. Available from: https://www.nature.com/articles/s41581-021-00454-y
132. Priest C, Tontonoz P (2019) Inter-organ cross-talk in metabolic syndrome. Nat Metab [Internet] 1(12):1177–1188. [cited 2025 Aug 14]. Available from: https://www.nature.com/articles/s42255-019-0145-5
133. Tokizane K, Imai S, ichiro. (2025) Inter-organ communication is a critical machinery to regulate metabolism and aging. Trends Endocrinol Metab [Internet] 36(8):756–766. [cited 2025 Aug 14]. Available from: https://www.cell.com/action/showFullText?pii=S1043276024003205

Chapter 10
ROS and Inflammation Interplay in Hypertension and Other CVDs

Debarshi Mondal, Mustak Ahamed, Pratyush Porel, and Khadga Raj Aran

10.1 Introduction

Cardiovascular diseases (CVDs) belong to one of the greatest health risks globally, holding the first position according to their morbidity and mortality rates. It is estimated that 19.8 million people died in 2022 due to CVDs, accounting for an estimated 32% of total global mortality. This has an unequal impact on low- and middle-income countries, where three-quarters of deaths are reported due to acute or chronic CVDs [1]. CVDs possess a wide range of disorders affecting various organs of our body, such as the heart and blood vessels, which lead to several minor to severe complications, like coronary heart disease, cerebrovascular disorder, and rheumatic heart disease [2]. Most CVDs have their pathophysiology based on atherosclerosis, which is a gradual accumulation of larger molecules, such as proteins and cholesterol, leading to the development of various complications like heart attack and stroke. The formation of atheroma is closely associated with several well-established behavioral and physiological risk factors, including an unhealthy diet, physical inactivity, tobacco use, hypertension, hypercholesterolemia, and obesity. At the core of the onset and progression of CVDs are these risk factors' physiological manifestations, which establish the foundation for the molecular abnormalities that drive CVDs. Two pivotal molecular processes, oxidative stress and inflammation, are now understood to be fundamental to the pathogenesis of CVDs, moving beyond their traditional role as mere consequences of the disease. Oxidative stress is defined as a state of disruption of homeostasis between the production of reactive oxygen species (ROS) and the capacity of the body's

D. Mondal · P. Porel
ISF College of Pharmacy and Research, Moga, Punjab, India

M. Ahamed · K. R. Aran (✉)
Department of Pharmacy Practice, ISF College of Pharmacy, Moga, Punjab, India

S. Ashique, N. Mishra (eds.), *Reactive Oxygen Species in Metabolic Inflammation*,
https://doi.org/10.1007/978-3-032-11752-6_10

endogenous antioxidant defense systems to minimize the harmful effects of ROS [3]. Although high levels of ROS induce considerable damage to lipids, proteins, and DNA, low levels of the same molecules are necessary or serve to regulate several physiological processes, including vascular tone and cellular signaling. This contradictory nature of ROS demonstrates the dual effects of ROS on cardiovascular health and diseases [4].

On the same note, inflammation is a complex physiological response to damage or infection. Although clinically acute inflammation is a healthy process critical to wound healing and pathogen resolution, in the chronic state, low-level, cellularly resolved inflammation is a harmful mechanism within the cardiovascular system. This persistent scenario is typified by the chronic retention of immunocytes and liberation of pro-inflammatory mediators, the latter of which play a central role in the pathophysiology of CVDs, especially as regards atherosclerotic disease [5].

The concurrent presence of both oxidative stress and chronic inflammation in the development of CVDs suggests a deep, underlying relationship between the two mechanisms. The core premise of this report is that oxidative stress and inflammation are not two parallel, independent pathologies but are inextricably interconnected through a synergistic, self-perpetuating positive feedback loop. Pro-inflammatory signaling pathways are likely to be activated by oxidative stress, and vice versa, ROS production may be significantly upregulated, while the inflammatory scenario imposes its mark on the system. Such cyclical multiplication is the driving force behind chronic disease and can cause progressive tissue destruction, which defines the inexorable character of atherosclerosis and other CVDs. Comprehension of this interrelated causality has far-reaching implications in the development of therapeutics [6].

This chapter aims to explore the molecular mechanisms of oxidative stress and inflammation in the development of hypertension and other CVDs. In this chapter, the role of individual sources and functions of ROS and inflammatory pathways has been reviewed systematically, and thereafter, a thorough overview of the cellular and molecular constituents of vascular inflammation has been provided. This chapter also discusses various therapeutic approaches that target the ROS-inflammation axis and are intricately involved in the management of several CVDs. The conclusion of this chapter emphasizes its future directions and provides a more thorough overview of the preclinical and clinical studies that indicate promise for clinical implementation.

10.2 Reactive Oxygen Species: Sources and Significance

10.2.1 Physiological and Pathological Sources of ROS

ROS is a family of highly reactive molecules and free radicals derived from the partial reduction of molecular oxygen in various physiological processes. The most well-known species include the superoxide anion ($O_2^{\bullet}-$) and its derivative, hydrogen peroxide (H_2O_2), along with the enormously reactive hydroxyl radical (OH) [7]. Although high levels of these species are harmful to the cell, low or physiological levels are essential in maintaining homeostasis and in regulating crucial molecular mechanisms, such as growth, adhesion, and differentiation. ROS are generated through various cellular and enzymatic systems, with the major sources in the cardiovascular system being the mitochondrial electron transport chain (ETC), the nicotinamide adenine dinucleotide phosphate (NADPH) oxidase (NOX) family of enzymes, xanthine oxidase (XO), and uncoupled endothelial nitric oxide synthase (eNOS) [8].

10.2.2 The Central Role of Mitochondria in ROS Production

It has long been known that mitochondria are the primary site for the synthesis of Adenosine Triphosphate (ATP), and also a major source of ROS. Superoxide is a normal by-product of aerobic respiration, formed when a small percentage of electrons leak from the ETC, particularly at Complexes I and III, and react with molecular oxygen. Under normal physiological conditions, the low levels of ROS synthesized are effectively balanced by endogenous antioxidant proteins, whereas in pathological conditions, such as myocardial ischemia-reperfusion injury, this homeostasis is disrupted. The ETC is a significant source of ROS during reperfusion, causing excess production far beyond the defenses of antioxidants [9]. This elevation of ROS during disease pathology degrades mitochondrial components to produce more ROS and resulting in a self-sustaining vicious cycle. This concept, articulated in the mitochondrial free radical theory of aging, explains how a localized cellular insult can trigger the cascade of oxidative damage, accelerate cellular dysfunctions, and lead to cardiomyocyte death. Consequently, protecting mitochondrial integrity or function is a crucial consideration for therapeutic strategies aimed at disrupting this self-destructive cycle and mitigating a broad range of CVD pathologies [10].

10.2.3 NADPH Oxidase (NOX) as a Key Vascular ROS Source

The nicotinamide adenine dinucleotide phosphate (NADPH) oxidase family of enzymes is considered a major source of ROS within the vascular system. This enzyme complex generates superoxide by transferring an electron from NADPH to molecular oxygen. The resulting superoxide is highly reactive and, in the vasculature, is known to quickly inactivate nitric oxide (NO), a crucial vasodilator, by forming peroxynitrite. This process is actively involved in endothelial dysfunction, a key initiating event in atherosclerosis, and it facilitates other reactions involving ROS that led to vascular disease progression and cardiac remodeling [11]. The significance of NOX has been demonstrated by its direct link to various conventional risk factors. An example of this would be hypertension, which not only promotes oxidative stress due to poor lifestyle but also conveys various cellular and molecular signaling. The phenomenon of high pressure on the circumferential wall activates the pathway through integrin-linked kinase 1, which results in the activation of NADPH oxidase and excess generation of ROS. This precise, stepwise mechanism illustrates how acute or chronic hypertension can translate into microscopic molecular damage, thereby creating a new layer of the vicious cycle and explaining the roots of these risk factors, those potently damaging the cardiovascular system [12].

10.3 Inflammation in Cardiovascular Pathophysiology

10.3.1 The Immune Response in Vasculature

Atherosclerosis is now widely regarded as a chronic inflammatory and partially autoimmune disease. Its lesions are characterized by a dense infiltration of both innate and adaptive immune cells. The atherogenic process is initiated by the accumulation and subsequent oxidative modification of low-density lipoproteins (LDL) in the subendothelial space. The modified LDL, known as oxidized LDL (oxLDL), is a potent pro-inflammatory signal. In response, monocytes and T cells are recruited from the bloodstream and adhere to the vascular wall [13]. Monocytes differentiate into macrophages, which avidly ingest oxLDL, becoming lipid-laden "foam cells." The sustained influx of cholesterol and the formation of intracellular cholesterol microcrystals activate a key molecular machinery known as the NOD-, LRR-, and pyrin domain-containing protein 3 (NLRP3) inflammasome. This activation leads to the cleavage of pro-interleukin-1β (pro-IL-1β) into its active form IL-1β, which is a primary inflammatory cytokine. The myeloid response is accompanied by the infiltration of B and T cells, which contribute to the lesion's growing cellular content. This complex immune infiltrate is a key determinant of plaque vulnerability and correlates with clinical complications and the risk of plaque rupture [14].

10.3.2 Cytokines and Chemokines as Molecular Drivers

Cytokines and chemokines are the central molecular messengers that orchestrate the inflammatory response. Chemokines are a specific class of cytokines that induce the directed migration, or chemotaxis, of immune cells to a site of injury or inflammation. Notable pro-inflammatory mediators corresponding to tumor necrosis factor-α (TNF-α) and IL-1β, which are released by macrophages, endothelial cells, among other inflammatory mediators, are deemed to be highly active in the promotion of atherosclerosis with instability of the atheroma [15]. Release of cytokines can elevate the plasma levels of C-reactive protein (CRP), a well-known clinical biomarker of inflammation and cardiovascular risk, such as IL-6. High involvement is also

Table 10.1 Key inflammatory mediators, along with their cellular sources and functions in CVDs

S. no.	Mediators	Cellular source	Primary functions	Role in CVD pathology	References
1.	IL-1α and IL-1β	Macrophages, monocytes	Activate the inflammasome and release cytokines	Plaque instability, atheroma formation, and myocardial infarction	[17]
2.	TNF-α	Macrophages, monocytes	Enhance expression of multiple inflammatory genes	Atherosclerosis, plaque instability, and hypertension	[18]
3.	IL-6	Macrophages, monocytes, and endothelial cells	Increase plasma fibrinogen and CRP	Atherosclerosis, systemic inflammation, CVDs risk stratification	[19]
4.	CXC family	Endothelial cells, smooth muscle cells, and leukocytes	Induce chemotaxis of immune cells	Atherosclerosis, myocardial infarction, cardiac fibrosis	[20]
5.	CXCL-1 and – 8	Fibroblasts, activated endothelial cells, macrophages, and neutrophils	Recruit neutrophils	Myocardial infarction, cardiac fibrosis, atherosclerosis	[21]
6.	CXCL4	Lymphocyte and myofibroblast	Recruit monocytes	Atherosclerosis, atrial fibrillation	[22]
7.	CXCL-9 and –10	Neutrophils, macrophages, lymphocytes, and endothelial cells	Recruit T cells; modulate fibroblast migration	Exacerbate tissue inflammation, tissue repair	[23]

seen in the chemokine family, and most notably the Cysteine-X-Cysteine chemokines (CXC) group. The recruitment of neutrophils by chemokines, including CXCL1 and CXCL8, occurs via the phosphoinositide 3-kinase/protein kinase B (PI3K/Akt) pathway to aggravate tissue inflammation. However, certain chemokines have been shown to have protective, antifibrotic effects and can attenuate CVDs by inhibiting fibroblast migration, highlighting the multifaceted and even paradoxical functions of these effectors in CVD [16]. As these mediators serve multiple functions, Table 10.1 summarizes their sources and primary functions, along with their role in CVD pathologies.

10.4 ROS and Inflammation: The Mechanistic Interplay

10.4.1 ROS as Activators of Inflammatory Pathways

ROS are not simply incidental metabolic by-products, but are involved in several intracellular signaling and inflammatory responses. One of the primary transcriptional factors, that is, nuclear factor-kappa B (NF-κB), is a key controller of the transcription of several genes associated with inflammation, immune response, and oxidative stress. NF-κB is sequestered by its inhibitory protein, inhibitory kappa B (IκB), in its inactive state in the cytoplasm. An overproduction of ROS initiates the activation of Inhibitor of kappa B kinase subunit (IKK) and the phosphorylation of IKK. The result of this phosphorylation is that IκB is marked to be degraded via the ubiquitin proteasome system, thus releasing the inhibition of NF-κB to move out of the cell into the nucleus, where it can begin to transcribe a large number of pro-inflammatory pathways, including IL-6 pathways and TNF-α [24].

The relationship between this ROS-NF-κB connection is, however, not so straightforward as a cause-and-effect relationship. A critical review of the literature will demonstrate that there is a major controversy as to the direct role of ROS as the second messengers to the activation of NF-κB. Several studies have indicated that putative antioxidants such as N-acetyl-l-cysteine (NAC) and pyrrolidine dithiocarbamate (PDTC) also inhibit NF-κB in manners that are not related to antioxidant mechanisms [25]. NAC, e.g., can block TNF-alpha signaling by reducing the affinity of the TNF receptor, whereas PDTC can block ubiquitin ligase activity in a test tube. These observations refute the very simple picture that ROS causes inflammation, and suggest that the anti-inflammatory activities of these compounds may not be because of their antioxidant properties. This indicates the necessity to have a more in-depth appreciation of cellular signaling and implies that the association between ROS and NF-κB activation is more complex and cell-type-specific than was previously assumed [26].

10.4.2 Redox-Sensitive Gene Expression and Cell Fate

In addition to the NF-κB pathway, long-term effects are broadly mediated by the redox state of the cell, which can modulate the long-term effects of gene expression and ultimately the cellular proteome. ROS can directly activate other redox-sensitive transcription factors, such as nuclear factor erythroid 2-related factor 2 (Nrf2). It is activated by lipid peroxidation and subsequently induces the expression of various antioxidant enzymes, providing a protective counter-response to oxidative stress [27]. An overload of ROS may also cause direct damage to important cellular structures and constituents. The nitrogenous base guanine in DNA can be oxidized by high levels of ROS, resulting in damage that triggers the onset of senescence and related signaling networks. This permanent injury can cause the formation of a state of chronic inflammation by means of stimulating the secretion of senescence-induced secretory phenotypes (SASPs) and other inflammatory factors [28]. On a more mechanistic level, excessive ROS can derail some basic proteins, including cardiac ion channels. This redox-related modification of the channel proteins may cause a decrease in repolarizing currents of potassium, an excess of the sodium and calcium inflow, which results in poor contractile strength, arrhythmic changes of the heart rate, and other electrophysiological anomalies [29].

10.4.3 Inflammation-Driven ROS Generation: Completing the Vicious Cycle

The reversed dependence between oxidative stress and inflammation is finalized by how inflammatory factors potentiate the production of ROS, completing the positive feedback loop. An important axis in this process is the pro-inflammatory cytokines that activate NADPH oxidase (NOX). It has been shown that several cytokines, including TNF-α and IL-1β, may directly stimulate ROS production through activation of NOX. As an example, TNF-α can activate the expression of the NOX subunits and stimulate certain kinases, namely cyclin-dependent kinase 5 (CDK5), which stimulates NOX-mediated ROS production [30].

This inflammation-mediated ROS is generated further in the mitochondria. Inflammatory signaling may enhance mitochondrial ROS generation via mechanisms that include proteins such as reactive oxygen species modulator 1 (Romo1) and Bcl-XL, which can decrease mitochondrial membrane potential and stimulate an ensuing explosion of ROS. This forms a potent positive feedback loop: inflammation induces ROS production by mitochondria, which subsequently causes more inflammation [31]. This loop of constant feedback is an expectant feature in chronic inflammatory conditions according to CVDs. In understanding how these factors all come together to form this reciprocal interplay, their mechanistic explanation is critical in providing a complete explanation as to the unique and unforgiving, progressive nature of diseases such as atherosclerosis, as well as the idea that breaking this cycle is the ultimate goal to be used in future treatment procedures [32].

10.5 ROS-Inflammation Axis in Hypertension

Hypertension is a complicated disease state characterized by consistent elevation of blood pressure, loss of vascular, renal functions, and immunological systems of the body, and is also associated with oxidative stress and inflammation [33]. On the other hand, several intracellular signaling pathways are disturbed by increased ROS levels and loss of redox homeostasis. It induces the production of ROS, resulting in oxidative damage, reduced availability of NO, endothelial dysfunction, renal damage, and an activated/inflammatory immune system. Thus, the multiple pathways collectively contribute to a self-sustaining process of inflammation and ROS production, which sustains elevated blood pressure and alters vascular architecture. It is important to recognize that the ROS-inflammation axis mechanistically contributes to hypertension and can be targeted therapeutically with anti-inflammatory or antioxidant strategies [34, 35].

10.5.1 ROS in Kidney Functions

The vascular endothelium plays a significant role in the regulation of vascular tone, remodeling, and inflammatory signaling. Hypertension is characterized by endothelial dysfunction that is directly related to the overproduction of ROS. Under physiological conditions, NO produced from endothelial nitric oxide synthase (eNOS) acts as a potent vasodilator. However, in hypertensive states, ROS, notably H_2O_2 and superoxide, decrease NO bioavailability by both directly inactivating NO and uncoupling eNOS, producing superoxide instead of NO. Together, vascular stiffness, peripheral resistance, and chronic hypertension are caused by the imbalance between vasodilation and vasoconstriction [36].

Pro-inflammatory cytokines and cell adhesion molecules, such as Vascular Cell Adhesion Molecule-1 (VCAM-1), Intercellular Cell Adhesion Molecule-1 (ICAM-1), and E-selectin, are produced by ROS and can act as signaling molecules to stimulate transcription factors that drive AP-1 and NF-κB to promote leukocyte adhesion, infiltration, and chronic vascular inflammation. These factors induce vascular remodeling characterized by decreased arterial compliance in part because of the proliferation of smooth muscle cells, neovasculogenesis, and matrix deposition due to oxidative stress. Since antioxidant supplements restore endothelium-dependent vasodilation as indicated by brachial artery flow-mediated dilatation, bench studies support the idea that ROS is causing a major contribution to endothelial dysfunction in hypertension [37].

10.5.2 ROS in Endothelial Functions

The main marker of hypertension, ROS, is consistently generated and eventually triggers processes that result in endothelial injury. The vascular endothelium regulates vascular tone, remodeling of the vasculature, and inflammatory signals. NO is a potent vasodilator, is consistently and reliably generated by eNOS. ROS, such as superoxide and hydrogen peroxide, can separate eNOS and directly eliminate NO in the pathological condition of hypertension. They solely generate superoxide instead of NO. Vascular stiffness, chronic hypertension, sustained hypertension, and heightened peripheral resistance result from augmented vasoconstriction and additional factors at the arterial wall level [38].

ROS can act as signaling molecules to activate transcription factors like NF-κB and AP-1, as well as pro-inflammatory cytokines and adhesion molecules. This leads to changes in leukocyte adhesion and retention as well as chronic vascular inflammation. Furthermore, extracellular matrix (ECM) deposition and smooth muscle cell growth are encouraged by oxidative stress, which ultimately leads to vasculature remodeling and decreased arterial compliance. The concept that ROS play a significant role in endothelial dysfunction in hypertension is supported by published experimental studies that demonstrated antioxidants restored endothelial-dependent vasodilation. Additionally, published clinical data demonstrate that hypertensive patients have elevated blood pressure-related biomarkers of oxidative stress and decreased endothelial function [39].

10.5.3 Immune-Mediated Dysfunctions

Hypertension has been demonstrated to engage the immune system in its pathogenesis, and ROS are amongst the most critical enhancers of immunological activation. MACs and DCs upon stimulation by oxidative stress produce pro-inflammatory cytokines IL-6, TNF-α, and IL-1β. They activate and recruit T cell activation, more so Th1 and Th17 cells that produce IL-17 and interferon-γ (IFN-γ). These mediators disrupt renal sodium excretion, enhance oxidative stress, and cause vascular inflammation [40].

ROS will also induce neoantigens through endogenous oxidative modification of lipids and proteins recognized by dendritic cells to sustain adaptive immune responses. Activated immune cells produce more ROS using the NADPH oxidases, maintaining a continual cycle of oxidative stress and inflammation. In hypertension, regulatory T cells (Tregs), the immune suppressors of immune activation, will be diminished or dysfunctional. Additionally, immune-ROS dynamics facilitate vascular injury and renal fibrosis, enhancing sympathetic nervous system tone, and reinforcing the hypertensive state [41].

10.6 ROS-Inflammation Axis in Other Cardiovascular Diseases

CVDs remain the major cause of morbidity and mortality globally, and the underlying mechanisms are being increasingly identified as interlinked with elevated oxidative stress and vascular inflammation. Apart from hypertension, other CVDs, like atherosclerosis, ischemic heart disease, cardiac arrhythmia, and heart failure, are also heavily reliant upon the coordination of ROS. In healthy individuals, levels of ROS are properly regulated and play a crucial role as essential signaling molecules for maintaining vascular homeostasis, but excessive ROS production ultimately overwhelms antioxidant defense, causing oxidative stress and the initiation of inflammatory mechanisms [42]. This section will discuss the ROS-inflammation axis in three of the most important conditions apart from hypertension: atherosclerosis-associated ischemic heart disease, heart failure, cardiac remodeling, and cardiac arrhythmia [43].

10.6.1 Atherosclerosis and Ischemic Heart Disease

Reactive oxygen species (ROS) contribute to atherosclerosis at every stage of the disease, from plaque formation to rupture, and it is now recognized as a chronic inflammatory disease of the vascular wall. NADPH oxidases, mitochondrial failure, xanthine oxidase, and uncoupled eNOS can all cause vascular oxidative stress. Endothelial dysfunction, the initial stage of atherogenesis, results from ROS's faster reactivity with NO, which lowers its bioavailability. The recruitment and trans-endothelial migration of monocytes will occur when the endothelium's vasodilatory function declines, despite an increase in adhesion molecules such as VCAM-1, ICAM-1, E-selectin, and cytokines [44].

On the other hand, oxLDL is another important initiator of immune response in the subendothelial space when LDL is oxidatively modified. Macrophages ingest oxLDL by means of scavenger receptors, resulting in the fatty streaks and foam cells that promote a more pro-inflammatory environment. ROS activate pro-inflammatory transcriptional factors, such as NF-κB, resulting in increased cytokine production and sustained vascular inflammation. Additionally, as lesions progress, ROS eventually stimulate matrix metalloproteinases (MMPs), contributing to thinning of the fibrous cap and collagen breakdown, which could provoke clot formation resulting in ruptured plaque and acute coronary syndromes [45].

Ischemia-reperfusion injury refers to the injury to the heart muscle in ischemic heart disease exacerbated by bursts of ROS with reperfusion. Excessive production of ROS damages mitochondria, causes mitochondrial permeability transition pore (mPTP) opening, and activates cardiomyocyte apoptosis. Collectively, these mechanisms demonstrate that inflammatory responses and oxidative damage are important contributing factors in the development of plaque and ischemic cardiac injury [46].

10.6.2 Heart Failure and Cardiac Remodeling

Heart failure is characterized by maladaptive remodeling, systemic inflammation, and an irreversible loss of contractile function. Heart failure is the ultimate end-stage condition uniting numerous cardiac pathologies. In cardiomyocytes, one of the main drivers of pathology is ROS generation from dysfunctional mitochondria and NADPH oxidases. Mitochondrial ROS compromise both excitation-contraction coupling and cardiac output through reduced calcium homeostasis and inhibition of ATP production [47].

The redox-sensitive signaling pathways that ROS can activate include AP-1, NF-κB, and Mitogen-Activated Protein Kinase (MAPKs). These pathways can promote the production of pro-inflammatory cytokines, including MCP-1, IL-6, and TNF-α, which facilitate the migration of neutrophils and macrophages to heart tissues, thereby promoting inflammation and ROS production. While the inflammatory mediators increase systemic inflammation and generate oxidative stress, chronic inflammation causes hypertrophy, promotes interstitial fibrosis, and leads to cardiomyocyte apoptosis, which results in maladaptive ventricular remodeling [48].

Fibroblast activation, ECM deposition, and altered collagen turnover are all components of structural remodeling in heart failure that are significantly influenced by ROS-mediated MMP activation, ultimately resulting in a decline in ejection fraction, increased ventricular stiffness, and impaired diastolic filling, all of which heighten the clinical manifestations of heart failure. Clinical evidence also supports these mechanisms: patients with severe heart failure have increased oxidative stress measures (e.g., oxLDL and 8-isoprostane) and inflammatory measures (e.g., CRP and TNF-α), both of which are correlated to worse prognosis, and thus, the ROS-inflammation axis acts more as an enabler of the development of heart failure rather than only a physiological mechanism [49].

10.6.3 Cardiac Arrhythmia and Electrophysiological Malfunction

The relationship between oxidative stress, inflammation, and cardiac arrhythmias, including atrial fibrillation and ventricular tachyarrhythmias, is becoming increasingly recognized. ROS produce electrophysiological instability in cardiomyocytes by directly altering ion channel function. For example, alterations to the ryanodine receptor (RyR2) and L-type calcium channels due to ROS lead to triggered activity, delayed afterdepolarizations, and calcium overload. Likewise, oxidative stress causes dysfunction of potassium channels (particularly IKr and Ik), resulting in prolonged action potential lengths and promoting early afterdepolarizations; sodium channel dysfunction slows conduction and induces reentrant arrhythmias [50].

The expression and localization of gap junctions are changed by cytokines such as TNF-α and IL-6, which create conduction heterogeneity and can further exacerbate arrhythmogenesis. Addressing the fact that chronic ROS and inflammation cause fibrosis, which directly forms a structural substrate for reentry circuits. Ischemia-reperfusion injury in ventricular arrhythmias may arise from bursts of ROS, leading to disruption of electrophysiological activity. Oxidative stress in atrial tissue in atrial fibrillation contributes to both electrical and structural remodeling. The importance of this axis is emphasized by experimental data, which demonstrate that anti-inflammatory or antioxidant therapies may reduce the risk of arrhythmia [51].

Pathological role of the ROS-inflammation axis in various CVDs, like hypertension, atherosclerosis, heart failure, fibrosis, and cardiac remodeling, is diagrammatically illustrated in Fig. 10.1.

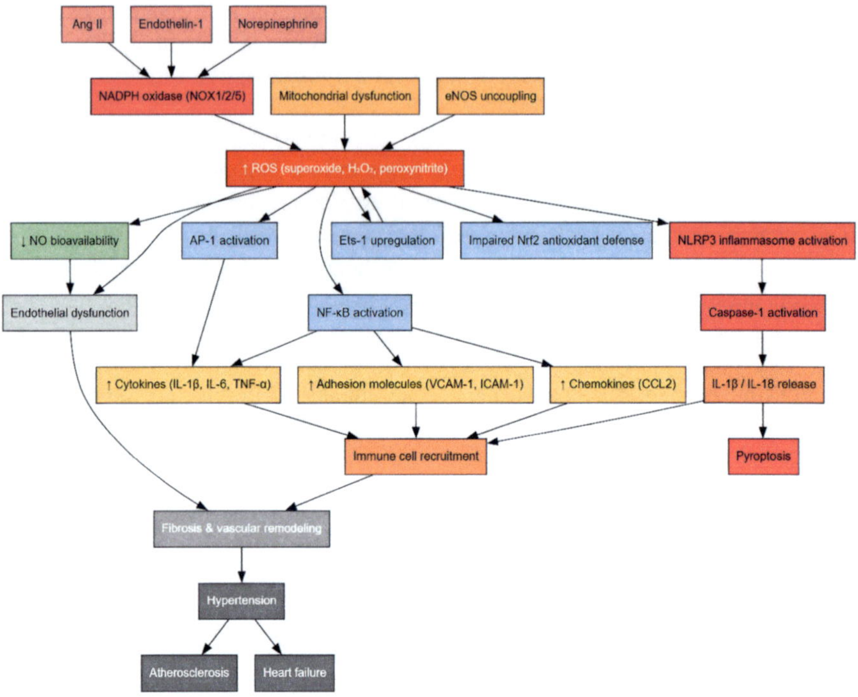

Fig. 10.1 Pathological role of the ROS-inflammation axis in various CVDs

10.7 ROS-Inflammation Axis: Target for Therapeutic Approaches

Treatments targeting the ROS-inflammation axis have become more appealing as oxidative stress and inflammation are recognized as key drivers of several CVDs. Current medications, such as statins and different groups of antihypertensives, indirectly modify redox homeostasis and inflammatory signaling; there is growing evidence that more direct interventions could provide additional benefits. The three main therapeutic approaches are phytomedicine and nutraceuticals, which use natural products with redox-modulating and immunoregulatory effects, anti-inflammatory therapies, and antioxidant therapies [52].

10.7.1 Antioxidant-Based Therapies

Antioxidant therapies are intended to restore the redox homeostasis by enhancing endogenous antioxidant protein expression or scavenging ROS. Quite promising effects were observed in earlier preclinical research showing reduced oxidative stress; however, large-scale clinical studies have demonstrated variable primary cardiovascular benefits, partly due to the ineffectiveness of non-specific antioxidants in counteracting spurious localized surges of ROS, poor absorption, and low specificity to tissues [53].

Targeted antioxidant methods show greater promise:

- *NADPH oxidase inhibitors*: Since the NOX family of NADPH oxidases is a significant source of ROccS in the kidneys and arteries, blocking them makes sense as a therapeutic target. In preclinical settings, compounds like GKT137831 (setanaxib, a dual NOX-1/−4 inhibitor) have demonstrated effectiveness in lowering vascular inflammation and fibrosis [54].
- *Antioxidants that target mitochondria*: Scavenging ROS at the main source of generation, agents such as MitoQ and SkQ1 are designed to accumulate in mitochondria. In trials on animals, they attenuate cardiac remodeling, decrease oxidative damage, and enhance mitochondrial function [55].
- *Endogenous enzyme mimetics*: By neutralizing superoxide and hydrogen peroxide, catalase analogs and superoxide dismutase (SOD) mimetics improve antioxidant defense. Tempol, a SOD mimic, has been shown to reduce blood pressure in studies on hypertension [56].
- *Upregulation of endogenous antioxidant pathways*: Glutathione peroxidase and heme oxygenase-1 are two examples of antioxidant enzymes whose transcription is increased by compounds that activate Nrf2. This indirect approach strengthens cells' resistance to oxidative damage [57].

Clinical translation of targeted antioxidants is still difficult, although they exhibit more promise than conventional vitamin supplementation. Optimal dose, tissue-specific distribution, and long-term safety are important concerns that must be addressed before broad clinical adoption.

10.7.2 Antioxidant-Inflammation-Based Therapies

Anti-inflammatory treatments represent another interesting approach due to the intertwined nature of ROS and inflammation. Anti-inflammatory treatments reduce ROS production and break the cycle of inflammatory signals. Different anti-inflammatory-based therapeutic approaches are discussed below:

- *Cytokine targeted therapies*: The development of monoclonal antibodies against pro-inflammatory cytokines has been studied, with a special emphasis on TNF-α, IL-1β, and IL-6. The CANTOS study provided evidence for the therapeutic potential of selective anti-inflammatory medication, showing dramatically decreased recurrent cardiovascular events in post-myocardial infarction patients with IL-1β inhibition (canakinumab), without alterations in lipid levels [58].
- *Inhibitors of NF-κB and inflammasome*: Inhibitors of these signaling pathways have potential because ROS promote the NF-κB signaling and subsequently the

activation of the NLRP3 inflammasome. Preclinical research findings have outlined reduced vascular remodeling and inflammation [59].

- *Immunomodulatory drugs*: Immunosuppressants such as colchicine and low-dose methotrexate have been repurposed for heart disease. Colchicine, in particular, has been associated with decreased rates of coronary events by blocking neutrophilic inflammation [60].
- *Traditional heart medications that inhibit inflammation*: Statins, angiotensin-converting enzyme (ACE) inhibitors, and angiotensin receptor blockers (ARBs) all have the dual benefit of reducing blood pressure or cholesterol while inhibiting oxidative and inflammatory signals. The difficulty with anti-inflammatories is finding the fine line between host-protective effects versus immune suppression. If patients are immunosuppressed for long periods of time, they may also become more susceptible to infections or neoplasms, signaling a need for precision-targeted therapeutics [61].

10.7.3 Phytomedicines and Nutraceuticals in ROS Management

Due to their pleiotropic action, accessibility, and favorable safety profile, phytomedicines and nutraceuticals are attractive adjuncts for mitigating the ROS-inflammation axis. Several plant-derived substances possess anti-inflammatory and antioxidant activities that can enhance the outcomes of the conventional approaches.

- *Polyphenols*: Products such as quercetin (apples, onions), curcumin (turmeric), and resveratrol (grapes, red wine) induce Nrf2, boost antioxidant defenses, and inhibit NF-κB. Curcumin inhibits vascular inflammation and fibrosis, while resveratrol improves endothelial function and enhances mitochondrial efficiency [62].
- *Flavonoids*: Epigallocatechin gallate (EGCG) and catechins from green tea scavenge ROS, modulate NO bioavailability, and inhibit vascular dysfunction [63].
- *Omega-3 fatty acids*: Present in oils from fish, omegas can improve endothelial function, inhibit pro-inflammatory cytokines, and reduce ROS [64].
- *Coenzyme Q10 (CoQ10)*: CoQ10 is considered a mitochondrial cofactor that has antioxidant capabilities. CoQ10 improves myocardial energetics and has been researched in heart failure, with a minor therapeutic effect [65].
- *Conventional herbal formulations*: Bioactive compounds in ginseng, ashwagandha, ginkgo biloba, and other Chinese and Ayurvedic remedies help regulate inflammation and oxidative stress [66].

Despite promising preclinical and observational findings, clinical translation of phytomedicines is complicated due to heterogeneity of formulation, lack of standardized dosing, and issues surrounding bioavailability. They also have attractions as adjunct therapies in addition to pharmacological interventions because of the history of usage and safety associated with their use.

10.8 Preclinical and Clinical Perspectives

In the past few decades, ROS and inflammation have been strongly implicated in the pathology of hypertension. The ROS-inflammation axis is an interesting potential therapeutic target, not only contributing to immunological activation, but also renal and vascular dysfunction. Antioxidant and anti-inflammatory modulators have been investigated in both pre-clinical and clinical studies to determine their efficacy in reducing blood pressure and preventing organ damage. This section summarizes the translational aspect of ROS-based interventions in the management of hypertension with consideration of the preclinical and clinical data.

Numerous animal models have been established to study the role of oxidative stress in hypertension, such as salt-sensitive hypertension, angiotensin II-infused rodent model, and spontaneously hypertensive rats (SHR). These studies have provided a series of preclinical evidence that indicates the excessive ROS generation leads to vascular remodeling, activation of pro-inflammatory pathways, and diminished presence of NO bioavailability [67]. In these models, several targeted therapies, like NADPH oxidase inhibition (apocynin and setanaxib), Nrf2 activators (sulforaphane and bardoxolone), SOD mimetics (tempol), and mitochondria-targeted antioxidants (MitoQ, SkQ1), have shown promising results in various validated preclinical models. Collectively, all data provide compelling support that damage associated with hypertension can be prevented or repaired by either supporting the body's own antioxidant defense systems or by diminishing sources of ROS [68–70].

As there are compelling preclinical data, it has been more difficult to translate to a clinical dose. Although some challenges and limitations need to be addressed before the successful implementation of ROS-targeted therapeutic approaches in clinical settings, like precise dose calculation, targeting pathological ROS while avoiding disruption of normal signaling, and interindividual heterogeneity in redox biology. Clinical research has been done using a variety of approaches, including mitochondrial antioxidants, NADPH oxidase inhibitors, traditional antioxidants (vitamin C, E, and β-carotene) alone and in combination therapies with ACE inhibitors, ARBs, and statins [71]. The outcomes of the preclinical research are discussed in detail in Table 10.2.

Table 10.2 Detailed discussion on the findings of various preclinical studies on ROS-targeted therapies

S. no.	Therapeutic groups	Drug name	Animal model	Mechanism of action	Key findings	References
1.	NADPH oxidase inhibitors	Morindone	Ang II-infused mouse; SHR rats	Reduce NADPH-derived superoxide; dampen redox-sensitive inflammation	Decreased vascular ROS, increased endothelial NO, decreased BP; attenuated vascular remodeling	[69]
2.	NOX blocker	Apocynin	SHR rats; dahl salt-sensitive rats	Prevents cytosolic subunit assembly of the NOX2 complex	Decreased renal oxidative stress, decreased proteinuria, and modest BP reduction	[72]
3.	NOX-1/−4 inhibitor	Setanaxib and GKT137831	Mouse models of vascular fibrosis, Ang II-hypertension	Selective NOX-1/−4 inhibition	Decreased vascular fibrosis and stiffness; improved endothelium-dependent relaxation	[73]
4.	Mitochondria-targeted antioxidant therapy	MitoQ	Ang II-infused mouse; SHR rats	Accumulates in mitochondria to scavenge ROS; protects mtDNA	Improved endothelial function, decreased aortic stiffness, and decreased cardiac hypertrophy	[55]
		SkQ1	Deoxycorticosterone Acetate (DOCA)-salt hypertensive rats	Targets the inner mitochondrial membrane to neutralize ROS	Decreased oxidative injury in vessels and the kidney; improved BP control	[55]
5.	SOD mimetic	Tempol	SHR rats; dahl salt-sensitive rats	Catalytic dismutation of superoxide to H_2O_2	Decreased systolic blood pressure, improved renal hemodynamics, and also suppressed vascular wall thickening	[74]
6.	SOD/catalase mimetic	MnTBAP	Ang II-induced hypertensive mouse	Scavenges superoxide and peroxynitrite	Restored acetylcholine-mediated vessel dilation	[75]
7.	Nrf2 activators	Sulforaphane, bardoxolone	Nuclear factor erythroid 2-related factor 2; Angiotensin II-induced hypertension	Upregulate HO-1, NQO1, SOD, catalase via Nrf2 signaling cascade	Enhanced antioxidant defense; decreased vascular inflammation; improved compliance	[76]

8.	Coenzyme Q10	Resveratrol	SHR rats	Mitochondrial electron carrier; antioxidant	Decreased oxidative biomarkers; improved endothelial-dependent relaxation; mild BP effect	[65]
9.	Polyphenol	Resveratrol	SHR rats; dahl salt-sensitive rats	SIRT1 activation; decreased Silent Information Regulator 2 homolog; NF-κB modulation	Decreased vascular inflammation; increased NO bioavailability; decreased blood pressure	[77]
		Curcumin	DOCA-salt rats	NF-κB inhibition; direct ROS scavenging	Decreased vascular oxidative stress, decreased cardiac hypertrophy, and fibrosis	[78]
10.	NF-κB modulators	MCC950 and parthenolide	Ang II–HTN; TAC; atherosclerosis models	Block IL-1β maturation or NF-κB transcription	Decreased vascular/cardiac inflammation; decreased remodeling; improved hemodynamics	[79]
11.	Endothelial cofactor restoration	Tetrahydrobiopterin (BH4)	eNOS-uncoupling models; SHR rats	Recouple eNOS to restore NO production	Decreased NO bioavailability; reduced blood pressure and oxidative markers	[80]
12.	Immunomodulators	Colchicine	Ang II-hypertension; post-MI animal models	Suppress neutrophil inflammasome/cytokine cascades	Decreased vascular inflammation, decreased remodeling, and arrhythmia protection in I/R	[81]

10.9 Conclusion

This chapter's conclusion emphasizes the crucial roles that inflammation and oxidative stress play in the pathogenesis of hypertension and other CVDs. A thorough examination of the most recent molecular and cellular research demonstrates that the production of ROS, mainly due to the overactivation of NOX, specifically NOX-1, −2, −4, and − 5, is a major cause of redox dysregulation in the vascular system. Persistently high ROS levels damage vascular endothelium, interfere with redox-sensitive signaling pathways, induce endoplasmic reticulum stress, and cause chronic inflammation. Along with aggravating systemic hypertension and hastening the development of CVDs, this pro-oxidant milieu also plays a role in endothelial dysfunction and vascular remodeling. Pro-inflammatory cytokines increase the generation of ROS, and ROS intensify inflammatory signaling, creating a vicious cycle that leads to organ malfunction and vascular damage. This feedback loop plays a key role in the pathophysiology of stroke, atherosclerosis, and other CVDs. There is still a lack of clinical translation despite preclinical research showing that utilizing ROS scavengers or blocking NOX enzymes can lower blood pressure and vascular damage. Various therapeutic approaches, such as NOX-specific inhibitors, and modulators of redox-sensitive transcription factors, such as Nrf2, have shown greater promise, yet antioxidant therapy has yielded uneven outcomes. According to genetic profiles and biomarkers, individualized redox-based treatments ought to be the main focus of future research. Providing novel paves to interfere with the ROS-inflammation axis and ameliorate CVDs, the integration of physiology and omics methods may enhance therapeutic outcomes.

References

1. Roth GA, Mensah GA, Johnson CO, Addolorato G, Ammirati E, Baddour LM, Barengo NC, Beaton AZ, Benjamin EJ, Benziger CP, Bonny A (2020) Global burden of cardiovascular diseases and risk factors, 1990–2019: update from the GBD 2019 study. J Am Coll Cardiol 76(25):2982–3021. https://doi.org/10.1016/j.jacc.2020.11.010
2. Frąk W, Wojtasińska A, Lisińska W, Młynarska E, Franczyk B, Rysz J (2022) Pathophysiology of cardiovascular diseases: new insights into molecular mechanisms of atherosclerosis, arterial hypertension, and coronary artery disease. Biomedicine 10(8):1938. https://doi.org/10.3390/biomedicines10081938
3. Jamison T (2024) Cardiovascular diseases and pathophysiology. In: mitochondrial transplantation and transfer. Academic Press, pp 327–341. https://doi.org/10.1016/B978-0-443-18858-9.00015-9
4. Sies H, Jones DP (2020) Reactive oxygen species (ROS) as pleiotropic physiological signalling agents. Nat Rev Mol Cell Biol 21(7):363–383. https://doi.org/10.1038/s41580-020-0230-3
5. Chen L, Deng H, Cui H, Fang J, Zuo Z, Deng J, Li Y, Wang X, Zhao L (2017) Inflammatory responses and inflammation-associated diseases in organs. Oncotarget 9(6):7204. https://doi.org/10.18632/oncotarget.23208

6. Mahmoud AM, Wilkinson FL, Sandhu MA, Lightfoot AP (2021) The interplay of oxidative stress and inflammation: mechanistic insights and therapeutic potential of antioxidants. Oxidative Med Cell Longev 2021(1). https://doi.org/10.1155/2021/9851914
7. de Almeida AJ, de Oliveira JC, da Silva Pontes LV, de Souza Júnior JF, Gonçalves TA, Dantas SH, de Almeida Feitosa MS, Silva AO, de Medeiros IA (2022) ROS: basic concepts, sources, cellular signaling, and its implications in aging pathways. Oxid Med Cell Longev 2022(1):1225578. https://doi.org/10.1155/2022/1225578
8. Jomova K, Raptova R, Alomar SY, Alwasel SH, Nepovimova E, Kuca K, Valko M (2023) Reactive oxygen species, toxicity, oxidative stress, and antioxidants: chronic diseases and aging. Arch Toxicol 97(10):2499–2574. https://doi.org/10.1007/s00204-023-03562-9
9. Harrington JS, Ryter SW, Plataki M, Price DR, Choi AM (2023) Mitochondria in health, disease, and aging. Physiol Rev 103(4):2349–2422. https://doi.org/10.1152/physrev.00058.2021
10. Palma FR, Gantner BN, Sakiyama MJ, Kayzuka C, Shukla S, Lacchini R, Cunniff B, Bonini MG (2024) ROS production by mitochondria: function or dysfunction? Oncogene 43(5):295–303. https://doi.org/10.1038/s41388-023-02907-z
11. Koju N, Qin ZH, Sheng R (2022) Reduced nicotinamide adenine dinucleotide phosphate in redox balance and diseases: a friend or foe? Acta Pharmacol Sin 43(8):1889–1904. https://doi.org/10.1038/s41401-021-00838-7
12. Tarafdar A, Pula G (2018) The role of NADPH oxidases and oxidative stress in neurodegenerative disorders. Int J Mol Sci 19(12):3824. https://doi.org/10.3390/ijms19123824
13. Raggi P, Genest J, Giles JT, Rayner KJ, Dwivedi G, Beanlands RS, Gupta M (2018) Role of inflammation in the pathogenesis of atherosclerosis and therapeutic interventions. Atherosclerosis 276:98–108. https://doi.org/10.1016/j.atherosclerosis.2018.07.014
14. Jiang H, Zhou Y, Nabavi SM, Sahebkar A, Little PJ, Xu S, Weng J, Ge J (2022) Mechanisms of oxidized LDL-mediated endothelial dysfunction and its consequences for the development of atherosclerosis. Front Cardiovasc Med 9:925923. https://doi.org/10.3389/fcvm.2022.925923
15. Turner MD, Nedjai B, Hurst T, Pennington DJ (2014) Cytokines and chemokines: at the crossroads of cell signalling and inflammatory disease. Biochimica et Biophysica Acta (BBA) Mol Cell Res 1843(11):2563–2582. https://doi.org/10.1016/j.bbamcr.2014.05.014
16. Sproston NR, Ashworth JJ (2018) Role of C-reactive protein at sites of inflammation and infection. Front Immunol 9:754. https://doi.org/10.3389/fimmu.2018.00754
17. Tapia VS, Daniels MJ, Palazon-Riquelme P, Dewhurst M, Luheshi NM, Rivers-Auty J, Green J, Redondo-Castro E, Kaldis P, Lopez-Castejon G, Brough D (2019) The three cytokines IL-1β, IL-18, and IL-1α share related but distinct secretory routes. J Biol Chem 294(21):8325–8335. https://doi.org/10.1074/jbc.RA119.008009
18. Ajoolabady A, Pratico D, Lin L, Mantzoros CS, Bahijri S, Tuomilehto J, Ren J (2024) Inflammation in atherosclerosis: pathophysiology and mechanisms. Cell Death Dis 15(11):817. https://doi.org/10.1038/s41419-024-07166-8
19. Choy E, Rose-John S (2017) Interleukin-6 as a multifunctional regulator: inflammation, immune response, and fibrosis. J Scleroderma Relat Disord 2(2_suppl):S1–S5. https://doi.org/10.5301/jsrd.5000265
20. Gao JH, Yu XH, Tang CK (2019) CXC chemokine ligand 12 (CXCL12) in atherosclerosis: an underlying therapeutic target. Clin Chim Acta 495:538–544. https://doi.org/10.1016/j.cca.2019.05.022
21. Luscinskas FW (2013) CXCL1 excess stops neutrophils in their tracks. Blood 122(23):3708–3710. https://doi.org/10.1182/blood-2013-10-531905
22. Affandi AJ, Carvalheiro T, Ottria A, de Haan JJ, Brans MA, Brandt MM, Tieland RG, Lopes AP, Fernández BM, Bekker CP, van der Linden M (2022) CXCL4 drives fibrosis by promoting several key cellular and molecular processes. Cell Rep 38(1). https://doi.org/10.1016/j.celrep.2021.110189
23. Poosti F, Soebadi MA, Crijns H, De Zutter A, Metzemaekers M, Berghmans N, Vanheule V, Albersen M, Opdenakker G, Van Damme J, Sprangers B (2022) Inhibition of renal fibro-

sis with a human CXCL9-derived glycosaminoglycan-binding peptide. Clin Transl Immunol 11(2):e1370. https://doi.org/10.1002/cti2.1370
24. Lennicke C, Cochemé HM (2021) Redox metabolism: ROS as specific molecular regulators of cell signaling and function. Mol Cell 81(18):3691–3707. https://doi.org/10.1016/j.molcel.2021.08.018
25. Nan N, Yan Z, Zhang Y, Chen R, Qin G, Sang N (2023) Overview of PM2. 5 and health outcomes: focusing on components, sources, and pollutant mixture co-exposure. Chemosphere 323:138181. https://doi.org/10.1016/j.chemosphere.2023.138181
26. Van Loo G, Bertrand MJ (2023) Death by TNF: a road to inflammation. Nat Rev Immunol 23(5):289–303. https://doi.org/10.1038/s41577-022-00792-3
27. Lingappan K (2018) NF-κB in oxidative stress. Curr Opin Toxicol 7:81–86. https://doi.org/10.1016/j.cotox.2017.11.002
28. Hong Y, Boiti A, Vallone D, Foulkes NS (2024) Reactive oxygen species signaling and oxidative stress: transcriptional regulation and evolution. Antioxidants 13(3):312. https://doi.org/10.3390/antiox13030312
29. Juan CA, Pérez de la Lastra JM, Plou FJ, Pérez-Lebeña E (2021) The chemistry of reactive oxygen species (ROS) revisited: outlining their role in biological macromolecules (DNA, lipids and proteins) and induced pathologies. Int J Mol Sci. 22(9):4642. https://doi.org/10.3390/ijms22094642
30. Bellanti F, Coda AR, Trecca MI, Buglio AL, Serviddio G, Vendemiale G (2025) Redox imbalance in inflammation: the interplay of oxidative and reductive stress. Antioxidants. 14(6):656. https://doi.org/10.3390/antiox14060656
31. Rimessi A, Previati M, Nigro F, Wieckowski MR, Pinton P (2016) Mitochondrial reactive oxygen species and inflammation: molecular mechanisms, diseases and promising therapies. Int J Biochem Cell Biol 81:281–293. https://doi.org/10.1016/j.biocel.2016.06.015
32. Dainichi T, Iwata M (2023) Inflammatory loops in the epithelial–immune microenvironment of the skin and skin appendages in chronic inflammatory diseases. Front Immunol 14:1274270. https://doi.org/10.3389/fimmu.2023.1274270
33. Zhang Z, Zhao L, Zhou X, Meng X, Zhou X (2023) Role of inflammation, immunity, and oxidative stress in hypertension: new insights and potential therapeutic targets. Front Immunol 13:1098725. https://doi.org/10.3389/fimmu.2022.1098725
34. Griendling KK, Camargo LL, Rios FJ, Alves-Lopes R, Montezano AC, Touyz RM (2021) Oxidative stress and hypertension. Circ Res 128(7):993–1020. https://doi.org/10.1161/CIRCRESAHA.121.318063
35. Chen Q, Wang Q, Zhu J, Xiao Q, Zhang L (2018) Reactive oxygen species: key regulators in vascular health and diseases. Br J Pharmacol 175(8):1279–1292. https://doi.org/10.1111/bph.13828
36. Ray A, Maharana KC, Meenakshi S, Singh S (2023) Endothelial dysfunction and its relation in different disorders: recent update. Health Sci Rev 7:100084. https://doi.org/10.1016/j.hsr.2023.100084
37. Oates JC, Russell DL, Van Beusecum JP (2022) Endothelial cells: potential novel regulators of renal inflammation. Am J Physiol Renal Physiol 322(3):F309–FF21. https://doi.org/10.1152/ajprenal.00371.2021
38. Félétou M, Köhler R, Vanhoutte PM (2010) Endothelium-derived vasoactive factors and hypertension: possible roles in pathogenesis and as treatment targets. Curr Hypertens Rep 12(4):267–275. https://doi.org/10.1007/s11906-010-0118-2
39. Harlacher E, Wollenhaupt J, Baaten CC, Noels H (2022) Impact of uremic toxins on endothelial dysfunction in chronic kidney disease: a systematic review. Int J Mol Sci 23(1):531. https://doi.org/10.3390/ijms23010531
40. da Silva CHND, Guedes IH, de Lima JCS, Sobrinho JMDR, dos Santos AA (2023) Responses triggered by the immune system in hypertensive conditions and repercussions on target organ damage: a review. Curr Cardiol Rev 19(2):56–65. https://doi.org/10.2174/1573403X18666220920090632

41. Lin W, Shen P, Song Y, Huang Y, Tu S (2021) Reactive oxygen species in autoimmune cells: function, differentiation, and metabolism. Front Immunol 12:635021. https://doi.org/10.3389/fimmu.2021.635021
42. Weseler AR, Bast A (2010) Oxidative stress and vascular function: implications for pharmacologic treatments. Curr Hypertens Rep 12(3):154–161. https://doi.org/10.1007/s11906-010-0103-9
43. Moris D, Spartalis M, Tzatzaki E, Spartalis E, Karachaliou G-S, Triantafyllis AS et al (2017) The role of reactive oxygen species in myocardial redox signaling and regulation. Ann Transl Med 5(16):324. https://doi.org/10.21037/atm.2017.06.17
44. Myszko M, Bychowski J, Skrzydlewska E, Łuczaj W (2025) The dual role of oxidative stress in atherosclerosis and coronary artery disease: pathological mechanisms and diagnostic potential. Antioxidants 14(3):275. https://doi.org/10.3390/antiox14030275
45. Liang Y, Ip MSM, Mak JCW (2019) (−)-Epigallocatechin-3-gallate suppresses cigarette smoke-induced inflammation in human cardiomyocytes via ROS-mediated MAPK and NF-κB pathways. Phytomedicine 58:152768. https://doi.org/10.1016/j.phymed.2018.11.028
46. Tang L, Zhang W, Liao Y, Wang W, Deng X, Wang C et al (2025) Autophagy: a double-edged sword in ischemia–reperfusion injury. Cell Mol Biol Lett 30(1):1–47. https://doi.org/10.1186/s11658-025-00713-x
47. Tsutsui H, Kinugawa S, Matsushima S (2011) Oxidative stress and heart failure. Am J Phys Heart Circ Phys 301(6):H2181–H2H90. https://doi.org/10.1152/ajpheart.00554.2011
48. Sorescu D, Griendling KK (2002) Reactive oxygen species, mitochondria, and NAD (P) H oxidases in the development and progression of heart failure. Congest Heart Fail 8(3):132–140. https://doi.org/10.1111/j.1527-5299.2002.00717.x
49. Aimo A, Castiglione V, Borrelli C, Saccaro LF, Franzini M, Masi S et al (2020) Oxidative stress and inflammation in the evolution of heart failure: from pathophysiology to therapeutic strategies. Eur J Prev Cardiol 27(5):494–510. https://doi.org/10.1177/2047487319870344
50. Adameova A, Shah AK, Dhalla NS (2020) Role of oxidative stress in the genesis of ventricular arrhythmias. Int J Mol Sci 21(12):4200. https://doi.org/10.3390/ijms21124200
51. Hool LC (2007) What cardiologists should know about calcium ion channels and their regulation by reactive oxygen species. Heart Lung Circ 16(5):361–372. https://doi.org/10.1016/j.hlc.2007.01.005
52. Pashkow FJ (2011) Oxidative stress and inflammation in heart disease: do antioxidants have a role in treatment and/or prevention? Int J Inflamm 2011(1):514623. https://doi.org/10.4061/2011/514623
53. Ashok A, Andrabi SS, Mansoor S, Kuang Y, Kwon BK, Labhasetwar V (2022) Antioxidant therapy in oxidative stress-induced neurodegenerative diseases: role of nanoparticle-based drug delivery systems in clinical translation. Antioxidants 11(2):408. https://doi.org/10.3390/antiox11020408
54. Teixeira G, Szyndralewiez C, Molango S, Carnesecchi S, Heitz F, Wiesel P et al (2017) Therapeutic potential of NADPH oxidase 1/4 inhibitors. Br J Pharmacol 174(12):1647–1669. https://doi.org/10.1111/bph.13532
55. Parker AM, Lees JG, Tate M, Phang RJ, Velagic A, Deo M et al (2025) MitoQ protects against oxidative stress-induced mitochondrial dysregulation in human cardiomyocytes. Journal of Molecular and Cellular Cardiology Plus:100469. https://doi.org/10.1016/j.jmccpl.2025.100469
56. Rosa AC, Corsi D, Cavi N, Bruni N, Dosio F (2021) Superoxide dismutase administration: a review of proposed human uses. Molecules 26(7):1844. https://doi.org/10.3390/molecules26071844
57. Alonso-Piñeiro JA, Gonzalez-Rovira A, Sánchez-Gomar I, Moreno JA, Durán-Ruiz MC (2021) Nrf2 and heme oxygenase-1 involvement in atherosclerosis related oxidative stress. Antioxidants 10(9):1463. https://doi.org/10.3390/antiox10091463
58. Rocha VZ, Rached FH, Miname MH (2023) Insights into the role of inflammation in the management of atherosclerosis. J Inflamm Res:2223–2239. https://doi.org/10.2147/JIR.S276982

59. Bai Y, Zhang L, Zheng B, Zhang X, Zhang H, Zhao A et al (2023) circACTA2 inhibits NLRP3 inflammasome-mediated inflammation via interacting with NF-κB in vascular smooth muscle cells. Cell Mol Life Sci 80(8):229. https://doi.org/10.1007/s00018-023-04840-6
60. Ferone E, Segev A, Tempo E, Gentile P, Elsanhoury A, Baggio C et al (2024) Current treatment and immunomodulation strategies in acute myocarditis. J Cardiovasc Pharmacol 83(5):364–376. https://doi.org/10.1097/FJC.0000000000001542
61. Montecucco F, Mach F (2009) Statins, ACE inhibitors and ARBs in cardiovascular disease. Best Pract Res Clin Endocrinol Metab 23(3):389–400. https://doi.org/10.1016/j.beem.2008.12.003
62. Pechanova O, Dayar E, Cebova M (2020) Therapeutic potential of polyphenols-loaded polymeric nanoparticles in cardiovascular system. Molecules 25(15):3322. https://doi.org/10.3390/molecules25153322
63. Mokra D, Adamcakova J, Mokry J (2022) Green tea polyphenol (−)-epigallocatechin-3-gallate (EGCG): a time for a new player in the treatment of respiratory diseases? Antioxidants 11(8):1566. https://doi.org/10.3390/antiox11081566
64. Bayram SŞ, Kızıltan G (2024) The role of Omega-3 polyunsaturated fatty acids in diabetes mellitus management: a narrative review. Curr Nutr Rep 13(3):527–551. https://doi.org/10.1007/s13668-024-00561-9
65. Martelli A, Testai L, Colletti A, Cicero AF (2020) Coenzyme Q10: clinical applications in cardiovascular diseases. Antioxidants 9(4):341. https://doi.org/10.3390/antiox9040341
66. Kadiri SK, Tiwari P (2025) Integrating traditional medicine with network pharmacology for Alzheimer's treatment. Health Sci Rev:100223. https://doi.org/10.1016/j.hsr.2025.100223
67. Carlstrom M, Montenegro MF (2019) Therapeutic value of stimulating the nitrate-nitrite-nitric oxide pathway to attenuate oxidative stress and restore nitric oxide bioavailability in cardiorenal disease. J Intern Med 285(1):2–18. https://doi.org/10.1111/joim.12818
68. Sylvester AL, Zhang DX, Ran S, Zinkevich NS (2022) Inhibiting NADPH oxidases to target vascular and other pathologies: an update on recent experimental and clinical studies. Biomolecules 12(6):823. https://doi.org/10.3390/biom12060823
69. Drummond GR, Selemidis S, Griendling KK, Sobey CG (2011) Combating oxidative stress in vascular disease: NADPH oxidases as therapeutic targets. Nat Rev Drug Discov 10(6):453–471. https://doi.org/10.1038/nrd3403
70. Quiroz Y, Ferrebuz A, Vaziri ND, Rodriguez-Iturbe B (2009) Effect of chronic antioxidant therapy with superoxide dismutase-mimetic drug, tempol, on progression of renal disease in rats with renal mass reduction. Nephron Exp Nephrol 112(1):e31–e42. https://doi.org/10.1159/000210577
71. Kang JH, Cook NR, Manson JE, Buring JE, Albert CM, Grodstein F (2009) Vitamin E, vitamin C, beta carotene, and cognitive function among women with or at risk of cardiovascular disease: the women's antioxidant and cardiovascular study. Circulation 119(21):2772–2780. https://doi.org/10.1161/CIRCULATIONAHA.108.816900
72. Graton ME, Potje SR, Troiano JA, Vale GT, Perassa LA, Nakamune AC, Tirapelli CR, Bendhack LM, Antoniali C (2019) Apocynin alters redox signaling in conductance and resistance vessels of spontaneously hypertensive rats. Free Radic Biol Med (134):53–63. https://doi.org/10.1016/j.freeradbiomed.2018.12.026
73. Nosalski R, Mikolajczyk T, Siedlinski M, Saju B, Koziol J, Maffia P, Guzik TJ (2020) Nox1/4 inhibition exacerbates age dependent perivascular inflammation and fibrosis in a model of spontaneous hypertension. Pharmacol Res 161:105235. https://doi.org/10.1016/j.phrs.2020.105235
74. Welch WJ, Mendonca M, Blau J, Karber A, Dennehy K, Patel K, Lao YS, Jose PA, Wilcox CS (2005) Antihypertensive response to prolonged tempol in the spontaneously hypertensive rat. Kidney Int 68(1):179–187. https://doi.org/10.1111/j.1523-1755.2005.00392.x
75. Batinić-Haberle I, Cuzzocrea S, Rebouças JS, Ferrer-Sueta G, Mazzon E, Di Paola R, Radi R, Spasojević I, Benov L, Salvemini D (2008) Pure MnTBAP selectively scavenges peroxynitrite over superoxide: comparison of pure and commercial MnTBAP samples to MnTE-2-PyP in two different models of oxidative stress injuries, SOD-specific E. coli model and

carrageenan-induced pleurisy. Free Radic Biol Med 46(2):192. https://doi.org/10.1016/j.freeradbiomed.2008.09.042
76. Lu W (2024) Sulforaphane regulates AngII-induced podocyte oxidative stress injury through the Nrf2-Keap1/ho-1/ROS pathway. Ren Fail 46(2):2416937. https://doi.org/10.1080/0886022X.2024.2416937
77. Dolinsky VW, Chakrabarti S, Pereira TJ, Oka T, Levasseur J, Beker D, Zordoky BN, Morton JS, Nagendran J, Lopaschuk GD, Davidge ST (2013) Resveratrol prevents hypertension and cardiac hypertrophy in hypertensive rats and mice. Biochimica et Biophysica Acta (BBA) Mol Basis Dis 1832(10):1723–1733. https://doi.org/10.1016/j.bbadis.2013.05.018
78. Iyer A, Chan V, Brown L (2010) The DOCA-salt hypertensive rat as a model of cardiovascular oxidative and inflammatory stress. Curr Cardiol Rev 6(4):291–297. https://doi.org/10.2174/157340310793566109
79. Li S, Li Q, Zhou Q, Li S, Wang S, Yao Q, Ouyang C, Liu C, Li M (2024) Attenuating atherosclerosis through inhibition of the NF-κB/NLRP3/IL-1β pathway-mediated Pyroptosis in vascular smooth muscle cells (VSMCs). Cardiovasc Ther 2024(1):1506083. https://doi.org/10.1155/2024/1506083
80. Faria AM, Papadimitriou A, Silva KC, Lopes de Faria JM, Lopes de Faria JB (2012) Uncoupling endothelial nitric oxide synthase is ameliorated by green tea in experimental diabetes by re-establishing tetrahydrobiopterin levels. Diabetes 61(7):1838–1847. https://doi.org/10.2337/db11-1241
81. Wang X, Chen L, Wei J, Zheng H, Zhou N, Xu X, Deng X, Liu T, Zou Y (2025) The immune system in cardiovascular diseases: from basic mechanisms to therapeutic implications. Signal Transduct Target Ther 10(1):166. https://doi.org/10.1038/s41392-025-02220-z

Chapter 11
Interplay of Reactive Oxygen Species (ROS) and Inflammatory Pathways in Diabetes

Bedanta Bhattacharjee, Sandhanam Kuppusamy, Ram Kumar Sahu, and Jiyauddin Khan

Abbreviations

Bcl-2	B-cell lymphoma 2
DM	Diabetes mellitus
DNA	Deoxyribonucleic acid
ER	Endoplasmic reticulum
IFN-γ	Interferon gamma
IKK	IκB kinase
IL-1β	Interleukin-1 beta
JNK	c-Jun N-terminal kinase
MODY	Maturity-onset diabetes of the young
NEFAs	Non-esterified fatty acids
OS	Oxidative stress
RNAs	Ribonucleic acids
ROS	Reactive oxygen species
T1DM	Type 1 diabetes mellitus
T2DM	Type 2 diabetes mellitus
TNF-α	Tumor necrosis factor alpha

B. Bhattacharjee
School of Pharmaceutical Sciences, Girijananda Chowdhury University, Tezpur Campus, Assam, India

S. Kuppusamy
Department of Pharmacology, SRM College of Pharmacy, SRMIST, Chennai, India

R. K. Sahu (✉)
Department of Pharmaceutical Sciences, Hemvati Nandan Bahuguna Garhwal University (A Central University), Chauras, Campus, Tehri Garhwal, Uttarakhand, India

J. Khan
School of Pharmacy, Management and Science University, Shah Alam, Selangor Darul Ehsan, Malaysia

S. Ashique, N. Mishra (eds.), *Reactive Oxygen Species in Metabolic Inflammation*,
https://doi.org/10.1007/978-3-032-11752-6_11

11.1 Introduction

Diabetes is a long-term metabolic disorder that causes blood glucose levels to elevate over time. Types of diabetes are as follows: type 1, type 2, gestational diabetes (occurs during pregnancy), maturity-onset diabetes of the young (MODY), neonatal diabetes (appears in newborns), and secondary diabetes (hormone problems, or medications like steroids) [1]. Diabetes mellitus type 1 (T1DM) is the most prevalent form in children and teenagers; body is unable to produce enough insulin, the hormone that helps in controlling the blood glucose levels. Diabetes mellitus type 2 (T2DM) is more frequent in adults and usually occurs when the body either doesn't respond properly to insulin or doesn't make enough of it. Both T1DM and T2DM types lead to high blood glucose [2]. Lifestyle factors like a poor diet, lack of exercise, and long-standing unhealthy habits contribute to T2DM. Diabetes is a growing health concern worldwide. According to the International Diabetes Federation, nearly 537 million people were living with diabetes in 2021. This disease is also affecting more people in Africa, about 24 million cases and a prevalence of 4.5% that same year, largely due to urbanization and changing lifestyles [3]. Diabetes is a complicated condition caused by a combination of genetic and environmental factors. Long-term high blood glucose called chronic hyperglycemia can damage the cells in the pancreas that release insulin, making it even harder for the body to regulate blood glucose. This can set off a harmful cycle, where persistent high glucose levels make metabolic problems worse. Blood glucose above 180 mg/dL is considered high, though the symptoms people experience can vary [4]. If glucose levels rise above 250 mg/dL, classic diabetes symptoms often appear, such as increased urination (polyuria) and excessive thirst (polydipsia). This happens because the kidneys can't reabsorb all the extra glucose, resulting in loss of water and more frequent trips to the bathroom [5].

Insulin resistance often develops due to an excess of fatty acids and inflammatory substances in the body. This makes it harder for cells to take up glucose, while fat breakdown increases. The body's response to insulin is weakened, or it can't produce enough insulin; it compensates by making too much glucagon, which in turn raises blood glucose levels even further [6]. Insulin resistance is a key feature of T2DM; the disease truly sets in when the pancreas can no longer make enough insulin to overcome this resistance. Long-term high blood glucose causes glucoses to attach to proteins and fats in the body, a process called glycation. This buildup can be measured with the HbA1c blood test [7]. Glycation particularly affects small blood vessels in the eyes, kidneys, and nerves, and higher blood glucose speeds up this damage. As a result, people with poorly controlled diabetes risk complications such as vision loss (diabetic retinopathy), kidney failure (diabetic nephropathy), and nerve damage (diabetic neuropathy), which can lead to blindness, the need for dialysis, or even limb amputation complications that can often be prevented with good blood glucose control [8].

Oxidative stress (OS) and inflammation are vital biological processes that help protect the body under normal conditions. Reactive oxygen species (ROS), a key component of OS, play an essential role in the immune system by attacking invading organisms and supporting tissue repair. Inflammation acts as a defense mechanism; it responds to infections or injuries, removes harmful substances, and aids in healing. These processes decrease their regulation, and they can contribute to the healing of a variety of diseases. Maintaining a careful balance of both inflammation and OS is crucial for good health and disease prevention [9]. ROS not only function as signaling molecules within cells, but they also help produce substances that drive inflammation. This relationship can become a vicious cycle; excess of ROS can increase inflammation, which in turn leads to the production of even more ROS. When there is too much ROS, they can damage important parts of our cells, such as DNA, fats, and proteins [10]. Due to high blood glucose (hyperglycemia), the body generates more free radicals and its natural antioxidant defenses become weaker, which leads to further oxidative imbalance. ROS come with different effects on our biology that, they are generally grouped into two main categories: radical and non-radical species. Radical ROS like alkoxyl, hydroxyl, superoxide, and peroxyl radicals are highly reactive and can cause significant biological effects [11]. Non-radical ROS, such as ozone, singlet oxygen, and hydrogen peroxide, also influence cell function, though they behave differently. Understanding how ROS work and how they are classified is important for developing treatments that target these processes. ROS are strongly linked to the control of metabolism and inflammation. When ROS levels stay high for a long time, they can disrupt how insulin works and is released, leading to problems like metabolic dysfunction and insulin resistance [12]. This highlights why it is important to study how ROS are involved in chronic inflammation and metabolic diseases. Both OS and inflammation are now widely recognized as major contributors to developing insulin resistance, a key feature of T2DM. These conditions activate certain stress-response proteins inside cells such as Jun N-terminal kinase (JNK) and IκB kinase (IKK) that interfere with insulin signaling after it binds to its receptor. OS and inflammation can damage fat tissue, which reduces the production of adiponectin, a hormone that helps keep the body sensitive to insulin [12]. Scientific research on diabetes in relation to ROS and inflammation during the period 2024–2025 is presented in Fig. 11.1. The dataset was obtained from PubMed, Scopus, and Web of Science.

Besides insulin resistance, OS is another key factor contributing to organ damage in diabetes. In diabetic kidney disease, persistently high blood glucose triggers harmful processes like the buildup of toxic fats, lack of oxygen in kidney tissues, and disrupted kidney filtration. Over time, these issues lead to ongoing kidney inflammation and a gradual decline in kidney function. The liver is also at risk, as constant exposure to ROS and inflammation can activate immune cells and cause oxidative injury [13]. This can result in scarring of liver tissue (fibrosis) and even progress to cirrhosis. Similarly, chronic OS and inflammation can damage the heart, increasing the risk of cardiovascular problems such as atherosclerosis or irregular heart rhythms like atrial fibrillation. Long-standing inflammation and OS in the brain may contribute to nerve cell damage and have been linked to both neurodegenerative and psychiatric conditions [14]. These wide-ranging effects highlight how diabetes-driven OS and inflammation can harm multiple body systems. This chapter

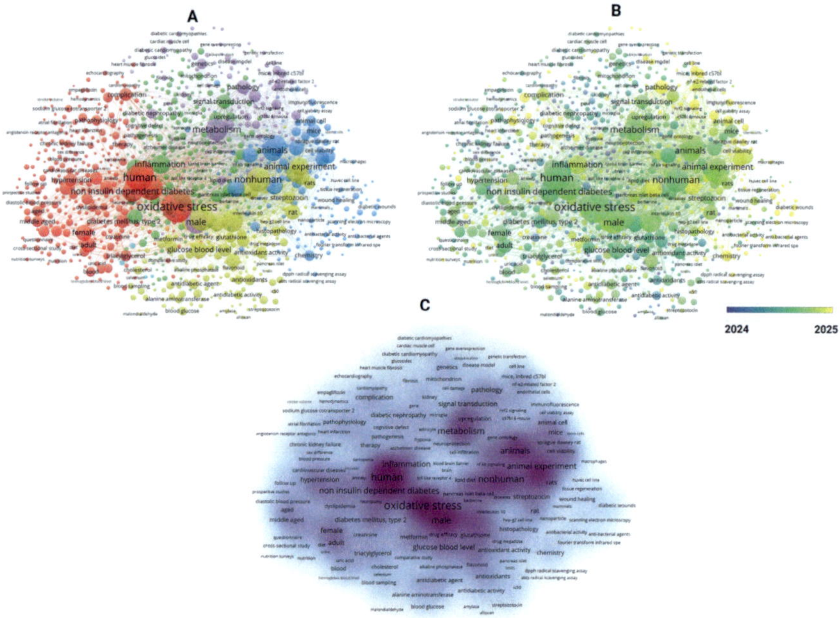

Fig. 11.1 A bibliometric network map of scientific research on diabetes related with ROS and inflammation. The network visualization based on four colored clusters (**a**) indicates the relatedness of the scientific journals and an overlay visualization (**b**) specifies a period of the occurrence of the keyword from 2024 (blue) to 2025 (yellow). This figure is created from VOSviewer (Jan van Eck and Waltman 2018), and collected data were obtained from Web of Science Core Collection and Scopus database utilizing keywords "diabetes" and "reactive oxygen species" and "inflammation" and a time range of 2024–2025. (**c**) Density heat map visualization of diabetes research

will explore in detail how diabetes, OS, and inflammation interact and will review current and emerging treatment strategies aimed at reducing inflammation and preventing diabetes complications.

11.2 Reactive Oxygen Species in Diabetes

It has been established that reactive oxygen species (ROS) were developed in numerous tissues in diabetes situations [15]. ROS may be produced in cells by a variety of mechanisms, including the membrane-mediated NADPH oxidase (NOX), the electron transport chain (ETC) in mitochondria, and non-enzymatic glycosylation process [16–18]. In preclinical diabetic model, glycation process is recognized in numerous tissue and organ levels, and multiple types of glycated proteins such as lens crystalline, albumin, and glycosylated hemoglobin are formed in a non-enzymatic way using the glycation process [16]. The reaction generates Amadori rearrangement product, Schiff base, and eventually advanced glycosylation end products (AGEs). Through this mechanism, ROS are also generated [16]. The ETC in mitochondria is additionally a key mechanism for ROS production. During diabetic situations, the ETC is stimulated, resulting in increased generation of ROS [17]. It has been demonstrated that membrane-mediated NOX is a significant generator of ROS. NOX is comprised of the membrane-mediated subunit p22 phox, and gp91 phox (Nox2)/Nox1/Nox4, as well as the oxidase regulatory domain and cytoplasmic components p67 phox and p47 phox. NOX is triggered by several stimuli such as angiotensin-II, insulin, and AGEs, each of which is potentially involved in diabetes development [18].

11.3 Inflammatory Pathways and Their Role in Diabetes

Multiple investigations have reported elevated levels of interleukins, chemokines, cytokines, and acute-phase proteins (i.e., haptoglobin, serum-amyloid A, and C-reactive protein [CRP]) in diabetic patients [19]. The high systemic levels of CRP, IL-6, and IL-1β are considered as important biomarkers of diabetes development. An increased concentration of inflammatory mediators can activate the natural immune response in diabetic individuals because of over-nutrition [20].

Moreover, such inflammatory mediators are unable to demonstrate the severity of inflammation in the various peripheral organs [21]. Over-nutrition is defined as excessive nutrient consumption above the level that is essential for regular body growth and development, that can be deleterious to one's health [21]. Nutrients are any substances, such as vitamins, fats, proteins, minerals, and carbohydrates, that are required for regular bodily processes and growth [22]. The increased levels of pro-inflammatory biomarkers in the bloodstream, as well as obvious organ inflammation, have been observed in diabetic patients. As a result, examining the quantity of these pro-inflammatory biomarkers in the blood simply could not be sufficient to forecast the type and intensity of inflammation in an organ level [22]. The evidence outlined above make it extremely apparent that inflammation is a key driver in the advancement of diabetes. Therefore, diabetes is a persistent auto-inflammatory conditions characterized by the production of IL-1β by β-cells, which causes β-cell impairment [23]. The below are the most significant inflammatory reactions that contribute to the pathophysiology of diabetes:

11.3.1 Gene Regulatory Cascades

There are multiple biological systems which generate insulin insensitivity in peripheral tissues. They induce inflammation and activate stress protein kinases including Jun N-terminal kinase (JNK) and IκB kinase-β (IKKβ) [24]. These protein kinases play a significant role in the progression of diabetes. IKKβ could accelerate the expression of nuclear factor-κB (NF-κB), which, further, activates the pro-inflammatory mediators like IL-1β and tumor necrosis factor alpha (TNF-α) in adipose tissue and liver. These mediators are known to cause insulin insensitivity in the peripheral tissues [24]. On the other hand, JNK stimulates ETS like-1 protein (ELK1) and activates transcription factor-2 (ATF2) [25]. Since the significance of JNK-mediated transcription factors is unknown, pre-clinical investigations reveal that JNK contributes an integral function in the inflammatory response as well as etiopathogenesis of diabetes [25]. IL-1β and TNF-α, generated by NF-κB triggering, further activate both JNK and NF-κB via feed-forward loops utilizing specific receptors [26]. In addition to the JNK and NF-κB cascades, AGEs and FFAs may activate receptors for advanced glycation end products (RAGE) and toll-like receptors (TLRs), which in turn may increase insulin insensitivity [26]. The extracellular signals trigger intracellular circuits such as NF-κB and JNK for binding to cell-specific receptors. Stimulation of such systems in adipose tissues and liver increases the generation of IL-6, IL-1β, and TNF-α [27]. Hyperglycemic conditions and excess IL-1β release by β-cells might stimulate NF-κB. The inhibition of NF-κB activity with the help of interleukin-1 receptor antagonist prevents β-cells from detrimental consequences. The incidence of inflammation may be lowered by inhibiting the expression of the JNK and NF-κB cascades, which are active in several tissues and are crucial for inflammation in tissues [27].

11.3.2 Chemokines

Adipocytes release chemokines, which attract macrophages and monocytes [28]. There is evidence that the adipose tissues of obese person release higher chemokines than those of non-obese person. The example of chemokines includes CCL2 or MCP-1, CCL3 or MIP-1α, CCL5, CCL6, CCL7, CCL8, and CCL9 [29]. Such chemokines are secreted by endothelial cells, pancreatic cells, and adipose tissue. In preclinical models, diabetes has been linked to an increase in chemokine secretion [30]. Several laboratory studies have shown that such chemokines contribute an essential role in the progression of diabetes together with pro-inflammatory cytokines; although, the specific mechanism involving these chemokines is yet to be established [30].

11.3.3 Cytokines

The role of many pro-inflammatory cytokines, including IL-6, TNF-α, and IL-1β in the multimodal pathophysiology for the spread of diabetes is especially intriguing [31]. Such cytokines, which are generated by adipose tissues, cause inflammation in the appropriate tissue as well as in the β-cells, which eventually results in insulin resistance [31, 32]. At first, it was proposed that inflammation in β-cells was caused by hyperglycemia-induced apoptosis. This suggested process were later identified by Maedler et al. who discovered that high glucose level can activates the β-cell's pro-apoptotic receptor Fas cell surface death receptor (FAS) and may cause the release of IL-1β [33]. IL-1β expression in β-cells is controlled by free fatty acids (FFAs), which enhance the release of IL-1β and other pro-inflammatory cytokines, in addition to glucose-induced stimulation [33]. When IL-1β is stimulated by FFAs or glucose, it controls its secretion and synthesis in β-cells via auto-stimulation. To inhibit the auto-stimulation of IL-1β, decrease NF-κB activity and/or reduce Interleukin-1 Receptor-1 (IL-1RI) activation using IL-1R antagonist. Thus, inhibiting IL-1RI activity could impede the stimulation of IL-1β driven by FFAs or glucose [33]. Additionally, β-cell malfunction and decreased insulin release could result from activation of IL-1β production, which concurrently promotes nitric oxide synthesis and lowers mitochondrial Adenosine triphosphate (ATP) level [33]. Apart from IL-1β, TNF-α also performs a crucial function by forming a bridge between inflammation, obesity, and insulin resistance. TNF-α is a critical element in the relationship between insulin resistance and inflammation [34]. It affects the JNK and NF-κB/IKKβ cascades responsible for insulin resistance. The excessive production of TNF-α in adipose tissues leads to insulin resistance in peripheral tissues, promoting inflammation and pancreatic β-cell death [34].

The function of IL-6 in diabetes is complicated and contentious. Although, experimental investigations have shown that IL-6 causes inhibition of cytokine trafficking proteins, apoptotic death of β-cells, and insulin resistance in peripheral tissues. In addition to the negative consequences, IL-6 has become a distinct risk component and works as a predictor and pathologic indicator for insulin resistance and development of diabetes [35].

11.3.4 Adipocytokines

Adipocytokines include adiponectin and leptin, which are often generated by adipocytes. Such adipocytokines exhibit immunomodulatory impacts on diabetes [36]. Leptin is classified as a pro-inflammatory cytokine because of its structural homology to other cytokines. Leptin triggers apoptosis in β-cells by increasing IL-1β production and suppressing IL-1R antagonists, impairing insulin release from β-cells [37]. Adiponectin serves a distinct role in diabetes compared to leptin. It reduces inflammation in endothelial cells by inhibiting NF-κB and TNF-α-mediated adhesion protein production, while activating IL-1R antagonists. In obese pre-clinical models, it enhances insulin sensitivity and lowers glucose concentrations [37].

11.3.5 Low Oxygen Levels

This condition, also known as hypoxia, develops when oxygen level is inadequate [38]. Angiogenesis is an adaptive process that releases angiogenic mediators as a substitute for the lack of oxygen in aggressively developing tissues (e.g., cancer tissues) [39]. A comparable process has been found in pre-clinical models of obesity and human adipose tissues, potentially causing tissue malfunction. Hypoxia may trigger the expression of pro-inflammatory proteins in macrophages [40]. Macrophages build up in hypoxic areas, triggering inflammation in quickly expanding fat tissues. Hypoxia can cause endoplasmic reticulum-stress in pancreatic β-cells by affecting redox potential and ATP concentration, leading to β-cell death [40].

11.4 Interplay Between ROS and Inflammation

Overnutrition can result in excessive levels of glucose and fats in the body, which, over time, contribute to obesity and the development of metabolic syndrome. This state leads to persistent, low-grade inflammation. Predominantly elevated nutrient levels affect insulin sensitivity, disrupting the insulin-signaling pathway by altering the phosphorylation of molecules like IRS1, PI3K, and AKT [41]. High fat intake also promotes inflammation through the creation of ceramides, while too much glucose increases OS in the body. As T2DM progresses, tissues that become resistant to insulin put extra pressure on insulin-producing β-cells, eventually exhausting them and resulting in decreased insulin release. The central role of pancreatic β-cells is to sense nutrient levels and adjust insulin secretion accordingly. When exposed to increases in nutrients such as glucose, fats, and certain amino acids, β-cells normally ramp up insulin production [42]. Short-term exposure to these nutrients helps maintain blood sugar; however, prolonged high levels of glucose and fats can be toxic to β-cells, leading to diminished insulin secretion and contributing to insulin resistance elsewhere in the body. Initially, this may cause β-cells to overwork and release too much insulin, but chronic exposure raises OS, which eventually impairs β-cell function or even causes their death. This reduction in insulin-producing cells worsens the harmful cycle of high blood sugar and fat seen in T2DM [43]. Excess glucose pushes β-cells to synthesize more insulin, overloading the endoplasmic reticulum (ER) and triggering an unfolded protein response. If ER stress persists, it can prompt cell death and release of inflammatory signals like IL-1β via activation of the inflammasome. The rise in local inflammatory cytokines activates pathways like NF-κB, which switch on pro-apoptotic genes, further driving β-cells toward apoptosis. NF-κB also increases inducible nitric oxide synthase (iNOS) activity, which contributes to both ER stress and cell death. Additionally, high levels of non-esterified fatty acids (NEFAs) aggravate lipotoxicity, causing further ER stress by boosting the production of reactive oxygen and nitrogen species, disrupting calcium

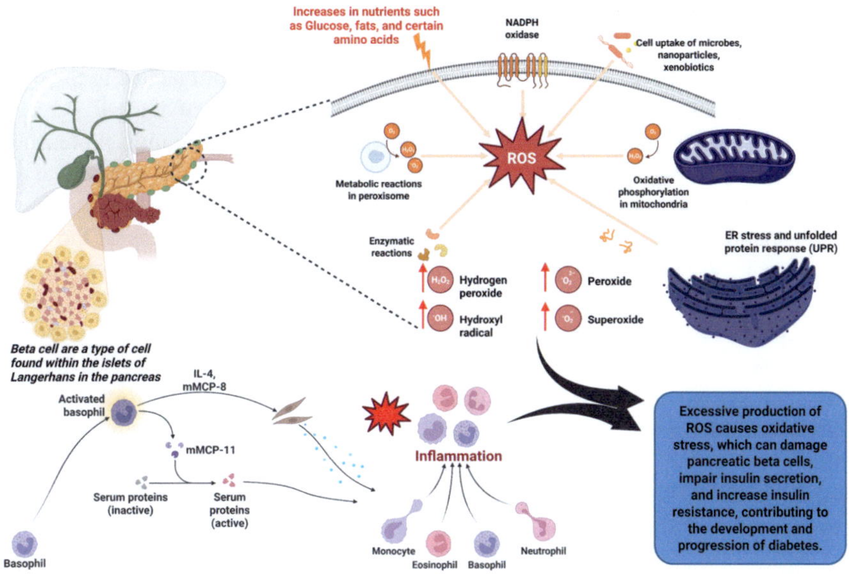

Fig. 11.2 ROS and inflammation contribute to the development and progression of diabetes. In the pancreas, within insulin-producing beta cells located in the islets of Langerhans, excessive production of ROS leads to oxidative stress. ROS are highly reactive molecules generated through several cellular processes, including metabolic reactions in organelles like mitochondria and peroxisomes, and during immune responses. ROS production increases the damage in pancreatic beta cells, impairing their ability to secrete insulin properly. ROS promote the activation and recruitment of immune cells such as monocytes, eosinophils, basophils, and neutrophils that lead to inflammation. This inflammation disrupts pancreatic function and increases insulin resistance. The combined effects of OS and chronic inflammation weaken insulin secretion and action, play an important role in both the onset and worsening of diabetes. (Created with Biorender.com [Premium Version])

balance, and interfering with protein trafficking within the cell is shown in Fig. 11.2 [44].

Unregulated OS is an important factor in the impairment of pancreatic β-cell function, prompting ongoing research into the use of antioxidant compounds to support β-cell insulin production and survival, an approach that shows promise for diabetes management, particularly with glutathione peroxidase mimetics like ebselen. OS, nitrosative and endoplasmic reticulum (ER) stress also play considerable roles in organs targeted by insulin, such as skeletal muscle and adipose tissue [45]. Inflammatory cytokines are central to β-cell deterioration. In T1DM, it has been well established that cytokines like TNF-α, IFN-γ, and IL-1β, which are secreted by infiltrating immune cells, contribute to insulitis and induce β-cell death. Research has shown that IL-1β on its own or combined with TNF-α and IFN-γ can significantly disrupt β-cell function and promote cell death. Moreover, high levels of glucose and non-esterified fatty acids (NEFAs) within the islets can stimulate local production of IL-1β, suggesting that cytokine-driven β-cell apoptosis isn't exclusive to autoimmune diabetes but is also relevant in metabolic syndrome and T2DM [46]. Inflammation pathway involved in this process is by activation of NF-κB in the β-cells. Under normal conditions, NF-κB is kept inactive in the cytoplasm, bound to inhibitory IκB proteins. Predominant stimulation of the IL-1 receptor, IκB is phosphorylated, marking it for breakdown by the ubiquitin-proteasome system. As IκB is degraded, NF-κB is released and moves to the nucleus, and it activates the expression of numerous genes affected by cytokines. These include genes such as iNOS as well as both anti- and proapoptotic members of the Bcl-2 family [47]. When iNOS is ramped up by NF-κB, it increases nitric oxide production and it has been linked directly to disrupted insulin secretion and death of β-cells in the presence of cytokines. Additionally, NF-κB enhances the expression of p53 upregulated modulator of apoptosis (PUMA), leading to the activation of Bax and Bak proteins which worsen β-cell apoptosis triggered by inflammatory cytokines [48]. Excess nutrients trigger OS and inflammation in pancreatic β-cells, ROS and inflammatory cytokines activate pathways like NF-κB, leading to β-cell dysfunction and apoptosis. This vicious cycle worsens insulin resistance and accelerates the progression of diabetes [49].

11.5 Therapeutic Implications and Future Directions

In recent years, researchers have been paying closer attention to OS and the inflammation contributing to the development and worsening of DM. This has led to growing interest in adding antioxidants and anti-inflammatory agents to traditional diabetes treatments. OS happens when harmful molecules called ROS damage our cells, which plays a big part in diabetes-related complications [50]. Antioxidants like vitamins E and C haven't consistently delivered strong results in clinical trials, newer strategies are focusing more precisely on mechanism going on inside the cells, especially in the mitochondria the tiny powerhouses that generate energy. Some of the most promising treatments mimic natural enzymes in our bodies that help neutralize these damaging molecules [51]. α-lipoic acid acts by boosting the body's own antioxidant defenses. Some medications previously prescribed for diabetes complications, such as statins and ACE inhibitors, may also have antioxidant properties, which help to reduce the risk of further issues [52].

Natural substances found in plants like certain flavonoids, anthocyanins, and puerarin are also showing encouraging effects. They seem to reduce OS and inflammation by improving mitochondrial health and blocking key inflammatory processes. This might protect against problems like diabetic kidney disease and heart conditions [9]. These natural substances can help the body fight off free radicals and lower the levels of toxic chemicals that cause inflammation and tissue damage. Some studies even suggest that some natural substances might have an effect on small molecules known as non-coding RNAs, which are critical for controlling inflammation and OS [53]. This makes novel treatments possible that are quite interesting. But there are still a lot of problems that need to be solved before these treatments can be used regularly to treat diabetes. Diabetes is a complex disease that affects people in different ways, so it's hard to show that antioxidant and anti-inflammatory medications will work well for everyone. Other problems include making sure these treatments are safe, work well, are always of excellent quality, and are given the right amount, as well as receiving permission from health authorities [54]. It's also very vital to test these treatments on a wide range of people to find out more about their pros and cons. The future of diabetes treatment probably lies in precision medicine, which tailor antioxidant and anti-inflammatory medicines to each person's individual genetics and metabolism. To achieve that objective, researchers must investigate the exact molecular targets implicated and conduct more clinical studies to validate these hypotheses. It's just as vital to include people from all areas of life in these research and look at the long-term impacts of natural chemicals and novel antioxidants [55]. Ultimately, combining these new methods with the routine diabetes care of today could help patients with diabetes live better lives and lower their risk of complications.

11.6 Conclusion

Diabetes is a complex disease that deteriorates progressively due to two primary factors: oxidative damage and inflammation. The cells generate too many harmful substances called ROS which damage the oxidative balance and stress. This damages the beta-cells in the pancreas that create insulin and makes it harder for the body to use insulin properly. The body's immune response keeps these harmful processes by making proteins that keep the inflammation going at a low level. This sets off a bad cycle in which OS and inflammation make each other worse, which damage more cells and make insulin problems worse. This can cause really severe health problems for persons with diabetes, like kidney disease, nerve damage, vision problems, and severe cardio vascular problems. Understanding these damaging processes work together helps us realize why diabetes is so challenging to treat and merely maintaining blood glucose levels stable isn't always enough. This knowledge highlights how crucial it is to make treatments that act on both OS and inflammation. Early research on antioxidants, anti-inflammatory drugs, and natural chemicals shows that they might be able to help minimize the oxidative stress effects. It is still an ongoing effort to turn these findings into safe and effective treatments for patients. In the future research, using these new treatment techniques along with regular diabetes care could make a big impact in helping individuals live better lives and avoid serious issues. Continued research into the molecular mechanisms underlying this crosstalk will be essential for developing more effective and multi-targeted treatments to improve clinical outcomes for diabetic patients.

References

1. Choudhury AA, Devi Rajeswari V (2021 Nov) Gestational diabetes mellitus – a metabolic and reproductive disorder. Biomed Pharmacother 143:112183
2. Galicia-Garcia U, Benito-Vicente A, Jebari S, Larrea-Sebal A, Siddiqi H, Uribe KB et al (2020) Pathophysiology of type 2 diabetes mellitus. IJMS 21(17):6275
3. Pappachan JM, Ashraf AP (2024) Global strategies for preventing type 2 diabetes: a public health perspective. Indian J Med Res 160:391–395
4. Goyal S, Rani J, Bhat MA, Vanita V (2023) Genetics of diabetes. World J Diabetes 14(6):656–679
5. Ramírez-Guerrero G, Müller-Ortiz H, Pedreros-Rosales C (2022) Polyuria in adults. A diagnostic approach based on pathophysiology. Rev Clín Esp (English Edition) 222(5):301–308
6. Lee SH, Park SY, Choi CS (2022) Insulin resistance: from mechanisms to therapeutic strategies. Diabetes Metab J 46(1):15–37
7. Zhao X, An X, Yang C, Sun W, Ji H, Lian F (2023) The crucial role and mechanism of insulin resistance in metabolic disease. Front Endocrinol 14:1149239
8. Li Y, Liu Y, Liu S, Gao M, Wang W, Chen K et al (2023) Diabetic vascular diseases: molecular mechanisms and therapeutic strategies. Sig Transduct Target Ther 8(1):152
9. Jomova K, Alomar SY, Valko R, Liska J, Nepovimova E, Kuca K et al (2025 May) Flavonoids and their role in oxidative stress, inflammation, and human diseases. Chem Biol Interact 413:111489
10. De Almeida AJPO, De Oliveira JCPL, Da Silva Pontes LV, De Souza Júnior JF, Gonçalves TAF, Dantas SH et al (2022) ROS: basic concepts, sources, cellular signaling, and its implications in aging pathways. Giustarini D, editor. Oxid Med Cell Longev 2022(1):1225578
11. Bhatti JS, Sehrawat A, Mishra J, Sidhu IS, Navik U, Khullar N et al (2022 May) Oxidative stress in the pathophysiology of type 2 diabetes and related complications: current therapeutics strategies and future perspectives. Free Radic Biol Med 184:114–134
12. Aranda-Rivera AK, Cruz-Gregorio A, Arancibia-Hernández YL, Hernández-Cruz EY, Pedraza-Chaverri J (2022) RONS and oxidative stress: an overview of basic concepts. Oxygen 2(4):437–478
13. Wang N, Zhang C (2024) Oxidative stress: a culprit in the progression of diabetic kidney disease. Antioxidants 13(4):455
14. El Hadi H, Di Vincenzo A, Vettor R, Rossato M (2020) Relationship between heart disease and liver disease: a two-way street. Cells 9(3):567
15. Iacobini C, Vitale M, Pesce C, Pugliese G, Menini S (2021) Diabetic complications and oxidative stress: a 20-year voyage back in time and back to the future. Antioxidants [Internet] 10(5):727. [cited 2025 Aug 14]. Available from: https://www.mdpi.com/2076-3921/10/5/727
16. Kang Q, Dai H, Jiang S, Yu L (2022) Advanced glycation end products in diabetic retinopathy and phytochemical therapy. Front Nutr [Internet] 9:1037186. [cited 2025 Aug 14]. Available from: https://www.frontiersin.org/articles/10.3389/fnut.2022.1037186/full
17. Saucedo R, Ortega-Camarillo C, Ferreira-Hermosillo A, Díaz-Velázquez MF, Meixueiro-Calderón C, Valencia-Ortega J (2023) Role of oxidative stress and inflammation in gestational diabetes mellitus. Antioxidants [Internet] 12(10):1812. [cited 2025 Aug 14]. Available from: https://www.mdpi.com/2076-3921/12/10/1812
18. Sun Y, Wu D, Hu Q (2025) $NADP^+$/NADPH in metabolism and its relation to cardiovascular pathologies. CMC [Internet] 32(23):4671–4687. [cited 2025 Aug 14]. Available from: https://www.eurekaselect.com/227059/article
19. Gart E, Snabel J, De Jong JCBC, Verschuren L, Van Den Hoek AM, Morrison MC et al (2025) Biomarkers of metabolism and inflammation in individuals with obesity and normal weight: a comparative analysis exploring sex differences. IJMS [Internet] 26(15):7576. [cited 2025 Aug 14]. Available from: https://www.mdpi.com/1422-0067/26/15/7576

20. Savulescu-Fiedler I, Mihalcea R, Dragosloveanu S, Scheau C, Baz RO, Caruntu A et al (2024) The interplay between obesity and inflammation. Life [Internet] 14(7):856. [cited 2025 Aug 14]. Available from: https://www.mdpi.com/2075-1729/14/7/856
21. Mao L, Gao B, Chang H, Shen H (2024) Interaction and metabolic pathways: elucidating the role of gut microbiota in gestational diabetes mellitus pathogenesis. Metabolites [Internet] 14(1):43. [cited 2025 Aug 14]. Available from: https://www.mdpi.com/2218-1989/14/1/43
22. Weinberg Sibony R, Segev O, Dor S, Raz I (2024) Overview of oxidative stress and inflammation in diabetes. J Diabetes [Internet] 16(10):e70014. Oct [cited 2025 Aug 14]. Available from: https://onlinelibrary.wiley.com/doi/10.1111/1753-0407.70014
23. Ratan Y, Rajput A, Pareek A, Pareek A, Kaur R, Sonia S et al (2024) Recent advances in biomolecular patho-mechanistic pathways behind the development and progression of diabetic neuropathy. Biomedicines [Internet] 12(7):1390. [cited 2025 Aug 14]. Available from: https://www.mdpi.com/2227-9059/12/7/1390
24. Püschel GP, Klauder J, Henkel J (2022) Macrophages, Low-grade inflammation, insulin resistance and hyperinsulinemia: a mutual ambiguous relationship in the development of metabolic diseases. JCM [Internet] 11(15):4358. [cited 2025 Aug 14]. Available from: https://www.mdpi.com/2077-0383/11/15/4358
25. Yan H, He L, Lv D, Yang J, Yuan Z (2024) The role of the dysregulated JNK signaling pathway in the pathogenesis of human diseases and its potential therapeutic strategies: a comprehensive review. Biomolecules [Internet] 14(2):243. [cited 2025 Aug 14]. Available from: https://www.mdpi.com/2218-273X/14/2/243
26. Zhao L, Hu H, Zhang L, Liu Z, Huang Y, Liu Q et al (2024) Inflammation in diabetes complications: molecular mechanisms and therapeutic interventions. MedComm [Internet] 5(4):e516. [cited 2025 Aug 14]. Available from: https://onlinelibrary.wiley.com/doi/10.1002/mco2.516
27. Burillo J, Marqués P, Jiménez B, González-Blanco C, Benito M, Guillén C (2021) Insulin resistance and diabetes mellitus in Alzheimer's disease. Cells [Internet]. 10(5):1236. [cited 2025 Aug 14]. Available from: https://www.mdpi.com/2073-4409/10/5/1236
28. Pan X, Kaminga AC, Wen SW, Liu A (2021) Chemokines in prediabetes and type 2 diabetes: a meta-analysis. Front Immunol [Internet] 12:622438. [cited 2025 Aug 14]. Available from: https://www.frontiersin.org/articles/10.3389/fimmu.2021.622438/full
29. Sandhanam K, Bhattacharjee B, Rynjah D, Kalita P, Sen S, Chakraborty R (2024) Deciphering the interlinked CXCR4-mediated feedback loop among signaling pathways in diabetic wound healing. CDR [Internet] 21. [cited 2025 Jan 6]. Available from: https://www.eurekaselect.com/236649/article
30. Chang TT, Chen C, Chen JW (2022) CCL7 as a novel inflammatory mediator in cardiovascular disease, diabetes mellitus, and kidney disease. Cardiovasc Diabetol [Internet] 21(1):185. [cited 2025 Aug 14]. Available from: https://cardiab.biomedcentral.com/articles/10.1186/s12933-022-01626-1
31. Donate-Correa J, Ferri CM, Sánchez-Quintana F, Pérez-Castro A, González-Luis A, Martín-Núñez E et al (2021) Inflammatory cytokines in diabetic kidney disease: pathophysiologic and therapeutic implications. Front Med [Internet]. 7:628289. [cited 2025 Aug 14]. Available from: https://www.frontiersin.org/articles/10.3389/fmed.2020.628289/full
32. Velikova TV, Kabakchieva PP, Assyov YS, Georgiev TA (2021) Targeting inflammatory cytokines to improve type 2 diabetes control. Infante M, editor. BioMed Res Int [Internet]. 2021(1):7297419. [cited 2025 Aug 14]. Available from: https://onlinelibrary.wiley.com/doi/10.1155/2021/7297419
33. Maedler K, Oberholzer J, Bucher P, Spinas GA, Donath MY (2003) Monounsaturated fatty acids prevent the deleterious effects of palmitate and high glucose on human pancreatic β-cell turnover and function. Diabetes [Internet] 52(3):726–733. [cited 2025 Aug 14]. Available from: https://diabetesjournals.org/diabetes/article/52/3/726/24032/Monounsaturated-Fatty-Acids-Prevent-the
34. Sheng X, Zhang C, Zhao J, Xu J, Zhang P, Ding Q et al (2024) Microvascular destabilization and intricated network of the cytokines in diabetic retinopathy: from the perspective of cellular

and molecular components. Cell Biosci [Internet]. 14(1):85. [cited 2025 Aug 14]. Available from: https://cellandbioscience.biomedcentral.com/articles/10.1186/s13578-024-01269-7

35. Kreiner FF, Kraaijenhof JM, Von Herrath M, Hovingh GKK, Von Scholten BJ (2022) Interleukin 6 in diabetes, chronic kidney disease, and cardiovascular disease: mechanisms and therapeutic perspectives. Expert Rev Clin Immunol [Internet] 18(4):377–389. [cited 2025 Aug 14]. Available from: https://www.tandfonline.com/doi/full/10.1080/1744666X.2022.2045952
36. Jiang Y, Fan H, Xie J, Xu Y, Sun X (2023) Association between adipocytokines and diabetic retinopathy: a systematic review and meta-analysis. Front Endocrinol [Internet]. 14:1271027. [cited 2025 Aug 14]. Available from: https://www.frontiersin.org/articles/10.3389/fendo.2023.1271027/full
37. Kim J, Oh CM, Kim H (2023) The interplay of adipokines and pancreatic beta cells in metabolic regulation and diabetes. Biomedicines [Internet] 11(9):2589. [cited 2025 Aug 14]. Available from: https://www.mdpi.com/2227-9059/11/9/2589
38. Catrina SB, Zheng X (2021) Hypoxia and hypoxia-inducible factors in diabetes and its complications. Diabetologia [Internet] 64(4):709–716. [cited 2025 Aug 14]. Available from: http://link.springer.com/10.1007/s00125-021-05380-z
39. Durrani IA, Bhatti A, John P (2021) The prognostic outcome of 'type 2 diabetes mellitus and breast cancer' association pivots on hypoxia-hyperglycemia axis. Cancer Cell Int [Internet] 21(1):351. [cited 2025 Aug 14]. Available from: https://cancerci.biomedcentral.com/articles/10.1186/s12935-021-02040-5
40. Luo Z, Tian M, Yang G, Tan Q, Chen Y, Li G et al (2022) Hypoxia signaling in human health and diseases: implications and prospects for therapeutics. Sig Transduct Target Ther [Internet] 7(1):218. [cited 2025 Aug 14]. Available from: https://www.nature.com/articles/s41392-022-01080-1
41. Mann V, Sundaresan A, Shishodia S (2024) Overnutrition and lipotoxicity: impaired efferocytosis and chronic inflammation as precursors to multifaceted disease pathogenesis. Biology 13(4):241
42. Spaggiari R, Angelini S, Di Vincenzo A, Scaglione G, Morrone S, Finello V et al (2024 Jul) Ceramides as emerging players in cardiovascular disease: focus on their Pathogenetic effects and regulation by diet. Adv Nutr 15(7):100252
43. Gerber PA, Rutter GA (2017 Apr) The role of oxidative stress and hypoxia in pancreatic Beta-cell dysfunction in diabetes mellitus. Antioxid Redox Signal 26(10):501–518
44. Yuan S, She D, Jiang S, Deng N, Peng J, Ma L (2024) Endoplasmic reticulum stress and therapeutic strategies in metabolic, neurodegenerative diseases and cancer. Mol Med 30(1):40
45. Dinić S, Arambašić Jovanović J, Uskoković A, Mihailović M, Grdović N, Tolić A et al (2022) Oxidative stress-mediated beta cell death and dysfunction as a target for diabetes management. Front Endocrinol 13:1006376
46. Lu J, Liu J, Li L, Lan Y, Liang Y (2020) Cytokines in type 1 diabetes: mechanisms of action and immunotherapeutic targets. Clin & Trans Imm 9(3):e1122
47. Pakjoo M, Ahmadi SE, Zahedi M, Jaafari N, Khademi R, Amini A et al (2024) Interplay between proteasome inhibitors and NF-κB pathway in leukemia and lymphoma: a comprehensive review on challenges ahead of proteasome inhibitors. Cell Commun Signal 22(1):105
48. Amen OM, Sarker SD, Ghildyal R, Arya A (2019) Endoplasmic reticulum stress activates unfolded protein response signaling and mediates inflammation, obesity, and cardiac dysfunction: therapeutic and molecular approach. Front Pharmacol 10:977
49. Chen J, Fei S, Chan LWC, Gan X, Shao B, Jiang H et al (2025 Jun) Inflammatory signaling pathways in pancreatic β-cell: new insights into type 2 diabetes pathogenesis. Pharmacol Res 216:107776
50. Caturano A, D'Angelo M, Mormone A, Russo V, Mollica MP, Salvatore T et al (2023) Oxidative stress in type 2 diabetes: impacts from pathogenesis to lifestyle modifications. CIMB 45(8):6651–6666
51. Jiang Q, Yin J, Chen J, Ma X, Wu M, Liu G et al (2020) Mitochondria-targeted antioxidants: a step towards disease treatment. Hou DX, editor. Oxidative Med Cell Longev 2020:1–18

52. Borghi C, Levy BI (2022 Apr) Synergistic actions between angiotensin-converting enzyme inhibitors and statins in atherosclerosis. Nutr Metab Cardiovasc Dis 32(4):815–826
53. Chandimali N, Bak SG, Park EH, Lim HJ, Won YS, Kim EK et al (2025) Free radicals and their impact on health and antioxidant defenses: a review. Cell Death Discov 11(1):19
54. Antar SA, Ashour NA, Sharaky M, Khattab M, Ashour NA, Zaid RT et al (2023 Dec) Diabetes mellitus: classification, mediators, and complications; a gate to identify potential targets for the development of new effective treatments. Biomed Pharmacother 168:115734
55. Sun D, Gao W, Hu H, Zhou S (2022) Why 90% of clinical drug development fails and how to improve it? Acta Pharm Sin B 12(7):3049–3062

Chapter 12
Connection Between Thyroid Hormones and ROS-Mediated Inflammation in Hyperthyroidism

Mohini Mondal, Rajarshi Nath, Syed Mustafizur Rahaman, Shreyasi Katari, Mirazuddin Mollick, Sushant Kumar Sharma, Touseef Begum, and Prashant Kumar Singh

12.1 Introduction

The thyroid hormone, especially thyroxine (T4) and triiodothyronine (T3) that has an essential role to play in maintaining metabolic balance, growth, and the developmental processes. These hormones also exert the central influence on energy expenditure through their effects on basal metabolic rate, thermogenesis, and mitochondrial performance. When levels of circulating thyroid hormone exceed the normal physiologic range, as in hyperthyroid states, there is an upregulated spillover of reactive oxygen species (ROS), a paradoxical metabolic effect [1]. The resulting oxidative imbalance has metabolic and inflammatory implications as a predisposing factor to tissue injury and nosologically acceleration of trajectories. ROS, as inherent

M. Mondal (✉) · R. Nath
Department of Pharmaceutical Technology, Bharat Technology, Uluberia, West Bengal, India

S. M. Rahaman
Bengal College of Pharmaceutical Sciences and Research, Durgapur, West Bengal, India

S. Katari
Derozio Pharma Institute Moyna, Purba Medinipur, West Bengal, India

M. Mollick
Momtaz Begum Pharmacy College, Rajarhat, West Bengal, India

S. K. Sharma
Institute of Bio-Medical Education and Research Department of Pharmacy, Mangalayatan University, Aligarh, Beswan, Uttar Pradesh, India

T. Begum
Department of Pharmaceutical Sciences, Ibn Sina National College for Medical Studies, Jeddah, Kingdom of Saudi Arabia

P. K. Singh
Faculty of Pharmaceutical Sciences Rama University, Mandhana Kanpur, Uttar Pradesh, India

S. Ashique, N. Mishra (eds.), *Reactive Oxygen Species in Metabolic Inflammation*,
https://doi.org/10.1007/978-3-032-11752-6_12

components of aerobic energy transduction are produced mainly in the mitochondrial compartment [2]. Given a homeostatic environment, their production is acted upon by a unified antioxidant defense system, thus maintaining a steady state between oxidation and reduction fluxes. Nevertheless, T4 and T3 increased circulation triggers the hypermetabolic state of hyperthyroidism, which potentiates electron transport and oxidative phosphorylation in the mitochondrial membrane, thus forefront in collective increase in ROS. In addition to the endogenous reserve of the antioxidant enzymes, the excess has an effect at the cellular homeostasis; the level which has been balanced by physiological regulators is crossed [3]. The oxidative stress then takes the form of a hostile cellular milieu, characterized by lipid peroxidation, breakup of DNA strands, and oxidation of proteins, all of which constitute destructive triads that could spread dysfunction into cellular and tissue territories [4]. In addition to their abilities to cause direct oxidant damages, the reactive oxygen species (ROS) act as communication mediators, coordinating the induction of the redox-responsive transcription components including the AP-1 complex and NF-kB. At the transcriptional level, these transcriptional networks become involved in an inflammatory cascade that results in the overexpression of cytokines and chemokines [5]. Through this signaling axis, a mechanistic dependency is created between the excess of thyroid hormone, an oxidative ambience out of proportion, and the initiation of the antecedent of a chronically inflammatory process. Delineation of this nexus inherits epistemological importance, as it clarifies previously obscure molecular aspects of hyperthyroid pathogenesis as well as identifying priori therapeutic targets at the level of oxidant and inflammatory pathways [6].

12.2 Thyroid Hormones and Metabolic Regulation

The interacting basal metabolic rate has the most crucial impetus of the thyroid hormones such as triiodothyronine (T3) and thyroxine (T4). Their effects can be seen in virtually every tissue where they alleviate low-levels of energy turnover, oxygen usage, and mitochondrial action. T4 is mainly transformed to biological active T3 at the target tissues, and T3 undergoes combination with thyroid hormone nuclear receptors (TRs) [7]. This type of ligand receptor complexes controls the expression of genes dealing with energy metabolism and oxidative phosphorylation and thermogenesis. The most characteristic of the activities of the thyroid hormones is the control of the basal metabolic rate (BMR). High T3 levels elicit mitochondrial biogenesis and improve the performance of respiratory chain multiplexes and the turnover of ATP [8]. It is this increased metabolism that is the basis of the weight loss, intolerance of heat, and the muscle wasting that happens to most hyperthyroid patients. The consequence of this faster-achieved oxidative phosphorylation is, however, unavoidable overproduction of ROS, mainly superoxide and hydrogen peroxide, which at unsuppressed levels instigate oxidative stress [9].

The thyroid hormones also regulate the energy metabolism in terms of modulating the use of carbohydrates as well as lipids. They also generate glycolysis, gluconeogenesis, and lipolysis and hence provide sustained energy to serve the demands of increased metabolism [10]. Moreover, T3 increases the activity of Na^+/K^+-ATPases and uncoupling proteins in mitochondria, which also increased the oxygen consumption and heat production [11]. Although such adaptations are beneficial to the physiology, there is a consequence of hyperthyroidism that leads to excessive oxygen demands and formation of free radicals. Therefore, thyroid hormones, essential to the continued normal metabolic state, turn into a double-edged sword in hyperthyroidism. The very mechanisms that enhance energy release are the same that makes tissues involved susceptible to oxidative damage, establishing redox imbalance paving way to inflammation [12].

12.3 Oxidative Stress and ROS in Thyroid Dysfunction

The reactive oxygen species (ROS) are inevitably by-products of aerobic metabolism and they are mostly produced during electron leakage injuring the mitochondrial respiratory chain. They are also produced by other cellular mechanisms, such as Nicotinamide Adenine Dinucleotide Phosphate (NADPH) oxidases, xanthine oxidase, and peroxisomal oxidases. Unrestricted ROS are associated with acute inflammation and stand at redox equilibrium below standard situations; nevertheless, the high ROS concentration is detected in the presence of inflammatory disorders [13]. An increase in the level of hormones of the thyroid gland in hyperthyroidism intensively activates the growth of oxidative phosphorylation in the mitochondrial process and oxygen consumption. This condition induces a hypermetabolic state with excessive ROS generation that exceeds the buffering capacity of endogenous antioxidant systems, thereby resulting in oxidative stress [14].

The overall imbalance is negative since uninhibited ROS may trigger the onset of lipid peroxidation, protein oxidation, and attack on DNA which ultimately end in disturbance to homeostasis and tissue functionality. Clinical and experimental evidence support this state of affairs. Hyperthyroid patients exhibit consistently elevated malondialdehyde (MDA) as one of the most accepted measures of lipid peroxidation, as well as protein carbonyls and oxidized nucleic acids. Simultaneously, decreasing activity of some of the antioxidant enzymes, such as superoxide dismutase (SOD), catalase, and glutathione peroxidase, enhances oxidative damage [15]. All these findings correlate to determine oxidative stress as one of the major mechanisms in thyroid dysfunction. In this regard, ROS cannot be viewed only as metabolic products but as the factors actively participating in the molecular pathology of hyperthyroidism [16].

12.4 ROS-Mediated Inflammatory Pathways in Hyperthyroidism

Other than their well-known direct cytotoxic effect, reactive oxygen species (ROS) also serve as important second messengers in intracellular communication. In high levels, the nuclear factor kappa B known as (NF-kB) and activator protein-1 known as (AP-1) are the most well-known redox-sensitive transcription factors that are activated when these are present in high concentrations [17]. When these transcription factors are turned on, they go into the nucleus and start the transcription of genes that code for pro-inflammatory mediators. The resultant effects include heightened synthesis of cytokines and chemokines such as interleukin-6 (IL-6), tumor necrosis factor-alpha known as (TNF-alpha), and interleukin-1beta known as (IL-1beta) [18]. These mediators cascade inflammation through the recruitment of leukocytes, the increase of vascular permeability as well as the increased production of further ROS. These processes lead to homeostatic imbalance of oxidative stress and inflammation, which causes progressive cellular and tissue injury [19]. In hyperthyroidism this mechanism plays a special role. The thyroid and peripheral tissues are always exposed to the effects of excess thyroid hormone-mediated metabolism and redox stress, and thus present an environment in which the pool of redox signaling is perpetuated by inflammatory signaling [20]. This paradigm is further supported by clinical and experimental evidence since the patients with hyperthyroidism regularly show higher inflammatory and biomarkers of oxidative damage to the tissues. Therefore, hyperthyroidism cannot be seen only as a metabolic disease but, in a broader view, as a state of low-grade inflammation, and oxidative signaling can be viewed as a unifying factor between phenotypic signatures of hormone excess, oxidative imbalances, and systemic manifestations [21].

12.5 Antioxidant Defense System in Hyperthyroidism

Cells possess an effective antioxidant defense that is efficient in removing oxidative damage by scavenging ROS. The major enzymatic antioxidants are superoxide dismutase, which transform superoxide ions into hydrogen peroxide, catalase which converts hydrogen peroxide into water (H_2O) and oxygen (O_2), and glutathione peroxidase which utilizes glutathione as a cosubstrate to reduce peroxides [22]. Studies have revealed that such antioxidant enzymes are less effective in persons with hyperthyroidism. This may be due to the fact that ROS inactivates the enzymes or usage of cofactors. Oxidative stress is further aggravated by lower levels of the non-enzymatic antioxidant glutathione. Consequently, hyperthyroid patients experience increased lipid and protein oxidation, i.e., the level of antioxidants required to effectively counteract the surplus of ROS being produced is not sufficient [23].

12.6 Clinical Evidence and Experimental Models

Clinical trials have also established an elevated level of increased oxidative stress and inflammatory markers in patients with hyperthyroidism. Higher levels of MDA, 8-Hydroxy-2′-deoxyguanosine (8-OHdG), IL-6, and TNF-alpha are found to be correlated with disease severity. Antithyroid medications/thyroidectomy diminish these biomarkers demonstrating how excessive thyroid hormone production, oxidative stress, and inflammation are all related [24]. There are similarities of increased reactive oxygen species (ROS) formation and upregulation of inflammation pathways as well as oxidative tissue injury in experimental animal models of hyperthyroidism induced by T3 or T4 administration. Opportunities to resolve oxidative stress and the associated inflammatory cytokine expression exist following antioxidant supplementation in these models, including vitamins C and E or enzyme mimetic supplementation, pointing to therapeutic potential [25].

12.7 Future Perspective and Research Gap

Thyroid hormone production (T3 and T4) plays a significant role in ROS generation by enhancing metabolic activity and increasing mitochondrial turnover and respiration rates. ROS not only enhance metabolic activity but also function as signaling molecules that activate inflammatory pathways. In hyperthyroidism, this can lead to a chronic inflammatory state, potentially contributing to various complications [26]. Thyroid hormone releases enzymes that catalyze ROS generation. ROS does not help produce thyroid hormones, acting as a buffer to neutralize free radicals and maintain whole-body homeostasis. An excess of free radicals causes structural cell damage, undermining genomic stability. In the future, we have to be focused on improving the target side of ROS factor production [27]. As a result, hyperthyroidism is characterized by increased oxidative metabolism and reactive oxygen species (ROS) generation, which causes tissue damage. The last few years have rapidly increased the cost-effectiveness of the treatment of thyroid. The main difficulty of this disease is not properly managing the disease, which continuously does not respond to existing treatment methods. So, all these limitations will be overcome with the help of targeted treatment of the thyroid, and improve the target site of ROS. Targeted treatment is investigated specifically for the abnormalities of the thyroid gland. Antioxidants and adjuvant treatment can produce better outcomes than targeted therapy. So, in the future perspective and research gap are developing the target side treatment, and with the advancements in the use of antioxidants as an adjuvant treatment, improving cost effectiveness and decreasing the adverse reaction [28].

12.8 Conclusion

Thyroid hormones play an important role in maintaining antioxidant levels. Oxidative stress can affect hyperthyroidism. The most recent studies suggest a close link between thyroid diseases and oxidative stress. Therefore, the conclusion strengthens that hyperthyroidism is not only a hormonal disorder but also involves significant oxidative stress and inflammation. Hyperthyroidism drives ROS production, and the subsequent ROS-mediated inflammation can further contribute to tissue dysfunction. The intricate interplay between thyroid hormones, oxidative stress, and inflammation in hyperthyroidism is paramount. This holistic approach is essential for optimizing patient outcomes and averting the long-term complications frequently linked to persistent oxidative stress and inflammation in hyperthyroidism.

References

1. Kawara RS, Moawed FS, Elsenosi Y, Elmaksoud HA, Ahmed ES, Abo-Zaid OA (2024) Melissa officinalis extract palliates redox imbalance and inflammation associated with hyperthyroidism-induced liver damage by regulating Nrf-2/Keap-1 gene expression in γ-irradiated rats. BMC Complement Med Ther 24(1):71
2. Rocco M, D'alessio C, Sánchez V, Romeo H, Laura M (2025) Hyperthyroidism-induced lymphoid cell activation in the lymph nodes and spleen of BALB/c mice. Biocell 49(4):629
3. Małujło-Balcerska E, Gałecka M, Pietras T (2025) Thyroid hormone pathways in respiratory diseases: mechanism and therapeutic potential. J Inflamm Res 18:10519–10531
4. Yoldas A, Bahtiyar N, Aydemir B, Toplan S (2025) Evaluation of selenoproteins and proinflammatory cytokines in L-thyroxine-induced hyperthyroid rats: effects of selenium supplementation. Arch Endocrinol Metab 69(2):e240444
5. Mahdi MF, Abdulkhaliq RJ (2025) Evaluation of pro-and anti-inflammatory cytokines IL-17 and IL-33 in autoimmune hyperthyroidism. Trends Immunother 9:140–150
6. Pashaei M, Hesari M, Shackebaei D, Godini A (2024) Cardioprotective effects of losartan and diminazene against the ischemia/reperfusion injury in hyperthyroidism rats. Can J Physiol Pharmacol 102(6):383–390
7. Sabatino L, Vassalle C (2025) Thyroid hormones and metabolism regulation: which role on Brown adipose tissue and Browning process? Biomolecules 15(3):361
8. Sinha RA, Yen PM (2024) Metabolic messengers: thyroid hormones. Nat Metab 6(4):639–650
9. Sagliocchi S, Restolfer F, Cossidente A, Dentice M (2024) The key roles of thyroid hormone in mitochondrial regulation, at interface of human health and disease. J Basic Clin Physiol Pharmacol 35(4–5):231–240
10. Sinha RA, Bruinstroop E, Yen PM (2025) Actions of thyroid hormones and thyromimetics on the liver. Nat Rev Gastroenterol Hepatol 22(1):9–22
11. Zwahlen J, Gairin E, Vianello S, Mercader M, Roux N, Laudet V (2024) The ecological function of thyroid hormones. Philos Trans R Soc Lond B Biol Sci 379(1898):20220511
12. Jo'rayeva G (2025) The role of thyroid hormones in child development. Modern Sci Res 4(2):990–995
13. Chourasiya S (2024) The role of oxidative stress in human pathology with a special attention to thyroid disorders. In: Frontiers in animal science: research and development, vol I, p 45
14. Ma C, Li H, Lu S, Li X (2024) Thyroid-associated ophthalmopathy: the role of oxidative stress. Front Endocrinol 15:1400869

15. Festus O, Dic-Ijiewere E, Ailemen F, Imafidon N, Ebo J, Nwankwo C, Iweka F (2025) Oxidative stress pattern of patients with abnormal thyroid function in Southern Nigeria. J Res Appl Basic Med Sci 11(1):85–97
16. Sabbar Jebur A, Mohammed Saleh BO, Nafea Al-Azzawi OF (2025) Status of serum levels of oxidative stress biochemical markers and total antioxidant capacity in primary hypothyroidism. Rep Biochem Mol Biol 14:1–9
17. Gu W, Tian S, Fu L, Wang L, Zhang L, Wang F (2025) Investigation of the molecular mechanism of Xiaoluo Wan in thyroid-associated ophthalmopathy: network analysis and in vivo study. Hereditas 162(1):137
18. Li B, Liu J, He W, Zhou Y, Zhao M, Xia C, Pan X, Ji Z, Duan R, Lian H, Xu K (2025) Inhibition of macrophage inflammasome assembly and pyroptosis with GC-1 ameliorates acute lung injury. Theranostics 15(6):2360
19. Shokri F, Zarei M, Komaki A, Raoufi S, Ramezani-Aliakbari F (2024) Effect of diminazene on cardiac hypertrophy through mitophagy in rat models with hyperthyroidism induced by levothyroxine. Naunyn Schmiedeberg's Arch Pharmacol 397(2):1151–1162
20. Gupta V, Srivastava R (2024) Ashwagandha diminishes hippocampal apoptosis induced by microwave radiation by acetylcholinesterase dependent neuro-inflammatory pathway in male Coturnix coturnix Japonica. Neurochem Res 49(7):1687–1702
21. Mishra P, Sahoo DK, Mohanty C, Samanta L (2024) Curcumin-loaded nanoparticles effectively prevent T4-induced oxidative stress in rat heart. Cell Biochem Funct 42(4):e4070
22. Somayaji H, Shetty S, Hegde P, Shetty V (2024) Association of antioxidants in thyroid dysfunction. Chin J Lab Med 6(1):203–211
23. Wu J, Niu J, Jia C, Yang Q, Li D, Dong X (2025) A higher composite dietary antioxidant index is associated with a decreased risk of subclinical hyperthyroidism in adults: evidence from epidemiological studies. Front Nutr 12:1613223
24. Huang Z, Hamblin MR, Zhang Q (2024) Photobiomodulation in experimental models of Alzheimer's disease: state-of-the-art and translational perspectives. Alzheimer's Res Ther 16(1):114
25. Zhou M, Theologis AA, O'Connell GD (2024) Understanding the etiopathogenesis of lumbar intervertebral disc herniation: from clinical evidence to basic scientific research. JOR Spine 7(1):e1289
26. Wang Z, Cao L, Danek B, Jin Q, Lu Z, Sun J (2025) Accelerating clinical evidence synthesis with large language models. NPJ Digit Med 8(1):509
27. Socała K, Żmudzka E, Lustyk K, Zagaja M, Brighenti V, Costa AM, Andres-Mach M, Pytka K, Martinelli I, Mandrioli J, Pellati F (2024) Therapeutic potential of stilbenes in neuropsychiatric and neurological disorders: a comprehensive review of preclinical and clinical evidence. Phytother Res 38(3):1400–1461
28. Alkubaisy MM, Khazaal WM, Shakir KH, Muhammad SR, Habelalmateen MI, Ayad H, Saadoun ON (2024) AI for neurodegenerative experimental models: advancements and applications. In: 2024 international conference on IoT, communication and automation technology (ICICAT). IEEE, pp 675–680

Chapter 13
ROS-Mediated Inflammation in Cancer Cell and Tumor Microenvironment

Aganta Chakraborty, Sanket Pramanik, Shrabasti Mullick, Mohini Mondal, Biplab Debnath, Debajit Dewan, Sumel Ashique, and Farzad Taghizadeh-Hesary

13.1 Introduction

Reactive oxygen species includes singlet oxygen ($^1O_2$), superoxide anions (O_2 •), hydroxyl radicals (•OH), and hydrogen peroxide (H_2O_2). These types of reactive oxygen species are a group of oxygen-centered chemicals that are highly reactive. The biological activities and characteristics of ROS represent a double-edged sword with respect to cellular signaling as well as creating cellular damage [1]. Research work has established that uncontrolled ROS is strongly reactive and is capable of destroying cells in the organism through oxidation of cellular elements, including nucleic acids, proteins, and lipids, and thus inflammation [2]. Successful cell stress response systems maintain homeostasis and shield against this damage by responding to departures from the steady-state set point of oxidative levels in the cell's compartments and then taking timely countermeasures [3]. Detoxifying harmful ROS is very important for all cells to survive. Organisms have come up with many ways to keep a balance between making and removing ROS, so they can live in an environment that has a lot of oxygen. When there is too much ROS produced and the cells cannot quickly get rid of it or fix the damage it causes, this situation is known as oxidative stress [4]. ROS is produced within the cell in diverse compartments like the subcellular organelles and cytoplasm like the peroxisomes, mitochondria, as well as the endo(sarco)plasmic reticulum. Mitochondria are the center of cell existence and regulate the progression of nearly all human disorders in

A. Chakraborty · S. Pramanik · S. Mullick · M. Mondal · B. Debnath · D. Dewan · S. Ashique
Department of Pharmaceutical Technology, Bharat Technology, Uluberia, West Bengal, India

F. Taghizadeh-Hesary (✉)
ENT and Head and Neck Research Center, The Five Senses Health Institute, School of Medicine, Iran University of Medical Sciences, Tehran, Iran

Radiation Oncology Department, Iran University of Medical Sciences, Tehran, Iran
e-mail: taghizadeh_hesary.f@iums.ac.ir

S. Ashique, N. Mishra (eds.), *Reactive Oxygen Species in Metabolic Inflammation*,
https://doi.org/10.1007/978-3-032-11752-6_13

addition to the cell's maturation [5]. ROS importantly influences cancer characteristics. Even though ROS acts primarily as an oxidative killing feature against cancer cells, it also strengths multiple secondary signaling pathways and the close relation between ROS and genetic instability, which may have implications for inducing mutations [6]. The role of reactive oxygen species (ROS) in the organism is complicated and multifaceted. It can influence biological aging, inflammation, and even cancer, by acting as an important modulator, but also a destructive demolisher. For example, in a recent in vivo study aimed at gut inflammation, it was shown that the copper chaperone antioxidant 1 (Atox1) increased the development of inflammation by regulating the ROS-NLRP3 inflammatory pathway and increasing amounts of pro-inflammatory cytokines in gut tissue. In an in vitro experiment, cadmium (Cd) induced oxidative stress by increasing ROS levels and decreasing the activity of total antioxidant capacity (T-AOC) under conditions of lipopolysaccharide (LPS) activation [7]. There has been strong evidence for anticancer activity when targeting ROS generation or increasing ROS in tumor cells. These approaches are highly promising as they can slow the growth of tumors and modify parts of the tumor microenvironment simultaneously [8]. Reactive oxygen species (ROS) can be produced by activated T cells as well as NK cells, and, in turn, facilitate the recruitment of neutrophils and macrophages, which can help eradicate cancer cells [9]. Oxidative stress is a complicated and important component in carcinogenesis, affecting not just the early steps of tumor formation, but also the tumor cell microenvironment, survival, and proliferation. To both prevent and treat breast cancer, a better understanding of the mechanisms involved in oxidative stress must occur [10].

13.2 ROS and Inflammation: Molecular Interplay

13.2.1 ROS as Signaling Molecules

Reactive oxygen species (ROS) such as superoxide and H_2O_2 are produced by numerous cellular processes, including the β-oxidation of fatty acids in peroxisomes, complexes I and III of the electron transport chain in mitochondria, NADPH oxidases (NOX) on cellular membranes, flavin-centered oxidases in peroxisomes, and the endoplasmic reticulum, and from external sources (toxins, irradiation, and pharmaceutical compounds) [11]. Redox-sensitive proteins utilize important cysteine residues oxidized by H_2O_2 in order to control the signaling processes [12]. By oxidizing important cell signaling proteins, such as tyrosine phosphatases, the ROS is produced from cells involved in the host-defense response (e.g., poly morpho nuclear neutrophils, PMN) and cause endothelial dysfunction. ROS is not solely a cell-signaling molecule, but it also triggers inflammation. These ROS can rapidly interact with NO at diffusion-limited rates ($k = 5–10·109\ M^{-1}\ s^{-1}$), creating reactive nitrogen species (RNS) such as peroxynitrite, for example, 3–4 times faster than the superoxide dismutation by superoxide dismutase (SOD) [13]. Stabilized HIFs initiate a large transcriptional program, which includes genes that have a vital role in angiogenesis. HIF upregulates vascular endothelial growth factor (VEGF), which allows for the proliferation of the endothelial cell that is required for angiogenesis. VEGF binds to its receptor, which activates key signaling pathways that are necessary for endothelial cell proliferation. Also, VEGF's mitogenic effects can be enhanced through activation of the extracellular signal-regulated kinase/mitogen-activated protein kinase (ERK/MAPK) pathway, which will enhance the proliferation capacity of the cells [9]. An environment that favors tumor growth and metastasis can be established via oxidative stress by enhancing angiogenesis, cell proliferation, and inhibiting cell death. Oxidative stress is associated with cancer pathogenesis; however, it alters a range of physiological processes including cell signaling and immune responses. A balance of ROS generation and the antioxidant defense systems in a cell is essential [14]. Sirtuin 1 (SIRT1), a NAD + -dependent deacetylase that controls the p53 and Forkhead box class O (FoxO) transcription factors, can be activated by ROS [11]. Redox regulation of proteins largely depends on their redox-reactive cysteine (Cys) residues, which undergo reversible oxidation by ROS and serve as key sensors at the oxidative interface. Active cysteines are oxidized to generate reactive sulfenic acids (-SOH), which can further oxidize to disulphide bond (-S-S-) and sulfinic (-SO2H) and sulfonic (-SO3H) acids with nearby cysteines. If nitrogen is nearby, sulfenic acid can generate a sulfenamide. There are structural and/or functional changes to the protein due to these oxidative changes. Disulfide bond formation typically begins with the oxidation of cysteine residues to sulfenic acid, and in some cases to sulfinic acid. These oxidized states can be reversed by cellular reducing systems such as thioredoxin and peroxiredoxin, restoring proteins to their original redox state. This reversibility is crucial because these oxidative modifications play an important role in redox sensing and signaling [15].

13.2.2 ROS and Inflammasome Activation (E.G., NLRP3)

Reactive oxygen species (ROS) consists of oxygen-reduced molecules that are produced in the intracellular environment. ROS is important not just for its ability to kill microbes but also for their possible inflammatory and metabolic regulatory properties as well. The second signal to inflammasome activation that cleaves pro-IL-1β and pro-IL-18 before they are secreted, as well as gasdermin-D that triggers pyroptosis, are associated with the pro-inflammatory properties of ROS [16]. ROS could be detected indirectly by cytoplasmic proteins mediating inflammasome activity, or could directly lead to inflammasome assembly. It is known that ATP-mediated ROS production activates the PI3K pathway and that pharmacological inhibition of PI3K prevents ATP-mediated caspase-1 activation, suggesting that PI3K may have a role in inflammasome activation after generation of ROS [17]. Thioredoxin-interacting protein (TXNIP) is triggered to release thioredoxin, which leads to TXNIP detached thioredoxin to bind to NLRP3 and facilitate its activation.

This is one of the currently known ways to activate the NLRP3 inflammasome [16]. It was recently revealed that NCF4, a constituent of the NADPH oxidase complex that primarily generates ROS in phagocytes, maintains the equilibrium between ROS generation and the inflammasome activation in macrophages. The puncta distribution and phosphorylation of NCF4 transfer from the membrane-bound NADPH complex to the perinuclear region and promote ASC specks formation and inflammasome activation, thereby preventing the progression of colorectal cancer (CRC) through the activation of downstream IL-18-IFN-γ signaling [18]. Without priming, human monocytes can activate NLRP3 inflammasome by using K+ and Cl − efflux [19]. Potassium (K+) efflux is usually linked to production of reactive oxygen species (ROS); as of now, we do not know how the two pathways interact, although it may be that low intracellular K+ concentrations lead to ROS production or vice versa. K+ efflux by compounds inducing ROS-mediated inflammasome activation for non-ATP agonists is not inhibited by P2X7 channel activity [20].

13.3 ROS-Induced Inflammation in Cancer Cells

13.3.1 Mutagenesis and Tumor-Promoting Inflammation

Reactive oxygen species (ROS) play an important role in chronic inflammation development and tumor metastasis. Inflammatory cells can generate ROS in a pathological inflammatory tumor microenvironment induced by chronic inflammation. ROS can directly damage DNA or affect cellular signaling pathways that then promote growth and metastasis in tumors, including colorectal, lung, liver, and breast cancers [7]. Increased metabolic rate, gene mutation, and relative hypoxia all contribute to increased production of reactivate oxygen species (ROS) in tumor cells. The cells will then limit the excess ROS with increased antioxidant enzymatic and nonenzymatic mechanisms. Moderate increases in ROS ares associated with multiple pathological diseases caused by ROS signaling and DNA mutation, including promoting tumor progression. Conversely, ROS can induce programmed cell death (PCD) [21]. Superoxide radicals can also generate a genotoxic effect through indirect action mechanisms by inducing clastogenic factors (CF), or chromosome breakage factors, a secondary clastogenic material that persists longer. Superoxide acts as a signal in the formation of CF and drives the competent cells to produce additional superoxide. The whole process leads to clastogenesis, which, as an ongoing plasma cell type, has the potential to grow out of the DNA repair mechanism and end in cancer. It has been established that conditions that involve CF development are linked with increased cancer risk. Antioxidants may be important as anticlastogens and thus collectively anticarcinogens, when reactive oxygen species (ROS) are shown to participate in the production and activities of cystic fibrosis [22].

13.3.2 Evasion of Apoptosis

Cells may utilize a stress-response mechanism of apoptosis evasion to protect themselves from noxious stimuli. Since most modern-day anticancer therapies (radiotherapy, chemotherapy, and immunotherapies) work under the premise of activating cell-death pathways, predominantly apoptosis pathways, within cancer cells, it can be posited that apoptosis resistance might support carcinogenesis, tumor proliferation, and poor treatment outcomes [23]. The level of ROS exceeds the level at which oncogenic cells can self-regulate production; in essence, ROS directly downregulates EGF and EGFR expression and greatly impairs EGFR phosphorylation, which subsequently inhibits the downstream signaling pathways important for cell proliferation (e.g., ERK and PI3K/Akt) [24]. High levels of ROS result in cell-cycle arrest and apoptosis. To protect themselves from excess intracellular reactive oxygen species (ROS), cancer cells adapt and respond to oxidative stress by producing more antioxidant enzymes. One thiol-based physiological sensor-effector system that (a) responds to an oxidant challenge and (b) assists with eukaryotic redox homeostasis is the NRF2-KEAP1 complex [25]. A new study reports that eIF5B acts as a regulatory node to facilitate the translation of pro-survival proteins from IRES mRNAs, allowing cancer cells to evade apoptosis [26].

13.3.3 Induction of Epithelial-Mesenchymal Transition (EMT)

When cells lose their polarity, the connections between them, and their epithelial markers, it is a sign that they are going through a process called epithelial-mesenchymal transition, or EMT. During this process, the cells start to take on mesenchymal features and become more mobile. Many studies have linked these changes to the start and spread of breast cancer. Some enzymes called matrix metalloproteinases, or MMPs, can trigger EMT in breast cells. Specifically, MMP-3 causes EMT in a way that leads to cancer development, and this process involves the creation of reactive oxygen species, or ROS [27]. ROS plays an important role in the growth of MCF-7 cancer cells by inducing EMT, the glycolytic switch, and mitochondrial suppression through the activation of the Dlx-2/Snail axis [28]. A recent study found that diosmetin decreases TGF-β1-induced EMT by inhibiting the MAPK and PI3K/Akt pathways and inhibiting the production of ROS [29]. Consequently, modifying the calcium-signaling pathways responsible for the formation of EMT in cancer cells may be a critical intervention to inhibit metastasis [30].

13.4 ROS and the Tumor Microenvironment (TME)

13.4.1 Crosstalk between Cancer Cells and TME Via ROS

Reactive oxygen species (ROS) can affect the tumor microenvironment or TME, specifically leukocytes that have infiltrated the tumor. Improved understanding of ROS biology in this context may allow therapeutic manipulation of ROS levels to alter the TME and enhance anti-tumor activity [31]. The selective distribution of CSCs in niche areas shares characteristics with a tumor microenvironment (TME). These manifestations of synergistic effects are shown by the close relationship between the TME and CSCs. As a result of the phenotypic diversity within CSCs and the spatial relationships within the surrounding tumor microenvironment, additional treatment problems arose. CSCs communicate with immune cells using certain immune checkpoint molecules that are capable of immunosuppressive effects to protect themselves from immunologic clearance. Additionally, CSCs also protect themselves from immune surveillance by altering the basis of the TME using growth factors, metabolites, cytokines, and extracellular vesicles (EVs). Thoughtful assessment is being made to include these interactions in the therapeutic development of anti-tumor therapies [32]. A diversity of cytokines produced by TAMs and additional immune cells in the tumor microenvironment is part of the link between inflammation, immunity, and carcinogenesis. Inflammation in response to TAMs, at multiple stages of tumor progression, is a prerequisite for tumors and related processes [33]. Recent studies demonstrate that AMPK and AKT have opposing activities when metabolic stress occurs. Metabolic stress-induced ROS signaling occurs at the intersection of redox homeostasis and metabolic reprogramming [34].

13.4.2 Cancer-Associated Fibroblasts (CAFs) and ROS

The most significant factors of the tumor microenvironment (TME) are cancer-associated fibroblasts (CAFs), a heterogeneous group of stromal cell populations with cell-of-origin, phenotypic, and functional heterogeneity. Activated CAFs can promote tumorigenesis, angiogenesis, invasion, and metastasis via many pathways. They can also improve extracellular matrix (ECM) remodeling and even chemoresistance [35]. CAFs promote EMT through nuclear factor-κB, cycloxygenase-2, and hypoxia-inducible factor-1. CAF-derived metalloproteases induce reactive oxygen species release in the cancer cells required for EMT, stemness, and the spread of metastatic cells. By targeting the proinflammatory signature via RNA interference, tumor growth is not only eliminated, but metastasis is also significantly inhibited, which further endorses the therapeutic targeting of the COX-2/nuclear factor-κB/hypoxia-inducible factor-1 circuit as a valid anti-metastatic approach in targeting malignancy of cancer cells [36]. Future anticancer strategies should target ligand receptors and redox-mediated pathways simultaneously, as CAFs and cancer cells exploit mitogenic signaling and redox signaling circuits to bolster their interaction [37]. ROS affects CAF activity by stimulating the transformation of fibroblasts to myofibroblasts, which promote the growth and metastasis of tumors [38].

13.4.3 Tumor-Associated Macrophages (TAMs) and ROS

Tumor-associated macrophages (TAMs) with proangiogenic and immune-suppressive characteristics similar to alternatively activated (M2) macrophages promote tumor growth [39]. The fatty acid oxidation (FAO) metabolic efficiencies of TAM subtypes differ. Reactive oxygen species (ROS) is produced in association with FAO and can contribute to oxidative stress and other chemical processes. There is current evidence that ROS and FAO can influence TAM recruitment, polarisation, and phagocytosis ability together and/or separately, and play a role in tumor growth [40]. Exosomal NOX1 stimulates the generation of ROS in cervical cancer, which advances TAM M2-polarization-mediated cancer progression [41]. Tumor-associated macrophages (TAMs) generate reactive oxygen species (ROS) throughout the metastatic cascade, contributing to a pro-metastatic microenvironment. Kupffer cells in the liver secrete reactive oxygen species (ROS) in response to LPS, causing damage to the endothelia and enabling circulating tumor cells to adhere [42]. Unharmed Nrf2 was released and translocated to the nucleus in a study in which itaconate was used to inhibit KEAP1-NRF2 complex degradation, which activated transcription of downstream genes. Therefore, ferroptosis and lipid peroxidation in macrophages were inhibited [40].

13.4.4 Immune Suppression Mediated by ROS

Reactive oxygen species (ROS) functions beyond mediators of oxidative stress and also help modulate immune function during tumor development [43]. The reasons why cancer cells can avoid the immune system are complex and involve several factors. These include the creation of an environment around the tumor that suppresses the immune response, the tumor's ability to resist immune attacks, reduced movement of immune cells like T cells to the tumor site, and issues with the immune cells that are meant to fight the tumor. Myeloid-derived suppressor cells (MDSCs) are a special type of immature immune cells that help cancer cells avoid detection by the immune system. They do this by suppressing immune responses and supporting the growth of an environment that protects the tumor. Studies have shown that the number of MDSCs found in the blood of cancer patients is closely linked to the cancer's progression, likelihood of spreading, and the stage of the disease [44]. Besides activating antioxidative pathways, ROS also activates transcriptional programs that dictate not only MDSC destiny but also their function [45]. Transforming growth factor beta 1 (TGF-β1), stromal cell-derived factor 1 (SDF-1), and other factors mainly mediate the change of fibroblasts to myofibroblasts. It is considered well established that TGF-β1 activation requires mitochondrial ROS. TGF-β1 expression levels decreased when fibroblasts were exposed to an inhibitor of mitochondrial ROS pharmacologically. Moreover, SDF-1 has the capability to induce the conversion of fibroblasts to myofibroblasts in a ROS-dependent manner [31]. Immunosuppressive cells generate ROS in the tumor microenvironment, which leads to T cell death and functional inhibition. Therefore, regulating ROS might be important for prolonging T cell survival and enhancing anticancer activity. A combination of T cell-based therapy and antioxidant therapy (for example, ROS scavenger) should be considered as a potent option for treating cancers with T cell-derived antitumor immunity [46].

13.5 Chronic Inflammation, ROS, and Tumor Progression

13.5.1 Role of Chronic Infections (e.g., H. pylori*, HBV/HCV) in ROS Production*

Cancers related to infections are also rapidly becoming notable global health hazards. This includes gastric cancer caused by *Helicobacter pylori* (*H. pylori*), liver cancer associated with either the hepatitis B or hepatitis C viruses, colon rectal cancer associated with Escherichia coli (E. coli), and cervical cancer caused by human papillomavirus (HPV) [47]. Neutrophil-derived ROS interact with DNA to produce mutagenic lesions that build up over time in gastric neoplasia associated with *H. pylori*. The oxidative stress caused by ROS also activates signaling pathways, including the PI3K-Akt and MAPK pathways, which further promote carcinogenesis [48]. All types of *H. pylori* contain mechanisms to neutralize ROS, whereby they produce enzymes like catalase, superoxide dismutase, and arginase to decrease NO production. In addition to this, they recruit macrophages and persist in host tissues by inducing a mitochondria-dependent apoptosis process in macrophages [49]. HBeAg's inhibition of LPS-induced NLRP3 inflammasome activation and IL-1β synthesis occurs via two mechanisms: blockade of NF-κB phosphorylation, which reduces NLRP3 and pro-IL-1β expression, and blockade of ROS generation, which inhibits caspase-1 activation and IL-1β maturation. This effect promotes immune tolerance and prolonged HBV infection [50]. Recent studies suggest that oxidative stress (OS) and chronic inflammation may factor into the pathophysiology of HPV infection and the carcinogenic process. OS occurs as a result of chronic inflammation, which drives the continuation of the inflammatory response, releasing various mediators that damage cells [51]. Several inflammatory signaling pathways, such as the TLR/MyD88, NF-κB, pro-inflammatory cytokines, as well as COX-2 pathways, have been implicated in the association between inflammation and carcinogenesis [52].

13.5.2 Link Between Chronic Inflammation and Cancer Initiation

Various studies have shown that inflammation associated with tumors has a detrimental effect that continues the progression of the tumor, and a chronic inflammatory state continues the development of cancer. Estimates have suggested that chronic inflammation or persistent infections may be responsible for up to 20% of tumors [53]. The epithelial NADPH oxidase (NOX) family proteins, which produce ROS, regulated by cytokines, are elevated in cancer and chronic inflammation. ROS modulate the inflammatory response in damaged tissues as effector molecules that facilitate host defense or as chemoattractants for leukocytes to the wound site. ROS can also act as signaling molecules that regulate tumor cell survival, proliferation, metabolism, angiogenesis, and metastasis, as well as modify chromosomal DNA, resulting in genomic instability [54]. The intrinsic and extrinsic pathways contribute to the connections between inflammation and cancer. Both pathways contribute to the accumulation of tumor-promoting factors in the tumor and microenvironment and to the activation of transcription factors, including NF-κB, STAT-3, and HIF-1. At multiple levels, STAT-3 interacts with NF-κB to increase tumor-associated inflammation, which may be immunosuppressive to anti-tumor immune responses. As well, these factors also promote tumor establishment, growth, and metastasis. TNF, PGHS-2, IL-1, and IL-6 are key mediators of an inflammatory environment through the modulation of the expression of tumor-promoting factors [55]. Certain essential cytokines, like TNF-α, promote angiogenesis, aiding the spread and metastasis of cancer itself. It will be useful to prove an uptick of other levels which could serve as predictive markers in liver cancer initiation and recurrence [56].

13.6 ROS and Cancer Immunity

13.6.1 Immunosuppressive Microenvironment via ROS

Cellular metabolism generates reactive oxygen species (ROS), which are important signaling messengers within the immune system. Tumor-exacerbated immunosuppression is associated with ROS levels that are elevated within the tumor microenvironment [46]. An excess of reactive oxygen species (ROS) can damage T cells because an immunosuppressive tumor microenvironment is developed and well established at inappropriate levels [57]. Tumor-associated macrophages (TAMs) and myeloid-derived suppressor cells (MDSCs) are two types of immune cells in the tumor environment that help the tumor avoid the body's defenses. These cells make it hard for cancer treatments that use the immune system to work effectively. Most medicines made with nanotechnology have not been able to target both of these cell types at the same time. In addition to helping MDSCs stop the immune response, reactive oxygen species (ROS) also play a role in changing TAMs into a type that supports tumor growth [58]. The proliferation of T cells directed by an antigen and dependent on dendritic cells is inhibited by antioxidants that suppress the production of reactive oxygen species (ROS) generated by dendritic cells and T cells during an antigen-presenting event [59]. The rise in macrophage infiltration and T cell inactivity observed in ROS was due to a reduction in their successful uptake of tumor exo-mir-155 and, hence, by overexpression of programmed death ligand 1 (PD-L1) [60]. Reducing ROS in the tumor microenvironment improved T lymphocyte infiltration and produced antitumor immunity, which resulted in a very strong antitumor effect [61].

13.6.2 ROS Impact on T Cells, NK Cells, and Dendritic Cells

Reactive oxygen species (ROS) and the voltage across the mitochondrial membrane (MMP) play a big role in how T cells work. T cells with a strong MMP show more activity in genes linked to fighting infections, use more sugar for energy, and produce more ROS compared to T cells with a weaker MMP. These weaker MMP T cells are better at fighting cancer and have more genes related to long-term memory. When there is a moderate amount of ROS, it helps turn on the gene for IL-2 and a protein called NFAT, which helps T cells become active and multiply. But when there is too much ROS, it stops these proteins from working and messes up signals from the T cell receptor, which stops T cells from working and lets the tumor escape the immune system [62]. Elevated levels of reactive oxygen species (ROS) and a lack of glucose and amino acids can influence the DNA methylation and histone modification of CD8+ T cells within the TME [63]. ROS seems to be quite detrimental to NK viability and function within the TME. Specifically, NK cell death is facilitated by ROS made in NKs by either lactate or the tryptophan catabolite kynurenine [64]. Physiological levels of ROS potentiate DCs' cross-presentation and enhance antitumor immunity. T-MPs are taken up by DCs into lysosomes, wherein the elevation of NOX2 levels raises the pH and prompts ROS generation. Increased pH enhances the production of tumor antigenic peptides, which in turn can make them more easily presented to CD8 T cells. Meanwhile, through lysosomal $Ca2+$ pathways and downstream transcription factor EB (TFEB), greater ROS levels concomitantly enhance the expression of CD80 and CD86, which provide costimulatory signals to improve T cell activation [65].

13.6.3 Redox Regulation of Immune Checkpoint Expression (PD-L1)

Due to ROS's ability to both elevate and decrease PD-L1 levels in cancer cells, increased ROS production typically increases PD-L1 expression, while ROS scavenging can inhibit PD-L1. However, notable exceptions exist—for example, certain drugs simultaneously increase ROS production while reducing PD-L1 expression, thereby acting through an opposing regulatory pathway. The expression of PD-L1 may be significantly affected by drugs that modulate ROS production, even though there are no clear or distinct relationships that can be established [66]. The differential effects of ROS-modulating drugs on PD-L1 were first reported by Bailly et al. It was found that NF-κB and HIF-1, two transcription factors regulated by redox processes, induce PD-L1. However, much like the antioxidants mentioned, these drugs likely have multi-target effects that could contribute to the variability of PD-L1 expression with these pro-oxidant drugs. Similarly, since N-acetyl-cysteine and other antioxidants likely have multi-target effects, hydrogen peroxide (H_2O_2) may be the ideal way to directly test the proof of principle for studying redox control of PD-L1. In this instance, PD-L1 protein on monocytes was also increased with H_2O_2 [67]. Oncogenic K-ras led to a significant increase in PD-L1 expression, which occurred through a redox-mediated mechanism. Activation of K-ras^{G12V} increased the generation of reactive oxygen species (ROS) and FGF receptor 1 (FGFR1) expression, which subsequently led to a significant upregulation of PD-L1 [68]. The complexity of ROS biology under the tumor microenvironment is illustrated by the differential PD-L1 response to ROS modulation. A clearer understanding of how ROS is involved in modulating the PD-(L)1 immunologic checkpoint is warranted [69].

13.7 Therapeutic Implications

13.7.1 Antioxidant Therapy

In many cancers, antioxidant enzymes are reduced and can inhibit both initiation and promotion stages of carcinogenesis. The Mn-SOD mitochondrial enzyme is the one most reduced. Therefore, speculation has arisen that Mn-SOD may be a new class of tumor-suppressor gene [70]. Phenethyl isothiocyanate (PEITC) works with glutathione (GSH) through a process where one part acts as an electrophile and the other as a nucleophile. This interaction uses up the GSH, which leads to more oxidative stress and harmful effects in cancer cells. PEITC also stops the activity of the enzyme GPX. This compound was found to cause death in certain cancer cells, such as those transformed by BCR-ABL in blood cells and HRAS (V12) in ovarian cancer cells. In experiments with mice, PEITC helped improve survival when cancer cells were implanted into them. Piperlongumine (PL), a natural compound from plants, was found through a test using cancer cells to trigger cell death in cancer cells but not in normal cells. PL interacts with an enzyme called glutathione transferase, changing how the cell handles reactive oxygen species (ROS), which increases ROS levels. PL has been shown to cause cell death in osteosarcoma, breast, bladder, and lung cancer cells. However, it does not have the same effect on normal cells that divide quickly. PL was also found to be effective in animal models of lung, bladder, and breast cancer [71]. L-ascorbic acid (vitamin C) and tocopherol (vitamin E), which are the most frequently used dietary antioxidants, are highly desirable for the treatment of cancer [72]. Phenolic compounds, which are the largest class of naturally occurring antioxidants found in the human diet, possess anti-inflammatory and antioxidant properties, both directly and indirectly, which subsequently assist in decreasing cellular oxidative stress [73].

13.7.2 Targeting ROS-Generating Enzymes (e.g., NOX Inhibitors)

NOX enzymes are one of the main ways the body makes ROS. Because ROS is very reactive chemically, it is no surprise that many diseases and problems are connected to it. Health issues that occur because of tissue damage caused by ROS, like inflammation, aging, and the gradual breakdown of cells and organs, are especially common. In short, it seems that nearly every part of the body can be affected by the harmful effects of ROS [74]. Reactive oxygen species (ROS) produced by NOX enzymes provide protection in autoimmune demyelinating disorders, most likely by reducing the specific immune response [75]. NOX is a multi-subunit complex that consists of four cytosolic subunits and two membrane-bound subunits. Consequently, researchers investigating therapies for acute and chronic disorders in the central nervous system (CNS), and their potential side effects, have directed a considerable amount of investigation toward the inhibition of NOX activity in the last several years. Increasing evidence indicates that inhibiting NOX in circulating immune cells can improve neurological disease states and that NOX inhibition has neuroprotective properties [76].

13.7.3 Anti-Inflammatory Strategies

In multiple solid tumors, several early-stage clinical trials are expected to begin soon to evaluate cytokine and chemokine inhibitors, as well as other therapeutic strategies designed to target tumor-associated macrophages (TAMs). While no overt toxicity was detected, a small number of individuals showed signs of partial clinical improvement and disease stabilization. These intriguing findings provide new perspectives for treatment regimens that aim to combat inflammatory environments that promote cancer and enhance the efficacy of anti-obstruction treatments [77]. The development of COX-2-selective inhibitors, including rofecoxib and celecoxib, has resulted in the production of protective prostaglandins while providing an anti-inflammatory effect [78]. MABp1 is a first-in-class anti-IL-1α monoclonal antibody developed for the treatment of systemic inflammation in cancer. Vitamin D has the capacity to modulate the host immune system, which is associated with improved immune responses and decreases in damaging inflammatory responses [79]. P10 has potential as an NRF2 activator and treatment for acute liver inflammation disorders, because it specifically disrupts the connection between β-TrCP1 and NRF2 [80].

13.7.4 Combination Therapies with Immunotherapy or Chemotherapy

Most chemotherapeutic agents have been designed mainly for their direct cytotoxic effects, without regard to their effects on the immune system. As shown in animal models, immunotherapy and chemotherapy can interact effectively: mice with healthy immune systems had significantly greater tumor responses following anthracycline administration. Adoptive cell therapy (ACT) and immune checkpoint blockade (ICB) are two immunotherapies that have transformed the treatment of cancer [81]. The simultaneous blockade of cytotoxic T-lymphocyte–associated antigen-4 (CTLA-4) and programmed death-1 (PD-1), both of which serve as critical negative regulators of T-cell activation [82]. Along with triple-negative breast cancer, patients diagnosed with recurrent or refractory small cell lung cancer (SCLC) receive atezolizumab single-agent therapy. Patients with severe SCLC may also receive atezolizumab along with first-line chemotherapy (carboplatin/cisplatin + etoposide), which significantly increases overall survival (OS) and progression-free survival (PFS). This indicates that the combination of immune checkpoint suppression with cytotoxic therapy may have therapeutic benefit and clinical utility [83].

13.8 Future Perspectives and Conclusion

ROS are important for cell communication and many chemical processes in the body. It can also change how genes are controlled, which affects cells over a long time. When there is too much ROS, it causes oxidative stress, which is linked to different types of cell death, like apoptosis, necrosis, and autophagy. Special proteins called transcription factors help control the balance of ROS, which is important for keeping cells healthy. In unhealthy conditions, ROS is still part of the body's inflammatory response. There is a cycle between ROS and inflammation—when inflammation happens, it causes an imbalance in the body's redox state, and this imbalance affects how inflammation is controlled. Because of this connection, ROS plays a key role in cancer development caused by inflammation. Studies show that ROS and inflammation both play big roles in the start of cancer. ROS can make DNA damage worse and increase genetic changes caused by inflammation, which helps cancer grow faster. Even after cancer forms, ROS continues to be important. It affects how cancer cells grow, use energy, spread, and move to other parts of the body. The role of ROS in cancer is complex and has many different effects. Cancer cells usually have more ROS than healthy cells. If ROS levels get too high and the body's ability to handle it is not enough, this can cause more damage to the DNA and lead to cell death in cancer cells. In these processes, the protein NRF2 is very important for both cancer development and keeping the redox balance steady.

Acknowledgments Authors are thankful to their respective institutions.

Conflict of Interest None.

Funding None.

References

1. Zhao W, Zhang Y, Chen J, Hu D (2025) Revolutionizing oral care: reactive oxygen species (ROS)-regulating biomaterials for combating infection and inflammation. Redox Biol 79:103451
2. Liu J, Han X, Zhang T, Tian K, Li Z, Luo F (2023) Reactive oxygen species (ROS) scavenging biomaterials for anti-inflammatory diseases: from mechanism to therapy. J Hematol Oncol 16(1):116
3. Sies H, Belousov VV, Chandel NS, Davies MJ, Jones DP, Mann GE et al (2022) Defining roles of specific reactive oxygen species (ROS) in cell biology and physiology. Nat Rev Mol Cell Biol 23(7):499–515
4. Bardaweel SK, Gul M, Alzweiri M, Ishaqat A, ALSalamat HA, Bashatwah RM (2018) Reactive oxygen species: the dual role in physiological and pathological conditions of the human body. Eurasian J Med 50(3):193
5. Kozlov AV, Javadov S, Sommer N (2024) Cellular ROS and antioxidants: physiological and pathological role. Antioxidants (Basel) 13:602
6. Okon IS, Zou M-H (2015) Mitochondrial ROS and cancer drug resistance: implications for therapy. Pharmacol Res 100:170–174
7. Chen A, Huang H, Fang S, Hang Q (2024) ROS: a "booster" for chronic inflammation and tumor metastasis. Biochimica et Biophysica Acta (BBA)-Rev Cancer 1879(6):189175
8. Attique I, Haider Z, Khan M, Hassan S, Soliman MM, Ibrahim WN et al (2025) Reactive oxygen species: from tumorigenesis to therapeutic strategies in cancer. Cancer Med 14(10):e70947
9. Yu Y, Liu S, Yang L, Song P, Liu Z, Liu X et al (2024) Roles of reactive oxygen species in inflammation and cancer. MedComm 5(4):e519
10. Dong R, Wang J, Guan R, Sun J, Jin P, Shen J (2025) Role of oxidative stress in the occurrence, development, and treatment of breast cancer. Antioxidants 14(1):104
11. Averill-Bates D (2024) Reactive oxygen species and cell signaling. Rev Biochim Biophys Acta, Mol Cell Res 1871(2):119573
12. Reczek CR, Chandel NS (2015) ROS-dependent signal transduction. Curr Opin Cell Biol 33:8–13
13. Mittal M, Siddiqui MR, Tran K, Reddy SP, Malik AB (2014) Reactive oxygen species in inflammation and tissue injury. Antioxid Redox Signal 20(7):1126–1167
14. Iqbal MJ, Kabeer A, Abbas Z, Siddiqui HA, Calina D, Sharifi-Rad J et al (2024) Interplay of oxidative stress, cellular communication and signaling pathways in cancer. Cell Commun Signal 22(1):7
15. Ray PD, Huang B-W, Tsuji Y (2012) Reactive oxygen species (ROS) homeostasis and redox regulation in cellular signaling. Cell Signal 24(5):981–990

16. Rosa CP, Belo TCA, de Melo Santos NC, Silva EN, Gasparotto J, Corsetti PP et al (2023) Reactive oxygen species trigger inflammasome activation after intracellular microbial interaction. Life Sci 331:122076
17. Martinon F (2010) Signaling by ROS drives inflammasome activation. Eur J Immunol 40(3):616–619
18. Li L, Xu T, Qi X (2025) Balanced regulation of ROS production and inflammasome activation in preventing early development of colorectal cancer. Immunol Rev 329(1):e13417
19. Paik S, Kim JK, Shin HJ, Park E-J, Kim IS, Jo E-K (2025) Updated insights into the molecular networks for NLRP3 inflammasome activation. Cell Mol Immunol 22:1–34
20. Tschopp J, Schroder K (2010) NLRP3 inflammasome activation: the convergence of multiple signalling pathways on ROS production? Nat Rev Immunol 10(3):210–215
21. Perillo B, Di Donato M, Pezone A, Di Zazzo E, Giovannelli P, Galasso G et al (2020) ROS in cancer therapy: the bright side of the moon. Exp Mol Med 52(2):192–203
22. Emerit I (1994) Reactive oxygen species, chromosome mutation, and cancer: possible role of clastogenic factors in carcinogenesis. Free Radic Biol Med 16(1):99–109
23. Fulda S (2010) Evasion of apoptosis as a cellular stress response in cancer. Int J Cell Biol 2010(1):370835
24. Huang R, Chen H, Liang J, Li Y, Yang J, Luo C et al (2021) Dual role of reactive oxygen species and their application in cancer therapy. J Cancer 12(18):5543
25. Verma P, Rishi B, George NG, Kushwaha N, Dhandha H, Kaur M et al (2023) Recent advances and future directions in etiopathogenesis and mechanisms of reactive oxygen species in cancer treatment. Pathol Oncol Res 29:1611415
26. Ross JA, Dungen KV, Bressler KR, Fredriksen M, Khandige Sharma D, Balasingam N et al (2019) Eukaryotic initiation factor 5B (eIF5B) provides a critical cell survival switch to glioblastoma cells via regulation of apoptosis. Cell Death Dis 10(2):57
27. Cichon MA, Radisky DC (2014) ROS-induced epithelial-mesenchymal transition in mammary epithelial cells is mediated by NF-κB-dependent activation of Snail. Oncotarget 5(9):2827
28. Lee SY, Ju MK, Jeon HM, Lee YJ, Kim CH, Park HG et al (2019) Reactive oxygen species induce epithelial-mesenchymal transition, glycolytic switch, and mitochondrial repression through the Dlx-2/Snail signaling pathways in MCF-7 cells. Mol Med Rep 20(3):2339–2346
29. Ge A, Ma Y, Liu Y-N, Li Y-S, Gu H, Zhang J-X et al (2016) Diosmetin prevents TGF-β1-induced epithelial-mesenchymal transition via ROS/MAPK signaling pathways. Life Sci 153:1–8
30. Davis FM, Azimi I, Faville RA, Peters AA, Jalink K, Putney J et al (2014) Induction of epithelial–mesenchymal transition (EMT) in breast cancer cells is calcium signal dependent. Oncogene 33(18):2307–2316
31. Weinberg F, Ramnath N, Nagrath D (2019) Reactive oxygen species in the tumor microenvironment: an overview. Cancer 11(8):1191
32. Wu B, Shi X, Jiang M, Liu H (2023) Cross-talk between cancer stem cells and immune cells: potential therapeutic targets in the tumor immune microenvironment. Mol Cancer 22(1):38
33. Wang J, Li D, Cang H, Guo B (2019) Crosstalk between cancer and immune cells: role of tumor-associated macrophages in the tumor microenvironment. Cancer Med 8(10):4709–4721
34. Zhao Y, Hu X, Liu Y, Dong S, Wen Z, He W et al (2017) ROS signaling under metabolic stress: cross-talk between AMPK and AKT pathway. Mol Cancer 16(1):79
35. Mao X, Xu J, Wang W, Liang C, Hua J, Liu J et al (2021) Crosstalk between cancer-associated fibroblasts and immune cells in the tumor microenvironment: new findings and future perspectives. Mol Cancer 20(1):131
36. Giannoni E, Bianchini F, Calorini L, Chiarugi P (2011) Cancer associated fibroblasts exploit reactive oxygen species through a proinflammatory signature leading to epithelial mesenchymal transition and stemness. Antioxid Redox Signal 14(12):2361–2371
37. JSK C, Tan MJ, Sng MK, Teo Z, Phua T, Choo CC et al (2018) Cancer-associated fibroblasts enact field cancerization by promoting extratumoral oxidative stress. Cell Death Dis 8(1):e2562

38. Arcucci A, Ruocco MR, Granato G, Sacco AM, Montagnani S (2016) Cancer: an oxidative crosstalk between solid tumor cells and cancer associated fibroblasts. Biomed Res Int 2016(1):4502846
39. Zhang Y, Choksi S, Chen K, Pobezinskaya Y, Linnoila I, Liu Z-G (2013) ROS play a critical role in the differentiation of alternatively activated macrophages and the occurrence of tumor-associated macrophages. Cell Res 23(7):898–914
40. Teng Y, Xu L, Li W, Liu P, Tian L, Liu M (2023) Targeting reactive oxygen species and fat acid oxidation for the modulation of tumor-associated macrophages: a narrative review. Front Immunol 14:1224443
41. Gu L, Feng C, Li M, Hong Z, Di W, Qiu L (2023) Exosomal NOX1 promotes tumor-associated macrophage M2 polarization-mediated cancer progression by stimulating ROS production in cervical cancer: a preliminary study. Eur J Med Res 28(1):323
42. Kennel KB, Greten FR (2021) Immune cell-produced ROS and their impact on tumor growth and metastasis. Redox Biol 42:101891
43. Kotsafti A, Scarpa M, Castagliuolo I, Scarpa M (2020) Reactive oxygen species and antitumor immunity—from surveillance to evasion. Cancer 12(7):1748
44. Lu J, Luo Y, Rao D, Wang T, Lei Z, Chen X et al (2024) Myeloid-derived suppressor cells in cancer: therapeutic targets to overcome tumor immune evasion. Exp Hematol Oncol 13(1):39
45. Ohl K, Tenbrock K (2018) Reactive oxygen species as regulators of MDSC-mediated immune suppression. Front Immunol 9:2499
46. Chen X, Song M, Zhang B, Zhang Y (2016) Reactive oxygen species regulate T cell immune response in the tumor microenvironment. Oxidative Med Cell Longev 2016(1):1580967
47. Zhang Y, Liu T, He W (2024) The application of organoids in cancers associated with pathogenic infections. Clin Exp Med 24(1):168
48. Ofoezie EF, Ogbonna CA, Olisakwe SC, Anunobi CJ, George ET, Babarinde S et al (2025) Role of infectious agents in cancer pathogenesis and therapy. The Microbe 6:100284
49. Chakraborty A, Guillen PV, Bhatta OP, Asfeen U, Anirudh MGP et al (2025) Gut microbiota and gastric cancer: the immunoregulatory mechanism and the preventive role of microbiota-derived short-chain fatty acids. Adv Gut Microbiome Res 2025(1):1095370
50. Yu X, Lan P, Hou X, Han Q, Lu N, Li T et al (2017) HBV inhibits LPS-induced NLRP3 inflammasome activation and IL-1β production via suppressing the NF-κB pathway and ROS production. J Hepatol 66(4):693–702
51. Georgescu SR, Mitran CI, Mitran MI, Caruntu C, Sarbu MI, Matei C et al (2018) New insights in the pathogenesis of HPV infection and the associated carcinogenic processes: the role of chronic inflammation and oxidative stress. J Immunol Res 2018(1):5315816
52. Kipanyula MJ, Etet PFS, Vecchio L, Farahna M, Nukenine EN, Kamdje AHN (2013) Signaling pathways bridging microbial-triggered inflammation and cancer. Cell Signal 25(2):403–416
53. Munn LL (2017) Cancer and inflammation. Wiley Interdiscip Rev Syst Biol Med 9(2):e1370
54. Wu Y, Antony S, Meitzler JL, Doroshow JH (2014) Molecular mechanisms underlying chronic inflammation-associated cancers. Cancer Lett 345(2):164–173
55. Multhoff G, Molls M, Radons J (2012) Chronic inflammation in cancer development. Front Immunol 2:98
56. Fernandes Q, Inchakalody VP, Bedhiafi T, Mestiri S, Taib N, Uddin S et al (2024) Chronic inflammation and cancer; the two sides of a coin. Life Sci 338:122390
57. Wang T, Xu H (2022) Multi-faced roles of reactive oxygen species in anti-tumor T cell immune responses and combination immunotherapy. Explor Med 3(1):77–98
58. Mo W, Liu S, Zhao X, Wei F, Li Y, Sheng X et al (2023) ROS scavenging nanozyme modulates immunosuppression for sensitized cancer immunotherapy. Adv Healthc Mater 12(21):2300191
59. Maemura K, Zheng Q, Wada T, Ozaki M, Takao S, Aikou T et al (2005) Reactive oxygen species are essential mediators in antigen presentation by Kupffer cells. Immunol Cell Biol 83(4):336–343

60. Li X, Wang S, Mu W, Barry J, Han A, Carpenter RL et al (2022) Reactive oxygen species reprogram macrophages to suppress antitumor immune response through the exosomal miR-155-5p/PD-L1 pathway. J Exp Clin Cancer Res 41(1):41
61. Deng H, Yang W, Zhou Z, Tian R, Lin L, Ma Y et al (2020) Targeted scavenging of extracellular ROS relieves suppressive immunogenic cell death. Nat Commun 11(1):4951
62. Yang M-Q, Zhang S-L, Sun L, Huang L-T, Yu J, Zhang J-H et al (2024) Targeting mitochondria: restoring the antitumor efficacy of exhausted T cells. Mol Cancer 23(1):260
63. Xia X, Yang Z, Lu Q, Liu Z, Wang L, Du J et al (2024) Reshaping the tumor immune microenvironment to improve CAR-T cell-based cancer immunotherapy. Mol Cancer 23(1):175
64. Miao L, Lu C, Zhang B, Li H, Zhao X, Chen H et al (2024) Advances in metabolic reprogramming of NK cells in the tumor microenvironment on the impact of NK therapy. J Transl Med 22(1):229
65. Chang Q, Zhang Y, Liu X, Miao P, Pu W, Liu S et al (2025) Oxidative stress in antigen processing and presentation. MedComm–Oncol 4(2):e70020
66. Allegra A, Murdaca G, Mirabile G, Gangemi S (2023) Redox signaling modulates activity of immune checkpoint inhibitors in cancer patients. Biomedicine 11(5):1325
67. Glorieux C, Xia X, Huang P (2021) The role of oncogenes and redox signaling in the regulation of PD-L1 in cancer. Cancer 13(17):4426
68. Glorieux C, Xia X, He Y-Q, Hu Y, Cremer K, Robert A et al (2021) Regulation of PD-L1 expression in K-ras-driven cancers through ROS-mediated FGFR1 signaling. Redox Biol 38:101780
69. Bailly C (2020) Regulation of PD-L1 expression on cancer cells with ROS-modulating drugs. Life Sci 246:117403
70. Matés JM, Sánchez-Jiménez FM (2000) Role of reactive oxygen species in apoptosis: implications for cancer therapy. Int J Biochem Cell Biol 32(2):157–170
71. Glasauer A, Chandel NS (2014) Targeting antioxidants for cancer therapy. Biochem Pharmacol 92(1):90–101
72. Luo M, Zhou L, Huang Z, Li B, Nice EC, Xu J et al (2022) Antioxidant therapy in cancer: rationale and progress. Antioxidants 11(6):1128
73. Zhang H, Tsao R (2016) Dietary polyphenols, oxidative stress and antioxidant and anti-inflammatory effects. Curr Opin Food Sci 8:33–42
74. Lambeth JD, Krause K-H, Clark RA (eds) (2008) NOX enzymes as novel targets for drug development. Semin Immunopathol 30(3):339–363
75. Sorce S, Krause K-H, Jaquet V (2012) Targeting NOX enzymes in the central nervous system: therapeutic opportunities. Cell Mol Life Sci 69(14):2387–2407
76. Barua S, Kim JY, Yenari MA, Lee JE (2019) The role of NOX inhibitors in neurodegenerative diseases. IBRO Rep 7:59–69
77. Marelli G, Sica A, Vannucci L, Allavena P (2017) Inflammation as target in cancer therapy. Curr Opin Pharmacol 35:57–65
78. Ricchi P, Zarrilli R, Di Palma A, Acquaviva A (2003) Nonsteroidal anti-inflammatory drugs in colorectal cancer: from prevention to therapy. Br J Cancer 88(6):803–807
79. Hou J, Karin M, Sun B (2021) Targeting cancer-promoting inflammation—have anti-inflammatory therapies come of age? Nat Rev Clin Oncol 18(5):261–279
80. García-Yagüe ÁJ, Cañizares-Moscato L, Encinar JA, Cazalla E, Fernández-Ginés R, Escoll M et al (2025) A novel β-TrCP1/NRF2 interaction inhibitor for effective anti-inflammatory therapy. J Biomed Sci 32(1):1–19
81. Zhu S, Zhang T, Zheng L, Liu H, Song W, Liu D et al (2021) Combination strategies to maximize the benefits of cancer immunotherapy. J Hematol Oncol 14(1):156
82. Drake C (2012) Combination immunotherapy approaches. Ann Oncol 23:viii41–vviii6
83. Yu W-D, Sun G, Li J, Xu J, Wang X (2019) Mechanisms and therapeutic potentials of cancer immunotherapy in combination with radiotherapy and/or chemotherapy. Cancer Lett 452:66–70

Chapter 14
ROS: Pathogenesis and Progression in Obesity

Danúbia Frasson Furtado, Valéria Ernestânia Chaves, and Maria Emilia Soares Martins dos Santos

14.1 Introduction/Overview

In addition to metabolic disorders and nutrient deprivation, oxidative stress (OS) can also arise in response to various physiological and environmental challenges, including exposure to chemical stressors, aging, inflammation, and psychological stress [1]. The progression of metabolic abnormalities—such as insulin resistance, type 2 diabetes mellitus (T2DM), and obesity—is closely linked to OS and inflammation [2]. Reactive oxygen species (ROS) and free radicals are routinely encountered in the environment, but they are also endogenously generated within living organisms, primarily during normal mitochondrial metabolic processes and immune responses against pathogens [3, 4]. The human body possesses several antioxidant defenses, creating an equilibrium between ROS generation and the action of antioxidant systems. When this redox balance is disrupted, it can lead to a condition known as OS.

Brown adipose tissue (BAT) contains a high density of mitochondria and therefore has a greater capacity for ROS production compared to white adipose tissue (WAT) [5, 6]. The cafeteria diet, a hypercaloric dietary pattern, increases thermogenesis in BAT—a process known as diet-induced thermogenesis—to maintain

D. F. Furtado
Federal University for Latin American Integration, Latin American Institute of Life and Natural Sciences, Foz do Iguaçu, Paraná, Brazil

V. E. Chaves
Federal University of São João Del Rei, Laboratory of Physiology, Divinópolis, Minas Gerais, Brazil

M. E. S. M. dos Santos (✉)
Federal University of Triângulo Mineiro, Department of Biochemistry, Pharmacology, and Physiology, Uberaba, Minas Gerais, Brazil
e-mail: mariaemilia.santos@uftm.edu.br

S. Ashique, N. Mishra (eds.), *Reactive Oxygen Species in Metabolic Inflammation*,
https://doi.org/10.1007/978-3-032-11752-6_14

energy homeostasis and prevent weight gain [7]. However, excessive energy intake leads to increased ROS generation, which can overwhelm BAT's antioxidant defenses.

Although WAT has a lower mitochondrial density, it contributes to energy homeostasis through lipolysis of stored triglycerides (TAG), releasing free fatty acids (FFA) into the bloodstream [5, 8]. These fatty acids (FA) serve as substrates for mitochondrial oxidation in other tissues. WAT lipolysis is particularly stimulated under energy-demanding conditions such as fasting and diabetes [5, 8].

In this chapter, we will examine the potential mechanisms by which ROS contribute to the onset and progression of obesity, focusing on their roles in adipogenesis, lipid metabolism (lipogenesis and lipolysis), mitochondrial activity, endoplasmic reticulum (ER) function, inflammation, and adipokine secretion.

14.2 Characterization of Adipose Tissue: Types and Functions

Traditionally, adipose tissue (AT) has been classified into WAT and BAT; however, beige adipose tissue is now also recognized as a distinct type [5, 6, 9]. WAT is distributed throughout the body and is found surrounding internal organs (visceral WAT) in the abdominal cavity and beneath the skin (subcutaneous WAT) [6, 9, 10]. BAT is primarily located in the supraclavicular, cervical, axillary, paraspinal, mediastinal, and abdominal regions [6]. In humans, WAT is the predominant AT type in adults, whereas BAT is more prominent in newborns [5]. In rodents, beige adipocytes are found mainly in subcutaneous fat depots and rarely in visceral depots [6].

The adipocyte is the principal functional cell type in AT. White adipocytes are unilocular, with a single large lipid droplet and a limited number of mitochondria, whereas brown adipocytes are multilocular and contain a high number of mitochondria enriched with uncoupling protein 1 (UCP1) in the inner mitochondrial membrane [5, 6]. Beige, or "brite," adipocytes originate from white adipocytes that have been induced to adopt a brown-like phenotype. These cells are also multilocular and express UCP1 [6]. Interestingly, beige adipocytes can revert to a white phenotype depending on environmental conditions, such as ambient temperature [6]. In addition to adipocytes, AT contains various other cell types collectively known as the stromal vascular fraction. These include immune cells (macrophages, lymphocytes, eosinophils, and mast cells), fibroblasts, adipocyte precursors, vascular cells, multipotent mesenchymal stem-like cells, and nerve fibers [5, 6].

Both WAT and BAT store excess energy in the form of TAG [5–8]. Upon lipolysis, WAT releases FFA into the bloodstream to supply energy during conditions such as fasting or diabetes [5, 8]. BAT, on the other hand, utilizes FFA released from its own lipid stores to fuel thermogenesis via UCP1, particularly in response to sympathetic activation, such as during non-shivering thermogenesis or diet-induced thermogenesis [5, 7, 9].

AT also has important endocrine functions, as it produces and secretes adipokines into the systemic circulation [5, 9, 11, 12]. Through this secretory activity, AT regulates the function of other organs and systems and orchestrates intricate metabolic mechanisms to sustain energy balance [9, 11, 12]. WAT releases multiple adipokines, such as tumor necrosis factor-alpha (TNF-α), interleukin-6 (IL-6), adiponectin, FA-binding protein 4 (FABP4), leptin, and plasminogen activator inhibitor-1 [10, 11]. BAT also secretes adipokines, such as fibroblast growth factor 21 (FGF21), retinol-binding protein 4, angiopoietin-like proteins, and neuregulin [11].

The two major adipokines secreted by white adipose tissue (WAT) are leptin and adiponectin [10, 13]. Leptin is a highly conserved peptide-based hormone encoded by the ob gene [13]. Mutations in this gene lead to increased adiposity and body weight, while increases in body fat mass elevate circulating leptin levels [10, 13]. Although adipocytes are the main source of leptin, it is also produced by the hypothalamus, pituitary gland, placenta, skeletal muscle, and intestinal epithelial cells [10, 13]. Leptin plays a key role in energy homeostasis by regulating appetite and energy expenditure [10].

Upon binding to its receptor (OB-R), leptin causes a structural rearrangement in the OB-R homodimer, activating Janus kinase (JAK)/signal transducers and activators of transcription (STAT) signaling pathway [10, 13]. The OB-Rb isoform of the leptin receptor is found in two neuronal populations of the arcuate nucleus in the hypothalamus, where it induces the release of anorexigenic neuropeptides and suppresses orexigenic neuropeptides [10, 13]. In addition to appetite regulation, leptin influences whole-body metabolism by increasing energy expenditure, inhibiting insulin secretion, and promoting FA mobilization [10, 13]. Furthermore, leptin has pleiotropic effects, influencing growth, immune function, stress response, angiogenesis, cardiovascular health, follicle maturation during the reproductive cycle, and maternal nutrition [13].

Adiponectin is encoded by the ADIPOQ gene (Adiponectin, C1Q and collagen domain containing) in humans [9, 12]. Following transcription, adiponectin undergoes glycosylation, hydroxylation, and other post-translational modifications essential for its biological activity [9, 12]. Although adiponectin exists in several oligomeric forms, the larger oligomers—especially high-molecular-weight (HMW) forms—are more biologically active, enhancing insulin sensitivity, promoting FA oxidation, and exerting anti-inflammatory effects [12]. Adiponectin exists in both full-length and globular forms. The globular form is found at low concentrations in human plasma and stimulates FA oxidation in muscle mitochondria [12]. This adipokine exerts its effects primarily through two receptor isoforms: AdipoR1 and AdipoR2, and it has a half-life of approximately 45–75 min [12]. AdipoR1 is ubiquitously expressed, with particularly high levels in skeletal muscle, whereas AdipoR2 is primarily expressed in the liver [9, 12, 14]. Adiponectin can also bind to other receptors, including T-cadherin and calreticulin [9].

AdipoR2 enhances hepatic insulin sensitivity by activating peroxisome proliferator-activated receptor alpha (PPAR-α) [9, 12, 14]. AdipoR1 and AdipoR2 belong to a recently identified family of seven-transmembrane domain receptors, structurally distinct from classical G-protein-coupled receptors [9, 12, 14]. This configuration suggests a signaling mechanism involving ceramidase activity, which generates sphingosine and FFA [9, 12]. Upon receptor binding, adiponectin regulates energy expenditure, insulin sensitivity, inflammatory responses, and lipid oxidation [9, 12, 14]. The synthesis of adiponectin by WAT is influenced by the body's metabolic status and modulated by factors such as diet, inflammation, and fasting [12].

14.3 Definition of Obesity: Main Causes and Related Pathologies

The human body is a highly refined machine composed of multiple systems that must work in concert to ensure optimal and harmonious functioning. Disruptions to this balance, known as homeostasis, can lead to various pathologies, including metabolic disorders. One of the most significant of these is obesity—a metabolic disease widely regarded as a silent pandemic affecting millions of people worldwide [6].

Since 1948, obesity has been recognized as a disease by the World Health Organization (WHO) and, more recently, was designated as a "chronic complex disease" with a specific ICD code (5B81) [15, 16]. According to the WHO, overweight and obesity are defined as abnormal or excessive fat accumulation that poses a risk to health [15]. Obesity is commonly assessed using the body mass index (BMI) [10, 15]. However, it is now well established that BMI is not the most accurate method for evaluating obesity. The main limitation of BMI is that it does not reflect an individual's health status—it may underestimate or overestimate adiposity and either underdiagnose or overdiagnose disease. Specifically, BMI does not distinguish between fat and lean mass, nor does it consider fat distribution [16].

Recently, experts from the Lancet Commission have proposed a new definition of obesity as a disease characterized by excessive adiposity, with or without abnormal fat distribution or dysfunction of AT [16]. Obesity is recognized as a multifactorial condition in which genetic, psychological, environmental, metabolic, and nutritional factors interact to disrupt the mechanisms that regulate normal body weight, fat distribution, and AT function. It significantly affects overall health and is associated with an increased risk for various comorbidities [16]. Excess adiposity contributes to pathogenic mechanisms such as inflammation, hormonal imbalance, and dysregulation of appetite and satiety, which are key arguments supporting its classification as a chronic disease [16].

A well-established genetic component contributes to obesity, involving alterations in gene expression, gene mutations, and polymorphisms. Obesity can be classified into three categories: monogenic, polygenic, and syndromic. Monogenic obesity results from mutations in a single gene—most commonly in the leptin–melanocortin signaling pathway. Although rare, this form presents as severe early-onset obesity (typically between 3 and 5 years of age), often accompanied by marked hyperphagia. Polygenic obesity develops due to the combined influence of multiple genetic variants. Syndromic obesity, in turn, occurs as part of broader genetic syndromes and is usually associated with additional clinical features, including intellectual disability, developmental delay, dysmorphic traits, and congenital malformations (e.g., Prader-Willi and Cohen syndromes) [17].

As previously mentioned, genetic types of obesity are strongly linked with dysfunction in the leptin–melanocortin pathway in the hypothalamus, which is essential for regulating appetite and energy balance. Under physiological conditions, leptin acts on the central nervous system to suppress appetite and increase energy expenditure. However, in individuals with obesity, persistent high leptin levels lead to leptin resistance, causing hypothalamic neurons to respond less effectively to leptin signaling. This impairs satiety responses, resulting in persistent hunger and increased food intake [17].

Beyond the hypothalamic neurocircuitry, appetite regulation involves a complex integration of peripheral signals. The hypothalamus receives hormonal and neural input from the gastrointestinal tract, AT, and the vagus nerve. Key gut-derived peptides—such as peptide YY (PYY), glucagon-like peptide-1 (GLP-1), cholecystokinin (CCK), and ghrelin—are secreted in response to food intake, with their secretion patterns influenced by macronutrient composition. These peptides, along with vagal nerve signaling, provide feedback to the central nervous system regarding satiety and nutrient status, contributing to appetite regulation [17].

In addition to biological mechanisms, several modifiable behavioral and environmental factors significantly contribute to obesity. These include physical inactivity, excessive caloric intake, gut dysbiosis, insufficient sleep, exposure to endocrine-disrupting chemicals, and chronic psychosocial stress [17]. Notably, some of these factors can induce epigenetic changes—alterations in gene expression that occur without modifying the DNA sequence and may influence the onset of obesity [17].

The consequences of obesity are profound. As previously discussed, it is associated with a wide range of comorbidities, including insulin resistance, T2DM, hypertension, dyslipidemia, metabolic-associated fatty liver disease (MAFLD), cancer, and atherosclerosis, conditions that collectively represent a substantial public health burden [17, 18].

In this context, and considering that the underlying causes of obesity remain incompletely understood, gaining deeper insight into the mechanisms that contribute to its onset or progression is essential. Such understanding can guide the development of more effective and targeted therapeutic strategies [17].

14.4 Definition of ROS and Oxidative Stress: Basic Information

Oxygen is a vital element required for the survival of most living organisms. While molecular oxygen itself is not a free radical, it is the source of several well-known ROS. The main ROS include superoxide anion ($O_2^{\cdot-}$), singlet oxygen ($^1O_2$), hydroxyl radical (HO·), and hydrogen peroxide (H_2O_2) [1]. Superoxide ($O_2^{\cdot-}$) and the hydroxyl radical (HO·) are considered free radicals as a result of the existence of unpaired electrons, which makes them highly unstable and prone to react with nearby molecules to achieve electron pairing. In general, ROS can participate in various oxidation reactions and induce cytotoxicity—either alone or through synergistic interactions—by attacking lipids, proteins, and nucleic acids, vital cellular components [19, 20].

ROS and free radicals are routinely produced in the environment around us. For instance, air pollutants such as nitrogen dioxide (NO_2) and nitric oxide (NO·), tobacco smoke, herbicides, ultraviolet and ionizing radiation, and certain drugs (e.g., anticancer medications, antibiotics) are substances that can effectively stimulate ROS production [1, 21–24]. It is crucial to highlight that, beyond external sources, free radicals are also endogenously generated within living organisms, notably during regular mitochondrial metabolic processes and immune responses against pathogens [3, 4].

Although the generation of ROS and free radicals is natural and occurs under normal physiological conditions—and despite their importance in various biological processes—it is also crucial to mention that the body possesses antioxidant substances. These antioxidants serve to preserve tissues and cells from oxidative injury induced by ROS and free radicals [25]. Antioxidants may act through various mechanisms: preventive action (e.g., glutathione peroxidase, catalase), by blocking the production of free radicals; scavenging action, by neutralizing free radicals once they are formed (e.g., vitamin E, superoxide dismutase); and through repair/regeneration systems that restore oxidatively damaged molecules [26].

As previously mentioned, organisms are constantly engaged in maintaining the balance between ROS generation and antioxidant system activation. However, under some conditions, this redox balance can be disrupted, leading to a condition known as OS. OS may arise in response to various physiological and environmental challenges, including exposure to chemical stressors, nutrient deprivation, aging, inflammation, psychological stress, and metabolic disturbances, among others [1].

The progression of metabolic disorders (e.g., obesity, diabetes, metabolic syndrome) is closely associated with OS and inflammation [2]. Not only is OS a consequence of these metabolic alterations, but it also has a central role in their pathogenesis—as the increased concentration of ROS can disrupt various biomolecules—including carbohydrates, proteins, and lipids [27] (Fig. 14.1).

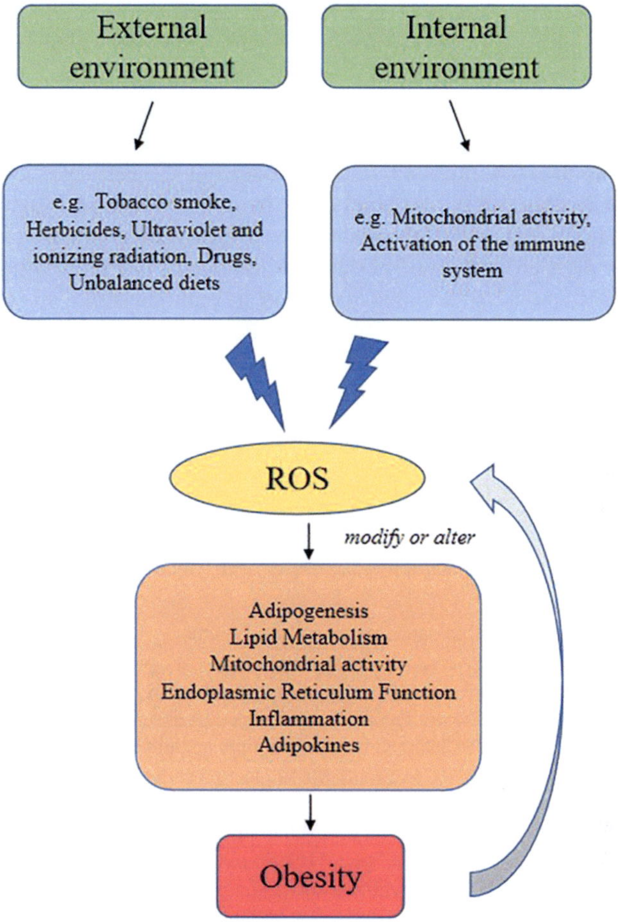

Fig. 14.1 Interplay between ROS and obesity

14.5 Potential Mechanisms Linking ROS/OS with the Onset and Progression of Obesity

AT is a metabolically active and adaptable organ, widely distributed throughout the body. In the context of the onset and progression of obesity, WAT is the most relevant [6]. The adipocyte—commonly referred to as the fat cell—is the central effector cell responsible for the energy storage and endocrine functions of AT. Energy is stored in adipocytes primarily in the form of TAG. These are synthesized through the esterification of glycerol with three FA in a process known as lipogenesis, which is a key mechanism for storing excess energy when caloric intake surpasses the body's immediate needs [28].

Changes in dietary patterns—such as the increased consumption of unbalanced diets and ultra-processed foods—combined with sedentary behavior and genetic predispositions have been identified as major contributors to the rise in body weight and obesity. According to the WHO, the global prevalence of obesity among adults has tripled since 1975. In 2022, approximately 2.5 billion adults were overweight, including more than 890 million living with obesity [15]. Estimates suggest that by 2050, approximately 3.8 billion adults will be affected by overweight or obesity, representing more than half of the anticipated global adult population [15, 29].

Given the high prevalence of obesity and its strong potential to promote other metabolic disorders, a better understanding of its pathophysiology is of significant clinical and public health interest. Chronic inflammation, genetic factors, and OS are well-established contributors to obesity and its associated comorbidities [30]. In line with the objectives of this chapter, we aim to outline the principal mechanisms through which ROS contribute to the initiation and progression of obesity (Table 14.1).

Table 14.1 Summary of the principal mechanisms through which ROS contribute to the initiation and progression of obesity

ROS influence the following biological pathways and systems		References
Adipogenesis	OS disrupts the activity of key transcriptional regulators responsible for maintaining lipid homeostasis and energy storage. This interference impairs adipocyte differentiation, exacerbates inflammation, and compromises adipocyte function	[31–39]
Lipogenesis	ROS enhance the incorporation of glucose-derived carbon into glyceride-bound fatty acid, increase NADPH availability, and upregulate key enzymes involved in lipogenesis	[40–42]
Lipolysis	ROS modulate the activity of key hormones and enzymes involved in the lipolytic pathway, exerting concentration-dependent effects that can either promote or inhibit lipid breakdown	[31, 41, 44–48]
Mitochondrial activity	Excessive ROS production can cause mitochondrial dysfunction, activate apoptotic pathways, and induce necrotic cell death. As a result, the function of one of the most important organelles for energy production is severely impaired	[49–55]
Endoplasmic reticulum	ER stress, primarily triggered by the secretion of pro-inflammatory cytokines in adipocytes, can contribute to the development of insulin resistance and obesity	[56–63]
Inflammation	In obesity, several pro-inflammatory cytokines are secreted that not only sustain local and systemic inflammation but also further elevate ROS production.	[27, 30, 64, 66–71]
Leptin	Hyperleptinemia contributes to OS by promoting the activation and recruitment of immune cells, particularly monocytes and macrophages, which subsequently secrete pro-inflammatory cytokines. Additionally, leptin directly stimulates key enzymes involved in ROS production while concurrently reducing the activity of antioxidant enzymes.	[31, 67, 68, 72–75]

14.5.1 Role of ROS in Adipogenesis

The interaction between OS and adipogenesis plays a key role in the onset of obesity and associated metabolic conditions. Adipogenesis is a precisely regulated process whereby mesenchymal stem cells develop into adipocytes, primarily governed by several transcription factors [31]. Elevated levels of ROS interfere with the activity of key transcriptional regulators, including CCAAT/enhancer-binding protein α (C/EBPα) [32, 33], peroxisome proliferator-activated receptor gamma (PPARγ) [32, 33], and sterol regulatory element-binding protein-1c (SREBP-1c) [33]. These transcription factors act together in a coordinated way to regulate lipid homeostasis and energy storage within AT. Oxidative stress disrupts this coordination, impairing adipocyte differentiation, exacerbating inflammation and insulin resistance, and ultimately compromising adipocyte functionality [32, 34]. The insulin mTOR pathway integrates metabolic cues to control adipogenesis by initiating a positive feedback loop via PPARγ, promoting both adipocyte differentiation and lipid accumulation. However, in the context of obesity, chronic activation of SREBP-1c leads to overproduction of lipids, which leads to the development of enlarged adipocytes and lipotoxicity, both of which negatively affect insulin sensitivity [35].

ROS also influence the mitogen-activated protein kinase (MAPK) pathway—a key controller of cell growth and differentiation—mainly by oxidatively modifying signaling molecules and inhibiting MAPK phosphatases [36]. This ROS-driven mechanism promotes abnormal adipocyte hypertrophy and excessive lipid accumulation, hallmark features of obesity [37]. Furthermore, NADPH oxidase (NOX) proteins, especially NOX4, have key functions in managing cell growth, differentiation, metabolism, immune responses, and apoptosis [31]. During adipocyte differentiation, preadipocytes release angiotensin II (Ang II), a peptide that promotes adipocyte growth and maturation, stimulates leptin secretion, and enhances prostaglandin I_2 (PGI_2) production in adipocytes. PGI_2 further promotes AT differentiation and increases TAG accumulation [38]. Notably, Ang II is a potent activator of NOX enzymes, and elevated NOX4 expression has been reported in adipocytes from both genetically and diet-induced obese animal models [39].

14.5.2 Role of ROS on Lipid Metabolism

14.5.2.1 ROS and Lipogenesis

As mentioned before, lipogenesis is a process which encompasses the synthesis of FA and TAG, occurring primarily in the liver and AT. The role of ROS in the stimulation of lipogenesis appears to be of considerable importance. Mice with elevated adipocyte ROS levels—through targeted depletion of glutathione—displayed limited AT expansion and increased ectopic fat accumulation, which indicated the inhibition of healthy expansion of AT [40, 41].

Among ROS, hydrogen peroxide (H_2O_2) has been suggested to play a key role in promoting lipogenesis. In rat adipocytes, H_2O_2 at concentrations ranging from 0.15 to 0.5 mM was demonstrated to enhance the incorporation of glucose-derived carbon into glyceride-FA [42]. H_2O_2 also stimulates lipogenesis by enhancing substrate transport, increasing NADPH availability, and activating pyruvate dehydrogenase [42].

ROS can further promote lipid synthesis by enhancing the utilization of glucose for lipogenesis. Experimental evidence indicates that increased ROS levels—induced by treatment with acetoacetate—stimulate de novo lipogenesis in human adipocytes by promoting the conversion of glucose into FA [40].

14.5.2.2 ROS and Lipolysis

Lipolysis is the metabolic process by which stored TAG are hydrolyzed into FFA and glycerol, primarily mediated by two enzymes, adipose tissue triglyceride lipase (ATGL) and hormone-sensitive lipase (HSL) [43]. Lipolysis in AT is always accompanied by the production of ROS and reactive nitrogen species (RNS), necessitating a strong antioxidant defense to preserve balanced redox signaling [41].

Activation of β-adrenergic receptors (β-ARs) by catecholamines results in higher intracellular ROS generation. Both $O_2^{\cdot -}$ and H_2O_2 can oxidize these receptors, sensitizing adipocytes to lipolytic stimuli and enhancing the lipolytic response [44, 45]. Similarly, the binding of growth hormone (GH) to its receptor also stimulates ROS production, particularly H_2O_2. This ROS generation activates downstream signaling pathways, including β-AR, adenylyl cyclase (AC), and protein kinases, such as PKA, ultimately promoting lipolysis [46]. Furthermore, ROS production is triggered upon activation of natriuretic peptide receptors A and C (NPR-A/C), which are expressed in AT. This activation enhances β-AR signaling, AC stimulation, and cyclic AMP (cAMP) production, further increasing lipolytic activity [47]. At higher concentrations, $O_2^{\cdot -}$ and H_2O_2 can also activate protein kinase C (PKC) through the promotion of diacylglycerol (DAG) release, contributing to lipolytic regulation [48].

In summary, ROS interact with key hormonal and enzymatic components of the lipolytic pathway, exerting concentration-dependent effects that can either stimulate or impair lipid breakdown. At physiological levels, ROS are normally produced during lipolysis and usually do not disturb the cell homeostasis. However, excessive ROS levels, as seen in obesity-related OS, may disrupt these mechanisms and contribute to metabolic dysregulation [31, 41].

14.5.3 Role of ROS in Mitochondrial Activity

Mitochondria are the primary source of cellular energy, responsible for generating over 90% of it through oxidative phosphorylation. However, an excessive supply of nutrients can disrupt mitochondrial function, affecting mitochondrial number, morphology, and dynamics. This imbalance may also impair mitochondrial biogenesis, increase ROS generation, and activate apoptotic pathways [49].

An unbalanced diet can induce elevated levels of ROS, contributing to the onset and progression of obesity. Mitochondria produce ATP via the oxidation of glucose, ketone bodies, and FA—with β-oxidation contributing up to 20% of total cellular energy production [50]. In the context of a Western diet, the overproduction of ROS is exacerbated by an abundant supply of dietary substrates, particularly FA and glucose. High-fat diets (HFD) lead to mitochondrial overload due to excessive FA influx, resulting in increased levels of acetyl-CoA. This, in turn, stimulates increased NADH production via the Krebs cycle, enhancing electron transport chain (ETC) activity and consequently promoting ROS overproduction [51]. Excessive ROS production, combined with disrupted calcium signaling, can trigger the opening of the mitochondrial permeability transition pore within the inner mitochondrial membrane. This process leads to proton leakage and the release of matrix solutes into the intermembrane space, resulting in mitochondrial dysfunction, initiation of apoptotic pathways, and, under severe conditions, necrotic cell death [52].

Additionally, Western diets typically contain roughly equal proportions of glucose and fructose. Although glucose is a key energy substrate, especially for the brain, substantial dietary intake is not essential, as the liver can synthesize it via gluconeogenesis. Nevertheless, excessive glucose consumption contributes to insulin resistance, dyslipidemia, and increased de novo lipogenesis [53].

Fructose metabolism, in particular, has been shown to significantly enhance ROS generation, especially superoxide anions. One major pathway involves the rapid production of NADPH and acetyl-CoA during fructose catabolism, fueling de novo lipogenesis and impairing mitochondrial function. This dysfunction increases electron leakage from the ETC, thereby amplifying ROS production [54].

Moreover, the phosphorylation of fructose to fructose-1-phosphate depletes intracellular ATP levels, which triggers the activation of xanthine oxidase. This enzyme catalyzes the conversion of hypoxanthine to xanthine and subsequently to uric acid, a process that generates superoxide radicals as metabolic byproducts [55].

14.5.4 Role of ROS in Endoplasmic Reticulum Function

The ER is responsible for orchestrating the proper folding and maturation of nascent polypeptides through the coordinated action of molecular chaperones and folding enzymes. When the protein-folding demand exceeds the ER's capacity, misfolded and unfolded proteins accumulate within the ER lumen, triggering a condition known as ER stress [56]. In reaction, cells trigger a well-preserved signaling pathway called the unfolded protein response (UPR), which seeks to reestablish endoplasmic reticulum (ER) balance by controlling multiple gene expression mechanisms [57].

Increasing data show that ER stress contributes to the cytosolic accumulation of ROS, thereby amplifying OS [58]. Notably, consumption of an HFD leads to a significant upregulation of ER stress markers, comprising eukaryotic initiation factor 2α (eIF2α), C/EBP homologous protein (CHOP), and protein kinase R-like ER kinase (PERK), while concurrently reducing the phosphorylation of key adaptive proteins such as binding immunoglobulin protein (BiP) and inositol-requiring enzyme 1 (IRE1). These modifications are intimately connected with the development of insulin resistance (IR) [59].

Moreover, HFD has been shown to significantly increase the expression of pro-inflammatory mediators, including tumor necrosis factor (TNF), Toll-like receptor 2 (TLR2), Toll-like receptor 4 (TLR4), and secreted phosphoprotein 1 (SPP1) [59]. Under obese conditions, immune cells are key contributors to ER stress within AT [60]. Recent studies have demonstrated that activated macrophages secrete pro-inflammatory cytokines that induce ER stress in adipocytes, contributing to the pathogenesis of insulin resistance [61]. UPR activation stimulates nuclear factor kappa-light-chain-enhancer of activated B cells (NF-κB), a central transcription factor in macrophages, leading to the upregulation of several pro-inflammatory genes such as TNF-α, interleukin-1β (IL-1β), IL-6, IL-12p40, and cyclooxygenase-2 (COX-2), thus reinforcing the inflammatory milieu that underlies metabolic dysfunction [57, 62, 63].

14.5.5 Role of ROS in Inflammation

There is a well-established interplay between OS and inflammation in the pathophysiology of obesity. In adipocytes from obese individuals, redox-sensitive transcription factors, particularly nuclear factor kappa B (NF-κB) and activator protein-1 (AP-1), become activated, promoting the expression and secretion of pro-inflammatory cytokines [30]. The main pro-inflammatory cytokines secreted are: TNF-α, involved in the regulation of immune cell activation, inflammatory response, lipid metabolism, adipocyte apoptosis, hepatic lipogenesis, insulin signaling, and OS [30, 64]; IL-1β, which transmits signals through multiple inflammatory kinases, including NF-κB and extracellular signal-regulated kinases (ERK), thereby enhancing the inflammatory response cascade [65] and IL-6, a pleiotropic cytokine that regulates energy homeostasis and promotes the synthesis of other pro-inflammatory mediators [66]. These cytokines contribute to ongoing local and systemic inflammation while simultaneously amplifying ROS production, establishing a self-perpetuating cycle of OS and inflammation [30, 67].

Nutritional factors are key modulators in this process. The overconsumption of lipids, carbohydrates, and especially saturated and trans-FA activates intracellular signaling cascades that culminate in excessive ROS generation [27]. During states of overnutrition, high glucose availability intensifies glycolytic flux and accelerates tricarboxylic acid (TCA) cycle activity, increasing NADH and $FADH_2$ production, which can promote electron leakage and ROS [68, 69]. In parallel, ROS inhibit the enzyme glyceraldehyde-3-phosphate dehydrogenase (GAPDH), diverting upstream glycolytic intermediates into alternative metabolic pathways. One byproduct of this shift is methylglyoxal, an early compound that leads to advanced glycation end products (AGEs) [68]. AGEs, in turn, activate NADPH oxidase (NOX), enhancing the production of both ROS and RNS. Simultaneously, the activation of NF-κB promotes the transcription of numerous pro-inflammatory genes, including TNF-α, adhesion molecules, inducible nitric oxide synthase (iNOS), IL-6, and specific microRNAs (miRs). These factors further promote adipogenesis, sustain inflammation, and increase OS, contributing to the pathogenesis of metabolic dysfunction [70, 71].

14.5.6 Role of ROS in Adipokines Secretion During Obesity

Adipokines are bioactive peptides secreted by AT that play critical roles in OS, metabolic regulation, and inflammation. Among the most studied adipokines are leptin and adiponectin, both of which are dysregulated in obesity and closely linked to redox imbalance.

Leptin, the first adipokine to be identified, is primarily secreted by adipocytes and is essential for maintaining energy homeostasis by modulating food intake and energy expenditure [72]. In obesity, leptin levels are significantly elevated (i.e., hyperleptinemia). Paradoxically, this is often accompanied by leptin resistance, whereby its central effects on satiety are blunted. Hyperleptinemia contributes to OS by promoting the stimulation and mobilization of immune cells, particularly monocytes and macrophages, which subsequently secrete pro-inflammatory cytokines such as IL-6 and TNF-α [68].

Furthermore, leptin directly stimulates NADPH oxidase (NOX) activity, enhancing the production of ROS, mainly H_2O_2 and ·OH [67]. Leptin also decreases the activity of paraoxonase-1 (PON-1), a key antioxidant enzyme associated with high-density lipoprotein (HDL), further compromising the cellular redox defense system [73].

Adiponectin is another major adipokine secreted by adipocytes; however, unlike leptin, its circulating levels decrease in obesity. It is widely recognized for its antioxidant and anti-inflammatory properties. Adiponectin suppresses the NF-κB signaling pathway, thereby reducing the expression and activity of pro-inflammatory cytokines such as TNF-α and IL-6, while also enhancing the production of the anti-inflammatory cytokine IL-10 [31, 74]. From a metabolic perspective, adiponectin enhances FA oxidation, decreases serum TAG levels, and improves insulin sensitivity, making it a protective factor against obesity-related metabolic disorders. Importantly, adiponectin also inhibits ROS generation, and its circulating levels are inversely correlated with markers of systemic OS [75]. Thus, the decline in adiponectin observed in obesity contributes to a pro-inflammatory and pro-oxidant state.

14.6 Strategies to Reduce ROS Levels and Improve Obesity Outcomes

As discussed above, OS plays a key role in the pathophysiology of obesity by disrupting energy homeostasis through its effects on hypothalamic neurons involved in appetite regulation and by exacerbating pro-inflammatory pathways within the AT microenvironment [31, 32].

Accordingly, antioxidants have gained attention as an effective method to counteract oxidative damage and restore cellular balance. Natural antioxidants—bioactive compounds abundantly found in various plant sources—demonstrate significant potential in modulating obesity-associated mechanisms. Major classes of antioxidants include: (1) polyphenols; (2) carotenoids; (3) capsaicinoids and capsinoids; (4) isothiocyanates; and (5) catechins [31, 32].

The antioxidant activity of these compounds is primarily attributed to their capacity to neutralize ROS, modulate signaling pathways linked to lipogenesis, and inhibit the activation of transcription, including PPARγ and NF-κB. These actions reduce the expression of pro-inflammatory genes and preserve crucial elements for normal adipocyte function [32].

Although the protective effects of antioxidants against OS and their role in maintaining adipocyte health are well documented, the interaction between endogenous antioxidants (e.g., glutathione, superoxide dismutase, and catalase) and those obtained from the diet has received less attention. Under conditions of obesity or poor nutrition, excessive ROS production may overwhelm endogenous defense systems, increasing cellular vulnerability to oxidative damage [32]. Therefore, the incorporation of dietary antioxidants—such as vitamins E and C, carotenoids, and polyphenols—represents a complementary and effective approach to mitigating OS and supporting adipocyte function in obesity-related conditions. Despite their promising potential, further well-controlled clinical trials are needed to confirm their therapeutic efficacy and elucidate their mechanisms of action within the context of metabolic disorders [31, 32].

14.7 Conclusions

ROS are byproducts of cellular metabolism that, at physiological concentrations, serve important signaling functions in various cellular processes. Under normal conditions, the antioxidant defense system tightly regulates ROS levels to maintain redox homeostasis. However, when antioxidant capacity is insufficient to counterbalance ROS production, a state of OS arises. Oxidative stress disrupts metabolic regulation in AT, impairs lipid metabolism, and promotes both local and systemic inflammation through interactions with immune cells and hormonal pathways.

Maintaining redox balance is considered one of the most effective strategies for preventing obesity and its associated metabolic disorders. As research progresses, a deeper understanding of the specific molecular mechanisms by which ROS contribute to the onset and progression of obesity—and how OS reduction (e.g., through anti-inflammatory and antioxidant-rich diets) might lessen these effects—could lead to the development of more targeted and effective interventions for managing obesity and related metabolic conditions (Table 14.1).

References

1. Yoshikawa T, You F (2024) Oxidative stress and bio-regulation. Int J Mol Sci 25(6):3360. https://doi.org/10.3390/ijms25063360
2. González-Domínguez Á, Belmonte T, González-Domínguez R (2023) Childhood obesity, metabolic syndrome, and oxidative stress: microRNAs go on stage. Rev Endocr Metab Disord 24(6):1147–1164. https://doi.org/10.1007/s11154-023-09834-0
3. Sun L, Wang X, Saredy J, Yuan Z, Yang X, Wang H (2020) Innate-adaptive immunity interplay and redox regulation in immune response. Redox Biol 37:101759. https://doi.org/10.1016/j.redox.2020.101759
4. Palma FR, Gantner BN, Sakiyama MJ, Kayzuka C, Shukla S, Lacchini R, Cunniff B, Bonini MG (2024) ROS production by mitochondria: function or dysfunction? Oncogene 43(5):295–303. https://doi.org/10.1038/s41388-023-02907-z
5. Metz S, Belanich JR, Claussnitzer M, Kilpeläinen TO (2025) Variant-to-function approaches for adipose tissue: insights into cardiometabolic disorders. Cell Genom 5(5):100844. https://doi.org/10.1016/j.xgen.2025.100844
6. Koenen M, Hill MA, Cohen P, Sowers JR (2021) Obesity, adipose tissue and vascular dysfunction. Circ Res 128(7):951–968. https://doi.org/10.1161/CIRCRESAHA.121.318093
7. Chaves VE, Frasson D, Martins-Santos MES, Navegantes LCC, Galban VD, Garófalo MAR et al (2008) Fatty acid synthesis and generation of glycerol-3-phosphate in brown adipose tissue from rats fed a cafeteria diet. Can J Physiol Pharmacol 86(7):416–423. https://doi.org/10.1139/y08-052
8. Chaves VE, Frasson D, Kawashita NH (2011) Several agents and pathways regulate lipolysis in adipocytes. Biochimie 93(10):1631–1640. https://doi.org/10.1016/j.biochi.2011.05.018
9. Shklyaev SS, Melnichenko GA, Volevodz NN, Falaleeva NA, Ivanov SA, Kaprin AD et al (2021) Adiponectin: a pleiotropic hormone with multifaceted roles. Probl Endokrinol (Mosk) 67(6):98–112. https://doi.org/10.14341/probl12827
10. Shehab MJ, Al-Mofarji ST, Mahdi BM, Ameen RS, Al-Zubaidi MM (2025) The correlation between obesity and leptin signaling pathways. Cytokine 192:156970. https://doi.org/10.1016/j.cyto.2025.156970
11. Rakib A, Mamun MAA, Mandal M, Sinha P, Singh UP (2025) Obesity-cancer axis crosstalk: molecular insights and therapeutic approaches. Acta Pharm Sin B 15(6):2930–2944. https://doi.org/10.1016/j.apsb.2025.04.029. Epub 2025 May 5. PMID: 40654338; PMCID: PMC12254798
12. Baldelli S, Aiello G, Mansilla Di Martino E, Campaci D, Muthanna FMS, Lombardo M (2024) The role of adipose tissue and nutrition in the regulation of adiponectin. Nutrients 16(15):2436. https://doi.org/10.3390/nu16152436
13. Marques-Oliveira GH, Silva TM, Lima WG, Valadares HMS, Chaves VE (2018) Insulin as a hormone regulator of the synthesis and release of leptin by white adipose tissue. Peptides 106:49–58. https://doi.org/10.1016/j.peptides.2018.06.007
14. Khoramipour K, Chamari K, Ahmadi Hekmatikar A, Ziyaiyan A, Taherkhani S, Elguindy NM et al (2021) Adiponectin: structure, physiological functions, role in diseases, and effects of nutrition. Nutrients 13(4):1180. https://doi.org/10.3390/nu13041180

15. World Health Organization (2021) Obesity and overweight [Internet]. [cited 2021 Jan 11]. Available from: https://www.who.int/news-room/fact-sheets/detail/obesity-and-overweight
16. Rubino F, Cummings DE, Eckel RH, Cohen RV, Wilding JPH, Brown WA et al (2025) Definition and diagnostic criteria of clinical obesity. Lancet Diabetes Endocrinol 13(3):221–262. https://doi.org/10.1016/S2213-8587(24)00316-4
17. Masood B, Moorthy M (2023) Causes of obesity: a review. Clin Med (Lond) 23(4):284–291. https://doi.org/10.7861/clinmed.2023-0168
18. Rohm TV, Meier DT, Olefsky JM, Donath MY (2022) Inflammation in obesity, diabetes, and related disorders. Immunity 55(1):31–55. https://doi.org/10.1016/j.immuni.2021.12.013
19. Cooke MS, Evans MD, Dizdaroglu M, Lunec J (2003) Oxidative DNA damage: mechanisms, mutation, and disease. FASEB J 17(10):1195–1214. https://doi.org/10.1096/fj.02-0752rev
20. Cadet J, Delatour T, Douki T, Gasparutto D, Pouget JP, Ravanat JL et al (1999) Hydroxyl radicals and DNA base damage. Mutat Res 424(1–2):9–21. https://doi.org/10.1016/s0027-5107(99)00004-4
21. Zi Y, Wang X, Zi Y, Yu H, Lan Y, Fan Y et al (2023) Cigarette smoke induces the ROS accumulation and iNOS activation through deactivation of Nrf-2/SIRT3 axis to mediate the human bronchial epithelium ferroptosis. Free Radic Biol Med 200:73–86. https://doi.org/10.1016/j.freeradbiomed.2023.03.002
22. Zimmermann R, Hertz-Schünemann R, Ehlert S, Liu C, McAdam K, Baker R et al (2015) Highly time-resolved imaging of combustion and pyrolysis product concentrations in solid fuel combustion: NO formation in a burning cigarette. Anal Chem 87:1711–1717. https://doi.org/10.1021/ac503512a
23. Bus JS, Aust SD, Gibson JE (1974) Superoxide- and singlet oxygen-catalyzed lipid peroxidation as a possible mechanism for paraquat (methyl viologen) toxicity. Biochem Biophys Res Commun 58(3):749–755. https://doi.org/10.1016/s0006-291x(74)80481-x
24. Allawzi A, Elajaili H, Redente EF, Nozik-Grayck E (2019) Oxidative toxicology of bleomycin: role of the extracellular redox environment. Curr Opin Toxicol 13:68–73. https://doi.org/10.1016/j.cotox.2018.08.001
25. McCord JM, Fridovich I (1969) Superoxide dismutase. An enzymic function for erythrocuprein (hemocuprein). J Biol Chem 244(22):6049–6055
26. Valgimigli L (2023) Lipid peroxidation and antioxidant protection. Biomolecules 13(9):1291. https://doi.org/10.3390/biom13091291
27. Masenga SK, Kabwe LS, Chakulya M, Kirabo A (2023) Mechanisms of oxidative stress in metabolic syndrome. Int J Mol Sci 24(9):7898. https://doi.org/10.3390/ijms24097898
28. Chung T-H, Kwon Y-J, Lee Y-J (2020) High triglyceride to HDL cholesterol ratio is associated with low testosterone and sex hormone-binding globulin levels in middle-aged and elderly men. Aging Male 23(2):93–97. https://doi.org/10.1080/13685538.2018.1501015
29. GBD 2021 Adult BMI Collaborators (2025) Global, regional, and national prevalence of adult overweight and obesity, 1990–2021, with forecasts to 2050: a forecasting study for the Global Burden of Disease Study 2021. Lancet 405(10481):813–838. https://doi.org/10.1016/S0140-6736(25)00355-1
30. Čolak E, Pap D (2021) The role of oxidative stress in the development of obesity and obesity-related metabolic disorders. J Med Biochem 40(1):1–9. https://doi.org/10.5937/jomb0-2465
31. Pérez-Torres I, Castrejón-Téllez V, Soto ME, Rubio-Ruiz ME, Manzano-Pech L, Guarner-Lans V (2021) Oxidative stress, plant natural antioxidants, and obesity. Int J Mol Sci 22(4):1786. https://doi.org/10.3390/ijms22041786
32. Fu M, Yoon K-S, Ha J, Kang I, Choe W (2025) Crosstalk between antioxidants and adipogenesis: mechanistic pathways and their roles in metabolic health. Antioxidants (Basel) 14(2):203. https://doi.org/10.3390/antiox14020203
33. Kim H-Y, Jang H-J, Muthamil S, Shin UC, Lyu J-H, Kim S-W et al (2024) Novel insights into regulators and functional modulators of adipogenesis. Biomed Pharmacother 177:117073. https://doi.org/10.1016/j.biopha.2024.117073

34. Lee H, Lee YJ, Choi H, Ko EH, Kim J-W (2009) Reactive oxygen species facilitate adipocyte differentiation by accelerating mitotic clonal expansion. J Biol Chem 284(16):10601–10609. https://doi.org/10.1074/jbc.M808742200
35. Jun I, Choi YJ, Kim B-R, Lee HK, Seo KY, Kim T-I (2024) Activation of the mTOR pathway enhances PPARγ/SREBP-mediated lipid synthesis in human meibomian gland epithelial cells. Sci Rep 14(1):28118. https://doi.org/10.1038/s41598-024-73969-6
36. Son Y, Kim S, Chung H-T, Pae H-O (2013) Reactive oxygen species in the activation of MAP kinases. Methods Enzymol 528:27–48. https://doi.org/10.1016/B978-0-12-405881-1.00002-1
37. Chang C-C, Sia K-C, Chang J-F, Lin C-M, Yang C-M, Lee I-T et al (2022) Participation of lipopolysaccharide in hyperplasic adipose expansion: involvement of NADPH oxidase/ROS/p42/p44 MAPK-dependent cyclooxygenase-2. J Cell Mol Med 26(14):3850–3861. https://doi.org/10.1111/jcmm.17419
38. Sharma AM (2004) Is there a rationale for angiotensin blockade in the management of obesity hypertension? Hypertension 44(1):12–19. https://doi.org/10.1161/01.HYP.0000132568.71409.a2
39. Vilchis-Landeros MM, Matuz-Mares D, Vázquez-Meza H (2020) Regulation of metabolic processes by hydrogen peroxide generated by NADPH oxidases. Processes 8(11):1424. https://doi.org/10.3390/pr8111424
40. Okuno Y, Fukuhara A, Hashimoto E, Kobayashi H, Kobayashi S, Otsuki M et al (2018) Oxidative stress inhibits healthy adipose expansion through suppression of SREBF1-mediated lipogenic pathway. Diabetes 67(6):1113–1127. https://doi.org/10.2337/db17-1032
41. Abou-Rjeileh U, Contreras GA (2021) Redox regulation of lipid mobilization in adipose tissues. Antioxidants (Basel) 10(7):1090. https://doi.org/10.3390/antiox10071090
42. Mukherjee SP (1980) Mediation of the antilipolytic and lipogenic effects of insulin in adipocytes by intracellular accumulation of hydrogen peroxide. Biochem Pharmacol 29(9):1239–1246. https://doi.org/10.1016/0006-2952(80)90280-4
43. Pérez-Torres I, Gutiérrez-Alvarez Y, Guarner-Lans V, Díaz-Díaz E, Manzano Pech L, Caballero-Chacón SC (2019) Intra-abdominal fat adipocyte hypertrophy through a progressive alteration of lipolysis and lipogenesis in metabolic syndrome rats. Nutrients 11(7):1529. https://doi.org/10.3390/nu11071529
44. Rambacher KM, Moniri NH (2020) The β2-adrenergic receptor-ROS signaling axis: an overlooked component of β2AR function? Biochem Pharmacol 171:113690. https://doi.org/10.1016/j.bcp.2019.113690
45. Burns RN, Moniri NH (2011) Agonist- and hydrogen peroxide-mediated oxidation of the β2 adrenergic receptor: evidence of receptor s-sulfenation as detected by a modified biotin-switch assay. J Pharmacol Exp Ther 339(3):914–921. https://doi.org/10.1124/jpet.111.185975
46. DeYulia GJ Jr, Cárcamo JM, Bórquez-Ojeda O, Shelton CC, Golde DW (2005) Hydrogen peroxide generated extracellularly by receptor-ligand interaction facilitates cell signaling. Proc Natl Acad Sci USA 102(14):5044–5049. https://doi.org/10.1073/pnas.0501154102
47. Tan CM, Xenoyannis S, Feldman RD (1995) Oxidant stress enhances adenylyl cyclase activation. Circ Res 77(4):710–717. https://doi.org/10.1161/01.res.77.4.710
48. Wang XT, McCullough KD, Wang XJ, Carpenter G, Holbrook NJ (2001) Oxidative stress-induced phospholipase C-gamma 1 activation enhances cell survival. J Biol Chem 276(30):28364–28371. https://doi.org/10.1074/jbc.M102693200
49. Qiu H, Schlegel V (2018) Impact of nutrient overload on metabolic homeostasis. Nutr Rev 76(9):693–707. https://doi.org/10.1093/nutrit/nuy023
50. Ebert D, Haller RG, Walton ME (2003) Energy contribution of octanoate to intact rat brain metabolism measured by 13C nuclear magnetic resonance spectroscopy. J Neurosci 23(13):5928–5935. https://doi.org/10.1523/JNEUROSCI.23-13-05928.2003
51. Lambertucci RH, Hirabara SM, Silveira LR, Levada-Pires AC, Curi R, Pithon-Curi TC (2008) Palmitate increases superoxide production through mitochondrial electron transport chain and NADPH oxidase activity in skeletal muscle cells. J Cell Physiol 216(3):796–804. https://doi.org/10.1002/jcp.21463

52. Marchi S, Guilbaud E, Tait SWG, Yamazaki T, Galluzzi L (2023) Mitochondrial control of inflammation. Nat Rev Immunol 23(3):159–173. https://doi.org/10.1038/s41577-022-00760-x
53. Phillips R (2025) Diet, mitochondrial dysfunction, vascular endothelial damage, and the microbiome: drivers of ocular degenerative and inflammatory diseases. Ophthalmol Ther 14(7):1429–1452. https://doi.org/10.1007/s40123-025-01160-9. Epub 2025 May 28. PMCID: PMC12167425
54. Todoric J, Di Caro G, Reibe S, Henstridge DC, Green CR, Vrbanac A et al (2020) Fructose stimulated de novo lipogenesis is promoted by inflammation. Nat Metab 2(10):1034–1045. https://doi.org/10.1038/s42255-020-0261-2
55. Velázquez AL, Vidal L, Alcaire F, Varela-Tomasco P, Ares G (2021) Significant sugar-reduction in dairy products targeted at children is possible without affecting hedonic perception. Int Dairy J 114:104937. https://doi.org/10.1016/j.idairyj.2020
56. Pagliassotti MJ, Kim PY, Estrada AL, Stewart CM, Gentile CL (2016) Endoplasmic reticulum stress in obesity and obesity-related disorders: an expanded view. Metabolism 65(9):1238–1246. https://doi.org/10.1016/j.metabol.2016.05.002
57. Ajoolabady A, Lebeaupin C, Wu NN, Kaufman RJ, Ren J (2023) ER stress and inflammation crosstalk in obesity. Med Res Rev 43(1):5–30. https://doi.org/10.1002/med.21921
58. Tu BP, Weissman JS (2004) Oxidative protein folding in eukaryotes: mechanisms and consequences. J Cell Biol 164(3):341–346. https://doi.org/10.1083/jcb.200311055
59. Zhao M, Zang B, Cheng M, Ma Y, Yang Y, Yang N (2013) Differential responses of hepatic endoplasmic reticulum stress and inflammation in diet-induced obese rats with high-fat diet rich in lard oil or bean oil. PLoS One 8(11):e78620. https://doi.org/10.1371/journal.pone.0078620
60. Hummasti S, Hotamisligil GS (2010) Endoplasmic reticulum stress and inflammation in obesity and diabetes. Circ Res 107(5):579–591. https://doi.org/10.1161/CIRCRESAHA.110.225698
61. Dasgupta A, Bandyopadhyay GK, Ray I, Bandyopadhyay K, Chowdhury N, De RK et al (2020) Catestatin improves insulin sensitivity by attenuating endoplasmic reticulum stress: in vivo and in silico validation. Comput Struct Biotechnol J 18:464–481. https://doi.org/10.1016/j.csbj.2020.02.005
62. Tam AB, Mercado EL, Hoffmann A, Niwa M (2012) ER stress activates NF-κB by integrating functions of basal IKK activity, IRE1 and PERK. PLoS One 7(10):e45078. https://doi.org/10.1371/journal.pone.0045078
63. Liu T, Zhang L, Joo D, Sun SC (2017) NF-κB signaling in inflammation. Signal Transduct Target Ther 2:17023. https://doi.org/10.1038/sigtrans.2017.23
64. Glass CK, Olefsky JM (2012) Inflammation and lipid signaling in the etiology of insulin resistance. Cell Metab 15(5):635–645. https://doi.org/10.1016/j.cmet.2012.04.001
65. Dinarello CA (2011) A clinical perspective of IL-1β as the gatekeeper of inflammation. Eur J Immunol 41(5):1203–1217. https://doi.org/10.1002/eji.201141550
66. Mohammadi M, Gozashti MH, Aghadavood M, Mehdizadeh MR, Hayatbakhsh MM (2017) Clinical significance of serum IL-6 and TNF-α levels in patients with metabolic syndrome. Rep Biochem Mol Biol 6(1):74–79
67. Savini I, Catani MV, Evangelista D, Gasperi V, Avigliano L (2013) Obesity-associated oxidative stress: strategies finalized to improve redox state. Int J Mol Sci 14(5):10497–10538. https://doi.org/10.3390/ijms140510497
68. Manna P, Jain SK (2015) Obesity, oxidative stress, adipose tissue dysfunction, and the associated health risks: causes and therapeutic strategies. Metab Syndr Relat Disord 13(10):423–444. https://doi.org/10.1089/met.2015.0095
69. Tangvarasittichai S (2015) Oxidative stress, insulin resistance, dyslipidemia and type 2 diabetes mellitus. World J Diabetes 6(3):456–480. https://doi.org/10.4239/wjd.v6.i3.456
70. Chaudhuri J, Bains Y, Guha S, Kahn A, Hall D, Bose N et al (2018) The role of advanced glycation end products in aging and metabolic diseases: bridging association and causality. Cell Metab 28(3):337–352. https://doi.org/10.1016/j.cmet.2018.08.014

71. Lai SW, Lopez Gonzalez ED, Zoukari T, Ki P, Shuck SC (2022) Methylglyoxal and its adducts: induction, repair, and association with disease. Chem Res Toxicol 35(10):1720–1746. https://doi.org/10.1021/acs.chemrestox.2c00160
72. Zhang Y, Chua S Jr (2017) Leptin function and regulation. Compr Physiol 8(1):351–369. https://doi.org/10.1002/cphy.c160041
73. Bełtowski J, Wójcicka G, Jamroz A (2003) Leptin decreases plasma paraoxonase 1 (PON1) activity and induces oxidative stress: the possible novel mechanism for proatherogenic effect of chronic hyperleptinemia. Atherosclerosis 170(1):21–29. https://doi.org/10.1016/s0021-9150(03)00236-3
74. Ouchi N, Kihara S, Arita Y, Okamoto Y, Maeda K, Kuriyama H et al (2000) Adiponectin, an adipocyte-derived plasma protein, inhibits endothelial NF-kappaB signaling through a cAMP-dependent pathway. Circulation 102(11):1296–1301. https://doi.org/10.1161/01.cir.102.11.1296
75. Choi HM, Doss HM, Kim KS (2020) Multifaceted physiological roles of adiponectin in inflammation and diseases. Int J Mol Sci 21(4):1219. https://doi.org/10.3390/ijms

Chapter 15
ROS and Inflammation in Mitochondrial Disorders

Amlan Bishal, Bratati Bandyopadhyay, Mohini Mondal, and Debajit Dewan

15.1 Introduction

Cellular metabolism plays an important role in the pathogenesis of human disease. Mitochondria are organelles that produce most of the energy needed for cells to function and promote cell metabolism. Understanding the relationship between metabolic and mitochondrial functions is a challenge due to variation in how mitochondrial function is measured and the heterogeneity of mitochondria, cells, tissues, and terminal organs. Mitochondrial dysfunction can be determined at both the cellular and tissue level using a variety of methods, including recent progression states involving new radiotracers in conjunction with PET imaging. Mitochondria are organelles with essential functions in almost all mammalian cells, including energy generation, the production of reactive oxygen species (ROS), coordination of cellular Ca2+ signals, regulation of autophagy, necrosis, and apoptotic cell death. Maintaining healthy mitochondria involves complex quality control systems in which misfolded proteins deteriorate, but damaged mitochondria are either repaired by fusion or removed by autophagy. Mitochondria are centrally important in several processes, including normal tissue height, metabolism, immunity, neurodegeneration, and infection. This leads to the transformation of mitochondrial functional and cellular signaling pathways into disordered cellular functions, which themselves manifest as diseases. Due to the important role of mitochondria in metabolism, mitochondrial dysfunction manifests itself as a heterologous multisystem disease, primarily involving OXPHOS-dependent tissues such as the brain, heart, and skeletal muscle [1–3].

ROS is not the same as superoxide radicals (O_2-), hydroxyl radicals ($-OH$), hydrogen peroxide (H_2O_2), or singlet oxygen ($^1O_2$). The small openings in short mitochondrial permeability transition pore (MPTPs) help keep mitochondria

A. Bishal (✉) · B. Bandyopadhyay · M. Mondal · D. Dewan
Department of Pharmaceutical Technology, Bharat Technology, Howrah, West Bengal, India

S. Ashique, N. Mishra (eds.), *Reactive Oxygen Species in Metabolic Inflammation*,
https://doi.org/10.1007/978-3-032-11752-6_15

healthy. The body's balance, or homeostasis, involves how mitochondria and other channels like MPTP and Inner Membrane Anion Channel (IMACs) respond to redox stress. When these channels open, they change the redox environment in a specific area, which causes ROS to be released. This process, where mitochondrial ROS is made and released, is known as ROS-induced ROS release (RIRR) [4–7].

15.2 Mitochondria as a Source of ROS

15.2.1 Electron Transport Chain (ETC) and ROS Generation

Mitochondrial electron transport chains, like complex I to IV, include electron transporters such as ubiquinone and cytochrome c. There are two main ways electrons move through these chains. One pathway starts with complex I, III, and IV. This allows NADH to be the main energy source, helping complexes II, III, and IV to work properly. As electrons pass through these complexes, they help create a proton gradient across the inner membrane. The energy from this gradient is used by complex V, called ATP synthase, to make ATP. The first part explains the structure and function of these complexes and how electrons are transferred. Some electrons go directly to oxygen, which can create reactive oxygen species, or ROS. The second part talks about where complex I is located and where ROS is made in different parts, like Complex II_F (Flavin site of Complex II), part of complex III, and complex IIIQO. It also explains how ROS is managed in healthy and disease conditions. ROS acts as a signaling molecule that helps cells grow, adapt to low oxygen, and decide the cell's fate. But too much ROS can cause damage and cell death. Many diseases are connected to high levels of ROS. Finally, it discusses proton leakage and the role of special proteins called uncoupling proteins (UCPs) in regulating mitochondrial energy efficiency. Proton leakage includes both regular and induced protons, with induced protons controlled by UCPs. There are five UCPs in mammalian cells: UCP1 to UCP5. UCP1 helps keep the body warm in the cold by making heat. UCP2 and UCP5 help reduce oxidative stress and protect cells in certain situations. All diseases related to oxidative stress are connected to UCPs [8, 9].

15.2.2 Mitochondrial Dynamics (Fusion/Fission) and Oxidative Stress

Neurodegenerative diseases (NDS) are a group of different disorders that slowly damage and destroy nerve cells, causing problems with the nervous system. Even though these diseases are different, they all share a common link with mitochondria. Mitochondria are important because they help make important chemicals and create energy in the form of ATP. This energy is made through a process called oxidative phosphorylation (OxPHOS), which also produces reactive oxygen species (ROS). If mitochondria work poorly, they can break apart and lose their membrane potential, which might lead to these diseases. Mitochondrial fusion and fission are key processes that help keep mitochondrial quality control in check. Mitochondrial quality control is very important. According to recent reports, mitochondrial division in mammalian cells can happen in two main ways: through midzones and surrounding areas. Midzone divisions occur in healthy cells and show that mitochondria are functioning well. This process does not cause unwanted changes like increased plasma membrane polarization or more ROS. This kind of division also happens after the cell divides, showing what happens to the organelles and the whole cell. Both parts that come from this division have working copies of mtDNA, which were copied from the midzone, ensuring both mitochondria work properly. This is different from the periphery, where parts of the mitochondria involved in energy production often become damaged or break down. In those cases, when the mitochondria's ends lose their membrane potential, the peripheral parts are more likely to be used. According to the periphery, when mtDNA becomes smaller, it shows that the mitochondria are dying and mitophagy is taking place. At the center of mitochondrial division is DRP1, a protein found in the cytoplasm that starts the process of splitting mitochondria. DRP1 forms a ring shape around the mitochondrion, helping to physically separate it. Some molecules on the outer part of the mitochondria help bring DRP1 in, like FIS1, MFF, MID9, and MID51 [10–14].

15.3 ROS-Mediated Inflammatory Signaling Pathways

Production of ROS is a physiological process, and mainly occurs at complexes I and III of the electron transport chain. In normal healthy conditions, the antioxidant defense system, including superoxide dismutase (SOD), glutathione peroxidase, and catalase, ensures a delicate redox balance [15]. Nonetheless, the mitochondrial malfunction, be it caused by mutations in the mtDNA or nuclear-coded mitochondrial genes, leads to leakage of electrons and excess production of ROS that overwhelms antioxidant capacity [16]. This oxidative/reduction imbalance leads to the pathway of inflammatory signaling, as illustrated in Fig. 15.1. Nuclear factor kappa-light-chain-enhancer of activated B cells (NF-κB) is one of the most characterized ROS-sensitive transcription factors. NF-κB IkB proteins trap NF-κB in the cytoplasm in resting cells [17]. The high levels of ROS stimulate the IkB kinases, causing IkB to be degraded and consequently the translocation of NF-κB into the nucleus. NF-κB binds to pro-inflammatory promoter regions on the inside, resulting in the increased transcription of cytokines (TNF-α, IL-1β, IL-6) and chemokines (MCP-1, CCL2). These intermediates increase the recruitment of immune cells and chronic inflammation [18]. Another major ROS-driven inflammatory axis involves

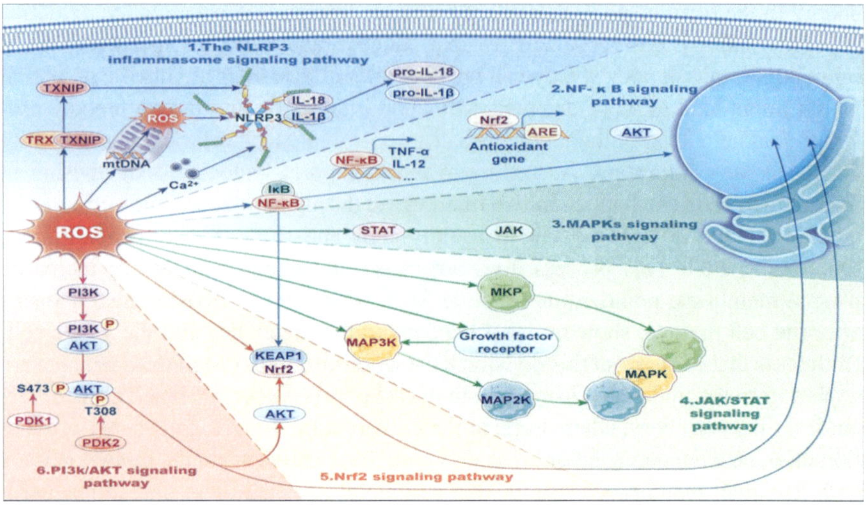

Fig. 15.1 Signal transduction of ROS in inflammation [23]

the NLRP3 inflammasome, a multiprotein complex composed of NLRP3, ASC (apoptosis-associated speck-like protein containing a CARD), and pro-caspase-1. Mitochondrial ROS serve as upstream activators of NLRP3, promoting its oligomerization and activation of caspase-1. Caspase-1 then processes pro-IL-1β and pro-IL-18 into their active forms, both of which are potent mediators of sterile inflammation [19]. Importantly, mtDNA released from damaged mitochondria can also act as a damage-associated molecular pattern (DAMP), synergizing with ROS to sustain inflammasome activation. ROS also alter mitogen-activated protein kinases (MAPK), such as ERK, JNK, and p38, which mediate the expression of cytokines and stress responses [20]. Likewise, JAK-STAT signaling is also affected by redox variations, which further boosts the expression of pro-inflammatory mediators. All these pathways lead to the production of a pro-inflammatory milieu, which continues to exist even without infectious stimuli. The end result of all these signaling processes is the release of cytokines and chemokines into the extracellular space [21]. High concentrations of IL-6, TNF-α, IL-1β, IL-18, and chemokines, including CXCL8 (IL-8) and MCP-1, not only recruit immune cells but also promote tissue damage locally. Increased circulating acute-phase proteins, such as C-reactive protein (CRP) are evidence of chronic systemic inflammation in patients with mitochondrial disease. In this way, ROS-mediated inflammatory signaling is a pivotal linkage between the malfunction of mitochondria and systemic pathophysiology [22].

15.4 Mitochondrial Disorders and Associated Inflammatory Profiles

Primary mitochondrial disorders are a heterogeneous category of diseases that occur due to mutations in either the mtDNA or nuclear genes that encode proteins in the mitochondrion. Although conventionally considered metabolic syndromes, an increasing body of evidence indicates that inflammation plays a major role in their pathology [24]. The following examples demonstrate how ROS-mediated signaling pathways appear as disease-specific inflammatory phenotypes. The m.3243A > G mutation in the mitochondrial tRNA 3leu (UUR) gene is most often linked to MELAS (Mitochondrial Encephalomyopathy, Lactic Acidosis, and Stroke-like Episodes) [25]. It is clinically manifested with stroke-like events, seizures, lactic acidosis, and progressive neuromuscular impairment. Its pathogenesis is based on inflammatory processes. Mitochondrial dysfunction in MELAS causes overproduction of ROS and triggers NF-κB signaling in vascular endothelial and neuronal cells. This stimulation triggers cytokine expression of IL-6, TNF-α, and IL-1β, which facilitates vascular inflammation, endothelial injury, and depleted nitric oxide bioavailability. These processes are part of the hallmark stroke-like episodes. Both serum and cerebrospinal fluid levels of inflammatory cytokines have been identified in MELAS patients and are associated with the severity of illness. Inflammation of MELAS, therefore, increases the metabolic crisis and worsens neuronal degeneration [26]. Mutations in the complex I subunits that are encoded by the mitochondrial DNA (LHON) are most commonly mutated to include the MT-ND4, MT-ND1, or MT-ND6 [27]. It is typified by acute or subacute loss of vision that is selective degeneration of retinal ganglion cells and optic nerve atrophy. In addition to energy failure, ROS-mediated inflammation is important in optic neuropathy. Overproduction of ROS stimulates microglial cells of the retina and optic nerve, which cause the release of pro-inflammatory cytokines such as IL-6 and TNF-α [28]. This neuroinflammatory environment increases the rate of ganglion cell apoptosis and degeneration of axons. NF-κB activation also increases adhesion molecules and chemokines, which increase the infiltration of immune cells [29]. Experimental LHON models and patient samples provide evidence of high levels of inflammatory markers, indicating that systemic and local inflammation are part of the disease progression. Anti-inflammatory interventions are thus being investigated as the adjunctive therapy in LHON [30]. Leigh syndrome is an early-onset, devastating, mitochondrial encephalopathy due to defects in complexes I, IV, or ATP synthase. It is characterized by developmental delay, lactic acidosis, respiratory dysfunction, and typical necrotic lesions in the basal ganglia and the brainstem. Neurodegeneration is a strong phenomenon that is facilitated by inflammatory mechanisms. The overproduction of ROS in neuronal and glial cells can activate the NLRP3 inflammasome, which results in caspase-1-dependent IL-1β and IL-18 maturation [31]. These cytokines mediate microglial activation and astrocytic reactivity, which encourages neuroinflammation and neuronal death. High systemic concentrations of TNF-α, IL-6, and IL-1β have been documented in patients and animal models, which promotes an interrelation between mitochondrial dysfunction and systemic inflammation. Prolonged low-grade inflammation is a contributing factor to multi-organ pathology, increasing the metabolic crisis of Leigh syndrome [32]. Systemic inflammatory features across mitochondrial diseases, MELAS, LHON, and Leigh syndrome, are clinically different disorders; however, they all display a similar signature of systemic inflammation. Raised levels of circulating cytokines, including IL-6, TNF-α, and C-reactive protein (CRP), are commonly found as a manifestation of chronic low-grade systemic inflammation [33]. This inflammatory state contributes to symptoms such as fatigue, muscle weakness, and progressive organ dysfunction. In addition, the emptying of mitochondrial DAMPs such as mtDNA, cardiolipin, and N-formyl peptides into the blood is a persistent stimulator of innate immune receptors. These signals sustain inflammation in the absence of infection, similar to the so-called inflammaging phenotype of age-related disorders [34]. In this way, mitochondrial diseases show how mitochondrial impairments may pair defects of metabolism with defects of immune regulation to generate complex inflammatory syndromes.

15.5 Oxidative Stress and Secondary Mitochondrial Dysfunction

Free radicals and other ROS are the prime source of oxidative stress, leading to cellular damage and mitochondrial dysfunction. Mitochondria are bilayer membrane structures that produce adenosine triphosphate (ATP) through oxidative phosphorylation. The formation of ATP comprises five respiratory chain enzyme complexes (I-V). Complex I-IV is the mitochondrial electron transport chain (mtETC) and complex V is ATP synthase. The mtETC oxidizes hydrogen-containing molecules to pyruvate and fatty acids with singlet oxygen and water. In the metabolic pathway, mitochondrial superoxide anion (O_2^-) and hydroxyl radical (OH^-) are generated at complex I and Complex III in mtETC. Thereby, complex I and III are the major sites of ROS formation. The highly reactive O_2^- and OH^- can damage biomolecules within mitochondria, including lipids, proteins, and DNA [35]. In cells, mitochondria are the only organelles besides the nucleus that have their own DNA, which is called mitochondrial DNA (mtDNA), and their own tools for making RNA and proteins [36]. Mitochondrial DNA constitutes roughly 1% of total cellular DNA, which might be damaged by ROS associated with oxidative stress. The persistence of mtDNA damage eventually leads to mutations in the mitochondrial genome, giving rise to increased mitochondrial dysfunction, which induces and aggravates the illnesses [37].

15.5.1 Inflammation-Induced ROS Aggravating Mitochondrial Damage

Inflammation is a disease characterized by immune cell accumulation (monocytes, macrophages, lymphocytes, polymorphonuclear neutrophils (PMN), and plasma cells) to the vascular wall, extravasation of immune cells to tissue, as well as production of ROS by these cells and tissue injury. ROS is mainly produced in PMN by the enzyme NADPH oxidase 2, which is activated by the formation of the catalytic subunit gp91phox (NOX2) [38]. Inflammation-induced reactive oxygen species (ROS) significantly aggravate mitochondrial damage, leading to a vicious cycle that perpetuates cellular injury and further inflammation. During inflammation, immune cell activation and mitochondrial dysfunction increase the production of ROS beyond the capacity of endogenous antioxidant defenses. This excess ROS results in oxidative stress, which indiscriminately targets mitochondrial membrane lipids, proteins, and mtDNA, leading to mitochondrial dysfunction and even cell death [39, 40]. The damaged mitochondria release excess ROS and mitochondrial macromolecules (mtDNA, peptides), which play a secondary messenger role to activate the inflammatory pathway like NLRP3 inflammasome [41].

15.5.2 Vicious Cycle Between ROS, Inflammation, and Mitochondrial Dysfunction

ROS are chemically reactive molecules generated primarily by mitochondria during oxidative phosphorylation. Several antioxidants' defenses system minimizes the excess ROS production. An excess amount of ROS generation in mitochondria seems to lead to cell death. The damaged mitochondria leak more electrons, mtDNA, and cardiolipin into the extracellular space. Moreover, damaged mitochondria release mitochondrial danger-associated molecular patterns (DAMPs), which trigger the pathogen-associated molecular patterns (PAMPs). These molecules activate several inflammatory pathways, including the TLR9, NLRP3 inflammasome, and NF-κB signaling [42, 43]. Inflammation is a protective immune response, but chronic or unresolved inflammation becomes harmful. Activated immune cells, such as macrophages and neutrophils, produce large amounts of ROS and reactive nitrogen species (RNS) as antimicrobial defenses. These molecules further damage host tissues and mitochondria, increasing oxidative stress. Furthermore, inflammatory cytokines (e.g., TNF-α, IL-1β, IL-6) disrupt mitochondrial biogenesis and impair antioxidant defenses, sustaining mitochondrial dysfunction. The vicious cycle between reactive oxygen species (ROS), inflammation, and mitochondrial dysfunction is a key driver in aging, neurodegeneration, vascular diseases, and inflammatory diseases [44].

15.6 Therapeutic Strategies Targeting ROS and Inflammation

Mitochondrial diseases, which may be caused by mutations in nuclear or mitochondrial DNA, are typified by inadequate oxidative phosphorylation, failed ATP production, and the buildup of reactive oxygen species (ROS). Overproduction of ROS, in relation to insufficient antioxidant defense mechanisms, triggers oxidative injury to mitochondrial and cellular constituents, such as proteins, lipids, and nucleic acids [45]. This injury is also enhanced by the pro-inflammatory cascades caused by mitochondrial dysfunction and results in a self-maintaining loop of oxidative stress and inflammation. The therapeutic interventions have, therefore, been mostly concerned with approaches that help in alleviating the burden of ROS and regulating the response of inflammation [46].

15.6.1 Antioxidants and ROS Scavengers

Coenzyme Q10 (CoQ10) is a lipid-soluble molecule of the mitochondrial electron transport chain (ETC) and an important electron carrier between complexes I/II and III. Its antioxidant effect is due to the capacity to serve as a redox buffer, which inhibits the spread of lipid peroxidation in mitochondrial and cellular membranes [47]. CoQ10 supplementation has been widely studied in mitochondrial diseases that include mitochondrial encephalomyopathy, lactic acidosis, and stroke-like episodes (MELAS), and Kearns-Sayre syndrome. Even though there are disparities in the clinical results, there are patients who show improvement in terms of exercise tolerance, fatigue, and cardiac functions [48]. Among the limitations are low bioavailability, inconsistent absorption, and low accumulation in the mitochondria. Mitochondria-targeted antioxidants have been created to circumvent the delivery issues associated with conventional antioxidants [49]. One of the most researched examples is MitoQ, a ubiquinone moiety attached to a lipophilic triphenyl phosphonium cation, enabling it to be concentrated in the mitochondrial matrix a number of times, owing to the negative membrane potential. MitoQ, once in, lowers the level of ROS, preserves mitochondrial DNA (mtDNA), and stabilizes mitochondrial membranes [50]. Preclinical models show that it can decrease oxidative injury, mitigate inflammation, and maintain cellular activity. There have been positive but small effects in clinical trials in conditions including Parkinson's disease, vascular dysfunction, and chronic kidney disease, but further research is required to confirm its long-term efficacy in primary mitochondrial disorders [51]. N-Acetylcysteine (NAC) is a proven thiol-based antioxidant which replenishes intracellular glutathione (GSH) stores as a cysteine donor. In addition to its effects in the scavenging of free radicals, NAC also has a heightened effect on redox homeostasis and the control of signaling pathways, including NF-κB, which have an indirect anti-inflammatory effect [52]. NAC has been investigated in mitochondrial myopathies, neurodegenerative diseases, and drug-induced mitochondrial toxicity, and evidence has indicated that it improves mitochondrial membrane potential, reduces the production of ROS, and increases cell survival. The positive safety profile of NAC predisposes it to attract an adjunct treatment, but further, highly rigorous, and long-term research is needed [53]. Other compounds are investigated, such as a-lipoic acid, vitamin E (tocopherols), and idebenone. The synthetic short-chain analog of CoQ10, idebenone, has shown better bioavailability and is approved in some locations to treat Leber hereditary optic neuropathy (LHON), with which it has improved visual function in some patients [54]. In the same way, a-lipoic acid plays a dual antioxidant and cofactor role in the mitochondrial enzyme complexes, whereas vitamin E can offer membrane-based protection against lipid peroxidation. Though such promising candidates exist, the heterogeneity of clinical outcomes indicates the complexity of oxidative stress in mitochondrial pathology [55].

15.6.2 Anti-Inflammatory Agents

ROS and mitochondrial dysfunction not only cause harm to biomolecules but also act as danger-associated molecular patterns (DAMPs) to provoke inflammatory pathways. The release of damaged mtDNA into the cytosol has the potential to trigger the NLRP3 inflammasome and release pro-inflammatory cytokines, including interleukin-1β (IL-1β) and interleukin-18 (IL-18). Persistent stimulation of these pathways works toward tissue degeneration in mitochondrial diseases, requiring measures related to curtailing inflammation [20]. Corticosteroids are the general anti-inflammatory medications that inhibit the production of cytokines, the activation of NF-κB, and the functionality of immune cells. They have been applied in mitochondrial myopathies and encephalopathies as a means of symptomatic relief. Their non-specific mode of action and side effects, including metabolic derangements, immunosuppression, and muscle wasting, however, restrict their use in the long term [56]. Targeted biologics are a more discriminatory method of modulating inflammation. TNF-α (e.g., infliximab, adalimumab), IL-6 (e.g., tocilizumab), or IL-1β (e.g., canakinumab, anakinra) directed monoclonal antibodies are undergoing research as potential mediators of cytokine-mediated inflammation in mitochondrial and neurodegenerative diseases. Such agents have the benefit of precision but are expensive, have the risk of infection, and have not been clinically validated in mitochondrial contexts [57]. The traditional NSAIDs and selective COX-2 inhibitors can have the symptomatic benefits of inflammation and oxidative stress, but there are gastrointestinal, renal, and cardiovascular risks of long-term use. New small molecules, including NLRP3 inflammasome activation inhibitors or mitochondrial dynamics modulators, are in preclinical development and provide promising avenues to therapeutic intervention in the future [58].

15.6.3 Combination Therapies and Adjunctive Approaches

Since there is a close interconnection between ROS and inflammation, combination therapies, which simultaneously focus on both processes, will tend to be more effective. As an example, a cytokine blocker can be combined with an antioxidant supplement to minimize oxidative and inflammatory damage [23]. Pharmacological interventions can also be combined with lifestyle interventions, including specific exercise programs, calorie restriction, and ketogenic diets, which can improve mitochondrial biogenesis and resilience. The nutraceutical polyphenols (resveratrol, curcumin) are under investigation as activators of sirtuins, enhancers of mitochondrial activity, and regulators of redox-inflammatory balance [59].

15.7 Future Perspectives and Challenges

Despite decades of research, effective therapies for mitochondrial disorders remain limited. Current strategies aimed at reducing ROS and inflammation are promising but face multiple challenges that must be addressed for meaningful clinical translation.

15.7.1 Biological and Clinical Heterogeneity

One of the most significant challenges of this understanding and treatment of mitochondrial disorders, in the context of ROS generation and inflammation, is biological and clinical heterogeneity. The mitochondrial diseases are caused by abnormalities of either the mitochondrial DNA (mtDNA) or nuclear DNA (nDNA) with each one of them making significant differences in disease onset, progression, and tissue distribution. Variations in the degree of heteroplasmy, tissue-specific energy needs and age-related penetrance add to the clinical appearance even more. Consequently, within the identical genetic defect, patients with mild myopathy might even have severe neurological or multisystem dysfunction. Such a heterogeneity also affects ROS generation and inflammation. The location and strength of ROS production can change with variations in respiratory chain defects, mitochondrial turnover, and metabolic condition, so it is hard to predict disease progression and therapeutic response. At the clinical level, it causes extensive phenotype variability, lack of diagnostic confidence and unreliable responses to treatment. Learning to cope with this complexity will be instrumental in the future of the therapeutic development since the more precise approaches to the issue, considering genotype, heteroplasmy distribution, and tissue specific mitochondrial dysfunction are needed to enhance patient outcomes [60].

15.7.2 Limitations of Current Antioxidant Therapies

Despite some promising results of the use of such antioxidants as CoQ10 and NAC, the clinical outcomes are not consistent. There are poor pharmacokinetics, ineffective delivery to mitochondria, and the risk that too much ROS scavenging could disrupt physiological redox signaling. ROS are not entirely harmful, they also play the role of signaling molecules in processes like autophagy and cell adaptation. Therefore, treatment should be directed to achieving redox balance and not unselective suppression [61].

15.7.3 Inflammation as a Double-Edged Sword

Though harmful in the chronic form, inflammatory reactions also take part in tissue repair and immune protection. Widespread inhibition of inflammation can therefore disrupt essential physiological functions. One of the most important issues is how to find a therapeutic window that minimizes pathological inflammation without undermining the positive immune activities. Biomarkers that can differentiate between harmful and adaptive inflammation will be needed in treatment guidance [62].

15.7.4 Emerging Therapeutic Platforms

In a bid to overcome shortcomings of the existing therapies in mitochondrial conditions, various novel therapeutic platforms are being evaluated that have the capacity to change the fate of patients. Nanoparticles, liposomes, and peptide-based carriers (also referred to as targeted drug delivery systems) have shown increased mitochondrial uptake of antioxidants in preclinical studies, including MitoQ and curcumin, resulting in decreased oxidative stress and cell survival in neurodegenerative and metabolic disease models [63]. Gene therapies are rapidly evolving; as a recent example, mitoTALENs and zinc-finger nucleases have been demonstrated in mouse models to selectively delete mutant mitochondrial DNA, restoring normal mitochondrial activity, with early CRISPR-based systems, albeit technically difficult due to the importation limitations of mitochondrial nucleases, having recently demonstrated proof-of-concept success in cell culture [64]. Mitochondrial replacement therapy (MRT), which has already been used in limited cases during birth, has been demonstrated to capture the transmission of mutations in the maternal mitochondrial DNA, and preliminary results have shown that children born through this method are healthy, although this method is still subject to scrutiny by ethics and regulatory agencies in most countries. Lastly, immunometabolism regulation is becoming a new approach, and research indicates that metabolic checkpoint regulation in immune cells, including mTOR or AMPK, can suppress chronic inflammation and restore mitochondrial homeostasis; experimental agonists that reestablish immune cell metabolic balance are currently in development in mitochondrial and other inflammatory diseases [65]. Together, these platforms transform a paradigm shift of symptomatic management to disease-modifying and potentially curative interventions, which is encouraged by the encouraging preclinical and early clinical results.

15.7.5 Need for Robust Biomarkers

The absence of valid biomarkers to precisely quantify oxidative stress, mitochondrial dysfunction, and inflammatory conditions in vivo is one of the major issues. Existing biomarkers like serum lactate, creatine kinase, or non-specific oxidative stress-indicators are insufficient. The creation of sensitive, non-invasive imaging techniques (e.g., redox-sensitive fluorescent probes, magnetic resonance spectroscopy), and molecular biomarkers (circulating mtDNA, cytokine panels) will play a pivotal role in early diagnosis, monitoring the patient, and assessing the efficacy of therapy [66].

15.7.6 Integrative and Personalized Medicine

Integrative methods combining the use of pharmacological agents with lifestyle modifications and individualized medicine are the future of mitochondrial therapy. Individualized exercise programs, dietary therapy, including ketogenic or intermittent fasting regimens, and stress control can be used in combination with specific antioxidant and anti-inflammatory medications [67]. Furthermore, natural intelligence and machine learning can be used to combine intricate omics information to forecast disease courses, optimize therapeutic schedules, and direct personalized treatment [68].

15.7.7 Ethical, Regulatory, and Economic Considerations

The translation of emerging therapies like gene editing, biologics, or mitochondrial replacement methods into clinical use brings up ethical and regulatory issues. There are long-term safety, intergenerational effects, and equitable access to high-cost therapies that have to be carefully considered [69]. Moreover, the scarcity of many mitochondrial diseases complicates the development of a large clinical trial, necessitating novel trial designs, including adaptive protocols, patient registries, and cross-border collaborations.

15.8 Conclusion

Mitochondria are at the crossroads of cellular energy metabolism, redox balance, and innate immune, and they play a pivotal role in the pathogenesis of mitochondrial disorders. Overproduction of reactive oxygen species (ROS), the major source of which is the electron transport chain, disturbs mitochondrial homeostasis and is a powerful promoter of inflammatory signaling, including activating NF-κB and assembling the NLRP3 inflammasome. This deviant cross-talking between inflammation and oxidative stress increases tissue damage, which leads to the development and clinical heterogeneity of primary mitochondrial diseases, such as MELAS, LHON, and Leigh syndrome. Moreover, the production of ROS through inflammation propagates secondary mitochondrial dysfunction, which leads to a vicious cycle that increases the severity of the disease. Treatment strategies that focus on interrupting this cycle, including the administration of antioxidants, mitochondrial-targeted ROS scavengers, and anti-inflammatory drugs, all of which have demonstrated promise, have a variable level of clinical effectiveness and depend upon the situation. The future needs to be precise in terms of the interventions that can control the production of ROS and inflammatory signaling without impairing the physiological redox and immune responses. Innovations in mitochondrial-targeted drug delivery, personalized medicine, and biomarker-based patient stratification could be the way to more effective treatment. On the whole, the interpretation of the complex relationship between ROS and inflammation in the context of mitochondrial diseases not only provides information on the pathophysiology of the disease but also demonstrates important areas of therapy. Further investigations in this field can make a difference in the management of these disabling conditions, and the management will be more focused on disease-modifying, rather than symptomatic management.

References

1. Ravagnan L, Roumier T, Kroemer G (2002) Mitochondria, the killer organelles and their weapons. J Cell Physiol 192(2):131–137
2. Lenaz G (1998) Role of mitochondria in oxidative stress and ageing. Biochimica Biophys Acta 1366(1–2):53–67
3. Lenaz G, Bovina C, D'aurelio M, Fato R, Formiggini G, Genova ML, Giuliano G, Pich MM, Paolucci UG, Castelli GP, Ventura B (2002) Role of mitochondria in oxidative stress and aging. Ann N Y Acad Sci 959(1):199–213
4. Zhang Y, Wong HS (2021) Are mitochondria the main contributor of reactive oxygen species in cells. J Exp Biol 224(5):jeb221606
5. Zorov DB, Juhaszova M, Sollott SJ (2014) Mitochondrial reactive oxygen species (ROS) and ROS-induced ROS release. Physiol Rev 94(3):909–950
6. Starkov AA (2008) The role of mitochondria in reactive oxygen species metabolism and signaling. Ann N Y Acad Sci 1147(1):37–52
7. Wikstrom M (1989) Identification of the electron transfers in cytochrome oxidase that are coupled to proton-pumping. Nature 338(6218):776–778

8. Lardy HA, Ferguson SM (1969) Oxidative phosphorylation in mitochondria. Annu Rev Biochem 38(1):991–1034
9. Williams RJ (1978) The history and the hypotheses concerning ATP-formation by energised protons. FEBS Lett 85(1):9–19
10. Chen H, Chan DC (2009) Mitochondrial dynamics–fusion, fission, movement, and mitophagy–in neurodegenerative diseases. Hum Mol Genet 18(R2):R169–R176
11. Green A, Hossain T, Eckmann DM (2022) Mitochondrial dynamics involves molecular and mechanical events in motility, fusion and fission. Front Cell Dev Biol 10:1010232
12. Ren L, Chen X, Chen X, Li J, Cheng B, Xia J (2020) Mitochondrial dynamics: fission and fusion in fate determination of mesenchymal stem cells. Front Cell Dev Biol 8:580070
13. Zhou H, Toan S (2020) Pathological roles of mitochondrial oxidative stress and mitochondrial dynamics in cardiac microvascular ischemia/reperfusion injury. Biomolecules 10(1):85
14. Chan DC (2012) Fusion and fission: interlinked processes critical for mitochondrial health. Annu Rev Genet 46(1):265–287
15. Kıran TR, Otlu O, Karabulut AB (2023) Oxidative stress and antioxidants in health and disease. J Lab Med 47(1):1–1
16. Kowalczyk P, Sulejczak D, Kleczkowska P, Bukowska-Ośko I, Kucia M, Popiel M, Wietrak E, Kramkowski K, Wrzosek K, Kaczyńska K (2021) Mitochondrial oxidative stress—a causative factor and therapeutic target in many diseases. Int J Mol Sci 22(24):13384
17. Saha S, Buttari B, Panieri E, Profumo E, Saso L (2020) An overview of Nrf2 signaling pathway and its role in inflammation. Molecules 25(22):5474
18. Guo Q, Jin Y, Chen X, Ye X, Shen X, Lin M, Zeng C, Zhou T, Zhang J (2024) NF-κB in biology and targeted therapy: new insights and translational implications. Signal Transduct Target Ther 9(1):53
19. Huang Y, Xu W, Zhou R (2021) NLRP3 inflammasome activation and cell death. Cell Mol Immunol 18(9):2114–2127
20. Zhao Y, Liu B, Xu L, Yu S, Fu J, Wang J, Yan X, Su J (2021) ROS-induced mtDNA release: the emerging messenger for communication between neurons and innate immune cells during neurodegenerative disorder progression. Antioxidants 10(12):1917
21. Sarapultsev A, Gusev E, Komelkova M, Utepova I, Luo S, Hu D (2023) JAK-STAT signaling in inflammation and stress-related diseases: implications for therapeutic interventions. Mol Biomed 4(1):40
22. Yi M, Li T, Niu M, Zhang H, Wu Y, Wu K, Dai Z (2024) Targeting cytokine and chemokine signaling pathways for cancer therapy. Signal Transduct Target Ther 9(1):176
23. Liu J, Han X, Zhang T, Tian K, Li Z, Luo F (2023) Reactive oxygen species (ROS) scavenging biomaterials for anti-inflammatory diseases: from mechanism to therapy. J Hematol Oncol 16(1):116
24. Heath O, Feichtinger RG, Achleitner MT, Hofbauer P, Mayr D, Merkevicius K, Spenger J, Steinbrücker K, Steindl C, Tiefenthaler E, Mayr JA (2024) Mitochondrial disorder diagnosis and management–what the pediatric neurologist wants to know. Eur J Paediatr Neurol 54:75–88
25. Ryytty S, Hämäläinen RH (2023) The mitochondrial m. 3243A> G mutation on the dish, lessons from in vitro models. Int J Mol Sci 24(17):13478
26. Na JH, Lee YM (2024) Diagnosis and management of mitochondrial encephalopathy, lactic acidosis, and stroke-like episodes syndrome. Biomolecules 14(12):1524
27. Lin X, Zhou Y, Xue L (2024) Mitochondrial complex I subunit MT-ND1 mutations affect disease progression. Heliyon 10(7):e28808
28. Kang EY, Liu PK, Wen YT, Quinn PM, Levi SR, Wang NK, Tsai RK (2021) Role of oxidative stress in ocular diseases associated with retinal ganglion cells degeneration. Antioxidants 10(12):1948
29. Duarte JN (2021) Neuroinflammatory mechanisms of mitochondrial dysfunction and neurodegeneration in glaucoma. J Ophthalmol 2021(1):4581909

30. Bouquet C, Clermont CV, Galy A, Fitoussi S, Blouin L, Munk MR, Valero S, Meunier S, Katz B, Sahel JA, Thomasson N (2019) Immune response and intraocular inflammation in patients with Leber hereditary optic neuropathy treated with intravitreal injection of recombinant adeno-associated virus 2 carrying the ND4 gene: a secondary analysis of a phase 1/2 clinical trial. JAMA Ophthalmol 137(4):399–406
31. Ruhoy IS, Saneto RP (2014) The genetics of Leigh syndrome and its implications for clinical practice and risk management. Appl Clin Genet 7:221–234
32. Muzio L, Viotti A, Martino G (2021) Microglia in neuroinflammation and neurodegeneration: from understanding to therapy. Front Neurosci 15:742065
33. Wei Y, Huang Y, Yang Y, Qian M (2021) MELAS/LS overlap syndrome associated with mitochondrial DNA mutations: clinical, genetic, and radiological studies. Front Neurol 12:648740
34. Kong C, Song W, Fu T (2022) Systemic inflammatory response syndrome is triggered by mitochondrial damage. Mol Med Rep 25(4):147
35. Guo C, Sun L, Chen X, Zhang D (2013) Oxidative stress, mitochondrial damage and neurodegenerative diseases. Neural Regen Res 8(21):2003–2014
36. Douarre C, Sourbier C, Dalla Rosa I, Brata Das B, Redon CE, Zhang H, Neckers L, Pommier Y (2012) Mitochondrial topoisomerase I is critical for mitochondrial integrity and cellular energy metabolism. PLoS One 7(7):e41094
37. Hollensworth SB, Shen CC, Sim JE, Spitz DR, Wilson GL, LeDoux SP (2000) Glial cell type-specific responses to menadione-induced oxidative stress. Free Radic Biol Med 28(8):1161–1174
38. Chelombitko MA (2018) Role of reactive oxygen species in inflammation: a minireview. Mosc Univ Biol Sci Bull 73(4):199–202
39. Patergnani S, Bouhamida E, Leo S, Pinton P, Rimessi A (2021) Mitochondrial oxidative stress and "mito-inflammation": actors in the diseases. Biomedicine 9(2):216
40. Jing W, Liu C, Su C, Liu L, Chen P, Li X, Zhang X, Yuan B, Wang H, Du X (2023) Role of reactive oxygen species and mitochondrial damage in rheumatoid arthritis and targeted drugs. Front Immunol 14:1107670
41. Yu Y, Liu S, Yang L, Song P, Liu Z, Liu X, Yan X, Dong Q (2024) Roles of reactive oxygen species in inflammation and cancer. MedComm 5(4):e519
42. Narala VR, Narala SR, Aiya Subramani P, Panati K, Kolliputi N (2024) Role of mitochondria in inflammatory lung diseases. Front Pharmacol 15:1433961
43. Jassim AH, Inman DM, Mitchell CH (2021) Crosstalk between dysfunctional mitochondria and inflammation in glaucomatous neurodegeneration. Front Pharmacol 12:699623
44. Dmytriv TR, Duve KV, Storey KB, Lushchak VI (2024) Vicious cycle of oxidative stress and neuroinflammation in pathophysiology of chronic vascular encephalopathy. Front Physiol 15:1443604
45. Nissanka N, Moraes CT (2018) Mitochondrial DNA damage and reactive oxygen species in neurodegenerative disease. FEBS Lett 592(5):728–742
46. Xu X, Pang Y, Fan X (2025) Mitochondria in oxidative stress, inflammation and aging: from mechanisms to therapeutic advances. Signal Transduct Target Ther 10(1):190
47. Gutierrez-Mariscal FM, de la Cruz-Ares S, Torres-Peña JD, Alcalá-Diaz JF, Yubero-Serrano EM, López-Miranda J (2021) Coenzyme Q10 and cardiovascular diseases. Antioxidants 10(6):906
48. Fan HC, Lee HF, Yue CT, Chi CS (2021) Clinical characteristics of mitochondrial encephalomyopathy, lactic acidosis, and stroke-like episodes. Life 11(11):1111
49. Jiang Q, Yin J, Chen J, Ma X, Wu M, Liu G, Yao K, Tan B, Yin Y (2020) Mitochondria-targeted antioxidants: a step towards disease treatment. Oxidative Med Cell Longev 2020(1):8837893
50. Sulaimon LA, Afolabi LO, Adisa RA, Ayankojo AG, Afolabi MO, Adewolu AM, Wan X (2022) Pharmacological significance of MitoQ in ameliorating mitochondria-related diseases. Adv Redox Res 5:100037

51. Leathem A, Ortiz-Cerda T, Dennis JM, Witting PK (2022) Evidence for oxidative pathways in the pathogenesis of PD: are antioxidants candidate drugs to ameliorate disease progression? Int J Mol Sci 23(13):6923
52. Sahasrabudhe SA, Terluk MR, Kartha RV (2023) N-acetylcysteine pharmacology and applications in rare diseases—repurposing an old antioxidant. Antioxidants 12(7):1316
53. Zhou J, Terluk MR, Orchard PJ, Cloyd JC, Kartha RV (2021) N-acetylcysteine reverses the mitochondrial dysfunction induced by very long-chain fatty acids in murine oligodendrocyte model of adrenoleukodystrophy. Biomedicine 9(12):1826
54. Ribeiro PV, Pari Mitre L, Gauza MM, Furukawa PH, Ferraz VE, Gomy I (2025) Therapeutic benefit of idebenone in Leber hereditary optic neuropathy: a systematic review and meta-analysis. Ophthalmic Genet 15:1–6
55. Abdelhalim MA, Qaid HA, Al-Mohy YH, Ghannam MM (2020) The protective roles of vitamin E and α-lipoic acid against nephrotoxicity, lipid peroxidation, and inflammatory damage induced by gold nanoparticles. Int J Nanomedicine 15:729–734
56. Ho GT, Theiss AL (2022) Mitochondria and inflammatory bowel diseases: toward a stratified therapeutic intervention. Annu Rev Physiol 84(1):435–459
57. Qin P, Sun Y, Li L (2024) Mitochondrial dysfunction in chronic neuroinflammatory diseases. Int J Mol Med 53(5):47
58. Taylor EB, Hall JE, Mouton AJ (2025) Current anti-inflammatory strategies for treatment of heart failure: from innate to adaptive immunity. Pharmacol Res 216:107761
59. Yip JM, Chiang GS, Lee IC, Lehming-Teo R, Dai K, Dongol L, Wang LY, Teo D, Seah GT, Lehming N (2025) Mitochondria and the repurposing of diabetes drugs for off-label health benefits. Int J Mol Sci 26(1):364
60. Mei J, Ding P, Gao C, Zhou J, Li Z, Zhang C, Gao J (2025) Mitochondrial diseases: molecular pathogenesis and therapeutic advances. MedComm 6(9):e70385. https://doi.org/10.1002/mco2.70385
61. Sinenko SA, Starkova TY, Kuzmin AA, Tomilin AN (2021) Physiological signaling functions of reactive oxygen species in stem cells: from flies to man. Front Cell Dev Biol 9:714370
62. Pahwa R, Goyal A, Jialal I (2025) Chronic inflammation. In: StatPearls. StatPearls Publishing, Treasure Island
63. Buchke S, Sharma M, Bora A, Relekar M, Bhanu P, Kumar J (2022) Mitochondria-targeted, nanoparticle-based drug-delivery systems: therapeutics for mitochondrial disorders. Life 12(5):657
64. Di Donfrancesco A, Massaro G, Di Meo I, Tiranti V, Bottani E, Brunetti D (2022) Gene therapy for mitochondrial diseases: current status and future perspective. Pharmaceutics 14(6):1287
65. Noohi F, Ravitsky V, Knoppers BM, Joly Y (2022) Mitochondrial replacement therapy: in whose interests? J Law Med Ethics 50(3):597–602
66. Hubens WH, Vallbona-Garcia A, De Coo IF, Van Tienen FH, Webers CA, Smeets HJ, Gorgels TG (2022) Blood biomarkers for assessment of mitochondrial dysfunction: an expert review. Mitochondrion 62:187–204
67. Zhong L, Hou X, Tian Y, Fu X (2025) Exercise and dietary interventions in the management of diabetic cardiomyopathy: mechanisms and implications. Cardiovasc Diabetol 24(1):159
68. 角田達彦. Advances in AI and machine learning for predictive medicine. J Hum Genet 2024
69. Zeps N, Lysaght T, Chadwick R, Erler A, Foo R, Giordano S, San Lai P, Schaefer GO, Xafis V, Chew WL, Sugarman J (2021) Ethics and regulatory considerations for the clinical translation of somatic cell human epigenetic editing. Stem Cell Rep 16(7):1652–1655

Chapter 16
ROS and Ferroptosis Impact on Inflammation-Mediated Hemochromatosis

Mirunalini Gobinath, Parikshit Roychowdhury, Poulami Sarkar, Naina Mohamed Pakkir Maideen, and Suvro Chatterjee

16.1 Introduction

Hemochromatosis, which is a metabolic disorder, occurs when excessive iron is present in the body due to increased iron accumulation or iron absorption. These deposits of iron are seen in the tissues of the liver, skin, joints, pancreas, heart, pituitary glands, etc., and often lead to iron toxicity [1]. In normal conditions, the red blood cells with hemoglobin act as oxygen transporters by carrying oxygen (4 heme molecules with iron) and provide oxygen to all tissues of the body [2]. Through this process, some iron is utilized and almost 1 mg is lost and gets excreted via sweat, GI tract, or skin cells. Even though the normal dietary intake of iron for an adult is around 10–20 mg/day, only 10% is absorbed from the GIT and around 1–2 mg/day is utilized and stored for normal physiological activities. But, in case of hemochromatosis, more iron absorption occurs, which results in almost 4–5 mg/day being absorbed and stored, leading to accumulation of iron and hemosiderosis [3, 4].

M. Gobinath
Department of Pharmacognosy & Phytopharmacy, JSS College of Pharmacy, JSS Academy of Higher Education & Research, Ooty, Tamil Nadu, India

P. Roychowdhury
Department of Pharmaceutics, JSS College of Pharmacy, JSS Academy of Higher Education & Research, Ooty, Tamil Nadu, India

P. Sarkar · S. Chatterjee (✉)
Department of Biotechnology, The University of Burdwan, Bardhaman, West Bengal, India

N. M. P. Maideen
College of Medicine, Mohammed Bin Rashid University of Medicine and Health Sciences, Dubai, UAE

Dubai Health, Dubai, UAE
e-mail: nmmaideen@dubaihealth.ae

S. Ashique, N. Mishra (eds.), *Reactive Oxygen Species in Metabolic Inflammation*,
https://doi.org/10.1007/978-3-032-11752-6_16

Unfortunately, when an excess of iron gets stored in the body, it leads to serious consequences such as free radicals' production, oxidative stress through the Fenton reaction, and serious damage. Reactive oxygen species (ROS) are produced, which act as a "double-edged sword" in the pathogenesis of hemochromatosis [5]. In addition, Fenton chemistry synthesizes more ferrous ions, which are free to interact with peroxides and generate ROS, which indirectly activate inflammatory signaling pathways and interlink oxidative stress with inflammation [6]. Though oxidative stress has a major role in pathophysiology, lipid peroxidation potentiates the same. Lipid peroxidation, defined as the oxidative degradation of lipids by free radicals, creates a chain reaction that catalyzes oxidative stress and inflammation. Thus, in hemochromatosis, the axis of ROS-ferroptosis-inflammation-lipid peroxidation is interconnected, which makes it intricate to rule out the disease. Through a thorough understanding of these crucial interactions, it is facile to develop novel therapeutic strategies, which not only target iron chelation, but also the multifaceted mechanism in hemochromatosis and other iron overload conditions [7].

16.2 Ferroptosis: Mechanistic Overview

Ferroptosis, as the name suggests, is an iron-dependent regulated non-apoptotic cell death pathway which is guided by lipid peroxidation and membrane damage due to the excessive overload of iron and ROS [8]. Over the past decades, cell death was a passive, non-resistant, deregulated mechanism, until the discovery of apoptosis, which is considered to be a form of regulated cell death pathway due to protease activation of caspase family (cysteine proteases). Later, other variants, such as necrosis, apoptosis, and autophagy, were found [9]. The reason why ferroptosis was highlighted is because of its non-apoptotic process, whereas apoptosis resistance, either intrinsic or acquired, is the limitation in cancer therapy. Subsequently, this states that the biochemical, genetic, and structural mechanism of ferroptosis is entirely distinct from other regulated-cell-death mechanisms such as pyroptosis, necrosis, necroptosis, and apoptosis [10]. Necroptosis, which is a form of regulated necrotic cell death, is primarily conciliated by mixed-lineage kinase domain-like protein (MLKL) phosphorylation and receptor-interacting protein kinase-3 (RIPK3), leading to membrane disruption and cell death. Pyroptosis is gasdermin-mediated programmed cell death, which is characterized by inflammation-mediated apoptosis, with morphological features including inflammation formation due to extracellular/intracellular activation, and flattened cytoplasm due to the leakage of the plasma membrane, IL-18, IL-1β release, and loss of membrane integrity [11].

The hallmarks of ferroptosis include features such as lipid peroxidation accumulation that executes ferroptosis, and leads to catastrophic membrane disruption of cells and organelles, which also depletes the antioxidant defenses of the body, and morphological characteristics that are similar to necrosis, which results in oncosis, which is a form of ischemic cell death due to rupture and swelling of cell components like cytoplasm, organelles, loss in plasma membrane integrity, and condensation of chromatin [12]. Changes in mitochondrial morphology of ferroptotic cells include swelling, a decrease in cristae, and an increased mitochondrial outer membrane. The interconnection of mitochondrial biology with ferroptosis is still controversial, but since the mitochondria is the hub for ROS, the mechanisms might interlink [13]. In contrast, some discoveries state that mitochondrial-mediated oxidative stress is not a prerequisite for ferroptosis, and it has been stated that it is similar to glutamate-induced excitotoxicity [8, 10].

Ferroptosis is further classified into exogenous pathways, that include the activation of lactotransferrin and serotransferrin iron transporters and glutamate or cystine inhibition of the reverse transport system (System Xc^{-}), whereas in endogenous pathways, there is inhibition of glutathione (GSH) peroxidase-4 (GPX4) and biosynthesis of intracellular antioxidant enzyme [14]. Alterations in these intracellular enzymes raise the levels of superoxide and cause ROS production. In previous research, it has been stated that ferroptosis suppressor protein-1 (FSP1) and GPX4 block and reduce the accumulation of ROS and the ferroptosis mechanism. Thus, the commencement and progression of ferroptosis are orchestrated with pro-ferroptotic pathways and cell defense mechanisms like antioxidant pathways [15]. This indicates that the key regulators are excessive iron metabolism, which leads to iron accumulation, and GSH depletion due to GPX4 inhibition, which results in the pathogenesis of various diseases and is a critical therapeutic target [16].

16.2.1 How Is Ferroptosis Mechanism Linked to Mitochondria?

Apart from being the powerhouses of the cell, mitochondria play an indispensable role in regulated cell deaths such as apoptosis, pyroptosis, necroptosis, and ferroptosis [17]. The classic mechanism that induces apoptosis through the mitochondria is by the enhanced permeability of the outer membrane of the mitochondria (OMM), leading to the leakage of cytochrome c into the intermembranous space which triggers apoptosis. In case of pyroptosis, permeability of OMM leaks cytochrome c into cytosol which potentiates caspase proteases and leads to subsequent cell death [18]. But in ferroptosis, mitochondria undergo various physical changes, such as decreased mitochondrial cristae, an increase in membrane density, and changes in energy metabolism, for instance ATP production, glycolysis, and oxidative phosphorylation are gradually decreased. In addition, ROS causes irreversible mitochondrial damage, which disrupts the integrity of the organelle, diminishes energy, and results in cell death. In other words, ferroptosis is closely linked with oxidative stress, mitophagy, mitochondrial fission and fusion, membrane potential, etc. [19, 20].

Mitochondria maintain the homeostasis of iron metabolism, which is important in performing bioenergetics and regulatory activities. When cells capture extracellular iron, it gets transported into mitochondrial membranes, and cofactors such as iron-sulfur and heme are fabricated. These cofactors are essential in redox reactions, DNA repair, synthesis, and other cellular processes like ATP production [21]. Mitoferrin and endosomes assist in the influx of iron into the cytoplasm, which is considered one of the most prevalent metals inside the mitochondria. Once iron is supplied into mitochondria, the OMM metabolizes it, and the ions and metabolites are interchanged between mitochondria and cytosol, which is regulated by voltage-dependent anion channels [22].

System Xc^-, which is known as the heterodimeric transporter of cystine and glutamine antiporter, comprises subunits of SLC3A2 and SLC7A11 (xCT). This imports cystine which is rapidly reduced to cysteine and exchanged for glutamate. The rate-limiting step for GSH synthesis is initiated by the substrate cysteine. Then, GSH and GPX are synthesized, which helps in the reduction of ROS. GPX4 is important in the reduction of lipid hydroperoxides (LOOH), such as cholesterol and esterified fatty acid hydroperoxides, for detoxification, by reducing them to their corresponding alcohols using glutathione as a cofactor [23]. In some cases, ferritin gets degraded when a specific autophagy mechanism gets initiated, which is known as ferritinophagy [24]. Furthermore, the tricarboxylic acid cycle and glutaminolysis trigger cells to become sensitized to ferroptosis [25].

Non-mitochondrial sites where ferroptosis occurs include organelles like the endoplasmic reticulum and peroxisomes. Phospholipid peroxidation happens in contact sites such as endoplasmic reticulum–mitochondria contact sites (EMCSs), which host pro-ferroptotic polyunsaturated lipids and potentiates ferroptosis without the help of mitochondria [26]. Besides, ferroptosis occurs in peroxisomes through the biosynthesis of plasmalogen. Alkylglycerone phosphate synthase (AGPS), glyceronephosphate O-acyltransferase (GNPAT), and fatty acyl-CoA reductase 1 (FAR1) are some peroxisomal enzymes that assist in the synthesis of plasmalogen, which are ether phospholipids, because in phospholipids, instead of ether, it utilizes ester and connects to glycerophospholipids. Further steps lead the cells to become more sensitive to ferroptosis by promoting lipid peroxidation, which is influenced by the number and activity of peroxisomes [27].

16.2.2 How Are Lipid Peroxidation and Ferroptosis Linked?

Earlier correlations between lipid peroxidation and cellular death originated in the 1960s, emerging from extensive research on carbon tetrachloride (CCl4)-induced hepatotoxicity studies [28]. It states that vitamin C, selenium, cysteine, and lipid peroxidation were recognized in a study showing that pretreating rats with selenium and vitamin C may mitigate CCl4-induced hepatotoxicity. Subsequently, the role of selenium was clarified when glutathione-dependent selenoenzyme was identified as the causative agent for inhibition of lipid peroxidation, especially in the phospholipid bilayers (PHGPx), which is phospholipid hydroperoxide glutathione peroxidase, currently known as GPX4 [29].

In lipid peroxidation, lipid gets deoxygenated and forms lipid peroxides and hydroperoxides, which are the key targets for ROS. An initiating free radical abstracts a labile hydrogen atom, which then combines with O_2 to form a peroxyl radical and propagates a chain reaction by continuing with another substrate moiety. At the end, two peroxyl radicals combine to produce a carbonyl compound, an alcohol molecule, and an oxygen molecule [30]. Enzymatic mechanisms of lipid peroxidation occur due to non-heme iron dioxygenases, lipoxygenases (LOX), and rarely by cyclooxygenases (COXs). The main substrates for lipid peroxidation include polyunsaturated fatty acids (PUFAs) and arachidonic acids, due to the presence of C-C double bonds, which makes them vulnerable to oxidative stress. Ferroptotic lipid peroxidation is commenced through acyl-CoA synthase long-chain family member-4 (ACSL4), lysophosphatidylcholine acyltransferase-3 (LPCAT3), and arachidonate lipoxygenases (ALOX) enzymatic cascade [31] (Fig. 16.1).

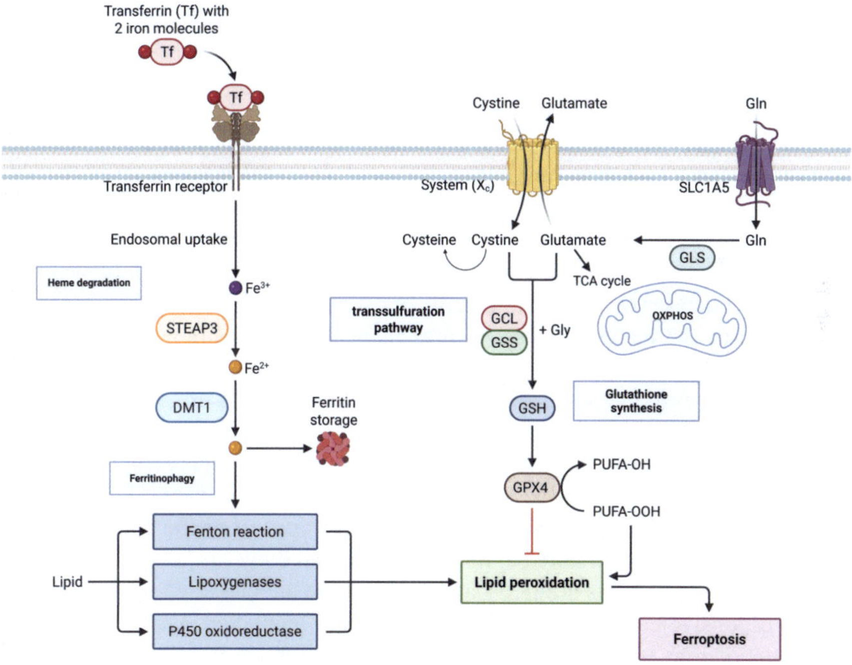

Fig. 16.1 Ferroptosis and its chemical biology

16.3 Physiology of Iron Homeostasis

Being one of the most important micronutrients required for biological processes, iron is absorbed through a sophisticated transportation mechanism involving various carriers and transporter proteins [32]. In the human body, iron is associated with the tasks of oxygen transport, DNA synthesis, respiration, electron transfer, etc. Iron can also be potentially toxic if it accumulates, and therefore it is regulated by a series of peptides controlling its absorption, distribution, storage, utilization, as well as excretion [33]. About 3–5 g of total iron are present in the human body, out of which 65–70% is transformed into hemoglobin in red blood cells. Myoglobin (10%), enzyme systems (5%), and storage forms that include ferritin and hemosiderin (15–20%) all include extra iron. The most metabolically active iron compartment has a tiny but crucial labile iron pool (LIP) of 2–4 mg that circulates bound to transferrin and turns over quickly (every 4 hours) [34].

Iron is absorbed in the human body through the duodenum and proximal parts of the jejunum. Iron absorbed by humans comes in two major forms: heme iron which makes up 10–15% of daily intake and is primarily sourced from hemoglobin and myoglobin present in meat products. They are particularly helpful due to their higher bioavailability of about 35%. The second source of iron is the non-heme type, the dietary portion including plants, fortified nutrition, and supplements makes up this type and is responsible for more than 85% of our daily intake of iron, this type of iron has a lower bioavailability of about 2–20% [3]. Absorption of the non-heme iron can be enhanced by using reducing agents such as vitamin C. Absorption of iron is a complicated process involving multiple chemical and physiological steps. Firstly, ferric ions are converted to ferrous ions by the enzyme duodenal cytochrome b reductase (Dcytb) located in the brush borders of enterocytes. This step is significant, as the metal transporters present in the gastric lining are specialized carriers of divalent metal ions rather than trivalent metal ions [4]. The ones responsible for iron transport are called divalent metal transporter 1 or DMT-1. There are four distinct types of DMT-1 in the human body, present in isoforms altering 3′ end and by the usage of promoter proteins. They are Exon 1A variants, which are mostly present in the duodenum and kidneys and contain an excess of 29–31 amino acids at the N terminus. Exon 1B variant is responsible for most of the endosomal iron transportation and is distributed in most of the human body [35].

The other two variants are IRE+ and IRE−, these are classified by the presence of ion-responsive elements for post-transcriptional regulation. Duodenum, as the site of major iron replenishment in the body, shows the presence of 1A-IRE isoforms that exhibit the highest iron regulation and are characterized by the expression of the highest iron responsiveness; therefore, they get triggered first for any iron deficiency in the blood. Post the extracellular absorption phase, the iron undergoes the intracellular absorption. Here, there are two major pathways that are possible. Either the iron enters the storage pathway, or it goes toward the export pathway. Iron is stored in ferritin; ferritin is a 24-protein subunit complex which can store up to 4500 iron atoms per complex [36]. Ferritin consists of a heavy chain and a light chain; the heavy chain provides ferroxidase activity which helps in oxidation of ferrous ions back to their original ferric form, which is optimal for the storage process; while the light chain helps in nucleation and enhances the storage stability of iron [37]. Ferroportin is the only known ion exporter which helps to transport iron via the basolateral membrane back to systemic circulation, this helps to maintain the iron homeostasis. Hephaestin and ceruloplasmin are two copper-containing iron oxidizing enzymes that oxidize iron to the ferric state in the systemic circulation, which is essential for it to be taken up by transferrin or macrophages, respectively. Enterocytes' apical membrane's heme carrier protein (HCP1) is involved in a unique absorption mechanism for heme iron. Heme oxygenase 1 (HMOX1) catalyzes the breakdown of heme within enterocytes, releasing biliverdin, carbon monoxide, and free iron. The released iron thereafter proceeds through the same intracellular processing pathways as iron that is not heme [38].

In iron homeostasis, cellular internalization, export, and transport play an important role, transferrin-mediated transport is the highest occurring process in the same, where transferrin receptors 1 and 2 form much of the uptake process, and GAPDH- and Cubilin-mediated transportation (TfR3 and TfR4) act as smaller alternate pathways. The primary uptake mechanism happens through steps in which transferrin 1 and 2 are both involved, transferrin 1 and 2 have similar structures, but transferrin 2 lacks the IRE regulation. Transferrin 1 is important in RBC synthesis and iron uptake, whereas 2 is significant in hepcidin regulation and other liver-specific iron regulation management [39]. The main pathway of iron uptake happens after diferric transferrin binds to transferrin receptors on the cell surface, it undergoes receptor binding, internalization, which results in the formation of clathrin-coated pits and vesicle internalization within the cell; acidification, which occurs when proton pumps lower endosomal pH to about 5.5, causes iron release because low pH encourages iron dissociation from transferrin; endosomal iron transport, in which DMT1 transports iron across the endosomal membrane to the cytosol; and finally, receptor recycling, where the Apotransferrin-receptor complex returns to the cell surface. Two different cellular iron uptake and regulation process take place when there is an iron deficiency in systemic circulation [40]. Glyceraldehyde-3-phosphate dehydrogenase (GAPDH) translocates to the plasma membrane and acts as a substitute transferrin receptor. Both clathrin-mediated and lipid raft-mediated endocytosis are involved in this mechanism. Cubilin acts as TfR4 in the kidney, promoting the reabsorption of iron from filtered transferrin [41]. Regarding the recycling of iron from senescent red blood cells, it is recycled by reticuloendothelial macrophages, specifically Kupffer cells in the liver and red pulp macrophages in the spleen. This mechanism, known as erythrophagocytosis, recovers roughly 20–25 mg of iron every day, which is significantly more than the 1–2 mg of iron that is absorbed from the diet [42].

16.3.1 Hepcidin-Ferroportin Axis

The understanding of iron metabolism and associated problems has entirely shifted since hepcidin was identified as the main controller of iron homeostasis. The circulating hormone hepcidin, a 25-amino acid antimicrobial peptide that is mainly produced by hepatocytes, regulates systemic iron homeostasis by targeting ferroportin. Hepcidin is an antimicrobial peptide that is classified as part of the defensin family. It has eight cysteine residues that create four disulfide bonds, which provide it with a stable structure. Furin and other proteases sequentially break an 84-amino acid precursor to produce the completed peptide [43]. Hepatocytes represent 95% of the synthesis of the systemic hepcidin but for local and organ-specific regulation of iron, the kidney and heart and, during inflammation, macrophages can produce the hormone as well. The only known cellular iron exporter in mammals is ferroportin (SLC40A1), a 12-transmembrane domain protein. Ferroportin is expressed in important tissues that handle iron. It is primarily found in duodenal enterocytes, which aid in the absorption of dietary iron, and in hepatocytes, which are involved in the storage and release of iron [44]. The protein, which is found in reticuloendothelial macrophages, controls iron recycling from erythrophagocytosis and aids in maternal-fetal iron transfer in placental syncytiotrophoblasts. Hepcidin binding to ferroportin causes quick internalization and destruction of the receptor in the hepcidin-ferroportin interaction, which is an example of a ligand-receptor system. The process goes as follows: Hepcidin attaches itself to ferroportin's extracellular domain, causing a conformational shift that initiates endocytosis. After that, internalized ferroportin undergoes ubiquitination, and within 1–2 hours, proteasomal destruction occurs [45, 46].

16.4 Understanding ROS and Its Role in Iron Overload Conditions

Free radicals are molecules that exist with an unpaired electron in their orbitals, due to which they are highly reactive and react with surface molecules, leading to a chain reaction. When in mild to moderate concentrations, free radicals are helpful in normal physiological functions, such as cellular signaling, host-pathogen defense, and metabolite production. The common free radical molecules are named reactive oxygen species (ROS) and reactive nitrogen species (RNS), which are either oxygen or nitrogen centric and categorized as radicals and non-radicals [47].

ROS are highly reactive molecules spawned naturally as the by-products of cellular metabolism, during the process such as oxidative phosphorylation in mitochondria. ROS is mainly composed of oxygen molecules, such as superoxide anion (O_2^-), hydrogen peroxide (H_2O_2), and hydroxyl radicals ($^{\cdot}OH$). ROS generation is proportionally dependent on iron level increment, especially free or labile iron. Iron is basically a "boon in disguise." These are also fabricated through the byproducts of aerobic metabolism [48].

A low to moderate level of ROS helps to maintain oxygen transport and enzymatic reactions. Whereas, overabundance of ROS can cause cell damage by triggering Fenton reaction [49], where hydroxyl radicals are produced from the reaction of ferrous ion (Fe^{2+}) interacting with hydrogen peroxide:

$$Fe^{2+} + H_2O_2 \rightarrow Fe^{3+} + {}^{\cdot}OH + OH^-$$

These hydroxyl ions are extremely reactive and toxic ROS and have the potential to cause damage to lipids, proteins, and DNA [50].

16.4.1 *Basic Concept of Redox Biology*

The oxidation-reduction reaction, which includes the transfer of electrons between molecules in the living system, is the key mechanism behind redox biology:

Oxidation: loss of electron (or hydrogen)/gain of oxygen
Reduction: gain of electron (or hydrogen)/loss of oxygen

These reactions are the core processes of life that control cellular signaling, metabolism, and energy production [51]. The balance between oxidants and antioxidants hinges on the redox reactions which administer cellular health. ROS serve as a signaling molecule that regulate the pathways like NF-κB (inflammation), Nrf2 (antioxidant gene expression), and MAPK (cell growth and stress response). Enzymatic antioxidants like superoxide dismutase (SOD), catalase, and GSH peroxidase (GPx), alongside non-enzymatic antioxidants including vitamin C and E, flavonoids, and carotenoids, also maintain redox equilibrium by influencing ROS levels and cellular integrity [52, 53]. Several pernicious diseases like cancer, cardiovascular diseases, and neurodegeneration are controlled by the lack of harmony of the redox balance. Reduction of ROS level promotes beneficial signaling and adaptation, while elevation of ROS levels promotes cell damage and diseases [54].

16.4.1.1 ROS Generation, Iron Overload, and Cellular Damage

The primary source of ROS production is the byproducts of normal cellular respiration, especially during aerobic respiration. Emergence of ROS depends upon mitochondrial respiration, enzymatic reactions, environmental stimuli, and photosensitizers. ROS mainly comprises oxygen molecules derived from diatomic oxygen (O_2), water, and hydrogen peroxide. Some common examples are superoxide anion (O_2^-), Hydroperoxide (H_2O_2), Hydroxyl radical (OH·), Singlet oxygen ($^1O_2$), Hypochlorous acid (HoCl), Peroxynitrite ($ONOO^-$), Ozone (O_3), Peroxyl radicals (LOO·), Alkoxyl radical (LO·), and Lipid hydroperoxide (LOOH) [55].

Each type of ROS has a distinct role in the biological system. Due to their physiological and pathological role, they are designated as the "double-edged sword." Iron overload occurs because of the labile (redox-active) iron increase, which sets off the Fenton reaction. Hydroxyl radicals(˙OH), one of the most destructive ROS, are generated when ferrous iron (Fe^{2+}) and hydrogen peroxide (H_2O_2) react through the Fenton reaction [56]. By initiating oxidative stress, cellular components like plasma membrane, mitochondria, DNA, proteins, and lipids are damaged by the ROS radicals. Cellular damage is formed due to the ROS in iron overload conditions, such as mitochondrial damage, endothelial cell injury, bone marrow suppression, and programmed cell death. These conditions trigger many pathological conditions like hemochromatosis, neurodegenerative diseases, cardiac injury, and osteoarthritis, which are linked with an iron overload condition [57, 58].

16.5 ROS and Inflammation

ROS plays a dual role in controlling inflammation. By defending against pathogens and regulating immune signaling, ROS helps at the physiological level, whereas at the pathological level, ROS production leads to oxidative stress, activating pro-inflammatory pathways like NF-κB and AP-1, which controls the increasing cytokine release (e.g., IL-1β, TNF-α) [59]. In addition, tissue damage can also trigger the NLRP3 inflammasome via TXNIP binding which leads to caspase-1 activation, IL-1β and IL-18 maturation, and pyroptosis [60]. Overproduction of ROS contributes to diseases such as diabetes, COPD, IBD, osteoarthritis, neurodegeneration, and angiogenesis [61].

16.6 Iron Overload and Oxidative Stress

Iron overload, medically known as hemochromatosis, is a pathological state characterized by the excessive accumulation of iron in the body, which plays a central role in causing toxicity and organ damage due to oxidative stress and ROS formation [62, 63]. Oxidative stress is exacerbated by the Fenton and Haber-Weiss reactions, which are the central processes of the ROS generation process with iron cycling between Fe^{2+} and Fe^{3+} states. Regulatory proteins such as ferritin and transferrin lead to the accumulation of labile iron during iron overload [64], which encourages the production of ROS and oxidative stress. The iron overload condition is induced by some key mechanisms which are induced by oxidative stress [65] (Table 16.1).

Iron overload and oxidative conditions have become a range of severe health issues which affect many parts of the human body, including vital organs and tissues, with many consequences. In the liver it manifests through hepatomegaly, cirrhosis, and liver cancer; in pancreas it can result in diabetes mellitus; in heart, it results in arrhythmias, cardiomyopathy, and heart failure; in the joints it manifests

Table 16.1 Key mechanisms of ROS, iron overload, and ferroptosis

Mechanisms	Triggered by	Consequences	References
Mitochondrial damage	Iron accumulation	Electron transport chain (ETC) dysfunction, ROS generation	[66]
Fenton reaction	Free Fe^{2+} reaction with H_2O_2	Hydroxyl radicals (·OH) formation	[67]
Iron homeostasis loss	Iron overproduction	Labile iron pool augmentation	[63, 68]
HO-1 activation	Oxidative stress	More free iron formation Feedback loop	[62]
Ferroptosis	Iron and lipid ROS reaction	Cell death due to lipid peroxidation	[8, 29]
Inflammation	ROS signaling	Tissue damage Cytokine release	[69]

as arthritis and joint pain; in the central nervous system and brain, it causes neurodegeneration (Alzheimer's, Parkinson's, Friedreich's ataxia), cognitive decline; whereas in the skin, it results in hyperpigmentation; and in the endocrine system, diseases like hypogonadism, infertility, and diabetes mellitus are seen. In the case of intracellular toxicities, mitochondria face the most problems with energy metabolism disruption and DNA damage [16, 70].

16.7 Pathophysiological Aspect

Hemochromatosis is a genetic condition primarily resulting from the mutations in the HFE gene, more accurately at C282Y and H63D homozygosity, which alters the polypeptide hepcidin expression in the liver therefore contributing to the disruption of the hepcidin ferroptosis axis, leading to an excessive absorption of iron from the duodenum as well as separation of iron from macrophages [71]. Dietary iron is absorbed through proteins that are present on the apical membrane of enterocytes, characterized as divalent metal transporters. When iron enters the cell, it is sequestered by ferritin to form an iron-ferritin complex. Systemic iron levels are generally monitored by iron-sensing chemoreceptors such as HFE. Hemojuvelin and transferrin I and II. Elevation of iron levels prompts the enterocyte cells to release it into the bloodstream, through unbinding from ferritin and binding to transferrin iron complexes which takes it to various tissues and then the rest is stored in hepatocytes, later being utilized in various other processes [72].

In case of mutated HFE, hepcidin hormone is reduced, this hormone effectively limits and controls the iron absorption, therefore the lowering of the hormone causes excessive absorption of iron from GIT. The reduced expression is mostly the result of the altered interaction between HFE protein and transferrin [73]. Iron accumulates in tissues everywhere in the body as a result of increased iron absorption. Since the liver plays a crucial role in iron metabolism, it is especially impacted. Ferritin, a protein that retains iron, is the first form of extra iron deposited in the liver's hepatocytes. But as iron overload worsens, hepatocytes are stressed out and unable to properly retain the extra iron [74]. As a result, iron begins to accumulate in other organs such as the endocrine glands, joints, pancreas, and heart. This accumulated iron impacts most organs through the production of ROS. When delivered to mitochondria for the synthesis of DNA, Cofactors, and damage repair, the excess amount of it can also harm it directly. This pathway ultimately results in tissue and organ damage, dysfunction, and inflammation [75].

16.8 In the Heart

Cardiomyocytes are the primary location of iron accumulation in the heart. Excessive iron deposition in cardiomyocytes impairs their capacity to generate contractions and interferes with their normal activity. Over-free iron impairs the sarco-endoplasmic reticulum calcium-ATPase 2 and sodium-calcium exchanger (NCX) as well as the excitation-contraction coupling. The contractile force of heart muscle cells is decreased when these transporters malfunction because elevated intracellular calcium and sodium result in reduced peak systolic and diastolic calcium [76]. The heart chambers widen as a result of this contractility deficit. The pathophysiology of DCM or Dilated Cardiomyopathy in hemochromatosis is additionally impacted by iron-mediated toxicity. Overproduction of iron promotes the production of harmful iron-based compounds, such as hydroxyl radicals, which cause direct harm to DNA, proteins, and lipids in cells. Cardiomyocytes' structural integrity and functionality are compromised by an accumulation of iron-induced harmful chemicals. The onset of cardiomyopathy in hemochromatosis is significantly influenced by inflammation in addition to oxidative stress and iron toxicity. Increased iron load in cardiomyocytes slowly contributes to their apoptosis over time, this further results in the development of iron-mediated fibrosis [77]. This reduces the cardiac contractility and thereby contributes to the progression of DCM further. However, the cardiac symptoms vary vastly among individuals based on their susceptibility, mutation extent, and quantitative burden of iron. When some individuals show symptoms of DCM early in life, others might stay asymptomatic for a longer period in their lifespan [78].

The most prominent direct implication of the iron overload and ferroptosis in the cardiac system is seen in the conduction system of the heart. In some cases, they are often correlated to atherosclerosis [79]. Irregular cardiac rhythms are caused by the iron accumulation in the conduction system, which consists of the His-Purkinje system, atrioventricular node, and sinoatrial node. This accumulation disrupts the normal transmission of electrical impulses. The regular operation of specialized cells involved in electrical conduction is disrupted by iron buildup within the cardiac conduction system. Ion-gated channels and membrane proteins that produce and conduct electrical signals are impacted by iron-induced toxicity, which causes oxidative stress and cell damage [78, 80].

16.9 In the Liver

In many liver disorders, oxidative stress and excessive iron are the two main factors that cause liver damage and disease progression. Ferroptosis, which is characterized by lipid peroxidation and iron excess, has been widely implicated in several liver disorders. The remainder of this study will concentrate on providing an overview of how ferroptosis contributes to the pathophysiology of a few prevalent liver conditions. IRI is one of the conditions where various organs show systemic pathogenic effects. It is believed that ferroptosis has key importance in the case of hepatic IRI. IRI-induced hepatocellular injury progresses through GPx4 inactivation-induced ferroptosis. Therefore, the diagnosis of hepatic IRI is often done by checking for iron accumulation, PTGS2 (which is an indicator of an increase in ferroptosis in the body) and lipid peroxidation [1].

Hemochromatosis-induced ferroptosis could become one of the major contributors to non-alcoholic fatty liver disease (NAFLD) in genetically compromised patients, there is an established pathological link between ferroptosis and NAFLD. The evidence from prior studies suggests that iron-induced lipid peroxide can be one of the biggest etiological factors for induction of NAFLD. There is verified evidence showcasing ferroptosis to be a major factor in regulated cell death initiation and inflammation in case of steatohepatitis, rather than necroptosis as it was assumed earlier. Clinical studies note a 90% or more patients are detected with enhanced levels of malondialdehyde and 4-hydroxynonenal, which are established markers for lipid peroxidation. RSL3 is a potent inhibitor of GPX4 that acts as a regulator of ferroptosis. By administering RSL3 to NASH-induced mice, it was observed that the symptoms increased multiple times; therefore, it was confirmed that a methionine/choline-deficient diet induced NASH can also progress through ferroptosis [81]. Thus, hemochromatosis can affect the liver in multiple possible mechanisms, including NAFLD, NASH, and IRI. Ferroptosis and alcoholic liver disease are also related by different pathogenic mechanisms, making individuals with hemochromatosis particularly vulnerable. This pathogenesis involves interplay between different organs and tissues, such as the gut, brain, lungs, and adipocytes, apart from liver, among all of these, liver-intestine and liver-adipocyte interactions are of highest prominence. The inhibition of sirtuin 1 and overexpression of adipose-specific lipin-1 have shown to be the contributing factors in aggravating liver conditions through ferroptosis and alcoholic damage in hemochromatosis [82, 83].

16.10 Therapeutic Implications and Emerging Strategies

Hemochromatosis is a complex disorder which leads to organ dysfunction and tissue damage due to iron overload and persistent inflammatory signaling. ROS and ferroptosis play a central role in this pathogenesis, and lipid peroxidation promotes cell death. Iron overload induces an imbalance in iron homeostasis and intensifies the inflammatory cascade by catalyzing the Fenton reaction, which produces ROS, exacerbates oxidative stress, and encourages ferroptotic cell death [84].

New development has shed light on recent developments potentially targeting ROS and ferroptosis to alleviate disease progression. Oxidative damage reduction and cellular integrity preservation are controlled by developed therapeutic strategies, such as: iron overload is controlled by iron chelators (e.g., deferoxamine, deferasirox), which help to mitigate the generation of ROS; oxidative stress is neutralized by antioxidants like N-acetylcysteine and vitamin E, and Lipid peroxidation is controlled by ferroptosis inhibitors (e.g., ferrostatin-1, liproxstatin-1) [85]. Additionally, iron homeostasis control, suppression of inflammation, and prevention of organ injury can be managed by hepcidin-ferroportin axis modulation, gene therapy, nanoparticle-based iron scavengers, and microbiome modulation. The promising approach to treating inflammation-mediated hemochromatosis involves focusing on ferroptosis and ROS [86].

16.10.1 ROS Inhibitors and Antioxidants

ROS inhibitors are the compounds that control the generation of ROS production by reducing them directly or indirectly in biological systems. Suppression of ROS formation includes blocking enzymatic inhibitors (NADPH oxidase), targeting pathways, and enhancing endogenous antioxidants (GSH synthesis boosting). Antioxidants are molecules that neutralize existing ROS to prevent cellular death. Antioxidants are categorized in two major types [87]: The major enzymatic antioxidants are superoxide dismutase which converts superoxide (O_2^-) into hydrogen peroxide (H_2O_2); catalase which helps in breaking down hydrogen peroxide (H_2O_2) into water (H_2O) and oxygen (O_2); and Glutathione peroxidase (GPx) which reduces hydrogen peroxide (H_2O_2) and lipid peroxides using GSH. Non-enzymatic antioxidants are molecules that are prone to oxidation; therefore, they act by absorbing oxidative stress before it can damage any cellular or tissue structures. The main examples of it are: vitamin C (ascorbic acid), which scavenges ROS and regenerates vitamin E; vitamin E (Tocopherol), which helps in protecting the cell membrane from lipid peroxidation; GSH which shows its action by detoxifying ROS, and polyphenols which reduce inflammation and modulate signaling [88].

16.10.2 *Ferroptosis Modulators*

Ferroptosis modulators are compounds that controls the inhibition of ferroptosis. It is a unique form of iron-dependent regulated cell death which is related to lipid peroxidation. These modulators have high value for their therapeutic potential in cancer, neurodegeneration, and other diseases. Several interconnected pathways, like the GPX4 pathway, iron metabolism, lipid metabolism, and the coq10-FSP1 axis, regulate ferroptosis. Ferroptosis modulators are working as either an inducer or inhibitor [89].

Some examples of ferroptosis inducers include erastin, which blocks system Xc^- and depletes intracellular GSH; RSL3, which directly inhibits GPX4; FIN56, which promotes GPX4 degradation and CoQ10 depletion; Brequinar which inhibits DHODH; and natural compounds such as Artemisinin, quercetin, and apigenin. Some instances of ferroptosis inhibitors are ferrostatin-1 and Liproxstatin-1 that protect against lipid peroxidation; vitamin E; β-carotene; Edaravone, which comes under the class of lipid peroxidase scavenger; melatonin; ENCG; and Curcumin, which show context-dependent action [90].

16.11 Novel Approaches for Hemochromatosis and Iron Chelation Therapy

Recent developments in the treatment of hemochromatosis and iron chelation therapy have introduced innovative therapeutic strategies beyond conventional methods like phlebotomy for hemochromatosis. Cardiac iron toxicity has been promisingly reduced by selenium-based compounds like ebselen, while early-stage diagnosis was performed by gene therapy. Modulation of hepcidin and ferroportin offers effective disease modification by targeted molecular therapies. Combination therapies and next-generation oral chelators show promising effectiveness in iron chelation [91] (Table 16.2).

16.12 Inflammatory Signaling Pathway Targeting for Hemochromatosis

Inflammatory signaling pathways targeted in hemochromatosis, which regulate hepcidin dysfunction driven by iron overloading. Gaining insight into these pathways is essential to discovering new therapeutic avenues in iron homeostasis control and organ damage reduction [95] (Table 16.3).

Table 16.2 Therapeutic approaches for hemochromatosis

Category	Approach	Key features	Benefits	References
Hemochromatosis therapies	Ebselen	Entry of iron into cardiac cells are inhibited by the selenium-based compounds	Demote cardiac iron toxicity, heart failure prevention	[92]
	Gene-based screening	Directed at specific genetic mutation and HFE testing	Genetic hemochromatosis inhibition, enable early-stage diagnosis	
	Targeted molecular therapies	Hepcidin and ferroportin modulation by drugs	Equipoise iron metabolism, potentially modifying targeted disease	
Iron chelation therapies	Combination chelation	Using multiple chelators (e.g., deferasirox + deferiprone, defritazole)	Reduce toxicity, specifically target organ	[93]
	Next-gen chelators	Better bioavailability, tolerability, specificity	Patient compliance improvement	
Supportive treatments	Phlebotomy	Reducing iron levels via routine blood removal	First-line treatment of hereditary hemochromatosis	[94]
	Lifestyle and dietary habits	Avoid iron containing foods and vitamin C supplements	Supports medical therapy	

Table 16.3 Therapeutic approaches for inflammation-mediated hemochromatosis

Pathways	Role in Hemochromatosis	Therapeutic benefits	References
JAK/STAT (e.g., STAT3)	Regulate hepcidin transcription of HFE function	Boost hepcidin to treat HFE-deficient patient	[96]
BMP/SMAD	Regulate hepcidin in hepatocyte	Normalize iron production and hepcidin level by targeting BMP6 or SMAD	[97]
NF-κB	Stimulate the production of inflammatory cytokine	Improve iron balance, manage pathway in acute inflammation	[98]
MAPK	Cytokine-induced, helps in the hepcidin generation	Hepcidin expression regulation, oxidative stress reduction	[63]

16.13 Conclusion

In the future, the treatment, monitoring, and diagnosis of hemochromatosis induced ferroptosis should include endophenotypes apart from HFE status, to integrate iron burden with co-morbid liver disease, alcohol usage, metabolic disease links, and status of inflammation. This will provide a better definition of ferroptosis-prone patient groups and help in creating a customized treatment regimen. There should be translatable biomarker discovery for iron handling, such as hepcidin ferritin ratio and plasma iron, lipid peroxidation products should also be given importance, and there should be quantifiable methods to identify the disease progression. Cell-free DNA or exosome markers to check ferroptosis damage need to be developed, then the future tracking of the disease can involve these techniques, merged with present-day MRI. The future of therapeutic research could revolve around evaluating proteins or small molecules that can effectively act as hepcidin analogues or agonists, higher antioxidant replenishment therapy, or membrane remodeling through dietary MUFA, etc. Ferroptosis inhibitors and lipid peroxidation chain-breaking molecules can be further explored, such as liproxatin-1. Prior to pursuing long-term results, the scientific community should evaluate time-limited, mechanism-anchored therapies that show biomarker reversal after shifting from correlative markers to causative, patient-specific maps of ferroptotic susceptibility. Suppressing pathologic ferroptosis in parenchymal organs without impairing physiologic ferroptosis that is important for tumor control is a difficult task that requires careful dose, duration, and population selection.

References

1. Adams PC, Jeffrey G, Ryan J (2023) Haemochromatosis. Lancet 401(10390):1811–1821
2. Delivoria-Papadopoulos M, McGowan JE (2004) Chapter 86 – Oxygen transport and delivery. In: Polin RA, Fox WW, Abman SH (eds) Fetal and neonatal physiology, 3rd ed [Internet]. W.B. Saunders [cited 2025 Aug 28], pp 880–889. Available from: https://www.sciencedirect.com/science/article/pii/B9780721696546500898
3. Geissler C, Singh M (2011) Iron, meat and health. Nutrients 3(3):283–316
4. Malik ZI, Ghafoor MU, Shah SHBU, Abid J, Farooq U, Ahmad AMR (2025) Unlocking iron: nutritional origins, metabolic pathways, and systemic significance. Front Nutr [Internet]. [cited 2025 Aug 28] 12. Available from: https://www.frontiersin.org/journals/nutrition/articles/10.3389/fnut.2025.1637316/full
5. Loveikyte R, Bourgonje AR, van Goor H, Dijkstra G, van der Meulen-de Jong AE (2023) The effect of iron therapy on oxidative stress and intestinal microbiota in inflammatory bowel diseases: a review on the conundrum. Redox Biol 68:102950
6. Chen HY (2019) Why the reactive oxygen species of the Fenton reaction switches from oxoiron(IV) species to hydroxyl radical in phosphate buffer solutions? A computational rationale. ACS Omega 4(9):14105–14113
7. Galaris D, Barbouti A, Pantopoulos K (2019) Iron homeostasis and oxidative stress: an intimate relationship. Biochim Biophys Acta Mol Cell Res 1866(12):118535
8. Tang D, Chen X, Kang R, Kroemer G (2021) Ferroptosis: molecular mechanisms and health implications. Cell Res 31(2):107–125
9. Kerr JFR, Wyllie AH, Currie AR (1972) Apoptosis: a basic biological phenomenon with wideranging implications in tissue kinetics. Br J Cancer 26(4):239–257
10. Dixon SJ, Lemberg KM, Lamprecht MR, Skouta R, Zaitsev EM, Gleason CE et al (2012) Ferroptosis: an iron-dependent form of nonapoptotic cell death. Cell 149(5):1060–1072
11. Yu P, Zhang X, Liu N, Tang L, Peng C, Chen X (2021) Pyroptosis: mechanisms and diseases. Signal Transduct Target Ther 6(1):128

12. Riegman M, Sagie L, Galed C, Levin T, Steinberg N, Dixon SJ et al (2020) Ferroptosis occurs through an osmotic mechanism and propagates independently of cell rupture. Nat Cell Biol 22(9):1042–1048
13. Gao M, Yi J, Zhu J, Minikes AM, Monian P, Thompson CB et al (2019) Role of mitochondria in ferroptosis. Mol Cell 73(2):354–363.e3
14. Chen X, Kang R, Kroemer G, Tang D (2021) Broadening horizons: the role of ferroptosis in cancer. Nat Rev Clin Oncol 18(5):280–296
15. Lei G, Zhuang L, Gan B (2022) Targeting ferroptosis as a vulnerability in cancer. Nat Rev Cancer 22(7):381–396
16. Zhang H, Li X (2025) Ferroptosis targeting: a novel therapeutic regimen in diabetic cardiomyopathy. Cell Signal 135:112053
17. Bock FJ, Tait SWG (2020) Mitochondria as multifaceted regulators of cell death. Nat Rev Mol Cell Biol 21(2):85–100
18. Li Q, Shi N, Cai C, Zhang M, He J, Tan Y et al (2021) The role of mitochondria in pyroptosis. Front Cell Dev Biol [Internet]. [cited 2025 Aug 28] 8. Available from: https://www.frontiersin.org/journals/cell-and-developmental-biology/articles/10.3389/fcell.2020.630771/full
19. Liu Y, Lu S, Wu L-L, Yang L, Yang L, Wang J (2023) The diversified role of mitochondria in ferroptosis in cancer. Cell Death Dis 14(8):519
20. Hirata Y, Iwasaki T, Makimura Y, Okajima S, Oh-hashi K, Takemori H (2019) Inhibition of double-stranded RNA-dependent protein kinase prevents oxytosis and ferroptosis in mouse hippocampal HT22 cells. Toxicology 418:1–10
21. Rouault TA, Tong WH (2005) Iron–sulphur cluster biogenesis and mitochondrial iron homeostasis. Nat Rev Mol Cell Biol 6(4):345–351
22. Shoshan-Barmatz V, De Pinto V, Zweckstetter M, Raviv Z, Keinan N, Arbel N (2010) VDAC, a multi-functional mitochondrial protein regulating cell life and death. Mol Asp Med 31(3):227–285
23. Parker JL, Deme JC, Kolokouris D, Kuteyi G, Biggin PC, Lea SM et al (2021) Molecular basis for redox control by the human cystine/glutamate antiporter system xc−. Nat Commun 12(1):7147
24. Hou W, Xie Y, Song X, Sun X, Lotze MT, Zeh HJ III et al (2016) Autophagy promotes ferroptosis by degradation of ferritin. Autophagy 12(8):1425–1428
25. Gao M, Monian P, Quadri N, Ramasamy R, Jiang X (2015) Glutaminolysis and transferrin regulate ferroptosis. Mol Cell 59(2):298–308
26. Sassano ML, Tyurina YY, Diokmetzidou A, Vervoort E, Tyurin VA, More S et al (2025) Endoplasmic reticulum–mitochondria contacts are prime hotspots of phospholipid peroxidation driving ferroptosis. Nat Cell Biol 27(6):902–917
27. Tang D, Kroemer G (2020) Peroxisome: the new player in ferroptosis. Signal Transduct Target Ther 5(1):273
28. Recknagel RO (1967) Carbon tetrachloride hepatotoxicity. Pharmacol Rev 19(2):145–208
29. Conrad M, Pratt DA (2019) The chemical basis of ferroptosis. Nat Chem Biol 15(12):1137–1147
30. Zheng Y, Sun J, Luo Z, Li Y, Huang Y (2024) Emerging mechanisms of lipid peroxidation in regulated cell death and its physiological implications. Cell Death Dis 15(11):859
31. Bai X, Kang J, Wei S, Wang Y, Liu Y, Yuan B et al (2025) A pH responsive nanocomposite for combination sonodynamic-immunotherapy with ferroptosis and calcium ion overload via SLC7A11/ACSL4/LPCAT3 pathway. Exploration 5(1):20240002
32. Abbaspour N, Hurrell R, Kelishadi R (2014) Review on iron and its importance for human health. J Res Med Sci 19(2):164–174
33. Zhao N, Enns CA (2012) Chapter three – Iron transport machinery of human cells: players and their interactions. In: Argüello JM, Lutsenko S (eds) Current topics in membranes [Internet]. Metal transporters, vol 69. Academic Press [cited 2025 Aug 29], pp 67–93. Available from: https://www.sciencedirect.com/science/article/pii/B9780123943903000033
34. Beard JL (2001) Iron biology in immune function, muscle metabolism and neuronal functioning. J Nutr 131(2):568S–580S

35. Roemhild K, von Maltzahn F, Weiskirchen R, Knüchel R, von Stillfried S, Lammers T (2021) Iron metabolism: pathophysiology and pharmacology. Trends Pharmacol Sci 42(8):640–656
36. Zhou ZD, Tan EK (2017) Iron regulatory protein (IRP)-iron responsive element (IRE) signaling pathway in human neurodegenerative diseases. Mol Neurodegener 12:75
37. Rochette L, Dogon G, Rigal E, Zeller M, Cottin Y, Vergely C (2023) Lipid peroxidation and iron metabolism: two corner stones in the homeostasis control of ferroptosis. Int J Mol Sci 24(1):449
38. Conway D, Henderson MA (2019) Iron metabolism. Anaesth Intensive Care Med 20(3):175–177
39. Galy B, Conrad M, Muckenthaler M (2024) Mechanisms controlling cellular and systemic iron homeostasis. Nat Rev Mol Cell Biol 25(2):133–155
40. Ru Q, Li Y, Chen L, Wu Y, Min J, Wang F (2024) Iron homeostasis and ferroptosis in human diseases: mechanisms and therapeutic prospects. Signal Transduct Target Ther 9(1):271
41. Venkataramani V (2021) Iron homeostasis and metabolism: two sides of a coin. In: Florez AF, Alborzinia H (eds) Ferroptosis: mechanism and diseases [Internet]. Springer International Publishing, Cham [cited 2025 Aug 29], pp 25–40. Available from: https://doi.org/10.1007/978-3-030-62026-4_3
42. Kumar S, Sheokand N, Mhadeshwar MA, Raje CI, Raje M (2012) Characterization of glyceraldehyde-3-phosphate dehydrogenase as a novel transferrin receptor. Int J Biochem Cell Biol 44(1):189–199
43. Nemeth E, Ganz T (2021) Hepcidin-ferroportin interaction controls systemic iron homeostasis. Int J Mol Sci 22(12):6493
44. Mohammad G, Matakidou A, Robbins PA, Lakhal-Littleton S (2021) The kidney hepcidin/ferroportin axis controls iron reabsorption and determines the magnitude of kidney and systemic iron overload. Kidney Int 100(3):559–569
45. Fisher AL, Phillips S, Wang CY, Paulo JA, Xiao X, Xu Y et al (2025) The hepcidin-ferroportin axis modulates liver endothelial cell BMP expression to influence iron homeostasis in mice. Blood 145(6):625–634
46. Liu S, Tsyplenkova S, Fillebeen C, Pantopoulos K (2025) Hypoferremic response to chronic inflammation is controlled via the hemojuvelin/hepcidin/ferroportin axis and does not involve hepcidin-independent regulation of Fpn mRNA. Am J Hematol 100(8):1323–1333
47. Gutteridge JMC, Halliwell B (2018) Mini-review: oxidative stress, redox stress or redox success? Biochem Biophys Res Commun 502(2):183–186
48. Mansoor S, Ali Wani O, Lone JK, Manhas S, Kour N, Alam P et al (2022) Reactive oxygen species in plants: from source to sink. Antioxidants 11(2):225
49. Xiao J, Guo S, Wang D, An Q (2024) Fenton-like reaction: recent advances and new trends. Chem Eur J 30(24):e202304337
50. Sies H, Belousov VV, Chandel NS, Davies MJ, Jones DP, Mann GE et al (2022) Defining roles of specific reactive oxygen species (ROS) in cell biology and physiology. Nat Rev Mol Cell Biol 23(7):499–515
51. Doelle HW (1975) 1 – Thermodynamics of biological reactions. In: Doelle HW (ed) Bacterial metabolism, 2nd edn. [Internet]. Academic Press [cited 2025 Aug 29], pp 1–37. Available from: https://www.sciencedirect.com/science/article/pii/B9780122193521500050
52. Zimmerman MC, Case AJ (2019) Redox biology in physiology and disease. Redox Biol 27:101267
53. Ray PD, Huang BW, Tsuji Y (2012) Reactive oxygen species (ROS) homeostasis and redox regulation in cellular signaling. Cell Signal 24(5):981–990
54. Dash UC, Bhol NK, Swain SK, Samal RR, Nayak PK, Raina V et al (2025) Oxidative stress and inflammation in the pathogenesis of neurological disorders: mechanisms and implications. Acta Pharm Sin B 15(1):15–34
55. Pooja G, Shweta S, Patel P (2025) Oxidative stress and free radicals in disease pathogenesis: a review. Discov Med 2(1):104

56. Jomova K, Alomar SY, Alwasel SH, Nepovimova E, Kuca K, Valko M (2024) Several lines of antioxidant defense against oxidative stress: antioxidant enzymes, nanomaterials with multiple enzyme-mimicking activities, and low-molecular-weight antioxidants. Arch Toxicol 98(5):1323–1367
57. Kowalczyk P, Sulejczak D, Kleczkowska P, Bukowska-Ośko I, Kucia M, Popiel M et al (2021) Mitochondrial oxidative stress—a causative factor and therapeutic target in many diseases. Int J Mol Sci 22(24):13384
58. Juan CA, Pérez de la Lastra JM, Plou FJ, Pérez-Lebeña E (2021) The chemistry of reactive oxygen species (ROS) revisited: outlining their role in biological macromolecules (DNA, lipids and proteins) and induced pathologies. Int J Mol Sci 22(9):4642
59. Li C, Deng C, Wang S, Dong X, Dai B, Guo W et al (2024) A novel role for the ROS-ATM-Chk2 axis mediated metabolic and cell cycle reprogramming in the M1 macrophage polarization. Redox Biol 70:103059
60. Leyane TS, Jere SW, Houreld NN (2022) Oxidative stress in ageing and chronic degenerative pathologies: molecular mechanisms involved in counteracting oxidative stress and chronic inflammation. Int J Mol Sci 23(13):7273
61. Yu M, Wang S, Lin D (2025) Mechanism and application of biomaterials targeting reactive oxygen species and macrophages in inflammation. Int J Mol Sci 26(1):245
62. Chen Y, Guo X, Zeng Y, Mo X, Hong S, He H et al (2023) Oxidative stress induces mitochondrial iron overload and ferroptotic cell death. Sci Rep 13(1):15515
63. Gordan R, Fefelova N, Gwathmey JK, Xie LH (2020) Iron overload, oxidative stress and calcium mishandling in cardiomyocytes: role of the mitochondrial permeability transition pore. Antioxidants 9(8):758
64. Xie T, Yao L, Li X (2024) Advance in iron metabolism, oxidative stress and cellular dysfunction in experimental and human kidney diseases. Antioxidants 13(6):659
65. Mo M, Pan L, Deng L, Liang M, Xia N, Liang Y (2024) Iron overload induces hepatic ferroptosis and insulin resistance by inhibiting the Jak2/stat3/slc7a11 signaling pathway. Cell Biochem Biophys 82(3):2079–2094
66. Yan F, Li K, Xing W, Dong M, Yi M, Zhang H (2022) Role of iron-related oxidative stress and mitochondrial dysfunction in cardiovascular diseases. Oxidative Med Cell Longev 2022(1):5124553
67. Wang W, Jing X, Du T, Ren J, Liu X, Chen F et al (2022) Iron overload promotes intervertebral disc degeneration via inducing oxidative stress and ferroptosis in endplate chondrocytes. Free Radic Biol Med 190:234–246
68. Mancardi D, Mezzanotte M, Arrigo E, Barinotti A, Roetto A (2021) Iron overload, oxidative stress, and ferroptosis in the failing heart and liver. Antioxidants 10(12):1864
69. Anderton H, Wicks IP, Silke J (2020) Cell death in chronic inflammation: breaking the cycle to treat rheumatic disease. Nat Rev Rheumatol 16(9):496–513
70. Nuñez MT, Chana-Cuevas P (2018) New perspectives in iron chelation therapy for the treatment of neurodegenerative diseases. Pharmaceuticals 11(4):109
71. Pietrangelo A (2010) Hereditary hemochromatosis: pathogenesis, diagnosis, and treatment. Gastroenterology 139(2):393–408.e2
72. Daniłowicz-Szymanowicz L, Świątczak M, Sikorska K, Starzyński RR, Raczak A, Lipiński P (2021) Pathogenesis, diagnosis, and clinical implications of hereditary hemochromatosis—the Cardiological point of view. Diagnostics (Basel) 11(7):1279
73. Nemeth E, Ganz T (2023) Hepcidin and iron in health and disease. Annu Rev Med 74:261–277
74. Canavesi E, Alfieri C, Pelusi S, Valenti L (2012) Hepcidin and HFE protein: iron metabolism as a target for the anemia of chronic kidney disease. World J Nephrol 1(6):166–176
75. Pietrangelo A (2011) Hepcidin in human iron disorders: therapeutic implications. J Hepatol 54(1):173–181
76. Ahmed S, Peterson SJ, Parikh MA, Frishman WH (2025) Cardiovascular manifestations of hemochromatosis: a review of pathophysiology, mechanisms, and treatment options. Cardiol Rev 33(4):359

77. Murphy CJ, Oudit GY (2010) Iron-overload cardiomyopathy: pathophysiology, diagnosis, and treatment. J Card Fail 16(11):888–900
78. Siri-Angkul N, Xie LH, Chattipakorn SC, Chattipakorn N (2018) Cellular electrophysiology of iron-overloaded cardiomyocytes. Front Physiol [Internet]. [cited 2025 Aug 29] 9. Available from: https://www.frontiersin.org/journals/physiology/articles/10.3389/fphys.2018.01615/full
79. Zhang M, Li J, Hu W (2024) The complex interplay between ferroptosis and atherosclerosis. Biomed Pharmacother 178:117183
80. Kirk EA, Heinecke JW, LeBoeuf RC (2001) Iron overload diminishes atherosclerosis in apoE-deficient mice. J Clin Invest 107(12):1545–1553
81. Hernaez R, Yeung E, Clark JM, Kowdley KV, Brancati FL, Kao WHL (2011) Hemochromatosis gene and nonalcoholic fatty liver disease: a systematic review and meta-analysis. J Hepatol 55(5):1079–1085
82. Barton JC, Barton JC, Acton RT (2023) Non-alcoholic fatty liver disease in hemochromatosis probands with iron overload and HFE p.C282Y/p.C282Y. BMC Gastroenterol 23(1):137
83. Teschke R (2024) Hemochromatosis: ferroptosis, ROS, gut microbiome, and clinical challenges with alcohol as confounding variable. Int J Mol Sci 25(5):2668
84. Wang H, An P, Xie E, Wu Q, Fang X, Gao H et al (2017) Characterization of ferroptosis in murine models of hemochromatosis. Hepatology 66(2):449–465
85. Li J, Cao F, Yin H-L, Huang Z-J, Lin Z-T, Mao N et al (2020) Ferroptosis: past, present and future. Cell Death Dis 11(2):88
86. Yan H-F, Zou T, Tuo Q-Z, Xu S, Li H, Belaidi AA et al (2021) Ferroptosis: mechanisms and links with diseases. Sig Transduct Target Ther 6(1):49
87. Patanè GT, Putaggio S, Tellone E, Barreca D, Ficarra S, Maffei C et al (2023) Ferroptosis: emerging role in diseases and potential implication of bioactive compounds. Int J Mol Sci 24(24):17279
88. Jena AB, Samal RR, Bhol NK, Duttaroy AK (2023) Cellular Red-Ox system in health and disease: the latest update. Biomed Pharmacother 162:114606
89. Młynarska E, Hajdys J, Czarnik W, Fularski P, Leszto K, Majchrowicz G et al (2024) The role of antioxidants in the therapy of cardiovascular diseases-a literature review. Nutrients 16(16):2587
90. Du Y, Guo Z (2022) Recent progress in ferroptosis: inducers and inhibitors. Cell Death Discov 8:501
91. Manolova V, Nyffenegger N, Flace A, Altermatt P, Varol A, Doucerain C et al (2019) Oral ferroportin inhibitor ameliorates ineffective erythropoiesis in a model of β-thalassemia. J Clin Invest 130(1):491–506
92. Ghazaiean M, Aliasgharian A, Karami H, Darvishi-Khezri H (2023) Ebselen: a promising therapy protecting cardiomyocytes from excess iron in iron-overloaded thalassemia patients. Open Med (Wars) 18(1):20230733
93. Entezari S, Haghi SM, Norouzkhani N, Sahebnazar B, Vosoughian F, Akbarzadeh D et al (2022) Iron chelators in treatment of iron overload. J Toxicol 2022:4911205
94. Zubair A (2014) Therapeutic phlebotomy. Clin Liver Dis (Hoboken) 4(5):102–106
95. Hübscher SG (2003) Iron overload, inflammation and fibrosis in genetic haemochromatosis. J Hepatol 38(4):521–525
96. Piperno A, Pelucchi S, Mariani R (2020) Inherited iron overload disorders. Transl Gastroenterol Hepatol 5:25
97. Camaschella C, Nai A, Silvestri L (2020) Iron metabolism and iron disorders revisited in the hepcidin era. Haematologica 105(2):260–272
98. Kowdley KV, Gochanour EM, Sundaram V, Shah RA, Handa P (2021) Hepcidin signaling in health and disease: ironing out the details. Hepatol Commun 5(5):723–735

Chapter 17
Nanomedicine-Driven Dual Delivery of ROS Scavengers and Anti-Inflammatory Agents in Metabolic Disorders

Abhishek Vijukumar, Gautam Singh, Mansha Aggarwal, Akanksha Sharma, and Navjot Kaur Sandhu

17.1 Introduction

Type 2 diabetes, obesity, and non-alcoholic fatty liver disease (NAFLD), which is now reclassified as metabolic dysfunction-associated steatotic liver disease (MASLD), are inextricably related disorders that occur because of insulin resistance and global metabolic maladaptation [1]. These disorders often coincide and represent different indicators of metabolic syndrome, which is fast gaining prominence in different parts of the world. At least 30 percent of the global population is already hit by MASLD alone, which is usually accompanied by obesity and diabetes, which are fuelled by visceral adiposity, sustained energy excesses, and inactive lifestyles [2, 3]. One key pathogenic mechanism shared by all disorders is the condition of oxidative stress, which is the result of an excess of reactive oxygen species (ROS) that overpower the innate antioxidant system. Over-production of ROS in metabolic syndrome is caused by numerous pathways, including mitochondrial abnormalities in insulin-sensitive organs, lipid peroxidation, protein oxidation, and poor reserve of antioxidants like glutathione. These oxidative stresses disrupt insulin signaling, the ability of the liver and muscle to absorb glucose and contribute to the development of insulin resistance [4]. Oxidative stress reflects the clinical importance of oxidative stress in individuals with metabolic syndrome, as there have been numerous reports of the elevation of indicators of oxidative damage in these individuals, such as lipid peroxides [5]. At the same time, low-grade inflammation is a major driver of metabolic deregulation. Obesity alters the immune cell profile in the adipose tissue by facilitating a pro-inflammatory environment through

A. Vijukumar · G. Singh · M. Aggarwal · A. Sharma
Department of Pharmacy Practice, ISF College of Pharmacy, Moga, Punjab, India

N. K. Sandhu (✉)
Department of Quality Assurance, ISF College of Pharmacy, Moga, Punjab, India

S. Ashique, N. Mishra (eds.), *Reactive Oxygen Species in Metabolic Inflammation*,
https://doi.org/10.1007/978-3-032-11752-6_17

the presence of M1 macrophages that secrete cytokines such as TNF-alpha and IL-6. These inflammatory mediators interfere with the insulin receptor signaling and stimulate the hepatic glucose production. In addition, inflammation is a general situation that extends beyond the adipose tissue to the liver, pancreas, and skeletal muscle, contributing to the β-cell dysfunction, endothelial damage, and a gradual decline in organ function [6]. Inflammation and oxidative stress are biologically related processes, working in an interrelated negative feedback loop. ROS can also stimulate redox-sensitive factors of transcription (such as NF-κB) and inflammasomes (such as NLRP3), even further enhancing the production of inflammatory cytokines. These inflammatory signals, in turn, promote the production of more ROS by initiating a self-propagating feedback loop that increases tissue damage and fibrosis [7]. This synergism is well evident in MASLD, whereby it contributes to progression from steatosis to steatohepatitis, culminating in cirrhosis. Oxidative stress, combined with inflammation, is a fundamental pathogenic pathway contributing to metabolic illness and is a major target of treatment with the goal of retarding or reversing disease progression [8]. It is essential that these intertwined mechanisms be addressed to develop effective interventions to meet the growing global disease burden of metabolic diseases.

17.2 Pathophysiological Basis

ROS are more than just by-products of metabolism, but they are fundamental players in communicating with the cells and maintaining immunological control. Within the context of metabolic diseases, such as obesity and type 2 diabetes, and MASLD, excessive ROS accumulation triggers and sustains a low-grade inflammation, creating a pathogenic feed-forward loop implicated in the development and evolution of the diseases [9]. Mitochondrial dysfunction, NADPH oxidase activation, endoplasmic reticulum stress, and xanthine oxidase activity are considered key factors that cause ROS [10]. Food excess and insulin resistance are worsened by these stresses, causing additional damage to mitochondrial integrity and an increase in ROS production in metabolic organs, including liver, adipose tissue, and skeletal muscles. Activation of nuclear factor-kappa B (NF-κB), which is a redox-sensitive transcription factor, is one of the main ways through which ROS stimulate inflammation [11]. Through the phosphorylation and degradation of IkB-alpha, ROS stimulate the release of NF-κB to move to the nucleus, where it enhances the synthesis of inflammatory cytokines such as TNF-α, IL-6, and MCP-1. These cytokines, subsequently, interfere with insulin signaling, attract immune cells to metabolic organs, and promote the production of more ROS [12, 13]. Simultaneously, mitochondrial ROS activate the NLRP3 inflammasome, a cytosolic protein complex that cleaves pro-IL-18 and pro-IL-18 to their active forms worsening, both local and systemic inflammation [14]. Such an activation of inflammasome is especially relevant within adipose tissue, where dysfunctional adipocytes and infiltrating immune cells (primarily M1 macrophages) create a pro-inflammatory local environment.

ROS are also of fundamental importance in the regulation of immune cells. They change the polarization of macrophages, promoting a switch to pro-inflammatory M1 phenotypes at the expense of the anti-inflammatory [15]. M2 phenotype leads to irreversible inflammation of adipose tissue. Besides, ROS suppress phagocytic activity of neutrophils and macrophages and antigen presentation, somatizing immunological resolution and perpetuating chronic inflammation. Besides their immunomodulatory actions, ROS are also mechanistically associated with metabolic dysregulation at many levels. ROS disrupt insulin signaling in insulin-responsive organs, eg, liver and muscle, by oxidizing and therefore inhibiting key proteins, like IRS-1 and AKT, and thereby reducing glucose uptake and glycogen synthesis [16]. This oxidative rearrangement is a trademark of insulin resistance as well as the primary issue in type 2 diabetes. Further, oxidative stress causes endothelial dysfunction by reducing nitric oxide (NO) bioavailability, raising vasoconstriction, leukocyte adhesion, and vascular inflammation, which are factors of hypertension and atherosclerosis, common comorbidities in metabolic syndrome.

From the metabolic perspective, repeated exposures to ROS alter cellular metabolic pathways. It modulates increased oxidative inactivation of glycolytic enzymes to favor a shift in metabolism toward the pentose phosphate pathway to generate NADPH that is significant in combating ROS [17]. Though adaptive initially, this alteration can adjust cellular energy balance and stimulate lipogenesis and steatosis. Chronic oxidative stress also results in the accumulation of oxidized lipids, proteins, and DNA that enhance cellular senescence, apoptosis, and fibrotic remodeling. This can be seen as the evolution in the liver, namely, the progression from a simple steatosis to steatohepatitis and ultimately to cirrhosis found in MASLD [18]. Similar mechanisms result in arterial stiffness and plaque formation in blood arteries.

Most importantly, oxidative stress and inflammation interact in a reciprocal fashion. ROS production by inducible nitric oxide synthase (iNOS) and NADPH oxidase is enhanced by inflammatory cytokines such as TNF-α and IL-6, and ROS keep promoting inflammatory transcription factors and inflammasomes [19]. This vicious circle facilitates tissue destruction across the body organs such as the liver, pancreas, adipose tissue, vasculatures, and skeletal muscles. As such, the correlation of the ROS generation and chronic inflammation is not a by-product but a leading cause in the causal path of metabolic disorders. Disrupting this feed-forward cycle, through a combined strategy of oxidative stress and inflammation targeting, is an important therapeutic alternative to slow the progression of these prevalent and debilitating diseases.

17.3 Nanomedicine in Targeted Drug Delivery

Nanomedicine shifted the paradigm of specific drug delivery, especially in addressing complex and multifactorial diseases such as metabolic disorders. The repeated inflammation, oxidative stress, and the pathophysiological systemic metabolic disturbances found in these disorders demand a variety of therapeutic solutions.

Nanocarriers have serious advantages over conventional delivery systems that render them highly appropriate to these challenges [20]. Among the most outstanding advantages is their ability to enhance the bioavailability of poorly soluble or unstable drugs. Nanocarriers have the potential to enhance solubility and extend the pharmacokinetic behavior of drugs by protecting the active substances in a nanoscale matrix against early degradation. This ensures that a higher level of the drug of therapy achieves systemic circulation and remains functioning over a prolonged period. Furthermore, surface modifications of nanocarriers can be performed by adding ligands in the form of peptides, antibodies, or smaller molecules to have a specific distribution in target tissue or cells [21]. The targeted approach not only increases efficacy by localizing the drug to the site of action, but it also reduces off-target activities and systemic toxicity.

The other characteristic feature of nanocarriers is controlled and sustained release of medication. Such systems can also be made time or stimuli-responsive, releasing their cargo on recognition of environmental cues like pH, temperature, or enzyme activity conditions, which are normally known to vary in inflamed or diseased tissues [22]. Nanocarriers help maximize the therapeutic results as they significantly reduce the frequency of the dosage by maintaining therapeutic drug concentrations over extended periods, increasing patient compliance. Also, their small size (typically 10–200 nm) ensures they have better permeability and retention (EPR) in diseased or malignant tissues to further aid in site-specific distribution. Other particle barriers, such as the gastrointestinal epithelium, hepatic filtration, and even the blood-brain barrier, which are frequently major challenges to traditional medication delivery, can also be traversed with nanocarriers [23].

A number of nanocarrier platforms are utilized as part of these objectives with various and different structural and functional attributes. Liposomes, made up of bilayers of phospholipids, are biocompatible and can entrap hydrophilic as well as hydrophobic drugs and are therefore versatile and widely used. Polymeric nanoparticles, fabricated by using biodegradable polymers, including PLGA or chitosan, have a high drug-loading capacity and tunable release profiles [24]. The highly branched architecture of dendrimers allows very precise tuning of molecule size and surface activity, and they are therefore very useful in conjugating with large numbers of therapeutic compounds or even targeting ligands. Self-assembled nanomicelles, formed through the aggregation of amphiphilic block copolymers, are especially applicable for solubilizing hydrophobic materials and release-controllable structuring. Carbon nanotubes and iron oxide nanoparticles, despite presenting distinct attributes, namely, the significant surface area and the capacity to respond to magnetic fields, respectively, must be thoroughly surface-functionalized to reduce the potential toxicity as well as enhance biocompatibility [25]. A similar type of structure is the polymer vesicle, related to the concept of the liposome but constructed with synthetic block copolymers with greater structural stability and the ability to be adapted toward encapsulating drugs simultaneously. The types of nanocarriers are mentioned in Table 17.1.

Together, the nanocarrier technologies are being actively researched toward a dual delivery usage, especially in co-delivery of ROS scavengers and

Table 17.1 Common nanocarriers used for dual delivery in metabolic disorders

Nanocarrier type	Structural features	Advantages	Therapeutic example
Liposomes	Phospholipid bilayer with aqueous core	Biocompatible, co-delivery of hydrophilic & hydrophobic drugs	Curcumin + NAC
Polymeric nanoparticles	Biodegradable polymer matrix (e.g., PLGA, chitosan)	High drug loading, tunable release	Resveratrol + Vitamin E
Core-shell nanoparticles	Distinct inner core and outer shell compartments	Spatial separation of agents, sequential release	CoQ10 (core) + Dexamethasone (shell)
Dendrimers	Tree-like branched structures	Precise size control, multifunctional surface	Aspirin + NAC
Nanomicelles	Self-assembled amphiphilic polymers	Ideal for hydrophobic drugs, controlled release	Tempol + NSAIDs
Layer-by-layer assemblies	Alternating layers of polymers or charged biomolecules	Precise control over sequential release	Sequential ROS scavenger + anti-inflammatory
Iron oxide nanoparticles	Magnetic core with modifiable surface	Magnetic targeting, imaging compatibility	ROS scavenger + Imaging probe

anti-inflammatory drugs to synergistically target their oxidative and inflammatory facets of metabolic syndromes. Nanocarrier properties such as size, charge, surface chemistry, and responsiveness to environmental stimuli can be customized to allow tight control of drug release kinetics and biodistribution profiles [26]. This level of control is essential to disorders, which affect many organs and pathways simultaneously.

17.4 ROS Scavengers Used in Nanocarriers

As a contrast therapeutic approach, the delivery of nanocarriers has evolved to enhance the use of ROS scavengers to treat and manage oxidative stress in metabolic diseases by improving stability, bioavailability, and delivery specificity of treatment compounds. These scavengers play an important role in countering a large amount of ROS that cause cellular damage, inflammation, and diseases like obesity, type 2 diabetes, and the development of non-alcoholic fatty liver disease [27]. Among the commonly used ROS scavengers, N-acetylcysteine (NAC) is unique in acting both directly as an ROS scavenger and indirectly, as a precursor to glutathione formation and hence restoring intracellular antioxidant defenses. Tempol is a nitroxide compound that acts as a superoxide dismutase (SOD) mimetic, and specifically, when superoxide anions are involved, it is useful in reducing mitochondrial and vascular oxidative stress. Another widely used chemical is Vitamin E (also known as 9-tocopherol), a lipophilic antioxidant that, when introduced in cell membranes, interrupts the chain of lipid peroxidation, protecting membrane integrity [28]. In the same regard, Coenzyme Q10 (ubiquinone), an important element of the mitochondrial electron transport chain, also possesses, significant antioxidant capabilities in that it provides protection against oxidative attacks on the mitochondria and the entire cell.

Nanozymes—nanoparticles produced with enzyme-like properties—have attracted increasing interest in recent years. Nanoparticles, including cerium oxide, copper oxide, and iron oxide, are of intrinsic catalytic activity and simulate natural antioxidant enzymes such as SOD, catalase, and peroxidase. These nanozymes provide a regenerative mechanism of action, which can facilitate reoccurring catalytic breakdown of ROS, including but not limited to superoxide, hydrogen peroxide, and hydroxyl radicals [29]. Their ability to maintain prolonged activity differentiates them from traditional antioxidants, which may frequently get consumed in the neutralization process. Moreover, natural dietary antioxidants such as polyphenols and flavonoids are promising in the management of metabolic diseases. They have, however, limited clinical application because of their low bioavailability. Nanocarrier formulations significantly increase their solubility, stability, and absorption, resulting in an improvement in their systemic antioxidant ability.

The mechanism of formation of these ROS scavengers resulting in the therapeutic effects is varied. Chemicals such as NAC, Tempol, and Vitamin E directly neutralize ROS by physically reacting with ROS to avoid damaging proteins, lipids, and DNA [30]. Nanozymes also have the capability of enzyme mimicry, which allows the decomposition of toxic species into harmless substances such as water and oxygen and helps to reduce the oxidative load without ingestion. Moreover, NAC and the like stimulate the synthesis of endogenous antioxidants, including glutathione, thereby re-establishing a cellular redox balance within cells. In addition to neutralization, a number of ROS scavengers inhibit inflammatory signaling pathways mediated by ROS. They alleviate metabolic diseases by preventing oxidative stress-induced activation of NF-κB and inflammasomes such as NLRP3, which is relevant to chronic inflammation [31]. Such a combination of antioxidants and anti-inflammatories is what makes ROS scavengers highly therapeutic.

In addition, lipid-soluble antioxidants, including vitamin E, specifically provide protection of the cellular membranes against oxidative damage, whereas nano-formulated chemicals are characterized by better targeting and pharmacokinetics so that therapeutic drugs accumulate in diseased tissues. The flexibility of nanocarrier technologies also means that it is possible to deliver ROS scavengers alongside anti-inflammatory or metabolic medicines, which target the interlinking pathophysiological processes that define metabolic disorders [32]. In this sense, ROS scavenging-oriented nanomedicine strategies lead the way toward developing better, more specific, and comprehensive treatments of oxidative-stress-related metabolic disease.

17.5 Anti-Inflammatory Agents in Nanoformulations

The field of anti-inflammatory therapy has been revolutionized by delivering medicines using nanocarriers, and in the treatment of metabolism and inflammatory diseases, this method has improved pharmacokinetics, bioavailability, and target delivery of drugs significantly. Nanocarrier technologies have been used to develop a wide range of anti-inflammatory drugs that are enhanced in terms of therapeutic effect and have a reduced systemic deleterious effect [33]. Among them, one of the exceptional cases is curcumin, the natural polyphenol that possesses considerable anti-inflammatory and antioxidant properties. Despite its favorable pharmacological properties, the limited solubility and rapid metabolism of curcumin limit its bioavailability. These drawbacks have been overcome through nanotechnology-based methodologies like the lipid-based nanoparticle, polymeric micelle, and the solid lipid nanoparticle that offer greatly increased solubility, stability, and tissue-specific delivery. Curcumin achieves its anti-inflammatory effects to a large extent through the activation of peroxisome proliferator-activated receptor gamma (PPAR-gamma), which reduces the secretion of pro-inflammatory cytokines like TNF-α [34]. It also inhibits the NF-κB signal pathway and decreases the generation of the inflammatory mediators COX-2 and iNOS.

Dexamethasone is another anti-inflammatory drug with wide application in nanoformulations; it is a potent synthetic corticosteroid. Nanocarriers administered with dexamethasone have better targeting and reduced systemic toxicity [35]. It binds to glucocorticoids, relocated in the nucleus, and regulates gene transcription positively, upregulating anti-inflammatory proteins and downregulating pro-inflammatory cytokines and chemokines. Dexamethasone also blocks NF-κB pathway and eicosanoid production by inhibiting phospholipase A2, leading to fewer bursts of immune cells and cytokines. Likewise, resveratrol, a naturally occuring polyphenol molecule, possesses anti-inflammatory and antioxidant effects yet is limited by infeasible water solubility and expedient elimination [36]. Nano-encapsulation significantly enhances stability and bioavailability, therefore leading to better therapeutic outcomes. Nanoformulation of non-steroidal anti-inflammatory medication (NSAIDs) such as ketoprofen, aspirin, and diclofenac has also been done to boost the distribution. Carbon nanotubes and polymeric nanoparticles nanocarriers help regulate drug release, enhance drug deposition at the site of injury, and reduce gastrointestinal and renal adverse outcomes that are common with systemic NSAIDs administration [37]. The major mechanism of action of NSAIDs is the inhibition of cyclooxygenase enzymes (COX-1 and COX-2) that cause a drop in the production of prostaglandins and inflammation. Nanoformulations facilitate controlled and selective distribution of medications, increasing treatment efficacy and curtailing adverse effects.

New advancements in the nanomedicine domain have managed to develop polymeric nanoparticles that possess inherent anti-inflammatory activity, like glycine-acryloyl chloride-based nanoparticles (GlyNPs) [38]. These novel nanoparticles have the ability to regulate inflammation processes without necessitating conventional drug-based medicines. Through studies, GlyNPs have been proven to interact with multiple inflammatory pathways, such as IL-6, IL-1β, TNF-alpha, TLR-4, STAT-1, MAPK-8, MAPK-14, NF-kappa-B, and the COX-2, with an adequate reduction of cytokines synthesized and the activation of immune cells both in vitro and in vivo. In some situations, they have been shown to be as effective or even more effective than conventional drugs such as aspirin. Anti-inflammatory medicaments incorporated into nanocarrier systems will be a groundbreaking paradigm of drug delivery [33]. These formulations not only surpass the pharmacological boundaries of conventional drugs but also allow the engineering of molecular pathways involved in inflammation to control it. Such a specific and sustained delivery system has a significant hope of treating chronic inflammatory diseases, especially those related to metabolic disturbances.

17.6 Dual Delivery Strategies

Dual delivery systems using nanocarriers have emerged as an effective method to co-deliver antioxidants (ROS scavengers) and anti-inflammatory agents to counter the intertwined pathophysiology of oxidative stress and inflammation in metabolic and chronic diseases [39]. These new delivery systems are created to maximize the pharmacokinetics, target efficacy, and therapeutic synergy of co-introduced medications, as well as to decrease harmful effects. Various nanocarrier designs have been proposed to allow this dual delivery, with each one offering certain advantages in drug entrapment, stability, and release properties, as shown in Fig. 17.1.

The commonly used solution is the core-shell nanocarrier, having a segmented structure that allows individual loadings of the various therapeutic chemicals.

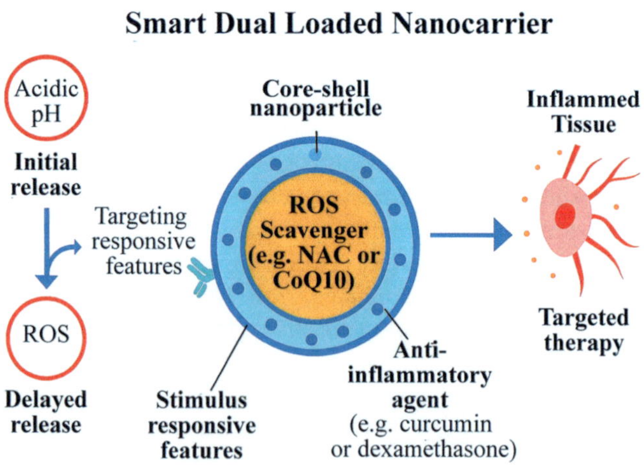

Fig. 17.1 Schematic diagram of dual delivery nanocarrier design

Hydrophilic drugs in these systems are typically adsorbed or conjugated to the shell, with hydrophobic drugs loaded into the core [40]. This spatial distance allows regulated release kinetics and protects the sensitive chemicals during the process of systemic circulation. The modified outermost layer of the core-shell nanocarriers can also be functionalized to increase targeting of inflamed or oxidatively damaged tissue and enhance cellular absorption.

The second flexible system to deliver dual drugs is through co-loaded liposomes. Such bilayer vesicles of lipid may have hydrophilic drugs in the aqueous interior and hydrophobic drugs in the lipid bilayer and hence are useful in the delivery of mixtures of compounds with some degree of solubility disparities [41]. In addition, they can design liposomes to target passively or actively and modify them for stimuli-responsive action. The capability to have a synchronized or differentiated release of medication through co-encapsulation within liposomes will allow different therapeutic applications, especially when the treatment needs a combination of ROS scavenger and anti-inflammatory drugs to be released in coordination with one another.

The layer-by-layer (LbL) assemblies enable significant release and delivery of medicinal substances. LbL nanocarriers can sequentially encapsulate the medicines within the alternating layers of polymers or other bioentities that have oppositely pertinent charges [42]. This design will permit precision regarding the order of drug release, with exterior layers disposing of their payload earlier than the interior-based ones. LbL carriers are especially useful in the development of sequential release systems that can take advantage of a phased treatment approach.

Nanoparticles and polymeric micelles are commonly used for dual delivery purposes as well. They may trap hydrophobic drugs inside the main core and subsequently conjugate the hydrophilic molecules to the surface or the polymeric film [43]. Their variable structure provides accurate control of degradation and release profiles of drugs. The systems based on polymeric materials can be developed in such a way that they react to specific environmental triggers (e.g., pH, redox potential, or enzyme activity), thereby allowing a controlled delivery in the diseased microenvironment. Nanocarriers can be designed to be dual-delivery systems, which can be consciously designed in such a way that they release their drugs either simultaneously or sequentially [44]. With simultaneous release systems, both therapeutic drugs are released at the same time. This approach is often used when the desired effect is to produce a synergistic effect within a specific time period, such as reducing oxidative stress and managing inflammation at the same time. These types of systems can be straightforward to design and are illustrated by co-loaded liposomes and polymeric nanoparticles, which enable the concurrent release of entrapped chemicals.

Sequential release systems, on the other hand, provide an additional layer of active medical control, as one medicine is given prior to the other, often triggered by a specific condition of the body [45]. Such stimuli can be pH shifts, redox potentials, enzyme action, or differences in temperature. Ordered release is achieved through building the nanocarrier with individual compartments or appending chemical connectors with different degradation properties. A different method uses stimuli-reactive substances, which specifically degrade or swell upon pathological conditions to offload medicines in an intermittent manner.

The sequential release technique features one drug as being released quickly, and the second in a dissolved prodrug form being activated afterward by nearby enzymes or redox states [46]. The strategy is especially useful when there are situations where pre-conditioning the target site—this would include reducing oxidative stress—would enhance the efficacy of the subsequent anti-inflammatory treatment. As an example, pre-release of a ROS scavenger can alleviate the oxidative environment, thereby reducing the functionality of anti-inflammatory drugs and facilitating a better cellular response [32].

Researchers have also shown that the sequencing and timing of release of medication in combination therapies have a significant impact on overall therapeutic efficacy. Sequential release strategies often present better results than simultaneous release, since they permit a more rational sequence of care—first diminishing oxidative harm to reduce tissue susceptibility, and afterward the elimination of inflammation to restore stability [47]. Additionally, sequential systems reduce drug-drug interactions within the target site and enable reduced doses of each therapy, thus reducing potential toxicity.

17.7 Challenges and Limitations

Clinical translation of dual-loaded nanocarriers, which aim at co-delivery of ROS scavengers and anti-inflammatory drugs, is faced with many significant challenges, primarily associated with their stability, scalability, and regulatory licensing [48]. Among the greatest challenges is the stability of nanocarriers during their lifetime, namely, formulation, storage, and in vivo administration. The nanocarriers should maintain structural integrity, avoid aggregation, and ensure retention of their therapeutic payloads up to the site of action. Nevertheless, the biggest challenge with most systems (including liposomes and other vesicular carriers) is that they are physicochemically unstable, releasing drugs prematurely or destroying them. Specific properties and susceptibilities, such as liposomes, show oxidative degradation of phospholipid components, bilayer fusion, low zeta potential, and Ostwald ripening, all promoting aggregation and loss of functionality [49]. Degradation of lipids via hydrolysis may also destabilize membrane integrity and cause leakage of encapsulated drugs, upon conversion to lysophosphatidylcholine. These stability issues not only degrade the therapeutic performance but also affect the shelf life, biocompatibility, and safety status of the formulation. To circumvent these challenges, the formulation alternatives, including modification of surfaces, addition of stabilizing reagents, and optimization of the synthesis, have been investigated, but reliable and generally applicable solutions still remain limited.

Another major challenge in upscaling these technologies out of laboratories into an industrial setting is associated with scalability. It is difficult to achieve homogeneous particle size, drug encapsulation efficiency, and shape when produced in large quantities [50]. The particularities of co-encapsulating two different pharmacologically active substances that have different solubility, stability, and release profiles render the manufacturing process immensely more complex. Approaches such as microfluidics and regulated mixing methods have demonstrated the potential to be scalable and repeatable in nanoparticle manufacture, but ensuring robustness and compliance with Good Manufacturing Practice (GMP), although at a higher cost, is a challenging task. Also, targeted ligands and surface alterations, which increase specificity and therapeutic effect, introduce degrees of complexity that hamper large-scale manufacturing and assaying of quality [51].

Dual-loaded nanocarriers pose additional challenges from a regulatory and translational perspective. Regulatory agencies like the FDA and EMA require comprehensive data about safety, toxicology, pharmacokinetics, and biodistribution of all ingredients used in the formulation. Excipients should possess properly defined safety profiles, and new ones are investigated thoroughly and over an extended period. The complexity of dual-drug systems increases the regulatory burden, as the safety and efficacy of both drugs must be demonstrated individually and in interaction [52]. Quality assurance is equally challenging, because dual-loaded systems need accurate control of drug ratios, encapsulation efficiency, and synchronized or staged release profiles, which need to be reproducible between production runs. Despite promising preclinical data, a relatively small number of nanomedicines have entered the clinic, due to high manufacturing costs, difficulty scaling production, the challenge of demonstrating unambiguous clinical utility compared to conventional therapies, and safety and immune concerns of possible long-term effects. Moreover, novel regulatory paths exclusive to nanomedicine are constantly being developed, necessitating timely and ongoing dialogue among scientists, manufacturers, and regulators to facilitate successful clinical translation [53]. Overall, the dual-loaded nanocarriers have great potential, but solving these stability, scalability, and regulatory issues will be important to realize clinical application.

17.8 Future Perspectives

The development of smart nanocarriers, highly sophisticated delivery vehicles designed to release medication in response to selected physiological or environmental stimuli, will reshape the future of nanomedicine. These stimuli-responsive nanocarriers provide one possible method of maximizing therapeutic delivery without the use of systemic toxicity and off-target effects, including the treatment of metabolic disorders [54]. Nanocarriers can respond to internal (endogenous) stimuli that are typical signatures of sick or inflamed tissues. As an illustration, pH-sensitive systems could utilize the low pH environment of inflamed regions to induce a release of drugs in that region, and enzyme systems can deliver drugs in response to elevated enzyme concentration that is characteristic of certain clinical conditions. Further control of the spatiotemporal characteristics of medication release is achieved through external (exogenous) stimuli, such as light, magnetic fields, temperature, and ultrasound, which allow a doctor to initiate medication release in spatiotemporally defined locations and time frames.

Mechanistically, the response of the smart nanocarriers to their intended stimuli is frequently characterized by chemical- or structural-transformation mechanisms, translating into a desired delivery of loaded pharmaceuticals [55]. These techniques, in addition to being highly versatile, can also be integrated with imaging agents to achieve real-time monitoring of drug distribution and release, and facilitate image-guided therapy. The major characteristics of these types of systems are higher therapeutic windows, minimizing the premature leak of the drug, better tissue accumulation in diseased tissues, and low systemic distribution. Additionally, they can surmount multidrug resistance and dynamically adjust to changing disease microenvironments, providing a sound basis in the management of complex, as well as chronically occurring diseases [56].

At the same time, the field is developing along the pathways of personalized nanomedicine, where nanocarriers are designed and tailored to the biological and clinical profile of the patient. Through patient-specific information, including genetics, proteomics, and metabolic deviations, scientists can tune chosen ligands, responsive elements, and drug cargo to be fitter and more successful in their treatment. Such customized systems can potentially provide multi-responsive features, thus being able to alter the kinetics of drug release as disease conditions/therapy windows change [57]. Moreover, the use of nanocarriers coupled to precision diagnostics, such as biosensing and imaging, can give real-time feedback on drug delivery, biodistribution, and therapeutic response, so that therapeutic regimes can be perfected over the duration of therapy. The combination of intelligent nanotechnology with personalized treatment models is opening up a truly vast promise to move toward precision medicine in the development of safe, efficacious, and adaptable therapeutics in metabolic diseases and beyond [58].

17.9 Conclusion

The use of nanotechnology in the formulation of therapy is one that promises change in managing inflammatory and metabolic diseases. Through the use of dual-loaded nanocarriers that cotransport ROS scavengers and anti-inflammatory agents, synergistic effects, enhanced specificity, and systemic toxicity can all be accomplished. Development of stimuli-responsive nanocarriers and patient-specific profile-based nanomedicine designs further boost the capabilities of nanomedicine, achieving target-specific, adaptable, and selectable drug delivery. Nonetheless, stability, production scale-up, regulatory clearance, and clinical translation have serious challenges. It will be necessary to solve these challenges via multidisciplinary cooperation, innovative smart carrier design, and reinforcing adaptation to changing regulatory environments. The intersection of smart delivery systems and personalized medicine is likely to lead to very effective, patient-centric treatments that will reshape the complexity of chronic diseases as we focus on their treatment.

References

1. Chan W-K, Chuah K-H, Rajaram RB, Lim L-L, Ratnasingam J, Vethakkan SR (2023) Metabolic dysfunction-associated steatotic liver disease (MASLD): a state-of-the-art review. J Obes Metab Syndr 32(3):197
2. Verma MK, Tripathi M, Singh BK (2024) Dietary determinants of metabolic syndrome: focus on the obesity and metabolic dysfunction-associated steatotic liver disease (MASLD). In: Metabolic syndrome-lifestyle and biological risk factors. IntechOpen
3. Ali H et al (2024) Non-pharmacological approach to diet and exercise in metabolic-associated fatty liver disease: bridging the gap between research and clinical practice. J Pers Med 14(1):61
4. Hajam YA et al (2022) Oxidative stress in human pathology and aging: molecular mechanisms and perspectives. Cells 11(3):552
5. Masenga SK, Kabwe LS, Chakulya M, Kirabo A (2023) Mechanisms of oxidative stress in metabolic syndrome. Int J Mol Sci 24(9):7898
6. Maurizi G, Della Guardia L, Maurizi A, Poloni A (2018) Adipocytes properties and crosstalk with immune system in obesity-related inflammation. J Cell Physiol 233(1):88–97
7. Chatterjee S (2016) Oxidative stress, inflammation, and disease. In: Oxidative stress and biomaterials. Elsevier, pp 35–58
8. Teodoro JS, Nunes S, Rolo AP, Reis F, Palmeira CM (2019) Therapeutic options targeting oxidative stress, mitochondrial dysfunction and inflammation to hinder the progression of vascular complications of diabetes. Front Physiol 9:1857
9. Mugaiga MI (2025) Adipose tissue inflammation in metabolic syndrome: molecular pathways and targeted interventions. INOSR APP SCI 13(2):122–127
10. Shin MH, Moon YJ, Seo J-E, Lee Y, Kim KH, Chung JH (2008) Reactive oxygen species produced by NADPH oxidase, xanthine oxidase, and mitochondrial electron transport system mediate heat shock-induced MMP-1 and MMP-9 expression. Free Radic Biol Med 44(4):635–645
11. Jabaut J, Ckless K (2012) Inflammation, immunity and redox signaling. In: Inflammation, chronic diseases and cancer–cell and molecular biology, immunology and clinical bases. InTech
12. Siebenlist U (2013) NF-KB: a mediator of pathogen and stress. In: Inducible gene expression, volume 1: environmental stresses and nutrients. Birkhäuser Boston Inc, p 93
13. Ahmed HS (2019) Upregulation of ROS detoxification genes by triterpenoid CDDO-Im in macrophages: protection against lipopolysaccharide-induced inflammatory injury. The University of North Carolina at Greensboro
14. Gupta S, Cassel SL, Sutterwala FS, Dagvadorj J (2025) Regulation of the NLRP3 inflammasome by autophagy and mitophagy. Immunol Rev 329(1):e13410
15. Pérez S, Rius-Pérez S (2022) Macrophage polarization and reprogramming in acute inflammation: a redox perspective. Antioxidants 11(7):1394
16. Khalid M, Alkaabi J, Khan MAB, Adem A (2021) Insulin signal transduction perturbations in insulin resistance. Int J Mol Sci 22(16):8590
17. Fuentes-Lemus E, Reyes JS, Figueroa JD, Davies MJ, López-Alarcón C (2023) The enzymes of the oxidative phase of the pentose phosphate pathway as targets of reactive species: consequences for NADPH production. Biochem Soc Trans 51(6):2173–2187
18. Ma Y, Wang J, Xiao W, Fan X (2024) A review of MASLD-related hepatocellular carcinoma: progress in pathogenesis, early detection, and therapeutic interventions. Front Med 11:1410668
19. Liu F-C, Lee H-C, Liao C-C, Chou A-H, Yu H-P (2023) Role of NADPH oxidase-derived ROS-mediated IL-6/STAT3 and MAPK/NF-κB signaling pathways in protective effect of corilagin against acetaminophen-induced liver injury in mice. Biology (Basel) 12(2):334
20. Ezike TC et al (2023) Advances in drug delivery systems, challenges and future directions. Heliyon 9(6):e17488
21. Kim CH, Lee SG, Kang MJ, Lee S, Choi YW (2017) Surface modification of lipid-based nanocarriers for cancer cell-specific drug targeting. J Pharm Investig 47(3):203–227

22. Karimi M et al (2016) Smart micro/nanoparticles in stimulus-responsive drug/gene delivery systems. Chem Soc Rev 45(5):1457–1501
23. Wu D, Chen Q, Chen X, Han F, Chen Z, Wang Y (2023) The blood–brain barrier: structure, regulation and drug delivery. Signal Transduct Target Ther 8(1):217
24. Lu B, Lv X, Le Y (2019) Chitosan-modified PLGA nanoparticles for control-released drug delivery. Polymers (Basel) 11(2):304
25. Salehirozveh M, Dehghani P, Mijakovic I (2024) Synthesis, functionalization, and biomedical applications of iron oxide nanoparticles (IONPs). J Funct Biomater 15(11):340
26. De R, Mahata MK, Kim K (2022) Structure-based varieties of polymeric nanocarriers and influences of their physicochemical properties on drug delivery profiles. Adv Sci 9(10):2105373
27. Zhang Z, Yang H, Han F, Guo P (2025) Reactive oxygen species as key molecules in the pathogenesis of alcoholic fatty liver disease and nonalcoholic fatty liver disease: future perspectives. Curr Issues Mol Biol 47(6):464
28. Iaria DL, Bitonti MB, Iorio G (2013) Structural and functional characterization of genes related to the quality of plant products in olive (Olea europaea L.). PhD diss
29. Singh N, Sherin GR, Mugesh G (2023) Antioxidant and prooxidant nanozymes: from cellular redox regulation to next-generation therapeutics. Angew Chem 135(33):e202301232
30. He L, He T, Farrar S, Ji L, Liu T, Ma X (2017) Antioxidants maintain cellular redox homeostasis by elimination of reactive oxygen species. Cell Physiol Biochem 44(2):532–553
31. Dwivedi DK, Jena GB (2020) NLRP3 inhibitor glibenclamide attenuates high-fat diet and streptozotocin-induced non-alcoholic fatty liver disease in rat: studies on oxidative stress, inflammation, DNA damage and insulin signalling pathway. Naunyn Schmiedeberg's Arch Pharmacol 393(4):705–716
32. Liu J, Han X, Zhang T, Tian K, Li Z, Luo F (2023) Reactive oxygen species (ROS) scavenging biomaterials for anti-inflammatory diseases: from mechanism to therapy. J Hematol Oncol 16(1):116
33. Placha D, Jampilek J (2021) Chronic inflammatory diseases, anti-inflammatory agents and their delivery nanosystems. Pharmaceutics 13(1):64
34. Li Q, Sun J, Mohammadtursun N, Wu J, Dong J, Li L (2019) Curcumin inhibits cigarette smoke-induced inflammation via modulating the PPARγ-NF-κB signaling pathway. Food Funct 10(12):7983–7994
35. Madamsetty VS et al (2022) Dexamethasone: insights into pharmacological aspects, therapeutic mechanisms, and delivery systems. ACS Biomater Sci Eng 8(5):1763–1790
36. Chen X et al (2025) Physiochemotherapy alginate-based hydrogel inhibiting lipogenesis through antioxidant and ant-inflammatory effects for obesity treatment. Int J Biol Macromol 304:140776
37. Zare H et al (2021) Carbon nanotubes: smart drug/gene delivery carriers. Int J Nanomedicine 16:1681–1706
38. Pareek D et al (2025) Mesoporous polymeric nanoparticles for effective treatment of inflammatory diseases: an in vivo study. J Mater Chem B 13(9):3094–3113
39. Nuñez-Selles AJ, Nuñez-Musa RA, Guillen-Marmolejos RA (2025) Linking oxidative stress biomarkers to disease progression and antioxidant therapy in hypertension and diabetes mellitus. Front Mol Biosci 12:1611842
40. Yang G et al (2019) Bioinspired core–shell nanoparticles for hydrophobic drug delivery. Angew Chem Int Ed 58(40):14357–14364
41. Al-Jipouri A, Almurisi SH, Al-Japairai K, Bakar LM, Doolaanea AA (2023) Liposomes or extracellular vesicles: a comprehensive comparison of both lipid bilayer vesicles for pulmonary drug delivery. Polymers (Basel) 15(2):318
42. Díez-Pascual AM, Rahdar A (2022) LbL Nano-assemblies: a versatile tool for biomedical and healthcare applications. Nanomaterials 12(6):949
43. Vrignaud S, Benoit J-P, Saulnier P (2011) Strategies for the nanoencapsulation of hydrophilic molecules in polymer-based nanoparticles. Biomaterials 32(33):8593–8604

44. Sun J, Luo C, Wang Y, He Z (2013) The holistic 3M modality of drug delivery nanosystems for cancer therapy. Nanoscale 5(3):845–859
45. Langer RS, Wise DL (2019) Medical applications of controlled release. CRC Press LLC, Boca Raton
46. Wong PT, Choi SK (2015) Mechanisms of drug release in nanotherapeutic delivery systems. Chem Rev 115(9):3388–3432
47. M. E. Coughlin, Transformative nursing in the NICU: trauma-informed, age-appropriate care. Springer Publishing Company, 2021
48. Rezvani M (2024) Oxidative stress-induced gastrointestinal diseases: biology and nanomedicines—a review. BioChem 4(3):189–216
49. Song F, Yang G, Wang Y, Tian S (2022) Effect of phospholipids on membrane characteristics and storage stability of liposomes. Innov Food Sci Emerg Technol 81:103155
50. Paliwal R, Babu RJ, Palakurthi S (2014) Nanomedicine scale-up technologies: feasibilities and challenges. AAPS PharmSciTech 15(6):1527–1534
51. Prajapati RN et al (2024) Recent advances in pharmaceutical design: unleashing the potential of novel therapeutics. Curr Pharm Biotechnol 25(16):2060–2077
52. Yandrapalli N (2024) Complex emulsions as an innovative pharmaceutical dosage form in addressing the issues of multi-drug therapy and polypharmacy challenges. Pharmaceutics 16(6):707
53. Mangla B, Kumar P, Javed S, Pathan T, Ahsan W, Aggarwal G (2025) Regulating nanomedicines: challenges, opportunities, and the path forward. Nanomedicine 20:1–17
54. Parvin N, Joo SW, Mandal TK (2025) Biodegradable and stimuli-responsive nanomaterials for targeted drug delivery in autoimmune diseases. J Funct Biomater 16(1):24
55. Nomikou N, Masood H, Patel S (2024) Stimulus-responsive nanoparticles for drug delivery. In: Fundamentals of pharmaceutical nanoscience. Springer, pp 389–422
56. Garg P, Malhotra J, Kulkarni P, Horne D, Salgia R, Singhal SS (2024) Emerging therapeutic strategies to overcome drug resistance in cancer cells. Cancers (Basel) 16(13):2478
57. Li W et al (2022) Recent advances in multiresponsive flexible sensors towards E-skin: a delicate design for versatile sensing. Small 18(7):2103734
58. Sengar A (2024) Precision in practice: nanotechnology and targeted therapies for personalized care. Int J Adv Nano Comput Anal 3:56–67

Chapter 18
Role of Iron in Metabolic Disorders via ROS Influence on Inflammation

Priya Chaudhary, Ayush Madan, Dipali Saxena, Komal Kriti, Satnam Singh, Manish R. Bhise, Mohini Mondal, Aganta Chakraborty, and Biplab Debnath

18.1 Introduction

Iron (Ferrum, Fe) is a metal that performs a vital function in the metabolism of the cellular system and human health. These roles mostly include the production of heme for molecules such as hemoglobin for transporting oxygen, redox catalytic processes requiring cytochromes, and iron-sulfur complexes for respiratory process

P. Chaudhary
Centre for Herbal Pharmacology and Environmental Sustainability, Chettinad Hospital and Research Institute, Chettinad Academy of Research and Education, Kelambakkam, Tamil Nadu, India
e-mail: priyachaudhary@care.edu.in

A. Madan
Department of Biotechnology, School of Research and Technology, People's University, Bhopal, Madhya Pradesh, India

D. Saxena
Department of Food and Nutrition, Shri Vaishanv Vidyapeeth Vishwavidyalaya, Indore, Madhya Pradesh, India

K. Kriti
Department of Pharmaceutical Science, Dhanbad College of Pharmacy and Research Institute, Dhanbad, Jharkhand, India

S. Singh
Department of Pharmacology, Adesh Institute of Medical Sciences and Research, Adesh University, Bathinda, India

M. R. Bhise
Department of Pharmaceutics, SGSPS, Institute of Pharmacy, Akola (MS), Affiliated to Sant Gadge Baba Amravati University, Amravati, India

M. Mondal · A. Chakraborty · B. Debnath (✉)
Department of Pharmaceutical Technology, Bharat Technology, Uluberia, West Bengal, India

S. Ashique, N. Mishra (eds.), *Reactive Oxygen Species in Metabolic Inflammation*,
https://doi.org/10.1007/978-3-032-11752-6_18

by the mitochondria. Both iron shortage and excessive levels of iron may have negative effects on wellness [1]. Due to its critical role, the metabolism of iron is closely monitored to meet the requirements for hemoglobin production, while avoiding the harmful effects of iron. Important characteristics of homeostasis of iron include different systems for systemic and cell-level control of iron intake, transportation, use, and storage, as well as the absence of a unique disposal process. A meticulously maintained equilibrium of iron is a physical requirement since unconstrained free iron contributes significantly to harmful effects as a cause of oxidative damage via the Fenton response, as well as its function in ferroptosis, an iron-based form of programmed cell fatality. Biochemical indications of elevated iron levels, as shown by high levels of serum ferritin, are a prevalent observation in roughly one-third of MASLD individuals [2]. Therefore, iron-induced damage to organs may play an important part in MASLD advancement. The following chapter will provide a clinical overview of iron's impact on oxidative stress production and ferroptosis in metabolic disorders.

18.2 Iron Metabolism and Homeostasis

18.2.1 Iron Absorption, Transport, and Storage

Duodenal absorption and discharge of iron: Iron is obtained by ingestion from food in both non-heme and heme varieties. Inorganic or non-heme iron, obtained initially from plant sources, is generally found in the ferric (Fe^{3+}) state. In the setting of stomach acid, a ferric reductase at the apical area of the duodenal surface, known as the duodenum cytochrome B (DcytB), converts Fe^{3+} to Fe^{2+} for intestinal absorption. Fe^{2+} is transferred to the cytosol via a metal divalent transporter 1 (DMT). The sole iron outflow protein ferroportin (FPN) exports Fe^{2+} to the basolateral border for subsequent systemic usage (mainly erythropoiesis) [3]. The additional dietary iron origin, heme iron, is primarily present in animal-based foods (meat and shellfish) in hemoprotein types such as myoglobin or hemoglobin. Unlike the digestion of non-heme iron, the integration of heme iron is not completely understandable. Both endocytosis via receptors and transport through membranes have been proposed for absorption into cells, but the precise processes of heme degradation and transportation remain unknown [4].

Extracellular transportation: Following FPN discharge, Fe^{2+} gets converted to Fe^{3+} by the ceruloplasmin, or ferroxidases and is deposited onto transferrin. Tf is the main metal carrier in outside of cells. Tf binds up to two Fe^{3+} atoms in the bloodstream, allowing them to be transported to the tissues. In normal settings, the concentration of transferrin with Fe^{3+} is approximately one-third of its overall capacity, providing a significant buffer. Outside the cell, Tf-bound iron constitutes about 0.1% (3 mg) of the entire iron content of the body (3–5 g). The percentage of dietary iron consumed provides about 1–2 mg each day, making up for non-specific everyday loss due to bleeding or cell elimination; thus, the iron necessary for erythropoiesis (the need of 25–30 mg/d) via flowing Tf is mainly caused by iron reused from elderly red blood cells through heme breakdown in macrophages [5].

Cell iron absorption, intracellular usage, and preservation: At the intended region, iron-comprised Tf interacts with the ubiquitous transferrin receptor 1 (TfR1). TfR1 activity is elevated in the erythrocyte progenitor region of bone marrow because of the constant requirement for iron during erythropoiesis. The variation Tf receptor 2 (TfR2) is located on hepatic and erythroid cells of precursors, where it plays an essential part in iron control and may act as a low-physiologically important alternative accumulation of iron pathway. Transferrin (Tf) binds to its receptor transferrin receptor 1 (TfR1) on the cell surface. This Tf–TfR1 complex is then internalized through clathrin-mediated endocytosis, allowing the cell to take up iron. Metalloreductase STEAP3 decreases Fe^{3+} within endosomes to Fe^{2+}. DMT1 transports decreased Fe^{2+} from the endosome to the cytosol, where it can be used or maintained, while TfR1 returns to the outer layer of the cell [6]. With the expulsion of Fe^{2+} from the endosomes into the cytosol, iron reaches what is known as the unstable iron reserve and is employed immediately for intracellular processes such as mitochondrial production of proteins via integration into an iron-sulfate or heme group, or it is retained inside the cell's membrane as ferritin. Ferritin is a molecule combination made up of 24 components that are both light and heavy chains. The heavy chains contain ferroxidase action, which oxidizes Fe^{2+} to beneficial Fe^{3+} types for stability [7].

18.2.2 Role of Key Regulators (Hepcidin, Ferroportin, Transferrin, Ferritin)

The average person has a few grams of iron. A great deal is used in erythrocytes for binding and carrying oxygen across the living system. Macrophages in the liver, bone marrow, and spleen recycle iron by absorbing and digesting dying erythrocytes, which give iron for mechanisms that include erythropoiesis [8]. The rest of the iron is kept in hepatocytes, which act as a regulating component to maintain circulatory iron levels. An alternate regulation measure for controlling body iron is to modify its digestion by enterocytes in the duodenum. In the case of enterocytes, iron is either retained to reduce the quantity of iron that circulates in the form of ferritin or transferred to the apical site and discharged into the circulatory system by Fpn, the sole recognized iron transporter at the level of the cell. Following the secretion, the iron is absorbed onto transferrin and securely disseminated through the body. Transferrin-bound iron is identified by the transferrin receptor (TfR), a membrane-bound protein that is extensively produced and frequently increased in cells with an excessive iron requirement, for example, erythroblasts [9].

Hepcidin, a 25-amino-acid-long peptide, has the primary role in regulating peripheral iron equilibrium. Hepcidin attaches to Fpn, causing it to be internalized and decomposed, decreasing the amount of iron in the circulatory system. A variety of substances, notably, bone morphogenetic protein (BMP), hemojuvelin (HJV), and human hemochromatosis protein (HFE), increase the activity of the HAMP gene, which produces hepcidin, by activating either the BMP receptor (BMPR) or the TfR in the liver [10]. Furthermore, cytokines associated with inflammation, such as interleukin 6 (IL-6), promote hepcidin generation and lower plasma concentration of iron to limit microbial development. To retrieve iron from its retention sites and enhance its accessibility for activities such as erythropoietin (EPO), erythropoiesis, it induces erythroblasts to discharge the protein erythroferrone (ERFE), which suppresses hepcidin synthesis by attaching to BMPs. Under deprived of oxygen situations, EPO production rises, facilitated by the hypoxia-inducible factor (HIF), which contributes to greater hepcidin decreased levels and, as a result, stimulates iron mobilization [11].

At the level of the cell, free iron is referred to as the unstable iron collection, which consists of ferrous iron linked by poor-affinity iron antioxidants. The unstable iron reserve controls internal iron flux via communicating with iron-responsive elements (IRE) and modulating the synthesis of mRNAs producing ferritin, TfR1, Fpn, and bivalent metal-ion transporter-1. Furthermore, HIF controls the gene expression of Fpn, DMT1, and TfR by attaching to their hypoxia-responsive element (HRE), acting as a strong transcriptional factor. Prolyl-4-hydroxylase (PHD), which modulates HIF destruction, is largely Fe^{2+} influenced, making HIF and PHD appealing targets for the therapy of iron metabolism-related diseases [12].

Overall, disturbances in any or all of its main actors disrupt iron equilibrium, resulting in either a surplus or a deficiency of circulatory iron. Both disorders are damaging to a person's general well-being. Intracellular iron management is basically aided by the hypoxia-inducible factor (HIF) and iron regulatory proteins (IRP) process, which affect the gene translation and the transcription process, whereas a particularly significant systemic mechanism for regulation is the hepcidin-ferroportin axis. Hepcidin is a type of peptide generated from the hepatocyte that provides adverse effects on iron levels in the bloodstream by inhibiting iron export from macrophages and enterocytes. It attaches to FPN and promotes absorption and destruction of the sole exporting protein, preventing iron outflow from the cell. Hepcidin production rises in the presence of iron to inhibit additional iron discharge, along with an irritation to diminish availability for metal-based pathogens. This inflammation process of hepcidin production is a major cause of anemia in inflammatory and chronic illnesses. Hepcidin production decreases in conditions associated with elevated iron demand, such as iron shortage, hemorrhaging, or ischemia [13].

18.2.3 Iron Overload Versus Deficiency: Systemic and Cellular Effects

Cellular iron homeostasis is tightly regulated, increasing iron uptake when cells are iron-deficient and restricting iron import while promoting iron retention when cells are iron-replete. This control can take place at multiple tiers, but the most comprehensible structure is the iron-based attachment of iron regulatory proteins (IRPs) to stem-loop buildings in the untranslated area of mRNAs that directs the synthesis of different iron-associated amino acids. When intracellular iron contents are low, IRPs adopt mRNA-binding configurations. IRPs attach to the 5′ UTR of ferritin mRNA, restricting translation and guaranteeing that minimal ferritin is generated when storage of iron is not needed. At the same time, the IRPs linked to IREs in the region of 3′ UTR of the TfR1 mRNA, preserving the information from destruction and allowing for increased TfR1 transcription on the cellular surface. This response, in return, allows the cellular system to increase its iron uptake. When intracellular iron contents escalate, IRPs do not attach to IREs, enabling ferritin mRNA transformation to take place while leaving TfR1 mRNA to demolition. These situations reduce intracellular iron intake and encourage retention. Other mRNAs with IREs include hypoxia inducible factor 2α, 5-aminolevulinic acid synthase 2, FPN1, and DMT1 [14]. The total impact of these alterations is to guarantee that cells respond appropriately to alterations in intracellular iron concentration. It is known that TfR1 and ferritin levels alter with iron status, which has remarkable clinical significance. As stated before, small quantities of ferritin evaporate from retention sites; hence, the percentage of ferritin in serum represents the quantity of iron deposited. Small levels of TfR1 are additionally observed in serum as the protein's extracellular region can be broken effectively at the cell surface. The outcome of soluble TfR corresponds to the proportion of cell surface TfR1 in the human system and serves as a helpful indicator of iron deficiency (ID), as TfR1 levels increase when iron levels are low. Other kinds of control have received fewer resources, although they are definitely worth investigating further. Hypoxia, cytokines, and hormones [15] are recognized translational factors that affect iron-related enzymes. Even iron regulates several iron-related genes [16].

The careful regulation of the equilibrium of iron at the concentration of cells is mirrored at the complete-body scale. If a person needs more iron, a greater amount of iron will be recruited from their own iron stores, resulting in increased intestinal absorption of the metal. These functions, nevertheless, will be decreased if the body is rich in iron. A great deal of study over the last 15 years has revealed that hepcidin, a tiny liver-derived peptide, mediates this control. Differences in body iron requirement are conveyed to the hepatocyte, which controls the generation of hepcidin (transcribed by the hepcidin antibacterial protein gene). Hepcidin travels through the bloodstream to its specific locations, where it interacts with the iron-exporting molecule FPN1. Thus, FPN1 functions as a hepcidin sensor. This compound is transformed, absorbed, and destroyed, reducing iron's ability to be discharged into circulation. Therefore, when the human body is iron packed, hepcidin levels are higher, and the transport of iron to the plasma is decreased; when iron needs are higher, hepcidin quantities fall, and major iron reaches the circulatory system. Important metal-transporting cells comprise macrophages, which recover heme-directed iron from aged hepatocytes, and RBCs, which store iron in large quantities, and intestine enterocytes, which aid in iron absorption through food. Yet, most cells can release iron via FPN1, making them suitable hepcidin receptors [16]. Hepcidin control has been extensively investigated, yet it is still unclear. The bone developmental protein-SMAD regulation process is essential for hepcidin control, and it is most favorably the mechanism by which iron-based hepcidin oversight occurs. However, cytokines from inflammation, hypoxia, and a variety of other variables may additionally stimulate signaling pathways that cause changes in the hepcidin transcriptional process. Increased stored iron and inflammatory process boost hepcidin expression analysis, respectively, while decreased iron reserves and hypoxia suppress it [17].

18.3 ROS Generation and Oxidative Stress

18.3.1 Sources of ROS in Metabolic Tissues

A single cell may generate a variety of ROS, with the kind of molecule generated depending on the cause of the event and its surrounding environment. As a result, different compartments within cells have developed to get rid of and/or promote the free radicals that they most frequently generate. The enzymatic reactions that comprise the ROS generation mechanism are separated and compartmentalized in the cell, much as specialized events in metabolism are. The endoplasmic reticulum, cytosol, peroxisomes, and mitochondria are the primary cell sections that create ROS, but they do so in diverse ways (Fig. 18.1):

18.3.1.1 Mitochondria

The generation of ROS occurs along the electron transport system in mitochondria. Electrons gained from derivatives of metabolism in the form of NADH reach the electron transport system at complex I, whereas electrons acquired as FADH2 reach to complex II, or the flavoproten if they originated from oxidation of extra-mitochondrial fats. In "forwardly electron transmission" the generation of ATP, electrons are transported to complex IV, cytochrome c, complex III, and ubiquinone in that order, providing energy for pumping hydrogen into the intermembrane gap. Nevertheless, in stressful or oxygen-limited situations, these molecules may be transported straight to the oxygen molecule at complex III, resulting in the generation of ROS superoxide regardless of the matrix of mitochondria or the space between the membranes. Furthermore, electrons originating from cytochrome c from ETF or complex II may be transported to complex I through "reverse electron transfer," resulting in the formation of superoxide. Superoxide can then be transformed to the more membrane diffusible H2O2 by Mn superoxide dismutase (SOD2) in the region of the matrix or Zn/Cu superoxide dismutase (SOD1) in the intermembrane region [18].

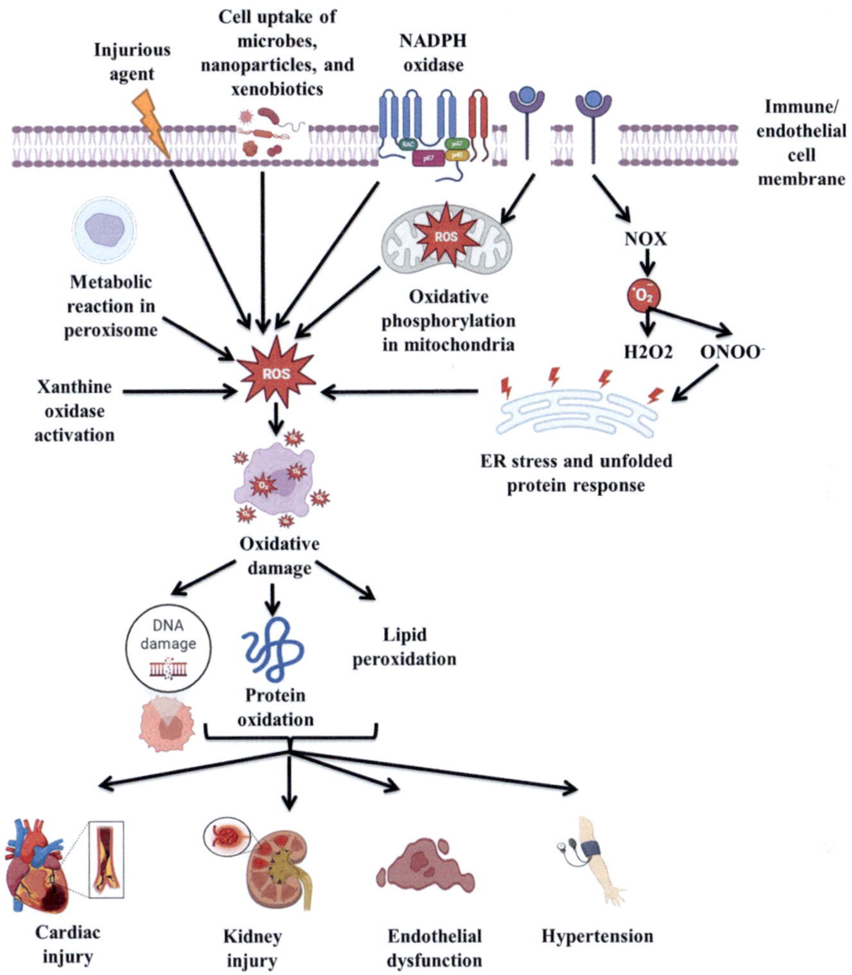

Fig. 18.1 ROS directed oxidative stress directed different metabolic disorders

18.3.1.2 Peroxisome

In the cell membrane, two dual oxidases (Duox) and NOX1–5 hydrolyze NADPH and transform O_2 to superoxide. Xanthine oxidase can produce various ROS, including superoxide from oxidizing an alkyl R-H group to hydrogen peroxide, $CO_3{\bullet}^-$ from acetaldehyde, and R-OH from converting hypoxanthine or xanthine to uric acid or xanthine, accordingly. The peroxisome produces H_2O_2 by triggering the reactions of glycolate oxidase, acyl-CoA oxidases, and flavin oxidases. In the meantime, in the organelle of endoplasmic reticulum, oxidized cysteine sites on isomerase of protein disulfide are decreased, providing the oxidative capacity for the degradation of cysteine residues on stretched proteins. This makes way for the creation of disulfide bonds, which provide both tertiary and secondary structures to foldable proteins. Lindsay and Rhodes [18] report that PDI is reoxidized by oxidized ERO-1α, resulting in the production of hydrogen peroxide.

18.3.1.3 Cytosolic ROS Sources

Dual oxidases (DUOX1/2) and NOX1–5 enzymes located in the cell membrane oxidize NADPH and reduce O_2 to form superoxide ($O_2•^-$). Xanthine oxidase also generates multiple ROS, including superoxide and hydrogen peroxide during hypoxanthine and xanthine oxidation, and carbonate radicals ($CO_3•^-$) during reactions with acetaldehyde. Collectively, many ROS-producing enzymes are distributed across the cell surface and other cellular compartments. NADPH Oxidase (NOX) is a membrane-related molecule whose active region faces the cell's cytoplasm. NOX converts NADPH and, less likely, NADH into superoxide. There are seven recognized mammalian NOX variants, which include naturally occurring NOX4 and NOX1–3, both of which require cytosolic monomer interaction for operation. At the same time, Ca2+ activates NOX5 and two different oxidase genes (Duox1 and Duox2) showcase potent interplay with other processes of cell signalling. A prime instance of this interaction emerges during thyroid-induced apoptosis in muscle tissue [19]. NOX variants bind to particular caveolins, lipid rafts, and membrane ruffles, allowing ROS-S to be localized in close contact to receptor-mediated signaling within cells. The translation in caveolin rafts of NOX4 with the AT1 GPCR, for instance, has been demonstrated to be important for the signalling of angiotensin II, extending the hypothesis that ROS-directing proteins are crucial to particular cellular communication machinery [20].

A well-studied cytosolic ROS-producing mechanism is xanthine oxidase (XO). XO can produce numerous distinct kinds of ROS based on its host. XO produces H_2O_2 through the modification of hypoxanthine to xanthine and then uric acid. However, if an amino acid or purine with hydrogen linked to an alcoholic group is oxidized, the enzyme produces superoxide [21]. In addition, XO has been shown to produce the extremely reactive molecule of carbonate radicals in the availability of bicarbonate. The carbonate free radical is a little-studied molecule in cell biology that continues within the cell for a long time and has been proposed to support bicarbonate's capacity to oxidize methionine leftovers, implying interest as a molecule that signals [22]. The range of ROS produced by XO action may provide valuable local data about the continually changing cellular biochemical environment.

18.3.1.4 Endoplasmic Reticulum

Superoxide dismutase (SOD) 1 and 2 (comprising Cu or Mn) transform two substances of superoxide atoms to H_2O_2 in the region of mitochondria and cytoplasm. In the meantime, in the mitochondrial region, peroxiredoxin III includes two groups of S-H that are oxidized to disulfide bonds by hydrogen peroxide. Excessive H_2O_2, on the other hand, inhibits Prx-III action by oxidizing its SH residue to SOOH. In the region of the endoplasmic reticulum, glutathione peroxidase's Se− is oxidized to Se-OH by H_2O_2. It is subsequently regenerated to Se− by oxidizing both substances of reduced glutathione (GSH) to oxidized, dimerized glutathione (GSSG). GSSG can be converted into two monomers of GSH in the cytoplasm by oxidizing two S-H groups on the glutathione reductase to a disulfide bond, which is subsequently converted back to S-H groups with the help of NADPH. Peroxisomes include the molecule catalase, which uses FeIII to transform hydrogen peroxide to H_2O in a cyclic mechanism [18].

18.3.2 Iron as a Catalyst in ROS Production (Fenton Reaction)

Iron's capacity to easily shift valence and transition between the Fe^{2+} and Fe^{3+} develops, supplying or absorbing electrons, maintains its favored position among living things as an intermediary of critical biological reactions. Yet, this chemical feature also generates iron, the main trigger for the formation of harmful reactive species in organisms that are aerobic. As a result, iron that circulates is securely linked to the iron transporter transferrin, preventing it from undergoing redox reactions. In contrast, the majority of intracellular iron is both extremely safe in the active regions of enzymes or properly deposited and eliminated in ferritin. Basic biological macromolecules, including proteins, nucleic acids, membrane lipids, and carbohydrates, cannot be oxidized by O_2 radical dot− or H_2O_2, as previously described. Yet, the existence of unprotected redox-active iron promotes Fenton reaction, requiring the interaction of H_2O_2 with Fe^{2+} to produce very active HO radical dots via perferryl/ferryl precursors [23].

Unprotected unstable iron, or NTBI, develops in plasma in pathological situations [24] and is easily absorbed by hepatocytes and other tissue parenchymal cells through the NTBI transporter ZIP14. Functional cells also have a tiny quantity of unprotected iron that is poorly understood and known as the "labile iron pool" (LIP) [25]. Unstable iron has a vital function in both H_2O_2-induced cytotoxicity and redox communication. Copper may initiate reactions with the same efficiency as iron, but its impact on H_2O_2-induced cytotoxicity is limited due to its low concentration in cells and organisms (<25-fold). HO radical dot, the main result of the Fenton response, is brief yet highly reactive, allowing it to oxidize any chemical group in close proximity (diffusion-controlled sensitivity). For example, iron bound to membrane phospholipids can initiate chain lipid-peroxidation reactions, which are closely associated with necrotic or ferroptotic forms of cell death. In addition, iron coupled with DNA causes abnormalities or single and double strand breaks, whereas iron freely bound to proteins promotes H_2O_2-dependent redox signals [26].

18.3.3 Mechanisms of Iron-Induced Oxidative Damage

Reactive oxygen species (ROS) are volatile and reactive substances generated by mitochondrial oxygen use, as well as by NADPH oxidase enzymes and cytochromes in cells. Anion superoxide (O_2-) is generated and converted to hydrogen peroxide (H_2O_2) by superoxide dismutase [27]. H_2O_2 is more stable than O_2- and can permeate the membranes of cells as a messenger molecule [28]. However, its levels must remain below a physiologically reasonable threshold, which is determined by enzymatic production. Yet, the reaction called Fenton (which requires a transitional metal) catalyzes H_2O_2 into hydroxyl radicals (HO), the most energetic ROS. Iron is an ubiquitous transitional metal that plays significant roles in metal-catalyzed oxidation processes [29].

Iron-catalyzed oxidation causes lipid peroxidation by reductively cleaving hydroperoxysides (ROOH) derived from phospholipids found in membranes, resulting in alkoxyl (RO.) and peroxyl ($RO_{2.}$) radicals. The production of lipid radicals is also important in the pathophysiology of ferroptosis, where iron-dependent peroxidation of lipids controls the procedure of controlling the death of cells. Superoxide dismutase enzyme (SOD) converts anion superoxide ($O_2\bullet-$) generated by oxidative phosphorylation (OXPHOS) to hydrogen peroxide (H_2O_2). Fe2+ catalyzes the reaction known as Fenton, which produces hydroxyl radicals (HO•), the most active ROS. Fe^{2+} and Fe^{3+} catalyze the reduction breakage of hydroperoxysides (ROOH), producing alkoxyl (RO•) and peroxyl (RO_2.) radicals [30].

18.4 Iron-Induced Inflammation

Although iron is essential for body processes, for instance, DNA replication, cellular respiration, and oxygen delivery, excessive iron redox activity can be harmful. The Fenton reaction, which produces hydroxyl radicals (•OH), strong reactive species that damage DNA, proteins, and lipids, is fueled by elevated labile iron pools, particularly ferrous Fe^{2+} [31].

18.4.1 Iron Overload Activates Inflammatory Pathways

Iron overload increases oxidative stress by generating ROS, which triggers stress-sensitive cascades such as MAPKs and IKK/NF-κB. These pathways direct the expression mechanism of pro-inflammatory factors. Iron-induced oxidative damage in metabolic organs (hepatic, adipose, pancreatic) attracts immune cells, including macrophages, and secretes cytokines, for example, IL-1β, IL-6, and TNF-α [32]. Recent research shows that iron buildup causes hepatocytes to generate extracellular RNA (eRNA), which acts as a damage-linked molecular pattern by engaging TLR3 and activating ROS generation and inflammatory signaling [33].

18.4.2 Role of NF-κB, NLRP3 Inflammasome, and Cytokine Release

Iron-mediated oxidative stress is a strong activator of NF-κB. Fe^{2+} activates IKK complex, causing IκB downgrade and nuclear relocation of NF-κB. This results in enhanced transcriptional analysis of IL-6, TNF-α, and IL-1β [34]. Elevated iron and ROS levels both enhance the start of the NLRP3 inflammasome. Iron toxicity can cause mitochondrial dysfunction and lysosomal instability, resulting in the direction of the NLRP3 complex, caspase 1 activation, and the production of IL-18 and IL-1β, which can increase inflammation and cause cell death..

18.4.3 Cross Talk Between Oxidative Stress, Ferroptosis, and Inflammatory Signaling

There is a bidirectional interaction between redox imbalance and inflammation. ROS activates NF-κB, AP-1, MAPKs, cGAS STING, and JAK STAT signaling. Inflammatory cytokines alter iron-regulatory gene expression, upregulating DMT1 and downregulating ferroportin, leading to iron buildup and oxidative damage [35]. Ferroptosis—a controlled, iron-dependent cell mortality mechanism determined by peroxidation of lipids—is inextricably related to inflammation. Iron overload and antioxidant system failure (e.g., GPX4, FSP1, BH_4 pathways) cause lipid hydroperoxide buildup, resulting in membrane damage and death. Ferroptosis promotes inflammatory signals, prolonging the vicious cycle.

18.4.4 Therapeutic Insights: Bridging Iron, ROS, and Inflammation

Therapeutic insights: Bridging Iron, ROS, and Inflammation.

Recent studies suggest that treatment methods should target ROS-driven inflammation and ferroptosis. In mouse animal models of iron-directed hepatic injury, chrysin (a natural flavonoid) reduced oxidative stress and NF-κB/NLRP3 activation [36]. General techniques include iron chelators, antioxidants, and inflammasome activation inhibitors.

18.5 Impact on Specific Metabolic Disorders

Excess iron deposition in metabolic tissues exacerbates oxidative stress, disrupts cellular homeostasis, and sustains chronic inflammation, thereby contributing to the onset and progression of diverse metabolic disorders [37]. The pathological effects of iron overload converge on three interconnected mechanisms: (1) enhanced generation of reactive oxygen species (ROS), (2) activation of inflammatory signaling cascades, and (3) dysregulation of iron trafficking at both systemic and cellular levels. The disease-specific impacts of iron overload are summarized in Table 18.1.

18.5.1 Obesity

In obesity, iron accumulation is commonly detected in adipose tissue macrophages (ATMs) and adipocytes, often accompanied by reduced iron export due to ferroportin (FPN) downregulation. This intracellular sequestration favors ROS accumulation and alters lipid droplet metabolism, leading to hypertrophy of adipocytes and increased release of free fatty acids (FFAs) [38]. Excess FFAs further stimulate Toll-like receptor 4 (TLR4) signaling, which works synergistically with iron-induced NF-κB activation to enhance pro-inflammatory cytokine production [39]. Chronic low-grade inflammation in obesity also modifies systemic iron homeostasis by elevating hepcidin, thereby perpetuating tissue iron retention.

18.5.2 Type 2 Diabetes Mellitus (T2DM)

Beyond β-cell damage, iron overload in T2DM impairs hepatic insulin clearance by disrupting hepatocyte insulin receptor signaling and IRS-1 phosphorylation balance. In skeletal muscle, iron-induced mitochondrial ROS reduce AMP-activated protein kinase (AMPK) activity, limiting glucose uptake and oxidation [40]. Furthermore, iron may contribute to advanced glycation end-product (AGE) accumulation in diabetic vasculature by catalyzing oxidative modifications, which exacerbate endothelial dysfunction. Elevated circulating non–transferrin-bound iron (NTBI) is increasingly recognized as a biomarker linking iron dysregulation to poor glycemic outcomes [41].

18.5.3 Non-Alcoholic Fatty Liver Disease (NAFLD)

In NAFLD, iron overload synergizes with lipotoxicity to impair mitochondrial β-oxidation, promoting ROS leakage and ATP depletion. Iron also modulates Kupffer cell polarization, favoring a pro-inflammatory M1 phenotype over the

Table 18.1 Pathophysiological impact of iron overload in major metabolic disorders

Metabolic disorder	Primary site(s) of iron deposition	Key mechanistic effects	Major inflammatory & oxidative pathways activated	Additional clinical implications	Source
Obesity	Adipocytes, ATMs	↓ FPN, ↑ intracellular ROS, lipid droplet hypertrophy, ↑ FFAs, TLR4–NF-κB synergy	NF-κB, MAPK, TLR4 signaling	Chronic inflammation, insulin resistance, elevated hepcidin perpetuating iron retention	[38, 39]
Type 2 diabetes mellitus (T2DM)	Pancreatic β-cells, skeletal muscle, liver	β-cell oxidative & ER stress, ↓ AMPK activity in muscle, impaired hepatic insulin clearance, ↑ AGEs	NF-κB, JNK, NLRP3, AMPK suppression	Hyperglycemia, endothelial dysfunction, NTBI as predictive biomarker	[40, 41]
Non-alcoholic fatty liver disease (NAFLD)	Hepatocytes, Kupffer cells, HSCs	Impaired β-oxidation, M1 Kupffer cell polarization, HFE-linked iron retention, sinusoidal capillarization	NF-κB, NLRP3, TGF-β/SMAD, hypoxia signaling	Rapid progression to NASH, fibrosis, hypoxia-aggravated liver injury	[42, 43]
Metabolic syndrome	Vascular endothelium, smooth muscle cells	oxLDL generation, EPC dysfunction, ↑ sympathetic tone, ↑ adhesion molecule expression	NF-κB, MAPK, VCAM-1/ICAM-1 upregulation	Hypertension, atherogenesis, increased carotid intima-media thickness	[44, 45]

Abbreviations: ↑ increased, ↓ decreased, *FPN* ferroportin, *FFAs* free fatty acids, *TLR4* Toll-like receptor 4, *ROS* reactive oxygen species, *NF-κB* nuclear factor kappa-light-chain-enhancer of activated B cells, *MAPK* mitogen-activated protein kinase, *JNK* c-Jun N-terminal kinase, *NLRP3* NOD-, LRR- and pyrin domain-containing protein 3, *AMPK* AMP-activated protein kinase, *AGEs* advanced glycation end-products, *NTBI* non-transferrin-bound iron, *HSCs* hepatic stellate cells, *TGF-β* transforming growth factor-beta, *VCAM-1* vascular cell adhesion molecule-1, *ICAM-1* intercellular adhesion molecule-1, *EPC* endothelial progenitor cells, *oxLDL* oxidized low-density lipoprotein

reparative M2 phenotype, which accelerates hepatocellular injury [42]. Dysregulated iron metabolism in NAFLD often coincides with genetic variants such as HFE mutations, which amplify iron retention in hepatocytes. In advanced disease, hepatic iron promotes sinusoidal capillarization and impairs oxygen delivery, aggravating hypoxia-driven fibrogenic signaling [43].

18.5.4 Metabolic Syndrome

In metabolic syndrome, iron's vascular effects extend to oxidative modification of low-density lipoprotein (oxLDL), which stimulates foam cell formation and accelerates atherogenesis. Iron-induced ROS impair endothelial progenitor cell (EPC) function, reducing vascular repair capacity [44]. At the systemic level, iron overload increases sympathetic nervous system activity, further contributing to hypertension. Elevated ferritin in metabolic syndrome has been independently associated with increased carotid intima-media thickness, linking iron status to subclinical atherosclerosis [45].

18.6 Therapeutic Perspectives

Given the central role of iron-induced oxidative stress and inflammation in the pathogenesis of metabolic disorders, therapeutic strategies have increasingly focused on modulating iron availability, reducing reactive oxygen species (ROS) burden, and interrupting pro-inflammatory signaling pathways [46]. These interventions aim not only to prevent further tissue injury but also to restore metabolic homeostasis. An overview of current and emerging therapeutic strategies is provided in Table 18.2.

18.6.1 Iron Chelation Therapy

Iron chelators reduce the labile iron pool (LIP) and limit Fenton reaction–mediated ROS production. Agents such as deferoxamine (DFO), deferiprone (DFP), and deferasirox (DFX) are clinically approved for iron overload syndromes and have demonstrated benefits in preclinical models of obesity, NAFLD, and type 2 diabetes mellitus (T2DM) [47]. Chelation therapy not only mitigates oxidative injury but also attenuates inflammatory responses by suppressing NF-κB and NLRP3 activation. However, concerns regarding potential depletion of essential iron-dependent enzymes and dose-related toxicity underscore the importance of careful patient selection and monitoring [48].

18.6.2 Antioxidant and Redox-Modulating Strategies

Antioxidants serve as adjuncts to reduce oxidative stress in iron overload–related metabolic disorders. N-acetylcysteine (NAC) replenishes glutathione (GSH) and supports glutathione peroxidase 4 (GPX4) activity, counteracting lipid peroxidation

Table 18.2 Current and emerging therapeutic strategies for iron-induced inflammation in metabolic disorders

Therapeutic approach	Primary target/ mechanism	Representative agents/strategies	Evidence level	Potential clinical outcomes	Source
Iron chelation therapy	Reduces labile iron pool (LIP), limits Fenton reaction	Deferoxamine, Deferiprone, Deferasirox	Clinical & preclinical	↓ ROS, ↓ NF-κB/NLRP3 activation, improved insulin sensitivity	[47, 48]
Antioxidant & redox modulation	Scavenges ROS, enhances antioxidant defenses, inhibits ferroptosis	N-acetylcysteine, curcumin, quercetin, Chrysin, Nrf2 activators	Clinical & Preclinical	↓ Lipid peroxidation, ↑ GPX4 activity, improved liver and vascular function	[49, 50]
Anti-inflammatory interventions	Inhibits cytokine production & inflammasome activation	NF-κB inhibitors (Parthenolide, Bardoxolone), NLRP3 inhibitors (MCC950), JAK–STAT modulators	Preclinical	↓ IL-1β, ↓ IL-6, ↓ TNF-α, reduced tissue inflammation	[51, 52]
Dietary & lifestyle modification	Lowers iron absorption, improves metabolic parameters	Reduced heme iron intake, polyphenol-rich foods, weight loss, coffee/tea consumption	Clinical	↓ Hepatic iron load, improved insulin sensitivity, reduced cardiovascular risk	[53, 54]
Gut microbiota–targeted approaches	Restores microbiome balance, limits pathogenic growth	Probiotics, prebiotics, Synbiotics, Lactoferrin	Preclinical & emerging clinical	↓ Endotoxemia, ↓ hepatic inflammation, improved gut barrier integrity	[55, 56]
Ferroptosis inhibition	Prevents lipid peroxidation & ferroptotic cell death	Ferrostatin-1, Liproxstatin-1, GPX4 activators, selenium	Preclinical	↓ Organ injury, ↓ inflammation, preserved cellular integrity	[57, 58]

Abbreviations: ↓ decreased, ↑ increased, *ROS* reactive oxygen species, *NF-κB* nuclear factor kappa-light-chain-enhancer of activated B cells, *NLRP3* NOD-, LRR-, and pyrin domain-containing protein 3, *GPX4* glutathione peroxidase 4, *LIP* labile iron pool, *IL* interleukin, *TNF-α* tumor necrosis factor-alpha

and ferroptosis [49]. Polyphenolic compounds (e.g., curcumin, quercetin, chrysin) exert dual effects by directly scavenging ROS and chelating redox-active iron. Nrf2 activators promote transcription of antioxidant response element (ARE)–driven genes, including heme oxygenase-1 (HO-1) and ferritin, enhancing cellular resilience to oxidative injury [50].

18.6.3 Anti-Inflammatory Interventions

Targeting inflammatory signaling downstream of iron-induced ROS is a promising therapeutic avenue. NF-κB inhibitors (e.g., parthenolide, bardoxolone methyl) reduce transcription of pro-inflammatory cytokines [51]. NLRP3 inflammasome inhibitors (e.g., MCC950) block caspase-1 activation and IL-1β/IL-18 maturation, effectively attenuating inflammation in metabolic tissues. JAK–STAT pathway modulators may further suppress cytokine-driven alterations in iron trafficking, disrupting the cycle of iron retention and inflammation [52].

18.6.4 Dietary and Lifestyle Interventions

Dietary modification represents a low-cost, sustainable strategy to modulate systemic iron load. Controlled iron intake through reduced consumption of heme iron–rich foods (e.g., red meat, organ meats) and enhanced intake of polyphenol-rich plant foods can limit iron absorption. Caloric restriction and weight loss improve insulin sensitivity and reduce hepatic iron deposition in NAFLD. Additionally, coffee and tea polyphenols have been shown to inhibit non-heme iron absorption [53, 54].

18.6.5 Gut Microbiota–Targeted Approaches

The gut–liver axis is increasingly recognized as a therapeutic target in iron-induced metabolic disorders. Excess luminal iron promotes the growth of pathogenic bacteria and disrupts intestinal barrier integrity, facilitating endotoxin translocation and hepatic inflammation [55]. Probiotics (e.g., Lactobacillus, Bifidobacterium), prebiotics, and synbiotics restore gut microbial balance and reduce systemic inflammation. Lactoferrin may limit luminal iron availability to harmful microbes while exerting immunomodulatory effects [56].

18.6.6 Targeting Ferroptosis

Given its central role in iron-driven tissue injury, ferroptosis inhibition is an emerging therapeutic strategy. Agents such as ferrostatin-1 and liproxstatin-1 prevent lipid peroxidation by scavenging lipid radicals and stabilizing membrane integrity. Upregulation of GPX4 activity through selenium supplementation or pharmacological activators can also counteract ferroptotic cell death [57, 58].

18.7 Fresh Perspectives and Future-Looking Strategies

Recent research studies indicate that iron actively contributes to inflammation in chronic metabolic circumstances for example, nonalcoholic fatty liver disease (NAFLD), type 2 diabetes, and obesity. Excess iron causes reactive species that degrade cells and provoke chronic inflammation by activating the NLRP3 and NF-κB inflammasome [2]. Researchers are currently focusing on iron-chelating therapies to treat iron overload and accompanying inflammation. Ferroptosis, a kind of cell mortality caused by iron and peroxidation of lipid, has been a key discovery in this field. Ferroptosis, unlike apoptosis, is highly inflammatory and causes tissue damage, particularly around the liver and pancreas [59]. This process is controlled by iron transporters such as the transferrin receptor (TfR1), antioxidant systems like GPX4 and glutathione, and enzymes that cause lipid peroxidation like ACSL4. Blocking ferroptosis using restrictors, for example, liproxstatin-1 or ferrostatin-1, has the possibility of shielding tissues from inflammation-induced damage.

Iron also affects immune-mediated cells, particularly macrophages, which induce inflammation in metabolic organs. High iron levels trigger macrophages to release pro-inflammatory "M1" chemokines like TNF-α and IL-1β, while reducing anti-inflammatory "M2" markers like arginase-1 [60]. However, under certain instances, iron-loaded macrophages could be acting protectively by storing excess iron in ferritin and reducing the formation of ROS. This illustrates that iron's immunological effects vary depending on the tissue environment and disease stage. The gut-liver axis is another rapidly expanding research field. Iron levels in the gut influence the composition of gut bacteria [61]. Excess iron triggers the proliferation of detrimental microbes, decreases the intestinal barrier, promotes endotoxin penetration, and induces liver inflammation. Probiotics and nutritional strategies that target the gut microbiota can assist in regulating iron levels while lowering inflammation. Gazing ahead, a variety of novel strategies are being developed. These include ferroptosis inhibitors, iron chelators, and immune-modulating drugs. Biomarkers such as non-transferrin-bound iron (NTBI), lipid peroxidation markers (e.g., malondialdehyde), GPX4 activity, and cytokines (e.g., IL-1β, IL-18) are being developed to improve the diagnosis and monitoring of iron-induced inflammation.

18.8 Conclusion

Iron plays a pivotal role in the development and progression of metabolic disorders by influencing oxidative stress and inflammation. While extra iron is required for many cellular activities, it also works as a strong pro-oxidant, increasing the creation of reactive oxygen species (ROS) that damage cellular components and activate important inflammatory pathways. Chronic low-grade inflammation causes obesity, insulin resistance, type 2 diabetes, and non-alcoholic fatty liver disease by activating NF-κB, the NLRP3 inflammasome, and by releasing pro-inflammatory cytokines. Finally, the complex interplay between iron metabolism, ROS, and inflammation provides a convincing paradigm for understanding the pathophysiology of metabolic disorders. Future research should focus on understanding tissue-specific iron management and defining safe treatment windows for adjusting iron levels to restore metabolic homeostasis.

References

1. Gozzelino R, Arosio P (2016) Iron Homeostasis in health and disease. Int J Mol Sci 17(1):130. https://doi.org/10.3390/ijms17010130
2. Gensluckner S, Wernly B, Datz C, Aigner E (2024) Iron, oxidative stress, and metabolic dysfunction-associated Steatotic liver disease. Antioxidants 13(2):208. https://doi.org/10.3390/antiox13020208
3. Vogt A-CS, Arsiwala T, Mohsen M, Vogel M, Manolova V, Bachmann MF (2021) On iron metabolism and its regulation. Int J Mol Sci 22(9):4591. https://doi.org/10.3390/ijms22094591
4. Mouleeswaran KS, Einstein D, Prathiba A (2022) Usefulness of reticulocyte Haemoglobin equivalent in the evaluation of iron deficiency status: a cross-sectional study. JCDR 16:EC10–EC12. https://doi.org/10.7860/JCDR/2022/56493.16491
5. Katsarou A, Pantopoulos K (2020) Basics and principles of cellular and systemic iron homeostasis. Mol Aspects Med 75:100866. https://doi.org/10.1016/j.mam.2020.100866
6. Kawabata H (2019) Transferrin and transferrin receptors update. Free Radic Biol Med 133:46–54. https://doi.org/10.1016/j.freeradbiomed.2018.06.037
7. Plays M, Müller S, Rodriguez R (2021) Chemistry and biology of ferritin. Metallomics 13(5):mfab021. https://doi.org/10.1093/mtomcs/mfab021
8. Ganz T (2013) Systemic iron homeostasis. Physiol Rev 93(4):1721–1741. https://doi.org/10.1152/physrev.00008.2013
9. Horton MA (1983) Expression of transferrin receptors during erythroid maturation. Exp Cell Res 144(2):361–366. https://doi.org/10.1016/0014-4827(83)90415-9
10. Gao J, Chen J, Kramer M, Tsukamoto H, Zhang A-S, Enns CA (2009) Interaction of the hereditary hemochromatosis protein HFE with transferrin receptor 2 is required for transferrin-induced hepcidin expression. Cell Metab 9(3):217–227. https://doi.org/10.1016/j.cmet.2009.01.010
11. Arezes J, Foy N, McHugh K, Sawant A, Quinkert D, Terraube V, Terraube V, Brinth A, Tam M, LaVallie ER, Taylor S, Armitage AE, Pasricha S_R, Cunningham AB, Lambert M, Draper SJ, Jasuja R, Drakesmith H (2018) Erythroferrone inhibits the induction of hepcidin by BMP6. Blood 132(14):1473–1477. https://doi.org/10.1182/blood-2018-06-857995
12. Jaakkola P, Mole DR, tian YM, Wilson MI, Gielbert J, Gaskell SJ, von Kriegsheim A, Hebestreit HF, Mukherji M, Schofield CJ, Maxwell PH, Pugh CW, Ratcliffe PJ (2001) Targeting of HIF-α to the von Hippel-Lindau Ubiquitylation complex by O2-regulated prolyl hydroxylation. Science 292(5516):468–472. https://doi.org/10.1126/science.1059796
13. Babitt JL, Huang FW, Wrighting DM, Xia Y, Sidis Y, Samad TA, Campagna JA, Chung RT, Schneyer AL, Woolf CJ, Andrews NC, Yin HY (2006) Bone morphogenetic protein signaling by hemojuvelin regulates hepcidin expression. Nat Genet 38(5):531–539. https://doi.org/10.1038/ng1777
14. Wilkinson N, Pantopoulos K (2014) The IRP/IRE system in vivo: insights from mouse models. Front Pharmacol 5:176. https://doi.org/10.3389/fphar.2014.00176
15. Harms K, Kaiser T (2015) Beyond soluble transferrin receptor: old challenges and new horizons. Best Pract Res Clin Endocrinol Metab 29(5):799–810. https://doi.org/10.1016/j.beem.2015.09.003

16. Anderson Gregory J, Frazer David M (2017) Current understanding of iron homeostasis. Am J Clin Nutri 106(6):15595–15665. https://doi.org/10.3945/ajcn.117.155804
17. Hou Y, Zhang S, Wang L, Li J, Qu G, He J, Rong H, Ji H, Liu S. Estrogen regulates iron homeostasis through governing hepatic hepcidin expression via an estrogen response element. Gene2012;511(2):398–403. https://doi.org/10.1016/j.gene.2012.09.060
18. Lindsay RT, Rhodes CJ (2025) Reactive oxygen species (ROS) in metabolic disease—don't shoot the metabolic messenger. Int J Mol Sci 26(6):2622. https://doi.org/10.3390/ijms26062622
19. Then AA, Goenawan H, Lesmana R, Christoper A, Sylviana N, Gunadi JW (2024) Exploring the potential regulation of DUOX in thyroid hormone autophagy signaling via IGF 1 in the skeletal muscle. Biomed Rep 22(3):39. https://doi.org/10.3892/br.2024.1917
20. Ushio-Fukai M, Zuo L, Ikeda S, Tojo T, Patrushev NA, Alexander RW (2005) cAbl tyrosine kinase mediates reactive oxygen species- and caveolin-dependent AT1 receptor signaling in vascular smooth muscle: role in vascular hypertrophy. Circ Res 97(8):829–836. https://doi.org/10.1161/01.res.0000185322.46009.f5
21. Galbusera C, Orth P, Fedida D, Spector T (2006) Superoxide radical production by allopurinol and xanthine oxidase. Biochem Pharmacol 71(12):1747–1752. https://doi.org/10.1016/j.bcp.2006.02.008
22. Medinas DB, Cerchiaro G, Trindade DF, Augusto O (2007) The carbonate radical and related oxidants derived from bicarbonate buffer. IUBMB Life 59(4–5):255–262. https://doi.org/10.1080/15216540701230511
23. Merkofer M, Kissner R, Hider RC, Brunk UT, Koppenol WH (2006) Fenton chemistry and iron chelation under physiologically relevant conditions: electrochemistry and kinetics. Chem Res Toxicol 19(10):1263–1269. https://doi.org/10.1021/tx060101w
24. Brissot P, Ropert M, Le Lan C, Loreal O (2012) Non-transferrin bound iron: a key role in iron overload and iron toxicity. Biochim Biophys Acta 1820(3):403–410. https://doi.org/10.1016/j.bbagen.2011.07.014
25. Cabantchik ZI (2014) Labile iron in cells and body fluids: physiology, pathology, and pharmacology. Front Pharmacol 5:45. https://doi.org/10.3389/fphar.2014.00045
26. Galaris D, Barbouti A, Pantopoulos K (2019) Iron homeostasis and oxidative stress: an intimate relationship. Biochimica et Biophysica Acta (BBA) 1866(12):118535. https://doi.org/10.1016/j.bbamcr.2019.118535
27. Valko M, Jomova K, Rhodes C, Kuča K, Musílek K (2016) Redox- and non-redox-metal-induced formation of free radicals and their role in human disease. Arch Toxicol 90(1):1–37. https://doi.org/10.1007/s00204-015-1579-5
28. Shadel G, Horvath T (2015) Mitochondrial ROS signaling in organismal homeostasis. Cell 163(3):560–569. https://doi.org/10.1016/j.cell.2015.10.001
29. Halliwell B (2007) Biochemistry of oxidative stress. Biochem Soc Trans 35(5):1147–1150. https://doi.org/10.1042/BST0351147
30. Yan F, Li K, Xing W, Dong M, Yi M, Zhang H (2022) Role of iron-related oxidative stress and mitochondrial dysfunction in cardiovascular diseases. Oxidative Med Cell Longev 2022:5124553. https://doi.org/10.1155/2022/5124553
31. Ru Q, Li Y, Chen L, Wu Y, Min J, Wang F (2024) Iron homeostasis and ferroptosis in human diseases: mechanisms and therapeutic prospects. Signal Transduct Target Ther 9(1):271. https://doi.org/10.1038/s41392-024-01969-z. PMID: 39396974; PMCID: PMC11486532
32. Sobieska K, Buczyńska A, Krętowski AJ, Popławska-Kita A (2024) Iron homeostasis and insulin sensitivity: unraveling the complex interactions. Rev Endocr Metab Disord. 25(5):925–939. https://doi.org/10.1007/s11154-024-09908-7. Epub 2024 Sep 17. PMID: 39287729; PMCID: PMC11470850
33. Raza S, Tewari A, Rajak S, Gupta P, Sinha RA (2025) Extracellular RNA mediates iron-induced toxicity and inflammatory signalling in hepatic cells. Toxicol Rep 14:102002. https://doi.org/10.1016/j.toxrep.2025.102002. PMID: 40162071; PMCID: PMC7617531
34. Bellanti F, Coda ARD, Trecca MI, Lo Buglio A, Serviddio G, Vendemiale G (2025) Redox imbalance in inflammation: the interplay of oxidative and reductive stress. Antioxidants

(Basel). 14(6):656. https://doi.org/10.3390/antiox14060656. PMID: 40563291; PMCID: PMC12189482
35. Chen Y, Fang ZM, Yi X, Wei X, Jiang DS (2023) The interaction between ferroptosis and inflammatory signaling pathways. Cell Death Dis 14(3):205. https://doi.org/10.1038/s41419-023-05716-0. PMID: 36944609; PMCID: PMC10030804
36. Alanazi ST, Salama SA, El-Ebiary AM, Altowairqi AK, Alharthi AT, Alzahrani SM, Althagafi SH, Alotaibi RA, Tammam AA (2024) Targeting SIRT1, NLRP3 inflammasome, and Nrf2 signaling with chrysin alleviates the iron-triggered hepatotoxicity in rats. Toxicology 504:153766. https://doi.org/10.1016/j.tox.2024.153766. Epub 2024 Mar 1. PMID: 38432408
37. Xie T, Yao L, Li X (2024) Advance in iron metabolism, oxidative stress and cellular dysfunction in experimental and human kidney diseases. Antioxidants (Basel) 13(6):659. https://doi.org/10.3390/antiox13060659. PMID: 38929098; PMCID: PMC11200795
38. Ruze R, Song J, Yin X, Chen Y, Xu R, Wang C, Zhao Y (2023) Mechanisms of obesity- and diabetes mellitus-related pancreatic carcinogenesis: a comprehensive and systematic review. Signal Transduct Target Ther 8(1):139. https://doi.org/10.1038/s41392-023-01376-w. PMID: 36964133; PMCID: PMC10039087
39. He L, Su Z, Wang S (2024) The anti-obesity effects of polyphenols: a comprehensive review of molecular mechanisms and signal pathways in regulating adipocytes. Front Nutr 11:1393575. https://doi.org/10.3389/fnut.2024.1393575. PMID: 39539361; PMCID: PMC11557335
40. Lee JH, Lee J (2022) Endoplasmic reticulum (ER) stress and its role in pancreatic β-cell dysfunction and senescence in type 2 diabetes. Int J Mol Sci 23(9):4843. https://doi.org/10.3390/ijms23094843. PMID: 35563231; PMCID: PMC9104816
41. Eguchi N, Vaziri ND, Dafoe DC, Ichii H (2021) The role of oxidative stress in pancreatic β cell dysfunction in diabetes. Int J Mol Sci 22(4):1509. https://doi.org/10.3390/ijms22041509. PMID: 33546200; PMCID: PMC7913369
42. Chen J, Deng X, Liu Y, Tan Q, Huang G, Che Q, Guo J, Su Z (2020) Kupffer cells in non-alcoholic fatty liver disease: friend or foe? Int J Biol Sci 16(13):2367–2378. https://doi.org/10.7150/ijbs.47143. PMID: 32760204; PMCID: PMC7378652
43. Gao H, Jiang Y, Zeng G, Huda N, Thoudam T, Yang Z, Liangpunsakul S, Ma J (2024) Cell-to-cell and organ-to-organ crosstalk in the pathogenesis of alcohol-associated liver disease. eGastroenterology 2(4):e100104. https://doi.org/10.1136/egastro-2024-100104. Epub 2024 Dec 9. PMID: 39735421; PMCID: PMC11674000
44. Levitan I, Volkov S, Subbaiah PV (2010) Oxidized LDL: diversity, patterns of recognition, and pathophysiology. Antioxid Redox Signal 13(1):39–75. https://doi.org/10.1089/ars.2009.2733. PMID: 19888833; PMCID: PMC2877120
45. Bidhu RP, Muraleedharan A, Daniel RA, Surya BN (2025) Prevalence and determination of metabolic syndrome among adults (18-60 years) in urban and rural South India: A community-based cross-sectional study. Cureus 17(5):e84384. https://doi.org/10.7759/cureus.84384
46. Dash UC, Bhol NK, Swain SK, Samal RR, Nayak PK, Raina V, Panda SK, Kerry RG, Duttaroy AK, Jena AB (2025) Oxidative stress and inflammation in the pathogenesis of neurological disorders: Mechanisms and implications. Acta Pharm Sin B. 15(1):15–34. https://doi.org/10.1016/j.apsb.2024.10.004. Epub 2024 Oct 16. PMID: 40041912; PMCID: PMC11873663
47. Entezari S, Haghi SM, Norouzkhani N, Sahebnazar B, Vosoughian F, Akbarzadeh D, Islampanah M, Naghsh N, Abbasalizadeh M, Deravi N (2022) Iron Chelators in treatment of iron overload. J Toxicol 2022:4911205. https://doi.org/10.1155/2022/4911205. PMID: 35571382; PMCID: PMC9098311
48. An Y, Zhang H, Wang C, Jiao F, Xu H, Wang X, Luan W, Ma F, Ni L, Tang X, Liu M, Guo W, Yu L (2019) Activation of ROS/MAPKs/NF-κB/NLRP3 and inhibition of efferocytosis in osteoclast-mediated diabetic osteoporosis. FASEB J. 33(11):12515–12527. https://doi.org/10.1096/fj.201802805RR. Epub 2019 Aug 28. PMID: 31461386; PMCID: PMC6902677
49. Wendlocha D, Kubina R, Krzykawski K, Mielczarek-Palacz A (2024) Selected flavonols targeting cell death pathways in cancer therapy: the latest achievements in research on apoptosis,

autophagy, necroptosis, pyroptosis, ferroptosis, and cuproptosis. Nutrients 16(8):1201. https://doi.org/10.3390/nu16081201. PMID: 38674891; PMCID: PMC11053927
50. Meng J, Lv Z, Zhang Y, Wang Y, Qiao X, Sun C, Chen Y, Guo M, Han W, Ye A, Xie T, Chu B, Shi C, Yang S, Chen C (2021) Precision redox: the key for antioxidant pharmacology. Antioxid Redox Signal. 34(14):1069–1082. https://doi.org/10.1089/ars.2020.8212. Epub 2020 Dec 2. PMID: 33270507; PMCID: PMC8080931
51. Ding W, Cai C, Zhu X, Wang J, Jiang Q (2022) Parthenolide ameliorates neurological deficits and neuroinflammation in mice with traumatic brain injury by suppressing STAT3/NF-κB and inflammasome activation. Int Immunopharmacol 108:108913. https://doi.org/10.1016/j.intimp.2022.108913. Epub 2022 Jun 1. PMID: 35729839
52. Markovics A, Rosenthal KS, Mikecz K, Carambula RE, Ciemielewski JC, Zimmerman DH (2021) Restoring the balance between pro-inflammatory and anti-inflammatory cytokines in the treatment of rheumatoid arthritis: new insights from animal models. Biomedicine 10(1):44. https://doi.org/10.3390/biomedicines10010044. PMID: 35052724; PMCID: PMC8772713
53. Romero-Gómez M, Aller R, Martín-Bermudo F (2022) Dietary recommendations for the management of Non-alcoholic Fatty Liver Disease (NAFLD): a nutritional geometry perspective. Semin Liver Dis 42(4):434–445. https://doi.org/10.1055/s-0042-1757711. Epub 2022 Oct 28. PMID: 36307105
54. Quetglas-Llabrés MM, Monserrat-Mesquida M, Bouzas C, García S, Mateos D, Casares M, Gómez C, Ugarriza L, Tur JA, Sureda A (2024) Effects of a two-year lifestyle intervention on intrahepatic fat reduction and renal health: mitigation of inflammation and oxidative stress, a randomized trial. Antioxidants (Basel) 13(7):754. https://doi.org/10.3390/antiox13070754. PMID: 39061823; PMCID: PMC11273830
55. Chowdhury M, Raj Chaudhary N, Kaur P, Goyal A, Sahu SK (2024) Different strategies targeting gut microbiota for the management of several disorders: a sustainable approach. Infect Disord Drug Targets 24(5):e160124225675. https://doi.org/10.2174/0118715265267536231121095634. PMID: 38317473
56. Long C, Zhou X, Xia F, Zhou B (2024) Intestinal barrier dysfunction and gut microbiota in non-alcoholic fatty liver disease: assessment, mechanisms, and therapeutic considerations. Biology (Basel) 13(4):243. https://doi.org/10.3390/biology13040243. PMID: 38666855; PMCID: PMC11048184
57. Fan BY, Pang YL, Li WX, Zhao CX, Zhang Y, Wang X, Ning GZ, Kong XH, Liu C, Yao X, Feng SQ (2021) Liproxstatin-1 is an effective inhibitor of oligodendrocyte ferroptosis induced by inhibition of glutathione peroxidase 4. Neural Regen Res 16(3):561–566. https://doi.org/10.4103/1673-5374.293157. PMID: 32985488; PMCID: PMC7996026
58. Sun JK, Huangfu XT, Yin X, Nie S, Deng YH, Ye ZY, Zhou XH, Wang X, Si YN (2025) Ferroptosis is involved in the IL-9-induced intestinal barrier injury in sepsis: an experimental animal and translational study. Int J Surg 111(7):4314–4324. https://doi.org/10.1097/JS9.0000000000002541. Epub 2025 May 20. PMID: 40391960
59. Chen F, Kang R, Tang D, Liu J (2024) Ferroptosis: principles and significance in health and disease. J Hematol Oncol 17(1):Article 41. https://doi.org/10.1186/s13045-024-01564-3
60. Ni S, Yuan Y, Kuang Y, Li X (2022) Iron metabolism and immune regulation. Front Immunol 13:816282. https://doi.org/10.3389/fimmu.2022.816282. PMID: 35401569; PMCID: PMC8983924
61. Hu Y, Gao M, Chenghuang J, Bao R (2025) Insights into the gut-liver axis: mechanisms and emerging therapies in hepatocellular carcinoma. Front Pharmacol 16:1595853. https://doi.org/10.3389/fphar.2025.1595853. PMID: 40458800; PMCID: PMC12127201

Chapter 19
Dietary Interventions in ROS-Mediated Metabolic Disorders

Prachi Jagdev, Jaipreet Singh, Sarita Sharma, Sumeet Gupta, Gaurav Kant Saraogi, and Raghavendra Pai

19.1 Introduction

Reactive oxygen species (ROS) are reactive molecules having oxygen that are synthesized during normal cellular metabolism, including free radicals, hydroxyl radicals, singlet oxygen, and hydrogen peroxide. Reactive oxygen species (ROS) have a dual role in human metabolism, acting as essential signaling mediators while also causing cellular damage and metabolic inefficiency when produced in excess. Excessive synthesis disrupts molecular signaling, adversely affecting DNA, proteins, and lipids, leading to metabolic disorders [1]. Reactive oxygen species (ROS) substantially influence metabolic dysregulation, requiring a comprehensive approach that encompasses dietary and lifestyle changes. A diet rich in antioxidants reduces excess reactive oxygen species, while the intake of highly processed foods and unhealthy fats exacerbates oxidative stress [2]. Targeted nutritional interventions are recommended. Lifestyle modifications such as physical activity, stress alleviation, smoking cessation, and weight control might enhance antioxidant defenses, diminish inflammation, and lower systemic oxidative stress, so strengthening the body's resistance to ROS-induced metabolic harm [3, 4]. Advancements

P. Jagdev · J. Singh · S. Gupta (✉)
Department of Pharmacology, M.M. College of Pharmacy, Maharishi Markandeshwar (Deemed to be University), Ambala, Haryana, India

S. Sharma
Department of Pharmacology, Amity Institute of Pharmacy, Amity University, Gwalior, Madhya Pradesh, India

G. K. Saraogi
Oriental Institute of Pharmacy and Research, Bhopal, Madhya Pradesh, India

R. Pai
Pharmacovigilance Department, India – South East Asia, Takeda Singapore, Singapore, Singapore

S. Ashique, N. Mishra (eds.), *Reactive Oxygen Species in Metabolic Inflammation*,
https://doi.org/10.1007/978-3-032-11752-6_19

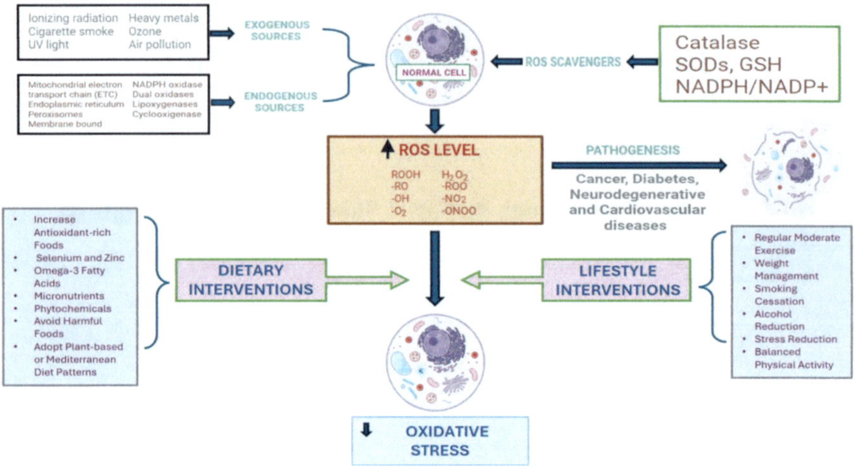

Fig. 19.1 Overview of dietary intervention in ROS-mediated metabolic disorder

in tailored diet and lifestyle treatment, using digital technologies and biomarker-based monitoring, provide precision and efficacy in regulating reactive oxygen species regulation and improving cardiometabolic health outcomes [5]. Researchers and clinicians are developing customized delivery methods and nano-formulations for medicines, despite challenges such as accurate ROS detection, variability in patient reactions, and limitations on long-term adherence. Advancements in molecular diagnostics, big data analytics, and lifestyle design are expected to advance the field of reactive oxygen species and metabolic illnesses, merging nutrition, lifestyle, and medical innovation [1, 6] (Fig. 19.1).

19.2 Role of ROS in Metabolic Disorders

A chronic inflammatory condition is brought on by an increase in the body's oxidative environment, which also serves as substantial risk determinant for the onset of several diseases, like diabetes, arthritis, cancer, and cardiovascular issues [7].

Overproduction of ROS can cause cellular dysfunction, including genetic mutations, altered transport pathways, energy metabolism loss, altered cell signaling, and a decline in biological activity, immunological activation, and inflammation [8]. Proteins, enzymes, and vitamins are examples of antioxidant defenses that organisms have, but overproduction or failure of the antioxidant system can cause ROS to accumulate, leading to the stage known as "oxidative stress" [9].

Protease activity, autophagy, and inflammatory signaling are among the metabolic processes that are impacted by oxidative stress, which leads to chronic inflammation and advancement of metabolic disorders. ROS impact the etiology and severity of major metabolic diseases by altering redox-sensitive signaling pathways linked to lipid metabolism, insulin resistance, and endothelial dysfunction [1].

19.2.1 How ROS Affect Metabolic Health?

The excessive production of ROS damages mitochondrial DNA, proteins, and lipids, which affects the production of cellular energy and produces harmful cycles [1, 10]. They break down polyunsaturated lipids, generating toxic lipid peroxides that compromise cell integrity and signaling. Oxidative alterations impair the function of proteins and nucleic acids, leading to cellular dysfunction and ultimately cell death [10, 11].

Excess ROS not only damages redox-sensitive transcription factors but also promotes inflammation, insulin resistance, and abnormal cell division [10, 11].

Moreover, oxidative stress brought on by dietary stress—especially that which is heavy in fat and carbohydrates—causes chronic illnesses and creates a pathogenic environment [7].

Ros certainly has dual roles:

- Essential roles of ROS in metabolism are signal transduction, regulation of metabolic pathways and homeostatic maintenance.
- Harmful roles of elevated ROS level are mitochondrial dysfunction, modification of biomolecules, impaired cell signaling and redox regulation, promotion of inflammatory pathways, and antioxidant system depletion.

19.2.2 ROS and Major Metabolic Disorders

The primary metabolic conditions that are strongly associated with ROS include obesity, atherosclerosis, type 2 diabetes mellitus, metabolic syndrome, insulin resistance, cardiovascular disease, and nonalcoholic fatty liver disease (NAFLD) (Table 19.1).

19.3 Dietary Interventions to Modulate ROS

Diet significantly affects the balance between oxidants and antioxidants; Western dietary patterns, marked by excessive saturated fats, processed carbohydrates, and animal proteins, coupled with inadequate plant-based fiber intake, lead to elevated oxidative stress and an increased susceptibility to chronic illnesses. These diets may also precipitate metabolic disorders including cardiovascular illnesses, hypertension, type 2 diabetes, gout, metabolic syndrome, obesity, and cognitive decline. The

Table 19.1 The role of ROS in metabolic disorders

Disorder	Influence of ROS	Reference
Obesity	Damages adipocyte function, encourages persistent inflammation in adipose tissue, and raises the risk of insulin resistance and associated metabolic disorders	[12]
Type 2 diabetes mellitus	NADPH oxidase activation and mitochondrial ROS generation due to hyperglycemia	[1]
Atherosclerosis	Endothelial dysfunction and oxidative alteration of LDL caused by ROS	[1]
Insulin resistance	ROS-driven disruption in insulin signaling pathways	[1]
Metabolic syndrome	Excess ROS due to chronic inflammation, insulin resistance, obesity, and dyslipidemia	[1]
Cardiovascular disease	ROS oxidizes LDL cholesterol, promotes the formation of arterial plaques, and damages the vascular endothelium	[1]
Nonalcoholic fatty liver disease (NAFLD)	Damages liver cells due to lipid peroxidation, promotes fibrosis, and speeds up the accumulation of hepatic fat	[12]

incidence of obesity in the USA is increasing, with 72.3% of American men and 64.1% of American women categorized as overweight or obese. Obese people have oxidative stress, marked by elevated reactive oxygen species (ROS) and diminished antioxidant defenses. Oxidative stress is linked to systemic inflammation, endothelial cell proliferation, apoptosis, and increased vasoconstriction, making it an essential factor in endothelial dysfunction. The processes driving oxidative stress associated with obesity remain incompletely elucidated; still, leptin, a hormone produced by adipocytes, is considered crucial. Elevated leptin levels induce oxidative stress in human endothelial cells and the endothelium of obese mouse models [13].

19.3.1 How Diet Influences ROS Antioxidant Defense

Diet contains various dietary antioxidants, including vitamins and polyphenols, which may neutralize reactive oxygen species (ROS) by donating electrons, hence diminishing free radical activity and mitigating oxidative damage to biological components. These chemicals stimulate cellular pathways that enhance the production of intrinsic antioxidant enzymes, including, glutathione peroxidase, superoxide dismutase, and catalase. Minerals such as zinc and selenium are essential cofactors for the production and function of these enzymes. Energy metabolism is affected by caloric consumption, macronutrient composition, and bioactive substances. Diets influencing glycolysis and the pentose phosphate pathway may impact a cell's ability to detoxify reactive oxygen species. Diverse dietary antioxidants facilitate the repair of oxidized molecules and avert more damage, hence preserving cellular structural integrity [14, 15].

19.3.2 Specific Dietary Components and Their ROS-Modulating Effects

19.3.2.1 Vitamin C, Vitamin E, Tocopherols, Tocotrienols, Vitamin D, and Beta Carotene

These are antioxidants with varied properties. Vitamin C neutralizes reactive oxygen species, replenishes vitamin E, and promotes detoxification. Vitamin E protects cell membranes from lipid peroxidation by interfering in the chain reaction of lipid breakdown caused by reactive oxygen radicals. Vitamin D modulates antioxidant enzyme activity and protects against oxidative damage in chronic illnesses. Carotenoids, found in carrots, sweet potatoes, and leafy greens, neutralize singlet oxygen and free radicals [14].

19.3.2.2 Selenium

It is a trace element that serves as a cofactor for glutathione peroxidase, an enzyme that mitigates hydrogen peroxide and lipid hydroperoxides. It additionally collaborates with vitamin E to diminish reactive oxygen species (ROS) and mitigate tissue damage. Zinc is essential for the action of superoxide dismutase [14].

19.3.2.3 Polyphenols and Flavonoids

These are present in fruits, tea, coffee, chocolate, red wine, apples, and onions; they directly scavenge ROS and chelate transition metal ions, thereby inhibiting the generation of hydroxyl radicals. They also enhance endogenous antioxidant enzymes by activating transcription factors such as Nrf2, exemplified by resveratrol [15, 16].

19.3.2.4 Omega-3 Fatty Acids

These are vital polyunsaturated fats, which are necessary for wellness. They are present in botanical sources such as flaxseeds, chia seeds, walnuts, and canola oil. They are also present in fatty fish, seafood, and docosahexaenoic acid (DHA). Omega-3 fatty acids are essential for cellular membrane integrity, energy metabolism, cardiovascular health, and the regulation of reactive oxygen species and oxidative stress. Omega-3 fatty acids are essential in mitigating reactive oxygen species (ROS) and oxidative damage. They activate antioxidant enzymes like GPx and SOD, counteracting surplus ROS. They directly inhibit ROS generation by diminishing membrane composition and enzymatic activity. They also modulate inflammatory pathways by inhibiting NF-κB and limiting pro-inflammatory eicosanoids derived from arachidonic acid. Furthermore, they stimulate AMPK and PPARγ pathways, enhancing mitochondrial fatty acid oxidation and further diminishing ROS production. These measures alleviate the detrimental impacts of oxidative stress. Also they inhibit the IL-23/IL-17 pathway, resulting in lower oxidative stress [17–20].

19.3.2.5 Functional Foods and Bioactive

These encompass cruciferous vegetables like broccoli and Brussels sprouts, which are rich in sulforaphane and isothiocyanates that stimulate the Nrf2 pathway; nuts, whole grains, and seeds, which supply micronutrients and phytochemicals that bolster antioxidant defenses and mitigate ROS burden; and glutathione, a principal intracellular antioxidant that plays an essential role in diminishing hydrogen peroxide and rectifying oxidative damage.

19.3.3 Dietary Patterns and Approaches in Oxidative Stress

A. DASH (Dietary Approaches to Stop Hypertension) Diet: This diet increases endogenous antioxidant levels, reduces lipid peroxidation, improves total antioxidant capacity (TAC), decreases NADPH oxidase activity, reduces Inflammatory and pro-oxidant pathways, also shows positive effects on endothelial function.

    - Examples: Whole grains, vegetables and fruits, fat-free dairy products, fish poultry, beans, and nuts [21].

B. Mediterranean Diet: This diet is abundant in polyphenols and micronutrients, diminishes ROS production, enhances antioxidant status, and mitigates lipid/DNA oxidative damage in both healthy persons and those with metabolic syndrome.

    - Examples: olive oil, fish, nuts, vegetables, whole grains [22].

C. Plant-Based/Vegetarian: The substantial use of natural antioxidants by plants may diminish systemic oxidative stress and inflammation, perhaps reducing ROS via the supply of ample phytonutrients.

    - Examples: Fruits, vegetables, grains, and legumes [4].

19.4 Lifestyle Interventions to Reduce ROS

In the context of reactive oxygen species, a lifestyle intervention focuses on altering lifestyle choices including food, exercise, and other aspects of daily living in order to reduce oxidative stress and enhance general health. Lifestyle changes may be able to impede or diminish the progression of disease and enhance patient outcomes by lowering reactive oxygen species (ROS) levels and enhancing antioxidant capacity.

ROS can be neutralized and oxidative stress indicators can be decreased by implementing plant-based or Mediterranean diets that are high in antioxidants [4]. The body's capacity to handle oxidative stress can be possibly improved by increasing physical exercise because it can improve endogenous antioxidant defenses and systemic redox equilibrium [23, 24]. Stress reduction and quitting smoking are two behavioral changes that can help lower excess reactive oxygen species (ROS) and support better physiological function [24, 25].

The following are the major lifestyle interventions that significantly reduce oxidative stress.

19.4.1 Physical Activity

Exercise may influence oxidative stress levels; however, the link is intricate and contingent upon the form, intensity, and duration of the activity. Consistent low to moderate intensity exercise may enhance endogenous antioxidant defenses and decrease oxidative stress levels [26]. Physical exercise enhances the formation of reactive oxygen species, increases the expression of antioxidant enzymes, and fortifies the cellular antioxidant defense system, therefore diminishing oxidative stress and chronic inflammation. Exercise reduces systemic meta-inflammation, a vital factor in age-related and metabolic illnesses, by enhancing the production of anti-inflammatory cytokines and modulating pro-inflammatory markers such as TNF-α, IL-6, and CRP, therefore breaking the cycle [27, 28]. Physical activity improves energy metabolism in skeletal muscle and tissues, promotes cellular health, resilience to metabolic stressors, and counteracts the negative effects of reactive oxygen species and inflammation. Exercise improves muscular strength, hypertrophy, and functioning, reduces the risks of frailty and sarcopenia, and rectifies immunological dysfunction in adipose tissue caused by obesity-related meta-inflammation [29, 30]. Intensive exercise or persistent overtraining may elevate detrimental reactive oxygen species (ROS) and inflammation without sufficient recovery, requiring tailored intervention strategies [31, 32].

19.4.2 Smoking Cessation

Smoking is a potent inducer of oxidative stress, resulting in the meta-inflammation, a loop of inflammation that is the underlying cause of many chronic illnesses. Cigarette smoke is recognized for releasing substantial amounts of reactive oxygen species (ROS) and free radicals, which can quickly increase oxidative stress in tissues. Smoke elevates the levels of ROS while simultaneously impairing the body's antioxidant mechanisms, leading to heightened oxidative stress. Reactive oxygen species (ROS) activate redox-sensitive signaling pathways and pro-inflammatory genes, resulting in cellular damage and the promotion of autophagy and inflammatory responses, which are prevalent in metabolic diseases and chronic conditions like chronic obstructive pulmonary disease (COPD). Research indicates that quitting cigarette smoking significantly reduces reactive oxygen species levels in lung tissue, diminishes autophagy and inflammatory indicators, and reinstates endothelial function in individuals. This results in enhanced nitric oxide bioactivity and reduced vascular inflammation [33, 34].

19.4.3 Calorie Restriction

Caloric restriction significantly reduces ROS-mediated meta-inflammation by decreasing ROS generation, boosting antioxidant defenses, and reducing systemic inflammatory markers linked to metabolic diseases and obesity. Caloric restriction boosts the regulation of intrinsic antioxidant systems, like superoxide dismutase and glutathione peroxidase, which reduce reactive oxygen species and improve cellular redox equilibrium. Caloric restriction reduces the production of ROS, especially inside mitochondria, by reducing metabolic rate and oxidative stress on cells. Calorie restriction decreases pro-inflammatory cytokines such as TNF-α, IL-6, CRP, and adipokines like leptin that are significant factors of meta-inflammation. Caloric restriction enhances insulin sensitivity, reduces body fat and weight, and fosters mitochondrial activity, hence alleviating oxidative and inflammatory stress at cellular and systemic levels. Overall, calorie restriction is an effective strategy for diminishing ROS-mediated meta-inflammation [35–38].

19.4.4 Stress Management

Psychological and physiological stress increases ROS and oxidative stress, exacerbates chronic inflammation via activating inflammatory pathways, elevating pro-inflammatory cytokines, and impairing cellular function, particularly mitochondrial ROS, which are significantly elevated in the NLRP3 inflammasome. Stress management strategies, including meditation, physical exercise, sleep enhancement, and psychiatric interventions, may reduce stress hormone pathways, such as cortisol and sympathetic nervous system activity, potentially exacerbating mitochondrial dysfunction and the production of reactive oxygen species. These strategies reduce the activation of pathways like the hypothalamic-pituitary-adrenal axis and the sympathetic nervous system, hence lowering mitochondrial ROS production and initiating inflammatory cascades. Effective stress management may mitigate excessive oxidative stress and meta-inflammation by reducing ROS production, inhibiting inflammatory pathways, and bolstering antioxidant defenses, thereby alleviating ROS overproduction and oxidative stress [24, 39, 40].

19.4.5 Dietary Plan

A nicely organized food plan significantly reduces ROS-mediated meta-inflammation by modulating oxidative stress and inflammation via many routes. Integrating fruits, vegetables, nuts, seeds, and olive oil into diets rich in antioxidants and omega-3 fatty acids reduces ROS, prevents oxidative damage, and lessens inflammatory signaling. Whole grains and legumes enhance gastrointestinal health, whereas higher fiber intake reduces inflammatory markers. The Mediterranean, DASH, and traditional diets may reduce inflammation over time. Reducing processed carbohydrates, saturated fats, and sugars alleviates chronic inflammation [41].

19.4.6 Limited Alcohol Consumption

Restricting alcohol intake may diminish ROS-mediated meta-inflammation by attenuating pathways that elevate ROS generation and foster chronic inflammation. Chronic alcohol use increases reactive oxygen species in the liver, diminishes antioxidant defenses, and rises free iron levels, leading to cellular injury and oxidative damage, hence intensifying inflammation [42]. Alcohol-induced oxidative stress triggers inflammatory signaling, enhances the generation of pro-inflammatory cytokines, increases gut–liver immunological contacts, induces mitochondrial dysfunction, and leads to chronic low-grade inflammation in metabolic diseases [43, 44]. Reduced alcohol intake promotes the recovery of antioxidant enzymes, restoring balance between the production and removal of ROS, while maintaining gut barrier integrity and interrupting the cycle of immune activation [45]. In essence, lifestyle interventions may reduce stress hormone activation, enhance mitochondrial function, strengthen antioxidant defenses, and inhibit pro-inflammatory signaling, thereby reducing oxidative stress, improving metabolic health, and mitigating chronic diseases.

19.5 Personalized Nutrition and Lifestyle Medicine

Personalized nutrition and lifestyle medicine is a modern approach, which considers genetic, metabolic, and environmental factors influencing an individual's oxidative stress burden. The procedure involves increasing antioxidant intake, correcting dietary deficiencies, and adopting beneficial lifestyle habits to reduce reactive oxygen species and improve general health [46, 47].

19.5.1 Personalized Nutrition

Personalized nutrition is a scientifically grounded approach that tailors' dietary recommendations according to an individual's genetic, physiological, metabolic, and behavioral information, with the objective of minimizing oxidative stress and improving health outcomes. Personalized nutrition enhances health outcomes by taking into account genetics, health status, lifestyle, and microbiome composition, employing modern technologies and emphasizing long-term health and chronic disease prevention [48].

Personalized Nutrition in the Context to ROS:

19.5.1.1 Macronutrient Adjustments

Macronutrient modifications entail decreasing simple carbohydrates, modifying omega-3/omega-6 ratios, and transitioning from animal to plant proteins for those experiencing elevated oxidative stress, diminishing anti-inflammatory compounds, and enhancing antioxidant consumption [57].

19.5.1.2 Antioxidant Intake and Micronutrient Personalization

Baseline antioxidant requirements among individuals differ according to genetic and lifestyle factors. Modifying dietary consumption according to inadequacies or genetic variations might enhance reactive oxygen species equilibrium and reduce the risk of oxidative stress, for instance, by increasing the intake of flavonoid-rich foods or supplementing with vitamin E [49, 50].

19.5.1.3 Gut Microbiome Modulation

Modifying the gut microbiota, such as assessing for dysbiosis or inadequate fiber fermentation ability, might augment antioxidant synthesis and diminish reactive oxygen species (ROS) through increased consumption of prebiotic fibers or fermented meals [51, 52].

19.5.1.4 Diet Patterns and Disease Risk

Dietary patterns influence disease risk, as Mediterranean diets diminish reactive oxygen species (ROS), low-FODMAP diets are advantageous for persons with irritable bowel syndrome, and weight control enhances antioxidant capacity [53].

19.5.1.5 Technological Tools in Personalized Development

Technological instruments in customized nutrition encompass genetic testing, biomarker surveillance, and digital platforms that utilize user data to enhance antioxidant levels and reduce the risk of reactive oxygen species [50].

19.5.1.6 Lifestyle Medicine

It employs evidence-based strategies like nutrition, physical exercise, sleep optimization, stress reduction, and toxin avoidance to prevent, manage, and reverse chronic diseases, with an emphasis on reducing reactive oxygen species.

Major lifestyle medicines include adopting a plant-based, antioxidant-rich diet comprising legumes, nuts, fruits, vegetables, and whole grains that can reduce signs of oxidative stress. The Mediterranean, vegetarian, vegan, and DASH diets are linked with enhanced oxidative stress indicators. Emphasize dark leafy greens, berries, nuts, and seeds, as well as orange and red fruits and vegetables. Restrict saturated fats, processed meals, and alcohol consumption. Novel nutraceuticals bolster antioxidant protection [4, 24]. Along with that consistent moderate-intensity physical activity, including walking, cycling, swimming, and resistance training, can improve antioxidant capacity and increase enzyme activity; nevertheless, one should exercise caution over excessive exercise [46]. In addition to this, sufficient sleep (7–9 h for adults) enables cellular repair mechanisms to eliminate accumulated reactive oxygen species (ROS). Prolonged sleep deprivation impairs antioxidant systems. Also stress management include mind-body techniques like mindfulness and social interactions to diminish oxidative burden and foster healthy biological responses, including the decline in oxidative stress [46]. To mitigate oxidative stress and inflammation, minimize exposure to ultraviolet radiation, tobacco smoke, industrial chemicals, and air pollution, while also restricting alcohol consumption and processed meals [4]. Holistic lifestyle programs that incorporate food, physical activity, stress management, and social well-being have demonstrated significant enhancements in oxidative indicators, improving antioxidant capacity, decreasing lipid peroxidation, and mitigating cellular damage [4, 46].

19.6 Delivery Methods and Formulation Strategies

Delivery methods are the main essential components in targeting and treating the ROS production in the human body. The following are some of the main key aspects where we can target ROS.

19.6.1 ROS-Responsive Drug Delivery Systems (DDSs)

ROS-responsive materials are intelligent biomaterials engineered to react to ROS, essential in biomedical applications like medication delivery, tissue engineering, and disease diagnostics [54].

Mechanism
ROS-responsive materials exhibit alterations in structure, chemical characteristics, and solubility with exposure to heightened amounts of reactive oxygen species. These alterations are initiated by ROS-induced solubility transition where polymers chemically convert from hydrophobic to hydrophilic, and ROS-mediated bond cleavage, which selectively disrupts chemical linkers within these materials, enabling targeted drug release or disassembly of delivery systems [55, 56].

Types of ROS Responsive Materials

(i) *Sulfur-based polymers*: Sulfur-based polymers, including thioether-containing PPS and poly(L-methionine), are oxidized by ROS, resulting in drug release via a hydrophobic-to-hydrophilic transition and the fragmentation of micelles or vesicles [54].
(ii) *Selenium and Tellurium-based polymers*: Exhibit enhanced sensitivity to reactive oxygen species (ROS) and selective reactivity; yet their usage is limited due to cost and stability issues [54].
(iii) *Boron/boronic acid-based materials*: These materials particularly cleaved by hydrogen peroxide, facilitate site-specific medication delivery in tumors or inflamed tissues with elevated H_2O_2 concentrations [57].
(iv) *Peroxalate*: It is a chemical that is cleaved by H_2O_2, is frequently utilized in diagnostic imaging and in triggered release systems for cargo delivery [54].
(v) *Hydrogels*: ROS-responsive hydrogels are cross-linked networks employed in wound dressings, tissue engineering, and controlled drug administration, owing to their sensitivity to localized reactive oxygen species [58].
(vi) *Nanoparticles, nanogels, and micelles*: These are nanoscale carriers that discharge their payload upon exposure to reactive oxygen species (ROS), facilitating spatially selective therapies, including anticancer nanomedicines [59].

19.6.2 Nanoparticle-Based Drug Delivery Systems

Nanoparticle-based delivery systems are advanced pharmaceutical delivery platforms that exploit elevated levels of ROS in pathological states, including cancer and inflammation. These approaches use the enhanced permeability and retention impact of nanoparticles, which accumulate in target tissues due to impaired vasculature and reduced lymphatic drainage. Fundamental processes of ROS responsiveness include ROS-induced bond breakage, solubility and hydrophobic-hydrophilic transitions, as well as multi-responsive systems that integrate ROS-triggered release with other sensitivity [54, 60].

Types of Nanoparticles Used in Drug Delivery Systems

- Polymeric nanoparticles: These particles are customized for ROS sensitivity by the incorporation of bonds or linkers that decompose in elevated ROS conditions, facilitating site-specific medication release.
- Lipid-based nanocarriers (liposomes and micelles): These encapsulate medications and release them in response to reactive oxygen species stimuli.
- Inorganic nanoparticles: It includes cerium oxide, gold, and silica nanoparticles, which may mimic antioxidant enzymes and directly neutralize reactive oxygen species (ROS).
- Hybrid nanomaterials: Combine properties of organic and inorganic compounds to enhance responsiveness and stability to reactive oxygen species [54, 61, 62].

Mechanism of Nanoparticles Used in Drug Delivery System
It acts via two methods:

- Stimulus-triggered release involves the localized provision of medication at the site of disease.
- ROS-scavenging nanocarriers and product-encapsulated nanoparticles.

Stimuli-responsive nanocarriers can release their therapeutic cargo in response to specific triggers such as pH, temperature, enzymes, or elevated ROS levels. Concurrently, these nanocarriers can help lower excessive reactive oxygen species (ROS), thereby enhancing the delivery and functional activity of antioxidant enzymes or their mimetics [54, 61, 62].

19.6.3 Gene Therapy for ROS Modulation

Gene therapy for the management of ROS is a biological approach aimed at alleviating oxidative stress, which affects ailments including cardiovascular diseases and dementia, by using modified DNA, RNA, or gene-editing technology.

19.6.3.1 Genes Involved in ROS Modulation

(a) *Antioxidant Genes*

- CATALASE GENE decomposes hydrogen peroxide, protects cardiac tissue from ischemia/reperfusion damage, and diminishes oxidative cell death.
- SUPEROXIDE DISMUTASE (SODs) are genes that encode enzymes responsible for eliminating superoxide radicals, therefore lowering blood pressure, minimizing infarct size, and safeguarding vascular and cardiac tissues from oxidative injury.

- OXR1(Oxidative stress resistance 1) is a crucial cellular antioxidant defense molecule, demonstrated to reduce ROS and cellular death in retinal degeneration models and may be beneficial in various neurodegenerative conditions.
- HEME OXYGENASE-1 (HO-1) catalyzes the degradation of pro-oxidant heme into biliverdin, iron ions, and carbon monoxide, all of which have anti-inflammatory and antioxidative properties. HO-1 gene therapy alleviates arterial damage, suppresses neointima development, and promotes cardiac healing after injury by converting pro-oxidant heme into biliverdin, iron ions, and carbon monoxide [63–65].

(b) *Direct ROS Scavenging and Neuroprotection in Neurodegeneration*

It acts by directly neutralizing ROS and preventing oxidative stress-induced cellular apoptosis; antioxidant gene therapy in the brain (e.g., superoxide dismutase, catalase, OXR1) can mitigate neuronal loss in disorders like Alzheimer's, Parkinson's, and postischemic injury [64].

19.6.3.2 Delivery Methods for Gene Therapy

- *Viral method*: Adenoviruses and adeno-associated viruses (AAVs) are often used for efficient gene transfer into somatic cells, particularly in neurological and cardiovascular regions [63].
- *Nonviral method*: To enhance safety and reduce immune reactions, novel technologies use nonviral plasmid DNA delivery and nanoparticles.

19.7 Research Evidences

Studies have shown that lifestyle and dietary modifications significantly diminish metabolic problems mediated by reactive oxygen species.

19.7.1 Preclinical Evidence

Preclinical research from animal models strongly indicates the effectiveness of lifestyle and dietary interventions in alleviating ROS-mediated metabolic disorders. Research on rodents demonstrates that obesity, high-fat diets, and hyperglycemia result in the production of ROS, oxidative damage, inflammation, and impaired insulin signaling. Obese mice consuming high-fat diets exhibit increased production of ROS and lipid peroxidation, which may be alleviated with regular resistance or aerobic exercise, resulting in decline of oxidative stress markers and pro-inflammatory cytokines. Dietary restriction reduces mitochondrial reactive oxygen species generation and enhances antioxidant enzymes, hence alleviating oxidative damage. Natural antioxidants, Mediterranean diets, and phytochemical nutrients improve mitochondrial function, reduce inflammation, and support metabolic health. These preclinical results suggest a correlation between lifestyle modifications, reduced oxidative stress, and the prevention or delay of ROS-induced metabolic and cardiovascular diseases [10, 66–68].

19.7.2 Clinical Evidence

Clinical studies indicate that structured lifestyle interventions, significantly reduce systemic oxidative stress markers in persons with metabolic syndrome, obesity, or type 2 diabetes and enhance endogenous antioxidant responses. Dietary patterns abundant in antioxidants, such as the Mediterranean, DASH, and vegetarian diets, augment antioxidant enzyme activity and reduce markers of lipid peroxidation, hence enhancing oxidative balance and modulating parameters of metabolic diseases. The combination of dietary improvements and increased physical activity leads to significant increases in antioxidant activity and greater reductions in oxidative damage and inflammation [69, 70].

Preclinical and clinical studies indicate that lifestyle and dietary modifications, including the intake of antioxidant-rich foods, regular physical exercise, and the reduction of processed and unhealthy fats, significantly reduce oxidative stress and improve outcomes in metabolic diseases [71, 72].

19.8 Challenges and Limitations

- Macronutrients may exhibit both pro-oxidant and antioxidant effects, depending upon the circumstances. Some lipids, particularly polyunsaturated fats from fish or olive oil, may provide more health advantages than saturated fats; nonetheless, diets high in fats or carbohydrates may elevate the formation of ROS and oxidative stress. The complex interconnections among dietary components complicate the design and interpretation of intervention [73].
- Personal attributes such as genetic predisposition, gender, age, metabolic condition, and preexisting health issues significantly influence the effects of dietary interventions on reactive oxygen species (ROS). Dietary restriction (DR) influences reactive oxygen species (ROS) production variably, contingent upon tissue type, species, and sex. The bulk of rodent research mostly focuses on males, resulting in a scarcity of data about females [67].
- Accurately detecting ROS and oxidative stress is challenging. Common markers, like lipid peroxidation products and protein carbonyls, may not sufficiently reflect dietary effects or encapsulate functional outcomes. Furthermore, there is a deficiency in studies about non-mitochondrial sources of reactive oxygen

species, and a thorough evaluation of oxidative damage to DNA, proteins, and lipids is necessary [67].

- Direct antioxidant supplementation, including vitamins C, E, and flavonoids, is the primary emphasis of several therapeutic approaches. Exogenous antioxidants may disrupt beneficial ROS-mediated signaling pathways and have not consistently shown significant benefits in clinical investigations [70].
- Reactive oxygen species (ROS) are not just detrimental; they are also crucial for appropriate physiological processes and cellular signaling. Consequently, indiscriminate inhibition of reactive oxygen species may jeopardize essential biological functions [73].
- Multiple studies indicate that dietary alterations marginally or inconsistently reduce markers of oxidative stress. Dietary restriction sometimes decreases mitochondrial ROS production; in rats, just 37% of studies indicated a reduction in ROS generation, while 62% showed no change. Moreover, comprehensive studies often report fewer positive outcomes, suggesting potential publication bias favoring good findings [67].
- There is a notable scarcity of comprehensive, long-term, controlled intervention studies in humans, with most material coming from animal models. Ethical considerations, dietary adherence, and practical limitations (such as the monitoring and regulation of actual intake over prolonged durations) impede human studies. Uncertainty persists about the safety and efficiency of dietary control of reactive oxygen species in humans [14].
- Individuals often disseminate research demonstrating positive effects, whereas they less commonly publicize studies yielding negative or null findings. This may lead individuals to perceive dietary modifications as more effective than they really are [67].

19.9 Future Perspectives

Modifications in diet and lifestyle in relation to ROS-mediated metabolic disorders have potential for both prevention and therapy. As research clarifies the molecular basis of oxidative stress, tailored dietary and lifestyle modifications will become essential elements. Advancements in metabolomics, redox biomarkers, and gut microbiome research will improve efficacy and safety. Recent research suggests that combining antioxidant-rich diets with regular physical activity, stress management, and improved sleep hygiene may boost metabolic resilience and promote disease remission, hence decreasing ROS production and systemic inflammation.

Moreover, the use of digital health technologies—such as wearable fitness devices, smartphone applications for food monitoring, and remote coaching—will enhance long-term compliance and individualized feedback, making lifestyle treatments more accessible and sustainable. Clinical studies will refine the intensity, duration, and dietary components of interventions for metabolic phenotypes, promoting a shift toward preventive treatment based on evidence-based lifestyle and dietary changes, potentially reducing ROS-mediated metabolic diseases and improving health outcomes [49, 74].

19.10 Conclusion

Modifications to food and physical exercise within a lifestyle intervention often enhance indicators of oxidative stress (SOD, CAT, GSH) and decrease MDA levels, hence improving overall cardiometabolic health. Consuming an abundance of plant-based foods, like nuts, legumes, vegetables, fruits, and whole grains, together with healthy fats, as seen in the Mediterranean, DASH, or vegetarian/vegan diets, is associated with significant decreases in oxidative stress and other metabolic disorders [70]. Systematic nutritional interventions coupled with consistent physical exercise lead to enhancements in weight, waist circumference, BMI, blood pressure, triglycerides, HDL cholesterol, and insulin sensitivity in individuals with metabolic syndrome. These enhancements are often associated with substantial decreases in several indicators of oxidative stress after several months of treatments. Nevertheless, some complex problems (e.g., cardiac autonomic neuropathy) may need prolonged or more extensive therapies for significant improvement [69]. Moreover, consuming foods that are inherently rich in antioxidants, rather than just relying on supplements, seems to be the most effective method for lowering oxidative stress and enhancing metabolic health. This may aid in halting or perhaps reversing the progression of disorders associated with reactive oxygen species (ROS). The benefits are evident across all demographics and dietary regimens; however, the outcomes may vary for individuals based on the duration of their adherence to the intervention [75]. The most effective and dependable method to mitigate ROS-induced damage and manage or prevent metabolic disorders is to integrate dietary modifications with increased physical activity. Prolonged commitment is crucial, since optimal outcomes manifest over an extended duration of months to years [4].

References

1. Forrester SJ, Kikuchi DS, Hernandes MS, Xu Q, Griendling KK (2018) Reactive oxygen species in metabolic and inflammatory signaling. Circ Res 122(6):877–902
2. Yang S, Lian G (2020) ROS and diseases: role in metabolism and energy supply. Mol Cell Biochem 467(1–2):1–12
3. Samoylenko A, Al Hossain J, Mennerich D, Kellokumpu S, Hiltunen JK, Kietzmann T (2013) Nutritional countermeasures targeting reactive oxygen species in cancer: from mechanisms to biomarkers and clinical evidence. Antioxid Redox Signal 19(17):2157–2196
4. Husain S, Hillmann K, Hengst K, Englert H (2023) Effects of a lifestyle intervention on the biomarkers of oxidative stress in non-communicable diseases: a systematic review. Front Aging 4

5. Hu D, Li Y, Li R, Wang M, Zhou K, He C et al (2024) Recent advances in reactive oxygen species (ROS)-responsive drug delivery systems for photodynamic therapy of cancer. Acta Pharm Sin B 14(12):5106–5131
6. Murphy MP, Bayir H, Belousov V, Chang CJ, Davies KJA, Davies MJ et al (2022) Guidelines for measuring reactive oxygen species and oxidative damage in cells and in vivo. Nat Metab 4(6):651–662
7. Rani V, Deep G, Singh RK, Palle K, Yadav UCS (2016) Oxidative stress and metabolic disorders: pathogenesis and therapeutic strategies. Life Sci 148:183–193
8. Yadav UCS, Rani V, Deep G, Singh RK, Palle K (2016) Oxidative stress in metabolic disorders: pathogenesis, prevention, and therapeutics. Oxid Med Cell Longev 2016(1)
9. Sies H, Berndt C, Jones DP (2017) Oxidative stress. Annu Rev Biochem 86(1):715–748
10. Masenga SK, Kabwe LS, Chakulya M, Kirabo A (2023) Mechanisms of oxidative stress in metabolic syndrome. Int J Mol Sci 24(9):7898
11. Afzal S, Abdul Manap AS, Attiq A, Albokhadaim I, Kandeel M, Alhojaily SM (2023) From imbalance to impairment: the central role of reactive oxygen species in oxidative stress-induced disorders and therapeutic exploration. Front Pharmacol 14
12. Akhigbe R, Ajayi A (2021) The impact of reactive oxygen species in the development of cardiometabolic disorders: a review. Lipids Health Dis 20(1):23
13. Ziaei R, Canas JA, Hajhashemy Z, Porta M Della, Foshati S, editors. Dietary and supplement strategies for the prevention and treatment of metabolic syndrome. Frontiers Media SA; 2024
14. Liu Z, Ren Z, Zhang J, Chuang CC, Kandaswamy E, Zhou T et al (2018) Role of ROS and Nutritional antioxidants in human diseases. Front Physiol 9
15. Lu Y, Wang K, Hu L (2025) Advancements in delivery systems for dietary polyphenols in enhancing radioprotection effects: challenges and opportunities. NPJ Sci Food 9(1):51
16. Rudrapal M, Khairnar SJ, Khan J, Dukhyil AB, Ansari MA, Alomary MN et al (2022) Dietary polyphenols and their role in oxidative stress-induced human diseases: insights into protective effects, antioxidant potentials and mechanism(s) of action. Front Pharmacol 13:806470
17. Kupczyk D, Bilski R, Szeleszczuk Ł, Mądra-Gackowska K, Studzińska R (2025) The role of diet in modulating inflammation and oxidative stress in rheumatoid arthritis, ankylosing spondylitis, and psoriatic arthritis. Nutrients 17(9):1603
18. Heshmati J, Morvaridzadeh M, Maroufizadeh S, Akbari A, Yavari M, Amirinejad A et al (2019) Omega-3 fatty acids supplementation and oxidative stress parameters: a systematic review and meta-analysis of clinical trials. Pharmacol Res 149:104462
19. Bayram SŞ, Kızıltan G (2024) The role of omega- 3 polyunsaturated fatty acids in diabetes mellitus management: a narrative review. Curr Nutr Rep 13(3):527–551
20. Muzammil K, Khaleel AQ, Merza MS, Kyada A, Ariffin IA, Verma S et al (2024) The effects of ω−3 fatty acids on inflammatory and oxidative stress markers in patients with type 2 diabetes mellitus: a systematic review and meta-analysis of controlled trials. Prostaglandins Other Lipid Mediat 175:106887
21. Asemi Z, Samimi M, Tabassi Z, Shakeri H, Sabihi SS, Esmaillzadeh A (2014) Effects of DASH diet on lipid profiles and biomarkers of oxidative stress in overweight and obese women with polycystic ovary syndrome: a randomized clinical trial. Nutrition 30(11–12):1287–1293
22. Marin C, Delgado-Lista J, Ramirez R, Carracedo J, Caballero J, Perez-Martinez P et al (2012) Mediterranean diet reduces senescence-associated stress in endothelial cells. Age (Omaha) 34(6):1309–1316
23. Helmink JH, Meis JJ, de Weerdt I, Visser FN, de Vries NK, Kremers SP (2010) Development and implementation of a lifestyle intervention to promote physical activity and healthy diet in the Dutch general practice setting: the BeweegKuur programme. Int J Behav Nutr Phys Act 7(1):49
24. Sharifi-Rad M, Anil Kumar NV, Zucca P, Varoni EM, Dini L, Panzarini E et al (2020) Lifestyle, oxidative stress, and antioxidants: Back and forth in the pathophysiology of chronic diseases. Front Physiol 11

25. Younge JO, Kouwenhoven-Pasmooij TA, Freak-Poli R, Roos-Hesselink JW, Hunink MM (2015) Randomized study designs for lifestyle interventions: a tutorial. Int J Epidemiol 44(6):2006–2019
26. Ye Y, Lin H, Wan M, Qiu P, Xia R, He J et al (2021) The effects of aerobic exercise on oxidative stress in older adults: a systematic review and meta-analysis. Front Physiol 12
27. Winn NC, Cottam MA, Wasserman DH, Hasty AH (2021) Exercise and adipose tissue immunity: outrunning inflammation. Obesity 29(5):790–801
28. El Assar M, Álvarez-Bustos A, Sosa P, Angulo J, Rodríguez-Mañas L (2022) Effect of physical activity/exercise on oxidative stress and inflammation in muscle and vascular aging. Int J Mol Sci 23(15):8713
29. Radak Z, Suzuki K, Higuchi M, Balogh L, Boldogh I, Koltai E (2016) Physical exercise, reactive oxygen species and neuroprotection. Free Radic Biol Med 98:187–196
30. Simioni C, Zauli G, Martelli AM, Vitale M, Sacchetti G, Gonelli A et al (2018) Oxidative stress: role of physical exercise and antioxidant nutraceuticals in adulthood and aging. Oncotarget 9(24):17181–17198
31. Powers SK, Deminice R, Ozdemir M, Yoshihara T, Bomkamp MP, Hyatt H (2020) Exercise-induced oxidative stress: friend or foe? J Sport Health Sci 9(5):415–425
32. Thomas DT, DelCimmuto NR, Flack KD, Stec DE, Hinds TD (2022) Reactive oxygen species (ROS) and antioxidants as immunomodulators in exercise: implications for heme oxygenase and bilirubin. Antioxidants. 11(2):179
33. Higashi Y (2023) Smoking cessation and vascular endothelial function. Hypertens Res 46(12):2670–2678
34. Morsch ALBC, Wisniewski E, Luciano TF, Comin VH, Silveira G d B, Marques S d O et al (2019) Cigarette smoke exposure induces ROS-mediated autophagy by regulating sestrin, AMPK, and mTOR level in mice. Redox Rep 24(1):27–33
35. Lopes Cantuária V, Rodrigues CM, Dias IR, Ottone V d O, Costa BO, Godinho LF et al (2025) Intense caloric restriction from birth protects the heart against ischemia/reperfusion injury and reduces reactive oxygen species in ovariectomized rats. Antioxidants 14(2):169
36. Lucas AMB, Palacio PB, Cunha PLO, Facundo HT (2022) Effects of calorie restriction on reactive oxygen species production by mitochondrial reverse electron transport, mitochondrial permeability transition pore, and beta-adrenergic stimulation during cardiac hypertrophy
37. Kanikowska D, Kanikowska A, Swora-Cwynar E, Grzymisławski M, Sato M, Bręborowicz A et al (2021) Moderate caloric restriction partially improved oxidative stress markers in obese humans. Antioxidants 10(7):1018
38. Dandona P, Ghanim H, Chaudhuri A, Dhindsa S, Kim SS (2010) Macronutrient intake induces oxidative and inflammatory stress: potential relevance to atherosclerosis and insulin resistance. Exp Mol Med 42(4):245
39. Mittal M, Siddiqui MR, Tran K, Reddy SP, Malik AB (2014) Reactive oxygen species in inflammation and tissue injury. Antioxid Redox Signal 20(7):1126–1167
40. García-Sánchez A, Miranda-Díaz AG, Cardona-Muñoz EG (2020) The role of oxidative stress in physiopathology and pharmacological treatment with pro- and antioxidant properties in chronic diseases. Oxidative Med Cell Longev 2020:1–16
41. Ramos-Lopez O, Martinez-Urbistondo D, Vargas-Nuñez JA, Martinez JA (2022) The role of nutrition on meta-inflammation: insights and potential targets in communicable and chronic disease management. Curr Obes Rep 11(4):305–335
42. Galicia-Moreno M, Gutiérrez-Reyes G (2014) The role of oxidative stress in the development of alcoholic liver disease. Revista de Gastroenterología de México (English Edition) 79(2):135–144
43. Tan HK, Yates E, Lilly K, Dhanda AD (2020) Oxidative stress in alcohol-related liver disease. World J Hepatol 12(7):332–349
44. Tharmalingam J, Gangadaran P, Rajendran RL, Ahn BC (2024) Impact of alcohol on inflammation, immunity, infections, and extracellular vesicles in pathogenesis. Cureus 16:e56923

45. Wang HJ (2010) Alcohol, inflammation, and gut-liver-brain interactions in tissue damage and disease development. World J Gastroenterol 16(11):1304
46. Husain S, Anand C, Weber RM, Hengst K, Gellner R, Bordewick-Dell U et al (2025) Lifestyle medicine as a pathway to longevity: the impact of the healthy lifestyle community program (cohort 2, HLCP-2) on oxidative stress biomarkers in the adult rural German population. Am J Lifestyle Med. https://doi.org/10.1177/15598276251348197
47. Moustakli E, Zikopoulos A, Katopodis P, Dafopoulos S, Paraschos VS, Zachariou A et al (2025) Dietary and lifestyle interventions to mitigate oxidative stress in male and female fertility: practical insights for infertility management—a narrative review. Meta 15(6):379
48. Roman S, Campos-Medina L, Leal-Mercado L (2024) Personalized nutrition: the end of the one-diet-fits-all era. Front Nutr 11
49. Margaritelis NV, Paschalis V, Theodorou AA, Kyparos A, Nikolaidis MG (2018) Antioxidants in personalized nutrition and exercise. Adv Nutr 9(6):813–823
50. Singar S, Nagpal R, Arjmandi BH, Akhavan NS (2024) Personalized nutrition: tailoring dietary recommendations through genetic insights. Nutrients 16(16):2673
51. Carlson DA, True C, Wilson CG (2024) Oxidative stress and food as medicine. Front Nutr 11
52. Jothi Marie Feula A, Saravanan M, Anandhalakshmi S, James Rajesh J (2024) Diet and oxidative stress: a way forward towards prevention of chronic diseases. In: Disease and health research: new insights, vol 10. BP International, pp 116–125
53. Ávila-Escalante ML, Coop-Gamas F, Cervantes-Rodríguez M, Méndez-Iturbide D, Aranda-González II (2020) The effect of diet on oxidative stress and metabolic diseases—clinically controlled trials. J Food Biochem 44(5):e13191
54. Liu J, Jia B, Li Z, Li W (2023) Reactive oxygen species-responsive polymer drug delivery systems. Front Bioeng Biotechnol 11
55. Ren X, Liu H, Wu X, Weng W, Wang X, Su J (2022) Reactive oxygen species (ROS)-responsive biomaterials for the treatment of bone-related diseases. Front Bioeng Biotechnol 9:820468
56. Zhang Y, Jiang M, Wang T (2024) Reactive oxygen species (ROS)-responsive biomaterials for treating myocardial ischemia-reperfusion injury. Front Bioeng Biotechnol 12
57. Zhou J, Fang C, Rong C, Luo T, Liu J, Zhang K (2023) Reactive oxygen species-sensitive materials: a promising strategy for regulating inflammation and favoring tissue regeneration. Smart Mater Med 4:427–446
58. Pu M, Cao H, Zhang H, Wang T, Li Y, Xiao S et al (2024) ROS-responsive hydrogels: from design and additive manufacturing to biomedical applications. Mater Horiz 11(16):3721–3746
59. Oh H, Jeong E, Lee JS, Kim J, Lee D, Kim BS et al (2023) ROS-responsive PEGylated ferrocene polymer nanoparticles with improved stability for tumor-selective chemotherapy and imaging. Mater Today Bio 22:100774
60. Liang J, Liu B (2016) ROS-responsive drug delivery systems. Bioeng Transl Med 1(3):239–251
61. Vasvani S, Vasukutty A, Bardhan R, Park IK, Uthaman S (2024) Reactive oxygen species driven prodrug-based nanoscale carriers for transformative therapies. Biomater Sci 12(17):4335–4353
62. Mauricio MD, Guerra-Ojeda S, Marchio P, Valles SL, Aldasoro M, Escribano-Lopez I et al (2018) Nanoparticles in medicine: a focus on vascular oxidative stress. Oxidative Med Cell Longev 2018(1)
63. Levonen AL, Vähäkangas E, Koponen JK, Ylä-Herttuala S (2008) Antioxidant gene therapy for cardiovascular disease. Circulation 117(16):2142–2150
64. Navarro-Yepes J, Zavala-Flores L, Anandhan A, Wang F, Skotak M, Chandra N et al (2014) Antioxidant gene therapy against neuronal cell death. Pharmacol Ther 142(2):206–230
65. Sahu B, Leon LM, Zhang W, Puranik N, Periasamy R, Khanna H et al (2021) Oxidative stress resistance 1 gene therapy retards neurodegeneration in the *rd1* mutant mouse model of retinopathy. Invest Opthalmol Vis Sci 62(12):8
66. Effting PS, Brescianini SMS, Sorato HR, Fernandes BB, Fidelis G dos SP, Silva PRL da et al (2019) Resistance exercise modulates oxidative stress parameters and TNF-a content in the heart of mice with diet-induced obesity. Arq Bras Cardiol 112(5):545–552

67. Walsh ME, Shi Y, Van Remmen H (2014) The effects of dietary restriction on oxidative stress in rodents. Free Radic Biol Med 66:88–99
68. Manzoor MF, Arif Z, Kabir A, Mehmood I, Munir D, Razzaq A et al (2022) Oxidative stress and metabolic diseases: relevance and therapeutic strategies. Front Nutr 9
69. Pennathur S, Jaiswal M, Vivekanandan-Giri A, White EA, Ang L, Raffel DM et al (2017) Structured lifestyle intervention in patients with the metabolic syndrome mitigates oxidative stress but fails to improve measures of cardiovascular autonomic neuropathy. J Diabetes Complicat 31(9):1437–1443
70. Jiang S, Liu H, Li C (2021) Dietary regulation of oxidative stress in chronic metabolic diseases. Foods 10(8):1854
71. Panicker LM, Zhang JH, Dagur PK, Gastinger MJ, Simonds WF (2010) Defective nucleolar localization and dominant interfering properties of a parafibromin L95P missense mutant causing the hyperparathyroidism–jaw tumor syndrome. Endocr Relat Cancer 17(2):513–524
72. Wang Y, Luo D, Jiang H, Song Y, Wang Z, Shao L et al (2023) Effects of physical exercise on biomarkers of oxidative stress in healthy subjects: a meta-analysis of randomized controlled trials. Open Life Sci 18(1)
73. Görlach A, Dimova EY, Petry A, Martínez-Ruiz A, Hernansanz-Agustín P, Rolo AP et al (2015) Reactive oxygen species, nutrition, hypoxia and diseases: problems solved? Redox Biol 6:372–385
74. Clemente-Suárez VJ, Martín-Rodríguez A, Redondo-Flórez L, López-Mora C, Yáñez-Sepúlveda R, Tornero-Aguilera JF (2023) New insights and potential therapeutic interventions in metabolic diseases. Int J Mol Sci 24(13):10672
75. Olalekan SO, Bakare OO, Okwute PG, Osonuga IO, Adeyanju MM, Olalekan RO (2024) The role of nutraceuticals in managing metabolic syndrome: a review of clinical studies. Egypt J Intern Med 36(1):108

Chapter 20
Role of Pro-, Pre-, and Post-Biotic Targeting Gut Microbes for ROS Inhibition

Navpreet Kaur, Shubham Kumar, Narendra Kumar Pandey, Dileep Singh Baghel, Saurabh Singh, Sumel Ashique, and Bimlesh Kumar

20.1 Introduction

Reactive oxygen species (ROS) has been defined as a class of endogenous, highly reactive, oxygen (and nitrogen)-bearing molecules [1]. About 2.5 billion years ago, cyanobacteria's oxygenic photosynthesis increased the amount of molecular oxygen (O2) in the atmosphere. This allowed aerobic respiration to evolve, which in turn led to the emergence of complex eukaryotic organisms [2, 3]. Molecular oxygen is a key component of cellular respiration for all aerobic species that are currently alive. However, some oxygen derivatives are extremely harmful to cells. Gerschman et al. postulated in the 1950s that free radicals containing oxygen were the cause of aerobic organisms' harmful effects [4, 5]. Over time, a new class of endogenous, highly reactive molecules that contain both oxygen and nitrogen has been defined by the terms ROS (reactive oxygen species), ROI (reactive oxygen intermediates), and RNS (reactive nitrogen species) [2].

ROS is described as oxygen-based radicals and non-radicals, generated inside the body organelles. Most of the ROS formed are short-lived and therefore they generate close proximity effect [3]. Nevertheless, ROS can even travel longer distances and show their impact at other locations in the cell as well. Due to reactive characteristics, ROS demonstrate a high damaging effect for certain biomolecules including lipid, proteins, carbohydrates, and nucleic acids, resulting in cellular damage and even death under severe conditions [6]. Overproduction of ROS has been evidently found toxic for living cells [3]. The most known examples of ROS are hydrogen peroxide (H2O2), superoxide anion (•O2 −), and hydroxyl radicals (•OH).

N. Kaur · S. Kumar · N. K. Pandey · D. S. Baghel · S. Singh · B. Kumar (✉)
School of Pharmaceutical Sciences, Lovely Professional University, Jalandhar, Punjab, India

S. Ashique
Department of Pharmaceutical Technology, Bharat Technology, Uluberia, West Bengal, India

S. Ashique, N. Mishra (eds.), *Reactive Oxygen Species in Metabolic Inflammation*,
https://doi.org/10.1007/978-3-032-11752-6_20

The subclass of ROS that contains nitrogen compounds is known as reactive nitrogen species (RNS) [7]. When physiologically present, ROS and RNS both play significant roles in regular cellular processes like preventing infection, controlling various intercellular signaling pathways, and promoting health [4]. Numerous substances, including heavy metals, tobacco, smoke, drugs, xenobiotics, pollutants, and radiation, can cause the production of ROS. According to a number of experimental investigations, ROS can either promote or suppress tumor growth. Through a sustained increase in cell cycle inhibition, the elevated ROS level can stop tumor growth [8, 9]. Both intrinsic and extrinsic pathways can cause apoptosis when ROS levels are elevated. Since the majority of chemotherapeutic agents produce ROS in order to trigger cell death, ROS is thought to be a tumor-suppressing agent [10]. ROS initiate apoptosis through their cytotoxic effects; however, when present in excessive amounts, they can contribute to carcinogenesis. Elevated ROS levels promote uncontrolled cell proliferation, ultimately leading to cancer and other related diseases [5]. The seven isoforms of the growing family of transmembrane NADPH oxidases (NOXs), a superoxide-generating system, are among the cytosolic enzyme systems that contribute to the production of ROS [11]. Another important source of intracellular ROS production is mitochondria. Following the transfer of a single electron to oxygen, mitochondrial superoxide radicals are primarily generated at two key sites within the electron transport chain: Complex I (NADH dehydrogenase) and Complex III (ubiquinone cytochrome c reductase) [12]. Under hyperbaric conditions, these two mitochondrial sites in vitro convert 1–2% of the oxygen molecules consumed into superoxide anions. Several intracellular enzymes, including the flavoenzyme ERO1 in the endoplasmic reticulum, xanthine oxidase, cyclooxygenases, cytochrome p450 enzymes, lipoxygenases, flavin-dependent demethylase, oxidases for polyamines and amino acids, and nitric oxide synthases that generate oxidants as part of their regular enzymatic function, are additional cellular sources of ROS production in addition to the mitochondria and NADPH oxidases [2, 13].

20.2 Impact of ROS on Human Health

ROS is a key molecule responsible for generation of inflammatory responses. Potentiating the ROS generation by polymorphonuclear neutrophils (PMNs) at the site of inflammation leads to endothelial dysfunction and tissue injury [14]. The vascular endothelium plays a vital role in regulating the passage of macromolecules and inflammatory cells from the bloodstream into surrounding tissues. Excessive ROS can disrupt endothelial integrity and damage neuronal and supporting cells, triggering inflammatory cascades, oxidative stress, and microvascular dysfunction. These events contribute not only to neoplastic, cardiovascular, and neurodegenerative diseases but also play a key role in the development of neuropathic pain by promoting nerve injury, sensitization, and abnormal neuronal signaling [15, 16]. ROS are highly reactive and cause multiple effects on organisms, either directly or through their derivatives, notably NO-derived radicals or RNS [17]. A common feature of different ROS is their nature to cause oxidative damage to proteins, DNA, and lipids. In other scenarios, humans can generate ROS as signaling molecules to control various processes including defense against pathogen and programmed cell death [18]. Accumulation of such information shows oxidative damage of tissue, and cellular components may act as a primary or secondary cause of ROS disease [19]. Studies claim that oxidative stress may alter the gastrointestinal motility in obesity, and even postinfectious disorders [20]. The study was conducted in order to investigate the impact of oxidative stress in gastrointestinal (GIT) pathologies, and how ROS in the signaling lead to alterations of the gastrointestinal tract (GIT). In addition, potential therapeutic use of antioxidants for the treatment of inflammatory gastrointestinal diseases are considered crucial [20, 21]. Therefore, further investigations are needed to clarify the mechanisms underlying the oxidative stress-related signaling can contribute to the alteration of the GIT mucosa in order to devise effective preventive and curative therapeutic strategies [20].

GIT track of the human body contains a variety of gut bacteria that are responsible for maintaining well-being. Gut microbiota are responsible for regulating the biological activity by generating a variety of metabolites. This REDOX equilibrium is controlled by the microbiota either directly or indirectly as it might affect the immune system by disrupting intracellular signaling and promoting inflammation [22, 23]. REDOX equilibrium mainly depends on the production of ROS inside and maintains the oxidative defense, thereby maintaining the oxidative damage of the cellular components [24]. Antioxidant defense of the body includes enzymatic and nonenzymatic antioxidants. The nonenzymatic antioxidant includes tripeptide glutathione and the proteins thioredoxin 1 and thioredoxin 2. On the other hand, enzymatic defense includes catalase (CAT), superoxide dismutase (SOD1/2/3), glutathione peroxidase (GPX), and glutathione reductase (GSR) [25]. Recent studies reported that enteric bacteria are also responsible for generating ROS. Several studies have described that intestinal bacteria are able to influence inflammation by manipulating the ubiquitin-proteosome pathway. Ubiquitination is a pathway responsible for regulating the biochemical events by preventing the degradation of protein molecules. The NF-κB also employs ubiquitination [26].

20.3 Need for Alternative Therapeutic Approaches: Why?

Failure of the conventional therapies leads to non-satisfaction in the patients suffering from the GIT disorders. Treatment options for functional gastrointestinal disorders (FGID) have limited efficacy to prevent the symptoms [27]. As a result, patients are looking for alternatives and nonconventional treatments. In a UK sample, 26% of patients with gastrointestinal symptoms willing to receive complementary and alternative medicine (CAM) therapy and 48% of those with irritable bowel syndrome (IBS) used CAM and found it satisfactory [28]. GIT is a complex organ, so maintenance of GIT organs is priority as it performs multiple functions [29]. GIT disturbances are quite common and lead to a health burden on the healthcare system. Pharmacological drugs produce symptomatic relief in the patients but are not effective for long-term treatments. Recently, the utilization of herbal treatment for GUT ailments have surged in developed as well as developing countries [30]. Dietary supplements like probiotics, prebiotics, and post-biotics helps to prevent and treat various diseases. Probiotics, prebiotics, and post-biotics are also responsible for modulating the gut microbiota. Understanding the effects and mechanisms of probiotics, prebiotics, and post-biotics for clinical utility is important. Prebiotics, post-biotics, and even probiotics are well-accepted for human health and need regulation to implement in the healthcare industry [31]. Management of GIT disorders even needs lifestyle changes along with conventional or alternative therapies. FGID are common but complex disorders need a healthy therapeutic relationship and a holistic approach to treat the patient [32].

20.4 Gut Microbiota and Homeostasis

Gut microbes include the community of small microorganisms present in the intestine and promote the human health by performing physiological functions. In physiological conditions, microbiota is responsible for maintaining the homeostasis; however, internal composition (dysbiosis) can lead to inflammatory bowel disease, ulcerative colitis, and even Crohn's disease [33, 34]. The occurrence of inflammatory conditions has rapidly increased in the last decade, becoming a significant problem for the healthcare system and a true challenge in finding novel therapeutic solutions. The issue is that, despite numerous studies, the etiopathogenesis of inflammatory bowel disease is not completely clear. Based on current knowledge, chronic intestinal inflammation occurs due to altered intestinal microbiota and environmental factors, as well as a complex interplay between the genetic predisposition of the host and an inappropriate innate and acquired immune response [35]. It is important to note that the development of biological and immunomodulatory therapy has led to significant progress in treating inflammatory bowel disease. Certain lifestyle changes and novel approaches—including fecal microbiota transplantation and nutritional supplementation with probiotics, prebiotics, and symbiotic—have offered solutions for dysbiosis management and paved the way toward restoring a healthy microbiome, with only minimal long-term unfavorable effects [34, 36].

The composition of gut bacteria in the host body determines the susceptibility to changes. The microbiota is also responsible for maintaining a normal host physiology, from nutritional health to behavior and even the body stress response [37]. Additionally, they play a crucial role in causing multiple diseases, affecting the body organs. The maintenance of a balanced gut microbial composition is essential for regulating various physiological responses. Disruption of this balance—or the absence of key microbial populations—can impair these responses. In conditions involving excessive ROS, such disturbances may contribute to hepatotoxicity and related abnormalities [38]. So, it's important to maintain homeostasis of the intestinal mucosa. The mechanisms by which microbiota exert detrimental influences or beneficial impact remain largely unclear. The advances in the analysis of gut microbiota will enhance the knowledge and provide the role in maintaining health and disease prevention [39]. The gut community is also responsible for maintaining normal human physiology, immune responses, and metabolism of nutrients So any disruption in the functioning of the gut bacteria leads to GIT conditions like IBD and even obesity. Initially, determining the composition of gut bacteria was difficult; but in recent years, the research based on culture-based techniques were identified. The European Metagenomics of the human intestinal tract (MetaHIT) and the US human Microbiome Project (HMP) have worked on large-scale sequencing and established the baseline healthy gut microbiota and reported its alteration in a disease state [40]. MetaHIT mainly work on investigating the relationship between the gut microbiome and intestinal pathologies, particularly obesity and IBD [41]. The HMP examined microbiota diversity in multiple body sites in healthy subjects, including the GI tract, to determine the baseline composition of healthy human gut. The intestinal microbiota generates numerous health benefits like, for instance, the pathogen protection, nutrition, host metabolism, and immune modulation [42].

20.4.1 Disruption of Gut Homeostasis by ROS

Oxidative signals play a vital role in activating the immune system. During this process, the body shifts from cellular respiration to glycolysis, leading to the production of reactive oxygen species (ROS). These ROS are crucial for regulating a proper immune response. Additionally, cells like monocytes, B cells, and neutrophils use ROS as part of their defense strategy against harmful microbes [43]. Normally, the body maintains a balance between pro- and anti-inflammatory responses. This balance is strongly influenced by the gut microbiota, which help regulate the REDOX system. Beneficial bacteria often have antioxidant effects and help reduce inflammation, while harmful bacteria can disrupt this balance, promoting inflammation and oxidative stress [25]. Microbial-induced reactive oxygen species (ROS) play a role in promoting cell growth, movement, and shaping the innate immune responses. This highlights how gut bacteria affect the regulatory systems of the intestinal lining [44]. These microbes thrive in a nutrient-rich environment and support the host by breaking down complex carbohydrates, producing vitamins and micronutrients, aiding immune system development, and preventing harmful bacteria from taking hold. ROS are short-lived, highly reactive molecules formed from the incomplete reduction of oxygen, including both radical types like superoxide (O_2^-) and non-radical forms like hydrogen peroxide (H_2O_2) [45]. In high concentrations, ROS are essential for immune defense, helping phagocytes destroy ingested pathogens. In nonimmune cells, however, ROS are often seen as harmful by-products of mitochondrial respiration, capable of damaging vital structures like membranes and DNA [46].

The intestinal lining consists of various specialized cell types that react differently to beneficial microbiota and harmful pathogens—promoting immune tolerance in the former case and triggering inflammatory responses in the latter [38]. Reactive oxygen species (ROS), generated by resident mucosal cells or newly recruited innate immune cells, play a critical role in defending against microbes and regulating signaling pathways, including those necessary for tissue repair. Deficient ROS production, often caused by genetic mutations affecting NADPH oxidases, has been linked to conditions such as Crohn's disease and pancolitis. Conversely, excessive ROS levels—resulting from increased oxidase activity or disrupted mitochondrial function—have been implicated in the development of ileitis and ulcerative colitis [46].

ROS in the gut are primarily generated by microorganisms, organelles, and enzymes, with NOX1 being the key enzyme in intestinal epithelial cells. Probiotics help maintain gut balance by regulating ROS, enhancing antioxidant levels, supporting immunity, and protecting the intestinal barrier. They are also beneficial in managing ROS-linked conditions like IBD and colon cancer. Approaches such as pro-biotics, post-biotics, and fecal microbiota transplantation (FMT) show promise in reshaping gut microbiota for personalized treatment of gastrointestinal disorders [47].

Redox signaling within the gastrointestinal mucosa operates through a finely regulated balance. In response to microbial threats, infiltrating immune cells—particularly neutrophils—deploy powerful antimicrobial strategies, including the production of reactive oxygen species (ROS), to defend compromised mucosal surfaces [48]. However, excessive or uncontrolled ROS release from neutrophils can cause unintended damage to nearby tissues. Therefore, preserving and restoring the integrity of the intestinal mucosal barrier is critical for host protection [49]. Traditionally, research on redox biology has emphasized its role in pathological conditions. More recently, emerging evidence highlights the beneficial effects of low, nontoxic levels of oxidative stress in maintaining mucosal equilibrium, modulating controlled inflammatory responses, and facilitating tissue repair [50]. Furthermore, resident gut microbiota has been found to activate redox signaling pathways in the mucosa, which involves enzymatic systems that resemble, but are not identical to, those found in professional phagocytes [23].

20.4.2 Interaction Between Gut Microbiota and Immune Systems

Gut microbiota and a host's innate and adaptive immune systems have a dynamic interaction and have a crucial impact on maintaining homeostasis and inhibiting inflammation. The gut microbiota helps to metabolize proteins as well as carbohydrates and even helps to form vitamins and produce metabolic products [51]. As a defense action, epithelial cells of the gut generate a mucosal barrier in order to

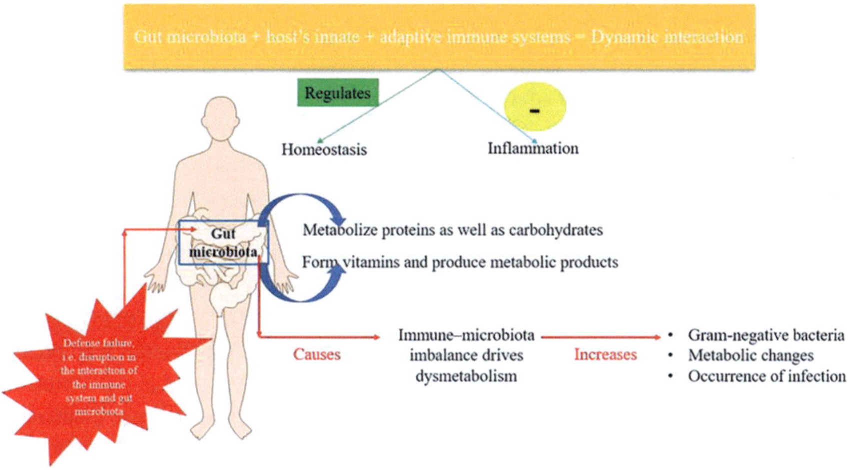

Fig. 20.1 Dynamics of gut microbiota, host's innate, and adaptive immune systems

segregate microbiota from host immune cells and reduce intestinal permeability. The disruption in the interaction of the immune system and gut microbiota can generate excessive gram-negative bacteria, leading to elevation in metabolic changes, host immunity, and potentiate the occurrence of infection [52].

Intestinal bacteria also help to maintain the health and disease-related physiology and pathophysiology in the host body. One of the mechanisms influenced by the gut microbiota is its interaction with the immune system. The presence or absence of specific immunological factors can determine disease susceptibility, and propose the identification and characterization of these bacteria in human. Studies also hypothesized that bacterial IgA-coating could be used to identify specific members of the microbiota that might interact with and impact the intestinal immune response and disease susceptibility [53]. Recent studies have demonstrated that not all commensals are equal in terms of their impact on the host immune system; indeed, specific commensals can exert dramatic effects on host immunity and disease susceptibility (Fig. 20.1).

We propose that these immunologically important commensals can be split into two broad categories based on their overall effects on the immune system [54].

20.4.3 Probiotics: Mechanism and Therapeutic Potential

Gut microbes generate a healthy environment and promote the well-being of the individual. Probiotics contain live organisms that, upon administration, promote gut health. Probiotic is a mixture of Latin and Greek words introduced in the year 1953 by German researcher Werner Kollath [55]. Prebiotics are categorized as nutrients and are degraded inside the gut by microbes. Prebiotics on degradation release short chain fatty acids, which are capable of providing health benefits to the GIT and other organs. Fructo-oligosaccharides and galacto-oligosaccharides are the prebiotics that exist in foods and produce numerous health benefits [56]. The main criteria used to classify a compound as a prebiotic are resistant of prebiotics to acidic pH of stomach, its fermentation by intestinal microbiota, and the growth inside intestine can improve host health. There are many types of prebiotics. Most of them are subtypes of carbohydrate groups and are oligosaccharide carbohydrates (OSCs). This category consists of inulin and fructo-oligosaccharide or oligofructose. Their structure is a linear chain of fructose with β (2 → 1) linkage. They usually have terminal glucose units with β (2 → 1) linkage. Inulin has DP of up to 60, while the DP of FOS is less than 10 [57]. Probiotics are mainly nonpathogenic microbes that provide health benefits on administered inside the human body. Molecular processes and applications of probiotics will be researched in recent years in order to evaluate its mechanism of actions [58]. Probiotics exert its mechanism by promoting the intestinal barriers. Probiotics inhibit binding of pathogens to intestinal mucosal cells and modulate the immunity of host health [59]. The actions of probiotics involve elevation of the gut bacteria that leads to reduction in nutritional intolerances, hence promoting the absorption of macro- and micronutrients and easing allergy symptoms in those who are sensitive. Probiotics can be taken by adding them to foods or drinks, including both dairy and nondairy options, or by using supplements Many fermented foods contain active microbes that are genetically similar to the strains used as probiotics. Studies show that fermented foods improve their nutritional value by changing substrates and creating bioactive and bioavailable end-products [60].

Probiotics structurally consist of complex carbohydrates (such as inulin and other fructo-oligosaccharides) that are utilized by the microorganisms inside the gastrointestinal tract. Prebiotics and probiotics are frequently combined and marketed as synbiotics in commercial products. Additionally, probiotics differ from post-biotics as post-biotics contain dead, whole, or fragmented microorganisms, along with or without their metabolites, and they provide health benefits to the host. Probiotics can be dispensed in different dosage form according to their mechanism and stability. Liquid formulations require to maintain the viability of the bacteria, hence they should be appropriately stored [61, 62].

Probiotics formulation containing *Escherichia coli* and lactic acid bacteria (LAB) helps to treat a wide variety of intestinal issues. Natural probiotics produce beneficial effects by generating immunomodulatory response on the host [63]. Lactic acid bacteria (LAB), such as *Lactococcus* and *Streptococcus*, are found in the host digestive system, including the ileum, jejunum, and colon. The most widely used LAB species for cytokine delivery is *Lactococcus lactis*. The presence of *Lactococcus* species is non-commensal, as they reside in the gut only transiently [64]. Most commonly used gram-negative strain of probiotics, *Escherichia coli* Nissle, was isolated from feces. The researchers evaluated the biosafety of EcN as probiotics in numerous trials to investigate the action. EcN was formulated in the drug called Mutaflor, which helps to treat infectious diarrheal diseases and IBD as well [65]. The formulation of probiotics is crucial as it requires to maintain the viability of the bacterial strain. Different methods have been included in order to enhance the stability of probiotics during industrial procedures. Modification in the production as well as formulation of products and microbiological approaches helps to reduce resistance. The adaptation and selection of the natural strain as probiotics has gained commercial interest. Stability of the probiotics can be altered by a number of factors includes including pH changes, temperature of storage, water content, and oxygen level of the formulation. Stability is a parameter to consider during manufacturing, and even the viability of the formulation should be maintained during gastrointestinal transit [66].

20.4.3.1 Antioxidant, Anti-Inflammatory, and Immunomodulatory Properties

Probiotics generate numerous health benefits when delivered in an adequate amount inside the human body. The protective actions of probiotic strains, including anti-inflammatory and antiproliferative effects against cancer cells, antagonism against pathogenic bacteria, cholesterol reduction, and antioxidant activity, were tested in the study [67]. Reactive oxygen species (ROS) are considered the common cause of gastrointestinal inflammation, even in cases of inflammatory bowel disease (IBD) and colorectal cancer (CRC) [68]. Some probiotic strains, such *Enterococcus faecium*, *Oenococcus oeni, Lactobacillus* spp., and *Bifidobacterium* spp., have strong antioxidant action by the reduction of 2,2-diphenyl-1-picrylhydrazyl radical (DPPH), antioxidant iron (FRAP), and scavenge O_2 radicals [69]. In a study, different strains of gram-negative and gram-positive strains were isolated. Subsequently, bacterial strains were tested for antioxidant and anti-inflammatory properties in vivo using chronic colitis model. The results reported the potential of LAB isolated from the agave fermentation stage in mezcal production. *Lactobacillus rhamnosus* LM07 and *Lactobacillus plantarum* LM17 strains represent potential candidates for developing new probiotic supplements to treat inflammatory bowel disease (IBD) [66].

Medicinal plants that produce lactic acid bacteria have positive health effects. Lactobacillus *agilis* isolated from plant origin show antibacterial, antioxidant, antidiabetic and anti-inflammatory properties [70]. Potential probiotic bacteria have limited adverse effects with potential to act as biotherapeutic agent, a sustainable substitute for antibiotics, and an antioxidant, anti-inflammatory, and antidiabetic agent [71]. *Lactobacillus* strains were isolated from the medicinal plants and the safety profile was evaluated. Out of 115 strains, 29 (25%) were found to be safe for human use [72]. The antibiotic assay methods were tested on the isolated strains using different antibiotics. The isolated strain of *Lactobacillus agilis* was tested by sequence analysis of 16 s rRNA gene and then in vitro antibacterial, antioxidant, antidiabetic, and anti-inflammatory agents were tested. Free radical scavenging activities and inhibition of α-amylase activities for *L. agilis* were found relative to standard drug values as 68% and 73% and 51.3% and 65.3%, respectively [70]. The in vitro anti-inflammatory assay showed 61.6% (*L. agilis*) while it showed 69%

(aspirin) activity for denaturation albumin protein. The results suggested that *L. agilis* can be used as a potential probiotic strain as well to produce nutraceuticals [73].

Nowadays, the interest toward products containing probiotics is growing due to their potential health benefits to the host and research is focusing on new probiotic microorganisms [74]. The present work has focused on the characterization of indigenous *Saccharomyces cerevisiae* strains, isolated from different food matrixes, with the goal to select strains with probiotic or health-beneficial potential [75]. A preliminary screening performed on 50 *S. cerevisiae* indigenous strains, in comparison to a commercial probiotic strain, allowed to individuate the most suitable ones for potential probiotic aptitude [76]. Fourteen selected strains were tested for survival ability in the gastrointestinal tract and finally, the strains characterized for the most important probiotic features were analyzed for health-beneficial traits, such as the content of glucan, antioxidants, and potential anti-inflammatory activities [77]. Three strains, 4LBI-3, LL-1, TA4–10, showing better attributes compared to the commercial probiotic *S. cerevisiae* var. *boulardii* strain, were characterized by interesting health-beneficial traits, such as high content of glucan, high antioxidant, and potential anti-inflammatory activities. Our results suggest that some of the tested *S. cerevisiae* strains have potential for probiotics and are candidates for different applications, such as dietary supplements and starter for the production of functional foods or as probiotics to be used therapeutically [77].

20.4.4 Prebiotics: Mechanism and Therapeutic Potential

Prebiotics belongs to the nutrients that undergo degraded process inside the gut by the gut microbiota. The benefits of using probiotics have gained interest in the recent years. The prebiotics help to feed the intestinal microbiota, hence on degradation short-chain fatty acids are produced that gain entry into the circulation to produce health benefits on the gastrointestinal and other organs [78]. Fructo-oligosaccharides and galacto-oligosaccharides are the types of commonly used prebiotics as they possess numerous health benefits. Health benefits are the main reason for considering the use of prebiotics in the diet; moreover, they also show better stability and storage when compared to the probiotics [57, 77]. The bacteria *Bifidobacterium* and *Lactobacillus* both are beneficial for human health and get amplified by prebiotics. The main mechanism of action of prebiotics includes maintenance of intestinal homeostasis after undergoing fermentation, production of fatty acids with non-carbohydrates molecules; fatty acids production maintains metabolic function and immune system. Prebiotic compounds must withstand the acidic pH of the stomach and resist hydrolysis. Other factors to be considered while adopting the prebiotics are that it should not be absorbed in the gastrointestinal tract and can be fermented by intestinal bacteria [57, 79].

Different types of prebiotics include fraction shaving insulin and fructo-oligosaccharide or oligofructose. Their structure is a linear chain of fructose with β (2 → 1) linkage. They usually have terminal glucose units with β (2 → 1) linkages

[56]. Inulin has DP of up to 60, while the DP of FOS is less than 10. Fructan was investigated to produce lactic acid bacteria and its chain is important in fermentation. Another prebiotics includes Galacto-oligosaccharides (GOS), which consist of the GOS with excess galactose at C_3, C_4 or C_6 and another GOS produced from lactose through enzymatic trans-glycosylation [80]. *Bifidobacteria* and *Lactobacilli* are stimulated inside the body of infants while *Enterobacteria*, *Bacteroidetes*, and *Firmicutes* are stimulated with the GOS. Resistant starch and glucose-derived oligosaccharides are also types of prebiotics that show beneficial effects [57].

20.4.4.1 Prebiotics in Hepatic and Metabolic Disorders

Short-chain fatty acids including acetic acid, propionic acid, and butyric acid are produced from carbohydrates during metabolism. Fermentation generated by the prebiotics produced energy for epithelial cells of the colon. SCFAs generates protection against gastrointestinal disorders, cardiovascular diseases, and cancers by inducing apoptosis [81]. Furthermore, prebiotics also helps to promote the immune system, produce vitamin B, inhibit pathogen growth, and lower blood ammonia. Moreover, SCFAs decrease glucagon levels in a dose-dependent manner, improve glucose tolerance, and activate glucagon-like peptide1 (GLP-1), which can stimulate the elevation of insulin production and increase insulin sensitivity [82]. In patients with type 2 diabetes, SCFAs continue to play crucial roles. Interestingly, some research has shown that individuals with type 2 diabetes had a significantly lower number of bacteria that produce SCFAs. These SCFAs exhibit several biological effects in addition to binding to G-protein-coupled receptors (GPCRs). For instance, SCFAs encourage the release of GLP-1, a key incretin hormone that is mostly produced by entero-endocrine L-cells [83].

Metabolic syndrome shows obesity and causes elevated level of glucose, hyperlipemia, and even hypertension. The incidence of metabolic disturbances is growing as it shows due to lifestyle and dietary changes [84]. MS is even increase the incidence of morbidity as well as mortality as a result of diabetes and hypertension. Recent evidences show that manipulation of the gut microbiota through prebiotics helps to maintain obesity, lipid level in the serum, and blood glucose level, which helps to lower the incidence of the type 2 DM. Endocannabinoid (eCB) system, bile salt hydrolase (BSH), metabolic endotoxemia, and short-chain fatty acids (SCFAs) play a crucial role in controlling the onset and progression of multiple sclerosis (MS) by normalizing adipogenesis and controlling insulin secretion, fat accumulation, energy homeostasis, and plasma cholesterol levels [85]. As a result, the gut microbiota could be a possible target for MS treatment. To improve our knowledge of gut microbiota manipulation and its role in MS prevention and treatment, more research is necessary [86].

The burden of nonalcoholic fatty liver disease (NAFLD) cases is increasing worldwide and it is the common cause linked to cardiovascular complications. Some studies reported the occurrence of liver disease is lower if there is improvement in the intestinal functions. Hence, the use of prebiotics along with probiotics and post-biotics is considered a positive approach for maintaining gut health. Some research studies evaluated the use of prebiotics and probiotics in NAFLD patients [87]. Research and review studies evaluated the impact of microbiota impact on liver injury, inflammation, and lipid levels in NAFD. Overall, supplementation with probiotic, prebiotic, or symbiotic had a positive effect on reducing liver enzymes, lipid profiles, and inflammatory cytokines in patients with NAFLD [88]. The use of prebiotics, probiotics, and synbiotics was evaluated and showed significant improvement in minimal hepatic encephalopathy [89].

20.5 Post-Biotics: Mechanism and Therapeutic Potential

Post-biotics are defined as metabolites of cell walls generated from probiotics. They do not have living cells, because of which they are safe to use. Post-biotics investigated have shown to have numerous health benefits to human such as anti-inflammatory, antibacterial, immunomodulatory, anti-aerogenic, anticarcinogenic, antioxidant, antihypertensive, antiproliferative, and hypocholesterolemia properties that promote the immune system and intestinal protection [90]. Common types of post-biotics, known as metabiotic, are para probiotics or "ghost probiotics" and fermented infant formula. Fermented infant formula is a starter or follow-on infant formula that has been fermented with lactic acid-producing or other bacteria, and, in most cases, it does not contain viable bacteria. The efficacy of post-biotics appears to depend on the interaction between the host and products generated by the microbiota. Post-biotics have the potential to provide new therapeutic approaches, increase the effectiveness of microorganisms, or provide a way to convert them into functional components [91].

Post-biotics are functional bioactive molecules that generate numerous health benefits and are generated from the fermentation process. Post-biotics comprise various metabolites like short-chain fatty acids, extracellular polysaccharides, cell lysates, and peptidoglycans. Post-biotics generate host health benefits in adults by relieving them from diarrhea, generating immunomodulatory effects, and improving adult dermatitis [92]. Post-biotics even treat the symptoms of an infant's colic. Post-biotics even enhance the potency of active microorganisms and convert them into functional molecules. Post-biotics can also be used in situations in which it is harder to control and maintain production and storage conditions, such as in developing countries. Moreover, it has been suggested that using post-biotics could be an attractive alternative for other "-biotics" in critically ill patients, young children, and premature neonates [93]. Post-biotics also exhibit antimicrobial, immunomodulatory, antioxidant, and anti-inflammatory activities. Its production requires a process that will ensure the safety and stability of the products. A study evaluating the in vivo tests of post-biotics for alopecia, acne, atopic dermatitis, and wound healing was investigated. The post-biotic use can also lead to some side effects when applied on the skin; so further research needs to be conducted to assess the potential of these compounds [94]. Moreover, investigation has shown that post-biotics demonstrate protection from inflammation by modulating the cytokines, elevating the immune response, and promoting epithelial barrier. Post-biotics' potential in treating inflammatory bowel diseases, autoimmune diseases, and metabolic syndrome was observed by regulating the immune mechanism. Post-biotics offer a promising approach in treatment of inflammation, with potential applications in functional foods and nutraceuticals [95].

20.5.1 Production and Metabolic Derivatives

In recent years, post-biotics are derived from the metabolic process and are biologically active compounds offering new research path and technological development for using it clinically [96]. Post-biotics promote the human health by showing unique mechanism and its potential use. The production of post-biotics involves fermentation procedure along with enzymatic conversion inside the body; various classes of post-biotics involve short chain fatty acids, peptides and proteins, bacteriocins, exopolysaccharides, and organic acids [97]. Post-biotics are produced from the inoculum obtained from the probiotics. The final production of post-biotics was done by the centrifugation of the fermented products. The most used approach for extracting post-biotics from microorganisms is to isolate metabolites produced by microbes using heat, enzymes, ultrasonication, ultraviolet treatment, centrifugation, and ultrafiltration procedures [98].

20.5.2 Role in Hepatic and Metabolic Disorders

Post-biotics shows localized effect on gut epithelial cells along with immunomodulatory, anti-inflammatory, and antibacterial properties; as anticarcinogenic and antiproliferative benefits as well as the prevention of celiac disease [99]. Post-biotics also play a crucial role in preventive therapy for wound healing and diabetes. Research studies have reported post-biotics promote the production of short chain fatty acids; as a result, it supports the integrity of the epithelial cell of the intestine [100]. Furthermore, the metabolites produced from *B. brev* are capable maintaining blood glucose levels, so post-biotics can be delivered as food and beverages. Post-biotics also exhibit excellent skin treatment properties, so are used in cosmetic industries for skin-related issues. Post-biotics have emerged as promising agents for improving agricultural practices and benefiting animal health [97] (Table 20.1).

Post-biotics can be used to treat skin wound healing and reduce the skin inflammation. The in vitro studies also show free radical scavenging activity. Certain post-biotics taken orally have been demonstrated to preserve microbiota balance and to strengthen the integrity of the skin barrier by inhibiting enzymes linked to the breakdown of extracellular matrix and exhibiting antioxidant activity [98, 116]. Post-biotics is capable of maintaining and restoring the host health in metabolic

Table 20.1 Application of post-biotics

S. No	Post-biotic name	Applications	References
1	*Bifidobacterium longum SPM1207*	Reduce total and LDL cholesterol	[101]
2	*B. longum CECT-7347*	Anti-inflammatory (reduce acute inflammatory response, gut-barrier disruption)	[102]
3	*B. animalis sub*sp. *lactis CECT 8145*	Anti-obesity (reduce nematode's fat content)	[103]
4	*B. longum W11*	Antioxidant (reduce ROS, increase RSH levels of H_2O_2-treated MRC-5)	[104]
5	*B. longum W11*	Immunomodulation (immune stimulatory in ConA-stimulated PBMCs)	[105]
6	*Lactobacillus fermentum, L. reuteri, and Bacillus subtilis natto*	Wound healing	[106]
7	*L. helveticus strain*	Induced skin photodamage	[107]
8	*L. plantarum (AN057), L. casei (AN177), and S. thermophilus*	Beneficial effects on pore size and wrinkle depth reduction, as well as skin moisture and elasticity increase	[108]
9	*Streptococcus thermophiles lysate*	Lipid barrier and reduced water loss	[109]
10	*L. plantarum K8*	Improve acne lesions	[110]
11	*Lactobacillus* sp.	Anti-inflammatory (decrease TNF-α, increase IL-10)	[111]
12	*L. plantarum*	Immunomodulation (attenuate the pro-inflammatory signaling in THP-1 cells)	[112]
13	*L. acidophilus, L. reuteri, L. plantarum*	Immunomodulation (inhibit LPS-induced TNF-α production)	[113]
14	*Bacillus subtilis* sp.	Decrease glucose levels in the blood increase insulin levels in the serum	[114]
15	*B. licheniformis*	Prevent diabetes complications (increase glycogen level, SOD, CAT decrease glucose level, TBARS, hepatic, and renal indices' toxicity)	[115]

disturbances. Post-biotic use required intense long-term studies to evaluate its potential for human use. Post-biotics show positive health results in metabolic syndromes, so they can cure the obesity and reduce hypertension and diabetes in patients [117]. Post-biotics are promising molecules that promotes health benefits and therapeutic effects when used clinically. Post-biotic use is attractive for altering the microbiota; however, further studies are needed to determine its efficacy and safety [118]. Post-biotics can also be released as metabolic product of liver secreted by the bacteria and considered as therapeutical agents having protection from metabolic dysfunction-associated steatosis liver diseases (MASLD). Exopolysaccharides, lipoteichoic acid, short-chain fatty acids, hydrogen sulfide, polyamines, tryptophan derivatives, and polyphenol metabolites show excellent anti-inflammatory and immunomodulatory potentials. These metabolites have shown potential in reducing liver infiltration, elevating fibrosis, and reducing inflammation during preclinical testing. The main mechanism of post-biotics is modulation of gut microbiota, promoting intestinal barrier functions, enhancing lipid metabolism and hepatoprotective action by hepatic steatosis [119, 120].

Post-biotics also inhibit the lipid accumulation inside the liver. Preventive effects of post-biotics prepared from *Lactobacillus paracasei* include prevention of nonalcoholic fatty liver disease (NAFLD) [120]. In addition, postbiotics (POST) regulated the gene expression in hepatocytes at the mRNA level, thereby regulating lipid metabolism. These findings suggest that POST plays a protective role against NAFLD and may exert its efficacy by modulating the gut microbiota and liver metabolism, and these findings may be applied to related functional foods [121]. Nonalcoholic fatty liver disease (NAFLD) The application of probiotics, as one of the most important strategies for the regulation of gut microbiota disorder, have been explored whether it is beneficial to gut-related diseases of intestine-distal organs. Some probiotics have showed to improve the liver parameters and phenotype in patients with NAFLD. The oral intake of them might become the effective management for the prevention and treatment of NAFLD [122].

20.6 Synergistic Effects of PPP with Conventional Drugs

Gut microbiota plays a crucial role in maintaining the host body balance. However, the present treatment can only relieve the symptoms for a short duration of time rather than improving the root cause of the disease [123]. Recent studies reported that probiotics, prebiotics, and post-biotics can improve and treat various diseases, but currently only be considered as dietary complement instead of medications. Numerous preclinical and clinical studies investigated the utilization of those probiotics, prebiotics, and post-biotics in various human health ailments [124]. It is expected that scientists can promote the application of probiotics, prebiotics, and post-biotics in disease treatment and utilize it in front-line treatment [31]. There is evidence that the gut microbiota plays a part in the pathophysiology of metabolic diseases, and using prebiotics, probiotics, synbiotics, and post-biotics to alter the gut microbiota has gained popularity [125].

Certain reviews have investigated the role of prebiotics, probiotics, synbiotic, and post-biotics—which referred to as "biotics"—in modulating the human gut microbiota and enhancing intestinal epithelial integrity [126]. Biotics also promotes health benefits with several mechanisms, including by promoting the growth of beneficial microbes, producing short-chain fatty acids (SCFAs), strengthening the gut barrier, and regulating immune responses. Each type of biotic have their own unique mechanism in preventing the inflammatory bowel disease, irritable bowel syndrome, obesity, constipation, and antibiotic-associated diarrhea [127].

20.6.1 Reducing Side Effects and Dosage of Conventional Therapies

The combination of the PPP shows promising health benefits. One of the major concerns inside the body for ROS generation is damage caused due to oxidative stress. Oxidative damage can lead to disturbance of the cellular component of the cell [128]. The generation of ROS inside the gut also causes deteriorating effect on the human body. Oxidative stress can lead to disturbance in the human health, lead to bowel inflammation and irritable bowel syndromes. Prebiotic, probiotic, post-biotic, and dietary polyphenols can improve the gut gut health by interacting with the gut microbiota [129]. It is reported that the combinations of these biotics can improve oxidative damage by reducing the production of ROS. In a study, antioxidant capacity of treatments with quercetin and inanimate *Lactobacillus acidophilus* were significantly evaluated and proved to have positive effect on CACO-2 cell of the intestine by reduction in the oxidative damage [130].

The clinical use of prebiotics and post-biotics gain interest as they exhibit antioxidant potential by preventing oxidative stress. In vivo and in vitro studies also reported the AO potential of these components. Probiotic bacteria and post-biotics are together marketed in the pharmaceuticals, nutraceuticals, and cosmetic products because of its antioxidant potentials [131]. Inflammatory bowel disease (IBD) is defined as an inflammatory condition, which includes ulcerative colitis and Crohn's disease, and is considered a major public health concern. Most of the conventional treatments only target the inflammation and the immune system, but holistic approaches targeting lifestyle and diet improvements are emerging. Pre-, pro-, syn-, and post-biotics are reported to prevent inflammation [132]. PPP along with synbiotic improves the intestinal barriers and integrity along with synthesis of metabolites. Researchers reported that PPP have potential to improve immunological function, alleviate symptoms associated with irritable bowel disease (IBD), decrease severity of allergies, and have antibacterial and anti-inflammatory effects [133].

Presently, prebiotics and probiotics are the products on the market; however, post-biotics are also gaining a great deal of attention. The numerous health advantages of post-biotic components may soon lead to an increase in consumer demand for post-biotic supplements. The most recent research aspects of post-biotics in the food and pharmaceutical industries are included in this review. The review encompasses a brief introduction, classification, production technologies, characterization, biological activities, and potential applications of post-biotics [134]. Celiac disease (CD) is an autoimmune inflammatory disease cause due to genetic factors, viral infection, or due to microbiota dysbiosis. The use of pro-, pre- and post-biotics in vitro and in vivo in CD was investigated. The study reported that probiotics and their products in CD could be interesting for observing their various effects on several different pathways, including anti-inflammatory properties [135].

20.6.2 Experimental Evidence

In vivo study of *Lactobacillus,* which is a probiotic, was conducted on neonatal animals. The effect of bacterial translocation (BT) was evaluated [136]. Neonatal rabbits receiving *Lacto* GG-supplemented formula exhibited a 25% decrease ($P < 0.05$) in small-bowel colonization by *Escherichia coli* K1A. The results conclude that the use of *Lactobacillus* show promising action on BT and sepsis. Probiotics and prebiotics both generate the effect on human by reducing the disease progression and pathogenesis by altering the circulatory cytokinin and gene expression [137].

Fermented milk supplemented with two probiotic strains, *Bifidobacterium lactis* Bi-07 and *Lactobacillus acidophilus* NCFM, and a prebiotic, isomalt oligosaccharide, was clinically tested on healthy adults. The same sample was provided to the experimental animals [138]. Fecal bacterial composition was evaluated and an increase in *bifidobacteria* and *lactobacilli* in both clinical and preclinical testing were observed. In conclusion, it was observed that prebiotic and probiotic both have the potential for the human gut health and immunity [139]. Some studies also reported that probiotics showed efficacy in improving psychiatric disorder-related behaviors including anxiety, depression, autism spectrum disorder (ASD), obsessive-compulsive disorder, and memory abilities, including spatial and nonspatial memory [140].

Clinical and preclinical studies of both pro- and prebiotics were tested and the results demonstrated that both are dependent on baseline numbers of lactic acid bacteria (LAB) and bifidobacteria (BIF). The survival of PRO was presumably enhanced by the use of prebiotic supplementation and advocates the use of particular combinations of pro- and prebiotics [141].

20.7 Future Perspectives and Challenges

Probiotics are considered as potential candidates for market use as they treat a wide variety of ailments. Some studies suggested that the release of metabolites by the bacteria are responsible for action not the viable bacteria. This evidence highlights that both the probiotic organisms and their bioactive metabolites—collectively known as postbiotics—are responsible for the observed health benefits. Therefore post-biotics show biological activities, such as antimicrobial, antioxidant, anti-inflammatory, anti-proliferative, and immunomodulation [142].

Numerous candidates of the probiotic and prebiotic substrates have been developed by scientists, creating greater demand for these products. The use of probiotic and prebiotic is developing, including synbiotic, post-biotics, microbial consortia, live biotherapeutic products, and genetically modified organisms, with renewed interest in polyphenols, fibers, and fermented foods. The incorporation of nutrition supplement and personalized treatment into the human lifestyle is gaining interest for its applications. Increasing demand and growing field discoveries required broad spectrum research in the industry. According to the reports, the annual growth of prebiotic is 12.7% and for probiotics it is 7% over the past 8 years. Even though the nutraceutical market is growing, there are certain gaps in establishing the best source of these in food sources and healthcare products. Certain microbiota groups consume the substrate, which promotes their growth and metabolic activity and mediates the traditional prebiotic effects. Providing specific bacterial groups with targeted substrates can indirectly influence the growth of pathogenic microorganisms[143]. Studies also reported that pre- and post-biotics can be used in metabolic disturbances as it maintains the insulin and cholesterol level, However the certain interaction of these molecules are still unclear and the challenge is in finding the appropriate data [144].

20.8 Regulatory Considerations

Regulatory specification data across the globe was a concern as we need to look up closely the safety and stability guidelines for probiotics. For market and clinical acceptance of probiotics, certain guidelines in production, safety, and uniform quality need to be adopted. Certain guidelines need to be established regarding the critical review of the products, which should be universally accepted [145]. Before probiotics can be considered for therapeutic use, their safety profile must be thoroughly evaluated. Further studies focusing on the probiotic strain's safety assessment parameters, as well as manufacturing, labeling, and delivery requirements are mandatory. The FDA should frame regulatory guidelines for probiotic marketing in a global context; the FSSAI's proposed regulations for probiotic manufacturing and sales in India; the approved list of probiotic cultures and brand licensing in current market trends; and future directions [146].

Post-biotics contain nonviable microbes that generate numerous positive physiological effects on administrations. The regulatory considerations should address important issues such as maintaining the quality, safety, and labeling standards of post-biotics. Based on the research, post-biotics are classified into four types based on commercial, technical, and commercial use. The guidelines need to be framed to ensure the quality, safety and also the efficacy of post-biotics. The growing demands of these products need further research to make it fit for consumers [147]. Post-biotics can be introduced in the pharmaceutical sector, in medicinal products and even for medical devices. The market needs regulation for products based on the research data available. Moreover, these products also developed into cosmetics for the external use on skin, for vaginal health, and as orally consumed products. The European Union for regulatory aspects was introduced along with regulations of other geographical areas [148].

Approaches to finding and creating new probiotics, along with their use in food, medicine, and pharmacology, are changing. The main reason has been the growing popularity and availability of latest technologies. In addition, the safety and efficacy of the probiotics products required legal as well as technical regulations [149]. Microbiome-based therapies involve modifying the gut microbiome to treat various diseases; however, proper regulatory oversight is essential during their development. So regulatory agencies like European Medicines Agency (EMA) and Food and Drug Administration (FDA) govern micro-biome therapeutics. Ensuring safety and effectiveness requires preclinical research, clinical trials, and consideration of negative effects on the microbiota and host physiology. Live biotherapeutic products (LBPs) follow good manufacturing practices (GMP) for making, storing, and distributing. It's important to determine the regulatory pathway, whether it is medicine, biologic, or medical device [150].

20.8.1 Integration into Therapeutic Regimens

Microbiome-based therapy can be effective in preoperative and postoperative therapies. The study revealed that prebiotics and probiotics, along with synbiotics, improve postoperative complications. The clinical study reported that the use of these reduces operation-related infections and sepsis and pneumonia. Hence, it is suggested to introduce synbiotic for adult patient's operative procedure to reduce the use of antibiotics and in order to prevent its resistance [151].

On the other flip side of the investigation the activity of prebiotic and post-biotics was evaluated in cancer patients. Studies conclude that the use of novel approach can reduce the GIT-related issues along with psychosocial health impact in cancer patients in improving the quality of life [152]. Inflammatory bowel diseases (IBD), which is Crohn's disease (CD) and ulcerative colitis (UC), associated with symptoms of chronic inflammation of the small intestine and colon lead to reduce the immune response in host health. The use of probiotics provide reduction in the oxidative stress and hence improve the safety and efficacy on IBD patients [117].

Metabolic diseases can also be regulated by the use of probiotics, prebiotics, synbiotic, and post-biotics. Probiotics, prebiotics, synbiotic, and post-biotics also show promising results in case of managing obesity and type 2 diabetes mellitus [117, 124]. The use of PPP has been promoted in case of obesity. Studies also reported that probiotics and synbiotic in obesity shows beneficial effects on weight reduction and other metabolic parameters via their involvement in gut microbiota modulation [124].

20.9 Conclusion

Prebiotics, probiotics, and post-biotics show beneficial potential in numerous diseases. Further clinical and preclinical studies need to be conducted in order to investigate the beneficial effects. Extensive research needs to be demonstrated to understand the mechanism of probiotics, prebiotics, and post-biotics in improving the gut health. Current research and clinical evidence suggest that probiotics, prebiotics, and post-biotics will play a bigger role in clinical treatments in the future. They may become the next generation of key medicines that will change how we treat and manage diseases. Advancement in personalized therapy and targeted therapy also enhances the market utilization of the PPP. Innovations such as genetically engineered probiotics and designer symbiotic are being explored to enhance functional outcomes, offering more precise modulation of the gut microbiome. Further regulatory guidelines need to be established in order to main the safety efficacy of PPP formulations. In conclusion, prebiotic, post-biotic and probiotics show excellent health benefits by promoting the maintenance of gut microbiota and strengthening the immune system. Hence, its utilization is promoted as nutraceutical products in the pharmaceutical industry. Post-biotics are poised to become a major player. Their stability, simple production, and absence of viability issues are crucial for people with weakened immune systems. Plus, they can be standardized, which makes them appealing for both pharmaceutical and industrial use.

References

1. Ahmed OM, Mohammed MT (2020) Oxidative stress: the role of reactive oxygen species (ROS) and antioxidants in human diseases. Plant Arch 20(2):4089–4095
2. Krumova K, Cosa G (2016) Overview of reactive oxygen species. In: Comprehensive series in photochemical & photobiological sciences, pp 1–21
3. Sachdev S, Ansari SA, Ansari MI (2023) Reactive oxygen species (ROS): an introduction. In: Reactive oxygen species in plants: the right balance. Springer, pp 1–22
4. Ahmad G et al (2017) Overview and sources of reactive oxygen species (ROS) in the reproductive system. In: Oxidative stress in human reproduction: shedding light on a complicated phenomenon. Springer, pp 1–16
5. Sahoo BM et al (2022) Reactive oxygen species (ROS): key components in cancer therapies. Anti-Cancer Agents Med Chem **22**(2):215–222
6. Juan CA et al (2021) The chemistry of reactive oxygen species (ROS) revisited: outlining their role in biological macromolecules (DNA, lipids and proteins) and induced pathologies. Int J Mol Sci 22(9):4642
7. Madkour LH (2019) Function of reactive oxygen species (ROS) inside the living organisms and sources of oxidants. Pharm Sci Anal Res J 2:180023
8. Cheung EC, Vousden KH (2022) The role of ROS in tumour development and progression. Nat Rev Cancer 22(5):280–297
9. Jomova K et al (2023) Reactive oxygen species, toxicity, oxidative stress, and antioxidants: chronic diseases and aging. Arch Toxicol 97(10):2499–2574
10. Mustafa M et al (2024) Apoptosis: a comprehensive overview of signaling pathways, morphological changes, and physiological significance and therapeutic implications. Cells 13(22):1838
11. Vermot A et al (2021) NADPH oxidases (NOX): an overview from discovery, molecular mechanisms to physiology and pathology. Antioxidants 10(6):890
12. Zhao RZ et al (2019) Mitochondrial electron transport chain, ROS generation and uncoupling. Int J Mol Med 44(1):3–15
13. Burtscher J et al (2022) Adaptive responses to hypoxia and/or hyperoxia in humans. Antioxid Redox Signal 37(13):887–912

14. Liu J et al (2023) Reactive oxygen species (ROS) scavenging biomaterials for anti-inflammatory diseases: from mechanism to therapy. J Hematol Oncol 16(1):116
15. Sheikh AM et al (2024) The role of the vascular system in degenerative diseases: mechanisms and implications. Int J Mol Sci 25(4):2169
16. Mittal M et al (2014) Reactive oxygen species in inflammation and tissue injury. Antioxid Redox Signal 20(7):1126–1167
17. Bergamini CM et al (2004) Oxygen, reactive oxygen species and tissue damage. Curr Pharm Des 10(14):1611–1626
18. Li Z et al (2017) Roles of reactive oxygen species in cell signaling pathways and immune responses to viral infections. Arch Virol 162(3):603–610
19. Islam MT, Shekhar HU (2014) Impact of oxidative stress on human health. In: Free radicals in human health and disease. Springer, pp 59–73
20. Vona R et al (2021) The impact of oxidative stress in human pathology: focus on gastrointestinal disorders. Antioxidants 10(2):201
21. Bhattacharyya A et al (2014) Oxidative stress: an essential factor in the pathogenesis of gastrointestinal mucosal diseases. Physiol Rev 94(2):329–354
22. Zhang Y-J et al (2015) Impacts of gut bacteria on human health and diseases. Int J Mol Sci 16(4):7493–7519
23. Campbell EL, Colgan SP (2019) Control and dysregulation of redox signalling in the gastro-intestinal tract. Nat Rev Gastroenterol Hepatol 16(2):106–120
24. He L et al (2017) Antioxidants maintain cellular redox homeostasis by elimination of reactive oxygen species. Cell Physiol Biochem 44(2):532–553
25. Kunst C et al (2023) The influence of gut microbiota on oxidative stress and the immune system. Biomedicine 11(5):1388
26. Jones RM, Mercante JW, Neish AS (2012) Reactive oxygen production induced by the gut microbiota: pharmacotherapeutic implications. Curr Med Chem 19(10):1519–1529
27. Cortez FGP (2019) The use of gastroscopy in the evaluation of treatment response of equine gastric ulcer syndrome (EGUS). Universidade de Lisboa (Portugal)
28. Usher L et al (2013) Factors associated with complementary and alternative medicine use in irritable bowel syndrome: a literature review. Psychol Community Health 2(3):346–361
29. Tillisch K (2006) Complementary and alternative medicine for functional gastrointestinal disorders. Gut 55(5):593–596
30. Yao Y et al (2022) Herbal therapies in gastrointestinal and hepatic disorders: an evidence-based clinical review. Front Pharmacol 13:962095
31. Ji J et al (2023) Probiotics, prebiotics, and postbiotics in health and disease. MedComm 4(6):e420
32. Fikree A, Byrne P (2021) Management of functional gastrointestinal disorders. Clin Med 21(1):44–52
33. Colella M et al (2023) Microbiota revolution: how gut microbes regulate our lives. World J Gastroenterol 29(28):4368
34. Talapko J et al (2022) Homeostasis and dysbiosis of the intestinal microbiota: comparing hallmarks of a healthy state with changes in inflammatory bowel disease. Microorganisms 10(12):2405
35. Basso P et al (2014) Association among genetic predisposition, gut microbiota, and host immune response in the etiopathogenesis of inflammatory bowel disease. Braz J Med Biol Res 47(9):727–737
36. Sayre CL et al (2023) Current use of complementary and conventional medicine for treatment of pediatric patients with gastrointestinal disorders. Front Pharmacol 14:1051442
37. Sirisinha S (2016) The potential impact of gut microbiota on your health: current status and future challenges. Asian Pac J Allergy Immunol 34(4):249–264
38. Pickard JM et al (2017) Gut microbiota: role in pathogen colonization, immune responses, and inflammatory disease. Immunol Rev 279(1):70–89
39. Sekirov I et al (2010) Gut microbiota in health and disease. Physiol Rev 90:859

40. Aasmets O et al (2022) Gut metagenome associations with extensive digital health data in a volunteer-based Estonian microbiome cohort. Nat Commun 13(1):869
41. Chanda D, De D (2024) Meta-analysis reveals obesity associated gut microbial alteration patterns and reproducible contributors of functional shift. Gut Microbes 16(1):2304900
42. Guinane CM, Cotter PD (2013) Role of the gut microbiota in health and chronic gastrointestinal disease: understanding a hidden metabolic organ. Ther Adv Gastroenterol 6(4):295–308
43. Martinvalet D, Walch M (2022) The role of reactive oxygen species in protective immunity. Front Immunol 12:832946
44. Jones RM, Neish AS (2017) Redox signaling mediated by the gut microbiota. Free Radic Biol Med 105:41–47
45. Al-Nu'airat J (2018) Implications of reactive oxygen species (ROS) in initiating chemical reactions. Murdoch University
46. Aviello G, Knaus U (2017) ROS in gastrointestinal inflammation: rescue or sabotage? Br J Pharmacol 174(12):1704–1718
47. Singh V et al (2022) Balancing reactive oxygen species generation by rebooting gut microbiota. J Appl Microbiol 132(6):4112–4129
48. Johnstone KF, Herzberg MC (2022) Antimicrobial peptides: defending the mucosal epithelial barrier. Front Oral Health 3:958480
49. Kong C et al (2024) Restore intestinal barrier integrity: an approach for inflammatory bowel disease therapy. J Inflamm Res 17:5389–5413
50. Rezvani M (2024) Oxidative stress-induced gastrointestinal diseases: biology and nanomedicines—a review. Biochemist 4(3):189–216
51. Kato LM et al (2014) The role of the adaptive immune system in regulation of gut microbiota. Immunol Rev 260(1):67–75
52. Yoo JY et al (2020) Gut microbiota and immune system interactions. Microorganisms 8(10):1587
53. DuPont HL et al (2022) Intestinal IgA-coated bacteria in healthy-and altered-microbiomes (Dysbiosis) and predictive value in successful fecal microbiota transplantation. Microorganisms 11(1):93
54. Palm NW, de Zoete MR, Flavell RA (2015) Immune–microbiota interactions in health and disease. Clin Immunol 159(2):122–127
55. Santacroce L, Charitos IA, Bottalico L (2019) A successful history: probiotics and their potential as antimicrobials. Expert Rev Anti-Infect Ther 17(8):635–645
56. Wilson B, Whelan K (2017) Prebiotic inulin-type fructans and galacto-oligosaccharides: definition, specificity, function, and application in gastrointestinal disorders. J Gastroenterol Hepatol 32:64–68
57. Davani-Davari D et al (2019) Prebiotics: definition, types, sources, mechanisms, and clinical applications. Foods 8(3):92
58. Tegegne BA, Kebede B (2022) Probiotics, their prophylactic and therapeutic applications in human health development: a review of the literature. Heliyon 8(6):e09725
59. Corr SC, Hill C, Gahan CG (2009) Understanding the mechanisms by which probiotics inhibit gastrointestinal pathogens. Adv Food Nutr Res 56:1–15
60. Latif A et al (2023) Probiotics: mechanism of action, health benefits and their application in food industries. Front Microbiol 14:1216674
61. Fernandes MA, Delgado L, Almeida AC (2024) The impact of probiotics on antibiotic resistance in newborns: a systematic review. J Neonatal-Perinatal Med 17(4):509–518
62. Wang G et al (2022) Characteristics of probiotic preparations and their applications. Foods 11(16):2472
63. Yousefi B et al (2019) Probiotics importance and their immunomodulatory properties. J Cell Physiol 234(6):8008–8018
64. Cano-Garrido O, Seras-Franzoso J, Garcia-Fruitos E (2015) Lactic acid bacteria: reviewing the potential of a promising delivery live vector for biomedical purposes. Microb Cell Factories 14(1):137

65. Behnsen J et al (2013) Probiotics: properties, examples, and specific applications. Cold Spring Harb Perspect Med 3(3):a010074
66. Gueimonde Fernández M, García BS (2012) Enhancing probiotic stability in industrial processes. Microb Ecol Health Dis 23. https://doi.org/10.3402/mehd.v23i0.18562
67. Nowak A, Paliwoda A, Błasiak J (2019) Anti-proliferative, pro-apoptotic and anti-oxidative activity of Lactobacillus and Bifidobacterium strains: a review of mechanisms and therapeutic perspectives. Crit Rev Food Sci Nutr 59(21):3456–3467
68. Bardelčíková A, Šoltys J, Mojžiš J (2023) Oxidative stress, inflammation and colorectal cancer: an overview. Antioxidants 12(4):901
69. Ghiasi F, Hashemi SMB, Abedi E (2023) Effective enhancement of food oxidative stability induced by Lactobacillus strains: in vitro activity. Food Control 153:109912
70. Khan AN et al (2021) Antagonistic, anti-oxidant, anti-inflammatory and anti-diabetic probiotic potential of Lactobacillus agilis isolated from the rhizosphere of the medicinal plants. Saudi J Biol Sci 28(11):6069–6076
71. Lacerda DC et al (2022) Potential role of Limosilactobacillus fermentum as a probiotic with anti-diabetic properties: a review. World J Diabetes 13(9):717
72. Gotteland M et al (2014) Probiotic screening and safety evaluation of Lactobacillus strains from plants, artisanal goat cheese, human stools, and breast milk. J Med Food 17(4):487–495
73. Virk MS et al (2024) The anti-inflammatory and curative exponent of probiotics: a comprehensive and authentic ingredient for the sustained functioning of major human organs. Nutrients 16(4):546
74. Grover S et al (2012) Probiotics for human health–new innovations and emerging trends. Gut pathogens 4(1):15
75. Kunyeit L, Rao RP, Anu-Appaiah K (2024) Yeasts originating from fermented foods, their potential as probiotics and therapeutic implication for human health and disease. Crit Rev Food Sci Nutr 64(19):6660–6671
76. Ansari JM et al (2019) Strain-level diversity of commercial probiotic isolates of Bacillus, Lactobacillus, and Saccharomyces species illustrated by molecular identification and phenotypic profiling. PLoS One 14(3):e0213841
77. Siesto G et al (2022) In vitro study of probiotic, antioxidant and anti-inflammatory activities among indigenous Saccharomyces cerevisiae strains. Foods 11(9):1342
78. Ashaolu TJ, Ashaolu JO, Adeyeye SA (2021) Fermentation of prebiotics by human colonic microbiota in vitro and short-chain fatty acids production: a critical review. J Appl Microbiol 130(3):677–687
79. Gulzar N et al (2019) Therapeutic potential of probiotics and prebiotics. In: Oral health by using probiotic products. IntechOpen
80. Smith ML (2018) Isolation and characterisation of novel enzymes producing fructo-and galacto-oligosaccharides through a metagenomic approach. Stellenbosch University, Stellenbosch
81. Fernández J et al (2016) Colon microbiota fermentation of dietary prebiotics towards short-chain fatty acids and their roles as anti-inflammatory and antitumour agents: a review. J Funct Foods 25:511–522
82. Ayesiga I et al (2024) Key regulatory aspects of prebiotics, probiotics, and synbiotics in the management of metabolic disorders. In: Synbiotics in metabolic disorders. CRC Press, pp 214–244
83. Yoo JY, Kim SS (2016) Probiotics and prebiotics: present status and future perspectives on metabolic disorders. Nutrients 8(3):173
84. Halpern A et al (2010) Metabolic syndrome, dyslipidemia, hypertension and type 2 diabetes in youth: from diagnosis to treatment. Diabetol Metab Syndr 2(1):55
85. Bisogno T, Lauritano A, Piscitelli F (2021) The endocannabinoid system: a bridge between Alzheimer's disease and gut microbiota. Life 11(9):934
86. He M, Shi B (2017) Gut microbiota as a potential target of metabolic syndrome: the role of probiotics and prebiotics. Cell Biosci 7(1):54

87. Carpi RZ et al (2022) The effects of probiotics, prebiotics and synbiotics in non-alcoholic fat liver disease (NAFLD) and non-alcoholic steatohepatitis (NASH): a systematic review. Int J Mol Sci 23(15):8805
88. Pan Y et al (2024) Efficacy of probiotics, prebiotics, and synbiotics on liver enzymes, lipid profiles, and inflammation in patients with non-alcoholic fatty liver disease: a systematic review and meta-analysis of randomized controlled trials. BMC Gastroenterol 24(1):283
89. Shukla S et al (2011) Meta-analysis: the effects of gut flora modulation using prebiotics, probiotics and synbiotics on minimal hepatic encephalopathy. Aliment Pharmacol Ther 33(6):662–671
90. Bourebaba Y et al (2022) Postbiotics as potential new therapeutic agents for metabolic disorders management. Biomed Pharmacother 153:113138
91. Hijová E (2024) Postbiotics as metabolites and their biotherapeutic potential. Int J Mol Sci 25(10):5441
92. Ozma MA et al (2022) Postbiotics as the key mediators of the gut microbiota-host interactions. Infez Med 30(2):180
93. Wegh CA et al (2019) Postbiotics and their potential applications in early life nutrition and beyond. Int J Mol Sci 20(19):4673
94. da Silva Vale A et al (2023) Production, formulation, and application of postbiotics in the treatment of skin conditions. Fermentation 9(3):264
95. Zdybel K et al (2025) Postbiotics formulation and therapeutic effect in inflammation: a systematic review. Nutrients 17(13):2187
96. Aggarwal S et al (2022) Postbiotics: from emerging concept to application. Front Sustain Food Syst 6:887642
97. Prajapati N et al (2023) Postbiotic production: harnessing the power of microbial metabolites for health applications. Front Microbiol 14:1306192
98. Paul Beulah B, Rajasekar T (2023) Preparation of postbiotics from Bacillus. In: Postbiotics. Springer, pp 75–79
99. Khani N, Bonyadi M, Soleimani RA, Raziabad RH, Ahmadi M, & Homayouni-Rad A (2025) Postbiotics: as a promising tools in the treatment of celiac disease. Probiotics and Antimicrobial Proteins, 17(3):1513–1522
100. Ragavan ML, Hemalatha S (2024) The functional roles of short chain fatty acids as postbiotics in human gut: future perspectives. Food Sci Biotechnol 33(2):275–285
101. Shin HS et al (2010) Hypocholesterolemic effect of sonication-killed Bifidobacterium longum isolated from healthy adult Koreans in high cholesterol fed rats. Arch Pharm Res 33(9):1425–1431
102. Martorell P et al (2021) Heat-treated Bifidobacterium longum CECT-7347: a whole-cell postbiotic with antioxidant, anti-inflammatory, and gut-barrier protection properties. Antioxidants 10(4):536
103. Silva Á et al (2020) An infant milk formula supplemented with heat-treated probiotic Bifidobacterium animalis subsp. lactis CECT 8145, reduces fat deposition in C. Elegans and augments acetate and lactate in a fermented infant slurry. Foods 9(5):652
104. Arsenijević D et al (2023) Bifidobacterium animalis and Laetiporus sulphureus Extract induce a strong increase in GSH levels in MRC-5 cells in response to oxidative stress. Biol Life Sci Forum 29(1):3
105. Inturri R et al (2017) Immunomodulatory effects of Bifidobacterium longum W11 produced exopolysaccharide on cytokine production. Curr Pharm Biotechnol 18(11):883–889
106. Leñini C et al (2023) Probiotic properties of Bacillus subtilis DG101 isolated from the traditional Japanese fermented food nattō. Front Microbiol 14:1253480
107. Rong J et al (2017) Skin resistance to UVB-induced oxidative stress and hyperpigmentation by the topical use of lactobacillus helveticus NS8-fermented milk supernatant. J Appl Microbiol 123(2):511–523

108. Huuskonen L et al (2022) Effects of Bifidobacterium animalis subsp. lactis Bl-04 on skin wrinkles and dryness: a randomized, triple-blinded, placebo-controlled clinical trial. Dermato 2(2):30–52
109. Min B et al (2020) Changes in cell membrane fatty acid composition of Streptococcus thermophilus in response to gradually increasing heat temperature. J Microbiol Biotechnol 30(5):739
110. Podrini C et al (2023) Topical administration of Lactiplantibacillus plantarum (SkinDuoTM) serum improves anti-acne properties. Microorganisms 11(2):417
111. Vincenzi A, Goettert MI, Volken de Souza CF (2023) Anti-inflammatory potential of lactic acid bacteria from southern Brazil by reducing TNF-α and increasing IL-10 synthesis: an in vitro study. J Appl Microbiol 134(8):lxad168
112. Kim W et al (2020) Lactobacillus plantarum-derived extracellular vesicles induce anti-inflammatory M2 macrophage polarization in vitro. J Extracell Vesicles 9(1):1793514
113. Zolfaghari SI, Khorasgani MR, Noorbakhshnia M (2021) The effects of lactobacilli *(L. rhamnosus, L. reuteri, L.* Plantarum) on LPS-induced memory impairment and changes in CaMKII-α and TNF-α genes expression in the hippocampus of rat. Physiol Behav 229:113224
114. Ghoneim MA et al (2016) Effect of polysaccharide from Bacillus subtilis sp. on cardiovascular diseases and atherogenic indices in diabetic rats. BMC Complement Altern Med 16(1):112
115. Wang D et al (2022) Dietary Bacillus licheniformis improves the effect of Astragalus membranaceus extract on blood glucose by regulating antioxidation activity and intestinal microbiota in InR [E19]/TM2 Drosophila melanogaster. PLoS One 17(7):e0271177
116. De Almeida CV, Antiga E, Lulli M (2023) Oral and topical probiotics and postbiotics in skincare and dermatological therapy: a concise review. Microorganisms 11(6):1420
117. Li H-Y et al (2021) Effects and mechanisms of probiotics, prebiotics, synbiotics, and postbiotics on metabolic diseases targeting gut microbiota: a narrative review. Nutrients 13(9):3211
118. Park M et al (2022) Role of postbiotics in diet-induced metabolic disorders. Nutrients 14(18):3701
119. Pan Z et al (2022) Postbiotics prepared using Lactobacillus paracasei CCFM1224 prevent nonalcoholic fatty liver disease by modulating the gut microbiota and liver metabolism. Int J Mol Sci 23(21):13522
120. Yilmaz Y (2024) Postbiotics as antiinflammatory and immune-modulating bioactive compounds in metabolic dysfunction-associated Steatotic liver disease. Mol Nutr Food Res 68(23):2400754
121. Yuan H et al (2024) Biomarkers for health functional foods in metabolic dysfunction-associated steatotic liver disorder (MASLD) prevention: an integrative analysis of network pharmacology, gut microbiota, and multi-omics. Nutrients 16(18):3061
122. Cao C et al (2023) Effects of probiotics on non-alcoholic fatty liver disease: a review of human clinical trials. Front Nutr 10:1155306
123. Khan I et al (2019) Alteration of gut microbiota in inflammatory bowel disease (IBD): cause or consequence? IBD treatment targeting the gut microbiome. Pathogens 8(3):126
124. Vallianou N et al (2020) Probiotics, prebiotics, synbiotics, postbiotics, and obesity: current evidence, controversies, and perspectives. Curr Obes Rep 9(3):179–192
125. Smolinska S, Popescu F-D, Zemelka-Wiacek M (2025) A review of the influence of prebiotics, probiotics, synbiotics, and postbiotics on the human gut microbiome and intestinal integrity. J Clin Med 14(11):3673
126. Liu Y, Wang J, Wu C (2022) Modulation of gut microbiota and immune system by probiotics, pre-biotics, and post-biotics. Front Nutr 8:634897
127. Meng X et al (2020) Gut dysbacteriosis and intestinal disease: mechanism and treatment. J Appl Microbiol 129(4):787–805
128. Ozougwu JC (2016) The role of reactive oxygen species and antioxidants in oxidative stress. Int J Res 1(8):1–8

129. Plamada D, Vodnar DC (2021) Polyphenols—Gut microbiota interrelationship: a transition to a new generation of prebiotics. Nutrients 14(1):137
130. Tan R, Loke WM (2022) Gut oxidative modulation of polyphenol, prebiotic, probiotic, and postbiotic in vitro. Curr Res Nutr Food Sci J 10(1):56–70
131. Blazheva D et al (2022) Antioxidant potential of probiotics and postbiotics: a biotechnological approach to improving their stability. Russ J Genet 58(9):1036–1050
132. Martyniak A et al (2021) Prebiotics, probiotics, synbiotics, paraprobiotics and postbiotic compounds in IBD. Biomolecules 11(12):1903
133. Al-Habsi N et al (2024) Health benefits of prebiotics, probiotics, synbiotics, and postbiotics. Nutrients 16(22):3955
134. Thorakkattu P et al (2022) Postbiotics: current trends in food and pharmaceutical industry. Foods 11(19):3094
135. Conte M et al (2021) Pro-pre and postbiotic in celiac disease. Appl Sci 11(17):8185
136. Frizzo LS et al (2010) Studies on translocation, acute oral toxicity and intestinal colonization of potentially probiotic lactic acid bacteria administered during calf rearing. Livest Sci 128(1–3):28–35
137. Lee D et al (2000) Evaluation of probiotic treatment in a neonatal animal model. Pediatr Surg Int 16(4):237–242
138. Rasinkangas P et al (2022) Bifidobacterium animalis subsp. lactis Bi-07 supports lactose digestion in vitro and in randomized, placebo-and lactase-controlled clinical trials. Am J Clin Nutr 116(6):1580–1594
139. Wang S et al (2012) Fermented milk supplemented with probiotics and prebiotics can effectively alter the intestinal microbiota and immunity of host animals. J Dairy Sci 95(9):4813–4822
140. Wang H et al (2016) Effect of probiotics on central nervous system functions in animals and humans: a systematic review. J Neurogastroenterol Motil 22(4):589
141. Martinez RCR, Bedani R, Saad SMI (2015) Scientific evidence for health effects attributed to the consumption of probiotics and prebiotics: an update for current perspectives and future challenges. Br J Nutr 114(12):1993–2015
142. Liang B, Xing D (2023) The current and future perspectives of postbiotics. Probiotics and Antimicrob proteins 15(6):1626–1643
143. Cunningham M et al (2021) Shaping the future of probiotics and prebiotics. Trends Microbiol 29(8):667–685
144. Yoo JiYoun YJ, Kim SungSoo KS (2016) Probiotics and prebiotics: present status and future perspectives on metabolic disorders. Nutrients 8(3):173
145. Arora M, Baldi A (2017) Comparative study of regulatory framework for probiotics: current status and future recommendations. Appl Clin Res Clin Trials Regul Aff 4(2):140–156
146. Misra S, Raghuwanshi S (2021) Safety concerns, regulatory guidelines, current market trends, and future directions toward the use of probiotics in gut-brain-skin axis. In: Probiotic research in therapeutics, Probiotics and gut skin axis–inside out and outside in, vol 3. Springer, pp 245–268
147. Guglielmetti S et al (2025) Commercial and regulatory frameworks for postbiotics: an industry-oriented scientific perspective for non-viable microbial ingredients conferring beneficial physiological effects. Trends Food Sci Technol 163:105130
148. Vinderola G et al (2023) Postbiotics in the medical field under the perspective of the ISAPP definition: scientific, regulatory, and marketing considerations. Front Pharmacol 14:1239745
149. Yunes R et al (2022) Lactobacilli: legal regulation and prospects for new generation drugs. Appl Biochem Microbiol 58(5):652–664
150. Waheed R, Farooq AZ, Asma L (2024) Regulatory considerations for microbiome-based therapeutics. In: Human microbiome: techniques, strategies, and therapeutic potential. Springer, pp 657–689

151. Kasatpibal N et al (2017) Effectiveness of probiotic, prebiotic, and synbiotic therapies in reducing postoperative complications: a systematic review and network meta-analysis. Clin Infect Dis 64(suppl_2):S153–S160
152. Deleemans JM et al (2021) The use of prebiotic and probiotic interventions for treating gastrointestinal and psychosocial health symptoms in cancer patients and survivors: a systematic review. Integr Cancer Ther 20:15347354211061733

Chapter 21
Artificial Intelligence, Deep Learning, and Machine Learning to Target ROS-Mediated Inflammation Inhibition in Metabolic Disorders

Arnab Seth, Satadeep Bandyapadhyay, Abiskar Palui, and Mainak Chatterjee

21.1 Introduction

A wide range of illnesses are classified as metabolic disorders, which are defined by aberrant biochemical processes that interfere with the body's natural metabolism of proteins, fats, and carbohydrates. These disorders cause metabolic intermediates to accumulate excessively or to be produced insufficiently, which results in cellular dysfunction and systemic problems. Obesity, metabolic syndrome, type 2 diabetes mellitus (T2DM), dyslipidemia, and nonalcoholic fatty liver disease (NAFLD) are among the most common metabolic illnesses and are all becoming more widely acknowledged as serious global public health issues [1, 2]. Globally, the prevalence of metabolic illnesses has increased due to a combination of genetic susceptibility, poor dietary practices, and sedentary lifestyles [3]. For example, obesity greatly raises the likelihood of insulin resistance, a defining feature of type 2 diabetes, and both illnesses are frequently linked to and can worsen one another [4]. Chronic low-grade inflammation and oxidative stress are key components of the pathogenesis of the majority of metabolic illnesses, and they are both intimately related to the over-production of reactive oxygen species (ROS) [5, 6]. According to Eckel et al., metabolic diseases encompass more than just metabolic imbalance; they also entail systemic changes that impact neurological, renal, and cardiovascular functioning [7]. Understanding the underlying molecular pathways of these complex and multi-faceted illnesses is essential for efficient diagnosis, treatment, and management.

Arnab Seth and Satadeep Bandyapadhyay contributed equally to this work.

A. Seth · A. Palui · M. Chatterjee (✉)
Department of Pharmaceutical Chemistry, Bharat Technology, Uluberia, West Bengal, India

S. Bandyapadhyay
Department of Pharmacology, Bharat Technology, Uluberia, West Bengal, India

S. Ashique, N. Mishra (eds.), *Reactive Oxygen Species in Metabolic Inflammation*,
https://doi.org/10.1007/978-3-032-11752-6_21

Precision medicine techniques have been made possible by recent developments in molecular biology, omics technologies, and computational tools that have deepened our understanding of the pathophysiology of metabolic illnesses [8]. The growing significance of artificial intelligence (AI) in understanding and addressing ROS-mediated inflammation as a major cause of metabolic dysfunction is the main objective of this chapter.

21.1.1 Role of ROS in Inflammation

Superoxide anion (O_2-), hydrogen peroxide (H_2O_2), and hydroxyl radical (OH-) are examples of chemically reactive molecules that contain oxygen and are known as reactive oxygen species (ROS). At the time of inflammation these molecules are created endogenously during cellular processes, mainly by mitochondrial respiration and enzymes like NADPH oxidase [9, 10]. In order to eliminate invasive infections and control inflammation, phagocytic immune cells like neutrophils and macrophages in particular create ROS during the inflammatory response [9, 11]. In addition to their microbicidal properties, ROS function as signaling molecules inside cells. By oxidizing redox-sensitive amino acid residues, they affect gene expression and the synthesis of inflammatory mediators by modulating the activity of proteins including kinases, phosphatases, and transcription factors (like NF-κB) [11]. By facilitating leukocyte recruitment and increasing the expression of adhesion molecules on endothelial cells, moderate ROS levels boost immunity by encouraging leukocyte migration to inflammatory areas [12]. Oxidative stress, which destroys biomolecules and is a contributing factor to chronic diseases like atherosclerosis, diabetes, and neurodegeneration, is caused when excessive ROS overwhelm cellular antioxidants. While natural ROS levels aid in wound healing, tissue regeneration, and immunological defense, high or prolonged levels can cause cellular damage, chronic inflammation, and even cancer. Important inflammatory pathways like NF-κB, NLRP3, JAK/STAT, and PI3K/AKT are all modulated by ROS. Antioxidants and ROS-scavenging biomaterials are being developed to resolve inflammation and restore redox equilibrium in order to prevent negative effects without compromising immunity [13].

21.2 Mechanistic Insights

Including both free radicals like superoxide anion (O_2^-) and hydroxyl radical (•OH) as well as non-radical species like hydrogen peroxide (H_2O_2), reactive oxygen species (ROS) are extremely reactive molecules that include oxygen [14]. Mostly produced as byproducts of mitochondrial oxidative phosphorylation, these species are particularly prevalent at complexes I and III of the electron transport chain (Fig. 21.1) [14]. NADPH oxidases (NOX enzymes), xanthine oxidases, cytochrome P450

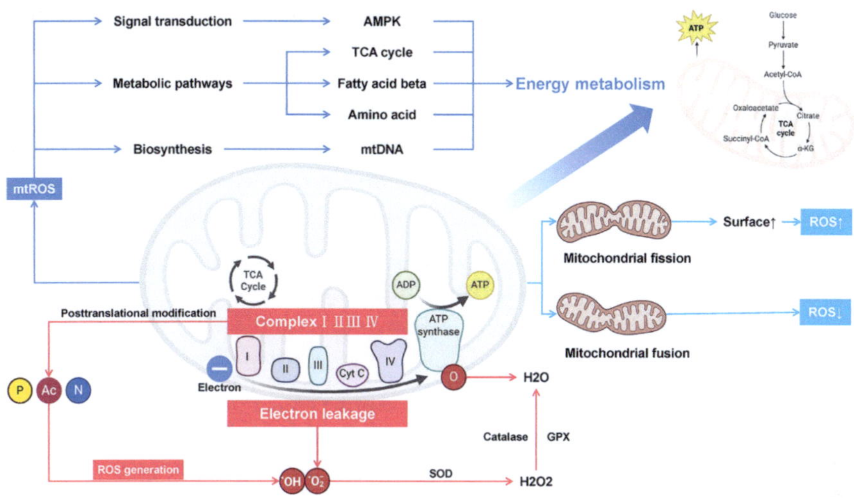

Fig. 21.1 Simple pathway of ROS generation [14]

enzymes, and peroxisomal oxidases are other producers of ROS [15]. Under normal circumstances, ROS at moderate concentrations serve as signaling molecules that control immunological responses, cell division, proliferation, and redox homeostasis [16]. The activity of redox-sensitive transcription factors, including NF-κB, AP-1, and Nrf2, which regulate the expression of genes related to survival, inflammation, and antioxidants, is modulated by ROS [17]. However, oxidative stress results when the creation of ROS surpasses the ability of the cellular antioxidant defense systems, which include enzymes such as glutathione peroxidase, catalase, and superoxide dismutase (SOD) [18]. Vital biomolecules such as proteins (oxidation and carbonylation), lipids (lipid peroxidation), and nucleic acids (strand breakage and base alterations) are all harmed by the result of this imbalance [19]. Numerous disease processes are significantly influenced by oxidative damage caused by ROS. It causes membrane damage, affects mitochondrial function, triggers pro-inflammatory pathways, and encourages cell death processes like ferroptosis, necrosis, and apoptosis [20, 21]. The pathophysiology of diseases like atherosclerosis, neurodegeneration, diabetes, cancer, and chronic inflammatory disorders is linked to persistent oxidative stress [22]. Furthermore, by triggering DNA damage responses and triggering the p53 and p21 pathways, high ROS levels aid in cellular senescence by stopping the cell cycle and fostering the senescence-associated secretory phenotype (SASP) [23]. Additionally, by interfering with calcium homeostasis and protein folding, ROS can cause endoplasmic reticulum (ER) stress, which activates the unfolded protein response (UPR) and contributes to metabolic and inflammatory diseases [24]. Autophagy is similarly influenced by ROS; low levels can promote autophagic removal of damaged organelles, whereas high levels hinder autophagic flow, resulting in the buildup of toxic aggregates and cellular dysfunction [25]. ROS control T-cell activation, antigen presentation, and cytokine synthesis in the immune system, which affects both innate and adaptive immunological responses [26]. Furthermore, ROS have a context-dependent function in angiogenesis. While high levels of ROS prevent angiogenesis by harming endothelial cells and causing vascular leakage, low levels encourage the expression of vascular endothelial growth factor (VEGF) and endothelial cell migration [27] (Fig. 21.1).

21.3 ROS-Mediated Signaling Pathways in Metabolic Diseases

The development and course of metabolic disorders such obesity, type 2 diabetes mellitus (T2DM), and nonalcoholic fatty liver disease (NAFLD) are significantly influenced by reactive oxygen species (ROS) [28]. ROS function as secondary messengers that control insulin signaling, glucose metabolism, and mitochondrial activity under typical physiological settings [29]. Redox equilibrium is disturbed by prolonged overnutrition and metabolic overload, which result in excessive ROS generation, mostly from mitochondria and NADPH oxidases. By encouraging the phosphorylation and degradation of IκB, reactive oxygen species (ROS) initiate the nuclear factor kappa B (NF-κB) signaling pathway. This results in the nuclear translocation of NF-κB and the transcriptional upregulation of pro-inflammatory cytokines like TNF-α and IL-6 [17]. Insulin resistance in skeletal muscle and adipose tissue is directly caused by this persistent inflammation, which is frequently referred to as "metaflammation" [30]. By encouraging the dissociation of thioredoxin-interacting protein (TXNIP), ROS activate the NLRP3 inflammasome. This allows TXNIP to connect to NLRP3 and then assemble with ASC and pro-caspase-1, releasing IL-1β and IL-18 [31]. An important anti-inflammatory, insulin-sensitizing adipokine, adiponectin, is secreted in adipose tissue but is suppressed by high ROS. This impairs insulin signaling and exacerbates insulin resistance (adiponectin levels are considerably lower in obese and insulin-resistant individuals) [32]. By oxidizing and deactivating protein-tyrosine phosphatases like PTEN and PTP1B, ROS further disrupt PI3K/AKT signaling, a pathway crucial for insulin sensitivity, which in turn affects downstream insulin-mediated signaling [33]. It has been considered that ROS generated primarily by NADPH oxidases and mitochondria function as vital second messengers that control key metabolic signaling pathways, such as HIF-1α, AMPK, NF-κB, PI3K/Akt, and MAPKs. However, excessive production that exceeds antioxidant buffering capacity causes oxidative stress and redox imbalance, which upsets metabolic homeostasis and promotes insulin resistance, chronic inflammation, endothelial dysfunction, and the development of metabolic diseases [14].

21.3.1 Targeting ROS and Inflammation

Increasingly, antioxidant and enzyme-modulating strategies to combat ROS and inflammation concentrate on triggering the Nrf2-ARE pathway, which strengthens endogenous defenses like HO-1, NQO1, and glutathione-related enzymes and reduces oxidative and inflammatory signaling. By efficiently inhibiting the production of pathogenic ROS, selective NOX inhibitors lower inflammation and vascular damage [34]. Superoxide dismutase, catalase, and peroxiredoxins are examples of antioxidant enzymes that function as important "ROS sinks," maintaining redox equilibrium [35]. It has been demonstrated that a number of natural substances, including dimethyl fumarate, resveratrol, and curcumin, activate the Nrf2 pathway and increase heme oxygenase-1 (HO-1), which has anti-inflammatory properties. In a variety of experimental settings, curcumin consistently increases HO-1 expression through Nrf2/ARE signaling to reduce inflammatory responses [36]. Similar to this, resveratrol activates Nrf2 in a variety of cell types and organisms to produce HO-1 [37]. In metabolic stress models, it has been shown that dimethyl fumarate (DMF) increases the production of HO-1 (and MnSOD) through Nrf2, which has anti-inflammatory and protective effects [38]. According to recent preclinical research, nanozymes that mimic SOD and catalase capabilities efficiently scavenge ROS, enhance mitochondrial function, and reduce inflammation [39]. A more targeted therapeutic strategy is provided by addressing redox imbalance, which acknowledges that oxidative and reductive stress both contribute to inflammatory dysfunction. In acute liver injury models, CA-Mn nanozymes exhibit ROS scavenging and Nrf2-Keap1 activation coupled with NF-κB suppression, indicating the great potential for precision ROS management by coordinated techniques combining Nrf2 activation, NOX inhibition, and targeted nanozyme therapy [40]. It has also been discovered that rebalancing NOX2 and Nrf2 in neural tissues inhibits the activation of the NLRP3 and NF-κB inflammasomes and prevents neuronal death [41].

21.3.2 Inflammatory Pathway Inhibitors

A variety of tactics are used to target ROS-linked inflammation, such as NLRP3 inflammasome inhibitors (DFV890, MCC950), which reduce inflammasome-mediated tissue damage by blocking caspase-1 activation and suppressing IL-1β/IL-18 production [42–44]. mPGES-1 modulators, such as KH176m, offer advantages in inflammatory diseases linked to mitochondria by concurrently lowering oxidative stress and prostaglandin E_2 synthesis. KH176m significantly normalizes high PGE_2 levels caused by inflammation and mitochondrial dysfunction by selectively inhibiting mPGES-1 activity and its expression in inflammatory models, according to recent mechanistic investigations [45]. Ziltivekimab and other anti-IL-6 monoclonal antibodies have been demonstrated to dramatically lower systemic inflammation in patients with chronic kidney disease and high cardiovascular risk experiencing a >93% decrease in hsCRP levels [46]. Pro-inflammatory transcriptional activation and ROS-amplified cytokine release are efficiently suppressed by NF-κB pathway inhibitors, which include both synthetic medications and substances originating from plants [46]. A mitochondria-focused anti-inflammatory strategy is provided by mitophagy enhancers that activate the PINK1–Parkin pathway, which reduces mitochondrial ROS generation and inhibits the activation of inflammasomes [47]. Additionally, recent data demonstrates dual-target medicines that simultaneously restore redox and decrease cytokine signaling by combining Nrf2 activation and NF-κB inhibition [48]. The exact control of ROS-driven inflammation may ultimately be achieved by combination techniques that incorporate transcription factor regulation, cytokine neutralization, inflammasome blockage, and mitochondrial redox modulation [49].

21.4 Artificial Intelligence in Biomedicine

Artificial intelligence (AI) has evolved as a crucial means in recent biomedicine, transforming drug discovery, personalized healthcare, prognostication, and diagnosis. Along with the enhanced incorporation of deep learning (DL) and machine learning (ML) methodologies, AI augments the analysis of medical images employing high-throughput, sophisticated radiomic profiling like computed tomography (CT), magnetic resonance imaging (MRI), and X-ray scans, in so doing assisting precise and early detection of disease [50]. Contemporary developments in multimodal AI systems that integrate with omics databases, clinical archives, and biomedical imaging have demonstrated their potential in providing personalized diagnostic insights across a wide range of applications [51]. In addition, LLMs (large language models) are becoming proficient in biomedical frameworks, including genetic sequencing, the interpretation of physiological data, and clinical notes, thereby enabling multifaceted reasoning across both clinical and molecular realms [52]. Despite this advancement, issues with the transparency of the algorithm used and regulatory acquiescence persist, warranting a trustworthy AI disposition. Emergent archetypes like XAI (explainable AI) and coalesced learning aim to address these apprehensions while upholding data concealment [53]. Furthermore, the ethical insinuations of prejudiced training datasets and inadequate demographic illustration are perilous. Recent studies have emphasized impartial AI moderation via the incorporation of assorted population data and extenuating structural discrepancies [54]. The continuous evolution of AI systems is embedding interdisciplinary collaboration to accelerate the biomedical landscape.

21.4.1 Overview of AI, ML, and DL Techniques

AI, including ML and DL, is reforming biomedicine by expediting evidence-based clinical high-throughput methodologies. ML algorithms like SVMs (support vector machine), RFs (random forest), and logistic regression have been employed for augmenting tasks such as classification of disease, discovery of biomarkers, and prediction of structured biomedical data. In distinction, the DL framework is based on CNNs (Convolutional Neural Networks) and U-Net variants. It emphasizes the modeling of indistinct and high-dimensional data like sequencing of genomes, natural language clinical notes, and medical imaging. For example, Islam et al. in 2022 have executed a CNN-derived model that can integrate MRI data with a synthetic CT image to advance brain tumor segmentation, accomplishing superior tumor boundary demarcation in comparison to conventional methodologies [55]. You et al. in 2023 have established a pulmonary nodule detection system based on DL that has shown sensitivity enhancement of specific locations throughout diverse zones of lungs, improving early recognition strategies of lung cancer [56]. Further in 2023, Bianconi et al. have technologically advanced a DL division algorithm with the help of data driven from postoperative MRI scans of glioblastoma patients. It has proven a suggestive enhancement of the quantification for tumor and has enabled better predictive assessment [57]. In 2024, Yunqiu and Min used Raman spectroscopy augmented with ML based on genome sequence, accomplishing more than 95% accurate classification for pathogenic organisms within 5 min. It has demonstrated the probability of label-free and real-time pathogen identification. These progressions accentuate the emerging amalgamation of AI into biomedical fields, along with a mounting prominence on data fusion, clinical translatability, and explainability [58].

21.4.2 Applications in Drug Discovery and Disease Modeling

AI is reconstructing drug discovery and disease modeling by integrating computational simulations and HTS (high-throughput screening) results to enhance therapeutic developments. In the realm of drug discovery, ML algorithms can be applied for the identification of targets and virtual screening of chemical libraries. For instance, VAEs (Variational Autoencoders) and GANs (Generative Adversarial Networks) have been used for small molecular development with specific binding interactions, improved ADMET profile, and reduced toxicity [55]. Remarkably, AI-simulated drug design for DDR1 kinase inhibitors shows a new paradigm in the recent era [59]. Fortification of learning workflows enables the iterative refinement of molecular structures based on synthetic convenience scores and predicted binding-free energies. AtomNet (an AI-based drug design platform) has been proven for drug modeling against infectious diseases, including SARS-CoV-2 [60]. AI enables the construction of in silico models in disease modeling, which act out disease development, molecular etiology, and diagnostic interferences. Neural networks are accomplished with a longitudinal patient database, which can envisage the trajectory of disease in neurodegenerative conditions like Alzheimer's. On the other hand, graph-based ML models can plot protein–protein interactions in oncology to predict possible druggable bulges [57]. AI-based approaches can accumulate clinical information with omics data to model orphan diseases, aiding virtual simulation of clinical trials and patient acceptance. Furthermore, mechanized phenotypical cell-based assay profiling through a combination of neural networking and high content imaging, enhanced genotype and phenotype correlation [58].

21.5 AI Approaches to ROS and Inflammation

AI has gradually advanced to decrypt the elaborate relationship between ROS and inflammatory pathways. For instance, in the case of TNBC (triple negative breast cancer), an amalgamated ROSig (ROS signature) engendered through an ML methodology including CosBoost and survival-SVM, which permits analytical stratifications and envisages therapeutic receptivity by differentiating immunological and molecular conglomeration [61]. Furthermore, AI-enabled DL algorithms alleviate noise and enhance the lesion detection, allowing improved imaging of inflammation dynamics, especially in low-grade inflammatory circumstances [62]. At the therapeutic edge, AI-based modeling of ROS scavenging and responsive nanomedicines enables target modification of redox conditions to lessen inflammation, showcasing potent anti-inflammatory activity in preclinical trials. Communally, these methodologies underline the growth of AI in enumerating ROS-mediated inflammatory pathways, filtering diagnostic assessments, and emphasizing precise nanotherapeutics.

21.5.1 ML Algorithms for Biomarker Discovery

ML methods for the discovery of biomarkers in ROS-mediated inflammation exploit omics datasets dealing with feature selection algorithms like LASSO, SVM-RFE, and filtration of mutual databases. Collaborative methods like XGBoost and Random Forest apply widely to exuberant contender biomarkers, whereas LIME and SHAP accelerate interpretations. For instance, ML algorithms based on metabolomic databases have been utilized to recognize oxidative stress and inflammatory symptoms in hypertensive OSA patients. Incorporation of multi-omics information with clinical constraints and vigorous cross-validation enables reproducible and robust biomarker pipelines for translational studies [63].

21.5.2 DL Models for Imaging and Omics Integration

DL methodologies, which amalgamate omics information and imaging, facilitate prevailing tools for the discovery of ROS-mediated inflammatory biomarkers. The CNNs excerpt longitudinal characteristics from imaging systems (like radiology, histopathology), which can be fused with omics-derived illustration employing attention-based gating or late fusion methodology for advanced prediction-based modeling [64]. On the other hand, end-to-end frameworks like Image2Omics describe the probability of forecasting proteomic and transcriptomic features unswervingly from cellular imaging, aiding noninvasive and rapid identification of inflammatory stimulus. These frameworks not only enhance investigative precision but also gather cross-modal relations, permeating explainable biologically standard inferences for oxidative stress-induced inflammations [65].

21.5.3 AI in Predicting ROS-Targeted Drug Efficacy

AI is progressively utilized in the prediction of ROS-targeted therapies by accumulating omics, chemical features, and pharmacological attributes. DL frameworks can model docking scores, molecular descriptors, and genetic information to approximate antioxidant efficacy, while ML models help in ranking molecular attributes and target selection. AI frameworks have been utilized for screening phytochemicals against ROS, predicting the antioxidant efficacy of nanomaterials, and evaluating drug-responsive genetic ailments in oxidative stress pathways. Combination of in silico methods with network pharmacology and system biology, aided by an understandable DL/ML tool (e.g., SHAP), confirms biological acceptability and model robustness. These approaches rationalize the ROS-targeted drug discovery by incorporating efficacy prediction, rapid VS (virtual screening), and ranking of promising chemicals [66, 67].

21.6 Case Studies and Applications

21.6.1 AI-Based Screening of Antioxidant Compounds

The use of artificial intelligence (AI) in drug development has made it possible to evaluate chemical libraries quickly, affordably, and in large quantities, which has greatly improved the screening of antioxidant chemicals. According to Lobo et al., antioxidants are essential for the prevention and treatment of a number of chronic illnesses, such as cancers, cardiovascular diseases, and neurological disorders [68]. They do this by scavenging reactive oxygen species (ROS) and preventing oxidative stress. Conventional screening techniques, including in vitro tests (DPPH, ABTS,

FRAP, etc.), need a lot of time, resources, and skilled manpower. On the contrary, the AI-driven tools especially machine learning and deep learning-based algorithms can effectively screen large databases by using the molecular structures as the key input for predicting antioxidant properties of various chemicals [69]. Besides prediction and screening, the AI-based tools are also effectively utilized for developing predictive models such as Quantitative Structure-Activity Relationships (QSARs), Read-across (RA), and Read-across Structure-Activity Relationship (RASAR) that helps identifying the important structural features responsible for antioxidant activity [70]. These features can further be used for designing new antioxidants and prioritizing hits for synthesis and further development. Accurate antioxidant potential prediction is made possible by the combination of algorithms like Random Forest, Support Vector Machines, and Graph Neural Networks with molecular descriptor-based QSAR (Quantitative Structure–Activity Relationship) models [71, 72]. Additionally, the discovery of new scaffolds with enhanced effectiveness and pharmacokinetic profiles—such as synthetic analogues and natural product derivatives—is made easier by the integration of AI with cheminformatics platforms and virtual screening pipelines [73]. De novo chemical production utilizing generative AI models is another recent development that can create completely new antioxidant compounds that are tuned for potency, safety, and drug-likeness [59]. The development of next-generation antioxidants with therapeutic potential against diseases linked to oxidative stress is expected to be accelerated by such AI-based screening.

21.7 Predictive Modeling in Type 2 Diabetes and NAFLD

The predictive modeling for ROS-induced T2D (Type 2 diabetes) and NAFLD (nonalcoholic fatty liver disease) conglomerates biochemical, clinical data with oxidative stress biomarkers to accelerate early diagnosis. DL models can amalgamate wearable-derived biological signals with standardized laboratory tests to evaluate the resistance risk of insulin in T2D [74]. In case of NAFLD, survival-based ML approaches (e.g., random survival forests) use feature selection methodologies (like LASSO) to modify significant predictor variables and accomplish high precision in long-term risk management. These AI frameworks permit dynamic, noninvasive, and insightful predictions, assisting precision medication in ROS-mediated metabolic disorders [75]. Table 21.1 gives an insight into the recently published reports on predictive modeling in T2D and NAFLD.

Table 21.1 Different published reports on predictive modeling in T2D and NAFLD

Disease	Model type	Data and features	Outcome and performance	References
T2D	ML models with 5 algorithms	32 molecular and 12 biochemical features Drug-treated T2D rats	Accuracy = approximately 80% AUC up to 0.93	[76]
	Explainable ML (SHAP + Extra Survival Trees)	Clinical group ($n = 554$) Mortality risk factors	C—Statistic 0.776 AUC = 0.86–0.80 for mortality prediction up to 16.8 years	[77]
	ML integrated with multi-omics data	Multi-omics database of human pancreatic islets of Langerhans	Identification of ROS-mediated predictive molecular networks leading to β-cell dysfunction	[78]
	Ensemble ML	Clinical and lifestyle data, along with oxidative stress-induced biomarkers	Improved risk assessment for T2D Accuracy = approximately 88%	[78]
	Risk scoring based on metabolites with ML classifiers	Conventional risk factors with circulating metabolites associated with oxidative stress	Development of a risk score with AUC >0.85 Validation done for the detection of early T2D	[78]
NAFLD	SVM-RFE on gene expression with LASSO	ROS and inflammation-linked gene (FOSB, GADD45B, AXUD1, SOCS2)	Diagnostic model AUC = 0.994 Validated through external mouse and human datasets	[79]
	FeRS using GBM and GLM boosting	ROS-induced Ferroptosis gene (FRG)	Identification of ZFP36 as key FRG	[80]
	WGCNA with ML	Oxidative stress biomarkers (for example, miRNAs, TF-miRNA-mRNA network)	Identification of specific biomarker	[81]
	XGBoost	Clinical database (e.g., neutrophil %, AST/ALT)	Internal AUC = 0.95 External AUC = 0.90	[82]
	Cascade ensemble with information fusion	Laboratory and anthropometric parameters	Accuracy = approximately 86% AUC = approximately 96%	[83]

21.8 Challenges and Future Perspectives

AI-based methods in antioxidant discovery hold promise, but there are still a number of obstacles to overcome. Due to variations in assay types, experimental conditions, and reporting standards, experimental findings often fluctuate between laboratories, resulting in heterogeneous and limited data, which is a significant bottleneck. The model's effectiveness and predictivity to generate new chemical scaffolds can be restricted by the bias of many datasets toward particular compound classes. Data imbalance can result in skewed predictions and decreased screening reliability when active chemicals greatly outnumber inactives or vice versa. Interpretability of modeled descriptors (structural features) is another important concern. It is sometime very difficult to understand the active and inactive predictions due to the "black box" design of deep learning models [84]. Transparency of models' architecture is an important aspect for regulatory acceptance and reliability in modern drug development and failing to do so may hinder the performance of AI and ML models. Discovering structure–activity correlations via the use of explainable AI (XAI) tools can support logical molecular design. Although, it is still difficult for AI models to be transferred between various chemical classes. For complicated phytochemical combinations or natural goods, models trained on synthetic substances might not function effectively. Standardized experimental procedures, open-access curated datasets, and hybrid modeling approaches that blend data-driven learning with mechanistic insights will be necessary to overcome these constraints.

21.9 Future Directions in AI-Aided Therapeutics

The combination of multi-omics data, real-time biosensor outputs, and sophisticated computer models to allow precision medicine techniques is key to the future of AI-aided treatments, especially in antioxidant drug development [71]. It is anticipated that explainable AI (XAI), which offers mechanistic insights into how structural characteristics affect antioxidant function, would advance AI platforms and enhance drug design interpretability and trust [84]. The identification of therapeutically relevant antioxidants will be accelerated by the creation of closed-loop systems that can quickly design, synthesize, test, and learn when in silico prediction models are combined with automated synthesis and high-throughput biological validation [59]. Furthermore, the creation of compounds that are concurrently optimized for potency, selectivity, bioavailability, and safety will be made possible by developments in generative AI, high-throughput computing, and molecular dynamics simulations [69, 73]. AI may eventually enable customized antioxidant treatment, adjusting therapy according to a person's genetic makeup and oxidative stress profile, increasing therapeutic effectiveness while reducing side effects [68].

21.10 Conclusion

The identification of antioxidant compounds may now be accelerated by AI-based screening, which provides quick, affordable, and scalable substitutes for conventional in vitro and in vivo testing. Researchers can estimate antioxidant potential, screen large chemical libraries, and find new molecular scaffolds with therapeutic significance by utilizing machine learning, deep learning, and QSAR modeling. These methods enable multiparameter optimization of potency, selectivity, and safety before experimental confirmation, in addition to making it easier to explore large chemical spaces. However, overcoming enduring obstacles is necessary to fully realize AI's potential in antioxidant therapies. Increasing interpretability through explainable AI (XAI) and incorporating mechanistic insights can improve trust, regulatory acceptance, and rational design. These difficulties include the lack and heterogeneity of high-quality datasets, the "black box" nature of complex predictive models, biases toward particular chemical classes, and variability in experimentations. Enhancing model repeatability and generalizability will also need the creation of curated open-access datasets and defined methodologies. Personalized antioxidant therapy might be revolutionized in the future by combining AI with multi-omics data, high-throughput screening, and automated synthesis in closed-loop pipelines. The development of next-generation antioxidants tailored to specific oxidative stress profiles may be accelerated by developments in generative AI, quantum computing, and molecular dynamics simulations. In order to ensure that AI-driven predictions result in clinically effective, safe, and targeted antioxidant therapies, success in this sector will ultimately depends on interdisciplinary cooperation, transparent model development, and data sharing.

References

1. Zimmet P, Alberti KGMM, Shaw J (2001) Global and societal implications of the diabetes epidemic. Nature 414:782–787
2. Younossi ZM, Koenig AB, Abdelatif D, Fazel Y, Henry L, Wymer M (2016) Global epidemiology of nonalcoholic fatty liver disease—Meta-analytic assessment of prevalence, incidence, and outcomes. Hepatology 64:73–84
3. Ng M, Fleming T, Robinson M, Thomson B, Graetz N, Margono C (2014) Global, regional, and national prevalence of overweight and obesity in children and adults during 1980–2013: a systematic analysis for the Global Burden of Disease Study 2013. Lancet 384:766–781
4. Kahn SE, Hull RL, Utzschneider KM (2006) Mechanisms linking obesity to insulin resistance and type 2 diabetes. Nature 444:840–846
5. Furukawa S, Fujita T, Shimabukuro M, Iwaki M, Yamada Y, Nakajima Y et al (2004) Increased oxidative stress in obesity and its impact on metabolic syndrome. J Clin Invest 114:1752–1761
6. Roberts CK, Sindhu KK (2009) Oxidative stress and metabolic syndrome. Life Sci 84:705–712
7. Eckel RH, Grundy SM, Zimmet PZ (2005) The metabolic syndrome. Lancet 365:1415–1428
8. Hasin Y, Seldin M, Lusis A (2017) Multi-omics approaches to disease. Genome Biol 18:83
9. Mittal M, Siddiqui MR, Tran K, Reddy SP, Malik AB (2014) Reactive oxygen species in inflammation and tissue injury. Antioxid Redox Signal 20:1126–1167

10. Jomova K, Raptova R, Alomar SY, Alwasel SH, Nepovimova E, Kuca K (2023) Reactive oxygen species, toxicity, oxidative stress, and antioxidants: chronic diseases and aging. Arch Toxicol 97:2499–2574
11. Liu J, Han X, Zhang T, Tian K, Li Z, Luo F (2023) Reactive oxygen species (ROS) scavenging biomaterials for anti-inflammatory diseases: from mechanism to therapy. J Hematol Oncol 16:1–34
12. Yu Y, Liu S, Yang L, Song P, Liu Z, Liu X et al (2024) Roles of reactive oxygen species in inflammation and cancer. MedComm 5:e519
13. Valko M, Leibfritz D, Moncol J, Cronin MTD, Mazur M, Telser J (2007) Free radicals and antioxidants in normal physiological functions and human disease. Int J Biochem Cell Biol 39:44–84
14. Liu S, Liu J, Wang Y, Deng F, Deng Z (2025) Oxidative stress: signaling pathways, biological functions, and disease. MedComm 6:e70268
15. Bedard K, Krause KH (2007) The NOX family of ROS-generating NADPH oxidases: physiology and pathophysiology. Physiol Rev 87:245–313
16. Ray PD, Huang B-W, Tsuji Y (2012) Reactive oxygen species (ROS) homeostasis and redox regulation in cellular signaling. Cell Signal 24:981–990
17. Hong Y, Boiti A, Vallone D, Foulkes NS (2024) Reactive oxygen species signaling and oxidative stress: transcriptional regulation and evolution. Antioxidants 13:312
18. Bhattacharyya A, Chattopadhyay R, Mitra S, Crowe SE (2014) Oxidative stress: an essential factor in the pathogenesis of gastrointestinal mucosal diseases. Physiol Rev 94:329–354
19. Juan CA, de la Lastra JMP, Plou FJ, Pérez-Lebeña E (2021) The chemistry of reactive oxygen species (Ros) revisited: outlining their role in biological macromolecules (dna, lipids and proteins) and induced pathologies. Int J Mol Sci 22:4642
20. Yang X, Li Y, Li Y, Ren X, Zhang X, Hu D et al (2017) Oxidative stress-mediated atherosclerosis: mechanisms and therapies. Front Physiol 8:1–16
21. Tang D, Chen X, Kang R, Kroemer G (2021) Ferroptosis: molecular mechanisms and health implications. Cell Res [Internet] 31:107–125. https://doi.org/10.1038/s41422-020-00441-1
22. Sharifi-Rad M, Anil Kumar NV, Zucca P, Varoni EM, Dini L, Panzarini E et al (2020) Lifestyle, oxidative stress, and antioxidants: back and forth in the pathophysiology of chronic diseases. Front Physiol 11:694
23. Dong Z, Luo Y, Yuan Z, Tian Y, Jin T, Xu F (2024) Cellular senescence and SASP in tumor progression and therapeutic opportunities. Mol Cancer [Internet] 23:181. https://doi.org/10.1186/s12943-024-02096-7
24. Malhotra JD, Kaufman RJ (2007) The endoplasmic reticulum and the unfolded protein response. Semin Cell Dev Biol 18:716–731
25. Scherz-Shouval R, Elazar Z (2011) Regulation of autophagy by ROS: physiology and pathology. Trends Biochem Sci 36:30–38
26. Chen X, Song M, Zhang B, Zhang Y (2016) Reactive oxygen species regulate T cell immune response in the tumor microenvironment. Oxid Med Cell Longev 2016:1580967
27. Kim Y-W, Byzova TV (2014) Oxidative stress in angiogenesis and vascular disease. Blood 123:625–631
28. Nascè A, Gariani K, Jornayvaz FR, Szanto I (2022) NADPH oxidases connecting fatty liver disease, insulin resistance and Type 2 diabetes: current knowledge and therapeutic outlook. Antioxidants 11:1131
29. Di MS, Iossa S, Venditti P (2017) Skeletal muscle insulin resistance: role of mitochondria and other ROS sources. J Endocrinol 233:R15–R42
30. Mukherjee S, Kundu R, Vidaković M (2024) Editorial: metaflammation in obesity and diabetes. Front Endocrinol (Lausanne) 15:1540999
31. Russell-Guzmán J, Silva LA-D, Cadagan C, Maturana M, Palomero J, Estrada M et al (2024) Activation of the ROS/TXNIP/NLRP3 pathway disrupts insulin-dependent glucose uptake in skeletal muscle of insulin-resistant obese mice. Free Radic Biol Med 222:187–198

32. Satish M, Saxena SK, Agrawal DK (2019) Adipokine dysregulation and insulin resistance with atherosclerotic vascular disease: metabolic syndrome or independent sequelae? J Cardiovasc Transl Res 12:415–424
33. Trinh VH, Nguyen Huu T, Sah DK, Choi JM, Yoon HJ, Park SC et al (2024) Redox regulation of PTEN by reactive oxygen species: its role in PHYSIOLOGICAL PROCESSES. Antioxidants 13:199
34. Nuñez-Selles AJ, Nuñez-Musa RA, Guillen-Marmolejos RA (2025) Linking oxidative stress biomarkers to disease progression and antioxidant therapy in hypertension and diabetes mellitus. Front Mol Biosci 12:1611842
35. Hunt M, Torres M, Bachar-Wikstrom E, Wikstrom JD (2024) Cellular and molecular roles of reactive oxygen species in wound healing. Commun Biol [Internet] 7:1534. https://doi.org/10.1038/s42003-024-07219-w
36. Funes SC, Rios M, Fernández-Fierro A, Covián C, Bueno SM, Riedel CA et al (2020) Naturally derived heme-oxygenase 1 inducers and their therapeutic application to immune-mediated diseases. Front Immunol 11:01467
37. Huenchuguala S, Segura-Aguilar J (2024) Natural compounds that activate the KEAP1/Nrf2 signaling pathway as potential new drugs in the treatment of Idiopathic Parkinson's disease. Antioxidants 13:1125
38. Lanza M, Basilotta R, Cuzzocrea S, Bulzomì M, Oddo S, Casili G et al (2023) An RNAi-Mediated reduction in transcription factor Nrf-2 blocks the positive effects of Dimethyl Fumarate on metabolic stress in Alzheimer's disease. Int J Mol Sci 24:11303
39. Mohsin M, Shams F, Li H, Alam A, Xia C, Fan L et al (2025) Nanozymes in neuropathic pain: strategies bridging oxidative stress, mitochondrial repair, and neuroimmune modulation for targeted therapy. J Neuroinflammation [Internet] 22:156. https://doi.org/10.1186/s12974-025-03456-w
40. Wang F, Yuan H, Shen J, Li Z, Li J, Luo P et al (2024) Nanozymes with broad-spectrum scavenging of reactive oxygen species (ROS) alleviate inflammation in acute liver injury. ACS Mater Lett 6:1304–1306
41. Ardizzone A, Capra AP, Repici A, Lanza M, Bova V, Palermo N et al (2024) Rebalancing NOX2/Nrf2 to limit inflammation and oxidative stress across gut-brain axis in migraine. Free Radic Biol Med 213:65–78
42. Cao L, Pi W, Zhang Y, Zheng C, Yong VW, Xue M (2025) Targeting the NLRP3-ROS axis: disrupting the oxidative-inflammatory vicious cycle in intracerebral hemorrhage. J Inflamm Res 18:9849–9870
43. Gatlik E, Mehes B, Voltz E, Sommer U, Tritto E, Lestini G et al (2024) Clinical translational Sci – 2024 – Gatlik – First-in-human safety tolerability and pharmacokinetic results of DFV890 an.pdf. Clin Transl Sci 17:e13789
44. Li H, Guan Y, Liang B, Ding P, Hou X, Wei W et al (2022) Therapeutic potential of MCC950, a specific inhibitor of NLRP3 inflammasome. Eur J Pharmacol 928:175091
45. Jiang X, Renkema H, Pennings B, Pecheritsyna S, Schoeman JC, Hankemeier T et al (2021) Mechanism of action and potential applications of selective inhibition of microsomal prostaglandin E synthase-1-mediated PGE2 biosynthesis by sonlicromanol's metabolite KH176m. Sci Rep [Internet] 11:880. https://doi.org/10.1038/s41598-020-79466-w
46. Wada Y, Jensen C, Meyer ASP, Zonoozi AAM, Honda H (2023) Efficacy and safety of interleukin-6 inhibition with ziltivekimab in patients at high risk of atherosclerotic events in Japan (RESCUE-2): a randomized, double-blind, placebo-controlled, phase 2 trial. J Cardiol [Internet] 82:279–285. https://doi.org/10.1016/j.jjcc.2023.05.006
47. Xu X, Pang Y, Fan X (2025) Mitochondria in oxidative stress, inflammation and aging: from mechanisms to therapeutic advances. Signal Transduct Target Ther [Internet] 10:190. https://doi.org/10.1038/s41392-025-02253-4
48. Bellanti F, Coda ARD, Trecca MI, Lo Buglio A, Serviddio G, Vendemiale G (2025) Redox imbalance in inflammation: the interplay of oxidative and reductive stress. Antioxidants 14:656

49. Saavedra-Torres JS, Pinzón-Fernández MV, Ocampo-Posada M, Nati-Castillo HA, Jiménez Hincapie LA, Cadrazo-Gil EJ et al (2025) Inflammasomes and signaling pathways: key mechanisms in the pathophysiology of sepsis. Cells 14:930
50. Parvin N, Joo S, Jung J, Mandal T (2025) Multimodal AI in biomedicine: pioneering the future of biomaterials, diagnostics, and personalized healthcare. Nanomaterials 15:895
51. Socol Y, Richardson A, Garali-Zineddine I, Grison S, Vares G, Klokov D (2023) Artificial intelligence in biology and medicine, and radioprotection research: perspectives from Jerusalem. Front Artif Intell 6:1291136
52. Bi Z, Dip S, Hajialigol D, Kommu S, Liu H, Lu M et al (2024) Ai for biomedicine in the era of large language models. arXiv Prepr 2403:15673
53. Basubrin O (2025) Current status and future of artificial intelligence in medicine. Cureus J Med Sci 17:77561
54. Kumar D, Malin BA, Vishwanatha JK, Wu L, Hedges JR (2024) AI in biomedicine—A forward-looking perspective on health equity. Int J Environ Res Public Health 21:1642
55. Islam KT, Wijewickrema S, O'Leary S (2022) A deep learning framework for segmenting brain tumors using MRI and synthetically generated CT images. Sensors 22:523
56. You S, Park JH, Park B, Shin H-B, Ha T, Yun JS et al (2023) The diagnostic performance and clinical value of deep learning-based nodule detection system concerning influence of location of pulmonary nodule. Insights Imaging 14:149
57. Bianconi A, Rossi LF, Bonada M, Zeppa P, Nico E, De MR et al (2023) Deep learning-based algorithm for postoperative glioblastoma MRI segmentation: a promising new tool for tumor burden assessment. Brain Informatics 10:26
58. Gao Y, Liu M (2024) Application of machine learning based genome sequence analysis in pathogen identification. Front Microbiol 15:1474078
59. Zhavoronkov A, Ivanenkov YA, Aliper A, Veselov MS, Aladinskiy VA, Aladinskaya AV et al (2019) Deep learning enables rapid identification of potent DDR1 kinase inhibitors. Nat Biotechnol 37:1038–1040
60. Gao C, Wu L, Wu W, Huang Y, Wang X, Sun Z et al (2024) Deep learning in pulmonary nodule detection and segmentation: a systematic review. Eur Radiol 35:255–266
61. Li J, Liang Y, Liang Y, Zhao X (2023) Integrating machine learning algorithms to systematically assess reactive oxygen species levels to aid prognosis and novel treatments for triple -negative breast cancer patients. Front Immunol 14:1196054
62. Schwenck J, Kneilling M, Riksen NP, la FC, Mulder DJ, Slart RJHA et al (2022) A role for artificial intelligence in molecular imaging of infection and inflammation. Eur J Hybrid Imaging 6:17
63. Du Z, Sun H, Du Y, Li L, Lv Q, Yu H et al (2022) Comprehensive metabolomics and machine learning identify profound oxidative stress and inflammation signatures in hypertensive patients with obstructive sleep Apnea. Antioxidants 11:1946
64. Ballard JL, Wang Z, Li W, Shen L, Long Q (2024) Deep learning-based approaches for multi-omics data integration and analysis. BioData Min [Internet] 17:1–29. https://doi.org/10.1186/s13040-024-00391-z
65. Mehrizi R, Mehrjou A, Alegro M, Zhao Y, Carbone B, Fishwick C et al (2023) Multi-omics prediction from high-content cellular imaging with deep learning. arXiv Prepr 2306:09391
66. Zhang X, Zheng Z, Xie L, Yang M, Wang J, Wang W et al (2025) Deep learning-enhanced development of innovative antioxidant liposomal drug delivery systems from natural herbs. Mater Horizons 12:7416–7424
67. Mirzaei M, Furxhi I, Murphy F, Mullins M (2023) Employing supervised algorithms for the prediction of nanomaterial's antioxidant efficiency. Int J Mol Sci 24:2792
68. Lobo V, Patil A, Phatak A, Chandra N (2010) Free radicals, antioxidants and functional foods: impact on human health. Pharmacogn Rev 4:118–126
69. Ekins S, Puhl AC, Zorn KM, Lane TR, Russo DP, Klein JJ et al (2019) Exploiting machine learning for end-to-end drug discovery and development. Nat Mater 18:435–441

70. Banerjee A, Roy K (2022) First report of q-RASAR modeling toward an approach of easy interpretability and efficient transferability. Mol Divers 26:2847–2862
71. Ghironi S, Viganò EL, Selvestrel G, Benfenati E (2025) QSAR models for predicting the antioxidant potential of chemical substances. J Xenobiotics 15:80
72. Idowu SO, Fatokun AA (2021) Artificial Intelligence (AI) to the rescue: deploying machine learning to bridge the biorelevance gap in antioxidant assays. SLAS Technol 26:16–25
73. Chen H, Engkvist O, Wang Y, Olivecrona M, Blaschke T (2018) The rise of deep learning in drug discovery. Drug Discov Today 23:1241–1250
74. Metwally AA, Heydari AA, McDuff D, Solot A, Esmaeilpour Z, Faranesh AZ et al (2025) Insulin resistance prediction from wearables and routine blood biomarkers. arXiv Prepr 2505:03784
75. Zhang H, Zhang L, Li N, Zhang Y, Zhang X, Wang D (2025) Machine learning survival models for non-alcoholic fatty liver disease based on a health checkup cohort. BMC Gastroenterol 25:518
76. Matboli M, Al-Amodi HS, Khaled A, Khaled R, Roushdy MMS, Ali M et al (2024) Comprehensive machine learning models for predicting therapeutic targets in type 2 diabetes utilizing molecular and biochemical features in rats. Front Endocrinol (Lausanne) 15:1384984
77. Vershinina O, Sabbatinelli J, Bonfigli AR, Colombaretti D, Giuliani A, Krivonosov M et al (2025) Explainable artificial intelligence model predicting the risk of all-cause mortality in patients with type 2 diabetes mellitus. arXiv Prepr 2507:23491
78. Kiran M, Xie Y, Anjum N, Ball G, Pierscionek B, Russell D (2025) Machine learning and artificial intelligence in type 2 diabetes prediction: a comprehensive 33-year bibliometric and literature analysis. Front Digit Health 7:1557467
79. Zhang Z, Wang S, Zhu Z, Nie B (2023) Identification of potential feature genes in non-alcoholic fatty liver disease using bioinformatics analysis and machine learning strategies. Comput Biol Med 157:106724
80. Dai L, Yuan W, Jiang R, Zhan Z, Zhang L, Xu X et al (2024) Machine learning-based integration identifies the ferroptosis hub genes in nonalcoholic steatohepatitis. Lipids Health Dis 23:23
81. Wang H, Cheng W, Hu P, Ling T, Hu C, Chen Y et al (2024) Integrative analysis identifies oxidative stress biomarkers in non-alcoholic fatty liver disease via machine learning and weighted gene co-expression network analysis. Front Immunol 15:1335112
82. Yan Y, Gan D, Zhang P, Zou H, Li MM (2024) A machine learning-based predictive model discriminates nonalcoholic steatohepatitis from nonalcoholic fatty liver disease. Heliyon 10:e38848
83. Delfan N, Moghadam PK, Khoshnevisan M, Chagahi MH, Hatami B, Asgharzadeh M et al (2024) AI-Driven non-invasive detection and staging of steatosis in fatty liver disease using a novel cascade model and information fusion techniques. arXiv Prepr 2412:04884
84. Samek W, Montavon G, Lapuschkin S, Anders CJ, Müller K-R (2021) Explaining deep neural networks and beyond: a review of methods and applications. Proc IEEE 109:247–278

Chapter 22
Nanocarrier-Based Targeted Approaches: To Scavenge ROS-Mediated Inflammation

Siuli Sen and Dipanjan Karati

22.1 Introduction

Inflammation is a fundamental defensive reaction that eliminates invading microbes and other entities to restore homeostasis [1]. Nonetheless, excessive stimulation of inflammatory responses is detrimental to the host and contributes to illnesses such as cancer, sepsis, and autoimmune disorders. This intricate biological process, which elicits clinical manifestations such as warmth, pain, erythema, and edema, is predominantly triggered by infections and tissue injury [2]. Inflammation is a fundamental pathogenic process that underpins the majority of human disorders [3]. For example, Coronavirus disease 2019 (COVID-19) has significantly disrupted worldwide healthcare systems and the economy; it is an inflammatory illness, with pneumonia as the most prevalent consequence [4]. Inflammation remains a complex medical challenge, making inflammatory mediators promising therapeutic targets for regulating inflammatory responses despite significant efforts.

The chemical entities produced by incomplete oxygen reduction are collectively termed "reactive oxygen species (ROS)" and serve as significant regulators of inflammation [5]. Superoxide anions (O_2^-), hydrogen peroxide (H_2O_2), singlet oxygen ($1O_2$), and hydroxyl radicals ($^{\cdot}OH$) are the main components of ROS [6]. Excess ROS is extremely reactive and can cause inflammation by oxidizing lipids, proteins, and nucleic acids, which can destroy cells in the body, according to studies [7]. Furthermore, ROS are signaling molecules that control different kinds of kinases and transcription factors that are involved in the start and progression of inflammation [8]. Many disorders, such as diabetes [9], inflammatory bowel disease, or IBD [10], chronic obstructive pulmonary disease or COPD [11], and osteoarthritis or OA [12], are thought to be exacerbated by ROS, which are thought to be involved in

S. Sen · D. Karati (✉)
Department of Pharmaceutical Technology, School of Pharmacy, Techno India University, Kolkata, West Bengal, India

S. Ashique, N. Mishra (eds.), *Reactive Oxygen Species in Metabolic Inflammation*,
https://doi.org/10.1007/978-3-032-11752-6_22

inflammatory reactions. Because ROS and inflammation are closely related, biomaterials that scavenge ROS have been developed to reduce inflammation and shield the host from harm brought on by aberrant inflammatory responses [13]. By either directly or indirectly balancing the production and elimination of ROS, ROS-scavenging biomaterials reduce inflammation [14]. Traditional anti-inflammatory treatments, such as nonsteroidal anti-inflammatory drugs and corticosteroids, can have negative effects on several organs, including the kidneys, the heart, and the gastrointestinal tract [15]. Biomedical nanotechnology has revolutionized the biotechnology and pharmaceutical sectors because of major developments in nanotechnology, especially in the areas of nanochemistry and nanomanufacturing. This has resulted in significant advancements for antioxidant therapy [16]. Numerous nanomaterials with distinct ROS-scavenging capabilities, including nanoparticulate C, Se, Ce, Pt, Cu, redox polymers, and polyphenols, have been created to reduce the inflammatory ROS response [17–19]. Nanotechnology, integrating material science with ROS biology and chemistry, has yielded innovative therapeutic approaches for modulating ROS levels across different biological contexts, thereby enhancing the treatment for ROS-related diseases. These tactics are very viable because of the regularity of chemical synthesis; nonetheless, some elements of the features defining nanomaterials and their interactions in biological environments remain ambiguous [20].

This chapter offers a thorough examination of the anti-inflammatory applications of nano-antioxidants by analyzing the molecular and pathophysiological processes of oxidative stress and its contribution to cellular and tissue damage. Then, for certain nanocarriers, we discuss the mechanisms governing ROS scavenging.

22.2 Reactive Oxygen Species (ROS): Double-Edged Swords

22.2.1 Sources and Types of ROS in Inflammatory Diseases

ROS are produced from both endogenous and exogenous sources in inflammatory diseases discussed in Table 22.1. NADPH oxidases, mitochondria, and various other cellular enzymes are examples of endogenous sources. Environmental elements such as pollution, radiation, and some medications are examples of exogenous sources. The human body's antioxidant defenses may be overpowered by excessive ROS production, which can result in oxidative stress and inflammatory disease pathology (Table 22.1).

Table 22.1 Endogenous and exogenous sources in inflammatory diseases

Source type	Source	ROS produced	Mechanism	Role in inflammation	References
Endogenous	Mitochondrial dysfunction	$O_2^{\cdot-}$, H_2O_2	Electron leakage from ETC complexes I/III during OXPHOS	Generates mtROS that activate NF-κB, MAPK, NLRP3 inflammasome; primes IL-1β, TNF; promotes chronic inflammation	[21]
	NADPH oxidases	$O_2^{\cdot-}$, H_2O_2	Transfer electrons from NADPH to O_2 via membrane-bound NOX enzymes	Superoxide/H_2O_2 drive vascular inflammation, endothelial dysfunction, atherogenesis, and pathogen defense	[22]
	Xanthine oxidase	$O_2^{\cdot-}$, H_2O_2	Oxidation of hypoxanthine/xanthine during purine catabolism	Supports reperfusion injury, triggers endothelial inflammation and oxidative stress	[23]
	Uncoupled NO synthase	$O_2^{\cdot-}$, $ONOO^-$	eNOS produces superoxide instead of NO when cofactors (BH_4) are deficient	Superoxide/peroxynitrite cause nitrosative stress, endothelial damage, amplify inflammatory signaling	[24]
	Endoplasmic reticulum	H_2O_2, $O_2^{\cdot-}$	ROS forming during protein folding and ER stress	ER stress-associated ROS induce UPR, boost NF-κB, cytokine release, and inflammation	[25]
	Cytochrome p450	$O_2^{\cdot-}$, H_2O_2	Electron leakage during xenobiotic metabolism	ROS contribute to oxidative stress in liver and vasculature; P450 in steroidogenic mitochondria also leak ROS	[26]
	Peroxidases	H_2O_2, hypochlorous acid (HOCl)	Produce ROS during immune response; DUOX generates H_2O_2	Pathogen killing but excessive activity causes tissue damage, lipid oxidation, and inflammation	[27]
	Cyclooxygenases	H_2O_2, lipid peroxides	Arachidonic acid oxidation yields ROS as by-products	ROS induce lipid peroxidation, enhance prostaglandin-mediated inflammatory signaling	[28]
	Lipid oxidases	Lipid hydroperoxides, $O_2^{\cdot-}$	Oxidation of polyunsaturated fatty acids	Drives lipid peroxidation, eicosanoid formation, and propagates oxidative inflammation	[29]

(continued)

Table 22.1 (continued)

Source type	Source	ROS produced	Mechanism	Role in inflammation	References
Exogenous	Ionizing and UV radiation	$^{\cdot}OH$, $O_2^{\cdot-}$, H_2O_2, singlet O_2	Water radiolysis by X-rays, γ-rays → hydroxyl radicals; UVA/UVB excite endogenous photosensitizers	DNA damage, lipid peroxidation → skin inflammation, premature aging, carcinogenesis	[30]
	Pollutants (PM, ozone), cigarette smoke	$O_2^{\cdot-}$, H_2O_2, $^{\cdot}OH$, $NO^{\cdot}$, HOCl	Direct ROS in smoke/pollution or via NADPH oxidase activation in lung cells	Respiratory inflammation, systemic oxidative stress, endothelial dysfunction	[31]
	Heavy and transition metals (Fe, Cu, Cd, Pb, As)	$^{\cdot}OH$ via	Metal-catalyzed H_2O_2 conversion to hydroxyl radicals	Causes DNA/lipid/protein damage, neuroinflammation, cardiovascular and renal inflammatory diseases	[32]
	Chemotherapeutics and X-ray therapy	$O_2^{\cdot-}$, H_2O_2, $^{\cdot}OH$	Redox cycling of chemo agents; radiation-induced water radiolysis	Tissue inflammation (e.g., cardiotoxicity, mucositis) as collateral damage in cancer therapy	[33, 34]
	Drugs, alcohol, smoked or processed foods	$O_2^{\cdot-}$, H_2O_2, lipid peroxides	Metabolic activation (ethanol via CYP2E1); food carcinogens (acrylamide, smoked meat)	GI and liver inflammation, increased colon cancer risk via lipid peroxidation and mucosal damage	[35, 36]
	Nanoparticles, industrial chemicals, solvents	$O_2^{\cdot-}$, H_2O_2, $^{\cdot}OH$	Particle uptake damages mitochondria, activates NOX, depletes GSH	Pulmonary and neuroinflammation, disruption of barrier function, chronic inflammatory conditions	[37, 38]

22.2.2 Cellular Mechanisms of ROS-Induced Tissue Damage

ROS are traditionally characterized as partly reduced oxygen metabolites with potent oxidizing properties. At elevated concentrations, they are harmful to cells; nevertheless, at lower concentrations (specific levels yet to be determined), they perform intricate signaling roles. They are detrimental since they oxidize proteins, lipids, cellular components, and cause DNA damage. At "physiological concentrations," ROS serve as signaling molecules that modulate cell proliferation, cell adhesion, differentiation, senescence, and cell death [39, 40]. The idea of sustained or chronic ROS generation is seen as important to the advancement of inflammatory diseases [41]. The superoxide anion (O_2), hydroxyl radical (OH), hydrogen peroxide (H_2O_2), and hypochlorous acid (HOCl) are the most studied and understood family members of physiologically relevant ROS [40]. Others may play key roles in signaling and illness, but their functions are unknown. The cytochrome P450 and the electron transport chain (ETC) in mitochondria generate ROS as by-products of cellular metabolism [41]. The NADPH oxidases found in a range of cells, particularly those in professional cells called phagocytes and the endothelial cells, are the other main source where ROS are not generated as by-products. These cells are essential to the development of the inflammatory response [41, 42].

O_2 is produced whenever O_2 is reduced by one electron through enzymatic catalysis by NADPH oxidase or xanthine oxidase (XO) or by the electron transfer processes in the mitochondrial ETC [40, 43, 44]. O_2^{-} has a half-life of 10^6 ns because it spontaneously dismutates to H_2O_2 ($k = 2 \times 10^5$ $M^{-1}s^{-1}$ at physiological circumstances) [45]. This process can be accelerated by a factor of 104 using enzyme SOD ($k = 1.6 \times 10^9$ $M^{-1}s^{-1}$) [40, 44]. When the transition metal iron is present, O2and H_2O_2 generate the very reactive OH and OH (Fig. 22.1) (Haber–Weiss reaction). O2 reacts with Fe^{3+} to form Fe^{2+} and O_2 at the start of this reaction. Nevertheless, this process is thermodynamically unfavorable under physiological circumstances [46].

This reaction's second step, also referred to as Fenton's reaction, takes place in biological conditions when Fe^{2+} combines with H_2O_2 to produce both OH and OH. OH has a very short half-life and is the most powerful oxidizing species of lipids and proteins found in biological membranes [40, 45, 47]. The myeloperoxidase enzyme, which is generally thought to be a PMN-specific enzyme, produces HOCl from H_2O_2 and chloride at inflammatory sites where PMNs are prevalent [48]. Negative charge of O_2 severely limits its ability to pass through biological membranes. Nonetheless, some transmembrane proteins, like the mitochondria's voltage-dependent anion channels (VDAC), permit the transmembrane passage of O_2 generated in ETC [48]. However, aquaporin channels like as AQP3 and AQP8 regulate cell H_2O_2 uptake, allowing H_2O_2 to pass across biological membranes and increase the probability that it will get into cells that are in touch with one another [49–51].

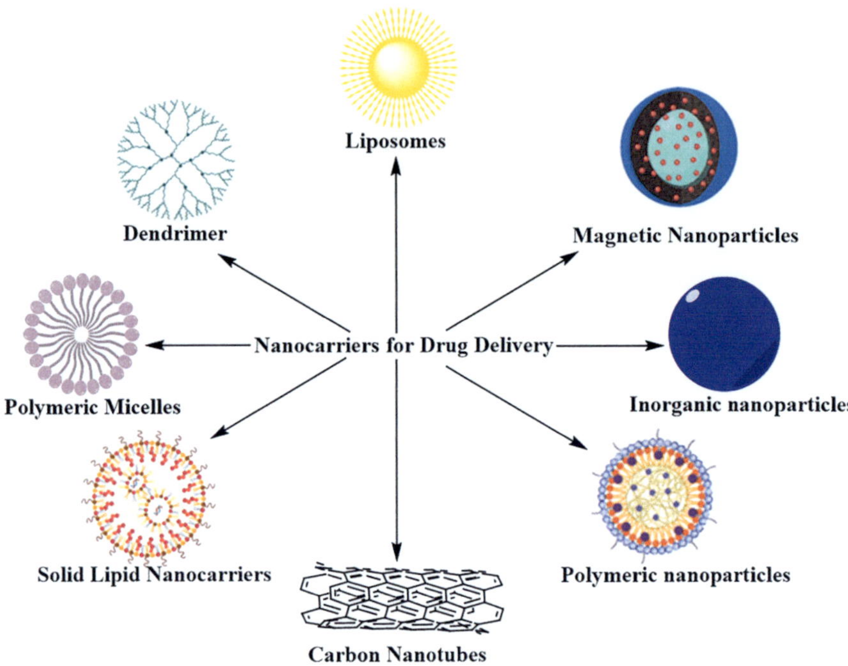

Fig. 22.1 Nanocarriers for drug delivery against inflammation

22.2.3 ROS as Therapeutic Targets in Inflammation

ROS play a crucial role in the etiology and development of inflammation. The generation of reactive oxygen species is essential for redox-signaling pathways and the immune system [39]. Excessive amounts lead to oxidative damage. Because of this, cellular damage, the triggering of transcription factors AP-1 and NF-κB, and an increase in inflammatory cytokines will contribute to the formation of more of such inflammatory cascades [52]. ROS should be targeted to combat inflammatory illnesses because of their significance. Scavenging ROS and tipping the redox balance in an inflammatory tissue can be done in a variety of ways.

22.2.3.1 Antioxidants

Synthetic antioxidants like N-acetylcysteine endeavor to neutralize ROS and reduce oxidative damage, while natural antioxidants like vitamin C, vitamin E, and polyphenols also do this. However, they never achieved clinical success due to their lack of selectivity and bioavailability [53].

22.2.3.2 Enzyme Therapy

It has been demonstrated that oxidative stress can be decreased by therapeutically modifying antioxidant enzymes. Inflammatory illnesses and disorders are also being studied in mimetics for these enzymes [54].

22.2.3.3 Enzyme Inhibitors

Inhibiting enzymes such as NADPH oxidases (like NOX1/4 inhibitors like GKT137831), xanthine oxidases (like allopurinol), and uncoupled nitric oxide synthase can be used to stop the production of ROS at the very beginning [55].

22.2.3.4 Mitochondrial Antioxidants

MitoQ and SS-31 are examples of mitochondrial antioxidants that preferentially accumulate in mitochondria to reduce the generation of mitochondrial ROS and maintain cellular bioenergetics. They also show promise in cases when inflammation is caused by mitochondrial malfunction [56].

22.2.3.5 Nanoparticle-Based Therapeutics

Novel nanotechnologies are improving therapeutic specificity and decreasing systemic toxicity by directly targeting inflamed tissue that can benefit from the process of antioxidants and ROS scavengers [57].

Despite these advancements, specificity to pathogen-like ROS remains a challenge to prevent disrupting healthy processes' redox signaling. Future technologies must focus on improving specificity, simplifying administration, and incorporating redox-based therapy to personalize therapy for inflammatory diseases with molecular diagnosis.

22.3 Nanocarriers in Modern Therapeutics

Nanocarriers have revolutionized contemporary drug delivery methods because they can enhance therapeutic drugs' bioavailability, targeted distribution, and controlled release [58]. Due to the low solubility, systemic toxicity, and off-target effects of traditional medications, nanocarriers can help the constraints of established therapeutic techniques in the treatment of inflammatory illnesses [59]. The utilization of nanotechnology in the production of these advanced therapeutic systems, which have a diameter of 1–1000 nm, enables for their exact transport and release in response to stimuli; this is especially useful in oxidative and inflammatory microenvironments (Fig. 22.1) [60, 61].

22.3.1 *Classification of Nanocarriers for Drug Delivery*

22.3.1.1 Solid Lipid Nanocarriers (SLNs)

Solid lipid nanoparticles (SLNs) can be made from solid lipids and enhanced with surfactants, purified triglycerides, complex glyceride mixtures, or waxes. These provide benefits over liposomes, such as improved tolerability and biocompatibility [62]. SLNs are employed in various therapeutic applications, including enhancing ocular bioavailability, targeting medications to the brain, and administering them via IV, pulmonary, and cutaneous routes. Stealth SLN enhances the accumulation of antineoplastic agents in tumors, while also allowing for the administration of tailored medicine [63]. Its diminutive dimensions, enhanced stability, and capacity to amalgamate hydrophilic and hydrophobic pharmaceuticals with a diverse array of ligands render it optimal for targeted administration. SLN has been utilized to generate targeted brain drug delivery for drugs such as cotrimoxazole, camptothecin, and vinpocetine [64]. SLN can be altered in size and charge on the surface to provide site-specific medication delivery. Good physical stability, enhanced drug bioavailability, higher drug loading capacity, and lower cytotoxicity are among the benefits [65].

22.3.1.2 Liposomes

Liposomes are nanocarriers created when lipids are dispersed in an aqueous environment. The structure comprises a core inside and a bilayer membrane, having hydrophilic medications encapsulated within the core and hydrophobic pharmaceuticals preferentially located in the surrounding bilayer [66]. Traditional liposomes face issues regarding stability, drug release, insufficient drug loading, and shorter blood circulation times. PEGylation helps liposomes escape the redox reaction, resulting in longer blood circulation time [67]. Liposomes can be grafted with monoclonal antibodies, proteins, peptides, carbohydrates, vitamins, antigen fragments, and glycoproteins to target cancer sites [68]. They are responsive to various stimuli and are used as model cells for the development of medicines, vaccines, cosmetics, and nutraceuticals. Advantages include good biocompatibility, nontoxicity, increased solubility, and improved pharmacokinetic properties. Disadvantages include low drug transport rate and poor stability [69].

22.3.1.3 Dendrimers

Dendrimers are branched macromolecules with various arms originating from the central core, produced by natural or synthetic components like nucleotides, amino acids, and sugars [70]. They have been chosen for drug delivery techniques due to their spherical morphologies, high branching, diverse molecular weights, and monodispersed polymers with a mean diameter of 1.4–1.45 nm [71]. Gene delivery, immunology, imaging with magnetic resonance, vaccinations, and the administration of antiviral, antimicrobial, and anticancer drugs are only a few of the biomedical applications for dendrimers [72]. By trapping hydrophobic medications in their intramolecular cleft and transferring it to their functional groups on the surface, they can improve their solubility and bioavailability. Benefits of dendrimers include their capacity to encapsulate and transport a variety of bioactive chemicals, their homogeneous particle size, and their favorable pharmacokinetic behavior [73].

22.3.1.4 Polymeric Micelles

Because of their inner core structure, which facilitates drug absorption and enhances stability and bioavailability, polymeric micelles, which are smaller than 100 nm, are a potential nanostructure for hydrophobic delivery of drugs [74, 75]. Mechanisms include increased permeability, holding effect stimuli, and complexing of certain targeting ligand molecules that can be used to target them. Advantages include biocompatibility, solubilization of lipophilic drugs, and conjugation of surface groups with targeting ligands. However, they have low drug loading capacity and are only used for lipophilic drugs [76].

22.3.1.5 Carbon Nanotubes

Carbon nanotubes, developed by Iijima in 1991, are hollow, tube-like carbon atom assemblies at the nanoscale [77]. They belong to the fullerene class and are constructed by folding sheets of graphene into a hollow tube [78]. Carbon nanotubes possess distinctive physicochemical and biological characteristics that render them effective drug delivery vehicles [79]. They can be altered to enhance circulation inside the body. When modified for solubility in aqueous human fluids, they exhibit little toxicity and are readily assimilated by cells [80]. The carbon nanotubes are among the most effective nanocarriers in cancer therapy, transporting plasmid DNA, as well as small interfering RNA (siRNA), complementary oligonucleotides, and aptamers [81]. They may be used as personalized instruments for preliminary cancer diagnoses and gene transfer. Their high absorption of light in the near-infrared range makes them a potential choice for photothermal ablation of malignant areas [82].

22.3.1.6 Polymeric Nanoparticles

Polymeric nanoparticles are solid, colloidal particles made of biodegradable polymers [83]. They can be dissolved, entrapped, adsorbed, encapsulated, or conjugated to the matrix in nanospheres. In nanocapsules, a polymeric shell surrounds an inner core, and the medication is typically dissolved in the cell center or adsorbed at the surface. These nanoparticles are suitable for brain-targeted drug delivery due to their low toxicity, biodegradability, better stability, and surface functional groups that modulate drug release rate [84, 85]. They are made from a range of natural and synthetic polymers, including albumin, collagen, chitosan, heparin, alginate, and dextran, which are both biocompatible and biodegradable. They provide improved physicochemical properties, longer drug circulation durations, more regulated drug delivery, homogeneous particle size distribution, enhanced drug payload, and improved storage stability. Surface modification using copolymers containing PEG or poly (ethylene oxide) can speed up plasma circulation and promote passive transport [86].

22.3.1.7 Magnetic Nanoparticles

Because magnetic nanoparticles can carry therapeutic chemicals on their surface or in the bulk, they can be utilized for the diagnosis and treatment of a variety of cancers [87]. Comprising three primary ferromagnetic elements (Fe, Ni, and Co), these nanoparticles can be employed as targeted carriers to transport medications to intratumor levels or to cause tumor necrosis [88]. The length of time that magnetic particles circulate through the bloodstream and are absorbed in the body is greatly influenced by their dimensions, charge, and surface composition [87]. The optimal size range is quite small to successfully distribute in certain tissues while avoiding the kidneys and penetrating small capillaries within bodily tissues [88]. In prostate and breast cancer cell lines, this formulation demonstrates dose-dependent antiproliferative efficacy and long-lasting intracellular drug retention. Its drawbacks include low dispersion capabilities, expensive synthesis material costs, and mobility that depends on environmental compatibility [89].

22.3.1.8 Inorganic Nanoparticles

Inorganic nanoparticles, consisting of a core and a shell, have been discovered for specialized medication administration [90]. These nanocarriers include inorganic components like gold, silica, or iron oxide, and are surrounded by an organic polymer shell. Inorganic nanoparticles having surface Plasmon resonance (SPR) include silica, gold, silver, and iron oxide being a property not found in liposomes, dendrimers, and micelles. These nanoparticles offer several advantages, including their high biocompatibility and versatility in surface functionalization [90, 91]. In contrast to organic materials, they are stable, hydrophilic, biocompatible, and nontoxic.

Because of their low toxicity, nonimmunogenic reactivity, and high absorption capacity, inorganic nanoparticles have drawn interest as drug delivery carriers [92]. Other inorganic nanoparticles like iron oxide and fullerenes have been considered as vehicles for drug delivery due to their nanoscale size in recent nanotechnology breakthroughs. The utility of these nanoparticles is increased as they could be absorbed by cells and transmit nucleic acids into living cells [93].

22.3.2 *Advantages of Nanocarriers in Inflammatory Disorders*

Chemotherapeutically active medications are transported using nanocarrier-based platforms (such as nanoparticles, nanocrystals, lipid complexes, nanocapsules, polymeric micelles, and dendrimers) using colloidal nanoparticles that are submicron in size (often less than 500 nm). Submicron-sized particles known as nanocarriers have a high surface-to-volume ratio, which speeds up the rate of dissolution. They are frequently used to provide medications that are poorly soluble to certain tissues, organs, or cells [94].

A recent development in nanotechnology called nanocrystals intends to increase the gastrointestinal absorption of BCS class II drugs and could assist BCS class IV drugs that have poor solubility and penetration [95]. The primary benefits of nanocrystals are higher dissolution rate, mucoadhesiveness, greater solubility, and improved fed/fasted state variation. These characteristics enhance the effectiveness and tolerability of anti-inflammatory medicines, resulting in a dramatic shift in the development of innovative products for diverse administration methods. Nanocrystals have been created for use in different administration routes, including oral, topical, ocular, and parenteral [96].

Proteins in biological fluids can be adsorbed by protein nanoparticles, creating a protein coating called a protein corona [97]. As a drug delivery technique, protein nanoparticles provide a number of benefits, such as stability, biodegradability, easy control over particle size, surface modification, and reduced toxicity issues, including immunogenicity [98]. The protein corona, often referred to as the bimolecular corona, is a coating of blood proteins that immediately coats nanomaterials when they are injected [99]. Several kinds of nanoparticles are being developed or are now in preclinical trials for the management of inflammation. The FDA has approved a few nanomedicines and numerous ongoing clinical studies (Table 22.2) [100].

22.4 Smart Nanocarrier Systems for ROS Scavenging

Smart nanocarrier systems are now being developed to deliver drugs to inflamed regions while also scavenging ROS. Nanocarrier systems can be categorized into four types: polymeric, lipid-based, inorganic, and hybrid/multifunctional

Table 22.2 List of clinical trials of several nanocarriers used against inflammation

Nanocarrier	Drug	Indication	Clinical phase	References
PEGylated liposomes	Prednisolone	Ilio-femoral atherosclerosis	Phase I	[161]
PEGylated liposomes	Prednisolone	Moderate to severe ulcerative colitis	Phase II	[162]
Liposome (lipogel)	Cyclosporine	Chronic plaque psoriasis	Phase II	[162]
Glutathione-PEGylated liposomes	Methyl prednisolone	Autoimmune/ Inflammatory diseases	Phase I	[163]
PLGA	Pitavastatin	Cardiovascular disease	Phase I	[164]
Nanoemulsion (oil-in-water)	Tretinoin	Acne vulgaris	Phase I	[165]
PEGylated silica-gold nanoshells	Photothermolysis (laser ultrasound)	Acne vulgaris	Phase I/II	[166]

nanocarriers. Each of these forms of nanocarriers has distinct therapeutic benefits for inflammatory disorders.

22.4.1 *Polymer-Based Nanocarriers*

Polymeric nanocarriers, such as polymeric nanogels, polymeric nanomicelles, polymeric nanoparticles, polymeric nanowires, and lipid-polymer hybrid nanoparticles (LPNs), are made from cationic polymers [101]. These nanocarriers have varying nucleic acid loading capacities and delivery efficiencies due to their unique three-dimensional network structure. The two main preparation methods for polymeric nanogels are cross-linking techniques and incorporating bifunctional monomers into a heterogeneous colloidal system [102].

In order to ensure steady and effective loading, nucleic acids are frequently added when the carrier preparation is finished [103]. In order to load nucleic acids onto nanogels, cationic polymers are often used, and different nucleic acids are added at the proper N/P ratio [104]. A number of factors affect the choice of the ideal N/P ratio, including the kinds of cationic polymers, concentration, length, and charge density of drugs [105].

Zhou et al. (2019) created a target gene vector, DMAPA-Amyp modified with RGD, that penetrates vascular endothelial cells in infarcted regions by binding to overexpressed integrins. In vivo investigations show that RGD-DMAPA-Amyp, coupled with a mutant hypoxia-inducible factor-1α (HIF-1α), efficiently treats cerebral infarction by improving neurological function recovery [106].

Yang et al. (2021) also developed an agonist (SAG) @ Pro-His-Ser-Arg-Asn (PHSRN)-HES with dual targeting function, leveraging amylopectin hydroxyethyl starch's capacity to delay breakdown and excretion in blood [107].

22.4.2 *Lipid-Based Nanoparticles*

In the last 10 years, nanomaterials' therapeutic potential as new drug delivery vehicles, in addition to classical pharmacology, has received widespread recognition. Lipid-based nanoparticles (LNPs) are one kind that has shown strong pharmacological activity and potential therapeutic advantages, prompting growing interest in both preclinical and clinical trials [108].

Phospholipids and cholesterol, cationic or ionizable lipids, and PEGylated lipids make up the four primary lipid components that make up LNPs. By interacting with negatively charged nucleic acids, these components promote particle stability and circulation time within the biological system and facilitate particle generation and stability [109]. LNPs facilitate endosomal escape and release therapeutic payload into the cytoplasm by readily traversing the cell membrane to distribute the contents of the nanoparticles to specific intracellular locations [110].

LNP properties are highly influenced by nanoparticle features such as size, surface charge, shape, stability, loading efficacy, and entrapment efficiency [111]. Smaller LNPs had greater cellular absorption and longer circulation duration, whereas bigger ones had more drug loading capacity but lower uptake [112]. Accurate assessments of these variables are critical for improving LNP manufacturing and drug delivery systems [113, 114].

Cui et al. (2017) studied the effects of thermally sensitive Danggui Buxue Extract-Loaded Liposomal Gel (DBLTG) on a dense dorsal excisional lesion. The DBLTG therapy greatly accelerated wound healing by lowering inflammation and increasing hydroxyproline levels in the granulation tissue. This resulted in the formation of newly formed granulation tissue, increased blood vessel density, and increased expression. The study assessed the effect of DBLTG on the VEGF/PI3K/Akt and TGF-/Smads Signaling Pathways [115].

Pires et al. (2020) created and described the Caryocar rasiliense Cambess (CBC) polymeric lipid nanocapsule (CBC-PLN) coated with chitosan (PLNCBC+) and without chitosan (PLNCBC) to aid in the healing of cutaneous wounds in male Wistar rats. The study found that by generating type 1 collagen, PLNCBC and PLNCBC+ both reduced cutaneous problems and aided in wound healing; however, PLNCCBC+ produced a higher output than PLNCBC [116].

22.4.3 Inorganic Nanocarriers

Recently, there has been a lot of interest in using inorganic materials as nanocarriers in systems for delivering drugs. Black phosphorus (BP), gold nanoparticles (GNPs), graphene oxide (GO), and mesoporous silica nanoparticles (MSNs) are the most often studied inorganic nanocarriers [117].

Inorganic nanoparticles used in tissue engineering (TME) feature chemical groups that may be modified and are coupled to medicinal drugs via stimuli-sensitive linkages [118]. When exposed to stimuli, these bonds release agents, such as pH-responsive bonds, GSH-responsive bonds, and ROS-responsive bonds [119, 120]. Tumor cells have a lower pH and greater ROS levels due to their quicker metabolism and biosynthetic processes [121].

Inorganic nanoparticles include carbon nanotubes and mesoporous silica nanoparticles [122].

Kunisaki et al. (2021) established that CNT fibers (cYarn) facilitate peripheral nerve regeneration. The researchers repaired a 15 mm lesion in the rat sciatic nerve using cYarns, resulting in the regeneration of axons that traversed the CNTs. Results indicated restoration of nerve function at 8 weeks of treatment. Histological assessment verified that axons traversed the CNT fibers, with a considerable increase in the proportion of axons spreading beyond the site of implantation compared to the silicone tube group [123].

In order to promote natural neutrophil apoptosis, control inflammation, and improve myocardial infarction recovery, Bao et al. (2021) created engineered neutrophil apoptotic bodies (eNABs). The neutrophil apoptotic body membrane (NABM) and eNABs may increase the release of TGF-β and IL-10 while decreasing the pro-inflammatory cytokines TNF-α and IL-6. MSNs loaded with hexyl 5-aminolevulinate hydrochloride (MSNHAL), which aids in reprogramming macrophages, promoting anti-inflammatory effects, and bolstering cardiac function, were fused with natural neutrophil apoptotic body membranes to create eNABs [124].

Dou et al. (2022) investigated the anti-inflammatory properties of Fe/Ce-MSN and polyethylene glycosylated Fe/Ce-MSN nanoparticles in RAW 264.7 macrophage cells. The nanoparticles enhanced superoxide dismutase activities and showed anti-inflammatory properties. They inhibited LPS-induced TNF-α and IL-1β levels and markedly reduced oxidative stress and inflammatory disorders. It was discovered that Fe/Ce-MSN was more efficient than Fe/Ce-MSNPEG. According to the findings, pH-responsive Fe/Ce-MSN and Fe/Ce-MSN-PEG prevent cell death and promote the function of anti-inflammatory nanotherapies [125].

22.5 Natural Antioxidants and Phytochemicals in Nanocarrier Systems

Skin conditions can result from oxidative stress, which is brought on by an imbalance between the body's natural antioxidant capacity and the creation of reactive oxygen species [126]. Nanocarriers, which improve drug solubility, skin penetration, and permeation, have revolutionized antioxidant delivery. Five types of nanocarrier systems include nanoemulsions, liposomes, SLNs, NLCs, and polymeric nanoparticles, which enhance antioxidant protection, improve skin permeation, and optimize the effectiveness of topical antioxidant therapies [127].

22.5.1 Curcumin

Curcumin[1,7-bis(4-hydroxy3-methoxyphenyl)-1,6-heptadiene-3,5-dione] is a powerful polyphenol antioxidant produced from *Curcuma longa*, which has a high lipophilicity and Light and oxygen sensitivity.

Suwannateep et al. (2012) used the self-assembly of polymer technology to construct nanospheres containing curcumin using ethyl cellulose or a combination of ethyl cellulose and methylcellulose to make it simpler to apply to the skin. Furthermore, curcumin administered via polymeric nanoparticles exhibited a greater antioxidant activity than free curcumin due to the enhanced skin penetration [128].

Studies have demonstrated curcumin's potential to promote wound healing by lowering ROS levels and promoting nerve regeneration and repair following tissue implantation [129]. Other research has lately looked at curcumin-loaded liposomes and curcumin-loaded PLGA nanoparticles (Nps-Cur) as ways to get around curcumin's poor oral bioavailability and low aqueous solubility. In AD, both of these methods showed promise in fostering neuroprotection against oxidative damage [130, 131].

Hydrogel-embedded curcumin (TM/PC), created by Qian F. et al. (2021), lowers ROS levels and enhances regeneration of nerves and recovery following tissue implantation [132]. Curcumin has a lot of potential for wound healing. Chen et al. (2021) developed a sandwich-like nanofibrous membrane (CSNM) covered with curcumin that exhibits strong antioxidant efficacy in DPPH radical scavenging investigations [133]. In diabetic full-thickness skin wounds, Hu et al. (2021) developed a hydrogel based on curcumin that can reduce inflammation during wound healing [134]. Zhang et al. (2022) created a hydrogel delivery method based on glycosaminoglycan that contains curcumin and has potent anti-inflammatory and ROS scavenging capabilities. By controlling the wound microenvironment, this technology speeds up the healing of chronic wounds. Curcumin can cure additional inflammatory disorders brought on by ROS and has been demonstrated to protect from ischemia-reperfusion damage [135].

Wu et al. (2008) found that encapsulating quercetin in a mixture of polyvinyl alcohol (PVA) and Eudragit E improved its drug release profile. The encapsulation effectiveness of the nanoparticles, which were made by the nanoprecipitation process, was discovered to be 99%. When compared to quercetin dissolved in water, the antioxidant activity was greatly increased by 883 times by using the combination of particle size reduction, crystalline transformation, and intermolecular hydrogen bonding. This implies that water-insoluble antioxidants can have their activity increased by polymeric nanoparticles [136].

22.5.2 *Resveratrol*

The polyphenol resveratrol is well-known for its antioxidant qualities and advantages for skin health [137]. When applied topically, resveratrol builds up in the epidermis' outermost layer, where it continues to scavenge free radicals. Resveratrol has been encapsulated utilizing a high-pressure homogenization technique into a solid lipid nanoparticle (SLN) by Gokce et al. (2012). With preferential deposition in the epidermis, the SLN enhances resveratrol transport into the skin. For depot systems, the slower absorption of alpha-lipoic acid into SLN systems with 1% Apifil could be beneficial since it ensures less variations in drug concentrations [138].

22.5.3 *Genistein*

Soy phytoestrogenic isoflavones include genistein, a strong antioxidant that has been demonstrated to fight against several age-related conditions, such as cancer, osteoporosis, obesity, type 2 diabetes, and neurological illnesses [139]. It has been demonstrated that genistein-loaded polymeric nanomicelles improve oral bioavailability, presumably as a result of increased dissolution and gastrointestinal release [140]. Compared to solutions or bulk powders, genistein loaded in solid lipid nanoparticles has also shown enhanced oral bioavailability. Incorporating genistein into tiny PEGylated silica nanoparticles could significantly improve its antioxidant properties, modifying endogenous CAT and SOD activity and H_2O_2 generation, according to studies conducted using HT29 human colon cancer cells [141]. Autophagy and apoptosis, two processes linked to cell death, are induced as a result, whereas free genistein only stimulates apoptosis.

22.5.4 *Epigallocatechin-3-Gallate*

Green tea contains a significant polyphenol called epigallocatechin-3-gallate (EGCG), which has potent antioxidant qualities as well as anti-inflammatory, anti-atherogenic, and anticancer effects. However, epidemiological studies frequently contradict the stability and bioavailability [142]. The bioavailability of EGCG has been increased by the use of novel nanodelivery methods. Solid lipid nanoparticles loaded with EGCG have demonstrated improved oral transport capability and greater stability [143]. In in vitro settings, these nanoparticles also exhibit antioxidant qualities by lowering intracellular ROS levels and lipid peroxidation, boosting keratinocyte survival, and exhibiting anti-inflammatory characteristics. Its low stability and bioavailability may be the cause of the uncertainty in epidemiological research and in vitro results [144].

22.6 Application in Disease Models

Immunosuppressive or anti-inflammatory medications are frequently used to treat chronic inflammatory illnesses; however, long-term use of these medications can have serious side effects and yield few therapeutic benefits. Immunogenicity, TB reactivation, liver and bone marrow toxicity, and vulnerability to infectious illnesses are some of the drawbacks of monoclonal antibodies. Immunosuppression, cardiovascular disorders, hypertension, and altered glucose metabolism can all result from long-term corticosteroid usage. There is a pressing need to create effective, selective, targeted strategies for chronic inflammatory disorders.

22.6.1 *Inflammatory Bowel Diseases (IBDs)*

IBDs are inflammatory conditions affecting the colon and small bowel, with Crohn's disease and Ulcerative Colitis being major types [145]. These diseases can cause significant damage to the bowel, leading to complications like malabsorption, malnutrition, and short bowel syndrome [146]. Pharmaceutical treatments focus on controlling chronic inflammation through immunosuppression, but have safety and long-term efficacy issues. Oral administration of nanotherapeutics, including small interfering RNAs (siRNAs) and nanoformulations of existing medications, has demonstrated promise [147]. In rat models, mesalazine (5-ASA) has been administered to the colon using PLGA nanoformulations following the production of colitis. Analysis of nonfunctionalized PLGA nanoparticle dispersion in people has revealed preferential accumulation at lesion locations [148].

22.6.2 *Plaque Psoriasis*

It is the most prevalent kind of psoriasis, a severe skin inflammatory condition that affects 2% of people worldwide [149]. It increases the risk of diabetes, arthritis, and cardiovascular illnesses. Inflammation in sick skin is maintained in large part by immunological variables such as Th17, Tc17, and γδ T cells [150, 151]. Animal models have demonstrated the efficacy of corticosteroid nanoformulations, while methotrexate, calcipotriol, and vitamin D3 loaded into nanostructured lipid carriers (NLPs) have demonstrated excellent skin infiltration and no skin irritation [152]. In animal models, NPs containing Tacrolimus or Cyclosporine A have demonstrated more effective anti-psoriatic actions [153]. An acid folate inhibitor called methotrexate (MTX) is used to treat severe psoriasis by inhibiting DNA synthesis and decreasing keratinocyte proliferation [154].

In Western nations, 0.1–3% of adults suffer with rheumatoid arthritis (RA), a chronic, systemic inflammatory disease. Nowadays, treatments use immunosuppressants or steroids to try to lessen synovial inflammation [155]. Finding novel, less harmful treatments is necessary, nonetheless, because of the serious side effects of existing ones. Since macrophages are passive targets of nanoparticles (NPs) because of their phagocytic activity, using NPs as medication carriers to target them at the diseased joints is one potential strategy [156].

NSAID-encapsulated liposomes have demonstrated strong anti-inflammatory properties, lowering inflammatory cytokines and accelerating RA recovery [157]. Nanoparticles (NPs) have been utilized to target active macrophages, transport siRNAs to inflammatory joints, postpone joint degeneration, and lower clinical scores for arthritis [158]. Compared to previous tacrolimus formulations, tacrolimus with albumin nanoformulations has demonstrated greater efficacy in lowering arthritis clinical scores [159].

22.7 Clinical Trials and Commercial Markets

A large number of nanomedicines—mostly liposomes—have entered clinical usage, and several more are undergoing clinical trials for a variety of imaging and medical uses. The FDA has only issued licenses for a tiny portion of nanomedicines, and hundreds of trials are now being conducted [160]. Only a small number of anti-inflammatory nanoparticles, meanwhile, have advanced to clinical trials (Table 22.2).

22.8 Challenges and Future Perspectives

Numerous targets for biological therapy against inflammatory illnesses have been identified as a result of recent developments in our understanding of the cellular and molecular aspects of the inflammatory response [167]. However, systemic toxicities and off-target side effects result from the extensive dispersion of inflammatory signaling pathways in various cell types and organs [168]. The development of nanocarriers to enhance the delivery of anti-inflammatory drugs has been spurred by this. The distribution of nanomedicines is dependent on their surface, whether they are coated with PEG or loaded with a ligand for active targeting [169]. Nanomedicines undergo a kinetic process of absorption, distribution, metabolization, and elimination.

Immune modulation may be influenced by the chemical makeup or physical characteristics of nanoparticles. New in vitro and in vivo preclinical models that more closely resemble the disease's pathogenesis in humans are desperately needed [170]. Various methods of delivery, including intravenous, topical, oral, and local, have been investigated using anti-inflammatory nanomedicines. The activation of the immune response can be affected by the delivery method, such as intravenous or direct injection into the site of inflammation [171].

Bio-inspired nanodevices represent a new generation of innovative nanomedicines with the aim to mimic natural circulatory cells. They can increase blood circulation time and improve the distribution of the loaded drug toward inflammatory cells and tissues. However, their immunogenicity, scale-up, and characterization remain important hurdles before starting clinical trials [172].

Reliable and standardized methodologies to obtain reproducible nanoparticles are required. There is often a gap between the academic environment where innovative nanocarriers are designed and the industrial context, where reproducible preparation and manufacturing are essential [173].

22.9 Conclusions

In order to address the challenges associated with administering fragile biomolecules, this chapter emphasizes the growing interest in enhancing the delivery of drug molecules or biological therapies for the resolution of inflammatory diseases. This presents an exceptional chance for nanoscaled drug delivery systems that can maximize both safety and effectiveness by means of improved biodistribution and pharmacokinetics.

NPs are being explored as a potential solution to autoimmune and inflammatory diseases, offering innovative solutions to improve the efficacy of current immunosuppressive therapies and overcome side effects. NP-based drug delivery has shown increased efficacy compared to current therapies, allowing for reduced dose and frequency of drug administration, and selective targeting of diseased tissues.

Understanding the anatomical barriers and innate immune mechanisms underlying these diseases is crucial for developing efficient NP-based therapies. If confirmed by clinical trials, nanomedicine promises a significant advance to current therapies and a real perspective for long-lasting disease control. The growing interest in enhancing drug delivery for inflammatory diseases presents an exceptional chance for nanoscaled drug delivery systems that can maximize safety and effectiveness through improved biodistribution and pharmacokinetics.

Acknowledgment Not applicable.

Author Contributions All the authors have made equal contributions in carrying out the work.

Funding Not applicable. Declarations

Conflict of Interest The authors declare no conflicting interests.

References

1. Chen L, Deng H, Cui H, Fang J, Zuo Z, Deng J, Li Y, Wang X, Zhao L (2017) Inflammatory responses and inflammation-associated diseases in organs. Oncotarget 9(6):7204
2. Megha KB, Joseph X, Akhil V, Mohanan PV (2021) Cascade of immune mechanism and consequences of inflammatory disorders. Phytomedicine 91:153712
3. Chaudhary R, Meher A, Krishnamoorthy P, Kumar H (2023) Interplay of host and viral factors in inflammatory pathway mediated cytokine storm during RNA virus infection. Curr Res Immunol 4:100062
4. Cascella M, Rajnik M, Aleem A, Dulebohn SC, Di Napoli R (2022) Features, evaluation, and treatment of coronavirus (COVID-19). In: StatPearls. StatPearls Publishing, Treasure Island
5. Liu J, Han X, Zhang T, Tian K, Li Z, Luo F (2023) Reactive oxygen species (ROS) scavenging biomaterials for anti-inflammatory diseases: from mechanism to therapy. J Hematol Oncol 16(1):116
6. Juan CA, Pérez de la Lastra JM, Plou FJ, Pérez-Lebeña E (2021) The chemistry of reactive oxygen species (ROS) revisited: outlining their role in biological macromolecules (DNA, lipids and proteins) and induced pathologies. Int J Mol Sci 22(9):4642
7. Dash UC, Bhol NK, Swain SK, Samal RR, Nayak PK, Raina V, Panda SK, Kerry RG, Duttaroy AK, Jena AB (2025) Oxidative stress and inflammation in the pathogenesis of neurological disorders: mechanisms and implications. Acta Pharm Sin B 15:15–34
8. Chatterjee S (2016) Oxidative stress, inflammation, and disease. In: Oxidative stress and biomaterials. Academic Press, pp 35–58
9. Rendra E, Riabov V, Mossel DM, Sevastyanova T, Harmsen MC, Kzhyshkowska J (2019) Reactive oxygen species (ROS) in macrophage activation and function in diabetes. Immunobiology 224(2):242–253
10. Zhu H, Li YR (2012) Oxidative stress and redox signaling mechanisms of inflammatory bowel disease: updated experimental and clinical evidence. Exp Biol Med 237(5):474–480
11. Lee IT, Yang CM (2012) Role of NADPH oxidase/ROS in pro-inflammatory mediators-induced airway and pulmonary diseases. Biochem Pharmacol 84(5):581–590
12. Ansari MY, Ahmad N, Haqqi TM (2020) Oxidative stress and inflammation in osteoarthritis pathogenesis: role of polyphenols. Biomed Pharmacother 129:110452

13. Huang X, He D, Pan Z, Luo G, Deng J (2021) Reactive-oxygen-species-scavenging nanomaterials for resolving inflammation. Mater Today Bio 11:100124
14. Yu M, Wang S, Lin D (2024) Mechanism and application of biomaterials targeting reactive oxygen species and macrophages in inflammation. Int J Mol Sci 26(1):245
15. Harirforoosh S, Jamali F (2009) Renal adverse effects of nonsteroidal anti-inflammatory drugs. Expert Opin Drug Saf 8(6):669–681
16. Malik S, Muhammad K, Waheed Y (2023) Nanotechnology: a revolution in modern industry. Molecules 28(2):661
17. Dai Y, Guo Y, Tang W, Chen D, Xue L, Chen Y, Guo Y, Wei S, Wu M, Dai J, Wang S (2024) Reactive oxygen species-scavenging nanomaterials for the prevention and treatment of age-related diseases. J Nanobiotechnol 22(1):252
18. Rosenkrans ZT, Sun T, Jiang D, Chen W, Barnhart TE, Zhang Z, Ferreira CA, Wang X, Engle JW, Huang P, Cai W (2020) Selenium-doped carbon quantum dots act as broad-spectrum antioxidants for acute kidney injury management. Adv Sci 7(12):2000420
19. Fan K, Xi J, Fan L, Wang P, Zhu C, Tang Y, Xu X, Liang M, Jiang B, Yan X, Gao L (2018) In vivo guiding nitrogen-doped carbon nanozyme for tumor catalytic therapy. Nat Commun 9(1):1440
20. Padmanaban S, Pully D, Samrot AV, Gosu V, Sadasivam N, Park IK, Radhakrishnan K, Kim DK (2023) Rising influence of nanotechnology in addressing oxidative stress-related liver disorders. Antioxidants (Basel) 12(7):1405
21. Venkatesan KB, Srinivasan MK, Prasad M, Jayaseelan N, Alamelu S, Panneerselvam C, Al-Duais MA, Alasmari A, Ajmal MR, Al-Aoh HA (2025) Pyrogallol induces apoptosis, oxidative stress and cell cycle arrest in C6 glioma cells: a potential therapeutic agent. Med Oncol 42(8):290
22. Yang XL, Chen YY, Zhao H, Zhang BY, Zhang XF, Li XJ, Yang XH (2025) Protective effect of aliskiren on renal injury in AGT-REN double transgenic hypertensive mice. Sheng Li Xue Bao 77(3):408–418. Chinese
23. Butts B, Herring C, Madariaga A, Graven LJ, Higgins MK, Dunbar SB (2025) Markers of inflammation are associated with symptoms and neighborhood deprivation in black adults with heart failure. medRxiv [Preprint]:2025.05.10.25327361
24. Zhang Y, Youn JY, Huang K, Zhang Y, Cai H (2025) Alleviation of accelerated diabetic atherogenesis in STZ-treated apoE/NOX1 DKO mice, apoE$^{-/-}$/tg-EC-DHFR mice, and by folic acid. Redox Biol 82:103570
25. Tahir R, He K, Yan H, Shrestha A, Li XH, Leji Z, Liu Q, Song K, Zhao L, Yang S (2025) Chronic abamectin induces hepatotoxicity and endoplasmic reticulum stress in largemouth bass (Micropterus salmoides): potential mitigation by curcumin and human health risk assessment. J Hazard Mater 495:138876
26. Onyena AP, Manohar CS, Irudayarajan L, Nkwoji JA, Chukwu LO (2025) Baseline oxidative stress responses and cytochrome P450 gene expression in Tympanotonos fuscatus from PAH-contaminated ecosystem in the Niger Delta, Nigeria. Environ Monit Assess 197(7):717
27. Caverzan A, Bonifacio A, Carvalho FE, Andrade CM, Passaia G, Schünemann M, Maraschin Fdos S, Martins MO, Teixeira FK, Rauber R, Margis R, Silveira JA, Margis-Pinheiro M (2014) The knockdown of chloroplastic ascorbate peroxidases reveals its regulatory role in the photosynthesis and protection under photo-oxidative stress in rice. Plant Sci 214:74–87
28. Zhang J, Wang W, Liu Y, Wan X, Zhang L (2025) Linarin relieves apoptosis, inflammation and oxidative stress in LPS-induced acute kidney injury by modulating COX2. Clin Exp Pharmacol Physiol 52(7):e70048
29. Rădulescu AL, Popescu RG, Balas M, Marinescu GC, Dinischiotu A (2025) Modulation of the antioxidant system of Caco-2 cells in the presence of aflatoxin B1, ochratoxin A, and ferulic acid. Toxins (Basel) 17(6):274
30. Rollo F, Martins GD, Gouveia AG, Ithurbide S, Servant P, Romão CV, Moe E (2023) Insights into the role of three Endonuclease III enzymes for oxidative stress resistance in the extremely radiation resistant bacterium *Deinococcus radiodurans*. Front Microbiol 14:1266785

31. Nel AE, Diaz-Sanchez D, Li N (2001) The role of particulate pollutants in pulmonary inflammation and asthma: evidence for the involvement of organic chemicals and oxidative stress. Curr Opin Pulm Med 7(1):20–26
32. Wang J, Cai R, Hu Z, Cai L, Wu J (2025) Study on the interaction effect of heavy metal cadmium in soil-plant system controlled by biochar and nano-zero-valent iron. Int J Mol Sci 26(9):4373
33. Fakhri KU, Sharma D, Fatma H, Yasin D, Alam M, Sami N, Ahmad FJ, Shamsi A, Rizvi MA (2025) The dual role of dietary phytochemicals in oxidative stress: implications for oncogenesis, cancer chemoprevention, and ncRNA regulation. Antioxidants (Basel) 14(6):620
34. Wexler B, Norris Q, Hoogland AI, Jim H, Huang J (2025) Radiation-induced cognitive impairment and its implications for anesthesia selection. J Clin Anesth 106:111917
35. Mulè S, Galla R, Ferrari S, Invernizzi M, Uberti F (2025) Molecular insights into the nociceptive modulation by palmitoylethanolamide and *Equisetum arvense* extract: an in vitro study across the blood-brain barrier. Nutrients 17(12):1998
36. Koukach D, Aljumaily M, Al-Attiyah N, Al-Amer R, Attia Y, Tayyem R (2025) From prevention to management: exploring the impact of diet on multiple sclerosis. Transl Neurosci 16(1):20250371
37. He Y, Lin J, Huang C, Bao Q, Hao Z, Zhao S, Li W, Zhu H, Zhai X (2025) Virus-inspired nanoparticles for intensified cancer therapy via cascade reinforcement of "nontoxicity-to-toxicity" transition and mitochondrial dysfunction. Adv Healthc Mater:e2405274
38. Tu J, Peng J, Wen L, Li C, Xiao Z, Wu Y, Xu Z, Hu Y, Zhong Y, Miao Y, Xiao J, Liu S (2025) Antioxidant peptides derived from woody oil resources: mechanisms of redox protection and emerging therapeutic opportunities. Pharmaceuticals (Basel) 18(6):842
39. Mittal M, Siddiqui MR, Tran K, Reddy SP, Malik AB (2014) Reactive oxygen species in inflammation and tissue injury. Antioxid Redox Signal 20(7):1126–1167
40. Thannickal VJ, Fanburg BL (2000) Reactive oxygen species in cell signaling. Am J Phys Lung Cell Mol Phys 279(6):L1005–L1028
41. Griffith B, Pendyala S, Hecker L, Lee PJ, Natarajan V, Thannickal VJ (2009) NOX enzymes and pulmonary disease. Antioxid Redox Signal 11(10):2505–2516
42. Pendyala S, Natarajan V (2010) Redox regulation of Nox proteins. Respir Physiol Neurobiol 174(3):265–271
43. Handy DE, Loscalzo J (2012) Redox regulation of mitochondrial function. Antioxid Redox Signal 16(11):1323–1367
44. Lambeth JD (2004) NOX enzymes and the biology of reactive oxygen. Nat Rev Immunol 4(3):181–189
45. Saran M, Bors W (1989) Oxygen radicals acting as chemical messengers: a hypothesis. Free Radic Res Commun 7(3–6):213–220
46. Kehrer JP (2000) The Haber–Weiss reaction and mechanisms of toxicity. Toxicology 149(1):43–50
47. Graf E, Mahoney JR, Bryant RG, Eaton JW (1984) Iron-catalyzed hydroxyl radical formation. Stringent requirement for free iron coordination site. J Biol Chem 259(6):3620–3624
48. Hampton MB, Kettle AJ, Winterbourn CC (1998) Inside the neutrophil phagosome: oxidants, myeloperoxidase, and bacterial killing. Blood J Am Soc Hematol 92(9):3007–3017
49. Bienert GP, Møller AL, Kristiansen KA, Schulz A, Møller IM, Schjoerring JK, Jahn TP (2007) Specific aquaporins facilitate the diffusion of hydrogen peroxide across membranes. J Biol Chem 282(2):1183–1192
50. Hara-Chikuma M, Chikuma S, Sugiyama Y, Kabashima K, Verkman AS, Inoue S, Miyachi Y (2012) Chemokine-dependent T cell migration requires aquaporin-3–mediated hydrogen peroxide uptake. J Exp Med 209(10):1743–1752
51. Miller EW, Dickinson BC, Chang CJ (2010) Aquaporin-3 mediates hydrogen peroxide uptake to regulate downstream intracellular signaling. Proc Natl Acad Sci 107(36):15681–15686
52. Guo Q, Jin Y, Chen X, Ye X, Shen X, Lin M, Zeng C, Zhou T, Zhang J (2024) NF-κB in biology and targeted therapy: new insights and translational implications. Signal Transduct Target Ther 9(1):53

53. Poljsak B, Šuput D, Milisav I (2013) Achieving the balance between ROS and antioxidants: when to use the synthetic antioxidants. Oxidative Med Cell Longev 2013(1):956792
54. Forman HJ, Zhang H (2021) Targeting oxidative stress in disease: promise and limitations of antioxidant therapy. Nat Rev Drug Discov 20(9):689–709
55. Zhang Y, Murugesan P, Huang K, Cai H (2020) NADPH oxidases and oxidase crosstalk in cardiovascular diseases: novel therapeutic targets. Nat Rev Cardiol 17(3):170–194
56. Jin H, Kanthasamy A, Ghosh A, Anantharam V, Kalyanaraman B, Kanthasamy AG (2014) Mitochondria-targeted antioxidants for treatment of Parkinson's disease: preclinical and clinical outcomes. Biochim Biophys Acta (BBA) Mol Basis Dis 1842(8):1282–1294
57. Huang Y, Guo X, Wu Y, Chen X, Feng L, Xie N, Shen G (2024) Nanotechnology's frontier in combatting infectious and inflammatory diseases: prevention and treatment. Signal Transduct Target Ther 9(1):34
58. Din FU, Aman W, Ullah I, Qureshi OS, Mustapha O, Shafique S, Zeb A (2017) Effective use of nanocarriers as drug delivery systems for the treatment of selected tumors. Int J Nanomedicine 5:7291–7309
59. Islam S, Ahmed MM, Islam MA, Hossain N, Chowdhury MA (2025) Advances in nanoparticles in targeted drug delivery-a review. Results Surf Interfaces:100529
60. Venturini J, Chakraborty A, Baysal MA, Tsimberidou AM (2025) Developments in nanotechnology approaches for the treatment of solid tumors. Exp Hematol Oncol 14(1):1–47
61. Shaik Nazma YP (2022) Nanocarriers and their types for targeted drug delivery. Int J Pharm Sci Rev Res 15:21–28
62. Müller RH, Rühl D, Runge S, Schulze-Forster K, Mehnert W (1997) Cytotoxicity of solid lipid nanoparticles as a function of the lipid matrix and the surfactant. Pharm Res 14:458–462
63. Cavalli R, Gasco MR, Chetoni P, Burgalassi S, Saettone MF (2002) Solid lipid nanoparticles (SLN) as ocular delivery system for tobramycin. Int J Pharm 238(1–2):241–245
64. Yang SC, Lu LF, Cai Y, Zhu JB, Liang BW, Yang CZ (1999) Body distribution in mice of intravenously injected camptothecin solid lipid nanoparticles and targeting effect on brain. J Control Release 59(3):299–307
65. Jain N, Jain R, Thakur N, Gupta BP, Jain DK, Banveer J, Jain S (2010) Nanotechnology: a safe and effective drug delivery system. Asian J Pharm Clin Res 3(3):159–165
66. Çağdaş M, Sezer AD, Bucak S (2014) Liposomes as potential drug carrier systems for Drug. In: Application of nanotechnology in drug delivery. IntechOpen, p 1
67. Noble GT, Stefanick JF, Ashley JD, Kiziltepe T, Bilgicer B (2014) Ligand-targeted liposome design: challenges and fundamental considerations. Trends Biotechnol 32(1):32–45
68. Sapra P, Allen TM (2003) Ligand-targeted liposomal anticancer drugs. Prog Lipid Res 42(5):439–462
69. Torchilin VP (2005) Recent advances with liposomes as pharmaceutical carriers. Nat Rev Drug Discov 4(2):145–160
70. Newkome GR, Moorefield CN, Vögtle F, Vögtle F, Vögtle F, Chemist G (2001) Dendrimers and dendrons: concepts, syntheses, applications. Wiley-VCH, Weinheim
71. Basu S, Sandanaraj BS, Thayumanavan S (2002) Molecular recognition in dendrimers. In: Encyclopedia of polymer science and technology. Wiley
72. Stiriba SE, Frey H, Haag R (2002) Dendritic polymers in biomedical applications: from potential to clinical use in diagnostics and therapy. Angew Chem Int Ed 41(8):1329–1334
73. Villalonga-Barber C, Micha-Screttas M, Steele BR, Georgopoulos A, Demetzos C (2008) Dendrimers as biopharmaceuticals: synthesis and properties. Curr Top Med Chem 8(14):1294–1309
74. Miyata K, Christie RJ, Kataoka K (2011) Polymeric micelles for nano-scale drug delivery. React Funct Polym 71(3):227–234
75. Xu W, Ling P, Zhang T (2013) Polymeric micelles, a promising drug delivery system to enhance bioavailability of poorly water-soluble drugs. J Drug Deliv 2013(1):340315
76. Li X, Tsibouklis J, Weng T, Zhang B, Yin G, Feng G, Cui Y, Savina IN, Mikhalovska LI, Sandeman SR, Howel CA (2017) Nano carriers for drug transport across the blood–brain barrier. J Drug Target 25(1):17–28

77. Iijima S (1991) Helical microtubules of graphitic carbon. Nature 354(6348):56–58
78. Bianco A (2004) Carbon nanotubes for the delivery of therapeutic molecules. Expert Opin Drug Deliv 1(1):57–65
79. Madani SY, Naderi N, Dissanayake O, Tan A, Seifalian AM (2011) A new era of cancer treatment: carbon nanotubes as drug delivery tools. Int J Nanomedicine 6:2963–2979
80. Iannazzo D, Piperno A, Pistone A, Grassi G, Galvagno S (2013) Recent advances in carbon nanotubes as delivery systems for anticancer drugs. Curr Med Chem 20(11):1333–1354
81. Wu W, Li R, Bian X, Zhu Z, Ding D, Li X, Jia Z, Jiang X, Hu Y (2009) Covalently combining carbon nanotubes with anticancer agent: preparation and antitumor activity. ACS Nano 3(9):2740–2750
82. Son KH, Hong JH, Lee JW (2016) Carbon nanotubes as cancer therapeutic carriers and mediators. Int J Nanomedicine 11:5163–5185
83. Lockman PR, Mumper RJ, Khan MA, Allen DD (2002) Nanoparticle technology for drug delivery across the blood-brain barrier. Drug Dev Ind Pharm 28(1):1–3
84. Sahoo SK, Labhasetwar V (2003) Nanotech approaches to drug delivery and imaging. Drug Discov Today 8(24):1112–1120
85. Panyam J, Labhasetwar V (2003) Biodegradable nanoparticles for drug and gene delivery to cells and tissue. Adv Drug Deliv Rev 55(3):329–347
86. Mishra BB, Patel BB, Tiwari S (2010) Colloidal nanocarriers: a review on formulation technology, types and applications toward targeted drug delivery. Nanomedicine 6(1):9–24
87. Babu A, Templeton AK, Munshi A, Ramesh R (2013) Nanoparticle-based drug delivery for therapy of lung cancer: progress and challenges. J Nanomater 2013(1):863951
88. Chouly C, Pouliquen D, Lucet I, Jeune JJ, Jallet P (1996) Development of superparamagnetic nanoparticles for MRI: effect of particle size, charge and surface nature on biodistribution. J Microencapsul 13(3):245–255
89. Chatterjee J, Haik Y, Chen CJ (2001) Modification and characterization of polystyrene-based magnetic microspheres and comparison with albumin-based magnetic microspheres. J Magn Magn Mater 225(1–2):21–29
90. Widder KJ, Senyei AE, Ranney DF (1980) In vitro release of biologically active adriamycin by magnetically responsive albumin microspheres. Cancer Res 40(10):3512–3517
91. Giner-Casares JJ, Henriksen-Lacey M, Coronado-Puchau M, Liz-Marzán LM (2016) Inorganic nanoparticles for biomedicine: where materials scientists meet medical research. Mater Today 19(1):19–28
92. Swierczewska M, Lee S, Chen X (2011) Inorganic nanoparticles for multimodal molecular imaging. Mol Imaging 10(1):7290–2011
93. Choi SJ, Lee JK, Jeong J, Choy JH (2013) Toxicity evaluation of inorganic nanoparticles: considerations and challenges. Mol Cell Toxicol 9:205–210
94. Abed SN, Deb PK, Surchi HS, Kokaz SF, Jamal SM, Bandopadhyay S, Tekade RK (2019) Nanocarriers in different preclinical and clinical stages. In: Basic fundamentals of drug delivery. Academic Press, pp 685–731
95. Peltonen L, Hirvonen J (2018) Drug nanocrystals–versatile option for formulation of poorly soluble materials. Int J Pharm 537(1–2):73–83
96. Sinha B, Müller RH, Möschwitzer JP (2013) Bottom-up approaches for preparing drug nanocrystals: formulations and factors affecting particle size. Int J Pharm 453(1):126–141
97. Schmid D, Park CG, Hartl CA, Subedi N, Cartwright AN, Puerto RB, Zheng Y, Maiarana J, Freeman GJ, Wucherpfennig KW, Irvine DJ (2017) T cell-targeting nanoparticles focus delivery of immunotherapy to improve antitumor immunity. Nat Commun 8(1):1747
98. Chu D, Dong X, Zhao Q, Gu J, Wang Z (2017) Photosensitization priming of tumor microenvironments improves delivery of nanotherapeutics via neutrophil infiltration. Adv Mater 29(27):1701021
99. Oh JY, Kim HS, Palanikumar L, Go EM, Jana B, Park SA, Kim HY, Kim K, Seo JK, Kwak SK, Kim C (2018) Cloaking nanoparticles with protein corona shield for targeted drug delivery. Nat Commun 9(1):4548

100. Dacoba TG, Olivera A, Torres D, Crecente-Campo J, Alonso MJ (2017) Modulating the immune system through nanotechnology. Semin Immunol 34:78–102. Academic Press
101. Zhang G, Niu A, Peng S, Jiang M, Tu Y, Li M, Wu C (2001) Formation of novel polymeric nanoparticles. Acc Chem Res 34(3):249–256
102. Chan JM, Valencia PM, Zhang L, Langer R, Farokhzad OC (2010) Polymeric nanoparticles for drug delivery. In: Cancer nanotechnology: Methods and protocols. Humana Press, Totowa, pp 163–175
103. Huang X, Zheng R, Ding F, Yang J, Xie M, Liu X, Li J, Feng J, Zhu X, Zhang C (2020) Efficient delivery of mRNA using crosslinked nucleic acid nanogel as a carrier. ACS Mater Lett 2(11):1509–1515
104. Sun Y, Jing X, Liu Y, Yu B, Hu H, Cong H, Shen Y (2023) A chitosan derivative-crosslinked hydrogel with controllable release of polydeoxyribonucleotides for wound treatment. Carbohydr Polym 300:120298
105. Zhang C, Gao S, Jiang W, Lin S, Du F, Li Z, Huang W (2010) Targeted minicircle DNA delivery using folate–poly (ethylene glycol)–polyethylenimine as non-viral carrier. Biomaterials 31(23):6075–6086
106. Deng L, Zhang F, Wu Y, Luo J, Mao X, Long L, Gou M, Yang L, Deng DY (2019) RGD-modified nanocarrier-mediated targeted delivery of HIF-1α-AA plasmid DNA to cerebrovascular endothelial cells for ischemic stroke treatment. ACS Biomater Sci Eng 5(11):6254–6264
107. Yang H, Luo Y, Hu H, Yang S, Li Y, Jin H, Chen S, He Q, Hong C, Wu J, Wan Y (2021) pH-sensitive, cerebral vasculature-targeting hydroxyethyl starch functionalized nanoparticles for improved angiogenesis and neurological function recovery in ischemic stroke. Adv Healthc Mater 10(12):2100028
108. Mehta M, Bui TA, Yang X, Aksoy Y, Goldys EM, Deng W (2023) Lipid-based nanoparticles for drug/gene delivery: an overview of the production techniques and difficulties encountered in their industrial development. ACS Mater Au 3(6):600–619
109. Cheng X, Lee RJ (2016) The role of helper lipids in lipid nanoparticles (LNPs) designed for oligonucleotide delivery. Adv Drug Deliv Rev 99:129–137
110. Tenchov R, Bird R, Curtze AE, Zhou Q (2021) Lipid nanoparticles─from liposomes to mRNA vaccine delivery, a landscape of research diversity and advancement. ACS Nano 15(11):16982–17015
111. Puri A, Loomis K, Smith B, Lee JH, Yavlovich A, Heldman E, Blumenthal R (2009) Lipid-based nanoparticles as pharmaceutical drug carriers: from concepts to clinic. Crit Rev Ther Drug Carrier Syst 26(6):523
112. Xu L, Wang X, Liu Y, Yang G, Falconer RJ, Zhao CX (2022) Lipid nanoparticles for drug delivery. Adv NanoBiomed Res 2(2):2100109
113. Mora-Huertas CE, Fessi H, Elaissari A (2010) Polymer-based nanocapsules for drug delivery. Int J Pharm 385(1–2):113–142
114. Rivas CJ, Tarhini M, Badri W, Miladi K, Greige-Gerges H, Nazari QA, Rodríguez SA, Román RÁ, Fessi H, Elaissari A (2017) Nanoprecipitation process: from encapsulation to drug delivery. Int J Pharm 532(1):66–81
115. Cui MD, Pan ZH, Pan LQ (2017) Danggui Buxue extract-loaded liposomes in thermosensitive gel enhance in vivo dermal wound healing via activation of the VEGF/PI3K/Akt and TGF-β/Smads signaling pathway. Evid Based Complement Alternat Med 2017(1):8407249
116. Pires J, Cargnin ST, Costa SA, Sinhorin VD, Damazo AS, Sinhorin AP, de Campos Bicudo R, Cavalheiro L, de Souza Valladão DM, Pohlmann AR, Guterres SS (2020) Healing of dermal wounds property of Caryocar brasiliense oil loaded polymeric lipid-core nanocapsules: formulation and in vivo evaluation. Eur J Pharm Sci 150:105356
117. Shi Z, Zhou Y, Fan T, Lin Y, Zhang H, Mei L (2020) Inorganic nano-carriers based smart drug delivery systems for tumor therapy. Smart Mater Med 1:32–47
118. Huo M, Yuan J, Tao L, Wei Y (2014) Redox-responsive polymers for drug delivery: from molecular design to applications. Polym Chem 5(5):1519–1528

119. Shi Z, Li Q, Mei L (2020) pH-sensitive nanoscale materials as robust drug delivery systems for cancer therapy. Chin Chem Lett 31(6):1345–1356
120. Wang X, Wang X, Jin S, Muhammad N, Guo Z (2018) Stimuli-responsive therapeutic metallodrugs. Chem Rev 119(2):1138–1192
121. Saravanakumar G, Kim J, Kim WJ (2017) Reactive-oxygen-species-responsive drug delivery systems: promises and challenges. Adv Sci 4(1):1600124
122. Sultana A, Zare M, Thomas V, Kumar TS, Ramakrishna S (2022) Nano-based drug delivery systems: conventional drug delivery routes, recent developments and future prospects. Med Drug Discov 15:100134
123. Kunisaki A, Kodama A, Ishikawa M, Ueda T, Lima MD, Kondo T, Adachi N (2021) Carbon-nanotube yarns induce axonal regeneration in peripheral nerve defect. Sci Rep 11(1):19562
124. Bao L, Dou G, Tian R, Lv Y, Ding F, Liu S, Zhao R, Zhao L, Zhou J, Weng L, Dong Y (2022) Engineered neutrophil apoptotic bodies ameliorate myocardial infarction by promoting macrophage efferocytosis and inflammation resolution. Bioact Mater 9:183–197
125. Dou Y, Zhang Y, Lin C, Han R, Wang Y, Wu D, Zheng J, Lu C, Tang L, He Y (2022) pH-responsive theranostic nanoplatform of ferrite and ceria co-engineered nanoparticles for anti-inflammatory. Front Bioeng Biotechnol 10:983677
126. Zazuli Z, Hartati R, Rowa CR, Asyarie S (2025) The potential application of nanocarriers in delivering topical antioxidants. Pharmaceuticals 18(1):56
127. Elmowafy M (2021) Skin penetration/permeation success determinants of nanocarriers: pursuit of a perfect formulation. Colloids Surf B Biointerfaces 203:111748
128. Suwannateep N, Wanichwecharungruang S, Haag SF, Devahastin S, Groth N, Fluhr JW, Lademann J, Meinke MC (2012) Encapsulated curcumin results in prolonged curcumin activity in vitro and radical scavenging activity ex vivo on skin after UVB-irradiation. Eur J Pharm Biopharm 82(3):485–490
129. Zhang ZY, Jiang M, Fang J, Yang MF, Zhang S, Yin YX, Li DW, Mao LL, Fu XY, Hou YJ, Fu XT (2017) Enhanced therapeutic potential of nano-curcumin against subarachnoid hemorrhage-induced blood–brain barrier disruption through inhibition of inflammatory response and oxidative stress. Mol Neurobiol 54:1–4
130. Doggui S, Sahni JK, Arseneault M, Dao L, Ramassamy C (2012) Neuronal uptake and neuroprotective effect of curcumin-loaded PLGA nanoparticles on the human SK-N-SH cell line. J Alzheimer's Dis 30(2):377–392
131. Fernandes M, Lopes I, Magalhães L, Sárria MP, Machado R, Sousa JC, Botelho C, Teixeira J, Gomes AC (2021) Novel concept of exosome-like liposomes for the treatment of Alzheimer's disease. J Control Release 336:130–143
132. Qian F, Han Y, Han Z, Zhang D, Zhang L, Zhao G, Li S, Jin G, Yu R, Liu H (2021) In situ implantable, post-trauma microenvironment-responsive, ROS depletion hydrogels for the treatment of traumatic brain injury. Biomaterials 270:120675
133. Chen K, Pan H, Ji D, Li Y, Duan H, Pan W (2021) Curcumin-loaded sandwich-like nanofibrous membrane prepared by electrospinning technology as wound dressing for accelerate wound healing. Mater Sci Eng C Mater Biol Appl 127:112245
134. Hu B, Gao M, Boakye-Yiadom KO, Ho W, Yu W, Xu X, Zhang XQ (2021) An intrinsically bioactive hydrogel with on-demand drug release behaviors for diabetic wound healing. Bioact Mater 6(12):4592–4606
135. Zhang X, Feng J, Feng W, Xu B, Zhang K, Ma G, Li Y, Yang M, Xu FJ (2022) Glycosaminoglycan-based hydrogel delivery system regulates the wound microenvironment to rescue chronic wound healing. ACS Appl Mater Interfaces 14(28):31737–31750
136. Wu TH, Yen FL, Lin LT, Tsai TR, Lin CC, Cham TM (2008) Preparation, physicochemical characterization, and antioxidant effects of quercetin nanoparticles. Int J Pharm 346(1–2):160–168
137. Alonso C, Martí M, Barba C, Carrer V, Rubio L, Coderch L (2017) Skin permeation and antioxidant efficacy of topically applied resveratrol. Arch Dermatol Res 309(6):423–431

138. Gokce EH, Korkmaz E, Dellera E, Sandri G, Bonferoni MC, Ozer O (2012) Resveratrol-loaded solid lipid nanoparticles versus nanostructured lipid carriers: evaluation of antioxidant potential for dermal applications. Int J Nanomedicine 7:1841–1850
139. Saha S, Sadhukhan P, C Sil P (2014) Genistein: a phytoestrogen with multifaceted therapeutic properties. Mini Rev Med Chem 14(11):920–940
140. Rassu G, Porcu EP, Fancello S, Obinu A, Senes N, Galleri G, Migheli R, Gavini E, Giunchedi P (2018) Intranasal delivery of genistein-loaded nanoparticles as a potential preventive system against neurodegenerative disorders. Pharmaceutics 11(1):8
141. Pool H, Campos-Vega R, Herrera-Hernández MG, García-Solis P, García-Gasca T, Sánchez IC, Luna-Bárcenas G, Vergara-Castañeda H (2018) Development of genistein-PEGylated silica hybrid nanomaterials with enhanced antioxidant and antiproliferative properties on HT29 human colon cancer cells. Am J Transl Res 10(8):2306
142. Botten D, Fugallo G, Fraternali F, Molteni C (2015) Structural properties of green tea catechins. J Phys Chem B 119(40):12860–12867
143. Mereles D, Hunstein W (2011) Epigallocatechin-3-gallate (EGCG) for clinical trials: more pitfalls than promises? Int J Mol Sci 12(9):5592–5603
144. Zhang XY, Zhang PY (2017) Polymersomes in nanomedicine-a review. Curr Nanosci 13(2):124–129
145. Baumgart DC, Carding SR (2007) Inflammatory bowel disease: cause and immunobiology. Lancet 369(9573):1627–1640
146. Baumgart DC, Sandborn WJ (2007) Inflammatory bowel disease: clinical aspects and established and evolving therapies. Lancet 369(9573):1641–1657
147. Coskun M (2014) Intestinal epithelium in inflammatory bowel disease. Front Med 1:24
148. Mahajan N, Sakarkar D, Manmode A, Pathak V, Ingole R, Dewade D (2011) Biodegradable nanoparticles for targeted delivery in treatment of ulcerative colitis. Adv Sci Lett 4(2):349–356
149. Brubaker SW, Bonham KS, Zanoni I, Kagan JC (2015) Innate immune pattern recognition: a cell biological perspective. Annu Rev Immunol 33(1):257–290
150. Laggner U, Di Meglio P, Perera GK, Hundhausen C, Lacy KE, Ali N, Smith CH, Hayday AC, Nickoloff BJ, Nestle FO (2011) Clinical relevance of pro-inflammatory skin homing CLA+ V gamma 9V delta 2 T cells in psoriasis. Br J Dermatol 164(4):907–907. Malden: Wiley-Blackwell
151. Laggner U, Di Meglio P, Perera GK, Hundhausen C, Lacy KE, Ali N, Smith CH, Hayday AC, Nickoloff BJ, Nestle FO (2011) Identification of a novel proinflammatory human skin-homing Vγ9Vδ2 T cell subset with a potential role in psoriasis. J Immunol 187(5):2783–2793
152. Baboota S, Alam MS, Sharma S, Sahni JK, Kumar A, Ali J (2011) Nanocarrier-based hydrogel of betamethasone dipropionate and salicylic acid for treatment of psoriasis. Int J Pharm Investig 1(3):139
153. Wan T, Pan W, Long Y, Yu K, Liu S, Ruan W, Pan J, Qin M, Wu C, Xu Y (2017) Effects of nanoparticles with hydrotropic nicotinamide on tacrolimus: permeability through psoriatic skin and antipsoriatic and antiproliferative activities. Int J Nanomedicine 12:1485–1497
154. Lakshmi PK, Devi GS, Bhaskaran S, Sacchidanand S (2007) Niosomal methotrexate gel in the treatment of localized psoriasis: phase I and phase II studies. Indian J Dermatol Venereol Leprol 73:157
155. Harris ED Jr (1990) Rheumatoid arthritis: pathophysiology and implications for therapy. N Engl J Med 322(18):1277–1289
156. Maeda H (2001) The enhanced permeability and retention (EPR) effect in tumor vasculature: the key role of tumor-selective macromolecular drug targeting. Adv Enzym Regul 41(1):189–207
157. Bernardi A, Zilberstein AC, Jäger E, Campos MM, Morrone FB, Calixto JB, Pohlmann AR, Guterres SS, Battastini AM (2009) Effects of indomethacin-loaded nanocapsules in experimental models of inflammation in rats. Br J Pharmacol 158(4):1104–1111

158. Kim MJ, Park JS, Lee SJ, Jang J, Park JS, Back SH, Bahn G, Park JH, Kang YM, Kim SH, Kwon IC (2015) Notch1 targeting siRNA delivery nanoparticles for rheumatoid arthritis therapy. J Control Release 216:140–148
159. Duan J, Dong J, Zhang T, Su Z, Ding J, Zhang Y, Mao X (2014) Polyethyleneimine-functionalized iron oxide nanoparticles for systemic siRNA delivery in experimental arthritis. Nanomedicine 9(6):789–801
160. Anselmo AC, Mitragotri S (2016) Nanoparticles in the clinic. Bioeng Transl Med 1(1):10–29
161. van der Valk FM, van Wijk DF, Lobatto ME, Verberne HJ, Storm G, Willems MC, Legemate DA, Nederveen AJ, Calcagno C, Mani V, Ramachandran S (2015) Prednisolone-containing liposomes accumulate in human atherosclerotic macrophages upon intravenous administration. Nanomedicine 11(5):1039–1046
162. Kumar R, Dogra S, Amarji B, Singh B, Kumar S, Vinay K, Mahajan R, Katare OP (2016) Efficacy of novel topical liposomal formulation of cyclosporine in mild to moderate stable plaque psoriasis: a randomized clinical trial. JAMA Dermatol 152(7):807–815
163. Kanhai KM, Zuiker RG, Stavrakaki I, Gladdines W, Gaillard PJ, Klaassen ES, Groeneveld GJ (2018) Glutathione-PEGylated liposomal methylprednisolone in comparison to free methylprednisolone: slow release characteristics and prolonged lymphocyte depression in a first-in-human study. Br J Clin Pharmacol 84(5):1020–1028
164. Nakano K, Matoba T, Koga JI, Kashihara Y, Fukae M, Ieiri I, Shiramoto M, Irie S, Kishimoto J, Todaka K, Egashira K (2018) Safety, tolerability, and pharmacokinetics of NK-104-NP a multicenter, randomized, placebo-controlled phase I investigator-initiated trial for intravenous administration of pitavastatin-loaded plga nanoparticles (Nk-104-NP) in healthy Japanese male subjects. Int Heart J 59(5):1015–1025
165. Sabouri M, Samadi A, Ahmad Nasrollahi S, Farboud ES, Mirrahimi B, Hassanzadeh H, Nassiri Kashani M, Dinarvand R, Firooz A (2018) Tretinoin loaded nanoemulsion for acne vulgaris: fabrication, physicochemical and clinical efficacy assessments. Skin Pharmacol Physiol 31(6):316–323
166. Paithankar D, Hwang BH, Munavalli G, Kauvar A, Lloyd J, Blomgren R, Faupel L, Meyer T, Mitragotri S (2015) Ultrasonic delivery of silica–gold nanoshells for photothermolysis of sebaceous glands in humans: nanotechnology from the bench to clinic. J Control Release 206:30–36
167. Dinarello CA (2010) Anti-inflammatory agents: present and future. Cell 140(6):935–950
168. Doshi N, Mitragotri S (2010) Macrophages recognize size and shape of their targets. PLoS One 5(4):e10051
169. Wibroe PP, Anselmo AC, Nilsson PH, Sarode A, Gupta V, Urbanics R, Szebeni J, Hunter AC, Mitragotri S, Mollnes TE, Moghimi SM (2017) Bypassing adverse injection reactions to nanoparticles through shape modification and attachment to erythrocytes. Nat Nanotechnol 12(6):589–594
170. Leong HS, Butler KS, Brinker CJ, Azzawi M, Conlan S, Dufès C, Owen A, Rannard S, Scott C, Chen C, Dobrovolskaia MA (2019) On the issue of transparency and reproducibility in nanomedicine. Nat Nanotechnol 14(7):629–635
171. Pradal J, Maudens P, Gabay C, Seemayer CA, Jordan O, Allémann E (2016) Effect of particle size on the biodistribution of nano-and microparticles following intra-articular injection in mice. Int J Pharm 498(1–2):119–129
172. Dormont F, Rouquette M, Mahatsekake C, Gobeaux F, Peramo A, Brusini R, Calet S, Testard F, Lepetre-Mouelhi S, Desmaële D, Varna M (2019) Translation of nanomedicines from lab to industrial scale synthesis: the case of squalene-adenosine nanoparticles. J Control Release 307:302–314
173. Zhou M, Hou J, Zhong Z, Hao N, Lin Y, Li C (2018) Targeted delivery of hyaluronic acid-coated solid lipid nanoparticles for rheumatoid arthritis therapy. Drug Deliv 25(1):716–722

Chapter 23
Complementary and Alternative Medicine Therapy to Inhibit ROS-Mediated Inflammatory Disorders

Aniqa Rehman, Teck Chwen Loh, Suet Lin Chia, and Hooi Ling Foo

23.1 Introduction

Oxidative stress is a fundamental feature of conditions driven by reactive oxygen species (ROS)-mediated inflammatory disorders, a pathological state marked by an abnormal accumulation of ROS, which are molecules containing oxygen that are chemically reactive, such as superoxide anion (O_2^-), hydrogen peroxide (H_2O_2), and hydroxyl radicals (OH) [1]. Although ROS at normal levels act as signaling

A. Rehman
Department of Bioprocess Technology, Faculty of Biotechnology and Biomolecular Sciences, Universiti Putra Malaysia, UPM Serdang, Selangor, Malaysia

T. C. Loh
Lactic Acid Bacteria Biota Technology Research Program, Research Laboratory of Probiotics and Cancer Therapeutics, UPM-MAKNA Cancer Research Laboratory (CANRES), Institute of Bioscience, Universiti Putra Malaysia, UPM Serdang, Selangor, Malaysia

Department of Animal Science, Faculty of Agriculture, Universiti Putra Malaysia, UPM Serdang, Selangor, Malaysia

S. L. Chia
Lactic Acid Bacteria Biota Technology Research Program, Research Laboratory of Probiotics and Cancer Therapeutics, UPM-MAKNA Cancer Research Laboratory (CANRES), Institute of Bioscience, Universiti Putra Malaysia, UPM Serdang, Selangor, Malaysia

Department of Microbiology, Faculty of Biotechnology and Biomolecular Sciences, Universiti Putra Malaysia, UPM Serdang, Selangor, Malaysia

H. L. Foo (✉)
Department of Bioprocess Technology, Faculty of Biotechnology and Biomolecular Sciences, Universiti Putra Malaysia, UPM Serdang, Selangor, Malaysia

Lactic Acid Bacteria Biota Technology Research Program, Research Laboratory of Probiotics and Cancer Therapeutics, UPM-MAKNA Cancer Research Laboratory (CANRES), Institute of Bioscience, Universiti Putra Malaysia, UPM Serdang, Selangor, Malaysia
e-mail: hlfoo@upm.edu.my

S. Ashique, N. Mishra (eds.), *Reactive Oxygen Species in Metabolic Inflammation*,
https://doi.org/10.1007/978-3-032-11752-6_23

molecules that help maintain cellular homeostasis, an overproduction of ROS can be detrimental. High levels of ROS can lead to damage within cells by oxidizing lipids, DNA, and proteins. Moreover, high ROS levels activate pro-inflammatory transcription factors such as NF-κB and AP-1, which play vital roles in initiating inflammatory responses [2]. This activation leads to increased expression of pro-inflammatory cytokine genes, which amplify the inflammatory response cascade. Consequently, the imbalance of ROS production not only disrupts cellular function but also perpetuates inflammation, potentially contributing to various inflammatory diseases [3].

Inflammation is a crucial and highly conserved biological response that functions as a protective barrier against pathogens, toxins, and cell damage. However, when unregulated, it can contribute to the development of various chronic diseases, including cancer, cardiovascular conditions, diabetes, and neurodegenerative disorders [4, 5]. This pathogenic connection between oxidative stress and the body's inflammatory response has prompted extensive research into therapeutic interventions that specifically target the reduction of reactive oxygen species (ROS), while simultaneously enhancing the intrinsic antioxidant defense mechanisms of the body. Such strategies aim to mitigate cellular damage and modulate inflammatory pathways by restoring redox balance [6]. Synthetic drugs like NSAIDs and corticosteroids are commonly used for symptom relief but pose challenges with prolonged use, due to side effects such as gastrointestinal issues, cardiovascular risks, and hormonal imbalances. Healthcare providers, therefore, seek safer and more effective long-term alternatives [7].

Growing scientific and clinical interest in Complementary and Alternative Medicine (CAM) therapies has emerged in response to the above limitations. CAM includes various remedies and products, such as herbal medicine, dietary supplements, acupuncture, mind–body techniques and nutraceuticals, which are used either in combination with or as alternatives to conventional pharmacological treatments [8]. Recent progress in system biology and metabolomics has further clarified and revealed the molecular mechanisms of CAM influences that mediate through redox signaling and immune modulation [9]. Additionally, growing clinical data support using CAM as an adjunctive measure to improve chronic inflammatory conditions, such as rheumatoid arthritis, inflammatory bowel disease, asthma, and even cancer [10]. Most CAM interventions exhibit excellent antioxidant, anti-inflammatory, and immunomodulatory effects, making them potential candidates for alleviating ROS-induced inflammation [11]. CAM modalities with antioxidant and anti-inflammatory effects provide a multifaceted approach to fighting oxidative stress and inflammation. This chapter discusses their rationale, mechanisms, and potential in reducing ROS-induced inflammation, setting the platform for integrative medicine's role in tackling a significant healthcare challenge.

23.2 Reactive Oxygen Species (ROS): Dual Roles in Cellular Signaling and Inflammation

Reactive oxygen species (ROS) are natural by-products of aerobic cellular metabolism that play dual roles in biological systems. At physiological levels, they serve as essential signaling mediators involved in various processes such as cellular proliferation, differentiation, and immune responses [12]. However, when their levels become dysregulated, ROS function as key instigators of oxidative stress, driving cell damage that contributes to disease pathogenesis [13]. Traditionally, mitochondrial oxidative phosphorylation was reflected as the primary origin of ROS, particularly through electron leakage in the mitochondrial respiratory chain [14]. Yet, recent insights have expanded this view, identifying significant contributions from enzyme systems, such as nicotinamide adenine dinucleotide phosphate (NADPH) oxidases, which can generate ROS in a regulated and stimulus-dependent manner [15].

Beyond mitochondrial metabolism, endogenous ROS production arises from multiple intracellular and tissue-specific sources. These include peroxisomal β-oxidation of fatty acids, enzymatic oxidation during metabolism of arginine, and the activity of various oxidases and oxygenases [16]. Furthermore, ROS generation is markedly amplified during immune responses, particularly through the respiratory burst that accompanies phagocytosis in response to microbial invasion [17]. Collectively, these diverse origins underscore the complexity of ROS biology and highlight their central role in both normal physiology and the progression of disorders related to oxidative stress [18].

A disproportion between reactive oxygen species (ROS) production and the cell's antioxidant defenses causes oxidative stress, which is a critical factor in the development of numerous inflammatory diseases. This condition arises when ROS are generated in excess, overwhelming the cell's ability to neutralize them, which causes damage to lipids, proteins, and nucleic acids [19]. Although ROS have historically been regarded as harmful, recent research shows they are essential signaling molecules regulating vital physiological processes [20] of cell proliferation, differentiation, migration, and apoptosis [21]. This understanding highlights the beneficial and essential roles that ROS play in maintaining normal biological functions, despite their reputation as merely harmful entities [22].

Unregulated release of reactive oxygen species (ROS) has been associated with the cause and development of many inflammatory situations, such as cardiovascular diseases [23], autoimmune diseases [24], and neurodegenerative diseases [25]. Understanding the link between ROS production and the cell's antioxidant defense is essential for developing effective therapies against these diseases [26]. While ROS was once thought to move freely and affect cells randomly, it is now clearly shown that they mediate highly specific actions [27].

23.2.1 ROS-Mediated Inflammatory Disorders

Inflammation is a complex immune response that helps defend the body from infection, injury, or tissue damage. Acute inflammation resolves once the threat passes, but chronic inflammation involves ongoing immune activation and links to many diseases [28]. A common feature of inflammatory disease is the overproduction of ROS produced by immune cells, such as eosinophils, neutrophils, and macrophages in affected tissues [29]. Disruption of redox signaling links to conditions like cardiovascular diseases, neurodegenerative disorders, cancer, and autoimmune diseases [30]. Produced by inflammatory cells, ROS levels are higher during chronic inflammatory conditions, such as ulcerative colitis, Barrett's esophagus (BE) and hepatitis [31]. BE is set forth as an esophagus cancer risk factor, with metaplastic changes progressing to adenocarcinoma [32]. In animal studies, reflux increases free radicals that damage esophageal cells. Clinically, BE patients show more ROS, H_2O_2 sensitivity, lipid peroxidation, fewer superoxide radicals, and chromosomal imbalance, especially in GERD and gastric acid exposure [33, 34]. Elevated levels of ROS can compromise the integrity of the endothelial barrier, resulting in a rise in vascular permeability, tissue damage, and eventually organ dysfunction [35, 36].

In addition, ROS has a significant role in autoimmune diseases, which emphasizes their pathologic significance [37]. IBD is a multifactorial disease involving gut microbiota dysbiosis, external factors like diet and lifestyle, and discrete variability. These factors increase ROS and immune activation in the colon, perpetuating chronic inflammation [38]. Similarly, in neurodegenerative diseases, including Alzheimer's [39, 40] and Parkinson's disease [41, 42], neuronal injury and disease development [43] as well as respiratory disorders, including asthma and chronic obstructive pulmonary disease (COPD). ROS have central roles in airway inflammation and lung injury [44]. The pathogenic mechanisms of these diseases typically involve complex regulation and signaling pathways governed by inflammatory mediators [45].

ROS play a dual role in cancer by functioning as both tumor suppressors and promoters, dependent on their levels and cellular context, a phenomenon referred to as the "redox paradox" [46]. Low to moderate levels of ROS encourage tumor growth by stimulating cell proliferation, forming new blood vessels, and promoting metastasis [47, 48], while higher concentrations induce oxidative stress, initiating apoptosis and senescence. Cancer cells have been found to display elevated ROS due to metabolic and mitochondrial alterations, but they strengthen antioxidant defense mechanisms to survive oxidative stress [49–51]. Understanding the role of antioxidants in managing oxidative stress has significant potential for developing effective preventive and therapeutic strategies against ROS-driven inflammatory diseases [52] (Table 23.1).

Table 23.1 Reactive oxygen species-mediated inflammatory disorders

Disorder	Role of ROS in pathogenesis	References
Cardiovascular diseases	ROS contribute to endothelial dysfunction, atherosclerosis, and progression to heart failure	[35, 53, 54]
Alzheimer's disease	ROS triggers neuronal apoptosis, protein aggregation, and neuroinflammation	[39, 40]
Parkinson's disease	ROS mediate mitochondrial dysfunction and dopaminergic neuronal dysfunction	[41, 42]
Chronic obstructive pulmonary disease (COPD)	ROS drives epithelial injury, airway inflammation and mucus hypersecretion.	[55, 56]
Asthma	ROS increased oxidative damage and the release of pro-inflammatory cytokines in the airways	[57, 58]
Inflammatory bowel disease (IBD)	Excess ROS damages the intestinal mucosa and enhances immune activation	[59, 60]
Rheumatoid arthritis	ROS promotes joint inflammation, cartilage degradation, and immune cell activation	[61]
Periodontal disease	ROS, along with the inflammasome dysregulation, contribute to the destruction of the gingival and alveolar bone	[62]
Autoimmune diseases	ROS enhance antigen presentation and pro-inflammatory cytokine release, worsening autoimmunity	[63]
Neuroinflammation	Chronic ROS production leads to neuronal death and cognitive impairment	[64, 65]
Colorectal cancer	Chronic inflammation and ROS-induced DNA damage promote carcinogenesis	[66]
Lung cancer	ROS from smoking and inflammation cause DNA mutations and support tumor progression	[67]
Gastric cancer	*Helicobacter pylori*-induced inflammation increases ROS, leading to mucosal damage and neoplastic transformation	[68]
Hepatocellular carcinoma	ROS contribute to liver inflammation, fibrosis, and DNA damage, facilitating hepatocarcinogenesis	[69]
Breast cancer	ROS regulate tumor microenvironment, activate pro-survival signaling pathways, and induce genomic instability	[70]

23.3 Current Treatment Strategies and Their Limitations

Conventional therapy for inflammatory diseases primarily relies on medications, including traditional pharmacotherapy, which encompasses aminosalicylates, corticosteroids, immunomodulators, nonsteroidal anti-inflammatory drugs (NSAIDs), and disease-modifying antirheumatic drugs (DMARDs). While these medications effectively reduce inflammation and symptoms, they often induce side effects, such as gastrointestinal issues, cardiovascular problems, and immunosuppression. Traditional treatments also face challenges, like limited specificity, modest effectiveness, and potential adverse effects. They may not adequately target the core causes, such as oxidative imbalance or may only provide symptom relief without addressing the root issues. The absence of reliable biomarkers further complicates the monitoring and delivery of personalized care. Additionally, many patients with chronic inflammatory conditions report insufficient symptom relief, prompting interest in alternative approaches. Lifestyle modifications, including dietary adjustments, physical activity, and stress management, are essential for overall well-being management and are often recommended as supplementary treatments; however, patient compliance and effectiveness can vary. The limitations of conventional therapies, paired with a longing for holistic, patient-focused options, have led to increased use of CAM for inflammation. Dissatisfaction with traditional medicine drives patients toward CAM alternatives, influencing clinical outcomes, healthcare delivery, research directions, and policy making [71].

23.4 Complementary and Alternative Medicine Therapies for Inflammatory Disorders

CAM encompasses a broad spectrum of healing and healthcare practices that fall outside conventional methods of Western medicine [72]. These practices have gained growing recognition regarding their possible role in the management of ROS-mediated inflammatory conditions. The terms "complementary" and "alternative" should not be used interchangeably: "Complementary" therapies are combined with conventional treatments, while "alternative" therapies replace them [73]. Some experts propose the term "integrative medicine" to describe the combination of mainstream and alternative approaches [74]. CAM is often integrated into the healthcare systems of various cultures, emphasizing holistic methods that seek to restore balance and boost the body's innate healing capabilities [75]. Numerous studies show that factors such as age, gender, education, income, and cultural background influence the likelihood of CAM adoption. Surveys in different countries reveal that people frequently turn to CAM for relief from chronic pain, fatigue, anxiety, and other symptoms linked to inflammatory conditions [76]. Consequently, many health systems worldwide increasingly incorporate CAM therapies to complement conventional care and enhance treatment effectiveness [77].

The use of CAM therapies for ROS-related inflammatory conditions relies on the idea that natural agents can influence the formation, neutralization, and signaling of ROS [78]. Alteration of these can have the potential to decrease oxidative stress, consequently reducing inflammation and attendant tissue injury. CAM methods can be categorized into two broad categories: natural products and mind–body interventions [79]. Natural products encompass a diverse range of compounds, including botanicals, vitamins, minerals, probiotics, and their derivatives, such as paraprobiotics and postbiotics. Comprehending the trends and implications of CAM utilization underscores the importance of a model to analyze CAM within a health service or resource framework [80] (Fig. 23.1).

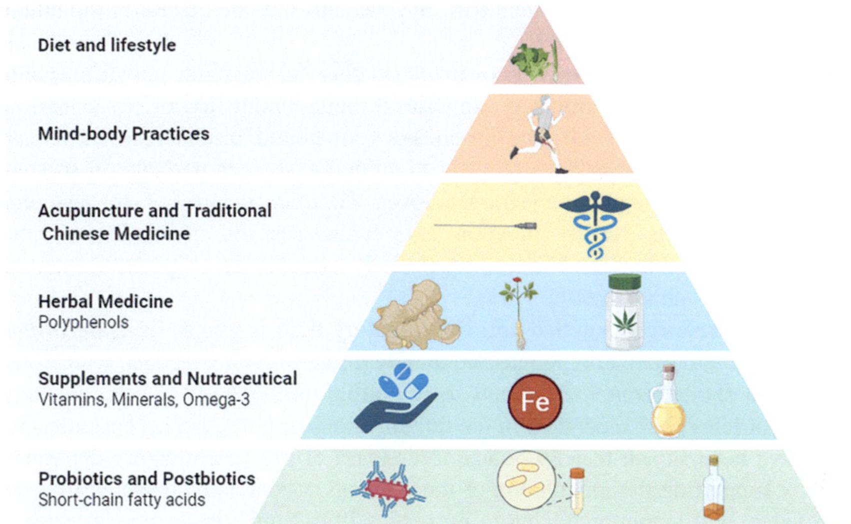

Fig. 23.1 Complementary and alternative medicine interventions for ROS-mediated inflammatory disorders

23.4.1 Herbal Medicine for Inflammatory Disorders

Historically, traditional medicine has been focused on the usage of herbal medicine, which is currently more widely accepted for its anti-inflammatory effects in modern evidence-based practice [81]. Polyphenols, flavonoids, and terpenoids are among the diverse bioactive compounds found in medicinal herbs that exert potent immunomodulatory, antioxidant, and anti-inflammatory effects by targeting key inflammatory signaling pathways [82]. Most noteworthy is that numerous such compounds are synergistic, whose overall activity when combined is greater than the activity of individual constituents and constitutes their amplified overall pharmacological profile [83]. In addition to their traditional uses, many medicinal plants are today being investigated for their varied pharmacological actions, including anti-inflammatory, analgesic, anti-ulcer, antidiabetic, and anticancer activities, further proclaiming their use in modern integrative therapeutic protocols [84].

Natural polyphenols represent a structurally diverse array of bioactive molecules characterized by one or more phenyl rings with appended hydroxyl groups [85]. Polyphenol's medicinal potential is primarily attributed to their capacity to regulate key molecular pathways involved in inflammation. Among these, curcumin, a polyphenolic molecule isolated from the rhizome of *Curcuma longa* (turmeric), has been the focus of extensive studies for its potent anti-inflammatory, antioxidant, and anticancer activities [86–88]. *Camellia sinensis,* or green tea, is rich in epigallocatechin gallate (EGCG), a bioactive polyphenol with established antioxidant and anti-inflammatory effects. EGCG produces its positive impact by inhibiting NF-κB signaling and lowering oxidative stress, mechanisms that are central to the pathogenesis of chronic diseases [89].

Similarly, *Panax ginseng* is known for its ginsenosides with potent immunomodulatory and anti-inflammatory activities through modulation of key pathways, including MAPK and NF-κB. The ginsenosides compounds also increase the release of anti-inflammatory cytokines (IL-10), highlighting their importance in reestablishing immune dysregulation [90]. Moreover, *Withania somnifera* (ashwagandha) profoundly inhibits NF-κB function and suppresses the generation of pro-inflammatory cytokines, and hence, is especially useful in treating neuroinflammatory and autoimmune diseases ([91].

Another extensively studied anti-inflammatory herb is ginger (*Zingiber officinale*), whose bioactive compounds are mainly gingerols and shogaols, which possess potent anti-inflammatory effects by inhibiting the synthesis of prostaglandins and leukotrienes, key mediators in the inflammatory response [92]. These metabolites have been shown to help reduce the release of pro-inflammatory cytokines, thereby supporting the modulation of the immune system and enhancing ginger's antioxidant and immunomodulatory properties [93]. Similarly, *Boswellia serrata*, or Indian frankincense, has demonstrated strong anti-inflammatory effects primarily through its active constituents, boswellic acids. These triterpenoids work by inhibiting 5-lipoxygenase, a key enzyme in the leukotriene biosynthetic pathway, thereby reducing leukotriene-mediated inflammatory processes [94, 95].

Another well-represented category of anti-inflammatory medications is cannabinoids, with cannabidiol (CBD) being a highly promising molecule because of its capacity to adjust immune responses and suppress key pro-inflammatory mediators, such as tumor necrosis factor-alpha (TNF-α) and prostaglandin E2 [96]. CBD exerts its effect by blocking the production and release of cytokines that promote inflammation and simultaneously inducing the expression of anti-inflammatory mediators like IL-1Ra, IL-4, and IL-10 and hence resulting in a balanced immunoregulatory process [97]. These findings further validate polyphenol's potential role as an add-on or alternative therapeutic regimen in treating chronic inflammatory disorders [98].

23.4.2 Acupuncture and Traditional Chinese Medicine

Acupuncture and Traditional Chinese Medicine (TCM) are ancient healing practices that have gained increasing interest due to their effectiveness in managing numerous inflammatory conditions [99]. Their acceptance is based on an integrative approach, which aims to balance inherent body imbalances rather than just treating external symptoms [100]. In TCM, health is viewed as the result of the dynamic interaction of fundamental elements like Qi, Blood, Yin, and Yang, and the free flow of energy through meridian pathways [101]. From this perspective, inflammatory diseases may occur because of pathogenic factors, such as "wind," "heat," "cold," "dampness," or "dryness," disrupting the body's internal balance and defense mechanisms [102]. This model contrasts with the Western biomedical approach, which focuses on molecular and cellular processes of inflammation, including cytokine signaling pathways and immune cell activation [103].

Acupuncture, an essential part of TCM, involves inserting fine needles into specific acupoints to stimulate sensory nerves and modulate physiological responses such as inflammation [104]. This neurostimulation is believed to activate brain pathways that control pain and immune responses, leading to the release of endorphins and prostaglandins, substances with pain-relieving and anti-inflammatory effects [105]. Its healing potential is based on its ability to induce neurochemical changes, including altering neurotransmitter levels and impacting the immune system [106]. Specifically, stimulation of acupoints has been shown to trigger the release of endogenous opioids responsible for analgesic and anti-inflammatory effects [107]. Additionally, acupuncture can modulate inflammatory processes by decreasing pro-inflammatory cytokines like TNF-α and IL-1, and increasing anti-inflammatory mediators, supporting its use as a nondrug approach for managing chronic inflammation [108].

23.4.3 Mind–Body Practices

Mind–body interventions use brain, body, and behavior interactions to boost physical and mental well-being. Methods like meditation, yoga, tai chi, and guided imagery involve cognitive, emotional, social, and spiritual elements to enhance health [109]. Such practices are believed to influence the body's systems, including the hypothalamic-pituitary-adrenal axis, autonomic nervous system, and immune mechanisms, thereby modulating inflammation and oxidative stress [110]. Frequent mind–body practice can reduce pro-inflammatory cytokines and increase anti-inflammatory mediators, aiding remission of chronic inflammatory diseases linked to oxidative stress [111]. These exercises also impact neuroendocrine function, the autonomic nervous system and stress hormones, which influence immune responses and improve health behaviors like nutrition, sleep, and emotional regulation [112].

This neurobiological dynamic interaction has been considered to underlie the improvement in immune function and psychological well-being seen with regular chronic practice of mind–body exercises [113]. In addition to modulating immune and stress pathways, these interventions may also enhance endogenous antioxidant protection and thereby facilitate redox homeostasis, and mitigate the toxic action of high ROS levels [114]. These physiological gains are often coupled with the acquisition of increases in self-efficacy and psychological resilience, both of which are requisite for effective self-management of chronic illnesses [115] and with enhanced emotional regulation [116]. By increasing an enhanced physiological and psychological state, mind–body practices may reduce systemic oxidative stress markedly, presenting a potentially valuable complementary therapeutic strategy for ROS-related disorders [117, 118].

23.5 Dietary Supplements and Nutraceuticals

Most supplements and nutraceuticals provide therapeutic effects mainly through antioxidant and anti-inflammatory mechanisms, counteracting ROS involved in chronic diseases [119]. These compounds reinforce the body's antioxidant system or detoxify ROS, restoring redox balance by supplying cofactors for antioxidant enzymes or inducing antioxidant gene expression [120]. This dual action reduces oxidative stress, a key factor in tissue damage and inflammation [121]. Diets like ketogenic and Mediterranean also influence oxidative stress, lipid metabolism, and inflammation markers [122]. Besides antioxidant roles, many supplements also have anti-inflammatory effects by modulating cellular pathways, downregulating cytokines, chemokines, and enzymes to suppress inflammation. Natural molecules like vitamins, minerals, polyphenols, and lipids often target ROS-mediated inflammation through synergistic, multi-targeted actions [123].

23.5.1 Vitamins

Micronutrients like vitamins C and E act as potent antioxidants through their ability to directly neutralize free radicals and prevent cellular damage, thereby reducing oxidative stress. Additionally, various B-complex vitamins play a role in metabolic processes that support the body's natural antioxidant defenses [124]. Vitamins A, C, D, and E are among the most well-studied antioxidants and have been linked to a lower risk of cardiovascular diseases in numerous extensive observational studies [125–127]. However, clinical trials on vitamin supplementation for primary and secondary prevention of cardiovascular disease have yielded mixed results, highlighting the complexity of nutrient interactions and individual differences in treatment response. Nonetheless, preclinical research consistently shows that polyphenol-based interventions reduce oxidative stress and inflammation markers, underscoring their potential as therapeutic agents [128].

23.5.2 Minerals

Major minerals like selenium, zinc, and copper serve as essential cofactors for various endogenous antioxidant enzymes, playing a key role in maintaining cell redox balance and controlling oxidative damage. For instance, zinc is indispensable for the catalytic activity of superoxide dismutase, an enzyme that detoxifies superoxide radicals [129]. Selenium is vital for glutathione peroxidases, enzymes that reduce hydrogen peroxide and organic hydroperoxides to water and alcohol, preventing lipid peroxidation and protecting cellular components and DNA [130]. Although magnesium does not directly participate in antioxidant enzyme reactions, it helps maintain cell integrity and exhibits anti-inflammatory effects that indirectly enhance protection against ROS damage. Iron, essential for oxygen transport and metabolism, can both support and induce oxidative stress: unchelated iron can generate highly reactive hydroxyl radicals via Fenton chemistry [131]. The interconnection of these trace elements underscores their crucial role in redox regulation and highlights the importance of mineral-rich diets in reducing oxidative stress and inflammation [132].

23.5.3 Omega-3 Fatty Acids

Omega-3 polyunsaturated fatty acids (PUFAs), especially eicosapentaenoic acid (EPA) and docosahexaenoic acid (DHA) from fish oils, are widely recognized for their potent anti-inflammatory and antioxidant properties. They help regulate ROS-driven inflammatory responses [133]. These fatty acids become part of cell membranes, affecting membrane fluidity and influencing signaling pathways related to inflammation and oxidative stress. Additionally, EPA and DHA serve as precursors to specialized pro-resolving mediators (SPMs) like resolvin and protectin, which actively promote the resolution of inflammation rather than just suppressing it [134]. This active process distinguishes omega-3 PUFAs from other anti-inflammatory agents and implies their ongoing benefit in managing chronic inflammatory conditions. Overall, these mechanisms suggest omega-3 fatty acids could be effective therapeutics for reducing ROS-induced inflammation by modulating redox balance and key inflammatory pathways. Their ability to lower pro-inflammatory gene expression and boost antioxidant defenses makes them promising dietary options for preventing and treating oxidative stress-related cellular damage and chronic inflammation [135].

23.5.4 Probiotics and Postbiotics

The rapidly evolving fields of probiotics and their derivatives of paraprobiotics and postbiotic metabolites attract considerable scientific attention because of their ability to alter the composition of the gut microbiota and their function, thereby influencing systemic responses related to inflammation and oxidative stress through the gut–brain and gut–immune axes [136]. Some probiotic strains can increase the production of short-chain fatty acids (SCFAs), which are acknowledged for their anti-inflammatory effects and help maintain intestinal barrier integrity by reducing permeability, which prevents pro-inflammatory microbial products from entering systemic circulation. Besides live microbes, paraprobiotics of inactivated probiotic microbial cells and postbiotics bioactive metabolites produced by probiotic cells, offer a stable, effective alternative that provides health benefits regardless of microbial viability [137]. Postbiotic metabolite treatments have been linked to immune regulation and improved antioxidant defenses.

For instance, probiotic strains like *Lactobacillus acidophilus* and *Bifidobacterium bifidum* can significantly lower pro-inflammatory cytokine levels and boost antioxidant capacity, evidenced by increased serum glutathione peroxidase activity and decreased lipid peroxidation [138, 139]. Probiotic strains *Lactobacillus casei* and *Lactiplantibacillus plantarum* inhibit myeloperoxidase activity, reducing neutrophil activation and inflammation [140, 141]. Manipulating the gut microbiome boosts intestinal health and offers systemic benefits by modulating immune responses and lowering oxidative stress. Supplementing with specific probiotics increases enzymes related to antioxidation, such as serum glutathione peroxidase, catalase, and superoxide dismutase, supporting internal defenses [142]. Restoring microbial balance with probiotics such as *Faecalibacterium prausnitzii*, *Propionibacterium freudenreichii*, and *Christensenella minuta* shows anti-inflammatory effects in preclinical models [143, 144]. Human evidence suggests some probiotic strains can reduce anxiety, lower stress, and enhance mood by connecting gut microbiota to neurological health through the gut–brain axis [145].

Postbiotic metabolites are therapeutic key bioactive compounds with potential in chronic inflammation management, notably IBD and IBS, which are typically characterized by recurrent inflammation, oxidative stress, and intestinal barrier function impairment [146]. Non-viable microbial metabolites, bioactive compounds like exopolysaccharides (EPS), show relevant anti-inflammatory and antioxidant properties that counteract the pathological processes that perpetuate these ailments [147]. For instance, *Lactiplantibacillus plantarum* was reported to produce EPS possessing good antioxidant activity that neutralizes ROS and prevents lipid peroxidation and, consequently, reduces oxidative stress-induced inflammation [148, 149]. Extensive research has also been conducted on the postbiotic metabolites [150] generated by these *L. plantarum* strains for their potential benefits in human and animal health [151–155]. Notably, these metabolites exhibit broad-spectrum inhibitory activity against a range of pathogens [156, 157]. Addition of such postbiotic metabolites offers a goal-oriented solution for enhancing gut inflammation, restoring the integrity of the intestinal barrier and alleviating oxidative damage. These sophisticated effects demonstrate the therapeutic potential of postbiotic metabolites as natural and efficient treatments for the prevention and management of inflammatory gastrointestinal diseases like IBD and IBS [158].

References

1. Sies H, Jones DP (2020) Reactive oxygen species (ROS) as pleiotropic physiological signalling agents. Nat Rev Mol Cell Biol 21(7):363–383
2. Mittal M, Siddiqui MR, Tran K, Reddy SP, Malik AB (2014) Reactive oxygen species in inflammation and tissue injury. Antioxid Redox Signal 20(7):1126–1167
3. Forrester SJ, Kikuchi DS, Hernandes MS, Xu Q, Griendling KK (2018) Reactive oxygen species in metabolic and inflammatory signaling. Circ Res 122(6):877–902

4. Medzhitov R (2021) The spectrum of inflammatory responses. Science 374(6571):1070–1075
5. Chen L, Deng H, Cui H, Fang J, Zuo Z, Deng J et al (2018) Inflammatory responses and inflammation-associated diseases in organs. Oncotarget 9(6):7204–7218
6. Reuter S, Gupta SC, Chaturvedi MM, Aggarwal BB (2010) Oxidative stress, inflammation, and cancer: how are they linked? Free Radic Biol Med 49(11):1603–1616
7. Gupta SC, Patchva S, Aggarwal BB (2013) Therapeutic roles of curcumin: lessons learned from clinical trials. AAPS J 15(1):195–218
8. Weber WJ, Hopp DC (2020) National center for complementary and integrative health perspectives on clinical research involving natural products. Drug Metab Dispos 48(10):963–965
9. Nair PP, Krishnakumar V, Nair PG (2024) Chronic inflammation: cross linking insights from Ayurvedic sciences, a silver lining to systems biology and personalized medicine. J Ayurveda Integr Med 15(4):101016
10. Spagnuolo C, Moccia S, Russo GL (2018) Anti-inflammatory effects of flavonoids in neurodegenerative disorders. Eur J Med Chem 153:105–115
11. Long AF (2014) Complementary and alternative medicine (CAM) and the public health: an innovative healthcare practice in supporting and sustaining health and well-being. Epidemiol Open Access 04(01). https://doi.org/10.4172/2161-1165.1000141
12. Khodr B, Khalil Z (2001) Modulation of inflammation by reactive oxygen species: implications for aging and tissue repair. Free Radic Biol Med 30(1):1–8
13. Rodrigues de Queiroz A, Hines C, Brown J, Sahay S, Vijayan J, Stone JM et al (2023) The effects of exogenously applied antioxidants on plant growth and resilience. Phytochem Rev 22(2):407–447
14. Zhao R, Jiang S, Zhang L, Yu Z (2019) Mitochondrial electron transport chain, ROS generation and uncoupling (review). Int J Mol Med 44:3–15
15. Bardaweel SK, Gul M, Alzweiri M, Ishaqat A, HA ALS, Bashatwah RM (2018) Reactive oxygen species: the dual role in physiological and pathological conditions of the human body. Eurasian J Med 50(3):193–201
16. Lassègue B, San Martín A, Griendling KK (2012) Biochemistry, physiology, and pathophysiology of NADPH oxidases in the cardiovascular system. Circ Res 110(10):1364–1390
17. Manoharan RR, Prasad A, Pospíšil P, Kzhyshkowska J (2024) ROS signaling in innate immunity via oxidative protein modifications. Front Immunol 15:1359600
18. Gupta SC, Hevia D, Patchva S, Park B, Koh W, Aggarwal BB (2012) Upsides and downsides of reactive oxygen species for cancer: the roles of reactive oxygen species in tumorigenesis, prevention, and therapy. Antioxid Redox Signal 16(11):1295–1322
19. Wang L, Gao C, Wang B, Wang C, Sagada G, Yan Y (2023) Methionine in fish health and nutrition: potential mechanisms, affecting factors, and future perspectives. Aquaculture 568:739310
20. Sies H (2015) Oxidative stress: a concept in redox biology and medicine. Redox Biol 4:180–183
21. Schieber M, Chandel NS (2014) ROS function in redox signaling and oxidative stress. Curr Biol 24(10):R453–R462
22. Checa J, Aran JM (2020) Reactive oxygen species: drivers of physiological and pathological processes. J Inflamm Res 13:1057–1073
23. Sies H (2007) Biological redox systems and oxidative stress. Cell Mol Life Sci 64(17):2181–2188
24. Kim GY, Jeong H, Yoon HY, Yoo HM, Lee JY, Park SH et al (2020) Anti-inflammatory mechanisms of suppressors of cytokine signaling target ROS via NRF-2/thioredoxin induction and inflammasome activation in macrophages. BMB Rep 53(12):640–645
25. Umeno A, Biju V, Yoshida Y (2017) *In vivo* ROS production and use of oxidative stress-derived biomarkers to detect the onset of diseases such as Alzheimer's disease, Parkinson's disease, and diabetes. Free Radic Res 51(4):413–427
26. Novo E, Parola M (2008) Redox mechanisms in hepatic chronic wound healing and fibrogenesis. Fibrogenesis Tissue Repair 1(1):5

27. Herb M, Gluschko A, Schramm M (2021) Reactive oxygen species: not omnipresent but important in many locations. Front Cell Dev Biol 9:716406
28. Lai J l, Liu Y h, Liu C, Qi M p, Liu R n, Zhu X f et al (2017) Indirubin inhibits LPS-induced inflammation via TLR4 abrogation mediated by the NF-kB and MAPK signaling pathways. Inflammation 40(1):1–12
29. Bach A (2017) Targeting oxidative stress in stroke. In: Neuroprotective therapy for stroke and ischemic disease. Cham: Springer International Publishing, pp 203–250
30. Ichiishi E, Li XK, Iorio EL (2016) Oxidative stress and diseases: clinical trials and approaches. Oxid Med Cell Longev 2016(1):3458276
31. Barnes RP, Fouquerel E, Opresko PL (2019) The impact of oxidative DNA damage and stress on telomere homeostasis. Mech Ageing Dev 177:37–45
32. Phillips WA, Lord RV, Nancarrow DJ, Watson DI, Whiteman DC (2011) Barrett's esophagus. J Gastroenterol Hepatol 26(4):639–648
33. Wetscher GJ, Perdikis G, Kretchmar DH, Stinson RG, Bagchi D, Redmond EJ et al (1995) Esophagitis in sprague-dawley rats is mediated by free radicals. Dig Dis Sci 40(6):1297–1305
34. Farhadi A, Fields J, Banan A, Keshavarzian A (2002) Reactive oxygen species: are they involved in the pathogenesis of GERD, barrett's esophagus, and the latter's progression toward esophageal cancer? Am J Gastroenterol 97(1):22–26
35. Incalza MA, D'Oria R, Natalicchio A, Perrini S, Laviola L, Giorgino F (2018) Oxidative stress and reactive oxygen species in endothelial dysfunction associated with cardiovascular and metabolic diseases. Vasc Pharmacol 100:1–19
36. Shaito A, Aramouni K, Assaf R, Parenti A, Orekhov A, Yazbi A El et al (2022) Oxidative stress-induced endothelial dysfunction in cardiovascular diseases. Front Biosci(Landmark Ed) 27(3):105
37. Chatterjee S (2016) Oxidative stress, inflammation, and disease. In: Oxidative stress and biomaterials. Elsevier, pp 35–58
38. Caetano MAF, Castelucci P (2022) Role of short chain fatty acids in gut health and possible therapeutic approaches in inflammatory bowel diseases. World J Clin Cases 10(28):9985–10003
39. Bhatt S, Puli L, Patil CR (2021) Role of reactive oxygen species in the progression of Alzheimer's disease. Drug Discov Today 26(3):794–803
40. Dumont M, Beal MF (2011) Neuroprotective strategies involving ROS in Alzheimer disease. Free Radic Biol Med 51(5):1014–1026
41. Weng M, Xie X, Liu C, Lim KL, Zhang C, wu, Li L (2018) The sources of reactive oxygen species and its possible role in the pathogenesis of Parkinson's disease. Parkinsons Dis 2018:1–9
42. Dias V, Junn E, Mouradian MM (2013) The role of oxidative stress in Parkinson's disease. J Parkinsons Dis 3(4):461–491
43. Gandhi S, Abramov AY (2012) Mechanism of oxidative stress in neurodegeneration. Oxid Med Cell Longev 2012:1–11
44. Zuo L, Wijegunawardana D (2021) Redox role of ROS and inflammation in pulmonary diseases. Adv Exp Med Biol 1304:187–204
45. Galvão I, Sugimoto MA, Vago JP, Machado MG, Sousa LP (2018) Mediators of inflammation. In: Immunopharmacology and inflammation. Springer International Publishing, Cham, pp 3–32
46. Chaiswing L, St. Clair WH, St. Clair DK (2018) Redox paradox: a novel approach to therapeutics-resistant cancer. Antioxid Redox Signal 29(13):1237–1272
47. Letafati A, Taghiabadi Z, Zafarian N, Tajdini R, Mondeali M, Aboofazeli A et al (2024) Emerging paradigms: unmasking the role of oxidative stress in HPV-induced carcinogenesis. Infect Agent Cancer 19(1):30
48. Fatma H, Jameel M, Siddique HR (2023) An update on phytochemicals in redox homeostasis: "virtuous or evil" in cancer chemoprevention? Chemistry (Easton) 5(1):201–222

49. Ray PD, Huang BW, Tsuji Y (2012) Reactive oxygen species (ROS) homeostasis and redox regulation in cellular signaling. Cell Signal 24(5):981–990
50. Ma Y, Zhang L, Rong S, Qu H, Zhang Y, Chang D et al (2013) Relation between gastric cancer and protein oxidation, DNA damage, and lipid peroxidation. Oxid Med Cell Longev 2013:1–6
51. Hileman EA, Achanta G, Huang P (2001) Superoxide dismutase: an emerging target for cancer therapeutics. Expert Opin Ther Targets 5(6):697–710
52. Ogrunc M (2014) Reactive oxygen species: the good, the bad, and the enigma. Mol Cell Oncol 1(3):e964033
53. Panth N, Paudel KR, Parajuli K (2016) Reactive oxygen species: a key hallmark of cardiovascular disease. Adv Med 2016:1–12
54. Daiber A, Kuntic M, Oelze M, Hahad O, Münzel T (2023) E-cigarette effects on vascular function in animals and humans. Pflugers Arch 475(7):783–796
55. McGuinness A, Sapey E (2017) Oxidative stress in COPD: sources, markers, and potential mechanisms. J Clin Med 6(2):21
56. Barnes PJ (2020) Oxidative stress-based therapeutics in COPD. Redox Biol 33:101544
57. Sahiner UM, Birben E, Erzurum S, Sackesen C, Kalayci Ö (2018) Oxidative stress in asthma: part of the puzzle. Pediatr Allergy Immunol 29(8):789–800
58. Qu J, Li Y, Zhong W, Gao P, Hu C (2017) Recent developments in the role of reactive oxygen species in allergic asthma. J Thorac Dis 9(1):E32–E43
59. Zhu H, Li YR (2012) Oxidative stress and redox signaling mechanisms of inflammatory bowel disease: updated experimental and clinical evidence. Exp Biol Med 237(5):474–480
60. Bronsart L, Nguyen L, Habtezion A, Contag C (2016) Reactive oxygen species imaging in a mouse model of inflammatory bowel disease. Mol Imaging Biol 18(4):473–478
61. Phull AR, Nasir B, Haq I u, Kim SJ (2018) Oxidative stress, consequences and ROS mediated cellular signaling in rheumatoid arthritis. Chem Biol Interact 281:121–136
62. Patil RT, Dhadse PV, Salian SS, Punse SD (2024) Role of oxidative stress in periodontal diseases. Cureus 16(5):e60779
63. Hoffmann MH, Griffiths HR (2018) The dual role of reactive oxygen species in autoimmune and inflammatory diseases: evidence from preclinical models. Free Radic Biol Med 125:62–71
64. Shabab T, Khanabdali R, Moghadamtousi SZ, Kadir HA, Mohan G (2017) Neuroinflammation pathways: a general review. Int J Neurosci 127(7):624–633
65. Simpson DSA, Oliver PL (2020) ROS generation in microglia: understanding oxidative stress and inflammation in neurodegenerative disease. Antioxidants 9(8):743
66. Sorolla MA, Hidalgo I, Sorolla A, Montal R, Pallisé O, Salud A et al (2021) Microenvironmental reactive oxygen species in colorectal cancer: involved processes and therapeutic opportunities. Cancers (Basel) 13(20):5037
67. ArulJothi KN, Kumaran K, Senthil S, Nidhu AB, Munaff N, Janitri VB et al (2022) Implications of reactive oxygen species in lung cancer and exploiting it for therapeutic interventions. Med Oncol 40(1):43
68. Wang Y, Xu J, Fu Z, Zhang R, Zhu W, Zhao Q et al (2024) The role of reactive oxygen species in gastric cancer. Cancer Biol Med 21:740–753
69. Xing L, Tang Y, Li L, Tao X (2023) ROS in hepatocellular carcinoma: what we know. Arch Biochem Biophys 744:109699
70. Oshi M, Gandhi S, Yan L, Tokumaru Y, Wu R, Yamada A et al (2022) Abundance of reactive oxygen species (ROS) is associated with tumor aggressiveness, immune response, and worse survival in breast cancer. Breast Cancer Res Treat 194(2):231–241
71. Shahjalal M, Chakma SK, Ahmed T, Yasmin I, Mahumud RA, Hossain A (2022) Prevalence and determinants of using complementary and alternative medicine for the treatment of chronic illnesses: a multicenter study in Bangladesh. PLoS One 17(1):e0262221
72. Ozioma Ezekwesili-Ofili J, Okechukwu Ogbonna A (2022) African traditional and complementary therapies. In: Complementary therapies. IntechOpen

73. de R SM (2012) Complementary therapies—considerations before recommend, tolerate or proscribe them. In: Complementary therapies for the contemporary healthcare. InTech
74. Chambliss LR (2001) Alternative and complementary medicine: an overview. Clin Obstet Gynecol 44(4):640–652
75. Naja F, Fadel RA, Alameddine M, Aridi Y, Zarif A, Hariri D et al (2015) Complementary and alternative medicine use and its association with quality of life among Lebanese breast cancer patients: a cross-sectional study. BMC Complement Altern Med 15(1):444
76. Tangkiatkumjai M, Boardman H, Walker DM (2020) Potential factors that influence usage of complementary and alternative medicine worldwide: a systematic review. BMC Complement Med Ther 20(1):363
77. Youn BY, Moon S, Mok K, Cheon C, Ko Y, Park S et al (2022) Use of traditional, complementary and alternative medicine in nine countries: a cross-sectional multinational survey. Complement Ther Med 71:102889
78. Michalczyk K, Pawlik J, Czekawy I, Kozłowski M, Cymbaluk-Płoska A (2021) Complementary methods in cancer treatment—cure or curse? Int J Environ Res Public Health 18(1):356
79. Smith JM, John Sullivan S, David Baxter G (2011) Complementary and alternative medicine: contemporary trends and issues. Phys Ther Rev 16(2):91–95
80. Fouladbakhsh JM, Stommel M (2007) Using the behavioral model for complementary and alternative medicine: the CAM healthcare model. J Complement Integr Med 4(1). https://doi.org/10.2202/1553-3840.1035
81. Shin S, Chen S, Xie K, Duhun SA, Ortiz-Cerda T (2025) Evaluating the anti-inflammatory and antioxidant efficacy of complementary and alternative medicines (CAM) used for management of inflammatory bowel disease: a comprehensive review. Redox Rep 30(1):2471737
82. Gu P, Feagins LA (2019) Dining with inflammatory bowel disease: a review of the literature on diet in the pathogenesis and management of IBD. Inflamm Bowel Dis 26:181
83. Che CT, Wang Z, Chow M, Lam C (2013) Herb-Herb combination for therapeutic enhancement and advancement: theory, practice and future perspectives. Molecules 18(5):5125–5141
84. Falci L, Shi Z, Greenlee H (2016) Multiple chronic conditions and use of complementary and alternative medicine among US adults: results from the 2012 National Health Interview Survey. Prev Chronic Dis 13:150501
85. Yeganeh PR, Leahy J, Spahis S, Patey N, Desjardins Y, Roy D et al (2018) Apple peel polyphenols reduce mitochondrial dysfunction in mice with DSS-induced ulcerative colitis. J Nutr Biochem 57:56–66
86. Rahaman MM, Rakib A, Mitra S, Tareq AM, Emran TB, Shahid-Ud-Daula AFM et al (2020) The genus curcuma and inflammation: overview of the pharmacological perspectives. Plants 10(1):63
87. Lv W, Zou K, Alouk I, Li X, Chen W, Miao S et al (2025) Unlocking curcumin's revolutionary: improvement of stability and elderly digestion by soybean oil bodies and soybean protein-chitosan complex based pickering emulsion. Int J Biol Macromol 284:138052
88. Pandey K (2018) Invited article herbs that heal: natural panacea for health promotion. Ann Phytomed Int J 7(2):6–12
89. Khan N, Mukhtar H (2007) Tea polyphenols for health promotion. Life Sci 81(7):519–533
90. Valdés-González JA, Sánchez M, Moratilla-Rivera I, Iglesias I, Gómez-Serranillos MP (2023) Immunomodulatory, anti-inflammatory, and anti-cancer properties of ginseng: a pharmacological update. Molecules 28(9):3863
91. Agarwal R, Diwanay S, Patki P, Patwardhan B (1999) Studies on immunomodulatory activity of Withania somnifera (Ashwagandha) extracts in experimental immune inflammation. J Ethnopharmacol 67(1):27–35
92. Sabina EP, Rasool M, Mathew L, EzilRani P, Indu H (2010) 6-shogaol inhibits monosodium urate crystal-induced inflammation—an in vivo and in vitro study. Food Chem Toxicol 48(1):229–235

93. Butt MS, Sultan MT (2011) Ginger and its health claims: molecular aspects. Crit Rev Food Sci Nutr 51(5):383–393
94. Ammon HPT (2010) Modulation of the immune system by Boswellia serrata extracts and boswellic acids. Phytomedicine 17(11):862–867
95. Umar S, Umar K, Sarwar AHMG, Khan A, Ahmad N, Ahmad S et al (2014) Boswellia serrata extract attenuates inflammatory mediators and oxidative stress in collagen induced arthritis. Phytomedicine 21(6):847–856
96. Nagarkatti P, Pandey R, Rieder SA, Hegde VL, Nagarkatti M (2009) Cannabinoids as novel anti-inflammatory drugs. Future Med Chem 1(7):1333–1349
97. Klein TW (2005) Cannabinoid-based drugs as anti-inflammatory therapeutics. Nat Rev Immunol 5(5):400–411
98. Yatoo MI, Gopalakrishnan A, Saxena A, Parray OR, Tufani NA, Chakraborty S et al (2018) Anti-inflammatory drugs and herbs with special emphasis on herbal medicines for countering inflammatory diseases and disorders – a review. Recent Pat Inflamm Allergy Drug Discov 12(1):39–58
99. Liu R, Luo Y, Ma J, Zhang Q, Sheng Y, Li J et al (2024) Traditional Chinese medicine for functional gastrointestinal disorders and inflammatory bowel disease: narrative review of the evidence and potential mechanisms involving the brain-gut axis. Front Pharmacol 15:1444922
100. Wang J, Zhao H, Shi K, Wang M (2023) Treatment of insomnia based on the mechanism of pathophysiology by acupuncture combined with herbal medicine: a review. Medicine 102(11):e33213
101. Li XT (2017) Introductory chapter: therapies based on kidney essence and Qi in Chinese medicine. In: Chinese medical therapies for diabetes, infertility, silicosis and the theoretical basis. InTech
102. Teppone M (2019) Medicine has always been "modern" and "scientific" from ancient times to the present day. J Integr Med 17(4):229–237
103. Ghasemian M, Owlia S, Owlia MB (2016) Review of anti-inflammatory herbal medicines. Adv Pharmacol Sci 2016:1–11
104. Chen B, Liu D, Li T, Zheng L, Lan L, Yang N et al (2023) Research hotspots and trends on acupuncture for anti-inflammation: a bibliometric analysis from 2011 to 2021. J Pain Res 16:1197–1217
105. Qin Z, Xiang K, Su DF, Sun Y, Liu X (2021) Activation of the cholinergic anti-inflammatory pathway as a novel therapeutic strategy for COVID-19. Front Immunol 11:595342
106. Huang J, Qin X, Cai X, Huang Y (2020) Effectiveness of acupuncture in the treatment of Parkinson's disease: an overview of systematic reviews. Front Neurol 11:917
107. Yu WY, Ma LX, Zhang Z, Mu JD, Sun TY, Tian Y et al (2021) Acupuncture for primary dysmenorrhea: a potential mechanism from an anti-inflammatory perspective. Evid Based Complement Alternat Med 2021:1–12
108. Kwon S, Lee B, Yeom M, Sur BJ, Kim M, Kim ST et al (2012) Modulatory effects of acupuncture on murine depression-like behavior following chronic systemic inflammation. Brain Res 1472:149–160
109. Bower JE, Irwin MR (2016) Mind–body therapies and control of inflammatory biology: a descriptive review. Brain Behav Immun 51:1–11
110. Moreno JJ (2024) Modulation of inflammatory response and pain by mind-body therapies as meditation. Brain Behav Immun Integr 5:100036
111. Hussain T, Tan B, Yin Y, Blachier F, Tossou MCB, Rahu N (2016) Oxidative stress and inflammation: what polyphenols can do for us? Oxid Med Cell Longev 2016:7432797
112. Greeson JM (2009) Mindfulness research update: 2008. Complement Health Pract Rev 14(1):10–18
113. Fang CY, Reibel DK, Longacre ML, Rosenzweig S, Campbell DE, Douglas SD (2010) Enhanced psychosocial well-being following participation in a mindfulness-based stress reduction program is associated with increased natural killer cell activity. J Altern Complement Med 16(5):531–538

114. Steven S, Frenis K, Oelze M, Kalinovic S, Kuntic M, Bayo Jimenez MT et al (2019) Vascular inflammation and oxidative stress: major triggers for cardiovascular disease. Oxida Med Cell Longev 2019:1–26
115. Muñoz-Vergara D, Burton W, Bain P, Rist PM, Khalsa SBS, Schreiber KL et al (2023) Understanding the dynamics of inflammatory mediators in response to mind-body movement therapies (MBMTs): a systematic review and meta-analysis of studies in healthy subjects. Brain Behav Immun Integr 2:100006
116. Grecucci A, Pappaianni E, Siugzdaite R, Theuninck A, Job R (2015) Mindful emotion regulation: exploring the neurocognitive mechanisms behind mindfulness. Biomed Res Int 2015:1–9
117. Burgoyne JR, Mongue-Din H, Eaton P, Shah AM (2012) Redox signaling in cardiac physiology and pathology. Circ Res 111(8):1091–1106
118. Scuto M, Rampulla F, Reali GM, Spanò SM, Trovato Salinaro A, Calabrese V (2024) Hormetic nutrition and redox regulation in gut–brain axis disorders. Antioxidants 13(4):484
119. Parian AM, Limketkai BN, Shah ND, Mullin GE (2015) Nutraceutical supplements for inflammatory bowel disease. Nutr Clin Pract 30(4):551–558
120. Verhoef MJ, Sutherland LR, Brkich L (1990) Use of alternative medicine by patients attending a gastroenterology clinic. CMAJ 142(2):121–125
121. Larussa T, Imeneo M, Luzza F (2017) Potential role of nutraceutical compounds in inflammatory bowel disease. World J Gastroenterol 23(14):2483
122. Lambadiari V, Katsimbri P, Kountouri A, Korakas E, Papathanasi A, Maratou E et al (2024) The effect of a ketogenic diet versus Mediterranean diet on clinical and biochemical markers of inflammation in patients with obesity and psoriatic arthritis: a randomized crossover trial. Int J Mol Sci 25(5):2475
123. Mijan M Al, Lim BO (2018) Diets, functional foods, and nutraceuticals as alternative therapies for inflammatory bowel disease: present status and future trends. World J Gastroenterol 24(25):2673–2685
124. van Herpen-broekmans WMR, IAA K-k, Michiel BL, Cornelis K, Hans P, Hendriks FJH et al (2004) Serum carotenoids and vitamins in relation to markers of endothelial. Eur J Epidemiol 19(10):915–921
125. Thurnham DI (2015) Inflammation and vitamin A. Food Nutr Bull 36(3):290–298
126. Singh U, Devaraj S, Jialal I (2005) Vitamin E, oxidative stress, and inflammation. Annu Rev Nutr 25(1):151–174
127. Guillot X, Semerano L, Saidenberg-Kermanac'h N, Falgarone G, Boissier MC (2010) Vitamin D and inflammation. Joint Bone Spine 77(6):552–557
128. Lee Y, Kim M, Choi K, Kim J, Bae W, Kim S et al (2011) Relationship between inflammation biomarkers, antioxidant vitamins, and bone mineral density in patients with metabolic syndrome. Nutr Res Pract 5(2):150
129. Hu P, Zhang S, Li H, Yan X, Zhang X, Zhang Q (2023) Association between dietary trace minerals and pelvic inflammatory disease: data from the 2015–2018 National Health and Nutrition Examination Surveys. Front Nutr 10:1273509
130. Ghishan FK, Kiela PR (2011) Advances in the understanding of mineral and bone metabolism in inflammatory bowel diseases. Am J Physiol Gastrointest Liver Physiol 300(2):G191–G201
131. Ghishan FK, Kiela PR (2017) Vitamins and minerals in inflammatory bowel disease. Gastroenterol Clin N Am 46(4):797–808
132. MLS B, Lima FS, Ganda EK, Foditsch C, EBS M, Machado VS et al (2014) Effect of trace mineral supplementation on selected minerals, energy metabolites, oxidative stress, and immune parameters and its association with uterine diseases in dairy cattle. J Dairy Sci 97(7):4281–4295
133. Simopoulos AP (2002) Omega-3 fatty acids in inflammation and autoimmune diseases. J Am Coll Nutr 21(6):495–505
134. Mori TA, Beilin LJ (2004) Omega-3 fatty acids and inflammation. Curr Atheroscler Rep 6(6):461–467

135. Calder PC (2010) Omega-3 fatty acids and inflammatory processes. Nutrients 2(3):355–374
136. Liu Y, Alookaran JJ, Rhoads JM (2018) Probiotics in autoimmune and inflammatory disorders. Nutrients 10(10):1537
137. Markowiak-Kopeć P, Śliżewska K (2020) The effect of probiotics on the production of short-chain fatty acids by human intestinal microbiome. Nutrients 12(4):1107
138. Li SC, Hsu WF, Chang JS, Shih CK (2019) Combination of Lactobacillus acidophilus and Bifidobacterium animalis subsp. Lactis shows a stronger anti-inflammatory effect than individual strains in HT-29 Cells. Nutrients 11(5):969
139. Peran L, Camuesco D, Comalada M, Bailon E, Henriksson A, Xaus J et al (2007) A comparative study of the preventative effects exerted by three probiotics, *Bifidobacterium lactis*, *Lactobacillus casei* and *Lactobacillus acidophilus*, in the TNBS model of rat colitis. J Appl Microbiol 103(4):836–844
140. Alipour B, Homayouni-Rad A, Vaghef-Mehrabany E, Sharif SK, Vaghef-Mehrabany L, Asghari-Jafarabadi M et al (2014) Effects of *Lactobacillus casei* supplementation on disease activity and inflammatory cytokines in rheumatoid arthritis patients: a randomized double-blind clinical trial. Int J Rheum Dis 17(5):519–527
141. Prete R, Dell'Orco F, Sabatini G, Montagano F, Battista N, Corsetti A (2024) Improving the antioxidant and anti-inflammatory activity of fermented milks with exopolysaccharides-producing Lactiplantibacillus plantarum strains. Foods 13(11):1663
142. Mack DR (2011) Probiotics in inflammatory bowel diseases and associated conditions. Nutrients 3(2):245–264
143. Lee NK, Jang HJ, Paik HD (2024) Non-lactic acid bacteria probiotics isolated from intestine or various circumstances. Food Sci Biotechnol 33(9):1997–2007
144. Ignatyeva O, Tolyneva D, Kovalyov A, Matkava L, Terekhov M, Kashtanova D et al (2024) Christensenella minuta, a new candidate next-generation probiotic: current evidence and future trajectories. Front Microbiol 14:1241259
145. Qi C, Wang P, Fu T, Lu M, Cai Y, Chen X et al (2021) A comprehensive review for gut microbes: technologies, interventions, metabolites and diseases. Brief Funct Genomics 20(1):42–60
146. Rafique N, Jan SY, Dar AH, Dash KK, Sarkar A, Shams R et al (2023) Promising bioactivities of postbiotics: a comprehensive review. J Agric Food Res 14:100708
147. Hosseinzadeh N, Asqardokht-Aliabadi A, Sarabi-Aghdam V, Hashemi N, Dogahi PR, Sarraf-Ov N et al (2025) Antioxidant properties of postbiotics: an overview on the analysis and evaluation methods. Probiotics Antimicrob Proteins 17(2):606–624
148. Yoon H, Ju J, Lee J, Park H, Lee J, Shin H et al (2013) The probiotic *Lactobacillus rhamnosus* BFE5264 and *Lactobacillus plantarum* NR74 promote cholesterol efflux and suppress inflammation in THP-1 cells. J Sci Food Agric 93(4):781–787
149. Wang J, Zhang W, Wang S, Wang Y, Chu X, Ji H (2021) *Lactobacillus plantarum* exhibits antioxidant and cytoprotective activities in porcine intestinal epithelial cells exposed to hydrogen peroxide. Oxid Med Cell Longev 2021(1):8936907
150. Danladi Y, Loh TC, Foo HL, Akit H, Md Tamrin NA, Naeem AM (2022) Effects of postbiotics and paraprobiotics as replacements for antibiotics on growth performance, carcass characteristics, small intestine histomorphology, immune status and hepatic growth gene expression in broiler chickens. Animals 12(7):917
151. Chang HM, Foo HL, Loh TC, Lim ETC, Abdul Mutalib NE (2021) Comparative studies of inhibitory and antioxidant activities, and organic acids compositions of postbiotics produced by probiotic Lactiplantibacillus plantarum strains isolated from Malaysian foods. Front Vet Sci 7:602280
152. Humam AM, Loh TC, Foo HL, Samsudin AA, Mustapha NM, Zulkifli I et al (2019) Effects of feeding different postbiotics produced by Lactobacillus plantarum on growth performance, carcass yield, intestinal morphology, gut microbiota composition, immune status, and growth gene expression in broilers under heat stress. Animals 9(9):644

153. Izuddin WI, Loh TC, Foo HL, Samsudin AA, Humam AM (2019) Postbiotic L. plantarum RG14 improves ruminal epithelium growth, immune status and upregulates the intestinal barrier function in post-weaning lambs. Sci Rep 9(1):9938
154. Thanh N, Chwen LT, HooiLing F, Hair-Bejo M, Azhar K (2010) Inhibitory activity of metabolites produced by strains Of Lactobacillus Plantarum isolated from Malaysian fermented food. Int J Probiotics Prebiotics 5(1):37–43
155. Thu TV, Loh TC, Foo HL, Yaakub H, Bejo MH (2011) Effects of liquid metabolite combinations produced by Lactobacillus plantarum on growth performance, faeces characteristics, intestinal morphology and diarrhoea incidence in postweaning piglets. Trop Anim Health Prod 43(1):69–75
156. Kareem KY, Hooi Ling F, Teck Chwen L, May Foong O, Anjas Asmara S (2014) Inhibitory activity of postbiotic produced by strains of Lactobacillus plantarum using reconstituted media supplemented with inulin. Gut Pathog 6(1):23
157. Lim YS (2003) Isolation of bacteriocinogenic lactic acid bacteria and purification of selected bacteriocins from traditional fermented foods. [Master's Thesis]. [Serdang]: Universiti Putra Malaysia, UPM Serdang, Malaysia
158. Żółkiewicz J, Marzec A, Ruszczyński M, Feleszko W (2020) Postbiotics—a step beyond pre- and probiotics. Nutrients 12(8):2189

Chapter 24
Clinical Perspectives on ROS-Mediated Inflammation in Metabolic Disorders: Clinical and Preclinical Study, Models, Etc.

Biplab Debnath, Rokeya Gazi, Ronit Deyashi, Mohammad Muztaba, Ansari Yaasir Ahmed, and Debajit Dewan

24.1 Introduction

ROS are chemically reactive molecules formed by the incomplete reduction of oxygen. They are naturally produced in cells and are essential for many biological processes [1]. When ROS (reactive oxygen species) levels are too high, they damage protein and fats in cells. This leads to problems in energy production, cell communication, and cell growth. It also weakens the immune system and causes inflammation. Unhealthy diets, especially those high in fat and sugar, increase oxidative stress by producing harmful molecules. This leads to obesity, which brings long-term oxidative stress and inflammation [2]. Obesity is a disease, where excessive fat is accumulated in the adipose tissue, leading to chronic health diseases. According to WHO, over one billion people are obese, including adults, adolescents, and children. The numbers are rising due to changes in lifestyle and diet. The COVID-19 pandemic has also increased obesity due to lockdowns and reduced physical activity [3]. Overproduction of ROS damage proteins, lipids, DNA, and cellular membranes. This contributes to many diseases, especially type 2 diabetes (T2DM). In diabetes, excess amount of blood sugar increase formation of cellular ROS. The ROS interferes with insulin signaling, leading to insulin resistance. Inflammation plays a key role in insulin resistance. Chronic low-grade inflammation is a silent, ongoing inflammation state. In obese or diabetic individuals, adipose tissue becomes inflamed and fat cells release inflammatory cytokines (e.g., TNF-ALPHA, IL-6,

B. Debnath · R. Gazi · R. Deyashi · D. Dewan (✉)
Department of Pharmaceutical Technology, Bharat Technology, Uluberia, West Bengal, India

M. Muztaba
Department of Pharmacology, Praduman Singh SPPS Pharmacy College, Basti, Uttar Pradesh, India

A. Y. Ahmed
Ali Allana College of Pharmacy, Akkalkuwa, Maharastra, India

S. Ashique, N. Mishra (eds.), *Reactive Oxygen Species in Metabolic Inflammation*,
https://doi.org/10.1007/978-3-032-11752-6_24

MCP-1). These cytokines attract immune cells like macrophage into the fat tissue, worsening inflammation. As a result, inflammatory molecules block insulin signaling pathways in the liver, muscle, and adipose tissue. This leads to systemic insulin resistance [4]. In insulin resistance and types 2 diabetes, ROS damages insulin signaling pathways and reduce GLUT4 (a glucose transporter), leading to high blood sugar [5]. Overproduction of ROS can also cause of nonalcoholic fatty liver disease. A type of cell death called pyroptosis involves cells swelling and bursting, releasing inflammatory substance like IL-1 beta and IL-18. This process is triggered by the NLRP3 inflammasome, which activates Caspase-1, leading to the production of GSDMD (Gasdermin D), a protein that forms holes in cell membranes, causing inflammation and cell death. Pyroptosis can worsen fat buildup in the liver [6]. Oxidative stress (OS) occurs when ROS levels exceed the body's antioxidant capacity. Oxidative stress and inflammation play central roles in the development of vascular diseases like atherosclerosis and heart complications. Vascular endothelium regulates blood vessels tone (via vasodilators like nitric oxide), blood clotting, inflammation, smooth muscle, and cell activity. Oxidative stress and inflammation give fuel to each other. This cycle causes nitric oxide inactivation, endothelial dysfunction, smooth muscle cell changes, plaque formations, and instability [7]. Many neurological disorders, such as Parkinson's, amyotrophic lateral sclerosis, Alzheimer's, multiple sclerosis, and ischemic stroke are associated with OS. In this chapter, we discuss different clinical and preclinical evidences associated with ROS generation and its associated metabolic diseases.

24.2 ROS and Inflammation: Molecular Basis

24.2.1 Generation and Types of ROS in Metabolic Tissues

Oxidative stress (OS) also termed as metabolic stress is defined as imbalance between formations of prooxidants and antioxidant molecules, resulting in damage of soft tissues, protein, lipids, and nucleic acids in the physiological system [8]. Most endogenous prooxidants contain oxygen and are referred to as reactive oxygen species (ROS). Other prooxidants may include nitrogen, or even halogens like chlorine, termed as RNS and RCS, respectively [9]. Reactive prooxidant species can be divided into two parts: (i) radical species ($O_2^{\bullet-}$, $OH^{\bullet}$, $RO_2^{\bullet}$, $NO^{\bullet}$, $NO_2^{\bullet}$, N_2O, and ClO^-), (ii) non-radical species (O_3, H_2O_2, HOCl, and $ONOO^-$) [10].

ROS can be produced as a by-product of cellular metabolism, enzymatic and nonenzymatic reactions, in mitochondrial matrix [11]. The electron transport chain (ETC) present in inner mitochondrial membrane (complex I to IV) basically helps in the formation of water molecules by reduction of oxygen [12]. In this process, the last step in complex IV, electrons react with molecular O_2 and form water. Nevertheless, there may be a "leakage" of O_2 in complexes I and III prior to electrons reaching complex IV, resulting in the production of $O_2^{\bullet-}$ rather than

H_2O. Superoxide radical ($O_2^{\bullet -}$) is produced in endothelial and inflammatory cells by the presence of lipoxygenases (LOX) and cyclooxygenases (COX) due to the metabolism of arachidonic acid [13]. The most powerful ROS oxidant is the hydroxyl radical OH•, which is produced in a very short time (10–9 s) by several reactions such as the Haber–Weiss reaction (A), the Fenton reaction (B), and decomposition of peroxynitrite (C), having high reactivity [14–16].The O_2 and ROS reduction are spontaneous reactions with a final product of water and with a negative Gibbs-free energy ($\Delta G^0 \leq 0$). A small percentage of the electrons in the ETC transport chain, in 0.1–2% of cases, passes along the chain to the mitochondrial matrix where they prematurely reduce molecular oxygen O_2 to superoxide ion, the precursor of most reactive oxygen species [17]. The endogenous and exogenous sources are mainly responsible for the formation of free radical. Endogenous free radical production is attributed to immune cell activation, inflammation, ischemia, infection, cancer, excessive exercise, mental stress, and aging. Whereas, environmental pollutants, heavy metals (such as Cd, Hg, Pb, Fe, and As), specific pharmaceuticals (including cyclosporine, tacrolimus, gentamycin, and bleomycin), chemical solvents, cooking methods, cigarette smoke, alcohol, and UV or X-ray radiation are the cause of exogenous free radical production. Exogenous compounds that enter the body undergo degradation or metabolism, resulting in the generation of free radicals as by-products [18].

ROS-mediated oxidative stress occurs at relatively moderate levels of O_2 and H_2O_2 in the mitochondria, constantly participating in redox signaling, in which there is a balance between the formation of ROS and the action of the antioxidant defense system. This leads to oxidative damage in cells, which can sharply increase in times of environmental stress and can impair the structural integrity of the cells. ROS cause three types of cell damage: (i) damage to DNA or RNA, (ii) lipid peroxidation of polyunsaturated fatty acids, including membrane phospholipids, and (iii) protein oxidation. The modified molecules lose their normal activities in the cytosol by irreversibly damaging DNA, lipids, and enzymes through oxidation [19–22]. Oxidative stress is another primary mediator of the senescence-associated secretary phenotype (SASP). Various studies reported that both telomere-dependent and telomere-independent DNA damage response (DDR) signaling promote the production of ROS and mitochondrial damage, leading to the promotion of SASP [23]. The mechanism of oxidative stress leading to different metabolic disorders is not properly clear but increasing ROS leads to cellular senescence, which is a physiological mechanism that blocks the cellular proliferation during replication [24]. Higher levels of oxidative stress generate different variables that are related with cellular damages, such as:

- The Regulation of mammalian rapamycin complex functions (mTOP pathway) [25].
- The formation of IL-1α creates a pro-inflammatory state, which increases nuclear factor kappa-B (NFκB) activity, resulting in epithelial-mesenchymal transition and tumor metastatic progression [25].

- The induction of the matrix metalloproteases (MMPs) expression, which is associated with age-related diseases such as atherosclerosis, cancer, osteoarthritis, Alzheimer's, and lung emphysema [25].
- Inhibiting Forkhead box (FOXO) proteins, which protect against oxidative damage mediated through the insulin/insulin-like growth factor (IGF-1) signaling cascade [25].
- Reduction in sarcoplasmic reticulum (SR), endoplasmic reticulum (ER) and calcium-dependent ATPase (Ca^{2+}-ATPase) activity leads to cardiac aging [26].
- Inhibits the activity of sirtuin, leading to (i) increased ROS formation due to inhibition of SOD; (ii) promotes anti-inflammatory state by inhibiting TNFα and NFκB; (iii) generates an oncogenic effect by inhibiting c-Jun and c-Myc [27].

Controlling of the $p16^{INK4a}$/pRB pathways, as well as p53/p21 pathways, leads to senescence [25]. Cellular senescence, oxidative stress, and SASP factors are responsible for damaging proteins, enzymes, lipids, DNAs, and RNAs, resulting in several acute and chronic disease, such as cardiovascular diseases, cancer, neurological diseases, diabetics, respiratory diseases, rheumatoid arthritis, kidney diseases, and sexual maturation. The production of an excess of $OH^{\bullet}$ ions and NO^{-} radical leads to the formation of malondialdehyde (MDA) and conjugated dienes, which are cytotoxic and mutagenic. This process is responsible for lipid peroxidation, impacting a significant number of lipid molecules [28]. Nishida et al. described that formation of 8-oxo-2′-deoxyguanosine (8-OHdG) is harmful for DNA mutagenesis due to impairment in CpG island [29, 30].

24.2.2 Link Between Mitochondrial Dysfunction (MD) and ROS Overproduction

The relationship between mitochondrial dysfunction (MD) and reactive oxygen (ROS) overproduction represents one of the most fundamental pathophysiological mechanisms underlying numerous diseases and aging processes. Mitochondria are simultaneously the primary sources of cellular ROS and the main targets of ROS-induced damage, creating a complex interplay that can lead to cellular dysfunction and death [31, 32]. Mitochondrial dysfunction (MD) refers to a failure in maintaining mitochondrial integrity, energy output, and cellular signaling. It can be caused directly via genetic mutations (primary mitochondrial diseases, PMDs) or be a secondary effect of others disease, such as types 2 diabetes, cardiovascular diseases, and neurodegeneration [33]. Defects in ETC complexes, particularly Complex I alterations, are the major sources of ROS generation in pathological conditions; impaired complex function leads to electron leakage to oxygen [31, 34]. MD often involves alterations in membrane potential. Hyperpolarization and depolarization increase ROS production by disrupting calcium handling, which in turn impairs antioxidant function [35, 36]. Metabolic intermediates like succinate can accumulate, leading to reverse electron transport through Complex I and massive ROS burst out [37]. Mitochondria rely heavily on glutathione (GSH) for ROS detoxification. MD decreases GSH level and impaired the function of glutathione peroxidase and reductase activities [38]. The superoxide dismutase 1 and superoxide dismutase 2 helps in converting superoxide to H_2O_2, and catalase and peroxiredoxins enzymes convert H_2O_2 into water. This function is disrupted due to mitochondrial dysfunctional. Hence, mitochondrial dysfunction is one of the leading causes of several metabolic disorders resulting in inflammation [39]. In this section we describe different metabolic disorders linked with mitochondrial dysfunction and ROS overproduction.

24.2.2.1 Cardiovascular Diseases

Cardiovascular diseases (CVDs) have multiple risk factors. Cholesterol, hypertension, smoking, diabetes, unbalanced nutrition, stress, and sedentary lifestyle contribute to CVDs [40–42]. Several studies suggest that oxidative stress may be the major or secondary cause of many CVDs [43]. Oxidative stress promotes atherosclerosis. It is widely accepted that early endothelial inflammation, which triggers local macrophages to produce ROS, is the cause of atheromatous plaque formation. Reactive oxygen species oxidize circulating LDL, forming foam cells and accumulating fat, resulting in the formation of an atherosclerotic plaque. Different in vivo and ex vivo studies demonstrate that oxidative stress is responsible for hypertension, cardiac hypertrophy, atherosclerosis, cardiomyopathy and congestive heart failure [44].

24.2.2.2 Neurological Disease

Oxidative stress is also associated with several neurological diseases such as Alzheimer's disease (AD), amyotrophic lateral sclerosis (ALS), and Parkinson's disease [45]. In case of Alzheimer's disease (AD), several researchers demonstrated that OS plays a vital role in neurological damages and leads to dementia [46]. β-amyloid is a toxic peptide often produced in AD patients' brain due to ROS and contributes to neurodegeneration during the disease's progression [47].

24.2.2.3 Respiratory Diseases

Numerous studies have shown that oxidative stress is connected to lung disorders like asthma and chronic obstructive pulmonary disease (COPD), which are characterized by systemic and local chronic inflammation [48]. Both cellular and environmental factors can contribute to the development of oxidative stress in COPD. The environmental oxidants such as, cigarette smoke stimulate macrophages and epithelial cells, which in turn causes the production of pro-inflammatory cytokines and chemokines that induce an immunological response. Releasing of protease degradation of connective tissues of lungs, leads to the development of emphysema or chronic bronchitis. Research studies have been demonstrated that excessive ROS generated by alveolar macrophages and polymorphonuclear leukocytes (PMNs) through oxidative bursts impede antiprotease mechanisms and hasten the deterioration of lung tissue [24]. OS is a crucial factor in the development of this illness in the elderly. The protein carbonylation has been demonstrated to alter the function of critical enzymes and structural proteins involved in muscle contractile performance [49].

24.2.2.4 Kidney Diseases

Oxidative stress plays a vital role in the development of chronic kidney disease (CKD) because it causes renal ischemia and glomerular damage, as well as indirectly through inflammation, hypertension, and endothelial dysfunction [50]. The CKD patients are in a chronic inflammatory state, which is characterized by the activation of PMNs and monocytes. This inflammatory cell increases the release of NADPH (nicotinamide adenine dinucleotide phosphate) oxidase and MPO (myeloperoxidase), which promote the generation of ROS [51]. During this phase, TNF-alpha and IL-1b play a key role as pro-inflammatory mediators, while NF-κB acts as a transcriptional factor to sustain inflammation. In the latter stage, TGF-beta production increases, leading to extracellular matrix synthesis. Hence, oxidative stress chronically stimulates kidney tissues, which causes inflammation and fibrotic tissue development, resulting in decreased organ function and renal failure [18]. Heavy metals (Cd, Hg, Pb, and As) and transition metals (Fe, Cu, Co, and Cr) cause oxidative stress, leading to nephropathy and certain malignancies [52].

24.2.2.5 Diabetics Mellitus

Diabetes mellitus (type 1 and 2) is a metabolic condition linked with higher levels of free radicals and lower antioxidant potential, leading to macrovascular and microvascular consequences. The specific process by which oxidative stress can accelerate the development of health issues in diabetes is only partially understood. Oxidative stress in type 2 diabetes (T2D) leads to prothrombotic responses, leading to cardiovascular disease [53]. The damage caused by diabetes can be considered as an oxidative tissue damaging effect of chronic hyperglycemia. Increased intracellular glucose causes ROS synthesis that surpassed the cell's antioxidant capacity to neutralize them [54]. ROS activate four key biochemical processes implicated in hyperglycemia-induced oxidative tissue: activation of protein kinase C (PKC), increased hexosamine pathway flux, increased AGEs, and increased polyol pathway flux [55]. Activation of the AGEs pathway can damage cells by altering gene transcription, signaling between cells, and binding blood proteins (e.g., albumin) to AGEs receptors on macrophages/mesangial cells. This leads to increased production of growth factors and pro-inflammatory cytokines [56]. The evidence above suggests that there is an enhanced oxidative stress in the elderly T2D patients, which can be partially combated by an improved antioxidant defense system in persons with impaired glucose tolerance (IGT) [57].

24.2.2.6 Nonalcoholic Fatty Liver (NAFLD)

NAFLD is an underappreciated and independent risk factor for atherosclerotic CVDs. NAFLD is a long-term liver condition where fat builds up in the liver without heavy alcohol use. It is becoming more common in China. People with NAFLD are more sensitive to liver damage caused by the drug acetaminophen (APAP). This is partly because their glutathione levels are lower, making them more vulnerable to oxidative stress. A type of cell death called pyroptosis has recently been studied. It involves cell swelling and bursting, releasing inflammatory substance like IL-1 beta and IL-18. This process is triggered by the NLRP3 inflammasome, which activities Caspase-1, leading to the production of GSDMD, protein that forms holes in cell membranes, causing inflammation. Pyroptosis can worsen fat buildup in the liver. The two-hit hypothesis of NAFLD, after fat accumulation (first hit) and a second hit like oxidative stress and inflammation, causes further damage. The thioredoxin (TRX) systems help control oxidative stress. But a protein called TXNIP blocks TRX and can activate NLRP3 when ROS levels are high, promoting pyroptosis. Caveolin-1 (CAV1) is a protein involved in fat metabolism and cell signaling. When CAV1 is missing or low, it can lead to more ROS, mitochondrial damage, and inflammation. It has been shown that CAV1 deficiency causes liver inflammation and injury. Activation of NLRP3 and CAV1 may help prevent pyroptosis by decreasing intracellular ROS levels [6].

24.2.2.7 Obesity

Obesity is a chronic medical condition characterized by excessive or abnormal accumulation of body fat that can impair health. Fat cells (adipocytes) enlarge and become inflamed. They produce more ROS due to stress and higher free fatty acid levels. Reactive oxygen species (ROS) promote NADPH oxidase activation while simultaneously reducing antioxidant enzyme levels. This shifts adipokine balance—decreases adiponectin (protective, anti-inflammatory) and increases TNF-alpha, IL-6, MCP-1 (pro-inflammatory cytokines). Obesity often leads to other chronic conditions, worsening health outcomes and increasing mortality. Like chronic kidney diseases (CKD), obesity leads to buildup in kidney, impaired filtration, and inflammation. Prevention includes weight control and healthy diets. Obesity causes cardiovascular disease by increasing blood pressure, bad cholesterol, vascular damage, and inflammation; the adipose tissue releases harmful substances that affect heart functions. Obesity also leads nonalcoholic fatty liver disease (NAFLD), caused by fat accumulation in the liver, strongly linked with T2D and obesity, which can progress to liver inflammation and cancer [5].

24.3 Preclinical Studies

24.3.1 Animal Models

Type 2 diabetes (T2D) and diet-induced obesity are very common metabolic disorders that can cause insulin resistance, hyperglycemia, and many other complications. It is very important to study the role of ROS and OS in the pathogenesis and progression of T2D. Excessive ROS production leads to destruction of cell homeostasis, impaired insulin secretion, and can cause diabetic complications such as endothelial dysfunction and inflammation. Animal models are primary tools for studying the pathophysiology of ROS-dependent type 2 diabetes and obesity [58–61]. Different animal models have been discussed below.

24.3.1.1 High-Fat Diet (HFD) Mice

High fat diet-induced obesity models are widely used to study metabolic dysfunction and ROS overproduction. In this model, the mice get high fat diet for 10–16 weeks before they are used in any experiment. In some cases, they are fed for only 4–8 weeks to get milder phenotypes for appropriate research [62]. The diet chart is discussed in Table 24.1.

24.3.1.2 Zucker Rats

These are genetically obese rats, best for studying early-onset obesity and ROS-related metabolic dysfunction. These rats show a mutation in the leptin receptor gene. This results in hyperphagia, hypometabolism, and progressive obesity. But unlike HFD mice, zucker rats are fed standard laboratory food, which help them to get a balanced nutrition. They get their 50–60 % calories from fat but they also follow a protein- and carbohydrate-rich diet. The feeding is continued for 2–16 weeks. Zucker rats develop obesity and diabetes naturally as they age. So, they are ideal for studies without any dietary manipulation [63, 64].

Table 24.1 The diet chart for high fat diet mice model

Ingredients	Amount (gram)
Casein	200
L-Cystine	3
Corn Starch	72.5
Sucrose	172.8
Cellulose	50
Soybean Oil	25
Lard Oil	177.5
Vitamin	10

24.3.1.3 Genetic Models (ob/ob, db/db Mice)

The ob/ob mice have a mutation in leptin gene resulting in leptin deficiency. They develop severe obesity and diabetes in an early age within 4–9 weeks. They are normally fed standard laboratory chow but they consume more than normal mice. Males eat 33% more and females eat about 25% more than usual. They are used as models in researches often at the age of 4–8 weeks. The db/db have a mutation leptin receptor that leads to leptin resistance. They are ideal for the study type 2 diabetes, insulin resistance, and hyperphagia. They develop obesity within 6–8 weeks. They are fed normal laboratory chow for 10–12 weeks depending upon the studies they are being used. These mouse models are used to study the role of ROS in genetic obesity and diabetes [63, 65].

24.3.1.4 Streptozotocin (STZ)-Induced Diabetes Models

Streptozotocin is a glucosamine-nitrosourea compound, used for creating both type 1 and type 2 diabetes models. STZ has structural similarity to glucose, so its uptake occurs via GLUT2 transporter in pancreatic β-cells. Inside the cell, STZ generates nitric oxide (NO) and ROS, which cause excessive alkylation and DNA fragmentation. This DNA damage triggers cell death, leading to severe insulin deficiency and hyperglycemia.

Single High-Dose Protocol

Doses of 150–200 mg/kg administered intraperitoneally induce diabetes within 48–72 h. This causes extensive β-cell destruction and severe hyperglycemia, leading to ROS overproduction and mitochondrial dysfunction.

Multiple Low-Dose Protocol

Daily injections of 40–60 mg/kg for 5–6 days helps to simulate autoimmune-mediated β-cell destruction. This helps in the study of oxidative stress and its role in β-cell dysfunction. This method is more relevant to human type 1 diabetes progression. Nicotinamide is used to partially protect β-cells from STZ toxicity, creating a type 2 diabetes-like phenotype. Nicotinamide prevents complete β-cell destruction but allows insulin resistance to develop (Table 24.2).

The mice are fasted for at least 4–8 h before STZ administration. This increases the effectiveness of STZ action. The STZ solution should be freshly prepared in sodium citrate buffer solution (pH = 4.5) before administration. Hyperglycemia may observe after 8–12 h after the administration [66].

Table 24.2 Different animal models used in T2D test and their outcome and disadvantages

Models	Disease	Test type	Test animals	Control animals	Outcomes	Disadvantages	References
HFD mice (C57BL/6J)	Diet-induced obesity	Glucose tolerance test, insulin tolerance test, body weight monitoring	$n = 8–12$ per group	$n = 8–12$ per group	Weight gain, glucose intolerance	Long induction time (generally 16–20 weeks)	[67]
HFD mice (C57BL/6J)	Type 2 diabetes	Glucose tolerance test, insulin tolerance test, HbA1c measurement	$n = 8–12$ per group	$n = 8–12$ per group	Hyper glycemia insulin resistance and β-cell dysfunction	Variable penetrance strain differences	[68]
Zucker rats (fa/fa)	Obesity	Glucose tolerance test, insulin tolerance test, hyperinsulinemic-euglycemic clamp studies	$n = 6–10$ per group	$n = 6–10$ per group	Obesity, hyperinsulinemia	Mild glucose intolerance	[69]
Zucker rats (fa/fa)	Type 2 diabetes	Glucose tolerance test, insulin tolerance test, hyperinsulinemic-euglycemic clamp studies	$n = 6–10$ per group	$n = 6–10$ per group	Mild glucose intolerance insulin resistance	Do not develop severe diabetes	[70]
Ob/ob mice	Obesity	Glucose tolerance test, insulin tolerance test, leptin gene therapy	$n = 6–10$ per group	$n = 6–10$ per group	Severe obesity hyperglycemia	Sterility leptin deficiency effects	[71]
Ob/ob mice	Type 2 diabetes	Glucose tolerance test, insulin tolerance test, leptin gene therapy	$n = 6–10$ per group	$n = 6–10$ per group	Insulin resistance glucose intolerance	Monogenic, not polygenic like humans	[71]
Db/db mice	Obesity	Glucose tolerance test, insulin tolerance test, HbA1c, pancreatic β-cell mass assessment	$n = 8–12$ per group	$n = 8–12$ per group	Morbid obesity. Hyperglycemia	Sterility requires heterozygous breeding	[72]

24.3.2 *In Vitro Models*

Cellular models allow precise investigation into several pathways linking ROS to cellular dysfunction.

24.3.2.1 3T3-L1 Cells

This cell line is used to study adipogenesis. Research shows that ROS level increases during differentiation, and acts as signaling molecules to promote adipogenesis through PPARγ. However, excess ROS are generated by treating mature adipocytes with agents like H_2O_2. This leads to impaired insulin signaling by activating stress-sensitive kinases (like JNK and p38 MAPK), which phosphorylate and inhibit insulin receptor substrate (IRS-1) [73].

HepG2 Cells

These cells are used to study hepatic steatosis (fatty liver). Treating HepG2 cells with excessive fatty acids or glucose induces mitochondrial ROS overproduction. This oxidative stress triggers inflammation, via activation of the NF-κB pathway, and apoptosis. It also contributes in the progression from simple steatosis to more severe liver injury [74] (Table 24.3).

24.3.2.2 ROS Inducers (H_2O_2, Tert-Butyl Hydroperoxide) and Antioxidants in Cell Studies

It is a widely used ROS inducer in both in vitro and in vivo models. It helps to study oxidative damage and cellular response like antioxidant defense system. The concentration of H_2O_2 ranges between 100 μM and 2 Mm. An exposure of 12–24 h to H_2O_2 triggers ROS overproduction in different cell lines usually. ROS induces lipid peroxidation and oxidative damage at higher concentration like ≥300–500 μM. But some differentiated adipocytes upregulate catalase, glutathione, thioredoxin-dependent systems to counteract the damage. H_2O_2 acts as a signaling molecule in early adipogenesis, but excessive ROS production can disrupt cell function [77]. This is an organic peroxide that is widely used as a ROS inducer in in vitro models to study oxidative stress, cellular antioxidant defense system, etc. The t-BHP produces ROS in the cells and cause oxidative damage to proteins and DNA. It also induces lipid peroxidation and leads to mitochondrial dysfunction. It induces apoptosis in some cells including hepatocytes and endothelial cells. The effect of this chemical is dose-dependent and can be measured by concentration of malondialdehyde in glutathione ratios and loss of the mitochondrial membrane potential. Concentration of t-BHP thus is typically used in the range of 100 μM–1 Mm for

Table 24.3 In vitro models to investigate several pathways linking ROS to cellular dysfunction

Cell line	Disease models studied	Type of test	Outcomes	Advantages	Disadvantages	Number of cell lines used	References
3t3-L1	Insulin resistance, obesity, diabetes, inflammation	Inflammatory cytokine expression ROS measurement differentiation assays	Increased ROS cause inflammation and increase adipocyte differentiation Upregulation of cytokines (TNF-A, IL-1B) Nutrient stress increases inflammatory response	Well characterized differentiation Inexpensive and easy to culture high differentiation efficiency	Mouse origin (not human) Cannot fully mimic human adipose tissue Homogeneous population limits tissue heterogeneity modeling	Typically, one subclone per study	[75]
Hepg2	Hepatocellular carcinoma Hepatoblastoma Liver metabolism toxicity	ROS production Viability assays Gene expression Drug metabolism	ROS elevation linked to mitochondrial dysfunction and cell death Oxidative stress marker changes (e.g., GSH, CAT, SOD1) Useful for cytotoxicity and metabolism studies	Human origin; Well-differentiated Secretes plasma proteins Suitable for liver metabolism and toxicity research	Tumor-derived (not normal hepatocytes) Low activity of some drug metabolizing enzymes Not fully representative of normal liver function	One cell line widely used in research	[76]

short-term exposures [78]. Antioxidants act as a protective barrier by neutralizing ROS and reducing oxidative damage. These compounds enhance insulin sensitivity and glucose metabolism by protecting β cells and stimulating insulin secretion. Some antioxidants like vitamin C, E, and α-lipoic acid help to reduce biomarkers of oxidative stress and improve insulin sensitivity [79].

24.3.3 Findings and Therapeutic Implications

Preclinical researches have established a clear mechanistic link between ROS over-production and metabolic disease progression.

24.3.3.1 Evidence of Oxidative Stress Biomarkers

OS in metabolic diseases is confirmed by measuring biomarkers in tissue and plasma. Examples of some common biomarkers are discussed below. In preclinical models, a wide range of antioxidant treatments (e.g., N-acetylcysteine, vitamins E and C, synthetic compounds) have successfully reduced markers of oxidative stress, decreased inflammation, and improved insulin sensitivity. Despite overwhelming success in animal models, results of large-scale human clinical trials with general antioxidants are disappointing and inconsistent in treating type 2 diabetes. This may be because ROS also play essential physiological roles. In addition, nonspecific antioxidant supplementation might interfere with these beneficial functions. Modern research is now focused on developing targeted antioxidant therapies that specifically act on pathological sources of ROS, like mitochondria-targeted antioxidants, to improve their efficacy and reduce off-target effects [80, 81].

24.4 Clinical Evidence

24.4.1 Observational Studies

24.4.1.1 Correlation of ROS Biomarkers (MDA, 8-OHdG) with Metabolic Parameters

A delicate balance exists between the production of reactive oxygen species and some metabolic parameters. This relationship is considered as a cornerstone cellular function and an important factor for developing numerous metabolic diseases. ROS biomarkers, which indicate the levels of these reactive oxygen species are the key indicators of metabolic dysregulation and some conditions like obesity, type 2 diabetes, and cardiovascular diseases [82].

Malondialdehyde (MDA). It is a very reactive three carbon dialdehyde that is a key biomarker of lipid peroxidation and oxidative cellular damage. It is the end-product of oxidation of polyunsaturated fatty acids by some reactive oxygen species. Malondialdehyde is teratogenic and cytotoxic also. This can lead to cellular damage and contribute to the pathogenesis of type 2 diabetes (T2D), cardiovascular disease, and other metabolic disorders [83, 84]. MDA is usually measured in the blood plasma, serum, and urine by thiobarbituric acid reactive substances (TBARS). MDA reacts with TBA under acidic condition to form a distinct pink color product (MDA-TBA_2), whose absorbance can be measured, typically around 532–535 nm. It is a less expensive method although more specific methods like HPLC, gas chromatography-mass spectrometry (GC-MS) are also used. Elevated MDA level is consistently observed in patients with obesity, metabolic disorders, and T2D. Plasma concentration is highly increased in obese patients. MDA levels are very high in T2D patients and also correlates positively with the disease, poor glycemic control (measured by HbA1c), and the presence of both micro- and macrovascular complications. Also, MDA levels show positive correlation with BODY MASS INDEX (BMI), total cholesterol, and LDL, but show negative correlation with HDL, firmly linking lipid peroxidation to the broader metabolic dysregulation seen in these conditions. Moreover, plasma MDA levels can predict endothelial dysfunction in hypertension and other cardiovascular diseases [85–88].

8-Hydroxy-2-deoxyguanosine (8-OHdG)- It is one of the most abundant and widely researched biomarkers of oxidative DNA damage. When some very highly reactive hydroxyl free radicals attack the c8 position of the guanine base of the DNA strand, it gives rise to 8-OHdG. It is further removed by the cellular base excision repair machinery and excreted by urine. Measuring urinary 8-OHdG provides a stable, noninvasive method to assess the rate of systemic, whole body oxidative damage [89, 90]. Measurement of 8-OHdG in urine, serum, or tissue DNA is performed using highly sensitive analytical techniques. Chromatographic methods coupled with mass spectroscopy is considered the best in terms of accuracy. But some immunological tests like ELISA are also used widely. High 8-OHdG levels are not only associated with aging and cancer but also increase the risk of several metabolic disease. It is observed that serum 8-OHdG is also very high in prediabetics, and are in highest levels in patients with severe diabetes. It indicates that oxidative DNA damage is an early event in diabetes progression. Researches show that there is a positive correlation between serum 8-OHdG and BMI in diabetic patients. In obese individuals, 8-OHdG levels correlate positively with BODY MASS INDEX, blood sugar (fasting), and triglycerides. Also, urinary 8-OHdG level is a very important marker in diabetic nephropathy [89, 91–94] (Table 24.4).

Table 24.4 Different biomarkers used for identification of metabolic disorders

Biomarker	Metabolic parameter	Direction of correlation	Significance and key findings	References
MDA	BMI	Positive (+)	Increased in obese patients. Positive correlation in T2D	[87]
	Glucose	Positive (+)	Indicates poor glycemic control	[94]
	HbA1c	Positive (+)	Significant positive correlation in T2DM	[87]
	Triglycerides	Positive (+)	Positive correlation with obesity	[94]
	LDL	Positive (+)	Positive correlation in obesity	[95]
	HDL	Negative (-)	Negative correlation in MetS	
8-OHdG	BMI	Positive (+)	Additive effect of obesity on DNA damage in T2D	[94]
	Glucose	Positive (+)	Increased risk of diabetes	[96]
	HbA1c	Positive (+)	Positive correlation in T2D patients	[94]
	Triglycerides	Positive (+) / weak positive	Weak correlation in diabetic nephropathy	
	LDL	No significant correlation	Not significantly associated with obesity	

24.4.1.2 Elevated Inflammatory Markers (CRP, IL-6) in Patients with T2DM and Obesity

T2DM and obesity both represent a complex metabolic dysfunction and inflammation. High levels of several inflammatory markers like C-reactive protein and interleukin-6 are notably observed in patients with T2DM and obesity. In addition, this scenario indicates maintenance of chronic low-grade inflammation, termed as "metabolic inflammation" [97, 98]. IL-6 is a pleiotropic cytokine, which has a major role in inflammation and metabolism in adipocytes, leukocytes, etc. Normal IL-6 level in healthy individuals is 2–20 pg/mL. In obese patients, altered adipose tissue becomes a primary source of IL-6, contributing up to 3% of total IL-6. Elevated level of total circulating IL-6 level is a well-established biomarker of the inflammation state associated with obesity and T2DM [99]. The direct link between IL-6 and T2DM is not fully understood yet. But it is observed that IL-6 can directly influence the cellular insulin resistance. This mechanism involves an IL-6-induced protein suppressor of cytokine signaling 3 (SOCS-3). This protein directly binds to the insulin receptor (IR) and insulin receptor substrate (IRS-1), and blocks the autophosphorylation of tyrosine residue. This whole cascade results in blocking of downstream pathway of insulin. Also, they can activate or trigger other inflammatory signaling pathways like JNK and NF-κB, which further result in insulin resistance [100, 101]. Many large-scale prospective studies and meta-analyses have proven that IL-6 is a very important predictor of future T2DM. A major meta-analysis was conducted with around 2000 participants, and it was concluded that, for every one log increase in IL-6 levels, the relative risk of developing T2DM increased to 31%. Circulating IL-6 has a correlation with insulin resistance (e.g., HOMA-IR), poor glycemic control (fasting glucose and HbA1c), and obesity [102, 103]. C-reactive protein is a very well-known and acute-phase protein and also an inflammatory biomarker. It rises rapidly in response to infection, inflammation, and tissue injury. CRP measurement gives a sensitive, nonspecific, and progressive indication of inflammation. CRP is synthesized majorly in the liver in response to pro-inflammatory cytokines like IL-6. This correlation between IL-6 and CRP is known as IL-6-CRP axis. Normal CRP level in healthy individuals is normally <3.0 mg/L, but this level can increase up to >1 gm/L during severe inflammation [103].

CRP is considered a key marker of inflammation driven by obesity. Many researches and studies have proven that high-sensitivity CRP (hs-CRP) can promote obesity, insulin resistance, and T2DM, but the relation between CRP and metabolic disorders are largely determined by BODY MASS INDEX. Recent studies and researches suggest that CRP can be an active participant in pathogenesis of various diseases rather than being a biomarker. An experiment was performed using transgenic rat models. That experiment demonstrated that chronic, low-grade CRP level can lead to development of early onset obesity [104, 105]. It is proved that high-sensitivity CRP is a powerful and independent factor for future cardiovascular complications like myocardial infraction and stroke. The processes of oxidative damage and inflammation is deeply interconnected. ROS can directly trigger some pro-inflammatory signaling pathways like NF-κB, while inflammatory cells, in turn, generate ROS as part of their function. This whole cascade creates a cycle where oxidative stress triggers inflammation and inflammation triggers more oxidative stress. Ultimately, this whole scenario increases the risk of diabetes and other complications [106, 107] (Table 24.5).

24.4.2 Clinical Trials

One of the main risk factors for heart disease and other health issues including diabetes and obesity is metabolic abnormalities. Oxidative stress and inflammation are directly linked to the onset of metabolic diseases. So, we can say oxidative stress contributes to the pathophysiology of several inherited metabolic diseases (IMDs). Antioxidants are chemical substances that are reducing in nature. These chemicals prevent the oxidation of several biological molecules and can be used to prevent and treat several major health crises. They can be eaten in normal diet as they are found in natural food sources and in supplements also. But the quality and extent of clinical evidence for the use of antioxidant to treat IMDs have not been fully understood yet. In this section, we describe the effect of antioxidants with some evidence, like vitamin E, N-acetylcysteine, and alpha-lipoic acid. Several experimental and clinical studies show that all of the above-mentioned antioxidants show a wide range of effects in protecting the human body, especially in MS patients. But these antioxidants should be used carefully and under proper supervision because their excess dose can be harmful or even fatal for humans.

Table 24.5 Different biomarkers identify for inflammation in T2DM and obesity

Biomarker	Metabolic parameter	Direction of correlation	Significance and key findings	References
IL-6	T2DM Risk	Positive (+)	Significant predictor of T2DM development (RR~1.31)	[104]
	Insulin resistance (HOMA-IR)	Positive (+)	Strong predictor of IL-6 levels, which can directly induce resistance via SOCS-3	[103]
	Glycemic control (FBG, HbA1c)	Positive (+)	Correlates with fasting plasma glucose and HbA1c	[108]
	Obesity (% body fat, BMI)	Positive (+)	Correlates with % body fat, adipose tissue is a major source	
	CRP	Positive (+)	Strong positive correlation (r = 0.78); IL-6 is the primary stimulus for CRP production	[103]
CRP	T2DM Risk	Positive (+)	Significant predictor of T2DM development (RR~1.26)	[109]
	Insulin resistance (HOMA-IR)	Positive (+)	Strongly associated, but relationship is largely mediated by BMI	[104]
	Glycemic control (FBG, HbA1c)	Positive (+)	Elevated CRP associated with higher HbA1c levels	[107]
	Obesity (BMI)	Positive (+)	Strongest determinant of CRP levels; may also have a causal role in obesity	[104]
	Cardiovascular events	Positive (+)	Powerful predictor of future cardiovascular events	[107]

Some researchers suggest that the effectiveness of antioxidant therapies is not universal, but is highly dependent on the specific antioxidant, the dose, the duration of treatment, and the clinical and genetic characteristics of the patient [110, 111].

24.4.2.1 Effect of Antioxidants in Metabolic Disorders

Vitamin E is a fat-soluble vitamin that has antioxidant properties. It contains eight isomers (α, β, γ, δ- tocopherol and tocotrienol). These compounds differ in their molecular arrangements. α-tocopherol is the most active isomer and has antioxidant and anti-inflammatory properties. Nuts, seeds, vegetable oil, soybean, canola, corn, and other green leafy vegetables are excellent sources of vitamin E. Vitamin E protects the cell membranes and lipoproteins from lipid peroxidation. It is proved that lipid peroxidation is one of the key characteristics of oxidative stress in type 2 diabetes and several cardiovascular diseases. Treatment with vitamin E has been seen to improve the overall metabolic health and mitigate the risk of vascular complications in the patients [112, 113]. The clinical studies of vitamin E are an example of complexities in antioxidant research. A meta-analysis was conducted in 2023 with 38 randomized controlled trials (RCTs) with over 2000 diabetic patients. This meta-analysis provided a significant clarification on the effects of vitamin E on glycemic control. In this analysis, scientists found that vitamin E supplementations have produced a significant reduction in HbA1c (mean difference of −0.21%), fasting insulin, and HOMA-IR. But it has no effectiveness in blood glucose level, except in short-term trials that last for less than 10 weeks. Some reviews suggests that benefits were confirmed to subgroups with poor glycemic control (HbA1c > 8%). A 2025 meta-analysis showed that vitamin E decreased total cholesterol and LDL-C. Moreover, greater benefits are seen at higher doses (>400 IU/day) and also for longer durations (>8 weeks) [113, 114]. The evidence of treatment with vitamin E on cardiovascular diseases has been very disappointing. Large-scale clinical trials, such as Women's Health Study did not find any benefit from vitamin E in cardiovascular disease in a large population of healthy women. However, this conclusion masked a crucial underlying factor. A major breakthrough emerged from the genetic sub-analysis of trials like the Heart Outcomes Prevention Evaluation (HOPE) study and the subsequent Israel Cardiovascular Events (I CARE) trial. These studies revealed that the effect of vitamin E significantly depends on the individual's haptoglobin (Hp) genotype. In the general population, vitamin E showed no benefit. But in specific subgroups of diabetic individuals with Hp 2-2 genotype, supplementation with 400 IU/day of vitamin E reduced the risk factors of major cardiovascular disease by a remarkable 50%. This analysis demonstrates that the failure of broad population trials could be due to a dilution effect, and this treatment has profound benefits in a genetically susceptible subgroup [114].

NAC is a nutritional supplement that primarily works as a precursor to the amino acid L-cysteine. L-cysteine is the rate-limiting substrate for the synthesis of glutathione (GSH), which is the most abundant and critical intracellular antioxidant in humans. In some oxidative stress and metabolic disease, intracellular GSH stores can become depleted. By providing sufficient amount of cysteine, NAC replenishes these stores, and helps to enhance the capacity of the cell to neutralize ROS and destroy harmful compounds [115, 116]. Study of NAC in type 2 diabetes mainly focused on diabetic nephropathy. A clinical trial—"Efficacy of N-Acetylcysteine in Treatment of Overt Diabetic Nephropathy" was conducted by Shiraz University of Medical Sciences to study and evaluate the effect of NAC in T2DM nephropathy. This study connects ROS overproduction to renal damage. This study helps to understand NAC's multifaceted mechanism, including its role as a GSH precursor and its ability to inhibit glycation and apoptosis as the basis for its potential therapeutic benefit in protecting the kidneys. The study was conducted with diabetic patients with significant proteinuria (defined as >500 mg in a 24-h urine sample) and a serum creatinine level below 3.0 mg/dL. Participants received 1200 mg of NAC per day (administered as two 600 mg effervescent tablets) for a duration of 3 months [117]. Another case study—"N-Acetylcysteine and Milk Thistle for Treatment of Diabetic Nephropathy (CGDN)" was sponsored by the VA Office of Research and Development. In this study, they used a more complex design to observe the effect of NAC in nephropathy. It was a two-center, double-blind, placebo-controlled, randomized pilot study with a modified-factorial design. The primary goal of this study was to observe the effect of NAC on both alone and in combination with another antioxidant, silibin (from milk thistle), on proteinuria and also markers of oxidative stress in type 2 diabetes patients with severe nephropathy. The study was conducted with five major characteristics. In addition, these are as follows: 1. placebo only, 2. NAC (600 mg twice daily) plus silibin placebo, 3. silibin only, 4. NAC plus silibin, 5. NAC plus a double dose of silibin, all administered for 3 months. This structure was used to study the individual effect as well as the synergistic effect of each antioxidant. The primary outcome measure was the change in the concentration of albumin excretion in urine, which is a key clinical marker of glomerular injury in severe diabetic kidney disease. Secondary outcomes were planned to analyze several markers of tubular injury and inflammation. But the study record shows that these were never measured, which suggests that the primary endpoint results did not want further investigation [118]. ALA is also known as thioctic acid. It is amphoteric in nature, so it can be solubilized in both fat and water. This is why it exerts its effects in both cytosolic and membranous cellular compartments. It shows a variety of mechanism in body, such as it directly neutralizes a wide range of free radicals, chelates (like iron and copper) that can initiate oxidative reactions and also it can regenerate other antioxidants such as, vitamin C, vitamin E, and coenzyme Q10. It also boosts the level of intracellular glutathione by increasing the cellular uptake of cysteine. This is why it is called as a universal antioxidant [119]. A randomized, double-blind, parallel-group clinical trial "Assessment of the Effect of Alpha Lipoic Acid Supplementation on Inflammatory Factors, Insulin Resistance, Glycemic Control and Anthropometric Indices in Patients with Metabolic Syndrome" was conducted to provide a comprehensive evaluation of ALA's effects in patients with severe metabolic syndrome. The trial was conducted with 44 individuals and they all were examined based on the NCEP: ATP III criteria, which ensures a well-defined study group. An important criterion was the "Lack of any cardiovascular disease," which suggests the study was specially designed to study the impact of ALA on the metabolic syndrome state, without any cardiovascular complications. The participants were given 600 mg ALA daily for 12 weeks, while the control group was given a matching placebo. This study showed a positive result in ALA's potential metabolic benefits. These included some key inflammatory markers like C-reactive protein (CRP), tumor necrosis factor-alpha, indicators of glycemic control (fasting blood sugar level, HbA1c, serum insulin concentration), lipid profiles, and other parameters such as Body Mass Index (BMI) and waist circumference. This whole experiment was performed to determine if ALA can modulate several interconnected pathways that are not properly regulated in metabolic syndrome [120].

A randomized controlled clinical trial, "Lipoic Acid and Prevention of Heart Disease" was conducted by Oregon State University, that was designed with a primary purpose of "Prevention of several cardiovascular disease." The main purpose of the trial was to determine if R-alpha lipoic acid (an isomer of alpha-lipoic acid) could mitigate the cardiovascular risk factors in healthy but at-risk population. The only criteria, where the participants should be obese and overweight (BMI 25.0–29.9 kg/m^2 or >30 kg/m^2, respectively) with elevated plasma triglyceride level (100–400 mg/dL). The participants were given a dose of 600 mg of R-alpha lipoic acid daily or a placebo. The results were primarily focused on some major cardiovascular risk factors like body weight and blood cholesterol levels. This trial confirmed that long-term use of R-alpha lipoic acid caused weight loss in the most of the participants, but it did not show any effects on elevated plasma triglyceride concentrations, which indicates that it has a specific effect on adiposity [121] (Table 24.6 and Fig. 24.1).

Table 24.6 Clinical trials study based on different antioxidants [115–121]

Antioxidant	Disease	Trial number	Study title	Phase	Therapeutic intervention	Study type	Status
Vitamin E	T2DM (with heart risk)	NCT00220831	Prevention of cardiovascular complication s in diabetic patients (I CARE)	Not specified	Vitamin E (400 IU/ day) vs. placebo in Hp 2-2 genotype patients	Interventional (randomized double-blind, placebo-controlled)	Terminated (due to benefit)
	Metabolic syndrome	NCT0560430 0	The role of vitamin E incorporated oat supplementation in metabolic syndrome	Not specified	Vitamin E incorporated oat supplement vs. oat supplement (placebo)	Interventional (randomized) double-blind, placebo-controlled)	Completed
	MetS/NAFLD	NCT0178759 1	Dairy fat as a mediator of vitamin E adequacy in individuals with metabolic syndrome	Not specified	A-tocopherol with fat-free, low-fat, milk	Interventional (randomized, crossover)	Completed
	NAFLD/NASH	NCT0179211 5	Treatment for nonalcoholic fatty liver with different doses of vitamins	Phase2	Vitamin E (200, 400, or 800 IU/day) + lifestyle modification	Interventional (randomized, open-label)	Completed
	PCOS (sterility)	NCT0330674 5	Micronutrient supplement in women with PCO-syndrome and sterility	Not specified	PRO fertile female (contains vitamin E 30 mg) vs. folic acid	Interventional (randomized, double-blind)	Recruiting
	PCOS (CC-resistant)	NCT0367438 5	Efficacy of vitamin E in PCOS resistant to clomiphene citrate	Phase2	Vitamin E + metformin + clomiphene citrate vs. metformin+ clomiphene citrate	Interventional (randomized, open-label)	Completed

N-acetyl-cys-teine(NAC)	T2DM (diabetic nephropathy)	NCT0126556 3	N-acetylcyst eine and milk thistle for treatment of diabetic nephropathy (CGDN)	Not specified	NAC (600 mg twice daily), Silibin, combination, or placebo	Interventional (randomized, double-blind, placebo-controlled)	Completed
	T2DM (beta-cell function)	NCT01394510	Effect of the antioxidant N-acetylcyste ine on beta-cell function	Not specified	NAC (600 mg then 1200 mg twice daily)	Interventional	Completed (single group, open-label, dose-finding)
	NAFLD		N-acetylcysteine improves liver function in patients with nonalcoholic fatty liver disease	Not specified	NAC (600 mg twice daily) vs. vitamin C (1000 mg twice daily)	Interventional (randomized)	Completed
	NASH	NCT0610506 0	A comparative study of N-acetyl cysteine, rosuvastatin, and vitamin E in the management of patients with nonalcoholic steatohepatitis	Phase1	Nac (1200 mg twice daily) vs. rosuvastatin vs. vitamin E	Interventional (randomized, double-blind, controlled)	Recruiting
	PCOS	NCT0683612 8	The combined effect of N - acetyl cysteine and metformin in polycystic ovary syndrome patients (PCOS)	Not specified	NAC+ metformin vs. metformin	Interventional (randomized)	Recruiting

(continued)

Table 24.6 (continued)

Antioxidant	Disease	Trial number	Study title	Phase	Therapeutic intervention	Study type	Status
Alpha-lipoic acid (ALA)	T2DM (diabetic neuropathy)	NCT02439879	Alpha lipoic acid for treatment of diabetic neuropathy	Not specified	Ala (1800 mg/day for 4 weeks, then 600 mg/day for16 weeks) vs. withdrawal	Interventional (randomized, open-label)	Completed
	T2DM (diabetic neuropathy)	NCT0432224 0	Alpha lipoic acid and diabetes mellitus: potential effects on peripheral neuropathy and different metabolic parameters	Not specified	ALA (600 mg/day for 3 months)	Interventional (prospective)	Completed
	Metabolic syndrome	NCT0358969 0	Assessment of the effect of alpha lipoic acid supplementation on inflammatory factors, insulin resistance. in patients with metabolic syndrome	Not specified	ALA (600 mg/day) vs. placebo	Interventional (randomized, double-blind, placebo-controlled)	Completed
	NAFLD (nondiabetic)	NCT0447527 6	Effect of alpha lipoic acid on nonalcoholic fatty liver diseases	Phase4	ALA (600 mg twice daily) vs. placebo + lifestyle modification	Interventional (randomized, placebo-controlled)	Recruiting
	PCOS (insulin resistance)	NCT00505427	Alpha lipoic acid and polycystic ovary syndrome	Not specified	ALA (800 mg twice daily for 16 weeks)	Interventional (single group, open-label)	Completed

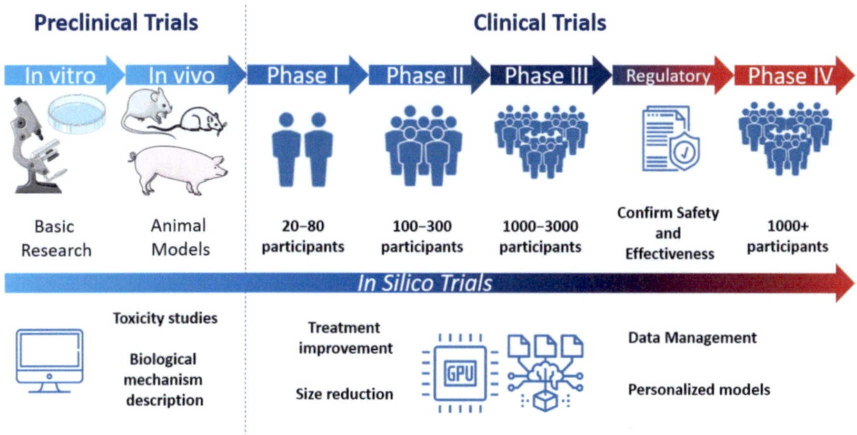

Fig. 24.1 Various phases of preclinical, clinical, and in silico study [122]

24.5 Conclusion

ROS-mediated inflammation has a central role in the pathogenesis and progression of metabolic disorders such as obesity, T2DM, NAFLD, and metabolic syndrome. Extensive research study from both preclinical and clinical data confirmed the intricate relationship between oxidative stress, chronic inflammation, and metabolic dysfunction. In metabolic tissues, excessive ROS production—often driven by mitochondrial dysfunction—triggers inflammatory signaling pathways, contributing to insulin resistance and organ damage. Preclinical models, including diet-induced and genetic animal models (e.g., ob/ob and db/db mice), as well as in vitro systems using adipocyte and hepatocyte cell lines, have demonstrated the detrimental effects of ROS on metabolic health. These studies consistently reveal elevated oxidative stress markers and support the beneficial effects of antioxidant therapies in reducing inflammation and improving insulin sensitivity. Clinical data further support these findings, showing strong associations between oxidative and inflammatory biomarkers and metabolic disease severity. Interventional trials using antioxidants such as vitamin E, N-acetylcysteine, and alpha-lipoic acid report promising outcomes, although results vary due to differences in patient profiles and treatment protocols. Overall, targeting ROS-mediated pathways represents a promising therapeutic strategy. However, further mechanistic insights and well-designed clinical trials are essential to optimize antioxidant-based therapies in metabolic disease management.

References

1. Jie Z, Liu J, Shu M, Ying Y, Yang H (2022) Detection strategies for superoxide anion: a review. Talanta 236:122892
2. Newsholme P, Cruzat VF, Keane KN, Warless R, de Bittencourt Jr PI (2016) Molecular mechanisms of ROS production and oxidative stress in diabetes. Biochem J 473(24):4527–4550
3. Yang M, Liu S, Zhang C (2022) The related metabolic diseases and treatment of obesity. Healthcare 10(9):1616
4. Godoy-Matos AF, Silva Junior WS, Valerio CM (2020) NAFLD as a continuum: from obesity to metabolic syndrome and diabetes. Diabetol Metab Syndr 12(1):60
5. Akhigbe R, Ajayi A (2021) The impact of reactive oxygen species in the development of cardiometabolic disorders: a review. Lipids Health Dis 20(1):23
6. Jiang X, Li Y, Fu D, You T, Wu S, Xin J, Wen J, Huang Y, Hu C (2023) Caveolin-1 ameliorates acetaminophen- aggravated inflammatory damage and lipid deposition in non-alcoholic fatty liver disease via the ROS/TXNIP/NLRP3 pathway. Int Immunopharmacol 114:109558
7. Higashi Y (2022) Roles of oxidative stress and inflammation in vascular endothelial dysfunction- related disease. Antioxidants 11(10):1958
8. Wu JQ, Kosten TR, Zhang XY (2013) Free radicals, antioxidant defense systems, and schizophrenia. Prog Neuro-Psychopharmacol Biol Psychiatry 46:200–206
9. Simioni C, Zauli G, Martelli AM, Vitale M, Sacchetti G, Gonelli A, Neri LM (2018) Oxidative stress: role of physical exercise and antioxidant nutraceuticals in adulthood and aging. Oncotarget 9(24):17181
10. Mueller MM, Castells-Roca L, Babu V, Ermolaeva MA, Müller RU, Frommolt P, Williams AB, Greiss S, Schneider JI, Benzing T, Schermer B (2014) DAF-16/FOXO and EGL-27/GATA promote developmental growth in response to persistent somatic DNA damage. Nat Cell Biol 16(12):1168–1179
11. Al-Gubory KH, Garrel C, Faure P, Sugino N (2012) Roles of antioxidant enzymes in corpus luteum rescue from reactive oxygen species-induced oxidative stress. Reprod Biomed Online 25(6):551–560
12. Juan CA, de la Lastra JM P, Plou FJ, Pérez-Lebeña E (2021) The chemistry of reactive oxygen species (ROS) revisited: outlining their role in biological macromolecules (DNA, lipids and proteins) and induced pathologies. Int J Mol Sci 22(9):4642
13. Halliwell B, Gutteridge JM (2015) Free radicals in biology and medicine. Oxford University Press
14. Kehrer JP (2000) The Haber–Weiss reaction and mechanisms of toxicity. Toxicology 149(1):43–50
15. Chen HY (2019) Why the reactive oxygen species of the Fenton reaction switches from oxoiron (IV) species to hydroxyl radical in phosphate buffer solutions? A computational rationale. ACS Omega 4(9):14105–14113
16. Fujiwara O, Fukuda S, Lopez E, Zeng Y, Niimi Y, DeWitt DS, Herndon DN, Prough DS, Enkhbaatar P (2019) Peroxynitrite decomposition catalyst reduces vasopressin requirement in ovine MRSA sepsis. Intensive Care Med Exp 7:1–4
17. Jiang H, Xue A, Wang Z, Xia R, Wang L, Tang Y, Wan P, Chen Y (2020) Electrochemical degradation of lignin by ROS. Sustain Chem 1(3):345–360
18. Pizzino G, Irrera N, Cucinotta M, Pallio G, Mannino F, Arcoraci V, Squadrito F, Altavilla D, Bitto A (2017) Oxidative stress: harms and benefits for human health. Oxidative Med Cell Longev 2017(1):8416763
19. Davalli P, Marverti G, Lauriola A, D'Arca D (2018) Targeting oxidatively induced DNA damage response in cancer: opportunities for novel cancer therapies. Oxidative Med Cell Longev 2018(1):2389523
20. Yan LL, Zaher HS (2019) How do cells cope with RNA damage and its consequences? J Biol Chem 294(41):15158–15171

21. Ito F, Sono Y, Ito T (2019) Measurement and clinical significance of lipid peroxidation as a biomarker of oxidative stress: oxidative stress in diabetes, atherosclerosis, and chronic inflammation. Antioxidants 8(3):72
22. Hawkins CL, Davies MJ (2019) Detection, identification, and quantification of oxidative protein modifications. J Biol Chem 294(51):19683–19708
23. Passos JF, Nelson G, Wang C, Richter T, Simillion C, Proctor CJ, Miwa S, Olijslagers S, Hallinan J, Wipat A, Saretzki G (2010) Feedback between p21 and reactive oxygen production is necessary for cell senescence. Mol Syst Biol 6(1):347
24. Liguori I, Russo G, Curcio F, Bulli G, Aran L, Della-Morte D, Gargiulo G, Testa G, Cacciatore F, Bonaduce D, Abete P (2018) Oxidative stress, aging, and diseases. Clin Interv Aging 13:757–772
25. Chandrasekaran A, Idelchik MD, Melendez JA (2017) Redox control of senescence and age-related disease. Redox Biol 11:91–102
26. Babušíková E, Lehotský J, Dobrota D, Račay P, Kaplán P (2012) Age-Associated changes in Ca 2+-ATPase and oxidative damage in sarcoplasmic reticulum of rat heart. Physiol Res 61(5):453–460
27. Ferrara N, Rinaldi B, Corbi G, Conti V, Stiuso P, Boccuti S, Rengo G, Rossi F, Filippelli A (2008) Exercise training promotes SIRT1 activity in aged rats. Rejuvenation Res 11(1):139–150
28. Frei B (1997) Reactive oxygen species and antioxidant vitamins, Linus Pauling Institute, Oregon State University. Retrieved May. 10:2019
29. Nishida N, Arizumi T, Takita M, Kitai S, Yada N, Hagiwara S, Inoue T, Minami Y, Ueshima K, Sakurai T, Kudo M (2013) Reactive oxygen species induce epigenetic instability through the formation of 8-hydroxydeoxyguanosine in human hepatocarcinogenesis. Dig Dis 31(5-6):459–466
30. Yasui M, Kanemaru Y, Kamoshita N, Suzuki T, Arakawa T, Honma M (2014) Tracing the fates of site-specifically introduced DNA adducts in the human genome. DNA Repair 15:11–20
31. Guo C, Sun L, Chen X, Zhang D (2013) Oxidative stress, mitochondrial damage and neurodegenerative diseases. Neural Regen Res 8(21):2003–2014
32. Kowalczyk P, Kaczyńska K, Kleczkowska P, Bukowska-Ośko I, Kramkowski K, Sulejczak D (2022) The lactoferrin phenomenon—a miracle molecule. Molecules 27(9):2941
33. Zong Y, Li H, Liao P, Chen L, Pan Y, Zheng Y, Zhang C, Liu D, Zheng M, Gao J (2024) Mitochondrial dysfunction: mechanisms and advances in therapy. Signal Transduct Target Ther 9(1):124
34. Bhatti JS, Bhatti GK, Reddy PH (2017) Mitochondrial dysfunction and oxidative stress in metabolic disorders- A step towards mitochondria based therapeutic strategies. Biochem Biophys Acta Mol Basis Dis 1863(5):1066–1077
35. Zorova LD, Popkov VA, Plotnikov EY, Silachev DN, Pevzner IB, Jankauskas SS, Babenko VA, Zorov SD, Balakireva AV, Juhaszova M, Sollott SJ (2018) Mitochondrial membrane potential. Anal Biochem 552:50–59
36. Gorospe CM, Carvalho G, Curbelo AH, Marchhart L, Mendes IC, Niedzwiecka K, Wanrooij PH (2023) Mitochondrial membrane potential acts as a retrograde signal to regulate cell cycle progression. Life Sci Alliance 6(12):e202302091
37. Okoye CN, Koren SA, Wojtovich AP (2023) Mitochondrial complex I ROS production and redox signaling in hypoxia. Redox Biol 67:102926
38. Apostolova N, Victor VM (2015) Molecular strategies for targeting antioxidants to mitochondria: therapeutic implications. Antioxid Redox Signal 22(8):686–729
39. Huang ML, Chiang S, Kalinowski DS, Bae DH, Sahni S, Richardson DR (2019) The role of the antioxidant response in mitochondrial dysfunction in degenerative diseases: cross-talk between antioxidant defense, autophagy, and apoptosis. Oxidative Med Cell Longev 2019(1):6392763
40. Bahorun T, Soobrattee MA, Luximon-Ramma V, Aruoma OI (2006) Free radicals and antioxidants in cardiovascular health and disease. Int J Med Update 1(2):25–41

41. Chatterjee M, Saluja R, Kanneganti S, Chinta S, Dikshit M (2007) Biochemical and molecular evaluation of neutrophil NOS in spontaneously hypertensive rats. Cell Mol Biol 53(1):84–93
42. Ceriello A (2008) Possible role of oxidative stress in the pathogenesis of hypertension. Diabetes Care 31(Suppl 2):S181–S184
43. Pacher P, Beckman JS, Liaudet L (2007) Nitric oxide and peroxynitrite in health and disease. Physiol Rev 87(1):315–424
44. Dröge W (2002) Free radicals in the physiological control of cell function. Physiol Rev 82(1):47–95
45. Halliwell B (2001) Role of free radicals in the neurodegenerative diseases: therapeutic implications for antioxidant treatment. Drugs Aging 18:685–716
46. Christen Y (2000) Oxidative stress and Alzheimer disease. Am J Clin Nutr 71(2):621S–629S
47. Allan BD (2002) Amyloid β-peptide (1-42)-induced oxidative stress and neurotoxicity: implications for neurodegeneration in Alzheimer's disease brain. A review. Free Radic Res 36(12):1307–1313
48. MacNee W (2001) Oxidative stress and lung inflammation in airways disease. Eur J Pharmacol 429(1-3):195–207
49. Barreiro E (2016) Role of protein carbonylation in skeletal muscle mass loss associated with chronic conditions. Proteome 4(2):18
50. Balasubramanian S (2013) Progression of chronic kidney disease: mechanisms and interventions in retardation. Apollo Med 10(1):19–28
51. Putri AY, Thaha M (2014) Role of oxidative stress on chronic kidney disease progression. Acta Med Indones 46(3):244–252
52. Valko MM, Morris H, Cronin MT (2005) Metals, toxicity and oxidative stress. Curr Med Chem 12(10):1161–1208
53. De Cristofaro R, Rocca B, Vitacolonna E, Falco A, Marchesani P, Ciabattoni G, Landolfi R, Patrono C, Davi G (2003) Lipid and protein oxidation contribute to a prothrombotic state in patients with type 2 diabetes mellitus. J Thromb Haemost 1(2):250–256
54. Brownlee M (2001) Biochemistry and molecular cell biology of diabetic complications. Nature 414(6865):813–820
55. Golbidi S, Alireza Ebadi S, Laher I (2011) Antioxidants in the treatment of diabetes. Curr Diabetes Rev 7(2):106–125
56. Brownlee M (2005) The pathobiology of diabetic complications: a unifying mechanism. Diabetes 54(6):1615–1625
57. Atli T, Keven K, Avci A, Kutlay S, Turkcapar N, Varli M, Aras S, Ertug E, Canbolat O (2004) Oxidative stress and antioxidant status in elderly diabetes mellitus and glucose intolerance patients. Arch Gerontol Geriatr 39(3):269–275
58. Buettner R, Schölmerich J, Bollheimer LC (2007) High-fat diets: modeling the metabolic disorders of human obesity in rodents. Obesity 15(4):798–808
59. Giacco F, Brownlee M (2010) Oxidative stress and diabetic complications. Circ Res 107(9):1058–1070
60. Leiter EH (1989) The genetics of diabetes susceptibility in mice. ASEB J 3(11):2231–2241
61. Herberg L, Coleman DL (1977) Laboratory animals exhibiting obesity and diabetes syndromes. Metabolism 26(1):59–99
62. King AJ (2012) The use of animal models in diabetes research. Br J Pharmacol 166(3):877–894
63. Forrester SJ, Kikuchi DS, Hernandes MS, Xu Q, Griendling KK (2018) Reactive oxygen species in metabolic and inflammatory signaling. Circ Res 122(6):877–902
64. Furukawa S, Fujita T, Shimabukuro M, Iwaki M, Yamada Y, Nakajima Y, Nakayama O, Makishima M, Matsuda M, Shimomura I (2017) Increased oxidative stress in obesity and its impact on metabolic syndrome. J Clin Invest 114(12):1752–1761
65. Furman BL (2015) Streptozotocin-induced diabetic models in mice and rats. Curr Protoc Pharmacol 70(1):5–47

66. He MQ, Wang JY, Wang Y, Sui J, Zhang M, Ding X, Zhao Y, Chen ZY, Ren XX, Shi BY (2020) High-fat diet-induced adipose tissue expansion occurs prior to insulin resistance in C57BL/6J mice. Chronic Dis Transl Med 6(3):198–207
67. Winzell MS, Ahrén B (2004) The High-Fat Diet–Fed mousea model for studying mechanisms and treatment of impaired glucose tolerance and type 2 diabetes. Diabetes 53(suppl_3):S215–S219
68. Shiota M, Printz RL (2012) Diabetes in Zucker diabetic fatty rat. In: Animal models in diabetes research. Humana Press, Totowa, pp 103–123
69. Peterson RG, Shaw WN, Neel MA, Little LA, Eichberg J (1990) Zucker diabetic fatty rat as a model for non-insulin-dependent diabetes mellitus. ILAR J 32(3):16–19
70. Lindström P (2007) The physiology of obese-hyperglycemic mice [ob/ob mice]. Sci World J 7(1):666–685
71. Chen H, Charlat O, Tartaglia LA, Woolf EA, Weng X, Ellis SJ, Lakey ND, Culpepper J, More KJ, Breitbart RE, Duyk GM (1996) Evidence that the diabetes gene encodes the leptin receptor: identification of a mutation in the leptin receptor gene in db/db mice. Cell 84(3):491–495
72. Lin Y, Berg AH, Iyengar P, Lam TK, Giacca A, Combs TP, Rajala MW, Du X, Rollman B, Li W, Hawkins M (2005) The hyperglycemia-induced inflammatory response in adipocytes: the role of reactive oxygen species. J Biol Chem 280(6):4617–4626
73. Wu WS, Tsai RK, Chang CH, Wang S, Wu JR, Chang YX (2006) Reactive oxygen species mediated sustained activation of protein kinase C α and extracellular signal-regulated kinase for migration of human hepatoma cell Hepg2. Mol Cancer Res 4(10):747–758
74. Kugo H, Sukketsiri W, Iwamoto K, Suihara S, Moriyama T, Zaima N (2021) Low glucose and serum levels cause an increased inflammatory factor in 3T3-L1 cell through Akt, MAPKs and NF-κB activation. Adipocyte 10(1):232–241
75. Sola-Leyva A, Jabalera Y, Chico-Lozano MA, Carrasco-Jiménez MP, Iglesias GR, Jimenez-Lopez C (2020) Reactive oxygen species (ROS) production in HepG2 cancer cell line through the application of localized alternating magnetic field. J Mater Chem B 8(34):7667–7676
76. Coyle CH, Martinez LJ, Coleman MC, Spitz DR, Weintraub NL, Kader KN (2006) Mechanisms of H2O2-induced oxidative stress in endothelial cells. Free Radic Biol Med 40(12):2206–2213
77. Yeh YC, Liu TJ, Lai HC (2020) Pathobiological mechanisms of endothelial dysfunction induced by tert-butyl hydroperoxide via apoptosis, necrosis and senescence in a rat model. Int J Med Sci 17(3):368
78. Casas AI, Dao VT, Daiber A, Maghzal GJ, Di Lisa F, Kaludercic N, Leach S, Cuadrado A, Jaquet V, Seredenina T, Krause KH (2015) Reactive oxygen-related diseases: therapeutic targets and emerging clinical indications. Antioxid Redox Signal 23(14):1171–1185
79. Veal EA, Day AM, Morgan BA (2007) Hydrogen peroxide sensing and signaling. Mol Cell 26(1):1–4
80. Wang J, Liao B, Wang C, Zhong O, Lei X, Yang Y (2022) Effects of antioxidant supplementation on metabolic disorders in obese patients from randomized clinical controls: a meta-analysis and systematic review. Oxidative Med Cell Longev 2022(1):7255413
81. Egea J, Fabregat I, Frapart YM, Ghezzi P, Görlach A, Kietzmann T, Kubaichuk K, Knaus UG, Lopez MG, Olaso-Gonzalez G, Petry A (2017) European contribution to the study of ROS: a summary of the findings and prospects for the future from the COST action BM1203 (EU-ROS). Redox Biol 13:94–162
82. Perez-Montero B, Fermin-Rodriguez ML, Portero-Fuentes M, Sarquis J, Caceres S, Portal JI, Juan LD, Miro G, Cruz-Lopez F (2025) Malondialdehyde (MDA) and 8-hydroxy-2′-deoxyguanosine (8-OHdG) levels in canine serum: establishing reference intervals and influencing factors. BMC Vet Res 21(1):161
83. Ayala A, Muñoz MF, Argüelles S (2014) Lipid peroxidation: production, metabolism, and signaling mechanisms of malondialdehyde and 4-hydroxy-2-nonenal. Oxidative Med Cell Longev 2014(1):360438

84. Teixeira de Lemos E, Oliveira J, Páscoa Pinheiro J, Reis F (2012) Regular physical exercise as a strategy to improve antioxidant and anti-inflammatory status: benefits in type 2 diabetes mellitus. Oxidative Med Cell Longev 2012(1):741545
85. Pallavi LC, SoUza UJ, Shivaprakash G (2017) Assessment of musculoskeletal strength and levels of fatigue during different phases of menstrual cycle in young adults. J Clin Diagn Res 11(2):CC11
86. Premkumar G, Bhagyalakshmi V, Sandhya S (2023) A correlation biomarker between BMI and lipid peroxidation in type 2 diabetes mellitus with and without other complications. Indian J Med Sci 74(3):122–125
87. Valavanidis A, Vlachogianni T, Fiotakis C (2009) 8-hydroxy-2′-deoxyguanosine (8-OHdG): a critical biomarker of oxidative stress and carcinogenesis. J Environ Sci Health Part C 27(2):120–139
88. Nishikawa T, Sasahara T, Kiritoshi S, Sonoda K, Senokuchi T, Matsuo T, Kukidome D, Wake N, Matsumura T, Miyamura N, Sakakida M (2003) Evaluation of urinary 8-hydroxydeoxyguanosine as a novel biomarker of macrovascular complications in type 2 diabetes. Diabetes Care 26(5):1507–1512
89. Giugliano D, Ceriello A, Paolisso G (1995) Diabetes mellitus, hypertension, and cardiovascular disease: which role for oxidative stress? Metabolism 44(3):363–368
90. Al-Aubaidy HA, Jelinek HF (2010) 8-Hydroxy-2-deoxy-guanosine identifies oxidative DNA damage in a rural prediabetes cohort. Redox Rep 15(4):155–160
91. Wu LL, Chiou CC, Chang PY, Wu JT (2004) Urinary 8-OHdG: a marker of oxidative stress to DNA and a risk factor for cancer, atherosclerosis and diabetics. Clin Chim Acta 339(1-2):1–9
92. Neelofar K, Arif Z, Arafat MY, Alam K, Ahmad J (2019) A study on correlation between oxidative stress parameters and inflammatory markers in type 2 diabetic patients with kidney dysfunction in north Indian population. J Cell Biochem 120(4):4892–4902
93. Majeed K, Ali H (2025) Association of low-grade inflammation and oxidative stress with metabolic dysfunction in healthy obese individuals. AIMS Allergy Immunology 9(2):56–69
94. Masenga SK, Kabwe LS, Chakulya M, Kirabo A (2023) Mechanisms of oxidative stress in metabolic syndrome. Int J Mol Sci 24(9):7898
95. Babakr A, Mukhtar M, Althubiti M, Al-Amodi H, Almaimani R, Nour Eldin MM, Elzubeir Abdalla M, Nasif W (2023) Investigation of hyperlipidemia associated with increased levels of oxidized low-density lipoproteins and 8-hydroxy-2´-deoxyguanosine. Diabetes Metab Syndr Obes 16:447–455
96. Nisar M, Iqbal MU (2021) Influence of hs-CRP, IL-6 and TNF-α and it's role in dyslipidemia and type 2 diabetes in population of Karachi, Pakistan. Pak J Pharm Sci 34:1217–1225
97. Mohammad IJ, Kashanian S, Rafipour R, Aljwaid H, Hashemi S (2024) Evaluation of the relationship of cytokines concentrations tumor necrosis factor-alpha, interleukin-6, and C-reactive protein in obese diabetics and obese non-diabetics: a comparative study. Biotechnol Appl Biochem 71(2):272–279
98. Wang X, Bao W, Liu J, OuYang YY, Wang D, Rong S, Xiao X, Shan ZL, Zhang Y, Yao P, Liu LG (2013) Inflammatory markers and risk of type 2 diabetes: a systematic review and meta-analysis. Diabetes Care 36(1):166–175
99. Rehman K, Akash MS, Liaqat A, Kamal S, Qadir MI, Rasul A (2017) Role of interleukin-6 in development of insulin resistance and type 2 diabetes mellitus. Crit Rev Eukaryot Gene Expr 27(3):229–236
100. Rehman K, Akash MS, Liaqat A, Kamal S, Qadir MI, Rasul A (2017) Role of interleukin-6 in development of insulin resistance and type 2 diabetes mellitus. Critical Reviews™ in Eukaryotic. Gene Expr 27(3):229–236
101. Akash MS, Rehman K, Chen S (2013) Role of inflammatory mechanisms in pathogenesis of type 2 diabetes mellitus. J Cell Biochem 114(3):525–531
102. Chakraborty S, Verma A, Garg R, Singh J, Verma H (2023) Cardiometabolic risk factors associated with type 2 diabetes mellitus: a mechanistic insight. Clin Med Insights Endocrinol Diabetes 16:11795514231220780

103. Kahn SE, Zinman B, Haffner SM, O'Neill MC, Kravitz BG, Yu D, Freed MI, Herman WH, Holman RR, Jones NP, Lachin JM (2006) Obesity is a major determinant of the association of C-reactive protein levels and the metabolic syndrome in type 2 diabetes. Diabetes 55(8):2357–2364
104. Fronczyk A, Molęda P, Safranow K, Piechota W, Majkowska L (2014) Increased concentration of C-reactive protein in obese patients with type 2 diabetes is associated with obesity and presence of diabetes but not with macrovascular and microvascular complications or glycemic control. Inflammation 37(2):349–357
105. King DE, Mainous AG III, Buchanan TA, Pearson WS (2003) C-reactive protein and glycemic control in adults with diabetes. Diabetes Care 26(5):1535–1539
106. Jialal I, Devaraj S (2003) Role of C-reactive protein in the assessment of cardiovascular risk. Am J Cardiol 91(2):200–202
107. Spranger J, Kroke A, Mohlig M, Hoffmann K, Bergmann MM, Ristow M, Boeing H, Pfeiffer AF (2003) Inflammatory cytokines and the risk to develop type 2 diabetes: results of the prospective population-based European Prospective Investigation into Cancer and Nutrition (EPIC)-Potsdam Study. Diabetes 52(3):812–817
108. Phosat C, Panprathip P, Chumpathat N, Prangthip P, Chantratita N, Soonthornworasiri N, Puduang S, Kwanbunjan K (2017) Elevated C-reactive protein, interleukin 6, tumor necrosis factor alpha and glycemic load associated with type 2 diabetes mellitus in rural Thais: a cross-sectional study. BMC Endocr Disord 17(1):44
109. Heimes F, Berendes LS, Hannibal L, Park J (2025) Antioxidant therapy in inborn metabolic diseases. Mol Genet Metab 145:109176
110. Martins Gregório B, Benchimol De Souza D, de Morais A, Nascimento F, Matta L, Fernandes-Santos C (2016) The potential role of antioxidants in metabolic syndrome. Curr Pharm Design 22(7):859–869
111. Nguyen NT, Jiang Y, McQuade JL (2024) Eating away cancer: the potential of diet and the microbiome for shaping immunotherapy outcome. Front Immunol 15:1409414
112. Asbaghi O, Nazarian B, Yousefi M, Anjom-Shoae J, Rasekhi H, Sadeghi O (2023) Effect of vitamin E intake on glycemic control and insulin resistance in diabetic patients: an updated systematic review and meta-analysis of randomized controlled trials. Nutr J 22(1):10
113. Milman U, Blum S, Shapira C, Aronson D, Miller-Lotan R, Anbinder Y, Alshiek J, Bennett L, Kostenko M, Landau M, Keidar S (2008) Vitamin E supplementation reduces cardiovascular events in a subgroup of middle-aged individuals with both type 2 diabetes mellitus and the haptoglobin 2-2 genotype: a prospective double-blinded clinical trial. Arterioscler Thromb Vasc Biol 28(2):341–347
114. Pedre B, Barayeu U, Ezeriņa D, Dick TP (2021) The mechanism of action of N-acetylcysteine (NAC): the emerging role of H2S and sulfane sulfur species. Pharmacol Ther 228:107916
115. https://clinicaltrials.gov/study/NCT01550432
116. https://clinicaltrials.gov/study/NCT00556465
117. https://clinicaltrials.gov/study/NCT01265563
118. https://clinicaltrials.gov/study/NCT03589690
119. https://clinicaltrials.gov/study/NCT00765310
120. https://clinicaltrials.gov/study/NCT02439879
121. https://clinicaltrials.gov/study/NCT06253429
122. Sánchez de la Nava AM, Gómez-Cid L, Ríos-Muñoz GR, Fernández-Santos ME, Fernández AI, Arenal Á, Sanz-Ruiz R, Grigorian-Shamagian L, Atienza F, Fernández-Avilés F (2022) Cardiovascular diseases in the digital health era: a translational approach from the lab to the clinic. Biotech 11(3):23

Chapter 25
Challenges and Opportunities in Translating ROS Research into Clinical Applications

Shankhadip Nandi, Srabani Podder, and Shaibal Chandra

Abbreviations

AD	Alzheimer's disease
AMPK	Adenosine monophosphate-activated protein kinase
BRB	Bilirubin derived biomaterials
BRSNP	Bilirubin-encapsulated silk fibrin nanoparticles
CAT	Catalase
CeO_2	Cerium oxide
CSCs	Cancer stem cells
CYP2E1	Cytochrome P450 2E1
DAMPs	Damage-associated molecular patterns
ER	Endoplasmic reticulum
FFA	Free fatty acids
GPx	Glutathione peroxidase
GSH	Glutathione
GSMM	Genome Scale Metabolic Model
H_2O_2	Hydrogen peroxide
HA	Hyaluronic acid
HO-1	Heme oxygenase-1
HT	HA-tyramine
IKKβ	Inhibitory kappa B kinase beta
JNK	c-Jun N-terminal kinase
LOXs	Lipoxygenases
MA	Malonic acid
MAPK	Mitogen-activated protein kinases

S. Nandi (✉) · S. Podder
Department of Pharmaceutics, Eminent College of Pharmaceutical Technology, Kolkata, West Bengal, India

S. Chandra
Department of Pharmacology, Bengal School of Technology, Chinsurah, West Bengal, India

S. Ashique, N. Mishra (eds.), *Reactive Oxygen Species in Metabolic Inflammation*,
https://doi.org/10.1007/978-3-032-11752-6_25

MTAs	Mitochondria-targeted antioxidants
mTOR	Mammalian target of rapamycin
NAC	N-acetylcysteine
NAFLD	Non-alcoholic fatty liver disease
NF-κB	Nuclear transcription factor-kappa B
NO	Nitric oxide
NOX	Nicotinamide adenine dinucleotide phosphate oxidases
Nrf2	Nuclear factor erythroid 2-related factor 2
PD	Parkinson's disease
PDT	Photodynamic therapy
PLGA	Poly (lactic-co-glycolic acid)
ROS	Reactive oxygen species
SDT	Sonodynamic therapy
sEH	Soluble epoxide hydrolase
SkQ1	Plastoquinonyl decyl triphenylphosphonium
SOD	Superoxide dismutase
SOD1	Copper-zinc superoxide dismutase
STAT	Signal transducer and activator of transcription
T2D	Type 2 diabetes
TA	Tannic acid
TGF-β1	Transforming growth factor beta 1
TRX	Thioredoxin
TrxR	Thioredoxin reductase
UVB	Ultraviolet-B radiation
XO	Xanthine oxidase
XOR	Xanthine oxidoreductase

25.1 Introduction

Reactive oxygen species (ROS) govern cellular homeostasis and serve as primary regulators of cellular disorder, resulting in disease pathogenesis. Several enzymatic activities in the cell compartments like cell membrane, endoplasmic reticulum (ER), cytoplasm, mitochondria, and peroxisome generate ROS as a by-product for basal metabolic processes [1]. An important enzymatic source of ROS production is the family of nicotinamide adenine dinucleotide phosphate oxidases (NOXs), xanthine oxidoreductases (XORs), lipoxygenases (LOXs), etc. [2]. In redox homeostasis, the delicate equilibrium remains intact within cells when endogenous antioxidant enzymes can catalyze the neutralization of ROS generated by the cells. The pathological condition termed oxidative stress is induced when the delicate equilibrium is disturbed, leading to excessive production of ROS and depletion of crucial macromolecules such as proteins, lipids, and DNA [3, 4]. At physiologically low levels,

ROS also play an important signaling function, a hypothesis fundamental to redox biology [5]. Hydrogen peroxide (H_2O_2) promotes redox signaling by reversibly oxidizing residues of cysteine in proteins [6]. Cysteine residues appear as thiolate anions at physiological pH and demonstrate greater susceptibility to oxidation compared to protonated cysteine thiols [7]. Redox signaling involves in the oxidation of thiolate anion to sulfenic species by H_2O_2, which leads to functional variations in the protein through allosteric modifications [8]. The reversed signal transduction process is necessary for several physiological and biological events, such as fibrosis, inflammation, apoptosis, autophagy, cell proliferation, mitosis, and immunological response [9]. Cells comprise specific enzymes, such as glutathione peroxidases, and peroxiredoxins, that are involved in preventing the formation of intracellular H_2O_2, thereby retaining ROS levels within the physiological range needed for signaling and avoiding the onset of unfavorable oxidative stress [3]. Oxidative stress has been recognized as an important driver in numerous chronic diseases, especially different types of cancer, neurological disorders like Alzheimer's disease (AD), Parkinson's disease (PD), as well as cardiovascular diseases. The cumulative impact of oxidative damage over the years is also acknowledged as an important factor in the aging process, leading to decreased cellular function and the emergence of age-related diseases [4]. The current research findings suggest that increased systemic oxidative stress is strongly correlated with metabolic disorder [10, 11]. The function of ROS signaling in regular physiological processes and its potential influence on abnormal responses like metabolic failure and inflammatory signaling are both characterized by the cell type, origin of ROS, and tissue environment. Diseases linked to increased inflammatory signaling and impaired metabolism, particularly obesity, diabetes mellitus, carcinogenesis, cardiovascular, nonalcoholic fatty liver disease (NAFLD), etc., are allied with a disrupted redox balance [1]. Several antioxidant enzymes like glutathione peroxidase (GPx), superoxide dismutase (SOD), and catalase (CAT) can be activated by ROS scavengers having low molecular weight [12]. Few studies have demonstrated that antioxidant enzymes can be successfully triggered by vitamins D, β-carotene, selenium, vitamin E, coenzyme Q, zinc, etc., to prevent oxidative chain reactions and accelerate the rate of ROS removal [13–15]. Nevertheless, arbitrarily suppressing ROS levels, without an understanding of their physiological functions, may adversely affect cellular activities and homeostasis [16]. If oxidative stress is fundamental to these pathologies, then therapies focused on restoring redox equilibrium or altering specific ROS pathways are obvious and prospective therapeutic approaches. The continued development of ROS scavenging biomaterials based on physiological concepts provides multiple opportunities for modulating ROS levels that is regarded as a viable therapeutic modality. Still, the physiological regulating platforms remain inadequately understood, restricting the capacity to create a systematic network [2]. This perspective encourages continued research endeavors in this delicate domain, despite existing barriers.

25.2 Role of ROS in Metabolic Inflammation

ROS are not the sole contributors of metabolic disorders; they act as vital and frequently concurrent components that considerably lead to the onset and development of obesity, type 2 diabetes (T2D), cardiovascular as well as neurodegenerative diseases, NAFLD, cancer, and other complications related to chronic metabolic inflammation. In obesity, excessive nutrient influx results in the hypertrophy and hyperplasia of adipocytes. Expanded fat deposits frequently become hypoxic, which together with nutrient excess stimulates the production mitochondrial ROS and activates NOX, thus ensuing local and systemic oxidative stress [17]. In T2D, chronic hyperglycemia and increased lipotoxicity deteriorates mitochondrial electron leakage and thus ROS is produced [18]. Excessive ROS activates pro-inflammatory/matrix protein-degrading stress kinases such as c-Jun N-terminal kinase (JNK) and inhibitory kappa B kinase beta (IKKβ) that lead to serine phosphorylation of insulin receptor substrates, thereby inhibiting insulin signaling and increasing peripheral insulin resistance [19]. Oxidative stress is particularly injurious to pancreatic β-cells that have low intrinsic antioxidant enzyme levels leading to DNA/protein damage, ER stress response, mitochondrial dysfunction ultimately resulting in apoptosis with reduction in islet cells [20]. These collected effects result in a combination of insulin resistance together with progressive β-cell failure, which underlies pathogenesis of T2D. NAFLD is caused by the accumulation of triglycerides in the hepatocyte when the rate of hepatic lipid inflow and synthesis exceeds that of outflow and disposal [21]. Accumulated free fatty acids (FFA) are preferentially subjected to β-oxidation within mitochondria, an organelle that also couples oxygen-derived phosphorylation to ATP production. The other major pathway for FFA processing, microsomal oxidation catalyzed by a member of the cytochrome P450 2E1 family (CYP2E1), engages molecular oxygen as a source of electrons during the breakdown of toxic xenobiotics [21]. Both CYP2E1-mediated microsomal oxidation and mitochondrial β-oxidation generate increased amounts of ROS, which in turn promote lipid peroxidation. ROS-induced lipid peroxidation products such as malondialdehyde and 4-hydroxynonenal constitute damage-associated molecular patterns (DAMPs) that activate Kupffer cells, driving inflammation and fibrogenesis via mediators such as transforming growth factor beta 1 (TGF-β1) [22]. The auto-amplification perpetuates hepatocellular damage through necrosis or apoptosis while also activating hepatic stellate cells, which can be viewed similarly to macrophages because they process extracellular lipids [23]. Atherosclerosis is a chronic inflammatory cardiovascular disease in which oxidative stress has been central to the pathogenesis. The major vascular sources of ROS are NOX isoforms, uncoupled endothelial nitric oxide synthase, and mitochondrial electron transport [24]. In addition, ROS can react with nitric oxide (NO) leading to peroxynitrite formation that further reduces NO bioavailability, impairs vasodilation, and promotes endothelial dysfunction [25]. Peripheral metabolic dysfunction characterized by insulin resistance, dyslipidemia, and chronic hyperglycemia induces significant oxidative stress in the brain [26]. In AD, amyloid-β aggregates induce ROS formation, lipid peroxidation, and microglial activation leading to chronic neuroinflammation. Lipid, protein, and DNA oxidation damages synaptic plasticity, disrupts neurotransmission, and eventually leads to cell death. The same mechanisms are seen in other metabolic-related neurodegenerative diseases like PD [27]. Persistence to excess ROS generation acts as a mechanistic link between metabolic dysregulation and associated oxidative stress that also promotes tumor growth among individuals at high risk for several cancers including hepatocellular carcinoma, pancreatic cancer, and colorectal cancer. Research studies have shown a novel nucleophilic mechanism of ROS-dependent epigenetic variations in tumor cells where enriched DNA methylation resulted in the silencing of tumor suppressor, antioxidant genes and augmenting the proliferation of cancer cells under the circumstance of oxidative stress [28].

25.3 ROS-Based Translational Research to Prevent Metabolic Inflammation

The strategy to abolish harmful oxidative stress includes attempting to restore the body's redox equilibrium. A clinical investigation found that the cardiovascular risk caused by obesity can be managed through losing weight, which eventually decreases indicators of oxidative stress and boosts the antioxidant system [29]. A number of studies attempted to understand the impact of ROS on particular signaling pathways, such as mitogen-activated protein kinases (MAPK), signal transducer and activator of transcription (STAT), and nuclear transcription factor-kappa B (NF-κB), which may enable the identification of novel therapeutic targets for the diseases linked to metabolic inflammation. Epidemiological research has demonstrated a decline in cancer incidence among diabetic individuals treated with metformin [30, 31]. Metformin suppresses mammalian target of rapamycin (mTOR) protein kinase-based biosynthesis by promoting the adenosine monophosphate-activated protein kinase (AMPK) pathway, an essential regulator of cellular energy state [32]. The administration of metformin in pancreatic cancer stem cells (CSCs), relying on oxidative metabolism and restricted metabolic flexibility, can result in mitochondrial inhibition, thus provoking shortage of energy and prompting apoptosis of CSCs [33].

The body's inherent ROS scavenging mechanism lacks the ability to prevent the detrimental consequences of excessive ROS generation in cases of ROS-mediated metabolic inflammation. Translational ROS research has currently focused on biomaterials with robust ROS scavenging capabilities as a promising therapeutic approach [34]. There are different categories of ROS scavenging biomaterials, each with a distinctive mode of action that catalyze the action of natural enzymes to speed up ROS elimination, directly react with ROS, or inhibit ROS sources which limits the generation of ROS [35].

25.3.1 Enzymatic Biomaterials

The agents that replicate enzymes or improve the function of natural enzymes in order to accelerate the removal of ROS via catalysis are referred as enzymatic biomaterials. Metabolic inflammation resulting from disruption of the ROS scavenging network can be effectively managed with natural enzymes that offer broad-spectrum capabilities [35]. Zhuang et al. fabricated SOD-loaded γ-PGAS/γ-PGA hydrogel (SOD-PGAS/PGA-H) through the process of cross-linking, which exhibited stronger ability to scavenge ROS [36]. Similarly, Dong et al. prepared a thermosensitive hydrogel, poly(N-isopropyl-acrylamide)/poly(gamma-glutamic acid), loaded with SOD, which showed the proficiency to inhibit ROS generation [37]. Altering natural enzyme biomaterials enhances the physiological catalytic ROS immune response, allowing the clearance of ROS and minimizing inflammation by increasing the function of natural ROS scavenging enzymes, including SOD, CAT, and GPx. One of the most likely targets for the modification of natural enzyme biomaterials is nuclear factor erythroid 2-related factor 2 (Nrf2), which activates a wide range of antioxidant enzymes. Based on the findings on various anti-inflammatory biomaterials that effectively activate Nrf2, Park et al. demonstrated that malonic acid (MA) extracted from *Pinus densiflora* boosts the key antioxidant enzymes like copper-zinc superoxide dismutase (SOD1) and heme oxygenase-1 (HO-1) by means of Nrf2 activation, leading to a significant degeneration in ultraviolet-B radiation (UVB)-induced ROS levels [38]. Using a dual-enzyme cross-linking process, Zhang et al. constructed an injectable hydrogel comprising gallic acid grafted hyaluronic acid (HA) and HA-tyramine (HT) that triggers the Nrf2 signaling pathway [39]. Further, it was shown by Qian et al. that peptides from seahorse hydrolysate activate Nrf2 and substantially decrease intracellular ROS concentrations [40]. Researchers have discovered that nanozymes can readily replicate the catalytic function of certain biological enzymes, owing to the rapid and major progress of nanotechnologies [41]. They have demonstrated an extensive range of benefits, like improved stability, unique ROS scavenging capabilities, multifunctionality, and programmable activity. Nanozymes composed of cerium oxide (CeO_2) have been reported by Yu et al. to diminish the proinflammatory agents by scavenging several ROS and inhibiting the signaling pathways like MAPK and NF-κB [42]. The efficient delivery of CeO_2 nanozymes conjugated micro-ribonucleic acid 146a (miRNA146a) over the trachea was observed by Niemiec et al. where NF-κB signaling pathway was successfully inhibited followed by prevention of acute lung injury through changes in leukocyte employment as well as lessening of inflammation and oxidative stress [43]. Using potassium permanganate oxidation surfactant templates, a novel manganese-loaded mesoporous silica nanozyme (MnMSN) was formed by Xiong et al. that could catalytically scavenge H_2O_2, hydroxyl radicals, and superoxide anion including extreme levels of ROS from inflammatory cells (M1 macrophages) [44].

25.3.2 Direct ROS Scavenging Biomaterials

These engineered materials can react directly with ROS to neutralize or eliminate ROS from the environment that can also alleviate oxidative stress and promote the healing course for injured tissues. Polyphenols have been frequently used as exogenous ROS scavenging substances and explored for their efficacy in preventing various ROS-related metabolic inflammations, owing to their aptitude to combine with free radicals, bind metal catalysts, trigger antioxidant enzymes, and prevent oxidases [2]. Metaflammation in the central nervous system has been controlled with curcumin, which suppresses the response to inflammation and oxidative stress [45]. Tannic acid (TA) is another example of plant-derived polyphenol characterized by its anti-inflammatory, anti-ROS, and antibacterial qualities. The extensive catechol groups enable TA as a naturally occurring free radical scavenging material [46]. Multidisciplinary hydrogels were utilized for linking TA-chelated iron-based molybdenum disulfide nanosheets, which were developed by Li et al. The ingredient possessed sufficient capability to destroy ROS and reduce the fabrication of inflammatory mediators, permitting the antioxidant system to remain in a condition of homeostasis [47]. Bilirubin-derived biomaterials (BRB) have also shown favorable outcomes in several metabolic inflammatory models so far. In their study, Kim et al. found that BRBs can mitigate Th2-mediated lung inflammation in mouse model with asthma [36]. This inhibition was mainly accomplished by eliminating ROS in CD4+ T cells [48]. Yao et al. synthesized bilirubin-encapsulated silk fibrin nanoparticles (BRSNP) for the management of acute pancreatitis, displaying excellent ROS scavenging activity [49].

25.3.3 ROS Generation-Blocking Biomaterials

Apocynin and other natural biomaterials have been found to prevent metabolic inflammatory circumstances in several animal models by inhibiting NOX activity or the production of ROS caused by NOX [50, 51]. Mehmood et al. isolated an extract from *Stevia rebaudiana* rich in flavonoids and chlorogenic acid, which revealed its beneficial impact on hyperuricemic mice model by inhibiting xanthine oxidase (XO) activity, an XOR variant that can be employed interchangeably, raising both oxidative stress and inflammatory responses [52]. Wan et al. showed that an ethanol extract from the mulberry leaves of *Morus alba* plant could significantly decrease XO activity in a dose-dependent way, leading to a decline in hypertension, hyperglycemia, and elevated cholesterol levels as well as exerting anti-inflammatory properties [53]. By means of indole compounds that can serve as dual inhibitors of 5-LOX and soluble epoxide hydrolase (sEH), Cerqua et al. confirmed the considerable anti-inflammatory assets in asthmatic mice [54].

25.4 Challenges in Translating ROS-Based Investigation into Clinical Interventions

The following obstacles may be faced in the translation of ROS-based investigations into viable treatment strategies.

25.4.1 Conflict of ROS as Signaling Molecule

The inherent inconsistency of ROS as crucial signaling molecules as well as harmful mediators poses an intense challenge to clinical interventions [3]. Although excessive ROS are associated with oxidative stress and pathology, full eradication of ROS would be detrimental to essential cellular processes and overall the state of equilibrium. A significant hurdle in this scenario is accurately specifying the onset at which ROS levels shift from favorable signaling molecules to harmful mediators resulting in oxidative stress [16]. The failure of a simple approach to ROS elimination in numerous clinical settings can be directly attributed to this intricacy. It further emphasizes the necessity of smart therapies that can differentiate between safe and hazardous ROS levels or sites that can lead to the formation of more specific and advanced interventions.

25.4.2 Limitations of Antioxidants in Clinical Trials

Rather than substantial evidence of their efficiency or specialized targeting abilities, many antioxidants employed in clinical studies were selected only because of their easy availability. However, when it involves preventing metabolic inflammation, multiple large-scale clinical investigations of systemic and broad-spectrum antioxidant therapies have often failed to establish consistent beneficial outcomes [55]. These trials indicated no valuable effect, while other studies suggested an elevated risk of mortality connected to the consumption of antioxidants like beta-carotene, vitamins A, and E [56, 57]. In the disorders linked with metabolic inflammation, the existing antioxidants cannot always target the fundamental enzymatic pathways and subcellular regions responsible for excessive ROS generation. The lack of discrimination action has the potential for disrupting important physiological ROS signals followed by inadequate scavenging of hazardous ROS. Systemic antioxidants are not competent to reach targeted tissues in adequate concentrations for exerting therapeutic benefits owing to variables like limited tissue penetration, fast metabolism, and poor bioavailability. For instance, constrained bioavailability of ferulic acid is a major roadblock to its clinical translation, although its potential anti-inflammatory effects were reported in preclinical studies [58]. Because of these challenges, therapeutic design requires to be reconsidered from the fundamental research, with a greater emphasis on improving current understanding of redox biology and developing more precise interventions.

25.5 Limitations of ROS Scavenging Biomaterials

The application of ROS scavenging biomaterials in the management of metabolic inflammation demands the resolution of many important and potentially challenging issues connected with oxidative stress. Shorter half-lives, poor stability, extreme susceptibility to production, etc., are the major obstacles to the efficient supply of natural enzymes utilizing biomaterials [59]. Natural biomaterials like celastrol and honokiol have been widely implicated in NOX inhibition. However, the exact mechanism of these natural biomaterials through which they inhibit ROS or reduce inflammation remains unclear. Further research into the ROS network must be continued for a deeper comprehension on numerous roles of ROS in inflammation; this, in turn, will facilitate the design of anti-inflammatory biomaterials and broaden the understanding regarding their mechanisms. For most of the ROS scavenging biomaterials, the selection and implementation of evaluation methods seem to be a major concern. The limited bioavailability in particular organs and the lack of tissue selectivity in newly developed biomaterials render them clinically inefficient. Additional research is also required to determine the approaches that can protect physiological functions of ROS from limiting its formation [2].

25.5.1 Insufficient Duration for Clinical Trial

The buildup of oxidative stress coupled with metabolic inflammation occurs through a prolonged duration. Considering an expected longer time frame needed for recovery from such detrimental effects, it will be challenging to establish any substantial benefit within a shorter intervention window compared to the ones used in existing clinical trial setups [56].

25.5.2 Lacking in Specificity

The functions of ROS in metabolic inflammation are still not comprehensively understood. Altered dosing of ROS, its wide range of sources, and distinct effects caused by its multiple areas of action, indicate that our understanding about ROS remains limited [60]. A significant hurdle toward developing prosperous ROS-targeted medications involves addressing the specificity due to the innate diversity of ROS and their various roles throughout the cell. The exact quantities at which ROS transform from appropriate signaling duties to harmful agents remain primarily unidentified, which limits the calculation of suitable therapeutic frames [16]. Moreover, a therapeutic approach intended to target a specific ROS may fail to properly mitigate the overall pathogenic consequence of other ROS, or it may unintentionally disrupt desirable signaling pathways regulated by another ROS. The regional and temporal delicacy of ROS formation and signaling essentially demands more sophisticated and customized delivery as well as regulated mechanisms to prevent undesirable outcomes.

25.5.3 Complexity of ROS Signaling Networks

ROS serve as vital constituents of highly sophisticated and interrelated physiological signaling networks that subsequently affect multiple biological activities by modulating the expression of crucial signaling molecules, notably kinases, phosphatases, transcription factors, etc. [61]. The biological cells generally offer remarkable ability to adapt for maintaining homeostasis. Upon exposure to increased quantities of ROS especially while experiencing metabolic inflammation, cells might initiate antioxidant pathways to eliminate free radicals and recover redox equilibrium. A prime instance of a recognized adaptive response is the Nrf2 pathway, which is accountable for stimulating antioxidant enzymes. This ability for adaptation can drastically decrease the efficiency of ongoing therapeutic intervention that aims to cause apoptosis in cancer cells or build drug resistance [62]. Conversely, treatments targeted for scavenging ROS may deliberately trigger compensating processes which undermine the purpose of the therapy.

25.5.4 Difficulties in Targeted Delivery

The safety and effectiveness of ROS-targeted interventions depend on their precise delivery to diseased sites in addition to specific organelles within cells like mitochondria [63]. Nonetheless, attaining this precisely focused delivery poses multiple serious difficulties. Targeting specified diseased domains, such as deeply submerged tumors or inflammatory tissues within systems, remained incredibly difficult owing to their restricted exposure. Creating nanocarriers that can preferentially respond to the differential pathogenic levels of ROS within heterogeneous damaged cells is also problematic [64]. The drug has to be effective and intact before it reaches its therapeutic target, and afterward it must be released at an adequate concentration in order to act. Numerous natural substances with antioxidant features exhibit low absorption and instabilities, resulting in inhibiting their clinical applicability in metabolic inflammation [65]. After reaching the target site, various biological barriers, including endosomal escape and cellular uptake mechanisms, generally create a fence for the drug delivery systems to deliver their load to the correct intracellular compartment, such as the mitochondria or cytoplasm, where the pathological ROS are generated [66]. These hurdles underscore the need for an interdisciplinary research approach, fostering collaboration among redox biologists, pharmacologists, and material scientists to design novel drug carriers as well as identify the specific molecular targets themselves.

25.6 Emerging Opportunities and Innovative Approaches for ROS-Targeted Research

Various creative techniques in alliance with plausible mechanisms underscore the developing opportunities, as shown in Table 25.1. This allows for targeted delivery for enhanced therapeutic success in metabolic inflammation, even though there are multiple challenges in converting ROS-based research. Mitochondria are substantial endogenous bases of pathological ROS in several ailments linked to metabolic inflammation, directly targeting these organelles offers a budding therapeutic plan [67]. Besides the ROS scavenging biomaterials, current studies have exposed that mitochondria-targeted antioxidants (MTAs), such as elamipretide, plastoquinonyl decyl triphenylphosphonium (SkQ1), mitoquinone, coenzyme Q10, nicotinamide mononucleotide, and melatonin, are competent to gather within mitochondria, where they can efficiently lessen ROS levels and stabilize the functions of mitochondria, thereby alleviating oxidative stress and downstream inflammatory signaling (Fig. 25.1). These antioxidants can endorse mitophagy that includes specific degradation and exclusion of injured mitochondria [68, 69]. The proficient removal method of these damaged organelles not only decreases oxidative stress but also supports to decline inflammation at its source [70]. A further targeted and well-defined therapeutic plan encompasses inhibiting the exact NOX isoforms that are accountable for generation of pathological ROS while conserving valuable redox signaling driven by other isoforms. This level of selectivity highlights the importance of detailed mechanistic understanding at the isoform level for successful translational research, rather than merely targeting a family of enzymes [71]. Additionally, the application of exogenous antioxidants or the inhibition of particular ROS-generating enzymes offers a promising therapeutic approach. It promotes the sustainability of the body's intrinsic and highly evolved antioxidant defense systems. These endogenous systems, like thioredoxin (TRX) and glutathione (GSH) systems, are essential to support cellular redox balance and are fundamental to cellular strength and homeostasis. The TRX system comprising thioredoxin reductase (TrxR), and peroxiredoxins, plays a dynamic role in governing cellular growth, gene expression, apoptosis, and overall antioxidant defense adjacent to all living cells. Controlling the TRX pathway has presented significant potential in inhibiting immune reactions in inflammatory disorders, with systemic management and transgenic overexpression of TRX elevating inflammation in several in vivo models [72]. Glutathione (GSH), a tripeptide consisting glycine, glutamine, and cysteine, is broadly accepted as the "master antioxidant" owing to its natural existence within all the cells. Increasing intracellular glutathione levels, for instance, through supplementation with N-acetylcysteine (NAC), a cysteine precursor, has demonstrated several benefits. NAC has been demonstrated to condense oxidative stress, expand insulin resistance in patients having obesity, and potentially support in the supervision of autoimmune diseases by defending cell mitochondria from free radicals [73]. This attitude of improving the body's extremely progressed and compartmentalized antioxidant defense systems presents a safer and potential approach to reestablishing redox balance than using exogenous scavengers.

Table 25.1 Emerging opportunities for translating ROS research into clinical applications

Emerging materials/ approaches	Targeted diseases	Compounds used	Mechanism of action	Efficiency
Mitochondria-targeted antioxidants (MTAs)	T2D, chronic kidney disease, cardiovascular diseases	Elamipretide, SkQ1, mitoquinone, coenzyme Q10, nicotinamide mononucleotide, melatonin, metformin	Reduction of ROS within mitochondria, stabilization of mitochondria, promotion of mitophagy, augmentation of mitochondrial function	Reduced oxidative damage, improved mitochondrial energetics, decreased inflammation, enhanced mitophagy
Endogenous antioxidant system enhancers	Obesity, T2D, autoimmune diseases, liver diseases	NAC, GSH, TRX	Advancement of intrinsic antioxidant defenses, improvement in GSH/TRX system activity, neutralization of ROS	Reduced oxidative stress, improved insulin sensitivity, anti-inflammatory effects, and cell protection
Specific ROS-generating enzyme inhibitors	Complications related to T2D, obesity-induced atrial fibrillation	NOX2 inhibitors	Specified inhibition of pathological ROS generation by utilizing specific enzymes	Lessened inflammation, reduced oxidative stress, defense against tissue damage
Sophisticated drug delivery systems	Inflammatory bowel diseases, different metabolic disorders with localized ROS, cancer	ROS-responsive linkers or moieties, such as thioether, thioketal, selenide, diselenide, telluride, and aryl boronic acids/esters sonosensitizers	Targeted drug release at the sites containing excessive ROS, superior therapeutic efficacy through localized action, improved synergistic effects associated with PDT/SDT	Precise drug delivery, reduced off-target effects, enhanced anti-inflammatory and anti-cancer effects
Personalized redox medicine approaches	Obesity, T2D, NAFLD, inflammatory diseases	Systems biology approach such as genome scale metabolic model (GSMM), genetic polymorphism-guided interventions	Tailoring the therapies based on distinct ROS profile, gut microbiome, genetic tendencies	Identification of definite targets for personalized interventions, improved risk stratification

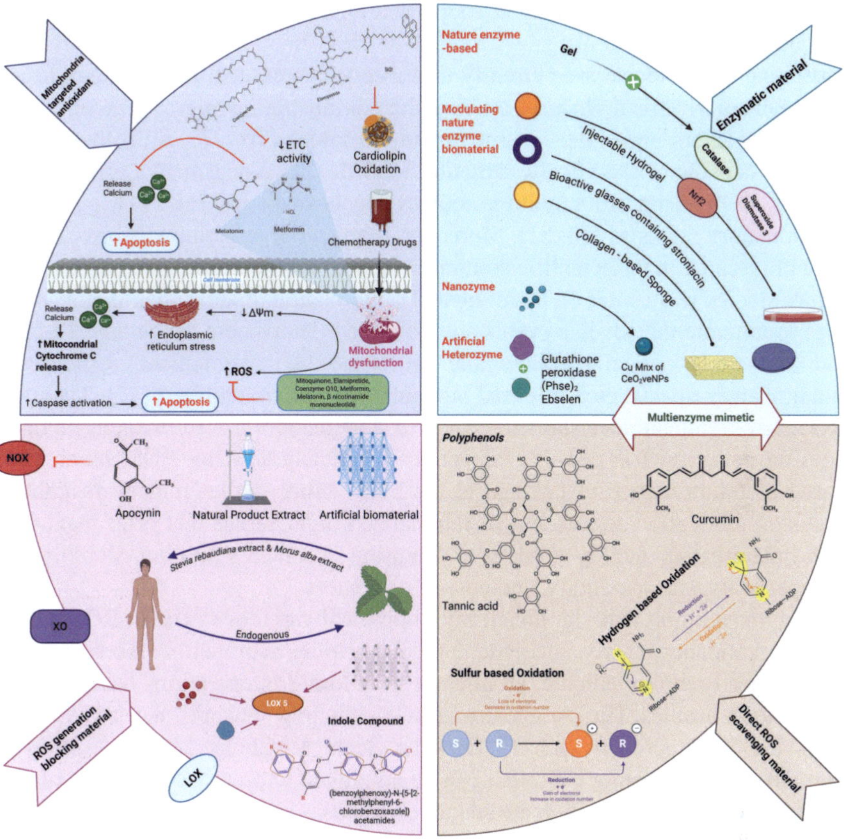

Fig. 25.1 Different categories of ROS scavenging materials used in translational research

To address the critical deficiencies in attaining therapeutic specificity and reducing off-target effects, development of few sophisticated drug delivery systems comprising ROS-responsive linkers or moieties, such as thioether, selenide, thioketal, diselenide, telluride, and aryl boronic acids/esters have been reported. These systems are capable of triggering the controlled and on-demand delivery of the encapsulated agent, confirming that the therapeutic agent is conveyed precisely where it is actually desired [63]. Besides, an expanding area of opportunity includes the incorporation of ROS-responsive systems with additional therapeutic modalities like photodynamic therapy (PDT) or sonodynamic therapy (SDT). In these synergistic methodologies, exogenously made ROS like light-activated photosensitizers in PDT, ultrasound-activated sonosensitizers in SDT, etc., can strengthen drug release from the emerging nanocarriers [64]. A number of novel nanocarriers have been recently fabricated to overcome the poor solubility and oral bioavailability of natural compounds used in ROS-based translational research. It is evidenced that few novel ROS scavenging biomaterials like curcumin-encapsulated poly (lactic-co-glycolic acid) (PLGA) nanoparticles and liposomes have been fabricated to boost neuroprotection against oxidative stress in AD [74, 75]. A number of intriguing technologies are currently under development that aim to target lesions by means of electrical charges, altering the administration approaches, and establishing initiators and other spatiotemporal variables [76, 77]. Still, biomaterials must be capable of specifically targeting ROS that have detrimental effects on the body, as ROS arising from multiple sources may have distinct roles in triggering the inflammatory response [78–81]. Moreover, biomaterial targeting is a breakthrough area of research in the scientific community at this moment. It is also predicted that future studies will center on targeting and optimization of anti-inflammatory therapies to mitigate the adverse events and enhance effectiveness of therapy. To facilitate numerous critical physiological functions, like metabolism, responses to inflammatory agents, etc., minimal amounts of ROS are required [77]. Therefore, ROS scavenging biomaterials should aim to store the levels of ROS to physiological equilibrium as much as possible rather than completely scavenge ROS, which would cause a certain degree of damage to the body. More studies remain desirable to focus on fabrication approaches of rationalized biomaterials that offer maximized anti-inflammatory facility while also featuring excellent solubility, stability, biocompatibility, site specificity, and cost-effectiveness.

The recognition that significant interindividual variations exist in ROS generation, antioxidant capacity, and genetic predispositions is propelling the field toward personalized redox medicine, an attempt in tailoring interventions based on individual ROS profiles. The complexity of ROS signaling networks necessitates a systems biology approach such as Genome Scale Metabolic Model (GSMM), a computational modeling [82], that can be employed to predict complexity of ROS signaling networks and anticipate adaptive responses [83]. This will guide the selection of the most appropriate therapy and dosage, representing a significant shift toward a more patient-centric model of care in metabolic diseases, ultimately leading to improved outcomes and more efficient resource utilization(Table 25.1).

25.7 Translational Gaps, Future Directions, and Clinical Impact

Indeed, many ROS-targeted treatments have shown encouraging preclinical results. There is frequently a big discrepancy between laboratory accomplishment and clinical reality. Although they are very useful for preliminary research, preclinical in vitro and animal experiments might not accurately represent the complex and varied nature of metabolic inflammation in humans. The design for clinical trials evaluating ROS-targeted remedies needs rigorous scrutiny of suitable consequences and criteria for choosing participants that can show substantial and enduring advantages. In order to remain patient safety, it is crucial that therapies should target abnormal ROS in particular, without affecting other critical functions of cells [84].

Bridging these translational gaps necessitates a consistent strategy that is crucial for transforming preclinical prospects into viable clinical therapies. Experimental studies are presently proceeding for substances such as coenzyme Q10 and nicotinamide riboside, which have exhibited minimal but encouraging outcomes in trials comprising human volunteers. In order to alleviate both macrovascular consequences like atherosclerosis-driven strokes, myocardial infarctions, coronary artery disease, and microvascular disorders (damage to retina, kidney, nerves, etc.) caused by metabolic inflammation, several pharmacological inhibitors of NOX enzymes are currently being examined [85]. In the course of time, individuals suffering from prevalent metabolic inflammatory diseases like obesity, type 2 diabetes, NAFLD, and related cardiovascular problems will be definitely benefitted from new treatment assumptions that restore redox homeostasis, successfully reduce chronic inflammation, and boost long-term results [1]. In the perspective of immense therapeutic constraints in the world today, this constitutes a major step ahead in clinical impact.

25.8 Conclusion

The complex and dual nature of ROS, functioning as fundamental signaling molecules as well as detrimental mediators of oxidative stress, is the key concept to metabolic inflammation. Several decades of investigation have demonstrated that imbalanced ROS plays a pathogenic function in numerous diseases, and yet there have been substantial barriers to translating these findings into clinical applications. The frequent flaws of existing systemic and broad-spectrum antioxidants illustrate the significant drawbacks in ambiguous interventions, especially their incapability to separate between the beneficial and detrimental ROS, lack in targeted efficiency, and intricate understanding of cellular adaptive responses. The field is currently going through a crucial evolution, from a fundamental mission of ROS scavenging to a multifaceted attempt for precise redox management. Several emerging prospects are propelled by a better molecular comprehension, prioritizing specifically tailored approaches. There are arrays of translational gaps that require to be solved in order for these innovations to accomplish their maximum clinical relevance. One of the key tasks that require to be performed precisely right now is to discover robust biomarkers that can detect oxidative stress, assist in patient categorization, and govern the therapeutic efficiency in vivo. It is also important to ensure the pharmacokinetic and pharmacodynamic integrity of novel substances as well as optimize preclinical models that more closely resemble human metabolic inflammatory diseases. Ultimately, there is an enormous opportunity in the extended and challenging journey from foundational ROS research to clinical practices intended for individuals with metabolic inflammation. Ongoing collaborative research, highlighting tailored approaches, and sophisticated delivery systems are really needed for translating these advancements into revolutionary clinical implementations, and thus curing the lives of millions suffering from metabolic inflammation.

References

1. Forrester SJ, Kikuchi DS, Hernandes MS, Xu Q, Griendling KK (2018) Reactive oxygen species in metabolic and inflammatory signaling. Circ Res 122(6):877–902
2. Liu J, Han X, Zhang T, Tian K, Li Z, Luo F (2023) Reactive oxygen species (ROS) scavenging biomaterials for anti-inflammatory diseases: from mechanism to therapy. J Hematol Oncol 16(1):116
3. Schieber M, Chandel NS (2014) ROS function in redox signaling and oxidative stress. Curr Biol 24(10):R453–R462
4. Hong Y, Boiti A, Vallone D, Foulkes NS (2024) Reactive oxygen species signaling and oxidative stress: transcriptional regulation and evolution. Antioxidants 13(3):312
5. Murphy MP, Bayir H, Belousov V, Chang CJ, Davies KJ, Davies MJ et al (2022) Guidelines for measuring reactive oxygen species and oxidative damage in cells and in vivo. Nat Metab 4(6):651–662
6. Nelson KJ, Bolduc JA, Wu H, Collins JA, Burke EA, Reisz JA et al (2018) H2O2 oxidation of cysteine residues in c-Jun N-terminal kinase 2 (JNK2) contributes to redox regulation in human articular chondrocytes. J Biol Chem 293(42):16376–16389
7. Roshanfekr A, Moeini Z, Golmohammadi F, Balalaie S, Seyedarabi A (2025) Antioxidative effects of phytoaromatic compounds on the cysteine of the SKCGS peptide as a critical aggregation domain of tau. Sci Rep 15(1):23206
8. Shields HJ, Traa A, Van Raamsdonk JM (2021) Beneficial and detrimental effects of reactive oxygen species on lifespan: a comprehensive review of comparative and experimental studies. Front Cell Dev Biol 9:628157
9. Bardaweel SK, Gul M, Alzweiri M, Ishaqat A, ALSalamat HA, Bashatwah RM (2018) Reactive oxygen species: the dual role in physiological and pathological conditions of the human body. Eurasian J Med 50(3):193
10. Reuter S, Gupta SC, Chaturvedi MM, Aggarwal BB (2010) Oxidative stress, inflammation, and cancer: how are they linked? Free Radic Biol Med 49(11):1603–1616
11. Furukawa S, Fujita T, Shimabukuro M, Iwaki M, Yamada Y, Nakajima Y et al (2017) Increased oxidative stress in obesity and its impact on metabolic syndrome. J Clin Invest 114(12):1752–1761
12. He L, He T, Farrar S, Ji L, Liu T, Ma X (2017) Antioxidants maintain cellular redox homeostasis by elimination of reactive oxygen species. Cell Physiol Biochem 44(2):532–553
13. Kim DH, Meza CA, Clarke H, Kim JS, Hickner RC (2020) Vitamin D and endothelial function. Nutrients 12(2):575
14. Traber MG, Atkinson J (2007) Vitamin E, antioxidant and nothing more. Free Radic Biol Med 43(1):4–15
15. Li MS, Adesina SE, Ellis CL, Gooch JL, Hoover RS, Williams CR (2017) NADPH oxidase-2 mediates zinc deficiency-induced oxidative stress and kidney damage. Am J Physiol Cell Physiol 312:C47–C55
16. Lindsay RT, Rhodes CJ (2025) Reactive oxygen species (ROS) in metabolic disease—don't shoot the metabolic messenger. Int J Mol Sci 26(6):2622
17. Netzer N, Gatterer H, Faulhaber M, Burtscher M, Pramsohler S, Pesta D (2015) Hypoxia, oxidative stress and fat. Biomolecules 5(2):1143–1150
18. Schrauwen P, Hesselink MK (2004) Oxidative capacity, lipotoxicity, and mitochondrial damage in type 2 diabetes. Diabetes 53(6):1412–1417
19. Rains JL, Jain SK (2011) Oxidative stress, insulin signaling, and diabetes. Free Radic Biol Med 50(5):567–575
20. Sun X, Pang H, Li J, Luo S, Huang G, Li X et al (2020) The NLRP3 inflammasome and its role in T1DM. Front Immunol 11:1595
21. Lu Q, Tian X, Wu H, Huang J, Li M, Mei Z et al (2021) Metabolic changes of hepatocytes in NAFLD. Front Physiol 12:710420

22. Ma Y, Lee G, Heo SY, Roh YS (2021) Oxidative stress is a key modulator in the development of nonalcoholic fatty liver disease. Antioxidants 11(1):91
23. Pu S, Pan Y, Zhang Q, You T, Yue T, Zhang Y et al (2023) Endoplasmic reticulum stress and mitochondrial stress in drug-induced liver injury. Molecules 28(7):3160
24. Montezano AC, Touyz RM (2012) Reactive oxygen species and endothelial function–role of nitric oxide synthase uncoupling and Nox family nicotinamide adenine dinucleotide phosphate oxidases. Basic Clin Pharmacol Toxicol 110(1):87–94
25. Chen JY, Ye ZX, Wang XF, Chang J, Yang MW, Zhong HH et al (2018) Nitric oxide bioavailability dysfunction involves in atherosclerosis. Biomed Pharmacother 97:423–428
26. Sultana MA, Hia RA, Akinsiku O, Hegde V (2023) Peripheral mitochondrial dysfunction: a potential contributor to the development of metabolic disorders and Alzheimer's disease. Biology 12(7):1019
27. Simpson DS, Oliver PL (2020) ROS generation in microglia: understanding oxidative stress and inflammation in neurodegenerative disease. Antioxidants 9(8):743
28. Afanas'ev I (2013) New nucleophilic mechanisms of ros-dependent epigenetic modifications: comparison of aging and cancer. Aging Dis 5(1):52
29. Bigornia SJ, Mott MM, Hess DT, Apovian CM, McDonnell ME, Duess MA et al (2010) Long-term successful weight loss improves vascular endothelial function in severely obese individuals. Obesity 18(4):754–759
30. Lee JH, Jeon SM, Hong SP, Cheon JH, Kim TI, Kim WH (2012) Metformin use is associated with a decreased incidence of colorectal adenomas in diabetic patients with previous colorectal cancer. Dig Liver Dis 44(12):1042–1047
31. Kim YH, Noh R, Cho SY, Park SJ, Jeon SM, Shin HD et al (2015) Inhibitory effect of metformin therapy on the incidence of colorectal advanced adenomas in patients with diabetes. Intest Res 13(2):145–152
32. Belda-Iniesta C, Pernía O, Simó R (2011) Metformin: a new option in cancer treatment. Clin Transl Oncol 13(6):363–367
33. Lonardo E, Cioffi M, Sancho P, Sanchez-Ripoll Y, Trabulo SM, Dorado J et al (2013) Metformin targets the metabolic achilles heel of human pancreatic cancer stem cells. PLoS One 8(10):e76518
34. Zhang J, Fu Y, Yang P, Liu X, Li Y, Gu Z (2020) ROS scavenging biopolymers for anti-inflammatory diseases: classification and formulation. Adv Mater Interfaces 7(16):2000632
35. Wang L, Zhu B, Deng Y, Li T, Tian Q, Yuan Z et al (2021) Biocatalytic and antioxidant nanostructures for ROS scavenging and biotherapeutics. Adv Funct Mater 31(31):2101804
36. Zhuang H, Hong Y, Gao J, Chen S, Ma Y, Wang S (2015) A poly (γ-glutamic acid)-based hydrogel loaded with superoxide dismutase for wound healing. J Appl Polym Sci 132(23). https://doi.org/10.1002/app.42033
37. Dong Y, Zhuang H, Hao Y, Zhang L, Yang Q, Liu Y et al (2020) Poly (N-isopropylacrylamide)/poly (γ-glutamic acid) thermo-sensitive hydrogels loaded with superoxide dismutase for wound dressing application. Int J Nanomedicine 15:1939–1950
38. Park C, Park J, Kim WJ, Kim W, Cheong H, Kim SJ (2021) Malonic acid isolated from Pinus densiflora inhibits UVB-induced oxidative stress and inflammation in HaCaT keratinocytes. Polymers 13(5):816
39. Zhang D, Ren Y, He Y, Chang R, Guo S, Ma S et al (2022) In situ forming and biocompatible hyaluronic acid hydrogel with reactive oxygen species-scavenging activity to improve traumatic brain injury repair by suppressing oxidative stress and neuroinflammation. Mater Today Bio 15:100278
40. Qian ZJ, Chen MF, Chen J, Zhang Y, Zhou C, Hong P et al (2021) Intracellular ethanol-mediated oxidation and apoptosis in HepG2/CYP2E1 cells impaired by two active peptides from seahorse (Hippocampus kuda bleeler) protein hydrolysates via the Nrf2/HO-1 and akt pathways. Food Sci Nutr 9(3):1584–1602
41. Li R, Hou X, Li L, Guo J, Jiang W, Shang W (2022) Application of metal-based nanozymes in inflammatory disease: a review. Front Bioeng Biotechnol 10:920213

42. Yu Y, Zhao S, Gu D, Zhu B, Liu H, Wu W et al (2022) Cerium oxide nanozyme attenuates periodontal bone destruction by inhibiting the ROS–NFκB pathway. Nanoscale 14(7):2628–2637
43. Niemiec SM, Hilton SA, Wallbank A, Azeltine M, Louiselle AE, Elajaili H et al (2021) Cerium oxide nanoparticle delivery of microRNA-146a for local treatment of acute lung injury. Nanomedicine 34:102388
44. Xiong Y, Zhang Y, Zhou C, Yu T (2023) ROS scavenging manganese-loaded mesoporous silica nanozymes for catalytic anti-inflammatory therapy. Adv Powder Technol 34(1):103886
45. Zhang ZY, Jiang M, Fang J, Yang MF, Zhang S, Yin YX et al (2017) Enhanced therapeutic potential of nano-curcumin against subarachnoid hemorrhage-induced blood–brain barrier disruption through inhibition of inflammatory response and oxidative stress. Mol Neurobiol 54(1):1–4
46. Sahiner N, Sagbas S, Aktas N, Silan C (2016) Inherently antioxidant and antimicrobial tannic acid release from poly (tannic acid) nanoparticles with controllable degradability. Colloids Surf B: Biointerfaces 142:334–343
47. Li Y, Fu R, Duan Z, Zhu C, Fan D (2022) Construction of multifunctional hydrogel based on the tannic acid-metal coating decorated MoS2 dual nanozyme for bacteria-infected wound healing. Bioact Mater 9:461–474
48. Kim DE, Lee Y, Kim M, Lee S, Jon S, Lee SH (2017) Bilirubin nanoparticles ameliorate allergic lung inflammation in a mouse model of asthma. Biomaterials 140:37–44
49. Yao Q, Jiang X, Zhai YY, Luo LZ, Xu HL, Xiao J et al (2020) Protective effects and mechanisms of bilirubin nanomedicine against acute pancreatitis. J Control Release 322:312–325
50. Barbieri SS, Cavalca V, Eligini S, Brambilla M, Caiani A, Tremoli E et al (2004) Apocynin prevents cyclooxygenase 2 expression in human monocytes through NADPH oxidase and glutathione redox-dependent mechanisms. Free Radic Biol Med 37(2):156–165
51. Kim SY, Moon KA, Jo HY, Jeong S, Seon SH, Jung E et al (2012) Anti-inflammatory effects of apocynin, an inhibitor of NADPH oxidase, in airway inflammation. Immunol Cell Biol 90(4):441–448
52. Mehmood A, Zhao L, Wang C, Hossen I, Nadeem M (2020) Stevia residue extract alone and combination with allopurinol attenuate hyperuricemia in fructose–PO-induced hyperuricemic mice. J Food Biochem 44(1):e13087
53. Wan L, Chen G, Jian S, Yin XJ, Zhu H (2018) Antioxidant and xanthine oxidase inhibitory properties and LC-MS/MS identification of compoundsof ethanolic extract from mulberry leaves. Acta Sci Pol Technol Aliment 17(4):313–319
54. Cerqua I, Musella S, Peltner LK, D'Avino D, Di Sarno V, Granato E et al (2022) Discovery and optimization of indoline-based compounds as dual 5-LOX/sEH inhibitors: in vitro and in vivo anti-inflammatory characterization. J Med Chem 65(21):14456–14480
55. Blagov AV, Summerhill VI, Sukhorukov VN, Zhigmitova EB, Postnov AY, Orekhov AN (2024) Potential use of antioxidants for the treatment of chronic inflammatory diseases. Front Pharmacol 15:1378335
56. Steinhubl SR (2008) Why have antioxidants failed in clinical trials? Am J Cardiol 101(10):S14–S19
57. Walker C (2008) Antioxidant supplements do not improve mortality and may cause harm. Am Fam Physician 78(9):1079–1080
58. Liu J, Guan Y, Yang L, Fang H, Sun H, Sun Y et al (2025) Ferulic acid as an anti-inflammatory agent: insights into molecular mechanisms, pharmacokinetics and applications. Pharmaceuticals 18(6):912
59. Abdel-Mageed HM, Abd El Aziz AE, Abdel Raouf BM, Mohamed SA, Nada D (2022) Antioxidant-biocompatible and stable catalase-based gelatin–alginate hydrogel scaffold with thermal wound healing capability: immobilization and delivery approach. 3 Biotech 12(3):73
60. Mijiritsky E, Gardin C, Ferroni L, Lacza Z, Zavan B (2020) Albumin-impregnated bone granules modulate the interactions between mesenchymal stem cells and monocytes under in vitro inflammatory conditions. Mater Sci Eng C 110:110678

61. Checa J, Aran JM (2020) Reactive oxygen species: drivers of physiological and pathological processes. J Inflamm Res 13:1057–1073
62. Attique I, Haider Z, Khan M, Hassan S, Soliman MM, Ibrahim WN et al (2025) Reactive oxygen species: from tumorigenesis to therapeutic strategies in cancer. Cancer Med 14(10):e70947
63. Liang J, Liu B (2016) ROS-responsive drug delivery systems. Bioeng Transl Med 1(3):239–251
64. Hu D, Li Y, Li R, Wang M, Zhou K, He C et al (2024) Recent advances in reactive oxygen species (ROS)-responsive drug delivery systems for photodynamic therapy of cancer. Acta Pharm Sin B 14(12):5106–5131
65. Ashok A, Andrabi SS, Mansoor S, Kuang Y, Kwon BK, Labhasetwar V (2022) Antioxidant therapy in oxidative stress-induced neurodegenerative diseases: role of nanoparticle-based drug delivery systems in clinical translation. Antioxidants 11(2):408
66. Azevedo C, Macedo MH, Sarmento B (2018) Strategies for the enhanced intracellular delivery of nanomaterials. Drug Discov Today 23(5):944–959
67. Xu X, Pang Y, Fan X (2025) Mitochondria in oxidative stress, inflammation and aging: from mechanisms to therapeutic advances. Signal Transduct Target Ther 10(1):190
68. Liu CJ, Wang LK, Tsai FM (2025) The application and molecular mechanisms of mitochondria-targeted antioxidants in chemotherapy-induced cardiac injury. Curr Issues Mol Biol 47(3):176
69. Chen C, Xiang A, Lin X, Guo J, Liu J, Hu S et al (2024) Mitophagy: insights into its signaling molecules, biological functions, and therapeutic potential in breast cancer. Cell Death Dis 10(1):457
70. Wang S, Long H, Hou L, Feng B, Ma Z, Wu Y et al (2023) The mitophagy pathway and its implications in human diseases. Signal Transduct Target Ther 8(1):304
71. Gola L, Bierhansl L, Csatári J, Schroeter CB, Korn L, Narayanan V et al (2023) NOX4-derived ROS are neuroprotective by balancing intracellular calcium stores. Cell Mol Life Sci 80(5):127
72. Muri J, Kopf M (2023) The thioredoxin system: balancing redox responses in immune cells and tumors. Eur J Immunol 53(1):2249948
73. Pedre B, Barayeu U, Ezeriņa D, Dick TP (2021) The mechanism of action of N-acetylcysteine (NAC): the emerging role of H2S and sulfane sulfur species. Pharmacol Ther 228:107916
74. Doggui S, Sahni JK, Arseneault M, Dao L, Ramassamy C (2012) Neuronal uptake and neuroprotective effect of curcumin-loaded PLGA nanoparticles on the human SK-N-SH cell line. J Alzheimers Dis 30(2):377–392
75. Fernandes M, Lopes I, Magalhães L, Sárria MP, Machado R, Sousa JC et al (2021) Novel concept of exosome-like liposomes for the treatment of Alzheimer's disease. J Control Release 336:130–143
76. Fokam D, Hoskin D (2020) Instrumental role for reactive oxygen species in the inflammatory response. Frontiers in Bioscience-Landmark 25(6):1110–1119
77. Pelletier M, Lepow TS, Billingham LK, Murphy MP, Siegel RM (2012) New tricks from an old dog: mitochondrial redox signaling in cellular inflammation. Semin Immunol 24(6):384–392
78. Yan M, Zhu L, Wu S, Cao Y, Mou N, Chi Q et al (2022) ROS responsive polydopamine nanoparticles to relieve oxidative stress and inflammation for ameliorating acute inflammatory bowel. Biomater Adv 142:213126
79. Huang X, He D, Pan Z, Luo G, Deng JJ (2021) Reactive-oxygen-species-scavenging nanomaterials for resolving inflammation. Mater Today Bio 11:100124
80. Wan WL, Tian B, Lin YJ, Korupalli C, Lu MY, Cui Q et al (2020) Photosynthesis-inspired H2 generation using a chlorophyll-loaded liposomal nanoplatform to detect and scavenge excess ROS. Nat Commun 11(1):534
81. Wan WL, Lin YJ, Chen HL, Huang CC, Shih PC, Bow YR et al (2017) In situ nanoreactor for photosynthesizing H2 gas to mitigate oxidative stress in tissue inflammation. J Am Chem Soc 139(37):12923–12926
82. Sertbas M, Ulgen KO (2020) Genome-scale metabolic modeling for unraveling molecular mechanisms of high threat pathogens. Front Cell Dev Biol 8:566702

83. Kolodkin AN, Sharma RP, Colangelo AM, Ignatenko A, Martorana F, Jennen D et al (2020) ROS networks: designs, aging, Parkinson's disease and precision therapies. NPJ Syst Biol Appl 6(1):34
84. Lei S, Fu Y, Zhang B, Yang H, Ji Z (2025) Knowledge graph and emerging trends in oxidative stress research on hepatic ischemia-reperfusion injury: a bibliometric analysis (1995–2024). Front Pharmacol 16:1587591
85. Urner S, Ho F, Jha JC, Ziegler D, Jandeleit-Dahm K (2020) NADPH oxidase inhibition: preclinical and clinical studies in diabetic complications. Antioxid Redox Signal 33(6):415–434

Index

S. Ashique, N. Mishra (eds.), *Reactive Oxygen Species in Metabolic Inflammation*, https://doi.org/10.1007/978-3-032-11752-6

MIX
Papier aus verantwortungsvollen Quellen
Paper from responsible sources
FSC® C105338

If you have any concerns about our products,
you can contact us on
ProductSafety@springernature.com

In case Publisher is established outside the EU,
the EU authorized representative is:
Springer Nature Customer Service Center GmbH
Europaplatz 3, 69115 Heidelberg, Germany

Printed by Libri Plureos GmbH
in Hamburg, Germany